Advances in Experimental Medicine and Biology

Volume 1304

Series Editors

Wim E. Crusio, Institut de Neurosciences Cognitives et Intégratives
d'Aquitaine, CNRS and University of Bordeaux, Pessac Cedex, France
Haidong Dong, Departments of Urology and Immunology,
Mayo Clinic, Rochester, MN, USA
Heinfried H. Radeke, Institute of Pharmacology & Toxicology,
Clinic of the Goethe University Frankfurt Main, Frankfurt am Main,
Hessen, Germany
Nima Rezaei, Research Center for Immunodeficiencies, Children's Medical
Center, Tehran University of Medical Sciences, Tehran, Iran
Junjie Xiao, Cardiac Regeneration and Ageing Lab, Institute of
Cardiovascular Science, School of Life Science, Shanghai University,
Shanghai, China

Advances in Experimental Medicine and Biology provides a platform for scientific contributions in the main disciplines of the biomedicine and the life sciences. This series publishes thematic volumes on contemporary research in the areas of microbiology, immunology, neurosciences, biochemistry, biomedical engineering, genetics, physiology, and cancer research. Covering emerging topics and techniques in basic and clinical science, it brings together clinicians and researchers from various fields.

Advances in Experimental Medicine and Biology has been publishing exceptional works in the field for over 40 years, and is indexed in SCOPUS, Medline (PubMed), Journal Citation Reports/Science Edition, Science Citation Index Expanded (SciSearch, Web of Science), EMBASE, BIOSIS, Reaxys, EMBiology, the Chemical Abstracts Service (CAS), and Pathway Studio.

2019 Impact Factor: 2.450   5 Year Impact Factor: 2.324

More information about this series at http://www.springer.com/series/5584

Yong-Xiao Wang
Editor

Lung Inflammation in Health and Disease, Volume II

 Springer

Editor
Yong-Xiao Wang
Department of Molecular & Cellular Physiology
Albany Medical College
Albany, NY, USA

Series Editors
Wim E. Crusio
Haidong Dong
Heinfried H. Radeke
Nima Rezaei
Junjie Xiao

ISSN 0065-2598   ISSN 2214-8019   (electronic)
Advances in Experimental Medicine and Biology
ISBN 978-3-030-68750-2   ISBN 978-3-030-68748-9   (eBook)
https://doi.org/10.1007/978-3-030-68748-9

This Springer imprint is published by the registered company Springer Nature Switzerland AG
The registered company address is: Gewerbestrasse 11, 6330 Cham, Switzerland

Preface

As previously discussed in Volume I, inflammation is a natural cellular process occurring in virtually all types of human body tissues, organs, and systems. This process can be acute or chronic. Acute inflammation is a healthy, immediate response to protect and repair the body from harmful stimuli. Usually it occurs within a couple of hours. Chronic inflammation is a lengthier cellular process that is not conducive to natural healing and may lead to pathological states including arthritis, asthma, and pulmonary hypertension.

Normally, inflammation can also be classified as systemic or localized. The former affects the entire human body, which is a pathogenetic component in numerous acute and chronic diseases including atherosclerosis, diabetes, sepsis, trauma, and others with a significant morbidity and mortality. The latter is localized as in a specific organ. For example, inflammation caused by asthma and pulmonary hypertension are localized in the lungs.

Lung diseases are very common and can also be very severe. It is well known that lung infections are the greatest single contributor to the overall global health burden. For instance, lung diseases are the most common causes of deaths of children under 5 years of age, which occur more than 9 million annually. Indeed, pneumonia is the leading killer for children worldwide. Asthma is the most common chronic disease, affecting about 14% of children globally and continuing to rise. Likewise, COPD is recognized to be the fourth leading cause of death in the world and the numbers are growing. The lung is not only the largest internal organ in the human body, but also the only internal organ that is exposed constantly to the external environment; as such, no other organ is more vital and vulnerable than the lung. This may explain the common morbidity and mortality of lung diseases.

Systemic inflammation may induce and even exacerbate local inflammatory diseases. Likewise, local inflammation can cause systemic inflammation. Indeed, there is increasing evidence of coexistence of systemic and local inflammation in patients with asthma, COPD, and other lung diseases. Moreover, the comorbidity of two and even multiple local inflammatory diseases often occurs. For instance, rheumatoid arthritis not only frequently happens together with but also promotes the development of pulmonary hypertension. The comorbidity of local and systemic as well as two or more inflammatory diseases significantly deteriorates the quality of life and may even exacerbate death in patients.

The current treatment options for lung diseases are neither always effective nor specific at all. The development of new therapeutics is earnestly needed. Equally desperately, the molecular mechanisms and physiological significance of lung diseases are still not fully understood. Apparently, this despondent fact is a major encumberment to creating new efficacious drugs in the treatment of lung diseases. This scenario is even worse in two and more lung diseases accompanied with other inflammatory diseases due to their complexity and diversity.

Despite the current state being unsatisfactory, great progresses have been made in many aspects of lung diseases from the molecular geneses to regulatory mechanisms to signaling pathways to cellular processes to basic and clinical technologies to new drug discoveries to clinical manifestations to laboratory and clinical diagnoses to treatment options to predictive prognosis. To the best of our knowledge, however, no one, cohesive book is available to present these state-of-the-art advances in the field. Thus, as one of the major aims, we compile this timely and much-needed book to provide a high-quality platform in which well-known scientists and emerging pioneers in basic, translational, and clinical settings can present their latest, exciting findings in the studies of lung inflammation in health and disease. The contents from multiple outstanding authors with unique expertise and skills of molecular and cell biology, biochemistry, physiology, pharmacology, biophysics, biotechnology, translational biomedicine, and medicine will provide new knowledge, concepts, and discoveries in the field. The second major aim is to help direct future research in lung diseases and other inflammatory diseases. The scope of the book includes nearly all new and important findings from very recent basic, translational, and clinical research in the studies of the molecular genesis, networks, microdomains, regulation, functions, elimination, and drug discoveries of inflammation in lung health and disease, which are involved in animal and human lung epithelial cells, smooth muscle cells, *endothelial* cells, adventitial cells, fibroblasts, neutrophils, *macrophages*, *lymphocytes*, and stem/progenitor cells. Lastly but importantly, the book will offer the latest and most promising results from clinical trials in terms of exploring interventions of local and systemic inflammation in the treatments of lung diseases.

This book features contributions from numerous basic, translational, and physician scientists in the fields of pulmonary vasculature redox signaling in health and disease, and as a result offers a widespread and comprehensive overview for academic and industrial scientists, postdoctoral fellows, and graduate students who are interested in redox signaling in health and disease and/or normal and pathological functions of the pulmonary vasculature. The book may also be valuable for clinicians, medical students, and allied health professionals.

We are sincerely grateful for the overwhelming support from leading scientists who contributed their expertise. Due to their contributions, we are pleased to share Volume II now. Similar to Volume I, the current volume is composed of 17 chapters from prominent investigators and clinicians covering novel fundamental roles and molecular mechanisms of inflammatory cellular responses in the development of acute respiratory distress syndrome, asthma, pulmonary hypertension, sarcoidosis, and other lung illnesses. Several articles principally deal with the interactions among inflammatory signaling with reactive oxygen species, calcium, sex, and other vital intracellular molecular signaling in lung diseases. We also share articles focused on the innovative diagnostic approaches and therapeutic treatment options in the aforementioned lung disorders. We are confident these reports detailing the most important basic, translational, clinical, and drug discovery studies will not only enrich our current knowledge, but will also serve to direct and promote future research in the field.

Albany, NY, USA   Yong-Xiao Wang

Acknowledgments

I want to express my wholehearted gratitude to all of the authors for their dedication and diligence in contributing book chapters, particularly during the challenging and unprecedented times of the global COVID-19 pandemic. Many of the authors in this book have not only performed exceptional roles as writer but also as reviewer. Their selfless contributions are sincerely appreciated. I also want to thank Ms. Alison Ball and Mr. Arjun Narayanan at Springer Nature for their assistance, patience, and enthusiasm in seeing this book to fruition.

Albany, NY, USA   Yong-Xiao Wang

Contents

1 **Can GPCRs Be Targeted to Control Inflammation in Asthma?**.............................. 1
   Pawan Sharma and Raymond B. Penn

2 **Cellular and Molecular Processes in Pulmonary Hypertension**............................. 21
   Vic Maietta, Jorge Reyes-García, Vishal R. Yadav, Yun-Min Zheng, Xu Peng, and Yong-Xiao Wang

3 **Inflammatory Pathways in Sarcoidosis**..................... 39
   Barbara P. Barna, Marc A. Judson, and Mary Jane Thomassen

4 **Innate Immune Responses and Pulmonary Diseases**.......... 53
   Tao Liu, Siqi Liu, and Xiaobo Zhou

5 **Interstitial Lung Disease Associated with Connective Tissue Diseases** 73
   Ruben A. Peredo, Vivek Mehta, and Scott Beegle

6 **Molecular Mechanisms of Vascular Damage During Lung Injury** 95
   Ramon Bossardi Ramos and Alejandro Pablo Adam

7 **Neurotrophin Regulation and Signaling in Airway Smooth Muscle**.................................. 109
   Benjamin B. Roos, Jacob J. Teske, Sangeeta Bhallamudi, Christina M. Pabelick, Venkatachalem Sathish, and Y. S. Prakash

8 **Novel Thoracic MRI Approaches for the Assessment of Pulmonary Physiology and Inflammation** 123
   Jonathan P. Brooke and Ian P. Hall

9 **Overview on Interactive Role of Inflammation, Reactive Oxygen Species, and Calcium Signaling in Asthma, COPD, and Pulmonary Hypertension** 147
   Lillian Truong, Yun-Min Zheng, Sharath Kandhi, and Yong-Xiao Wang

10 Protein S-Palmitoylation and Lung Diseases 165
Zeang Wu, Rubin Tan, Liping Zhu, Ping Yao, and Qinghua Hu

**11 Redox Role of ROS and Inflammation
in Pulmonary Diseases** 187
Li Zuo and Denethi Wijegunawardana

**12 Semaphorin3E/plexinD1 Axis in Asthma:
What We Know So Far!** 205
Latifa Koussih and Abdelilah S. Gounni

**13 Serine Protease Inhibitors to Treat Lung
Inflammatory Diseases** 215
Chahrazade El Amri

14 Sex and Gender Differences in Lung Disease 227
Patricia Silveyra, Nathalie Fuentes, and Daniel Enrique
Rodriguez Bauza

15 Sex Hormones and Lung Inflammation 259
Jorge Reyes-García, Luis M. Montaño, Abril Carbajal-García,
and Yong-Xiao Wang

**16 Synopsis of Clinical Acute Respiratory Distress
Syndrome (ARDS)** 323
Archana Mane and Naldine Isaac

**17 Redox and Inflammatory Signaling, the
Unfolded Protein Response, and the Pathogenesis
of Pulmonary Hypertension** 333
Adiya Katseff, Raed Alhawaj, and Michael S. Wolin

Index   375

Contributors

Alejandro Pablo Adam Department of Molecular and Cellular Physiology, Albany Medical College, Albany, NY, USA

Department of Ophthalmology, Albany Medical College, Albany, NY, USA

Raed Alhawaj Department of Physiology, New York Medical College, Valhalla, NY, USA

Department of Physiology, Faculty of Medicine, Kuwait University, Safat, Kuwait

Chahrazade El Amri Sorbonne Université, Faculty of Sciences and Engineering, IBPS, UMR 8256 CNRS-UPMC, ERL INSERM U1164, Biological Adaptation and Ageing, Paris, France

Barbara P. Barna Program in Lung Cell Biology and Translational Research, Division of Pulmonary and Critical Care Medicine, East Carolina University, Greenville, NC, USA

Scott Beegle Division of Pulmonary & Critical Care Medicine, Albany Medical College, Albany, NY, USA

Sangeeta Bhallamudi Department of Pharmaceutical Sciences, North Dakota State University, Fargo, ND, USA

Ramon Bossardi Ramos Department of Molecular and Cellular Physiology, Albany Medical College, Albany, NY, USA

Jonathan P. Brooke Department of Respiratory Medicine, University of Nottingham, Queens Medical Centre, Nottingham, UK

Abril Carbajal-García Departamento de Farmacología, Facultad de Medicina, Universidad Nacional Autónoma de México, CDMX, Mexico City, Mexico

Nathalie Fuentes National Institute of Allergy, Asthma, and Infectious Diseases, Bethesda, MD, USA

Abdelilah S. Gounni Department of Immunology, Max Rady College of Medicine, Rady Faculty of Health Sciences, University of Manitoba, Winnipeg, MB, Canada

Ian P. Hall Department of Respiratory Medicine, University of Nottingham, Queens Medical Centre, Nottingham, UK

Qinghua Hu School of Basic Medicine, Tongji Medical College, Huazhong University of Science and Technology, Wuhan, China

Naldine Isaac Department of Anesthesiology, Albany Medical Center, Albany, NY, USA

Marc A. Judson Division of Pulmonary and Critical Care Medicine, Albany Medical College, Albany, NY, USA

Sharath Kandhi Department of Molecular and Cellular Physiology, Albany Medical College, Albany, NY, USA

Adiya Katseff Department of Microbiology and Immunology, New York Medical College, Valhalla, NY, USA

Latifa Koussih Department of Immunology, Max Rady College of Medicine, Rady Faculty of Health Sciences, University of Manitoba, Winnipeg, MB, Canada

Department des sciences experimentales, Universite de Saint Boniface, Winnipeg, Manitoba, Canada

Siqi Liu Channing Division of Network Medicine, Brigham and Women's Hospital and Harvard Medical School, Boston, MA, USA

Tao Liu Channing Division of Network Medicine, Brigham and Women's Hospital and Harvard Medical School, Boston, MA, USA

Vic Maietta Department of Molecular & Cellular Physiology, Albany Medical College, Albany, NY, USA

Archana Mane Department of Anesthesiology, Albany Medical Center, Albany, NY, USA

Vivek Mehta Rheumatology, Alaska Native Medical Center, Anchorage, AK, USA

Luis M. Montaño Departamento de Farmacología, Facultad de Medicina, Universidad Nacional Autónoma de México, CDMX, Mexico City, Mexico

Christina M. Pabelick Department of Anesthesiology and Perioperative Medicine, Mayo Clinic, Rochester, MN, USA

Department of Physiology and Biomedical Engineering, Mayo Clinic, Rochester, MN, USA

Xu Peng Department of Medical Physiology, College of Medicine, Texas A&M University, College Station, TX, USA

Raymond B. Penn Center for Translational Medicine, Division of Pulmonary, Allergy, & Critical Care Medicine, Jane & Leonard Korman Respiratory Institute, Sidney Kimmel Medical College, Thomas Jefferson University, Philadelphia, PA, USA

Ruben A. Peredo Division of Rheumatology, Department of Medicine, Albany Medical College, Albany, NY, USA

Y. S. Prakash Department of Anesthesiology and Perioperative Medicine, Mayo Clinic, Rochester, MN, USA

Department of Physiology and Biomedical Engineering, Mayo Clinic, Rochester, MN, USA

Jorge Reyes-García Departamento de Farmacología, Facultad de Medicina, Universidad Nacional Autónoma de México, CDMX, Mexico City, Mexico

Department of Molecular and Cellular Physiology, Albany Medical College, Albany, NY, USA

Daniel Enrique Rodriguez Bauza Clinical Simulation Center, The Pennsylvania State University College of Medicine, Hershey, PA, USA

Benjamin B. Roos Department of Anesthesiology and Perioperative Medicine, Mayo Clinic, Rochester, MN, USA

Venkatachalem Sathish Department of Pharmaceutical Sciences, North Dakota State University, Fargo, ND, USA

Pawan Sharma Center for Translational Medicine, Division of Pulmonary, Allergy, & Critical Care Medicine, Jane & Leonard Korman Respiratory Institute, Sidney Kimmel Medical College, Thomas Jefferson University, Philadelphia, PA, USA

Patricia Silveyra Department of Environmental and Occupational Health, Indiana University Bloomington, Bloomington, IN, USA

Rubin Tan School of Basic Medicine, Tongji Medical College, Huazhong University of Science and Technology, Wuhan, China

School of Basic Medicine, Xuzhou Medical University, Xuzhou, China

Jacob J. Teske Department of Anesthesiology and Perioperative Medicine, Mayo Clinic, Rochester, MN, USA

Mary Jane Thomassen Program in Lung Cell Biology and Translational Research, Division of Pulmonary and Critical Care Medicine, East Carolina University, Greenville, NC, USA

Lillian Truong Department of Molecular and Cellular Physiology, Albany Medical College, Albany, NY, USA

Yong-Xiao Wang Department of Molecular & Cellular Physiology, Albany Medical College, Albany, NY, USA

Denethi Wijegunawardana Department of Pathology, Yale School of Medicine, New Haven, CT, USA

Michael S. Wolin Department of Physiology, New York Medical College, Valhalla, NY, USA

Zeang Wu School of Public Health, Tongji Medical College, Huazhong University of Science and Technology, Wuhan, China

First Affiliated Hospital, School of Medicine, Shihezi University, Shihezi, China

School of Basic Medicine, Tongji Medical College, Huazhong University of Science and Technology, Wuhan, China

Vishal R. Yadav Department of Molecular & Cellular Physiology, Albany Medical College, Albany, NY, USA

Ping Yao School of Public Health, Tongji Medical College, Huazhong University of Science and Technology, Wuhan, China

Yun-Min Zheng Department of Molecular and Cellular Physiology, Albany Medical College, Albany, NY, USA

Xiaobo Zhou Channing Division of Network Medicine, Brigham and Women's Hospital and Harvard Medical School, Boston, MA, USA

Liping Zhu School of Basic Medicine, Tongji Medical College, Huazhong University of Science and Technology, Wuhan, China

Li Zuo College of Arts and Sciences, Molecular Physiology and Biophysics Lab, University of Maine, Presque Isle Campus, Presque Isle, ME, USA

Interdisciplinary Biophysics Graduate Program, The Ohio State University, Columbus, OH, USA

Can GPCRs Be Targeted to Control Inflammation in Asthma?

Pawan Sharma and Raymond B. Penn

Abstract

Historically, the drugs used to manage obstructive lung diseases (OLDs), asthma, and chronic obstructive pulmonary disease (COPD) either (1) directly regulate airway contraction by blocking or relaxing airway smooth muscle (ASM) contraction or (2) indirectly regulate ASM contraction by inhibiting the principal cause of ASM contraction/bronchoconstriction and airway inflammation. To date, these tasks have been respectively assigned to two diverse drug types: agonists/antagonists of G protein-coupled receptors (GPCRs) and inhaled or systemic steroids. These two types of drugs "stay in their lane" with respect to their actions and consequently require the addition of the other drug to effectively manage both inflammation and bronchoconstriction in OLDs. Indeed, it has been speculated that safety issues historically associated with beta-agonist use (beta-agonists activate the beta-2-adrenoceptor (β_2AR) on airway smooth muscle (ASM) to provide bronchoprotection/bronchorelaxation) are a function of pro-inflammatory actions of β_2AR agonism. Recently, however, previously unappreciated roles of various GPCRs on ASM contractility and on airway inflammation have been elucidated, raising the possibility that novel GPCR ligands targeting these GPCRs can be developed as anti-inflammatory therapeutics. Moreover, we now know that many GPCRs can be "tuned" and not just turned "off" or "on" to specifically activate the beneficial therapeutic signaling a receptor can transduce while avoiding detrimental signaling. Thus, the fledging field of *biased agonism pharmacology* has the potential to turn the β_2AR into an anti-inflammatory facilitator in asthma, possibly reducing or eliminating the need for steroids.

Keywords

GPCR · Beta-2 agonists · Asthma · Inflammation · Bronchodilator · Obstructive lung disease · COPD · Biased agonism

Abbreviations

AHR	Airway hyperresponsiveness
ASM	Airway smooth muscle
AERD	Aspirin-exacerbated respiratory disease

P. Sharma · R. B. Penn (✉)
Center for Translational Medicine, Division of Pulmonary, Allergy, & Critical Care Medicine Jane & Leonard Korman Respiratory Institute, Sidney Kimmel Medical College Thomas Jefferson University, Philadelphia, PA, USA
e-mail: Raymond.Penn@jefferson.edu

© The Author(s), under exclusive license to Springer Nature Switzerland AG 2021
Y. -X. Wang (ed.), *Lung Inflammation in Health and Disease, Volume II*, Advances in Experimental Medicine and Biology 1304, https://doi.org/10.1007/978-3-030-68748-9_1

COPD	Chronic obstructive pulmonary disease
CaSR	Calcium-sensing receptor
CysLT	Cysteinyl leukotriene
IL	Interleukin
GPCR	G protein-coupled receptor
OLD	Obstructive lung diseases
mAChR	Muscarinic acetylcholine receptor
mPGES-1	microsomal prostaglandin E synthase-1
LABA	Long-acting beta-2 agonist
PG	Prostaglandin
LPS	Lipopolysaccharide
PAR-2	Protease-activated receptor-2
PMNT	Phenylethanolamine-N-mthyltransferase
PKA	Protein kinase A
SABA	Short-acting beta-2 agonist
TGFβ1	Transforming growth factor beta 1
TAS2R	Type II taste receptors
TNF-α	Tumor necrosis factor alpha
TRPV4	Transient receptor potential vanilloid 4
β_2AR	beta-2 adrenoceptor

The management of obstructive lung diseases (OLDs), asthma, and chronic obstructive pulmonary disease (COPD) is predicated on the importance of controlling excessive bronchoconstriction that increases airway resistance. Increased airway resistance manifests in the inability to breathe, which is not only potentially fatal but also impacts the quality of life [1–3]. Accordingly, drugs managing OLDs either directly regulate airway constriction by blocking or relaxing airway smooth muscle (ASM) contraction or indirectly regulate ASM contraction by inhibiting the principal cause of ASM contraction/bronchoconstriction and airway inflammation. To date, these tasks have been respectively assigned to two diverse drug types: agonists/antagonists of G protein-coupled receptors (GPCRs) and inhaled or systemic steroids.

Although it is important to recognize that non-allergic/nonatopic asthma is also an important health concern, this review will focus on allergic asthma, which has a rich history of research and drug discovery efforts dedicated to understanding and managing the disease. Herein, we will review the logic underlying allergic asthma management, the approaches undertaken to date to manage the two major features of the disease (bronchoconstriction and airway inflammation), and the ability of current and future GPCR-targeting drugs to go beyond their ability to directly regulate ASM contraction and manage airway inflammation [4].

1.1 Allergic Asthma Pathobiology and Attempts to Manage It

It is widely recognized that asthma is a complex disease, often labeled a syndrome, in which various factors can result in an exaggerated immune response to an allergen in the lung that results in obstruction to airflow. Although an increase in airway mucus contributes to this increased obstruction, airway narrowing caused by ASM contraction can greatly impede airflow, and the use of rapidly acting bronchodilators that work by relaxing ASM is usually sufficient to manage an acute asthmatic attack. Conceivably, preventing the *cause* of bronchoconstriction (airway inflammation) should be sufficient to manage asthma, but this is difficult to achieve in most asthmatics; thus, asthmatics are typically managed by either of the following: (1) prophylactic inhaled corticosteroids to control inflammation with the use of an inhaler of short-acting beta-agonist (SABA, acting on beta-2 adrenoceptors (β_2ARs) on ASM) bronchodilator to reverse bronchospasm when needed (mild asthmatics) or (2) daily prophylactic inhaled corticosteroids in combination with a long-acting beta-agonist (LABA) to help prevent (i.e., bronchoprotect)

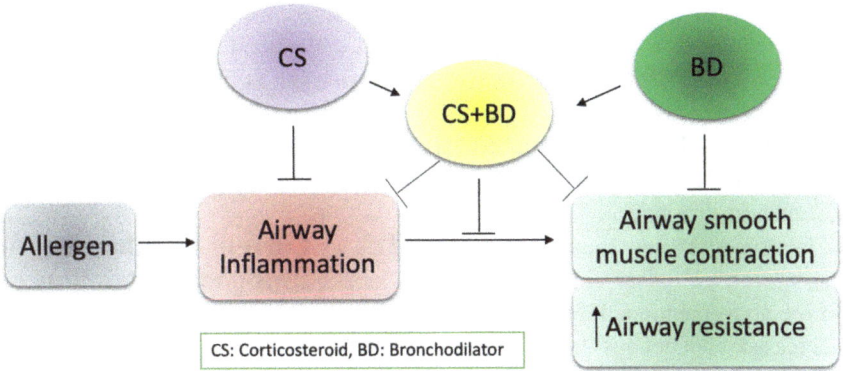

Fig. 1.1 Control of inflammation and bronchospasm in asthma. Corticosteroids (CS) when given alone can block allergic inflammation, while their effectiveness in asthma control is increased by concomitant administration of a bronchodilator (BD) such as β_2AR agonists which alone are not efficacious in controlling inflammation but are highly effective in preventing bronchospasm by targeting ASM. Of note, a modest cooperative effect in controlling inflammation and bronchospasm has been asserted by various studies

bronchospasm (mild, moderate, and some severe asthmatics). Schematic illustrating control of inflammation and bronchospasm by these drugs is depicted in Fig. 1.1.

1.2 The Relationship Between Airway Inflammation and Airway Contraction

The exaggerated immune response to allergen in the lung of asthmatics results in the production of multiple factors that cause ASM contraction. These factors include GPCR agonists that directly act on ASM to either effect contraction or sensitize the ASM to pro-contractile agents. Multiple GPCRs that can mediate ASM contraction are expressed on the plasma membrane of ASM cells [5–10]. Among the best characterized GPCRs in airway/asthma biology are the m3 muscarinic acetylcholine receptor (m3mAChR), H1 histamine receptor (H1HR), and cysteinyl leukotriene type 1 receptor (CysLT1R). Cognate ligands (acetylcholine, histamine, and CysLTs, respectively) for each of these receptors tend to be upregulated in expression in the allergen-exposed airway of asthmatics. Numerous other well-established GPCR agonists, including, but not limited to, endothelin (acting on ET-1R on ASM to induce ASM contraction), thromboxane (TP receptor), prostaglandin E2 (PGE$_2$; acting on EP1, EP2, EP3, and EP4 receptor with variable effects), and adenosine (acting on A1, A2a, and A2b adenosine receptors with variable effects), are induced during allergic lung inflammation and act either directly or indirectly on ASM to cause bronchoconstriction [9, 11–15]. Moreover, recent studies in human lung suggest that, whereas EP2 receptors dominate mast cell stabilizing effects of PGE$_2$, EP4 receptors dominate bronchodilation [16]. Recent studies have also identified (in both murine and human ASM) various GPCRs on ASM with the capacity to regulate (either contract or relax) ASM cells, ASM tissue ex vivo, or airways in vivo [17]. Numerous GPCRs have been identified as capable of mediating relaxation of contracted ASM cells, with the β_2AR being the principal GPCR targeted in asthma management for over the last 50 years [9].

Additionally, certain GPCRs that have little or limited capacity to directly stimulate pro-contractile or pro-relaxant signaling can stimulate the production of autocrine or paracrine GPCR agonists that in turn directly regulate ASM contractile state. Some GPCR agonists acting on (non-ASM) resident or infiltrating airway cells can stimulate the local release of cytokines that regulate ASM contractile state. The complexity of this intercellular communication and regulation can be evidenced in a recent study by Bonvini et al., in which trypsin-activated protease-activated receptor-2 (PAR-2) on ASM cells gates *transient receptor potential vanilloid 4* (TRPV4) channels to release ATP into the extracellular space and activate P2XY purinergic channels on mast cells, which in turn release CysLT to activate CysLT1Rs on ASM and cause contraction [18].

Many of the abovementioned GPCRs, stimulated by increased endogenous levels of their cognate ligands, likely contribute in some degree to the ASM contraction and airway narrowing caused by allergic inflammation in asthma. With the exception of the m3mAChR (therapeutically targeted in asthma/COPD with the m2/m3 mAChR antagonist tiotropium [19, 20]), however, pharmacological blockade of these receptors is not sufficient to reverse acute bronchoconstriction and manage an acute asthma attack [21].

Other inflammatory agents that are typically not GPCR agonists promote increased ASM contraction by sensitizing ASM to other more direct contractile stimuli. Such agents include cytokines such as tumor necrosis factor alpha (TNF-α) and interleukin (IL)-1β, as well as transforming growth factor beta 1 (TGFβ1), IL-4, IL-13, and IL-8 [22–26]. The mechanisms by which this sensitization occurs are well established and primarily involve increasing the ASM cell's ability to contract to a given level of intracellular calcium that is released in response to contractile GPCR activation (i.e., *calcium sensitization*) [27–36].

An additional mechanism by which inflammatory agents increase ASM contraction is via their enhancement of cholinergic discharge in the airway [37–39]. Physiological contraction of ASM occurs via the release of acetylcholine from sympathetic nerves innervating ASM, to activate m3mAChRs on ASM [40, 41]. In animal models of allergic inflammation, particularly rodent models, it is clear that increased cholinergic discharge contributes significantly to the airway hyperresponsiveness (AHR) associated with allergic lung inflammation [41–43]. In human asthmatics, the relative contribution of

excessive cholinergic discharge to AHR is not well understood (and is not necessary for AHR to exist) but is likely important given the therapeutic effectiveness of m3mAChR antagonists in at least a subpopulation of asthmatics [19, 20].

1.3 Current Treatment of Acute Bronchoconstriction

The life-threatening nature of an acute asthmatic attack requires rapid reversal of airway narrowing via direct regulation of ASM contraction. To date, such treatment has been almost exclusively limited to short-acting beta-agonists (SABAs) targeting the β_2AR on ASM and to a lesser extent to m3mAChR antagonists. Because multiple pro-contractile agonists often exist in the inflamed, asthmatic airway, specific m3mAChR antagonism may not be sufficient to reverse bronchoconstriction; thus, SABAs are the drug of choice for an asthmatic attack. Beta-agonists relax contracted airway smooth primarily by antagonizing intracellular pathways that are stimulated by contractile agonists and their cognate receptors (see [44–46] for a comprehensive review). Thus, beta-agonists tend to be effective in reversing ASM contraction caused by variable, and multiple, contractile stimuli.

1.4 Prophylactic Management of Bronchoconstriction by Controlling Inflammation

Not surprisingly, limiting or preventing the airway inflammation that causes bronchoconstriction is an excellent, and preferred, approach to manage asthma. For decades, inhaled corticosteroids have been the primary drug of choice for the management of mild asthma (with SABA inhalation when needed) [47]. However, majority of asthmatics today are managed by prophylactic, maintenance drugs consisting of an inhaled GPCR ligand (β_2AR agonist or m3mAChR antagonist) that directly targets ASM contraction, in a formulation that combines an inhaled steroid. Initially, the combination of an inhaled long-acting beta-agonist (LABA) salmeterol along with an inhaled steroid (fluticasone) dominated the market. However, over time, other combinations of LABAs plus steroids along with long-acting muscarinic receptor antagonists (LAMAs) combined with steroids emerged, although the latter combination was and is more applicable to COPD management. The prevention/mitigation of inflammation by steroids has remained the cornerstone of the combination approaches, while the addition of more direct bronchorelaxation agents represents an additional arm of prophylactic control that could translate into better disease management including reduced exacerbations [48–50].

1.5 Control of Allergic Airway Inflammation by Steroids: Why We Keep Using the Sledgehammer Instead of a Scalpel

For the majority of asthmatics, inhaled corticosteroids are effective in controlling lung inflammation, despite the fact that they are often administered during ongoing inflammation and not necessarily prophylactically [51]. Some asthmatics, including many severe asthmatics, are *steroid-resistant* (to both inhaled and oral steroids; the cause of which is a subject of intense research and debate) and consequently have difficulty managing their disease [52–54]. Notwithstanding steroid-resistant asthmatics, steroids are powerful in suppressing allergic inflammation based on their ability to greatly suppress the pro-inflammatory function of multiple resident and infiltrating cell types in the lung and halt the progression of inflammation at multiple steps [47, 55–57]. The principal mechanism of these actions of corticosteroids involves the glucocorticoid receptor-mediated suppression of inflammatory mediator genes and to a lesser extent glucocorticoid receptor-mediated induction of other genes [47, 56, 58–65]. However, although this "sledgehammer" effect on gene regulation and multiple cell functions typically results in effective control of allergic lung inflam-

mation, the effects of steroids are not exclusively anti-inflammatory. Numerous "off-target" effects of steroids exist, contributing to various clinical side effects associated with both inhaled and oral steroid use [53, 65]. The complement of steroid actions on inflammatory processes includes antithetical/counterproductive actions that limit therapeutic efficacy (i.e., certain anti-inflammatory processes maybe inhibited, along with pro-inflammatory processes being stimulated) [66, 67]. Ideally, a more refined approach that avoids off-target and antithetical effects would be preferable, *if* such targeting could *sufficiently* reduce inflammation to manage asthma features. Until recently, however, a more precise targeting of inflammatory mechanisms that drive asthma pathobiology did not exist. Even with the early promising results of certain biologics (typically antibodies targeting specific inflammatory mediators including IgE, IL-5, IL-5Rα, IL-4Rα, and IL-13) [68, 69], it is unclear whether targeting specific cytokines or inflammatory steps will be a superior or equivalent means of managing inflammation when compared with steroids for the majority of asthmatics. Moreover, the administration of biologics is both difficult (typically *s.c.* or *i.v.* injection) and expensive and justified only in severe asthmatics whose asthma is otherwise unmanageable [70, 71]. For now, the sledgehammer remains the most effective tool for the most asthmatics. Is there another solution?

1.6 Limitations of GPCR Ligands in the Control of Asthma and Inflammation

Arguably, the optimal asthma drug would be able to both prevent and neutralize inflammation while also directly inhibiting ASM contraction. To date, this drug does not exist, although as discussed recent studies have demonstrated the ability of certain GPCRs to regulate both ASM contraction and inflammation in murine models of asthma. CysLT1R antagonists have the capacity to both inhibit ASM contraction and airway inflammation, but their efficacy as a direct bronchodilator is minimal and as an anti-inflammatory agent is limited, perhaps being best in those patients whose lung inflammation is driven significantly by CysLT generation in the airway. Given the absence of any GPCR possessing efficacy as both a bronchodilator and anti-inflammatory, combination therapies of (GPCR ligand bronchodilator) LABAs/LAMAs and steroids are currently required for sufficient asthma control of both inflammation and bronchospasm and are overwhelming prescribed for most asthmatics.

However, this does not mean that inflammation control in asthma by GPCR pharmacology is not possible. Because multiple GPCRs participate in inflammation development, including many only recently appreciated receptors and others with but a nascent pharmacology, it is conceivable that the targeting of certain GPCRs may be sufficient to manage allergic lung inflammation. In addition, unlike biologics, GPCR agonists are small molecular drugs with inherent properties favorable for drug development, administration, and patient adherence [4, 72].

1.7 Recently Appreciated GPCRs Whose Targeting May Help in Managing Allergic Airway Inflammation

Advances in basic science capabilities in molecular and cell biology and receptor pharmacology have aided the discovery of previously unappreciated roles of various GPCRs in airway and asthma biology. Although many of these recent studies stemmed from attempts to find novel bronchodilators, a by-product of these studies has been the discovery of novel mechanisms by which GPCRs regulate allergic lung inflammation. Below, we discuss each of these GPCRs implicated in inflammation control and the potential of these receptors as asthma therapeutic targets.

1.8 Bitter Tastant Receptors

Evolutionarily bitter taste receptor signaling evolved as a mechanism to avoid potentially toxic food often bitter in taste. This function was primarily imparted by the bitter taste receptors (belonging to type II taste receptors, TAS2R), a family of seven transmembrane GPCRs expressed on the taste buds [73]. In the gastrointestinal system, these highly specialized chemosensory cells contribute to the innate host defense mechanism [74, 75]. It is now well established that TAS2Rs are expressed on variety of cell-types including ASM cells in the airways [76]. The first evidence to demonstrate the beneficial effect of bitter taste receptor signaling in providing effective bronchorelaxation was shown using human ASM cells where agonists of TAS2Rs were able to induce localized calcium and reverse airway obstruction to contractile agonists [76]. These observations were then verified in other species, and soon, it was established and recognized that TAS2R agonism may be a viable target to promote airway relaxation. Studies also demonstrated that the beneficial signaling activated by TAS2Rs in the airway smooth muscle was distinct and was not reliant on protein kinase A (PKA) activation, unlike β2 agonists [76–78]. Moreover, chronic treatment with TAS2R ligands does not lead to receptor desensitization in ASM, thereby preserving the beneficial bronchorelaxation effect [79]. Since the original characterization of bitter tastant receptors in the lung and ASM, it has been established that bitter tastants can provide effective bronchodilatory effects by promoting relaxation of ASM in multiple species and in animal models of asthma [76, 79–83]

It is also now apparent that bitter tastant receptor ligands can also mitigate other pathological features of asthma such as airway inflammation and airway remodeling, thereby providing a comprehensive asthma control [83]. The beneficial effects of bitter ligands have been shown in both prophylactic and treatment models where these agents acted on multiple levels in asthma pathology and prevented allergen-induced influx of immune cells into the airways and blocked key inflammatory cytokines that drive asthma pathogenesis that leads to airway remodeling and AHR [83]. Though bitter tastants are potentially an effective alternative to beta-agonists in terms of their bronchodilatory effects, it still remains to be seen whether these agents will be safe and equally effective in the clinic as the biggest challenge in their development is the identification of a specific TAS2R subtype that is highly relevant in asthma and translation of preclinical studies to humans [84].

1.9 Calcium-Sensing Receptor

The calcium-sensing receptor (CaSR) is best known for its role in regulating calcium homeostasis in the body. CaSRs on the parathyroid gland survey circulating calcium levels, which involves calcium binding to and activating the CaSR which initiates intracellular signals that suppress the release of parathyroid hormone. Interestingly, the CaSR can be activated by numerous other molecules, including polyvalent cations, amino acids, and virus elements, and is expressed on multiple cell types, including those in the lung. In Yarova et al., a prominent role of CaSRs in mediating the development of the asthma phenotype was revealed. The capacity of CaSRs to promote ASM contraction was demonstrated by a loss of CaSR-stimulated contraction in ASM cells and tissue in which the CaSR gene was ablated and by pretreatment with CaSR antagonists known as calcilytics. Importantly, calcilytics could also reverse the hyperresponsiveness and inflammation induced in vivo in a mixed allergen model of murine asthma. Relevance of CaSRs to human asthma was suggested by data demonstrating expression of CaSR in human ASM, with greater levels observed in ASM from asthmatics. The ability of CaSRs to regulate inflammation is likely due to its expression on invading inflammatory cells (eosinophils, macrophages). Interestingly, inflammation was shown to increase CaSR expression in both human and mouse tissues [85].

One of the most intriguing aspects of CaSR in asthma is that (CaSR antagonists) calcilytics have real potential as asthma drugs. They are small molecules that are readily deliverable by inhalation, and their efficacy is favored by the ability to target multiple cell types and mechanisms that contribute to the asthma phenotype.

Moreover, various calcilytics have already undergone clinical trials for safety and efficacy in diseases such as osteoporosis and autosomal dominant hypocalcemia (reviewed in [86]). Thus, despite the promiscuous nature of the CaSR, it might ultimately prove to be a useful asthma therapeutic target.

1.10 EP Receptors

The EP receptor family is activated by the ubiquitous inflammatory agent prostaglandin E2 and to a lesser extent other prostanoids [87]. The four members of the EP receptor family (EP1, EP2, EP3, and EP4) couple to different G proteins, signal to different pathways, and have variable expression in multiple cell types in the lung. With respect to allergic lung inflammation and asthma, PGE_2 through EP receptors thus regulates multiple cellular functions that serve different and often competing functions that control both inflammation and ASM contraction. For example, in humans, EP3 receptors in ASM cause contraction, whereas EP4 receptors mediate ASM relaxation [88]. Control of inflammation by EP receptors is complex. The net effect on PGE_2 activating multiple EP receptor subtypes in the allergen-challenged mice demonstrated that EP2, EP3, and EP4 agonists all could inhibit certain indices of allergen-induced inflammation in mice lacking mPGES [89]. In another study, employing three different airway disease models (including a more chronic ovalbumin (OVA) sensitization/challenge), in each of the EP knockout mice, demonstrated that the EP4R (and not EP1–EP3) was responsible for the anti-inflammatory effect of PGE_2 in each model [90]. Collectively, studies to date suggest that PGE_2 is largely beneficial, with the capacity to inhibit many pathological features of asthma. In both animal models and cell-based assays, PGE_2 inhibits multiple indices of allergic inflammation [89–93], inhibits proliferation of cultured ASM cells [94–99], and relaxes contracted airways ex vivo [100–103] while promoting bronchoprotection/bronchorelaxation in vivo [101, 103]. Moreover, these effects are conserved across species and most importantly are evident in human subjects.

The *potential* of PGE$_2$ as an asthma therapeutic, at least with respect to its bronchorelaxant properties, has been recognized for years. The bronchodilator effects of PGE$_2$ have been demonstrated in a range of patients (normal, asthmatic, and chronic bronchitis) [104]. This effect has been shown in other studies using healthy and asthmatic subjects, respectively [105, 106]. PGE$_2$ also protects against exercise-induced [105] and aspirin-induced bronchoconstriction in subjects with aspirin-exacerbated respiratory disease (AERD) [107, 108]. PGE$_2$ also prevents early and late allergen-induced bronchoconstrictor responses when given before allergen challenge [109, 110] and is protective against bronchoconstrictors such as histamine and methacholine [106, 111]. With respect to inflammation, PGE$_2$ also blocks the recruitment of eosinophils and basophils to the bronchial mucosa during allergen-induced late-phase responses and attenuates the release of mast cell-derived products. Thus, PGE$_2$ has validated functions as an anti-inflammatory and bronchoprotective agent in asthmatics [110, 112] In that regard, it is most established among potential asthma therapeutics for its ability to directly bronchodilate and to suppress inflammation.

However, despite these benefits of inhaled PGE$_2$, the development of prostanoid agonists for the treatment of asthma has been hindered as inhaled PGE$_2$ has repeatedly been shown to produce reflex cough in humans [104, 112, 113]. PGE$_2$ has been shown to excite airway afferent nerves [114], which concurs with the cough seen in both healthy and asthmatic patients during studies with inhaled PGE$_2$. Recent studies using cell, tissue, and in vivo models strongly implicate the EP3 receptor subtype in mediating cough induced by PGE$_2$ across all species tested, including human [115–120].

Clearly, the answer to harnessing the pro-relaxant and anti-inflammatory properties of PGE$_2$ as an asthma treatment or prophylaxis relies on specific targeting of EP receptor subtypes, with a primary goal of avoiding EP3 receptor agonism. Unfortunately, the development of ligands with sufficiently high specificity for each

of the EP subtypes has been difficult to date. Moreover, this solution first requires a clear understanding of the role of EP receptor subtypes in the many cell types participating in airway inflammation and pathobiology in asthma. In airway epithelial cells, PGE$_2$ modulates many functions including an increase in ciliary beat frequency [121] and Cl$^-$ channel conductance [122], whereas both induction [123] and inhibition [124] of mucin production have been reported. In vivo administration of EP agonists has been shown to inhibit LPS- (EP2) and allergen-induced (EP2/EP3/EP4) mucous cell metaplasia in rat nasal epithelium [124, 125]. EP receptors also play an immunomodulatory role in the epithelium; EP2/EP4 agonists increase IL-6 release [126], whereas EP4 inhibits IL-8 release [127]. A recent report shows that human airway epithelial (HAE) cell migration is promoted by PGE$_2$ and selective EP agonists (EP1–EP4), but upon undergoing TGFβ-induced Epithelial mesenchymal transition (EMT), the response to PGE$_2$ and EP2 and EP4 agonists becomes inhibitory, indicating adaptation of EP responses to remodeling in the lung [128].

Human mast cells have been shown to express EP2, EP3, and EP4 receptors. PGE$_2$ suppressed the generation of cytokines and cysteinyl leukotrienes primarily by eliciting signaling through EP2 receptors (although a suppressive effect was evident at high doses) [91]. Others have noted regulatory effects of PGE$_2$ on other inflammatory cell types including eosinophils [129], T cells [130], T regs [131], alveolar macrophages [132], and neutrophils [133] using cells from various species. Whereas some of these studies have suggested that EP2 and EP4 are the principal EP subtypes capable of inhibiting the inflammatory functions of these cells, definitive insight has been limited [134] due the limitations of subtype selective ligands.

Collectively, the above studies all point to a high potential of EP receptor subtype targeting for regulating both bronchospasm and airway inflammation in asthma. One caveat is that most of our understanding of EP receptor subtype function in the lung and lung cells during allergic inflammation comes from animal studies, and (1) the nature of, and control of, of allergen-induced inflammation in animals (particularly mice) differs (often significantly) from that of humans, and (2) species differences in cellular EP receptor subtype function have been identified, one striking difference being in the control of ASM contraction/relaxation [118]. However, the current pace of research in this area is encouraging, supported by ongoing industry efforts in EP subtype-selective drug discovery [135] aided by the increasing ability of both structural biology and receptor modeling science to enhance these efforts.

1.11 Protease-Activated Receptor 2 (PAR2)

PAR2 signaling in the lung has complex effects reflecting the diverse functions of the various lung cell types that express PAR2. Initial studies demonstrated that airway delivery of anti-PAR2 antibodies, or a cell permeable peptide inhibitor of PAR2 signaling, prevented allergen-induced AHR and airway inflammation in mice [136]. Moreover, using a mouse OVA model for PAR(2)-modulated airway inflammation, genetic ablation of β-arrestin2 (βarr2) decreased leukocyte recruitment, cytokine production, and mucin production in OVA-treated mice, yet PAR(2)-mediated PGE_2 production and the associated and decreased baseline airway resistance were unaffected by βarr2 knockout. Subsequent studies using OVA, cockroach extract, or *Alternaria alternata* to induce lung inflammation further confirmed a protective effect of PAR2 activity acting via PGE_2-mediated relaxation of ASM and a pathogenic effect of PAR2 mediated through PAR2 activation (and dependent on β-arrestin2) most likely on inflammatory cells [137, 138]. Collectively, these studies suggest that PAR2 is a potential useful GPCR target for controlling allergic lung inflammation, and a strategy enabling specific targeting of PAR2-β-arrestin2 signaling would be optimal, enabling the protective effect of PAR2-PGE_2 signaling axis to be retained.

1.12 Any Finally … the $β_2$AR? How Biased Agonism Pharmacology May Turn a Problem into a Solution

The inability of current beta-agonists to control inflammation almost certainly underlies the need to treat asthmatics with steroids. Although a handful of studies examining the effects in cell-based models have attributed anti-inflammatory properties to beta-agonists [139–145], there is little if any evidence that beta-agonist use reduces allergic inflammation in the lung [146, 147]. The cooperative effect of LABA and steroids in managing asthma is likely a function of each drug addressing an individual disease feature: LABA prophylactic inhibition of airway constriction (bronchoprotection) and steroid prophylactic inhibition of inflammation. Certain studies have identified some interesting cooperative mechanisms at the cellular level [48, 58, 148–150], but the predominate means by which these two drugs work together well appears to be that beta-agonists directly manages ASM contraction, while steroids limit inflammation. Both the cause (inflammation) and the effect (bronchoconstriction) are addressed to the extent that synergy at the cellular level is probably not required for this combination to be effective in most asthmatics.

A long history of safety issues also suggests that beta-agonists alone are inadequate to manage the disease in many asthmatics, perhaps due to the inability to control inflammation. "Epidemics" in which beta-agonist use was associated with high levels of asthma mortality occurred in the 1960s with high dose of inhaled isoproterenol (a nonspecific beta-agonist) use in the United Kingdom, followed by the use of a potent SABA, fenoterol, which also increased asthma-related mortality in New Zealand [151–157]. In the USA, a statistically insignificant increase in asthma-related deaths and safety concerns were reported with the use of LABAs. It was later noted that these life-threatening adverse events with the use of LABA therapy were limited by suboptimal study designs. Follow-up prospective clinical trials and meta-analyses consistently demonstrated their effectiveness and safety, although these trials did not contain sufficient power to address these safety concerns definitively [158, 159]. Moreover, clinical data further suggest that chronic beta-agonist use is associated with a reduction in bronchoprotective effect [160, 161], increase in AHR [162], and loss of asthma control [162–164]; whether this is due to a loss of effectiveness of beta-agonists caused by receptor desensitization or a failure to address the underlying cause (inflammation) of the disease is unclear. However, when the most recent (and highly publicized) beta-agonist safety issue arose during the SMART trial examining the efficacy of salmeterol monotherapy, it was widely speculated that the significant ability of salmeterol to relax ASM *masked* its inability to control airway inflammation [165, 166], thus leaving patients significantly at risk.

pathogenesis with β_1AR antagonism proved to be therapeutically beneficial [170].

In a series of elegant studies, Bond and colleagues demonstrated (highlights summarized in Fig. 1.2) that pharmacological blockade, or genetic ablation of the β_2AR or systemic epinephrine (the only endogenous ligand for the β_2AR), improved the asthma phenotype and that agonist-induced activation of the β_2AR was necessary for full development of the lung inflammation and the asthma phenotype [171–177]. Moreover, results from a pilot clinical trial determined that, in 8 out of 10 mild asthmatics, 9 weeks of treatment with the "beta-blocker" nadolol produced a significant, dose-dependent increase in PC20 as well as a significant reduction in FEV1 [178, 179].

1.13 Evidence that Beta-Agonists and the β_2AR Actually Promote Inflammation and Asthma Development

There is a paucity of meaningful clinical data assessing the effects of beta-agonist use on inflammation in asthma. A handful of underpowered clinical studies suggest limited anti-inflammatory effects of beta-agonists in reducing IL-8, eosinophils, mast cells in Bronchoalveolar lavage (BAL), and airway mucosa samples after LABA therapy [167–169]. In contrast, a considerable amount of research into the effects of beta-agonists and the β_2AR in murine models of allergic lung inflammation exists. These studies were pioneered by the Bond lab, who initially proposed that, as in heart failure, a "βAR paradox" exists for asthma. With heart failure, a loss of pump function occurs as cardiac myocytes and their β_1ARs become less responsive to endogenous norepinephrine. Although, theoretically, the use of exogenous beta-agonists as therapy might help stimulate the hypodynamic heart, it turns out that βAR *blockers* are the more effective treatment. This is because, as ultimately determined, β_1AR activation is actually a pathogenic driver of heart failure, and preventing this β_1AR-driven

1.14 Was It as Simple as β_2AR Signaling Was Actually Bad and Fostered Asthma Development?

Not exactly. During this time, the GPCR biology field discovered that certain GPCRs, including β_2ARs, were capable of stimulating diverse and sometimes functionally antithetical signaling pathways, and most of these signaling events were dependent on arrestin proteins [180, 181]. Arrestins were originally identified as regulatory proteins that promote β_2AR desensitization but were subsequently found capable of mediated G protein-independent signaling by binding the β_2AR and helping form a distinct signalosome. Strategies that inhibited the ability of arrestin proteins to bind to the β_2AR (e.g., arrestin knockdown) could inhibit specific signaling (e.g., ERK1/ERK2 signaling) while preserving other signaling (cAMP/PKA) [182]. In addition, drugs classically known as βAR antagonists, based on their ability to block βAR-stimulated cAMP production, could cause arrestin recruitment to βARs and stimulate signals that appeared independent of G proteins [182, 183]. These studies ushered in the exciting new age of "biased ligand pharmacology," and the race was on to develop new drugs that could "tune" receptors to preferen-

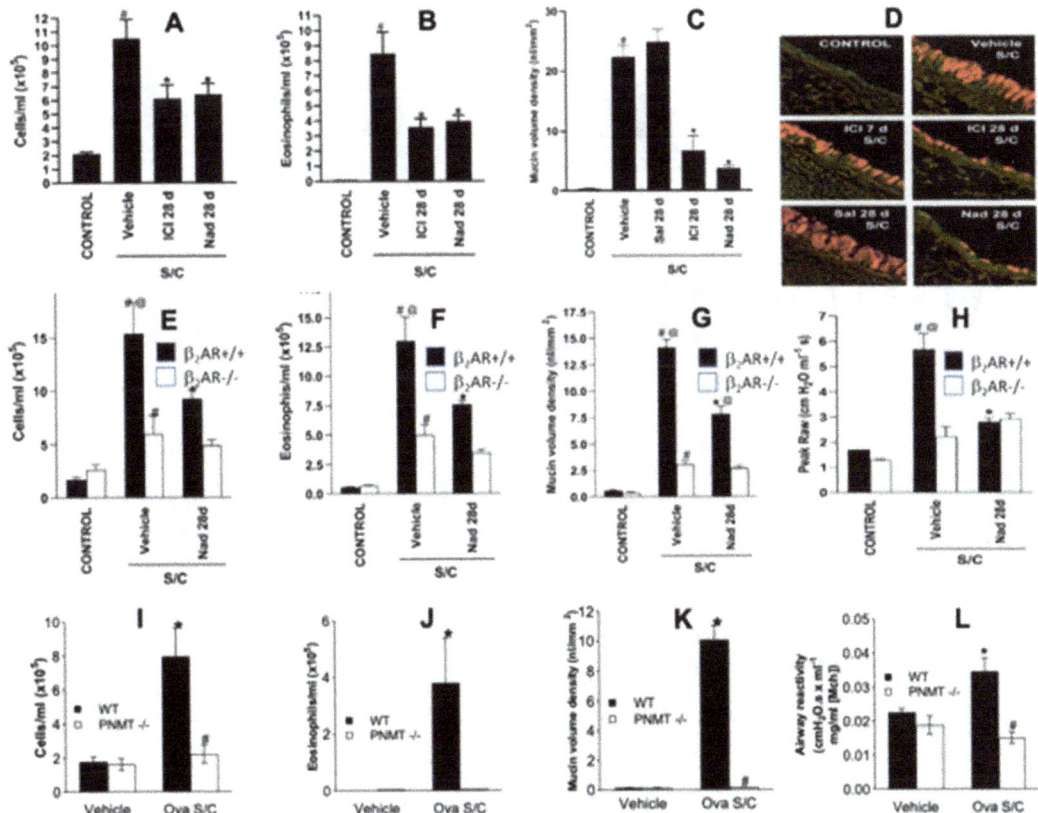

Fig. 1.2 Pharmacological, genetic inhibition of β_2AR signaling improves the asthma phenotype. *A-D.* In the classic ovalbumin (OVA) model of murine allergic lung inflammation, chronic co-administration of inverse agonists ICI118551 and nadolol, both of which block both canonical β_2AR signaling as well as arrestin binding to β_2AR, inhibits the OVA-induced increases in BALF total cellularity (**a**), eosinophils (**b**), and mucin (**c**). Moreover, chronic salbutamol treatment augments OVA-induced mucin (**c**). *D.* Periodic Acid-Schiff-stained airways from mice treated as per (**c**). *E-F.* Genetic ablation of the β_2AR similarly inhibits OVA-induced increases in lung cellularity (**e**), eosinophils (**f**), and mucin (**g**), as well as increases in methacholine-stimulated airway resistance (**h**). *I-J.* Genetic ablation of phenylethanolamine N-phenylethanolamine N-methyltransferase (PMNT, the enzyme catalyzing the final step in the synthesis of epinephrine) similarly inhibits OVA-induced increases in lung cellularity (**i**), eosinophils (**j**), and mucin (**k**), and methacholine-stimulated airway resistance (**h**). Data from Nguyen et al. *Am J Respir Cell Mol Biol* 2006 (**a–d**), Nguyen et al. Proc Natl Acad Sci USA 2009 (**e–h**), and Thanawala et al. *Am J Respir Cell Mol Biol* 2013 (**i–j**)

tially activate specific pathways, instead of simply turning receptors on or off.

Intrigued by this fledgling field of biased ligand pharmacology, Bond and colleagues considered whether the permissive effect of beta-agonists on asthma pathobiology was linked to arrestin effects on β_2ARs and possible arrestin-dependent signaling. The modest albeit encouraging effect of nadolol in the clinical pilot study led them to consider whether the use of nadolol,

which blocks both G protein and arrestin-dependent signaling by the β_2AR, might be to some extent "throwing the baby out with the bath water." *Might a better strategy be to preserve the β_2AR-Gs-cAMP-PKA signaling while blocking arrestin-dependent effects?* They therefore launched a series of studies that demonstrated that certain "beta-blockers" (e.g., carvedilol, propranolol) that blocked β_2AR-Gs signaling yet failed to inhibit β_2AR-arrestin binding (and stimulated

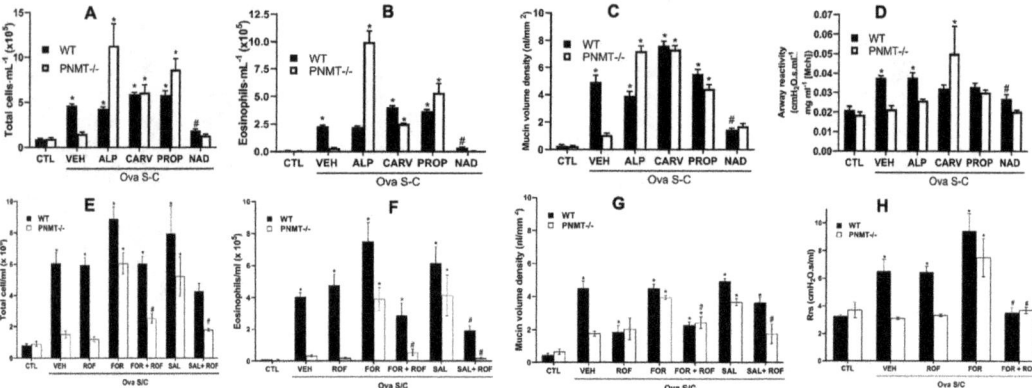

Fig. 1.3 Asthma pathology is influenced by the nature, degree of biased β_2AR signaling. *A-D*. Systemic depletion of epinephrine caused by PMNT deletion inhibits OVA-induced BALF total cellularity (**a**), eosinophils (**b**), mucin (**c**), and airway hyperresponsiveness (AHR) (**d**), while replacement of systemic epinephrine with a balanced β_2AR partial agonist alprenolol (ALP) or arrestin-biased carvedilol (CAR) or propranolol (PROP) helps restore each of these features of asthma. Conversely, nadolol (NAD), which blocks both β_2AR-Gs signaling as well as arrestin binding to the β_2AR does not. *E-F*. Replacement of systemic epinephrine with bal-anced β_2AR agonists formoterol (FOR) or salmeterol (SAL) in PMNT −/− mice restores OVA-induced increases in lung cellularity (**e**), eosinophils (**f**), and mucin (**g**), as well as increases in methacholine-stimulated airway resistance (**h**). Moreover, co-treatment with the PDE4 inhibitor roflumilast (ROF), which increases β_2AR-Gs-stimulated cAMP accumulation, partially reverses the deleterious effects of chronic treatment with balanced agonists formoterol and salmeterol on total lung cellularity, eosinophils, mucin, and AHR. Data from Thanawala et al. *Br J Pharmacol* 2015 (**a–d**) and Forkuo et al. *Am J Respir Cell Mel Biol* 2016 (**e, f**)

ERK1/ERK2 signaling in 293 cells as per [182]) (i.e., arrestin-biased ligands) were not effective in mitigating allergen-induced inflammation and AHR and in fact exacerbated the condition, especially in PMNT−/− (epinephrine-deficient) mice [177] (Fig. 1.3a–d). Without a Gs-biased β_2AR ligand in hand, Bond and colleagues biased signaling toward the Gs/cAMP/PKA pathway through use of phosphodiesterase (PDE) inhibitors which specifically increased intracellular cAMP in lung cells. Both PDE4 inhibitors roliram (not shown) and roflumilast (Fig. 1.3e–h) significantly reversed the adverse effects of formoterol and salmeterol on allergen-induced bronchoalveolar lavage fluid (BALF) cellularity and eosinophils, mucin, and AHR in both wild-type and PMNT−/− mice [173].

Additional studies demonstrated that the loss of IL-13-induced asthma phenotype caused by global genetic ablation of β_2AR was rescued by transgenic expression of the β_2AR in only airway epithelia [184], suggesting that β_2AR-arrestin signaling regulates the immunomodulatory function of airway epithelia and is critical to the development of the asthma phenotype. Should future studies clarify this as true, the two critical questions that remain are as follows: (1) is this the case with human asthma and (2) can Gs-biased β_2AR agonists or allosteric modulators that are both effective and safe in asthmatics be developed? To date (unlike arrestin-biased β_2AR ligands), identifying or developing such drugs has been a challenge, but with the rapid advances in structural biology and computer modeling that enable drug development, it is only matter of time before such drugs are known or developed.

1.15 What the Future Holds: The Promise of Biased Agonism Pharmacology

To date, the failure to identify a GPCR target, and a therapeutic ligand for it, capable of successfully managing allergic lung inflammation in human stems from the limitations of GPCR biology and pharmacology in basic research and the limitations of drug discovery. GPCR biology and phar-

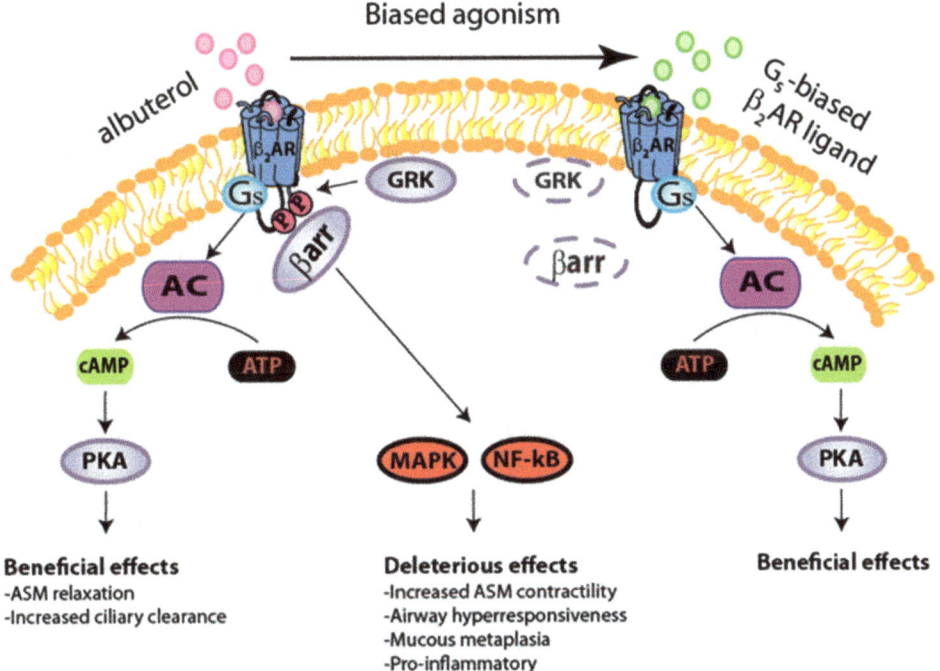

Fig. 1.4 Biased agonism pharmacology in ASM. Illustration of signaling mechanisms using principles of biased agonism to promote beneficial therapeutic effects in asthma

macology research in asthma is currently exploding, aided by increasing powerful genetic, molecular, cell biology, and computational tools and by the increasing rate of drug discovery, itself aided by advances in structural biology, computer modeling, and better screening tools and strategies. Moreover, the recent appreciation of qualitative signaling properties by GPCRs, and the realization that receptors can be "tuned" and not just turned "off" or "on" to specifically activate the beneficial therapeutic signaling a receptor can transduce, suggests that biased agonism pharmacology will develop the drugs we need to optimally control airway inflammation and asthma, as shown in Fig. 1.4 [185]. In addition, because GPCR ligands are small molecules and can be continuously refined to improve specificity of targeting while minimizing off-target effects, we should ultimately have an asthma therapy that supplants steroids and has a superior efficacy and safety profile.

1.16   Why Do We Care About Developing GPCR Ligands Capable of Managing Allergic Lung Inflammation?

As mentioned above, the properties of small-molecule GPCR ligands make them attractive therapies. In addition, it would be advantageous to control inflammation without the numerous side effects associated with corticosteroid treatment. The major issue remains whether control of inflammation by such drugs is truly sufficient to manage allergic inflammation and whether such a scalpel can do the job currently performed by a sledgehammer.

Acknowledgments Dr. Penn is supported by National Institutes of Health (NIH) grants HL58506, AI110007, AI161296, HL136209, HL145392, and HL114471. The authors thank Dr. Richard Bond for permission to utilize previously published figures.

References

1. Kaminsky DA. What does airway resistance tell us about lung function? Respir Care. 2012;57(1):85–96. discussion -9
2. Porsbjerg C, Rasmussen L, Nolte H, Backer V. Association of airway hyperresponsiveness with reduced quality of life in patients with moderate to severe asthma. Ann Allergy Asthma Immunol. 2007;98(1):44–50.
3. Cisneros C, Garcia-Rio F, Romera D, Villasante C, Giron R, Ancochea J. Bronchial reactivity indices are determinants of health-related quality of life in patients with stable asthma. Thorax. 2010;65(9):795–800.
4. Wendell SG, Fan H, Zhang C. G protein-coupled receptors in asthma therapy: pharmacology and drug action. Pharmacol Rev. 2020;72(1):1–49.
5. Billington CK, Penn RB. Signaling and regulation of G protein-coupled receptors in airway smooth muscle. Respir Res. 2003;4(1):2.
6. Penn RB. Embracing emerging paradigms of G protein-coupled receptor agonism and signaling to address airway smooth muscle pathobiology in asthma. Naunyn Schmiedeberg's Arch Pharmacol. 2008;378(2):149–69.
7. Billington CK, Penn RB. m3 muscarinic acetylcholine receptor regulation in the airway. Am J Respir Cell Mol Biol. 2002;26(3):269–72.
8. Deshpande DA, Penn RB. Targeting G protein-coupled receptor signaling in asthma. Cell Signal. 2006;18(12):2105–20.
9. Pera T, Penn RB. Bronchoprotection and bronchorelaxation in asthma: new targets, and new ways to target the old ones. Pharmacol Ther. 2016;164:82–96.
10. Sharma P, Ghavami S, Stelmack GL, McNeill KD, Mutawe MM, Klonisch T, et al. Beta-Dystroglycan binds caveolin-1 in smooth muscle: a functional role in caveolae distribution and Ca2+ release. J Cell Sci. 2010;123(Pt 18):3061–70.
11. Pera T, Penn RB. GPCRs in airway smooth muscle function and obstructive lung disease. In: Trebak M, Early S, editors. CRC methods and signaling transduction – experimental approaches for signal transduction of smooth muscles. Boca Raton: CRC Press; 2019. in press.
12. Wilson CN, Nadeem A, Spina D, Brown R, Page CP, Mustafa SJ. Adenosine receptors and asthma. Handb Exp Pharmacol. 2009;193:329–62.
13. Druey KM. Regulation of G-protein-coupled signaling pathways in allergic inflammation. Immunol Res. 2009;43(1–3):62–76.
14. Capra V. Molecular and functional aspects of human cysteinyl leukotriene receptors. Pharmacol Res. 2004;50(1):1–11.
15. Brown RA, Spina D, Page CP. Adenosine receptors and asthma. Br J Pharmacol. 2008;153(Suppl 1):S446–56.
16. Safholm J, Manson ML, Bood J, Delin I, Orre AC, Bergman P, et al. Prostaglandin E inhibits mast cell-dependent bronchoconstriction in human small airways through the E prostanoid subtype 2 receptor. J Allergy Clin Immunol. 2015;136(5):1232–9.e1.
17. Wright D, Sharma P, Ryu MH, Risse PA, Ngo M, Maarsingh H, et al. Models to study airway smooth muscle contraction in vivo, ex vivo and in vitro: implications in understanding asthma. Pulm Pharmacol Ther. 2013;26(1):24–36.
18. Bonvini SJ, Birrell MA, Dubuis E, Adcock JJ, Wortley MA, Flajolet P, et al. Novel airway smooth muscle-mast cell interactions and a role for the TRPV4-ATP axis in non-atopic asthma. Eur Respir J. 2020;56(1)
19. Aalbers R, Park HS. Positioning of long-acting muscarinic antagonists in the management of asthma. Allergy Asthma Immunol Res. 2017;9(5):386–93.
20. Murphy KR, Chipps BE. Tiotropium in children and adolescents with asthma. Ann Allergy Asthma Immunol. 2020;124(3):267–76.e3.
21. Chari VM, McIvor RA. Tiotropium for the treatment of asthma: patient selection and perspectives. Can Respir J. 2018;2018:3464960.
22. Ohta Y, Hayashi M, Kanemaru T, Abe K, Ito Y, Oike M. Dual modulation of airway smooth muscle contraction by Th2 cytokines via matrix metalloproteinase-1 production. J Immunol. 2008;180(6):4191–9.
23. Ding S, Zhang J, Yin S, Lu J, Hu M, Du J, et al. Inflammatory cytokines tumour necrosis factor-alpha and interleukin-8 enhance airway smooth muscle contraction by increasing L-type Ca(2+) channel expression. Clin Exp Pharmacol Physiol. 2019;46(1):56–64.
24. Ojiaku CA, Cao G, Zhu W, Yoo EJ, Shumyatcher M, Himes BE, et al. TGF-beta1 evokes human airway smooth muscle cell shortening and Hyperresponsiveness via Smad3. Am J Respir Cell Mol Biol. 2018;58(5):575–84.
25. Kaur D, Gomez E, Doe C, Berair R, Woodman L, Saunders R, et al. IL-33 drives airway hyperresponsiveness through IL-13-mediated mast cell: airway smooth muscle crosstalk. Allergy. 2015;70(5):556–67.
26. Amrani Y, Krymskaya V, Maki C, Panettieri RA Jr. Mechanisms underlying TNF-alpha effects on agonist-mediated calcium homeostasis in human airway smooth muscle cells. Am J Phys. 1997;273(5):L1020–8.
27. Koopmans T, Anaparti V, Castro-Piedras I, Yarova P, Irechukwu N, Nelson C, et al. Ca2+ handling and sensitivity in airway smooth muscle: emerging concepts for mechanistic understanding and therapeutic targeting. Pulm Pharmacol Ther. 2014;29(2):108–20.
28. Penn RB, Benovic JL. Regulation of heterotrimeric G protein signaling in airway smooth muscle. Proc Am Thorac Soc. 2008;5(1):47–57.
29. Somlyo AP, Somlyo AV. Ca2+ sensitivity of smooth muscle and nonmuscle myosin II: modulated by G

proteins, kinases, and myosin phosphatase. Physiol Rev. 2003;83(4):1325–58.

30. Somlyo AP, Himpens B. Cell calcium and its regulation in smooth muscle. FASEB J. 1989;3(11):2266–76.

31. Gong MC, Fujihara H, Somlyo AV, Somlyo AP. Translocation of rhoA associated with Ca2+ sensitization of smooth muscle. J Biol Chem. 1997;272(16):10704–9.

32. Croxton TL, Lande B, Hirshman CA. Role of G proteins in agonist-induced Ca2+ sensitization of tracheal smooth muscle. Am J Phys. 1998;275(4):L748–55.

33. Gosens R, Schaafsma D, Grootte Bromhaar MM, Vrugt B, Zaagsma J, Meurs H, et al. Growth factor-induced contraction of human bronchial smooth muscle is Rho-kinase-dependent. Eur J Pharmacol. 2004;494(1):73–6.

34. Schaafsma D, Boterman M, de Jong AM, Hovens I, Penninks JM, Nelemans SA, et al. Differential Rho-kinase dependency of full and partial muscarinic receptor agonists in airway smooth muscle contraction. Br J Pharmacol. 2006;147(7):737–43.

35. Deshpande DA, White TA, Dogan S, Walseth TF, Panettieri RA, Kannan MS. CD38/cyclic ADP-ribose signaling: role in the regulation of calcium homeostasis in airway smooth muscle. Am J Physiol Lung Cell Mol Physiol. 2005;288(5):L773–88.

36. Lanner JT, Georgiou DK, Joshi AD, Hamilton SL. Ryanodine receptors: structure, expression, molecular details, and function in calcium release. Cold Spring Harb Perspect Biol. 2010;2(11):a003996.

37. Gosens R, Zaagsma J, Meurs H, Halayko AJ. Muscarinic receptor signaling in the pathophysiology of asthma and COPD. Respir Res. 2006;7:73.

38. Fryer AD, Jacoby DB. Muscarinic receptors and control of airway smooth muscle. Am J Respir Crit Care Med. 1998;158(5 Pt 3):S154–60.

39. Coulson FR, Fryer AD. Muscarinic acetylcholine receptors and airway diseases. Pharmacol Ther. 2003;98(1):59–69.

40. Roffel AF, Davids JH, Elzinga CR, Wolf D, Zaagsma J, Kilbinger H. Characterization of the muscarinic receptor subtype(s) mediating contraction of the Guinea-pig lung strip and inhibition of acetylcholine release in the Guinea-pig trachea with the selective muscarinic receptor antagonist tripitramine. Br J Pharmacol. 1997;122(1):133–41.

41. ten Berge RE, Santing RE, Hamstra JJ, Roffel AF, Zaagsma J. Dysfunction of muscarinic M2 receptors after the early allergic reaction: possible contribution to bronchial hyperresponsiveness in allergic Guinea-pigs. Br J Pharmacol. 1995;114(4):881–7.

42. Buels KS, Jacoby DB, Fryer AD. Non-bronchodilating mechanisms of tiotropium prevent airway hyperreactivity in a Guinea-pig model of allergic asthma. Br J Pharmacol. 2012;165(5):1501–14.

43. Meurs H, Dekkers BG, Maarsingh H, Halayko AJ, Zaagsma J, Gosens R. Muscarinic recep-

tors on airway mesenchymal cells: novel findings for an ancient target. Pulm Pharmacol Ther. 2013;26(1):145–55.

44. Billington CK, Penn RB, Hall IP. beta2 Agonists. Handb Exp Pharmacol. 2017;237:23–40.

45. Pera T, Penn RB. Crosstalk between beta-2-adrenoceptor and muscarinic acetylcholine receptors in the airway. Curr Opin Pharmacol. 2014;16:72–81.

46. Walker JKL, Penn RB, Hanania NA, Dickey BF, Bond RA. New perspectives regarding β(2)-adrenoceptor ligands in the treatment of asthma. Br J Pharmacol. 2011;163(1):18–28.

47. Boardman C, Chachi L, Gavrila A, Keenan CR, Perry MM, Xia YC, et al. Mechanisms of glucocorticoid action and insensitivity in airways disease. Pulm Pharmacol Ther. 2014;29(2):129–43.

48. Newton R, Giembycz MA. Understanding how long-acting beta2 -adrenoceptor agonists enhance the clinical efficacy of inhaled corticosteroids in asthma – an update. Br J Pharmacol. 2016;173(24):3405–30.

49. Chung KF, Caramori G, Adcock IM. Inhaled corticosteroids as combination therapy with beta-adrenergic agonists in airways disease: present and future. Eur J Clin Pharmacol. 2009;65(9):853–71.

50. Kew KM, Dahri K. Long-acting muscarinic antagonists (LAMA) added to combination long-acting beta2-agonists and inhaled corticosteroids (LABA/ICS) versus LABA/ICS for adults with asthma. Cochrane Database Syst Rev. 2016;1:CD011721.

51. Barnes PJ. Glucocorticosteroids. Handb Exp Pharmacol. 2017;237:93–115.

52. Barnes PJ. Mechanisms and resistance in glucocorticoid control of inflammation. J Steroid Biochem Mol Biol. 2010;120(2–3):76–85.

53. Barnes PJ. Corticosteroid resistance in patients with asthma and chronic obstructive pulmonary disease. J Allergy Clin Immunol. 2013;131(3):636–45.

54. Adcock IM, Barnes PJ. Molecular mechanisms of corticosteroid resistance. Chest. 2008;134(2):394–401.

55. Louis R, Schleich F, Barnes PJ. Corticosteroids: still at the frontline in asthma treatment? Clin Chest Med. 2012;33(3):531–41.

56. Barnes PJ. How corticosteroids control inflammation: quintiles prize lecture 2005. Br J Pharmacol. 2006;148(3):245–54.

57. Durham A, Adcock IM, Tliba O. Steroid resistance in severe asthma: current mechanisms and future treatment. Curr Pharm Des. 2011;17(7):674–84.

58. Newton R, Leigh R, Giembycz MA. Pharmacological strategies for improving the efficacy and therapeutic ratio of glucocorticoids in inflammatory lung diseases. Pharmacol Ther. 2010;125(2):286–327.

59. Barnes PJ. Corticosteroid effects on cell signalling. Eur Respir J. 2006;27(2):413–26.

60. Barnes PJ. Targeting cytokines to treat asthma and chronic obstructive pulmonary disease. Nat Rev Immunol. 2018;18(7):454–66.

61. Barnes PJ, Adcock IM. Glucocorticoid resistance in inflammatory diseases. Lancet (London, England). 2009;373(9678):1905–17.

62. Barnes PJ, Ito K, Adcock IM. Corticosteroid resistance in chronic obstructive pulmonary disease: inactivation of histone deacetylase. Lancet (London, England). 2004;363(9410):731–3.

63. Chanez P, Bourdin A, Vachier I, Godard P, Bousquet J, Vignola AM. Effects of inhaled corticosteroids on pathology in asthma and chronic obstructive pulmonary disease. Proc Am Thorac Soc. 2004;1(3):184–90.

64. Ito K, Getting SJ, Charron CE. Mode of glucocorticoid actions in airway disease. ScientificWorldJournal. 2006;6:1750–69.

65. Amrani Y, Panettieri RA, Ramos-Ramirez P, Schaafsma D, Kaczmarek K, Tliba O. Important lessons learned from studies on the pharmacology of glucocorticoids in human airway smooth muscle cells: too much of a good thing may be a problem. Pharmacol Ther. 2020;213:107589.

66. Flaster H, Bernhagen J, Calandra T, Bucala R. The macrophage migration inhibitory factor-glucocorticoid dyad: regulation of inflammation and immunity. Mol Endocrinol. 2007;21(6):1267–80.

67. Cruz-Topete D, Cidlowski JA. One hormone, two actions: anti- and pro-inflammatory effects of glucocorticoids. Neuroimmunomodulation. 2015;22(1–2):20–32.

68. McGregor MC, Krings JG, Nair P, Castro M. Role of biologics in asthma. Am J Respir Crit Care Med. 2019;199(4):433–45.

69. Rogliani P, Calzetta L, Matera MG, Laitano R, Ritondo BL, Hanania NA, et al. Severe asthma and biological therapy: when, which, and for whom. Pulm Ther. 2020;6(1):47–66.

70. Dean K, Niven R. Asthma phenotypes and Endotypes: implications for personalised therapy. BioDrugs. 2017;31(5):393–408.

71. Anderson WC 3rd, Szefler SJ. Cost-effectiveness and comparative effectiveness of biologic therapy for asthma: to biologic or not to biologic? Ann Allergy Asthma Immunol. 2019;122(4):367–72.

72. Insel PA, Sriram K, Gorr MW, Wiley SZ, Michkov A, Salmeron C, et al. GPCRomics: An approach to discover GPCR drug targets. Trends Pharmacol Sci. 2019;40(6):378–87.

73. Workman AD, Palmer JN, Adappa ND, Cohen NA. The role of bitter and sweet taste receptors in upper airway immunity. Curr Allergy Asthma Rep. 2015;15(12):72.

74. Wu SV, Rozengurt N, Yang M, Young SH, Sinnett-Smith J, Rozengurt E. Expression of bitter taste receptors of the T2R family in the gastrointestinal tract and enteroendocrine STC-1 cells. Proc Natl Acad Sci U S A. 2002;99(4):2392–7.

75. Luo XC, Chen ZH, Xue JB, Zhao DX, Lu C, Li YH, et al. Infection by the parasitic helminth Trichinella spiralis activates a Tas2r-mediated signaling pathway in intestinal tuft cells. Proc Natl Acad Sci U S A. 2019;116(12):5564–9.

76. Deshpande DA, Wang WC, McIlmoyle EL, Robinett KS, Schillinger RM, An SS, et al. Bitter taste receptors on airway smooth muscle bronchodilate by localized calcium signaling and reverse obstruction. Nat Med. 2010;16(11):1299–304.

77. An SS, Liggett SB. Taste and smell GPCRs in the lung: evidence for a previously unrecognized widespread chemosensory system. Cell Signal. 2018;41:82–8.

78. Sharma P, Panebra A, Pera T, Tiegs BC, Hershfeld A, Kenyon LC, et al. Antimitogenic effect of bitter taste receptor agonists on airway smooth muscle cells. Am J Physiol Lung Cell Mol Physiol. 2016;310(4):L365–76.

79. An SS, Wang WC, Koziol-White CJ, Ahn K, Lee DY, Kurten RC, et al. TAS2R activation promotes airway smooth muscle relaxation despite beta(2)-adrenergic receptor tachyphylaxis. Am J Physiol Lung Cell Mol Physiol. 2012;303(4):L304–11.

80. Zhang CH, Lifshitz LM, Uy KF, Ikebe M, Fogarty KE, ZhuGe R. The cellular and molecular basis of bitter tastant-induced bronchodilation. PLoS Biol. 2013;11(3):e1001501.

81. Tan X, Sanderson MJ. Bitter tasting compounds dilate airways by inhibiting airway smooth muscle calcium oscillations and calcium sensitivity. Br J Pharmacol. 2014;171(3):646–62.

82. Pulkkinen V, Manson ML, Safholm J, Adner M, Dahlen SE. The bitter taste receptor (TAS2R) agonists denatonium and chloroquine display distinct patterns of relaxation of the Guinea pig trachea. Am J Physiol Lung Cell Mol Physiol. 2012;303(11):L956–66.

83. Sharma P, Yi R, Nayak AP, Wang N, Tang F, Knight MJ, et al. Bitter taste receptor agonists mitigate features of allergic asthma in mice. Sci Rep. 2017;7:46166.

84. Conaway S Jr, Nayak AP, Deshpande DA. Therapeutic potential and challenges of bitter taste receptors on lung cells. Curr Opin Pharmacol. 2020;51:43–9.

85. Yarova PL, Stewart AL, Sathish V, Britt RD, Thompson MA, Lowe AP, et al. Calcium-sensing receptor antagonists abrogate airway hyperresponsiveness and inflammation in allergic asthma. Sci Transl Med. 2015;7(284):284ra60.

86. Penn RB. Physiology. Calcilytics for asthma relief. Science. 2015;348(6233):398–9.

87. Kawahara K, Hohjoh H, Inazumi T, Tsuchiya S, Sugimoto Y. Prostaglandin E2-induced inflammation: relevance of prostaglandin E receptors. Biochim Biophys Acta. 2015;1851(4):414–21.

88. Guo M, Pascual RM, Wang S, Fontana MF, Valancius CA, Panettieri RA Jr, et al. Cytokines regulate beta-2-adrenergic receptor responsiveness in airway smooth muscle via multiple PKA- and EP2 receptor-dependent mechanisms. Biochemistry. 2005;44(42):13771–82.

89. Liu T, Laidlaw TM, Katz HR, Boyce JA. Prostaglandin E2 deficiency causes a phenotype of aspirin sensitivity that depends on platelets and

cysteinyl leukotrienes. Proc Natl Acad Sci U S A. 2013;110(42):16987–92.

90. Birrell MA, Maher SA, Dekkak B, Jones V, Wong S, Brook P, et al. Anti-inflammatory effects of PGE2 in the lung: role of the EP4 receptor subtype. Thorax. 2015;70(8):740–7.

91. Feng C, Beller EM, Bagga S, Boyce JA. Human mast cells express multiple EP receptors for prostaglandin E2 that differentially modulate activation responses. Blood. 2006;107(8):3243–50.

92. Lazzeri N, Belvisi MG, Patel HJ, Yacoub MH, Fan Chung K, Mitchell JA. Effects of prostaglandin E2 and cAMP elevating drugs on GM-CSF release by cultured human airway smooth muscle cells. Relevance to asthma therapy. Am J Respir Cell Mol Biol. 2001;24(1):44–8.

93. Lundequist A, Nallamshetty SN, Xing W, Feng C, Laidlaw TM, Uematsu S, et al. Prostaglandin E(2) exerts homeostatic regulation of pulmonary vascular remodeling in allergic airway inflammation. J Immunol. 2010;184(1):433–41.

94. Belvisi MG, Saunders M, Yacoub M, Mitchell JA. Expression of cyclo-oxygenase-2 in human airway smooth muscle is associated with profound reductions in cell growth. Br J Pharmacol. 1998;125(5):1102–8.

95. Billington CK, Kong KC, Bhattacharyya R, Wedegaertner PB, Panettieri RA, Chan TO, et al. Cooperative regulation of p70S6 kinase by receptor tyrosine kinases and G protein-coupled receptors augments airway smooth muscle growth. Biochemistry. 2005;44:14595–605.

96. Guo M, Pascual RM, Wang S, Fontana MF, Valancius CA, Panettieri RA, Jr., et al. Cytokines regulate beta-2-adrenergic receptor responsiveness in airway smooth muscle via multiple PKA- and EP2 receptor-dependent mechanisms. Biochemistry 2005;44(42):13771–13782.

97. Misior AM, Yan H, Pascual RM, Deshpande DA, Panettieri RA Jr, Penn RB. Mitogenic effects of cytokines on smooth muscle are critically dependent on protein kinase A and are unmasked by steroids and cyclooxygenase inhibitors. Mol Pharmacol. 2008;73(2):566–74.

98. Pascual RM, Billington CK, Hall IP, Panettieri RA Jr, Fish JE, Peters SP, et al. Mechanisms of cytokine effects on G protein-coupled receptor-mediated signaling in airway smooth muscle. Am J Physiol Lung Cell Mol Physiol. 2001;281(6):1425–35.

99. Yan H, Deshpande DA, Misior AM, Miles MC, Saxena H, Riemer EC, et al. Anti-mitogenic effects of beta-agonists and PGE2 on airway smooth muscle are PKA dependent. FASEB J. 2011;25(1):389–97.

100. Birrell MA, Maher SA, Buckley J, Dale N, Bonvini S, Raemdonck K, et al. Selectivity profiling of the novel EP(2) receptor antagonist, PF-04418948, in functional bioassay systems: atypical affinity at the Guinea pig EP(2) receptor. Br J Pharmacol. 2013;168(1):129–38.

101. Buckley J, Birrell MA, Maher SA, Nials AT, Clarke DL, Belvisi MG. EP(4) receptor as a new target for bronchodilator therapy. Thorax. 2011;66(12):1029–35.

102. Morgan SJ, Deshpande DA, Tiegs BC, Misior AM, Yan H, Hershfeld AV, et al. beta-agonist-mediated relaxation of airway smooth muscle is protein kinase A-dependent. J Biol Chem. 2014;289(33):23065–74.

103. Tilley SL, Hartney JM, Erikson CJ, Jania C, Nguyen M, Stock J, et al. Receptors and pathways mediating the effects of prostaglandin E2 on airway tone. American journal of physiology lung cellular and molecular. Physiology. 2003;284(4):599–606.

104. Kawakami Y, Uchiyama K, Irie T, Murao M. Evaluation of aerosols of prostaglandins E1 and E2 as bronchodilators. Eur J Clin Pharmacol. 1973;6(2):127–32.

105. Melillo E, Woolley KL, Manning PJ, Watson RM, O'Byrne PM. Effect of inhaled PGE2 on exercise-induced bronchoconstriction in asthmatic subjects. Am J Respir Crit Care Med. 1994;149(5):1138–41.

106. Walters EH, Bevan C, Parrish RW, Davies BH, Smith AP. Time-dependent effect of prostaglandin E2 inhalation on airway responses to bronchoconstrictor agents in normal subjects. Thorax. 1982;37(6):438–42.

107. Sestini P, Armetti L, Gambaro G, Pieroni MG, Refini RM, Sala A, et al. Inhaled PGE2 prevents aspirin-induced bronchoconstriction and urinary LTE4 excretion in aspirin-sensitive asthma. Am J Respir Crit Care Med. 1996;153(2):572–5.

108. Szczeklik A, Mastalerz L, Nizankowska E, Cmiel A. Protective and bronchodilator effects of prostaglandin E and salbutamol in aspirin-induced asthma. Am J Respir Crit Care Med. 1996;153(2):567–71.

109. Pasargiklian M, Bianco S, Allegra L. Clinical, functional and pathogenetic aspects of bronchial reactivity to prostaglandins F2alpha, E1, and E2. Adv Prostaglandin Thromboxane Res. 1976;1:461–75.

110. Pavord ID, Wong CS, Williams J, Tattersfield AE. Effect of inhaled prostaglandin E2 on allergen-induced asthma. Am Rev Respir Dis. 1993;148(1):87–90.

111. Manning PJ, Lane CG, O'Byrne PM. The effect of oral prostaglandin E1 on airway responsiveness in asthmatic subjects. Pulm Pharmacol. 1989;2(3):121–4.

112. Gauvreau GM, Watson RM, O'Byrne PM. Protective effects of inhaled PGE2 on allergen-induced airway responses and airway inflammation. Am J Respir Crit Care Med. 1999;159(1):31–6.

113. Costello JF, Dunlop LS, Gardiner PJ. Characteristics of prostaglandin induced cough in man. Br J Clin Pharmacol. 1985;20(4):355–9.

114. Coleridge HM, Coleridge JC, Ginzel KH, Baker DG, Banzett RB, Morrison MA. Stimulation of 'irritant' receptors and afferent C-fibres in the lungs by prostaglandins. Nature. 1976;264(5585):451–3.

115. Birrell MA, Maher SA, Dekkak B, Jones V, Wong S, Brook P, et al. Anti-inflammatory effects of PGE2 in

the lung: role of the EP4 receptor subtype. Thorax. 2015;70(8):740–7.

116. Jones VC, Birrell MA, Maher SA, Griffiths M, Grace M, O'Donnell VB, et al. Role of EP2 and EP4 receptors in airway microvascular leak induced by prostaglandin E2. Br J Pharmacol. 2016;173(6):992–1004.

117. Maher SA, Birrell MA, Belvisi MG. Prostaglandin E2 mediates cough via the EP3 receptor: implications for future disease therapy. Am J Respir Crit Care Med. 2009;180(10):923–8.

118. Buckley J, Birrell MA, Maher SA, Nials AT, Clarke DL, Belvisi MG. EP4 receptor as a new target for bronchodilator therapy. Thorax. 2011;66(12):1029–35.

119. Birrell MA, Maher SA, Buckley J, Dale N, Bonvini S, Raemdonck K, et al. Selectivity profiling of the novel EP2 receptor antagonist, PF-04418948, in functional bioassay systems: atypical affinity at the Guinea pig EP2 receptor. Br J Pharmacol. 2013;168(1):129–38.

120. Patel HJ, Birrell MA, Crispino N, Hele DJ, Venkatesan P, Barnes PJ, et al. Inhibition of Guinea-pig and human sensory nerve activity and the cough reflex in Guinea-pigs by cannabinoid (CB2) receptor activation. Br J Pharmacol. 2003;140(2):261–8.

121. Schuil PJ, Ten Berge M, Van Gelder JM, Graamans K, Huizing EH. Effects of prostaglandins D2 and E2 on ciliary beat frequency of human upper respiratory cilia in vitro. Acta Otolaryngol. 1995;115(3):438–42.

122. Seto V, Hirota C, Hirota S, Janssen LJ. E-ring Isoprostanes stimulate a Cl conductance in airway epithelium via prostaglandin E2-selective Prostanoid receptors. Am J Respir Cell Mol Biol. 2008;38(1):88–94.

123. Kook Kim J, Hoon Kim C, Kim K, Jong Jang H, Jik Kim H, Yoon JH. Effects of prostagladin E(2) on gel-forming mucin secretion in normal human nasal epithelial cells. Acta Otolaryngol. 2006;126(2):174–9.

124. Hattori R, Shimizu S, Majima Y, Shimizu T. EP4 agonist inhibits lipopolysaccharide-induced mucus secretion in airway epithelial cells. Ann Otol Rhinol Laryngol. 2008;117(1):51–8.

125. Hattori R, Shimizu S, Majima Y, Shimizu T. Prostaglandin E2 receptor EP2, EP3, and EP4 agonists inhibit antigen-induced mucus hypersecretion in the nasal epithelium of sensitized rats. Ann Otol Rhinol Laryngol. 2009;118(7):536–41.

126. Tavakoli S, Cowan MJ, Benfield T, Logun C, Shelhamer JH. Prostaglandin E(2)-induced interleukin-6 release by a human airway epithelial cell line. Am J Physiol Lung Cell Mol Physiol. 2001;280(1):L127–33.

127. Pelletier S, Dube J, Villeneuve A, Gobeil F Jr, Yang Q, Battistini B, et al. Prostaglandin E(2) increases cyclic AMP and inhibits endothelin-1 production/secretion by Guinea-pig tracheal epithelial cells through EP(4) receptors. Br J Pharmacol. 2001;132(5):999–1008.

128. Li YJ, Kanaji N, Wang XQ, Sato T, Nakanishi M, Kim M, et al. Prostaglandin E2 switches from a stimulator to an inhibitor of cell migration after epithelial-to-mesenchymal transition. Prostaglandins Other Lipid Mediat. 2015;116-117:1–9.

129. Sturm EM, Schratl P, Schuligoi R, Konya V, Sturm GJ, Lippe IT, et al. Prostaglandin E2 inhibits eosinophil trafficking through E-prostanoid 2 receptors. J Immunol. 2008;181(10):7273–83.

130. Jarvinen L, Badri L, Wettlaufer S, Ohtsuka T, Standiford TJ, Toews GB, et al. Lung resident mesenchymal stem cells isolated from human lung allografts inhibit T cell proliferation via a soluble mediator. J Immunol. 2008;181(6):4389–96.

131. Baratelli F, Lin Y, Zhu L, Yang SC, Heuze-Vourc'h N, Zeng G, et al. Prostaglandin E2 induces FOXP3 gene expression and T regulatory cell function in human CD4+ T cells. J Immunol. 2005;175(3):1483–90.

132. Canning BJ, Hmieleski RR, Spannhake EW, Jakab GJ. Ozone reduces murine alveolar and peritoneal macrophage phagocytosis: the role of prostanoids. Am J Phys. 1991;261(4 Pt 1):L277–82.

133. Armstrong RA. Investigation of the inhibitory effects of PGE2 and selective EP agonists on chemotaxis of human neutrophils. Br J Pharmacol. 1995;116(7):2903–8.

134. Peters T, Henry PJ. Protease-activated receptors and prostaglandins in inflammatory lung disease. Br J Pharmacol. 2009;158(4):1017–33.

135. Markovic T, Jakopin Z, Dolenc MS, Mlinaric-Rascan I. Structural features of subtype-selective EP receptor modulators. Drug Discov Today. 2017;22(1):57–71.

136. Nichols HL, Saffeddine M, Theriot BS, Hegde A, Polley D, El-Mays T, et al. beta-Arrestin-2 mediates the proinflammatory effects of proteinase-activated receptor-2 in the airway. Proc Natl Acad Sci U S A. 2012;109(41):16660–5.

137. Asaduzzaman M, Nadeem A, Arizmendi N, Davidson C, Nichols HL, Abel M, et al. Functional inhibition of PAR2 alleviates allergen-induced airway hyperresponsiveness and inflammation. Clin Exp Allergy. 2015;45(12):1844–55.

138. Yee MC, Nichols HL, Polley D, Saifeddine M, Pal K, Lee K, et al. Protease-activated receptor-2 signaling through beta-arrestin-2 mediates Alternaria alkaline serine protease-induced airway inflammation. Am J Physiol Lung Cell Mol Physiol. 2018;315(6):L1042–L57.

139. Chiu JC, Hsu JY, Fu LS, Chu JJ, Chi CS. Comparison of the effects of two long-acting beta2-agonists on cytokine secretion by human airway epithelial cells. J Microbiol Immunol Infect. 2007;40(5):388–94.

140. Hung CH, Chu YT, Hua YM, Hsu SH, Lin CS, Chang HC, et al. Effects of formoterol and salmeterol on the production of Th1- and Th2-related chemokines by monocytes and bronchial epithelial cells. Eur Respir J. 2008;31(6):1313–21.

141. Chorley BN, Li Y, Fang S, Park JA, Adler KB. (R)-albuterol elicits antiinflammatory effects in human airway epithelial cells via iNOS. Am J Respir Cell Mol Biol. 2006;34(1):119–27.

142. Kleine-Tebbe J, Frank G, Josties C, Kunkel G. Influence of salmeterol, a long-acting beta 2-adrenoceptor agonist, on IgE-mediated histamine release from human basophils. J Investig Allergol Clin Immunol. 1994;4(1):12–7.
143. Chong LK, Cooper E, Vardey CJ, Peachell PT. Salmeterol inhibition of mediator release from human lung mast cells by beta-adrenoceptor-dependent and independent mechanisms. Br J Pharmacol. 1998;123(5):1009–15.
144. Lewis RJ, Chachi L, Newby C, Amrani Y, Bradding P. Bidirectional Counterregulation of human lung mast cell and airway smooth muscle beta2 adrenoceptors. J Immunol. 2016;196(1):55–63.
145. Duffy SM, Cruse G, Lawley WJ, Bradding P. Beta2-adrenoceptor regulation of the K+ channel iKCa1 in human mast cells. FASEB J. 2005;19(8):1006–8.
146. Roberts JA, Bradding P, Britten KM, Walls AF, Wilson S, Gratziou C, et al. The long-acting beta2-agonist salmeterol xinafoate: effects on airway inflammation in asthma. Eur Respir J. 1999;14(2):275–82.
147. Green RH, Brightling CE, McKenna S, Hargadon B, Neale N, Parker D, et al. Comparison of asthma treatment given in addition to inhaled corticosteroids on airway inflammation and responsiveness. Eur Respir J. 2006;27(6):1144–51.
148. Giembycz MA, Kaur M, Leigh R, Newton R. A Holy Grail of asthma management: toward understanding how long-acting beta(2)-adrenoceptor agonists enhance the clinical efficacy of inhaled corticosteroids. Br J Pharmacol. 2008;153(6):1090–104.
149. Rider CF, Altonsy MO, Mostafa MM, Shah SV, Sasse S, Manson ML, et al. Long-acting beta2-adrenoceptor agonists enhance glucocorticoid receptor (GR)-mediated transcription by gene-specific mechanisms rather than generic effects via GR. Mol Pharmacol. 2018;94(3):1031–46.
150. Holden NS, Bell MJ, Rider CF, King EM, Gaunt DD, Leigh R, et al. beta2-adrenoceptor agonist-induced RGS2 expression is a genomic mechanism of bronchoprotection that is enhanced by glucocorticoids. Proc Natl Acad Sci U S A. 2011;108(49):19713–8.
151. Ortega VE, Peters SP. Beta-2 adrenergic agonists: focus on safety and benefits versus risks. Curr Opin Pharmacol. 2010;10(3):246–53.
152. Stolley PD. Asthma mortality. Why the United States was spared an epidemic of deaths due to asthma. Am Rev Respir Dis. 1972;105(6):883–90.
153. Grainger J, Woodman K, Pearce N, Crane J, Burgess C, Keane A, et al. Prescribed fenoterol and death from asthma in New Zealand, 1981-7: a further case-control study. Thorax. 1991;46(2):105–11.
154. Crane J, Pearce N, Flatt A, Burgess C, Jackson R, Kwong T, et al. Prescribed fenoterol and death from asthma in New Zealand, 1981-83: case-control study. Lancet (London, England). 1989;1(8644):917–22.
155. Pearce N, Grainger J, Atkinson M, Crane J, Burgess C, Culling C, et al. Case-control study of prescribed fenoterol and death from asthma in New Zealand, 1977–81. Thorax. 1990;45(3):170–5.
156. Pearce N, Beasley R, Crane J, Burgess C, Jackson R. End of the New Zealand asthma mortality epidemic. Lancet (London, England). 1995;345(8941):41–4.
157. Pearce N, Burgess C, Crane J, Beasley R. Fenoterol, asthma deaths, and asthma severity. Chest. 1997;112(4):1148–50.
158. Nelson HS, Weiss ST, Bleecker ER, Yancey SW, Dorinsky PM, Group SS. The salmeterol multicenter asthma research trial: a comparison of usual pharmacotherapy for asthma or usual pharmacotherapy plus salmeterol. Chest. 2006;129(1):15–26.
159. Pearlman DS, Chervinsky P, LaForce C, Seltzer JM, Southern DL, Kemp JP, et al. A comparison of salmeterol with albuterol in the treatment of mild-to-moderate asthma. N Engl J Med. 1992;327(20):1420–5.
160. Bhagat R, Kalra S, Swystun VA, Cockcroft DW. Rapid onset of tolerance to the bronchoprotective effect of salmeterol. Chest. 1995;108(5):1235–9.
161. Newnham DM, Grove A, McDevitt DG, Lipworth BJ. Subsensitivity of bronchodilator and systemic beta 2 adrenoceptor responses after regular twice daily treatment with eformoterol dry powder in asthmatic patients. Thorax. 1995;50(5):497–504.
162. Taylor DR. The β-agonist Saga and its clinical relevance: on and on it goes. Am J Respir Crit Care Med. 2009;179(11):976–8.
163. Salpeter S, Buckley N, Ormiston T, Salpeter E. Meta-analysis: effect of long-acting {beta}-agonists on severe asthma exacerbations and asthma-related deaths. Ann Intern Med. 2006;144:904–12.
164. Sears MR. Adverse effects of [beta]-agonists. J Allergy Clin Immunol. 2002;110(6, Part 2):S322–S8.
165. Sears MR, Taylor DR, Print CG, Lake DC, Li QQ, Flannery EM, et al. Regular inhaled beta-agonist treatment in bronchial asthma. Lancet (London, England). 1990;336(8728):1391–6.
166. Sears MR. Adverse effects of beta-agonists. J Allergy Clin Immunol. 2002;110(6 Suppl):S322–8.
167. Jeffery PK, Venge P, Gizycki MJ, Egerod I, Dahl R, Faurschou P. Effects of salmeterol on mucosal inflammation in asthma: a placebo-controlled study. Eur Respir J. 2002;20(6):1378–85.
168. Wallin A, Sandstrom T, Soderberg M, Howarth P, Lundback B, Della-Cioppa G, et al. The effects of regular inhaled formoterol, budesonide, and placebo on mucosal inflammation and clinical indices in mild asthma. Am J Respir Crit Care Med. 1999;159(1):79–86.
169. Wallin A, Sandstrom T, Cioppa GD, Holgate S, Wilson S. The effects of regular inhaled formoterol and budesonide on preformed Th-2 cytokines in mild asthmatics. Respir Med. 2002;96(12):1021–5.
170. Martinez-Milla J, Raposeiras-Roubin S, Pascual-Figal DA, Ibanez B. Role of Beta-blockers in cardiovascular disease in 2019. Rev Esp Cardiol (Engl Ed). 2019;72(10):844–52.

171. Al-Sawalha N, Pokkunuri I, Omoluabi O, Kim H, Thanawala VJ, Hernandez A, et al. Epinephrine activation of the beta2-adrenoceptor is required for IL-13-induced mucin production in human bronchial epithelial cells. PLoS One. 2015;10(7):e0132559.

172. Callaerts-Vegh Z, Evans KL, Dudekula N, Cuba D, Knoll BJ, Callaerts PF, et al. Effects of acute and chronic administration of beta-adrenoceptor ligands on airway function in a murine model of asthma. Proc Natl Acad Sci U S A. 2004;101(14):4948–53.

173. Forkuo GS, Kim H, Thanawala VJ, Al-Sawalha N, Valdez D, Joshi R, et al. PDE4 inhibitors attenuate the asthma phenotype produced by beta-adrenoceptor agonists in PNMT-KO mice. Am J Respir Cell Mol Biol. 2016;55:234–42.

174. Nguyen LP, Lin R, Parra S, Omoluabi O, Hanania NA, Tuvim MJ, et al. Beta2-adrenoceptor signaling is required for the development of an asthma phenotype in a murine model. Proc Natl Acad Sci U S A. 2009;106(7):2435–40.

175. Nguyen LP, Omoluabi O, Parra S, Frieske JM, Clement C, Ammar-Aouchiche Z, et al. Chronic exposure to beta-blockers attenuates inflammation and mucin content in a murine asthma model. Am J Respir Cell Mol Biol. 2008;38(3):256–62.

176. Thanawala VJ, Forkuo GS, Al-Sawalha N, Azzegagh Z, Nguyen LP, Eriksen JL, et al. beta2-adrenoceptor agonists are required for development of the asthma phenotype in a murine model. Am J Respir Cell Mol Biol. 2013;48(2):220–9.

177. Thanawala VJ, Valdez DJ, Joshi R, Forkuo GS, Parra S, Knoll BJ, et al. β-Blockers have differential effects on the murine asthma phenotype. Br J Pharmacol. 2015;172(20):4833–46.

178. Hanania NA, Mannava B, Franklin AE, Lipworth BJ, Williamson PA, Garner WJ, et al. Response to salbutamol in patients with mild asthma treated with nadolol. Eur Respir J. 2010;36(4):963–5.

179. Hanania NA, Singh S, El-Wali R, Flashner M, Franklin AE, Garner WJ, et al. The safety and effects of the beta-blocker, nadolol, in mild asthma: an open-label pilot study. Pulm Pharmacol Ther. 2007;21(1):134–41.

180. Urban JD, Clarke WP, von ZM, Nichols DE, Kobilka B, Weinstein H, et al. Functional selectivity and classical concepts of quantitative pharmacology. J Pharmacol Exp Ther. 2007;320(1):1–13.

181. Vaidehi N, Kenakin T. The role of conformational ensembles of seven transmembrane receptors in functional selectivity. Curr Opin Pharmacol. 2010;10(6):775–81.

182. Wisler JW, DeWire SM, Whalen EJ, Violin JD, Drake MT, Ahn S, et al. A unique mechanism of beta-blocker action: carvedilol stimulates beta-arrestin signaling. Proc Natl Acad Sci U S A. 2007;104(42):16657–62.

183. Drake MT, Violin JD, Whalen EJ, Wisler JW, Shenoy SK, Lefkowitz RJ. beta-arrestin-biased agonism at the beta2-adrenergic receptor. J Biol Chem. 2008;283(9):5669–76.

184. Nguyen LP, Al-Sawalha NA, Parra S, Pokkunuri I, Omoluabi O, Okulate AA, et al. beta2-adrenoceptor signaling in airway epithelial cells promotes eosinophilic inflammation, mucous metaplasia, and airway contractility. Proc Natl Acad Sci U S A. 2017;114(43):E9163–E71.

185. Penn RB, Bond RA, Walker JK. GPCRs and arrestins in airways: implications for asthma. Handb Exp Pharmacol. 2014;219:387–403.

Cellular and Molecular Processes in Pulmonary Hypertension

2

Vic Maietta, Jorge Reyes-García, Vishal R. Yadav, Yun-Min Zheng, Xu Peng, and Yong-Xiao Wang

Abstract

Pulmonary hypertension (PH) is a progressive lung disease characterized by persistent pulmonary vasoconstriction. Another well-recognized characteristic of PH is the muscularization of peripheral pulmonary arteries. This pulmonary vasoremodeling manifests in medial hypertrophy/hyperplasia of smooth muscle cells (SMCs) with possible neointimal formation. The underlying molecular processes for these two major vascular responses remain not fully understood. On the other hand, a series of very recent studies have shown that the increased reactive oxygen species (ROS) seems to be an important player in mediating pulmonary vasoconstriction and vasoremodeling, thereby leading to PH. Mitochondria are a primary site for ROS production in pulmonary artery (PA) SMCs, which subsequently activate NADPH oxidase to induce further ROS generation, i.e., ROS-induced ROS generation. ROS control the activity of multiple ion channels to induce intracellular Ca^{2+} release and extracellular Ca^{2+} influx (ROS-induced Ca^{2+} release and influx) to cause PH. ROS and Ca^{2+} signaling may synergistically trigger an inflammatory cascade to implicate in PH. Accordingly, this paper explores the important roles of ROS, Ca^{2+}, and inflammatory signaling in the development of PH, including their reciprocal interactions, key molecules, and possible therapeutic targets.

Vic Maietta, Jorge Reyes-García and Vishal R. Yadav contributed equally with all other contributors.

V. Maietta · V. R. Yadav · Y.-M. Zheng (✉) ·
Y.-X. Wang (✉)
Department of Molecular & Cellular Physiology,
Albany Medical College, Albany, NY, USA
e-mail: zhengy@amc.edu; wangy@amc.edu

J. Reyes-García
Department of Molecular & Cellular Physiology,
Albany Medical College, Albany, NY, USA

Departamento de Farmacología, Facultad de
Medicina, Universidad Nacional Autónoma de
México, Ciudad de México, Mexico

X. Peng (✉)
Department of Medical Physiology, College of
Medicine, Texas A&M University,
College Station, TX, USA
e-mail: xp23@tamu.edu

Keywords

Pulmonary hypertension · Vascular hypertrophy · Vascular remodeling · Mitochondria · Reactive oxygen species · Calcium signaling · Ion channel

Abbreviations

ACE	Angiotensin-converting enzyme
Ang II	Angiotensin II
Apaf1	Apoptotic protease-activating factor 1
AT_1	Angiotensin II type 1 receptor
ATP	Adenosine triphosphate
Bax	Bcl-2-like protein
CaMKII	Calcium/calmodulin kinase II
CaSR	Calcium sensing receptor
Ca_v	Voltage-dependent Ca^{2+} channel
c-MYC	Proto-oncogene encoding bHLH transcription factor protein
Coenzyme Q	Ubiquinol-cytochrome c reductase
Cyt	Cytochrome
DAG	1,2-Diacylglycerol
DNA	Deoxyribonucleic acid
DRP1	Dynamin-related protein 1
Duox	Dual oxidase
ER	Endoplasmic reticulum
ETC	Electron transport chain
FAD	Flavin adenine dinucleotide
Gpx1	Glutathione peroxidase 1
HIF-1	Hypoxia-inducible factor 1
HUVEC	Human umbilical vein endothelial cells
ICAM-1	Intercellular adhesion molecule 1
$IP_3(R)$	Inositol 1,4,5-trisphosphate (receptor)
K_V	Voltage-dependent potassium channel
LDL	Low-density lipoprotein
MCP-1	Monocyte chemoattractant protein
mPTP	Mitochondrial permeability transition pore
mRNA	Messenger ribonucleic acid
NADH	Nicotinamide adenine dinucleotide
NADPH	Nicotinamide adenine dinucleotide phosphate
NOX	NADPH oxidase
PASMCs	Pulmonary artery smooth muscle cells
PDK1	Gene coding for pyruvate dehydrogenase kinase 1
PH	Pulmonary hypertension
PHD	Prolyl hydroxylase
PKCε	Protein kinase C epsilon
PLC	Phospholipase C
RISP	Rieske iron-sulfur protein
RNAi	RNA interference
ROS	Reactive oxygen species
RyR	Ryanodine receptor
siRNA	Small interfering RNA
Smad3	Gene coding for Smad protein, transducers for TGF-β
SOD	Superoxide dismutase
SR	Sarcoplasmic reticulum
TGF-β	Transforming growth factor beta
TRPC	Transient receptor potential cation channels
VEGF(R)	Vascular endothelial growth factor (receptor)

2.1 Introduction

Pulmonary hypertension (PH) is a progressive and multifactorial disease which remains mostly undiagnosed with a poor prognosis and a 5-year mean survival rate under 70 percent in affected population [1]. This ailment is characterized by persistent pulmonary vasoconstriction progressing to proliferation of endothelial and smooth muscle cells that lines the arteries and arterioles of the pulmonary vasculature. This pulmonary vascular remodeling (PVR) results in elevated pulmonary vascular resistance and pressure leading to right ventricular hypertrophy and failure and ultimately death [2]. Mean pulmonary arterial pressure greater than 25 mmHg at rest or greater than 30 mmHg during exercise is indicative of pulmonary hypertension [3]. A histological view denotes muscularization of peripheral pulmonary arteries involving hypertrophy and hyperplasia, loss of small pre-capillary arteries, and neointimal formation [3]. Moreover, plexiform lesions (complex vascular lesions are aberrant channels in the obliterated vessel lumen and in the adventitia) form/arise at later stages of the disease due to either the clonal expansion of apoptosis-resistant endothelial cells (ECs) or the deposition of circulating endothelial progenitor cells at injury sites [3]. Untreated PH can eventually lead to right heart failure [4, 5]. Potential causes are linked to mitochondrial dysfunction. For instance, alterations in reactive oxygen species (ROS) homeostasis in pulmonary artery smooth muscle cells (PASMCS) and ECs are involved in the development of PVR seen in pulmonary arterial hypertension [6]. Moreover, changes in ROS intracellular concentration ($[ROS]_i$) may influence the vascular reactivity dependent on the intracellular Ca^{2+} concentration ($[Ca^{2+}]_i$) [7]. Cross-talk among these forces may initiate and enhance PH [3]. This article aims to review and discuss the molecular mechanisms underlying PH, primarily the role of ROS in the development of this lung disease and their cross-talk with several intermediate Ca^{2+} signaling molecules.

2.2 ROS

Growing evidence suggests that changes in $[ROS]_i$ mainly from mitochondrial dysfunction in pulmonary artery ECs (PAECs) and PASMCs contribute to the development of PH. Mitochondrial function plays a critical role in cell Ca^{2+} homeostasis, ATP and ROS generation, inflammation, progression of the cell cycle, and apoptosis [8]. ROS are reactive chemical entities, including oxygen-based free radicals (superoxide [$\cdot O_2^-$] and hydroxyl [$OH\cdot$]) and non-radical derivatives of molecular oxygen (H_2O_2) [9]. Free radicals are species or molecules containing one or more unpaired electrons and are capable of independent existence. The number of unpaired electrons renders them unstable and short-lived by imparting the reactive capabilities. Nonradical derivatives are less reactive and more stable [10]. The different sites of O_2^- and H_2O_2 distribution activate distinctive signaling pathways and functional responses in cells [11, 12]. In the mitochondria, ROS are the byproducts of chemical reactions along the electron transport chain (ETC). Electrons are transferred from NADH dehydrogenase enzymatic complex to other protein complexes in the process of reducing molecular oxygen to produce ATP. In the process, the formation of reduced and highly reactive

metabolites occurs, as is simplified in Fig. 2.1 [13]. These chemicals are normally considered as toxic byproducts of aerobic metabolism causing damage to lipids, proteins, and DNA [13, 14]. Cells contain antioxidant enzymes, like superoxide dismutase (SOD), catalase, and glutathione peroxidase, that turn ROS into less reactive forms, thereby preventing subsequent oxidative damage. The dismutation of $\cdot O_2^-$ in mitochondria, cytosol, and extracellular matrix is the main source of H_2O_2, a diffusible signaling molecule [13, 14]. Oxidative stress occurs when the overproduction of oxidants overwhelms the antioxidant capacity of the cell [14]. There is increasing evidence that ROS serve as signaling molecules in addition to being toxic oxidants. It is understood that nitric oxide (NO·) acts as a potent vasodilator when produced by the endothelial NO synthase (eNOS) [15]. NO· is also known for being a toxic oxidant in macrophages targeting bacteria [16]. This dual characteristic is surely not limited to this chemical species. It has been proposed that the mitochondrion acts as a hypoxic sensor that initiates ROS generation [17]. Furthermore, the hypoxic environment is responsible for elevated reactive oxygen species in PASMCs [1, 17, 18]. ROS are produced along the mitochondrial electron transport chain, both enzymatically and non-enzymatically [19]. The enzymatic pathway involves the action of mitochondrial complexes I (NADH: ubiquinone oxidoreductase), II (succinate dehydrogenase), and III (cytochrome bc_1 complex) as denoted in Fig. 2.1. NADH dehydrogenase in complex I, succinate dehydrogenase in complex II, and ubiquinol-cytochrome c reductase (coenzyme Q) in complex III have been found to leak electrons that produce superoxide anions, the precursors of most reactive oxygen species [19]. Experiments using inhibitor molecules to block key components of each mitochondrial respiratory chain complex provide evidence that ROS generation in the mitochondria arises from these proteins [17]. In this context, some researchers, like Archer and colleagues, showed that rotenone and antimycin A, the complex I inhibitor and complex III postubisemiquinone site inhibitor, respectively, mimic and eventually block the decrease in ROS generation induced by acute

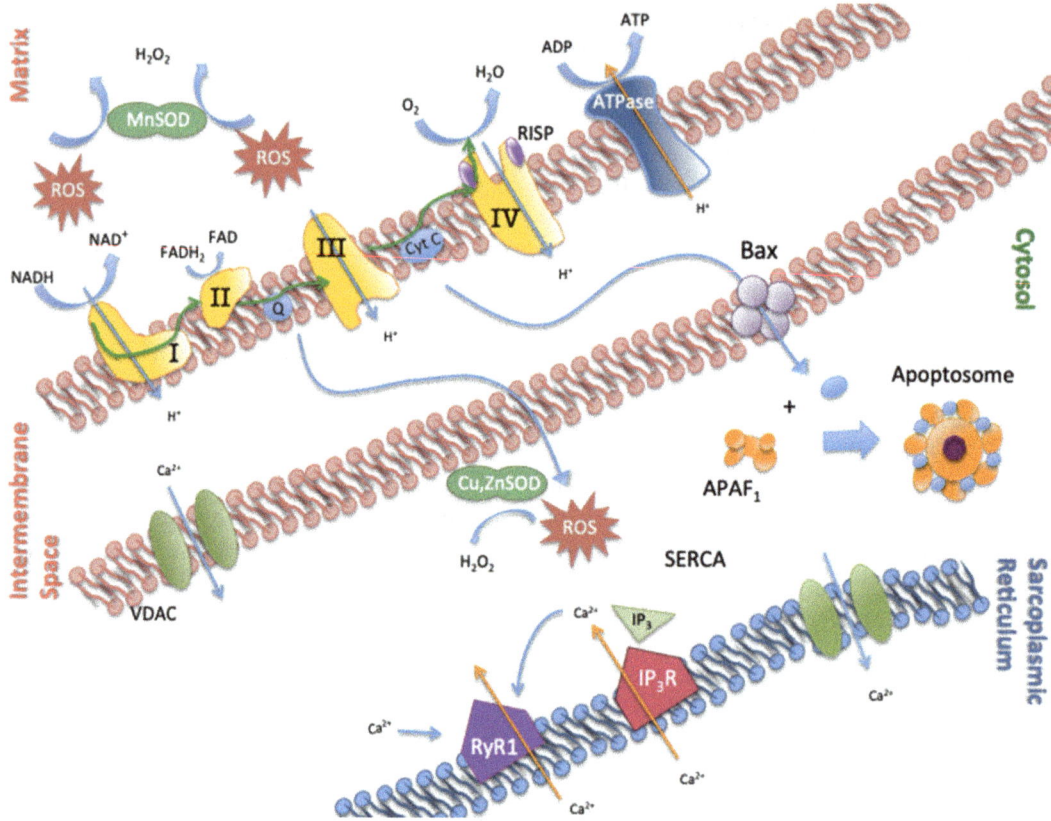

Fig. 2.1 A schematic diagram of the reactive oxygen species (ROS) leakage from various stages of the mitochondrial electron transport chain (ETC) in a vascular smooth muscle cell. Superoxide dismutase enzymes (SOD) convert these species into less reactive hydrogen peroxide molecules (H_2O_2) that can escape the mitochondria as signaling molecules, which interact with several cell proteins. Factors leading to Ca^{2+} accumulation in the cytosol, cell stress, or mitochondrial membrane depolarization trigger formation of the permeability transition pore (mPTP, not shown) between the inner and outer membranes. Bax and the voltage-dependent anion channels (VDACs) are the primary regulators of the mPTP. Loss of cytochrome c (Cyt c) from the ETC combined with apoptotic protease-activating factor 1 ($APAF_1$) complexes in the cytosol forming apoptosomes, a commitment to cell death. Inositol 1,4,5-trisphosphate (IP_3) will then trigger Ca^{2+} release from the sarcoplasmic reticulum, which in turn triggers a phenomenon known as Ca^{2+}-induced Ca^{2+} release from ryanodine receptor 1 (RyR1). Finally, Ca^{2+} will trigger contraction and other responses of the cell

hypoxia in isolated lungs and PASMCs from rats [18, 20]. Nevertheless, other works using rat PASMCs as well proved that rotenone and myxothiazol, a complex III preubisemiquinone site inhibitor, block, but do not mimic, the responses to hypoxic stimuli [21–23]. Correspondingly, studies made by our research group elucidated that the inhibitor actions of rotenone and methylphenylpyridinium iodide on complex I, nitropropionic acid and thenoyltrifluoroacetone on complex II, and myxothiazol on complex III block ROS generation in freshly isolated mouse

PASMC [7, 17, 24]. Furthermore, the inhibition of complex III postubisemiquinone site and complex IV failed to block a hypoxic-elicited ROS response. These insights provide strong evidence that mitochondrial ETC complexes I and II and preubisemiquinone site in complex III operate as a functional unit, which is responsible for ROS production in PASMCs.

Hypertension and metabolic syndrome are related ailments sharing an oxidative stress component. Zhang and colleagues found in isolated A10 cells (under metabolic syndrome conditions, i.e., incubated with 10 nmol/l insulin or 1 μmol/l dexamethasone) that ROS activity of complex I increased at 24 and 48 h with a decrease after 72 h, while the ROS activity of complex III declined across the full 72 h of the experiment . Moreover, they confirmed that complex I is essential for ROS overproduction since ROS generation is highly decreased by rotenone, whereas Apo, the NADPH oxidase (Nox) inhibitor, and AMA, the complex III inhibitor, failed to abolish ROS production [25]. Regardless of which complex contributes the most to $[ROS]_i$ increase, the use of MitoTracker and ROS-sensitive fluorescent dye H_2DCF indicates that mitochondria-dense areas of isolated mouse PASMCs show acute hypoxic increases in ROS concentrations ahead of nonmitochondrial areas of the cells [17]. This evidence points to the mitochondria as the intracellular source of ROS in response to hypoxic conditions.

ROS may act as a preconditioning adaptive response to brief episodes of ischemia that lower necrosis in myocardiocytes during subsequent extended periods of ischemia [26]. Vanden Hoek and colleagues demonstrated that 10 minutes of hypoxia prior to simulated ischemia and reperfusion challenge reduced cell death and potentiated the return of contraction in cultured chick myocardiocytes. In the same work, the use of NOS inhibitor (N-nitro-L-arginine) and NADPH oxidase inhibitor (diphenyleneiodonium, DPI), reported to eliminate superoxide formation from NOS, failed to attenuate the ROS generation during hypoxic preconditioning, excluding NOS as a source of superoxide. Conversely, myxothiazol diminished the ROS generation during hypoxic preconditioning, pointing mitochondrial complex III as the responsible of ROS production. During univalent electron transfer to O_2, usually at the ubisemiquinone site, the resulting superoxide may subsequently be converted to H_2O_2 in the cytosol by copper, zinc-SOD (Cu, Zn-SOD) [19, 26]. This assumption led to Vanden Hoek and coworkers to treat the cells with exogenous H_2O_2, which evokes preconditioning-like protection and markedly lowers apoptosis. Moreover, SOD inhibition which augments superoxide generation relative to H_2O_2 formation annuls the protec-

tive effects of preconditioning. Collectively, the hydrogen peroxide is more directly responsible for the protective effects during ischemia, and complex III is the main source of superoxide. In this context, Adesina and colleagues elegantly demonstrated that targeting mitochondrial H_2O_2 in pulmonary artery endothelial cells ameliorates pulmonary hypertension pathogenesis [2]. Altogether, these studies point out that reducing mitochondrial H_2O_2 may prevent hypoxia-induced pulmonary hypertension.

It has been found that more H_2O_2 is produced by PASMC mitochondria than systemic arterial SMC mitochondria. These mitochondria have lowered respiratory rates, are more depolarized, and contain more mitochondrial SOD [20]. Furthermore, the H_2O_2 is able to activate voltage-dependent potassium (K_v) channels.

2.3   Role of NADPH Oxidase

Another source of ROS is the NAD(P)H oxidase family of enzymes located in the outer membrane of cells and inner mitochondrial membranes. These enzymes are implicated in the regulation of vascular tone through the production of H_2O_2 and reducing the availability of NO by $\cdot O_2^-$ quenching (through $ONOO^-$ formation) [27, 28]. Nicotinamide adenine dinucleotide (NADH) and nicotinamide adenine dinucleotide phosphate (NAD(P)H) oxidase (Nox) are responsive to different cytokines, angiotensin II (Ang II), mechanical forces (e.g., shear stress), and metabolic factors [15, 29]. Hypertensive patients and normotensive subjects with a family history of hypertension (genetic risk) exhibit elevated H_2O_2 in plasma compared with blood pressure-matched normotensive subjects [30]. This suggests a genetic basis for hypertension, which is linked to ROS generation.

The Nox family of proteins includes Nox1–5 and Duox 1 and 2. The phagocyte Nox (Nox2) was the first characterized isoform comprised of six subunits that interact to form the active enzyme. The two integral membrane proteins, p22phox and gp91phox (α and β subunits, respectively), form the heterodimer flavocytochrome b558 (Cyt b558). The cytosolic regulatory complex is comprised of the three subunits p40[phox], p47[phox], and p67[phox].[22] Only upon stimulation and phosphorylation of p47[phox] will the cytosolic complex associate with Cyt b558 to form the active enzyme. Complete formation of the active oxidase requires Rac2 and Rap1 guanine nucleotide-binding proteins [31]. Once completed, electrons from the substrate make their way to oxygen forming O_2^- which is quickly converted to peroxide.

Nox4 was identified in kidney in 2000 by Geiszt and coworkers [32]. Primary expression of Nox4 was displayed in the medial layer of pulmonary blood vessels of mice and humans using anti-Nox4 antibodies and in situ hybridization by Mittal and coworkers [33]. Nox4 mRNA is the most abundant transcript in blood vessels (1000 times greater than Nox1 and Nox2) [34]. This enzyme is constitutively active in cardiovascular tissues and is involved in oxygen sensing, vasomotor control, cellular proliferation, differentiation, migration, apoptosis, senescence, fibrosis, and angiogenesis [35]. Unlike Nox1–3, Nox4 does not depend on cytosolic protein binding for ROS generation [36]. Transfer of electrons from NADPH to FAD occurs continuously due to the unique C-terminus [37].

Nox4 is expressed primarily in the mitochondria of cardiac myocytes. Ago and colleagues determined that the expression of Nox4 is upregulated in response to hypertrophic stimuli, e.g., pressure overload by imposing transverse aortic constrictions in mice, leading to oxidative stress in mitochondria of the heart and subsequently mitochondrial dysfunction and cardiac cell death [38]. Later on, Kuroda and colleagues showed that increases in Nox4 expression and O_2 production in mitochondria are blocked in cardiac-specific deletion of *Nox4* (c-*Nox4* $^{-/-}$) mice. Moreover, c-*Nox4* $^{-/-}$ displayed decreased fibrosis, cardiac hypertrophy, and apoptosis and improved cardiac function compared with wild-type (WT) mice. The authors also found that increased expression of Nox4 enhances oxidative stress and cardiac dysfunction [39]. These findings point to ROS generated by Nox4, as well as other areas of the ETC, which contribute to mito-

chondrial permeability transition-matrix swelling, outer membrane rupture, and the release of cytochrome *c* (CytC) into the cellular matrix [39, 40]. These insights point out Nox4 as the main contributor to oxidative stress in mitochondria of cardiomyocytes, driving mitochondrial and heart dysfunction during pressure overload.

Factors that have upregulated Nox4 activity include transforming growth factor beta (TGF-β), Ang II, tumor necrosis factor alpha (TNF-α), ER stress, shear stress, hypoxia, and ischemia [41–47]. It has been found that Nox4 expression is elevated in the intimal lesions of coronary arteries underlying atherosclerosis in humans [48]. Both shear stress and Ang II promote ROS generation from Nox1/NADPH located in the plasma membrane of cells along arterial bends, which oxidizes low-density lipoproteins (LDL) [35, 45]. The minimally modified LDL will induce monocyte chemoattractant protein (MCP-1), thereby eliciting an immune response and recruiting leukocytes to the area [45].

Nox4 mediates the growth of hypoxia-induced growth of PASMCs and induces vascular endothelial growth factor (VEGF) secretion and angiogenesis [49, 50]. Nox4 activity may be lowered by inhibitory agents of ROS generation, as well as the processes that are triggered by hypoxic activity, as shown in Fig. 2.2. PKCε translocation peptide inhibitor and glutathione peroxidase 1 (Gpx1) overexpression may be used to disrupt PKCε activation of Nox4 and lower intracellular ROS concentrations, respectively, in order to further reduce ROS [7]. Directly targeting Nox4 with apocynin effectively lowers ROS generation and additionally reduces the intracellular Ca^{2+} concentration in the cells that promote vasoconstriction and proliferation (to be discussed later) [7].

2.4 Hypoxia-Inducible Factor 1

During hypoxia, the transcription of several gene products is induced in order to supply O_2 to cells and tissues [51]. Most of the cellular responses to the hypoxic environment are regulated by hypoxia-inducible factors (HIFs), a family of

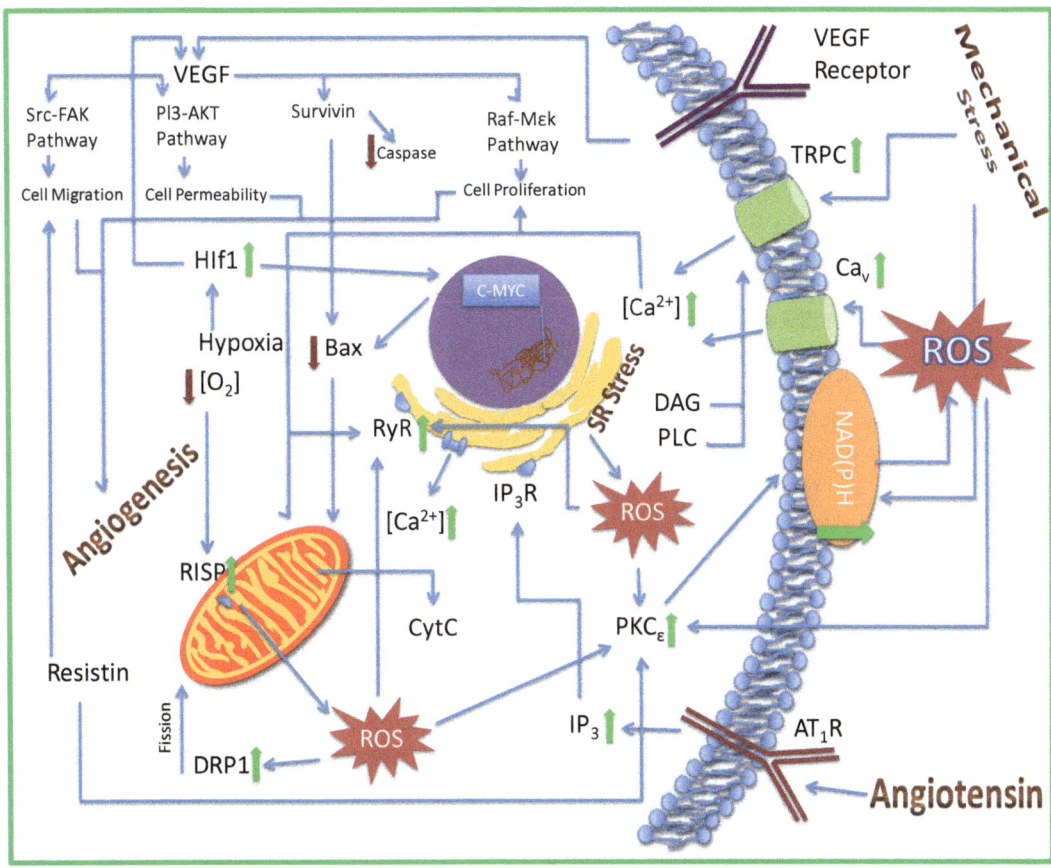

Fig. 2.2 A diagram of the underlying processes implicated in pulmonary hypertension. Plummeting O_2 concentrations in the cytoplasm can activate the hypoxia-inducible factor 1 (HIF-1), which can activate the fetal gene program and alter cellular metabolism. ROS generated by NADPH and the mitochondrial ETC are important signaling molecules that contribute to myocyte dysfunction and respiratory illnesses. ROS can activate proteins and channels that release Ca^{2+} into the cell, which elicits proliferation responses as well as apoptosis. Vascular endothelial growth factor (VEGF) and angiotensin may trigger and promote proliferation, migration, or permeability responses

DNA binding proteins. These transcription factors mediate the induction of genes related to the formation of new blood vessels guaranteeing the O_2 delivery and promoting cell survival [51]. Interestingly, the cellular adaptation to hypoxic conditions allows the growth and rapid development of certain cells such as embryos and solid tumors [52]. Moreover, HIFs have been associated with the pathogenesis of pulmonary arterial hypertension [53, 54]. HIFs are heterodimers formed by an O_2-regulated α subunit and a hydrocarbon receptor nuclear translocator (ARNT) β subunit [55, 56]. Studies about the synthesis of erythropoietin (EPO) as response to hypoxia led to the characterization of HIF-1 [56]. EPO is a peptide hormone that triggers erythropoiesis, subsequently increasing O_2 supply [52]. HIF-1α subunit is a master transcriptional regulator that is constitutively expressed and under normoxic conditions is rapidly degraded by the ubiquitin-proteasome system [51]. Under hypoxic conditions, whereby ROS generated in the mitochondria escape as H_2O_2, prolyl hydroxylase (PHD) is inhibited, which is normally responsible for HIF-1α degradation. PHD under normal conditions hydroxylates HIF-1 in proline and asparagine residues, targeting it for protein degradation. Once PHD is inhibited, the HIF-1 α-ARNT heterodimer is formed and binds to hypoxia-responsive element of several genes in the nucleus of the cell [57, 58].

HIF-1 can regulate O_2 mitochondrial homeostasis and biogenesis [59, 60]. This transcription factor facilitates the transition from aerobically derived ATP to anaerobically derived ATP by inducing glycolytic enzymes, pyruvate kinase M, and glucose transporters, helping cells to produce energy in hypoxic conditions [61, 62]. More specifically, HIF-1 upregulates pyruvate dehydrogenase kinase 1 (PDK1), which is the kinase responsible for inactivating pyruvate dehydrogenase (PDH). This process prevents the oxidative decarboxylation of pyruvate to acetyl-CoA, thereby interfering with pyruvate metabolism via the tricarboxylic acid (TCA) cycle [63, 64]. Furthermore, HIF-1 regulates cytochrome c oxidase (COX, complex IV) in hypoxic cells. This complex in the inner mitochondrial membrane is a dimer comprising 13 subunits each (COX1–COX13). Fukuda and coworkers showed in mammalian cells that HIF-1 switches the COX4–1 isoform to COX4–2 which optimizes COX activity under hypoxic conditions, exhibiting higher metabolic rates and ATP production as compared to cells without the COX4–1 isoform [65].

HIF-1 has also been implicated in the metabolic reprogramming and tumorigenesis in renal carcinoma by inhibition and degradation of c-MYC (a regulator of biogenesis and O_2 consumption). The inhibition of c-MYC leads to a decrease in respiration [66]. Moreover, c-MYC is involved in the downregulation of anti-apoptotic Bcl-2 family members such as Bax (Bcl-2-like protein 4) [67]. Bax, a channel-forming protein in the outer mitochondrial membrane, facilitates the release of cytochrome c to the cytosol [68]. The release of cytochrome c is illustrated in Fig. 2.1 through the Bax association with the mitochondrial outer membrane. That in itself is a commitment to cell apoptosis. Furthermore, the loss of cytochrome c and the Rieske iron-sulfur protein (RISP) center of complex III terminates the ROS signal from the mitochondria [69]. This downregulation mechanism may involve the reduced activity of Bax in order to protect the mitochondria and the cell from apoptosis (Fig. 2.2).

2.5 Vascular Endothelial Growth Factor

As stated before, HIF-1 is activated within hypoxic microenvironments, i.e., surrounding tumors, which eventually leads to vascular endothelial cell proliferation and angiogenesis in order to restore normoxic conditions to a specific location [70, 71]. During angiogenesis, the sprouting of new blood vessels from pre-existing vasculature is mediated by regulator factors such as fibroblast growth factor (FGF), platelet-derived growth factor (PDGF), epidermal growth factor (EGF), angiopoietins, and the vascular endothelial growth factor (VEGF), which is particularly upregulated as the result of HIF-1 activation [72]. During hypoxia, HIF-1 binds the regulatory region of the VEGF gene, inducing its transcription and protein synthesis in ECs. These cells contribute to the formation of new blood vessels, supplying oxygenated blood to deprived tissues and restoring the normoxic state [62].

VEGF exerts its biological actions through two receptors, the VEGF receptor 1 (VEGFR1) and the VEGF receptor 2 (VEGFR2), both members of the family of receptor tyrosine kinases (RTKs). VEGF and VEGFR2 are abundant in lung tissue and are essential for its development and maintenance [73]. In fact, patients suffering from severe PH show elevated levels of VEGF and soluble VEGFR2 in their blood plasma as well as in plexiform lesions [74]. The structural similarity between VEGFR and other RTKs makes its biology complex and involved. Angiogenesis is highly regulated and controlled by its own expression levels as well as by multiple proteins that can interact with the endogenous ligand. Positive regulators of VEGF mRNA expression include FGF, transforming growth factor alpha (TGF-α) and beta (TGF-β), and PDGF [75]. Moreover, it has been proposed that VEGFR1 is capable of binding placenta growth factor (PLGF) and VEGFB, but does not bind VEGFR2 [76, 77]. The overall ratio of these different ligands competing for the VEGFR binding site may potentiate the effects of VEGF. Currently, it is well accepted that VEGFR2 is the principal mediator of mitogenic and angiogenic effects of

VEGF, as well as microvascular permeability. Additionally, several isoforms of VEGF display different properties such as diffusibility and mitogenic activity [75]. Interestingly, some VEGF isoforms can bind to numerous proteins in the extracellular cell matrix (ECM), promoting integrin-dependent cell spreading, migration, and cell survival [78–80]. Additionally, VEGF also interact with the Notch signaling to regulate the formation of new vessels. Notch is a key signaling pathway that mediates the development of several cell types including ECs [81, 82].

VEGFR2 stimulation may activate several different signaling pathways, as illustrated in Fig. 2.2, including PI3K (phosphatidylinositol 3-kinase)-Akt (protein kinase B) pathway, PI3K-mTORC2 (mammalian target of rapamycin complex 2) pathway, Raf-MEK (mitogen-activated or extracellular signal-regulated protein kinase kinase)-MAPK (mitogen-activated protein kinase) pathway, and the Src-FAK (focal adhesion kinase) pathway. The mTORC2 pathway promotes cell survival and vascular permeability while inhibiting the Bcl-2-associated death promoter and caspases [83]. The Raf-MEK-MAPK pathway activates cell proliferation and the Src-FAK pathway elevates cell motility [84]. The relationship between mitochondrial function and VEGF was explored by Guo and colleagues. They found in human umbilical endothelial vein cells (HUVECs) that ROS generation decreased upon VEGF treatment. They also showed that oxidative phosphorylation and ATP levels increased, as well as catalase and Gpx1 expression levels, both part of the ROS defense system. VEGF was also able to activate mTOR through phosphorylation of the ribosomal S6 protein. The cells stimulated with VEGF accumulated in the S and G_2 phases, forming tube-like structures, and this phenomenon was abolished with the use of rapamycin (the mTOR inhibitor). The authors concluded that VEGF protects the ECs by enhancing mitochondrial function [85].

The evidence points out that hypoxia triggers the upregulation of VEGF and its receptors, leading to angiogenesis. In this context, the role of VEGF in PH has been widely explored in animal models through pharmacological and genetic approaches. It was found through experimental overexpression of VEGFA (the most abundant isoform) that the development of hypoxia-induced PH was blunted [86]. Then Farkas and colleagues demonstrated in experimental lung fibrosis rats that VEGF ameliorated PH via inhibition of endothelial apoptosis, while obstructing VEGF worsened it [87]. Endothelial cell survival is mainly regulated by VEGFR1, while VEGFR2 signaling cascade is involved in vascular differentiation and capillary-like tube formation [88]. VEGF induces production of the anti-apoptotic protein survivin (Fig. 2.2), which is transcriptionally regulated by mTOR, PI3K-Akt, and Bcl-2/ERK pathways [72, 89]. Survivin has been found in the pulmonary arteries of patients with PH and is associated with plexiform lesions [90, 91]. The inhibition of surviving expression elicits pulmonary vascular apoptosis and reverses PH, suggesting a novel therapeutic strategy [90]. Pharmacological inhibition of VEGFR1 and VEGFR2 by the small-molecular RTK inhibitor SU5416 in combination with chronic hypoxia in a murine model evokes angio-obliterative PH (by cell death-dependent pulmonary endothelial cell proliferation), vascular remodeling, and right ventricular hypertrophy and failure [92, 93]. It is well understood that, in the Su5416/hypoxia model, the triggering signal is the inhibition of VEGFR kinase activity. However, there is no conclusive evidence implicating the blockade of VEGF signaling as the modulator of PH in humans [92]. A possible exception is the PH displayed in patients suffering from chronic myelogenous leukemia and treated with the tyrosine kinase inhibitor dasatinib [94]. The existence of an endogenous molecule capable of inhibiting the VEGFR activity and subsequently triggering PH should be further explored.

Angio-obliterative PH may be described as vascular wound healing runaway train. Alterations in O_2 levels induce neointimal lesions and subsequent right ventricular remodeling. In order to accomplish angiogenesis in Su5416/hypoxia PH models or human PH, apoptosis-resistant cells resulting from the pharmacological or the autocrine VEGF2 blockade lead to a proliferate response via elevated levels of other growth factors such as FGF and PDGF. Moreover, VEGF signaling may continue uninhibited by the action of VEGFR3 or to the integrin $\alpha v\beta 3$ (expressed in the membrane of activated ECs) [92].

PH is a usual hemodynamic impediment in heart failure. In several pathophysiological circumstances such as ischemia, hypertrophy, and hypoxia, the postnatal heart activates the fetal gene program, which in turn affects protein synthesis, excitation-contraction coupling, intracellular Ca^{2+} signaling, and apoptosis [95]. The fetal gene program involves the activation of the PI3K-Akt pathway, which raises glycogen levels and protects against ischemic damage [95–97].

2.6 Other Proteins Involved in ROS Signaling

VSMCs contribute to hypertension development through the increase in vascular resistance via SMC growth and contraction. The renin-angiotensin system, mainly by the action of Ang II, increases arterial blood pressure through Ang II type 1 (AT_1) receptors [98]. An elevated protein expression and signaling of AT_1 in the pulmonary vasculature of patients with PH has been shown [99]. Ang II is produced in lung and endothelial cells by angiotensin-converting enzyme (ACE) acting on angiotensin I [98, 100]. Schelling and colleagues observed that Ang II activated the phospholipase C (PLC)-$\beta 1$ isoform, which is consistent with other signaling cascades induced by vasoconstrictor hormones [101]. Moreover, it has been demonstrated that PLC can interact with transient receptor potential canonical (TRPC) channels in the plasma membrane [102, 103], as depicted in Fig. 2.2. PLC principally hydrolyzes phosphatidylinositol 4,5-biphosphate (PIP_2) into the secondary messenger inositol 1,4,5-trisphosphate (IP_3) and diacylglycerol (DAG) [101]. DAG remains in the plasma membrane where it can activate TRPCs which allow Ca^{2+} entry into the cell [104]. IP_3 diffuses across the cytosol to bind the IP_3 receptors (IP_3Rs) on the surface of the sarcoplasmic reticulum (SR) [105]. Ca^{2+} flooding the cytoplasm triggers secondary cascade events wherein additional protein

kinase C (PKC) and Ca^{2+}/calmodulin kinase II (CaMKII) are activated [106]. Ca^{2+} influx also activates ryanodine receptors (RyRs) on the SR membrane, which in turn release more stored Ca^{2+}, a process known as Ca^{2+}-induced Ca^{2+} release (CICR) (shown in Fig. 2.1) [107]. Ca^{2+} is essential in smooth muscle contraction, elevating blood pressure, and in some cases provoking apoptosis. As portrayed in Fig. 2.2, Ang II can increase Nox1 activity and ROS generation in VSMCs, thereby mediating the development and progression of cardiovascular diseases [35, 108]. In addition, Matsuno and colleagues demonstrated using Nox1-deficient mice that ROS derived from Nox1/NADPH are critical for the pressor response to Ang II by reducing the availability of NO [34].

The dynamin-related protein 1 (DRP1) activity in mitochondrial fission may contribute to the quasi-malignancy of ECs and hyperproliferation of VSMCs involved in PH [35, 109]. DRP1 is a GTPase that mediates mitochondrial fission and autophagy. As can be seen in Fig. 2.2, this protein is located in the cytosol, but may be recruited to the outer mitochondrial membrane with calcium in response to stress and ROS [8, 110]. DRP1-dependent autophagy is triggered to remove damaged mitochondria during ischemia events [111]. In this regard, Zhang and colleagues showed how the non-specific ROS inhibitor, N-acetylcysteine (NAC), decreased DRP1 activity. Moreover, TEMPO, the mitochondrial ROS inhibitor, significantly lowered hypoxia-induced DRP1 expression. The genomic inhibition of DRP1 by siRNA also decreased ROS, while overexpression of DRP1 increased them. Furthermore, silencing DRP1 led to elevated rates of mitochondria fragmentation in cultured PASMCs, while overexpression of DRP1 partly inhibited fragmentation [8]. These findings point out that ROS are generated principally in mitochondria, mediating the fission of this organelle in PASMCs contributing to pulmonary vascular remodeling. Studies have found suppression of DRP1 to positively affected myocardiocytes and other cell types during ischemia-reperfusion events by preserving mitochondrial structure and distribution, lowering oxygen consumption, and reducing apoptosis rates [111]. However, Shirakabe and colleagues found that mitochondrial dysfunction, hypertrophy, and heart failure developed rapidly in DRP1 KO mice, suggesting a protective role of this protein [112]. The complexity of DRP1 may not be completely understood to fully map its involvement in mitochondrial dysfunction as it relates to PH.

In another study, glioblastoma U251 cells under hypoxia showed elevated expression of DRP1 through either ROS action or upregulation of HIF-1α. HIF-1α was inhibited using echinomycin, which reduced DRP1 expression and attenuated mitochondrial fission [113]. It was found that blocking the FIS1 fission protein and DRP1 prevented the fragmentation of mitochondria and abated the loss of CytC, which effectively protected cells from apoptosis [114]. Cytochrome c oxidase is the transmembrane protein imbedded in the cardiolipin of complex IV of the ETC. When elevated, ROS, the cardiolipin-cytochrome c complex, undergoes a peroxidase function, thereby releasing the heme protein. Zhang and colleagues were able to visualize the release of CytC through COX IV staining microscopy [8]. CytC is extruded through pores and eventually makes its way to the cytoplasm, as shown in Fig. 2.1.

Cytochrome c interacts with IP$_3$ on ER causing Ca^{2+} release [115]. Higher rates of CytC release from the mitochondria into the cytoplasm were observed in cells treated with siDRP1 (opposite to DRP1 overexpression). Knockdown of DRP1 even prevented mitochondrial fragmentation and CytC loss despite the translocation of Bax to the outer mitochondrial membrane [116]. Bax resides in the cytosol until apoptotic signaling (H$_2$O$_2$, heat, changes in pH, and mitochondrial membrane remodeling) causes it to bind to the outer mitochondrial membrane [117]. It is likely that Bax interacts with voltage-dependent anion channels and fission/fusion machinery in order to imbed itself into the outer membrane of the mitochondria [118]. CytC is then lost through voltage-dependent anion channels. Figure 2.1 displays how CytC binds to the apoptotic protease-activating factor 1 (Apaf1) forming the apoptosome and initiates programmed cell death [118]. In support, Thomas and Jacobson found delayed CytC release in A549 lung epithelial cells. These cells had impaired Drp1 mitochondrial recruitment and decreased Drp1-dependent fission. Consequently, the authors observed less mitochondrial fission and a resistance to apoptosis [119].

Finally, it is understood that Rac is an important signaling molecule in cardiovascular systems. Rac is a cytosolic low-molecular-weight guanine nucleotide-binding protein, which is a member of the Rho family of small GTPases. It was found that this GTPase promotes hypercholesterolemia in mice, which includes NADH-derived ROS, impaired vasorelaxation, elevated macrophage infiltration, and enhanced plaque rupture [120]. Rac1-Nox-mediated ROS have been shown to inhibit NO production and promote low-density lipoprotein (LDL) production as well as the production of inflammatory mediators like intracellular adhesion molecule 1 (ICAM-1) and vascular cell adhesion molecule 1 (VCAM-1), therefore progressing atherosclerotic lesions [121].

2.7 Ca^{2+} Signaling

There is cross-talk between cellular Ca^{2+} and ROS [122]. Extracellular ROS from NADPH oxidase and endogenous H$_2$O$_2$ were shown to stimulate the L (long-lasting)- and T (transient)-type voltage-dependent Ca^{2+}channels which favor the entry of this ion into vascular smooth muscle cells [123, 124]. The opening of these channels may be influenced by phosphorylation processes via protein kinases activated by ROS [123]. PLC pathway is known to activate TRPCs, in association with DAG (Fig. 2.2). TRPC1, TRPC3, and TRPC6 genes are upregulated in models of cardiovascular disease, which mediates smooth muscle proliferation, and thus contribute to hypertrophy. Also, the inhibition of these channels mitigates the associated pathophysiology [125]. Ca^{2+} influx through voltage-dependent Ca^{2+} channels (VDCC, Ca$_v$) is implicated in the proliferation of PASMCs [126]. Pharmacological blockade and genomic interference of the Ca$_v$3.1

subtype have been proved to inhibit PASMC proliferation and the entry to cell cycle [127]. Moreover, a G protein-coupled receptor (GPCR), the Ca^{2+} sensing receptor (CaSR), along with the PI3K and the MEK1/ERK1/2 pathways is involved in the hypoxia-evoked proliferation of PASMCs [128]. Additionally, Yamamura and colleagues found in a murine model of pulmonary hypertension induced by the exposure to hypoxia that NPS2143 (an antagonist of CaSR) prevented right ventricular hypertrophy and vascular remodeling [129].

Ca^{2+} is crucial to cell homeostasis and the maintenance of a proper vascular tone. A vascular smooth muscle cell at rest maintains an intracellular Ca^{2+} concentration ($[Ca^{2+}]_i$) around 100 nM. An agonist-stimulated VSMC displays an elevated $[Ca^{2+}]_i$ that oscillates between 500 nM and 1 mM [130]. One of the biggest Ca^{2+} stores in SMCs is the sarcoplasmic reticulum (SR), which contains roughly between 2 and 5 mM of Ca^{2+} [131]. The SR uptakes Ca^{2+} using sarco-/endoplasmic reticulum Ca^{2+} ATPases (SERCAs). It is in turn released through RyRs and IP_3Rs (Fig. 2.1). It is well known that the RyR can be stimulated by mitochondria-derived ROS, which oxidize the thiol groups present in the receptor [132, 133], as seen in Fig. 2.2. This modification enhances the channel activity and increases cytosolic Ca^{2+}. It is no surprise that mice with hypoxia-induced pulmonary hypertension also display increased activity of RyR in PASMCs [134]. Furthermore, RyR2 knockout mice do not show RyR activity or the associated elevation of intracellular Ca^{2+} and do not develop pulmonary hypertension followed by hypoxic exposure. The endogenous inhibitor molecule, FK506 binding protein 12.6 (FKBP12.6), forms a complex with RyR2 in a closed conformation. Disruption of this complex is proposed to underlie increased channel activity and elevated Ca^{2+} release into the cytosol [135]. Additionally, it has been remarkably proposed that alterations in $[ROS]_i$ in PASMCs can modify the activity of ion channels, evoking a large increase in $[Ca^{2+}]_i$. In this regard, hypoxia-induced changes in $[ROS]_i$ have been shown to inhibit K^+ currents (mediated by K_v channels) [18, 20, 136]. Moreover, the blockade

of these channels lead to membrane depolarization and the opening of Ca_v channels with the subsequent large increase in $[Ca^{2+}]_i$. Furthermore, PH patients with downregulated K_v proteins exhibit elevated Ca^{2+} influx via L-type voltage-dependent Ca^{2+} channels (L-VDCCs), which promote vasoconstriction and cell proliferation [20].

Appropriate Ca^{2+} uptake into the mitochondria supports metabolic processes, such as dehydrogenase activity. However, overwhelming Ca^{2+} stimulus may initiate apoptosis. Ca^{2+} efflux and influx across the mitochondria are regulated by distinctive proteins. The mitochondrial permeability transition pore (mPTP) is sensitive to voltage and Ca^{2+} and mediates the mitochondrial permeability transition (MPT). MPT is elicited by a Ca^{2+} overload in matrix Ca^{2+} concentration that results in organelle swelling. Interestingly, this process seems to play a critical role in cell death. On the other hand, Ca^{2+} import across the mitochondria occurs through the voltage-dependent anion channel (VDAC) (simplified in Fig. 2.1a). A portion of the outer mitochondrial membrane of this protein associates with other proteins (NADH, metabolic enzymes, chaperones) occurring in the inner mitochondrial membrane which modulate VDAC gating properties. The opening of this channel, usually initiated by fast or elevated Ca^{2+} ion uptake, causes depolarization, ETC inhibition and ROS generation, and loss of antioxidants like glutathione [25, 137]. Under ischemia, the opening of mPTP assists the cell by reducing Ca^{2+} overload in the cytosol, but this causes the mitochondria to swell [138]. Reperfusion causes additional ionic imbalances by reactivating energy transduction, contractions, and ROS generation [139]. Protein overload events in the SR may lead to excessive ROS generations and hyperoxidation in the SR. Accumulation of mis-/unfolded polypeptides give rise to ER stress and initiate Ca^{2+} release through RyRs and IP_3Rs into the cytosol [140].

Protein kinase C epsilon (PKCε) is abundantly expressed in adult cardiomyocytes. It regulates muscle contraction at a sarcomeric protein level, modulates intracellular metabolism through interactions with mitochondria, and plays a key role in protecting cells against ischemic injury through its involvement in hypertrophy. Acute hypoxia has been shown to activate PKCε by 3.2-fold in PASMC [24]. Through use of ETC inhibitors as well as Gpx1 overexpression mouse models, Rathore and colleagues demonstrated how ROS subsequently increases PKCε activity. Exogenous H_2O_2 was able to elevate PKCε activity even with blockaded ETC. By inhibiting PKCε, hypoxia-induced intracellular Ca^{2+} increases were shown reduced by 40 percent [24].

It has been found that ROS generated from the mitochondria elevate the activity of PKCε as well as Nox and propagate ROS generation [141]. Recent studies implicate more specifically the RISP of complex III as the key player for mitochondrial ROS production. Isolated mitochondria that were siRNA treated to silence the RISP activity had no hypoxic ROS generation, while those with RISP overexpression produced a greater ROS response [142].The inhibition of the RISP activity has been shown to abolish the increase in $[Ca^{2+}]_i$ in PASMCs and hypoxic vasoconstriction in isolated PAs [142] .

PKCε is also an upstream modulator of resistin-evoked VSMC migration [143]. Resistin is an adipokine mainly expressed in cells of monocyte/macrophage lineage in humans [144]. Elevated serum levels of resistin promote vascular cell dysfunction and are linked to inflammation associated with atherosclerosis and myocardial infarction [145]. Interestingly, this adipokine upregulates the expression of Nox*4* and *p22phox*, but only with active PKCε as the mediator. ROS may be halted through blockading PKCε. PKCε involvement in cardiovascular diseases is further demonstrated by its role in facilitating neointimal hyperplasia and luminal narrowing in mouse models. Moreover, resistin-treated mice only showed neointimal formations with unrestrained PKCε, while blocking PKCε showed no change [145].

2.8 MicroRNAs

Non-coding genomic transcripts are also regulators of biological and pathological processes. Evaluating the role microRNAs play in heart disease is essential for developing therapeutic targets and treatment. Reddy and coworkers found that miR-99a were associated with cardiomyocyte survival and growth at preliminary stages of pulmonary hypertension, while miR-208b was activated later [146]. The miR-208 is associated with the fetal gene program. There is an asymmetry of microRNA alterations between left and right ventricular remodeling. MicroRNAs 34a, 28, 148a, and 93 were upregulated in right ventricle remodeling and downregulated in left ventricle remodeling. In the right ventricle, there was downregulation in HIF-1α, a target of miR-199a. This microRNA also targets Smad3 and may impact Ca^{2+} concentration and NO release. There is a pathogenic role of miR-126. Lowered levels of miR-126 in right ventricular remodeling led to lowered activation of the Raf-Mek-MAPK pathway. This has been experimentally treated using miR-126 mimics that resulted in better cardiac vascular density and function [147]. With laser-assisted microdissection using microRNA analysis, it was found that plexiform lesions displayed an upregulation of miR-126 and mir-21 and a downregulation of miR-204 [148].

The loss of miR-145 (genetically and pharmacologically) in PH mouse models reduced right ventricle systolic pressure [149]. MicroRNA expression has been found to be significantly different in isolated pulmonary arteries than from normal arteries in PH patients [150]. Whether the expression is altered in several microRNAs in the right heart under hypoxia [151], further research is imperative to elucidate the role of miRNAs in the initiation and development of PH.

2.9 Conclusion

PH is a widespread lung disease affecting millions of people throughout the world. The underlying causes are multifaceted and interconnected. Progressively, we and other researchers are unraveling and piecing together the interplay of ROS with intracellular Ca^{2+} and inflammatory signaling in mediating complex pathophysiological processes in PA involved by numerous proteins and genes. Abnormal function in mitochondria as a result of cellular stress triggers multiple signaling cascades that may eventually lead to alteration in cellular homeostasis and apoptosis. Particularly, RISP in mitochondria acts as a primary hypoxia sensor in the pulmonary artery. Moreover, hypoxic environments promote ROS generation mainly from mitochondria leading to the activation of important signaling molecules such as PKCε and Nox. Both pathways result in the increase of $[Ca^{2+}]_i$ and the subsequent SMC contraction. Collectively, these mechanisms contribute to the development and progress of PH. Researchers are increasingly able to identify specific molecules in pathogenic processes in PH and thus create novel therapeutic targets to better treat this devastating illness.

References

1. Benza RL, et al. Predicting survival in pulmonary arterial hypertension: insights from the Registry to Evaluate Early and Long-Term Pulmonary Arterial Hypertension Disease Management (REVEAL). Circulation. 2010;122:164–72.
2. Adesina SE, et al. Targeting mitochondrial reactive oxygen species to modulate hypoxia-induced pulmonary hypertension. Free Radic Biol Med. 2015;87:36–47.
3. Aggarwal S, Gross CM, Sharma S, Fineman JR, Black SM. Reactive oxygen species in pulmonary vascular remodeling. Compr Physiol. 2013;3:1011–34.
4. Cao JY, Wales KM, Cordina R, Lau EMT, Celermajer DS. Pulmonary vasodilator therapies are of no benefit in pulmonary hypertension due to left heart disease: a meta-analysis. Int J Cardiol. 2018;273:213–20.
5. Spiekerkoetter E, Kawut SM, de Jesus Perez VA. New and emerging therapies for pulmonary arterial hypertension. Annu Rev Med. 2019;70:45–59.
6. Suresh K, Shimoda LA. Endothelial cell reactive oxygen species and Ca(2+) signaling in pulmonary hypertension. Adv Exp Med Biol. 2017;967:299–314.
7. Rathore R, et al. Hypoxia activates NADPH oxidase to increase [ROS]i and [Ca2+]i through the mitochondrial ROS-PKCepsilon signaling axis in pulmonary artery smooth muscle cells. Free Radic Biol Med. 2008;45:1223–31.
8. Zhang L, et al. Reactive oxygen species effect PASMCs apoptosis via regulation of dynamin-related protein 1 in hypoxic pulmonary hypertension. Histochem Cell Biol. 2016;146:71–84.
9. Droge W. Free radicals in the physiological control of cell function. Physiol Rev. 2002;82:47–95.
10. Phaniendra A, Jestadi DB, Periyasamy L. Free radicals: properties, sources, targets, and their implication in various diseases. Indian J Clin Biochem. 2015;30:11–26.
11. Zhang J, et al. ROS and ROS-mediated cellular signaling. Oxidative Med Cell Longev. 2016;2016:4350965.
12. Fridovich I. Superoxide anion radical (O2-.), superoxide dismutases, and related matters. J Biol Chem. 1997;272:18515–7.
13. Thannickal VJ, Fanburg BL. Reactive oxygen species in cell signaling. Am J Physiol Lung Cell Mol Physiol. 2000;279:L1005–28.
14. Paravicini TM, Touyz RM. NADPH oxidases, reactive oxygen species, and hypertension: clinical implications and therapeutic possibilities. Diabetes Care. 2008;31(Suppl 2):S170–80.
15. Wilcox CS, et al. Nitric oxide synthase in macula densa regulates glomerular capillary pressure. Proc Natl Acad Sci U S A. 1992;89:11993–7.
16. MacMicking JD, et al. Altered responses to bacterial infection and endotoxic shock in mice lacking inducible nitric oxide synthase. Cell. 1995;81:641–50.
17. Wang YX, Zheng YM. ROS-dependent signaling mechanisms for hypoxic Ca(2+) responses in pulmonary artery myocytes. Antioxid Redox Signal. 2010;12:611–23.
18. Archer SL, Huang J, Henry T, Peterson D, Weir EK. A redox-based O2 sensor in rat pulmonary vasculature. Circ Res. 1993;73:1100–12.
19. Turrens JF. Mitochondrial formation of reactive oxygen species. J Physiol. 2003;552:335–44.
20. Michelakis ED, et al. Diversity in mitochondrial function explains differences in vascular oxygen sensing. Circ Res. 2002;90:1307–15.
21. Waypa GB, Chandel NS, Schumacker PT. Model for hypoxic pulmonary vasoconstriction involving mitochondrial oxygen sensing. Circ Res. 2001;88:1259–66.
22. Waypa GB, et al. Increases in mitochondrial reactive oxygen species trigger hypoxia-induced calcium responses in pulmonary artery smooth muscle cells. Circ Res. 2006;99:970–8.
23. Waypa GB, et al. Mitochondrial reactive oxygen species trigger calcium increases during hypoxia in pulmonary arterial myocytes. Circ Res. 2002;91:719–26.
24. Rathore R, et al. Mitochondrial ROS-PKCepsilon signaling axis is uniquely involved in hypoxic increase in [Ca2+]i in pulmonary artery smooth muscle cells. Biochem Biophys Res Commun. 2006;351:784–90.
25. Zhang X, et al. A mechanism underlying hypertensive occurrence in the metabolic syndrome: cooperative effect of oxidative stress and calcium accumulation in vascular smooth muscle cells. Horm Metab Res. 2014;46:126–32.

26. Vanden Hoek TL, Becker LB, Shao Z, Li C, Schumacker PT. Reactive oxygen species released from mitochondria during brief hypoxia induce preconditioning in cardiomyocytes. J Biol Chem. 1998;273:18092–8.

27. Kajiya M, et al. Impaired NO-mediated vasodilation with increased superoxide but robust EDHF function in right ventricular arterial microvessels of pulmonary hypertensive rats. Am J Physiol Heart Circ Physiol. 2007;292:H2737–44.

28. Matoba T, Shimokawa H. Hydrogen peroxide is an endothelium-derived hyperpolarizing factor in animals and humans. J Pharmacol Sci. 2003;92:1–6.

29. Touyz RM, et al. Expression of a functionally active gp91phox-containing neutrophil-type NAD(P)H oxidase in smooth muscle cells from human resistance arteries: regulation by angiotensin II. Circ Res. 2002;90:1205–13.

30. Lacy F, Kailasam MT, O'Connor DT, Schmid-Schonbein GW, Parmer RJ. Plasma hydrogen peroxide production in human essential hypertension: role of heredity, gender, and ethnicity. Hypertension. 2000;36:878–84.

31. Diebold BA, Bokoch GM. Molecular basis for Rac2 regulation of phagocyte NADPH oxidase. Nat Immunol. 2001;2:211–5.

32. Geiszt M, Kopp JB, Varnai P, Leto TL. Identification of renox, an NAD(P)H oxidase in kidney. Proc Natl Acad Sci U S A. 2000;97:8010–4.

33. Mittal M, et al. Hypoxia-dependent regulation of nonphagocytic NADPH oxidase subunit NOX4 in the pulmonary vasculature. Circ Res. 2007;101:258–67.

34. Matsuno K, et al. Nox1 is involved in angiotensin II-mediated hypertension: a study in Nox1-deficient mice. Circulation. 2005;112:2677–85.

35. Chen F, Haigh S, Barman S, Fulton DJ. From form to function: the role of Nox4 in the cardiovascular system. Front Physiol. 2012;3:412.

36. Martyn KD, Frederick LM, von Loehneysen K, Dinauer MC, Knaus UG. Functional analysis of Nox4 reveals unique characteristics compared to other NADPH oxidases. Cell Signal. 2006;18:69–82.

37. Nisimoto Y, Jackson HM, Ogawa H, Kawahara T, Lambeth JD. Constitutive NADPH-dependent electron transferase activity of the Nox4 dehydrogenase domain. Biochemistry. 2010;49:2433–42.

38. Ago T, et al. Upregulation of Nox4 by hypertrophic stimuli promotes apoptosis and mitochondrial dysfunction in cardiac myocytes. Circ Res. 2010;106:1253–64.

39. Kuroda J, et al. NADPH oxidase 4 (Nox4) is a major source of oxidative stress in the failing heart. Proc Natl Acad Sci U S A. 2010;107:15565–70.

40. Baines CP, et al. Loss of cyclophilin D reveals a critical role for mitochondrial permeability transition in cell death. Nature. 2005;434:658–62.

41. Anilkumar N, Weber R, Zhang M, Brewer A, Shah AM. Nox4 and nox2 NADPH oxidases mediate distinct cellular redox signaling responses to ago-nist stimulation. Arterioscler Thromb Vasc Biol. 2008;28:1347–54.

42. Basuroy S, Bhattacharya S, Leffler CW, Parfenova H. Nox4 NADPH oxidase mediates oxidative stress and apoptosis caused by TNF-alpha in cerebral vascular endothelial cells. Am J Physiol Cell Physiol. 2009;296:C422–32.

43. Cucoranu I, et al. NAD(P)H oxidase 4 mediates transforming growth factor-beta1-induced differentiation of cardiac fibroblasts into myofibroblasts. Circ Res. 2005;97:900–7.

44. Hwang J, et al. Pulsatile versus oscillatory shear stress regulates NADPH oxidase subunit expression: implication for native LDL oxidation. Circ Res. 2003;93:1225–32.

45. Lu X, Murphy TC, Nanes MS, Hart CM. PPAR{gamma} regulates hypoxia-induced Nox4 expression in human pulmonary artery smooth muscle cells through NF-{kappa}B. Am J Physiol Lung Cell Mol Physiol. 2010;299:L559–66.

46. Pedruzzi E, et al. NAD(P)H oxidase Nox-4 mediates 7-ketocholesterol-induced endoplasmic reticulum stress and apoptosis in human aortic smooth muscle cells. Mol Cell Biol. 2004;24:10703–17.

47. Rajagopalan S, et al. Angiotensin II-mediated hypertension in the rat increases vascular superoxide production via membrane NADH/NADPH oxidase activation. Contribution to alterations of vasomotor tone. J Clin Invest. 1996;97:1916–23.

48. Sorescu D, et al. Superoxide production and expression of nox family proteins in human atherosclerosis. Circulation. 2002;105:1429–35.

49. Griffith B, et al. NOX enzymes and pulmonary disease. Antioxid Redox Signal. 2009;11:2505–16.

50. Zhang M, et al. NADPH oxidase-4 mediates protection against chronic load-induced stress in mouse hearts by enhancing angiogenesis. Proc Natl Acad Sci U S A. 2010;107:18121–6.

51. Chandel NS, et al. Reactive oxygen species generated at mitochondrial complex III stabilize hypoxia-inducible factor-1alpha during hypoxia: a mechanism of O2 sensing. J Biol Chem. 2000;275:25130–8.

52. Dengler VL, Galbraith M, Espinosa JM. Transcriptional regulation by hypoxia inducible factors. Crit Rev Biochem Mol Biol. 2014;49:1–15.

53. Fijalkowska I, et al. Hypoxia inducible-factor1alpha regulates the metabolic shift of pulmonary hypertensive endothelial cells. Am J Pathol. 2010;176:1130–8.

54. Lei W, et al. Expression and analyses of the HIF-1 pathway in the lungs of humans with pulmonary arterial hypertension. Mol Med Rep. 2016;14:4383–90.

55. Semenza GL. Hypoxia-inducible factors in physiology and medicine. Cell. 2012;148:399–408.

56. Semenza GL, Wang GL. A nuclear factor induced by hypoxia via de novo protein synthesis binds to the human erythropoietin gene enhancer at a site required for transcriptional activation. Mol Cell Biol. 1992;12:5447–54.

57. Murphy MP. How mitochondria produce reactive oxygen species. Biochem J. 2009;417:1–13.

58. Schofield CJ, Ratcliffe PJ. Oxygen sensing by HIF hydroxylases. Nat Rev Mol Cell Biol. 2004;5:343–54.
59. Hirota K. Hypoxia-inducible factor 1, a master transcription factor of cellular hypoxic gene expression. J Anesth. 2002;16:150–9.
60. Okamoto A, et al. HIF-1-mediated suppression of mitochondria electron transport chain function confers resistance to lidocaine-induced cell death. Sci Rep. 2017;7:3816.
61. Aragones J, et al. Deficiency or inhibition of oxygen sensor Phd1 induces hypoxia tolerance by reprogramming basal metabolism. Nat Genet. 2008;40:170–80.
62. Ziello JE, Jovin IS, Huang Y. Hypoxia-Inducible Factor (HIF)-1 regulatory pathway and its potential for therapeutic intervention in malignancy and ischemia. Yale J Biol Med. 2007;80:51–60.
63. Kim JW, Tchernyshyov I, Semenza GL, Dang CV. HIF-1-mediated expression of pyruvate dehydrogenase kinase: a metabolic switch required for cellular adaptation to hypoxia. Cell Metab. 2006;3:177–85.
64. Papandreou I, Cairns RA, Fontana L, Lim AL, Denko NC. HIF-1 mediates adaptation to hypoxia by actively downregulating mitochondrial oxygen consumption. Cell Metab. 2006;3:187–97.
65. Fukuda R, et al. HIF-1 regulates cytochrome oxidase subunits to optimize efficiency of respiration in hypoxic cells. Cell. 2007;129:111–22.
66. Zhang H, et al. HIF-1 inhibits mitochondrial biogenesis and cellular respiration in VHL-deficient renal cell carcinoma by repression of C-MYC activity. Cancer Cell. 2007;11:407–20.
67. Hoffman B, Liebermann DA. Apoptotic signaling by c-MYC. Oncogene. 2008;27:6462–72.
68. Eskes R, et al. Bax-induced cytochrome C release from mitochondria is independent of the permeability transition pore but highly dependent on Mg2+ ions. J Cell Biol. 1998;143:217–24.
69. Guzy RD, Schumacker PT. Oxygen sensing by mitochondria at complex III: the paradox of increased reactive oxygen species during hypoxia. Exp Physiol. 2006;91:807–19.
70. Carmeliet P, et al. Role of HIF-1alpha in hypoxia-mediated apoptosis, cell proliferation and tumour angiogenesis. Nature. 1998;394:485–90.
71. Laderoute KR, et al. 5′-AMP-activated protein kinase (AMPK) is induced by low-oxygen and glucose deprivation conditions found in solid-tumor microenvironments. Mol Cell Biol. 2006;26:5336–47.
72. Sanhueza C, et al. The twisted survivin connection to angiogenesis. Mol Cancer. 2015;14:198.
73. Voelkel NF, Vandivier RW, Tuder RM. Vascular endothelial growth factor in the lung. Am J Physiol Lung Cell Mol Physiol. 2006;290:L209–21.
74. Hirose S, Hosoda Y, Furuya S, Otsuki T, Ikeda E. Expression of vascular endothelial growth factor and its receptors correlates closely with formation of the plexiform lesion in human pulmonary hypertension. Pathol Int. 2000;50:472–9.
75. Ferrara N, Gerber HP, LeCouter J. The biology of VEGF and its receptors. Nat Med. 2003;9:669–76.
76. Olofsson B, et al. Vascular endothelial growth factor B (VEGF-B) binds to VEGF receptor-1 and regulates plasminogen activator activity in endothelial cells. Proc Natl Acad Sci U S A. 1998;95:11709–14.
77. Park JE, Chen HH, Winer J, Houck KA, Ferrara N. Placenta growth factor. Potentiation of vascular endothelial growth factor bioactivity, in vitro and in vivo, and high affinity binding to Flt-1 but not to Flk-1/KDR. J Biol Chem. 1994;269:25646–54.
78. Ashikari-Hada S, Habuchi H, Kariya Y, Kimata K. Heparin regulates vascular endothelial growth factor165-dependent mitogenic activity, tube formation, and its receptor phosphorylation of human endothelial cells. Comparison of the effects of heparin and modified heparins. J Biol Chem. 2005;280:31508–15.
79. Chen TT, et al. Anchorage of VEGF to the extracellular matrix conveys differential signaling responses to endothelial cells. J Cell Biol. 2010;188:595–609.
80. Mahabeleshwar GH, Feng W, Reddy K, Plow EF, Byzova TV. Mechanisms of integrin-vascular endothelial growth factor receptor cross-activation in angiogenesis. Circ Res. 2007;101:570–80.
81. Hellstrom M, Phng LK, Gerhardt H. VEGF and Notch signaling: the yin and yang of angiogenic sprouting. Cell Adhes Migr. 2007;1:133–6.
82. Thomas JL, et al. Interactions between VEGFR and Notch signaling pathways in endothelial and neural cells. Cell Mol Life Sci. 2013;70:1779–92.
83. Zhuang G, et al. Phosphoproteomic analysis implicates the mTORC2-FoxO1 axis in VEGF signaling and feedback activation of receptor tyrosine kinases. Sci Signal. 2013;6:ra25.
84. Claesson-Welsh L. VEGF receptor signal transduction – a brief update. Vasc Pharmacol. 2016;86:14–7.
85. Guo D, Wang Q, Li C, Wang Y, Chen X. VEGF stimulated the angiogenesis by promoting the mitochondrial functions. Oncotarget. 2017;8:77020–7.
86. Partovian C, et al. Adenovirus-mediated lung vascular endothelial growth factor overexpression protects against hypoxic pulmonary hypertension in rats. Am J Respir Cell Mol Biol. 2000;23:762–71.
87. Farkas L, et al. VEGF ameliorates pulmonary hypertension through inhibition of endothelial apoptosis in experimental lung fibrosis in rats. J Clin Invest. 2009;119:1298–311.
88. Zhang Z, Neiva KG, Lingen MW, Ellis LM, Nor JE. VEGF-dependent tumor angiogenesis requires inverse and reciprocal regulation of VEGFR1 and VEGFR2. Cell Death Differ. 2010;17:499–512.
89. Meng L, et al. Survivin is critically involved in VEGFR2 signaling-mediated esophageal cancer cell survival. Biomed Pharmacother. 2018;107:139–45.
90. McMurtry MS, et al. Gene therapy targeting survivin selectively induces pulmonary vascular apoptosis

and reverses pulmonary arterial hypertension. J Clin Invest. 2005;115:1479–91.

91. Rai PR, et al. The cancer paradigm of severe pulmonary arterial hypertension. Am J Respir Crit Care Med. 2008;178:558–64.

92. Nicolls MR, et al. New models of pulmonary hypertension based on VEGF receptor blockade-induced endothelial cell apoptosis. Pulm Circ. 2012;2:434–42.

93. Taraseviciene-Stewart L, et al. Inhibition of the VEGF receptor 2 combined with chronic hypoxia causes cell death-dependent pulmonary endothelial cell proliferation and severe pulmonary hypertension. FASEB J. 2001;15:427–38.

94. Montani D, et al. Pulmonary arterial hypertension in patients treated by dasatinib. Circulation. 2012;125:2128–37.

95. Taegtmeyer H, Sen S, Vela D. Return to the fetal gene program: a suggested metabolic link to gene expression in the heart. Ann N Y Acad Sci. 2010;1188:191–8.

96. Jonassen AK, Mjos OD, Sack MN. p70s6 kinase is a functional target of insulin activated Akt cell-survival signaling. Biochem Biophys Res Commun. 2004;315:160–5.

97. Sack MN, Yellon DM. Insulin therapy as an adjunct to reperfusion after acute coronary ischemia: a proposed direct myocardial cell survival effect independent of metabolic modulation. J Am Coll Cardiol. 2003;41:1404–7.

98. Yang R, Smolders I, Dupont AG. Blood pressure and renal hemodynamic effects of angiotensin fragments. Hypertens Res. 2011;34:674–83.

99. de Man FS, et al. Dysregulated renin-angiotensin-aldosterone system contributes to pulmonary arterial hypertension. Am J Respir Crit Care Med. 2012;186:780–9.

100. Peach MJ. Renin-angiotensin system: biochemistry and mechanisms of action. Physiol Rev. 1977;57:313–70.

101. Schelling JR, Nkemere N, Konieczkowski M, Martin KA, Dubyak GR. Angiotensin II activates the beta 1 isoform of phospholipase C in vascular smooth muscle cells. Am J Phys. 1997;272:C1558–66.

102. Rohacs T. Regulation of transient receptor potential channels by the phospholipase C pathway. Adv Biol Regul. 2013;53:341–55.

103. Song T, Hao Q, Zheng YM, Liu QH, Wang YX. Inositol 1,4,5-trisphosphate activates TRPC3 channels to cause extracellular Ca2+ influx in airway smooth muscle cells. Am J Physiol Lung Cell Mol Physiol. 2015;309:L1455–66.

104. Numata T, Kiyonaka S, Kato K, Takahashi N, Mori Y. In Zhu MX, editors. TRP Channels. Boca Raton, FL, 2011.

105. Hughes AR, Putney JW Jr. Inositol phosphate formation and its relationship to calcium signaling. Environ Health Perspect. 1990;84:141–7.

106. Liu Z, Khalil RA. Evolving mechanisms of vascular smooth muscle contraction highlight key targets in vascular disease. Biochem Pharmacol. 2018;153:91–122.

107. Collier ML, Ji G, Wang Y, Kotlikoff MI. Calcium-induced calcium release in smooth muscle: loose coupling between the action potential and calcium release. J Gen Physiol. 2000;115:653–62.

108. Griendling KK, Minieri CA, Ollerenshaw JD, Alexander RW. Angiotensin II stimulates NADH and NADPH oxidase activity in cultured vascular smooth muscle cells. Circ Res. 1994;74:1141–8.

109. Marsboom G, et al. Dynamin-related protein 1-mediated mitochondrial mitotic fission permits hyperproliferation of vascular smooth muscle cells and offers a novel therapeutic target in pulmonary hypertension. Circ Res. 2012;110:1484–97.

110. Ikeda Y, et al. Endogenous Drp1 mediates mitochondrial autophagy and protects the heart against energy stress. Circ Res. 2015;116:264–78.

111. Hu C, Huang Y, Li L. Drp1-dependent mitochondrial fission plays critical roles in physiological and pathological progresses in mammals. Int J Mol Sci. 2017;18:144.

112. Shirakabe A, et al. Drp1-dependent mitochondrial autophagy plays a protective role against pressure overload-induced mitochondrial dysfunction and heart failure. Circulation. 2016;133:1249–63.

113. Wan YY, et al. Involvement of Drp1 in hypoxia-induced migration of human glioblastoma U251 cells. Oncol Rep. 2014;32:619–26.

114. Suen DF, Norris KL, Youle RJ. Mitochondrial dynamics and apoptosis. Genes Dev. 2008;22:1577–90.

115. Hanson CJ, Bootman MD, Roderick HL. Cell signalling: IP3 receptors channel calcium into cell death. Curr Biol. 2004;14:R933–5.

116. Karbowski M, et al. Spatial and temporal association of Bax with mitochondrial fission sites, Drp1, and Mfn2 during apoptosis. J Cell Biol. 2002;159:931–8.

117. Westphal D, Kluck RM, Dewson G. Building blocks of the apoptotic pore: how Bax and Bak are activated and oligomerize during apoptosis. Cell Death Differ. 2014;21:196–205.

118. Dewson G, Kluck RM. Mechanisms by which Bak and Bax permeabilise mitochondria during apoptosis. J Cell Sci. 2009;122:2801–8.

119. Thomas KJ, Jacobson MR. Defects in mitochondrial fission protein dynamin-related protein 1 are linked to apoptotic resistance and autophagy in a lung cancer model. PLoS One. 2012;7:e45319.

120. Warnholtz A, et al. Increased NADH-oxidase-mediated superoxide production in the early stages of atherosclerosis: evidence for involvement of the renin-angiotensin system. Circulation. 1999;99:2027–33.

121. Violi F, Basili S, Nigro C, Pignatelli P. Role of NADPH oxidase in atherosclerosis. Futur Cardiol. 2009;5:83–92.

122. Gordeeva AV, Zvyagilskaya RA, Labas YA. Crosstalk between reactive oxygen species and calcium in living cells. Biochemistry (Mosc). 2003;68:1077–80.

123. Sag CM, Wagner S, Maier LS. Role of oxidants on calcium and sodium movement in healthy and diseased cardiac myocytes. Free Radic Biol Med. 2013;63:338–49.

124. Zimmerman MC, et al. Activation of NADPH oxidase 1 increases intracellular calcium and migration of smooth muscle cells. Hypertension. 2011;58:446–53.

125. Rowell J, Koitabashi N, Kass DA. TRP-ing up heart and vessels: canonical transient receptor potential channels and cardiovascular disease. J Cardiovasc Transl Res. 2010;3:516–24.

126. Rodman DM, et al. The low-voltage-activated calcium channel CAV3.1 controls proliferation of human pulmonary artery myocytes. Chest. 2005;128:581S–2S.

127. Rodman DM, et al. Low-voltage-activated (T-type) calcium channels control proliferation of human pulmonary artery myocytes. Circ Res. 2005;96:864–72.

128. Li GW, et al. The calcium-sensing receptor mediates hypoxia-induced proliferation of rat pulmonary artery smooth muscle cells through MEK1/ERK1,2 and PI3K pathways. Basic Clin Pharmacol Toxicol. 2011;108:185–93.

129. Yamamura A, et al. Enhanced Ca(2+)-sensing receptor function in idiopathic pulmonary arterial hypertension. Circ Res. 2012;111:469–81.

130. Monteith GR, Kable EP, Chen S, Roufogalis BD. Plasma membrane calcium pump-mediated calcium efflux and bulk cytosolic free calcium in cultured aortic smooth muscle cells from spontaneously hypertensive and Wistar-Kyoto normotensives rats. J Hypertens. 1996;14:435–42.

131. Wray S, Burdyga T. Sarcoplasmic reticulum function in smooth muscle. Physiol Rev. 2010;90:113–78.

132. Kourie JI. Interaction of reactive oxygen species with ion transport mechanisms. Am J Phys. 1998;275:C1–24.

133. Zima AV, Blatter LA. Redox regulation of cardiac calcium channels and transporters. Cardiovasc Res. 2006;71:310–21.

134. Truong L, Zheng YM, Wang YX. Mitochondrial Rieske iron-sulfur protein in pulmonary artery smooth muscle: a key primary signaling molecule in pulmonary hypertension. Arch Biochem Biophys. 2020;683:108234.

135. Cornea RL, Nitu FR, Samso M, Thomas DD, Fruen BR. Mapping the ryanodine receptor FK506-binding protein subunit using fluorescence resonance energy transfer. J Biol Chem. 2010;285:19219–26.

136. Firth AL, Yuill KH, Smirnov SV. Mitochondria-dependent regulation of Kv currents in rat pulmonary artery smooth muscle cells. Am J Physiol Lung Cell Mol Physiol. 2008;295:L61–70.

137. Rizzuto R, et al. Ca(2+) transfer from the ER to mitochondria: when, how and why. Biochim Biophys Acta. 2009;1787:1342–51.

138. Gorlach A, Bertram K, Hudecova S, Krizanova O. Calcium and ROS: a mutual interplay. Redox Biol. 2015;6:260–71.

139. Webster KA. Mitochondrial membrane permeabilization and cell death during myocardial infarction: roles of calcium and reactive oxygen species. Futur Cardiol. 2012;8:863–84.

140. Chaube R, Werstuck GH. Mitochondrial ROS versus ER ROS: which comes first in myocardial calcium dysregulation? Front Cardiovasc Med. 2016;3:36.

141. Song T, Zheng YM, Wang YX. Cross talk between mitochondrial reactive oxygen species and sarcoplasmic reticulum calcium in pulmonary arterial smooth muscle cells. Adv Exp Med Biol. 2017;967:289–98.

142. Korde AS, Yadav VR, Zheng YM, Wang YX. Primary role of mitochondrial Rieske iron-sulfur protein in hypoxic ROS production in pulmonary artery myocytes. Free Radic Biol Med. 2011;50:945–52.

143. Kawanami D, et al. Direct reciprocal effects of resistin and adiponectin on vascular endothelial cells: a new insight into adipocytokine-endothelial cell interactions. Biochem Biophys Res Commun. 2004;314:415–9.

144. Vozarova de Courten B, Degawa-Yamauchi M, Considine RV, Tataranni PA. High serum resistin is associated with an increase in adiposity but not a worsening of insulin resistance in Pima Indians. Diabetes. 2004;53:1279–84.

145. Raghuraman G, Zuniga MC, Yuan H, Zhou W. PKCepsilon mediates resistin-induced NADPH oxidase activation and inflammation leading to smooth muscle cell dysfunction and intimal hyperplasia. Atherosclerosis. 2016;253:29–37.

146. Reddy S, et al. Dynamic microRNA expression during the transition from right ventricular hypertrophy to failure. Physiol Genomics. 2012;44:562–75.

147. Potus F, et al. Downregulation of MicroRNA-126 contributes to the failing right ventricle in pulmonary arterial hypertension. Circulation. 2015;132:932–43.

148. Bockmeyer CL, et al. Plexiform vasculopathy of severe pulmonary arterial hypertension and microRNA expression. J Heart Lung Transplant. 2012;31:764–72.

149. Caruso P, et al. Dynamic changes in lung microRNA profiles during the development of pulmonary hypertension due to chronic hypoxia and monocrotaline. Arterioscler Thromb Vasc Biol. 2010;30:716–23.

150. Drake JI, et al. Molecular signature of a right heart failure program in chronic severe pulmonary hypertension. Am J Respir Cell Mol Biol. 2011;45:1239–47.

151. Shi L, et al. miR-223-IGF-IR signalling in hypoxia- and load-induced right-ventricular failure: a novel therapeutic approach. Cardiovasc Res. 2016;111:184–93.

Inflammatory Pathways in Sarcoidosis

Barbara P. Barna, Marc A. Judson,
and Mary Jane Thomassen

Abstract

Concepts regarding etiology and pathophysiology of sarcoidosis have changed remarkably within the past 5 years. Sarcoidosis is now viewed as a complex multi-causation disease related to a diverse collection of external environmental or infectious signals. It is generally accepted that the cause of sarcoidosis is unknown. Moreover, concepts of the inflammatory pathway have been modified by the realization that intrinsic genetic factors and innate immunity may modify adaptive immune responses to external triggers. With those potential regulatory pathways in mind, we will attempt to discuss the current understanding of the inflammatory response in sarcoidosis with emphasis on development of pulmonary granulomatous pathology. In that context, we will emphasize that both macrophages and T lymphocytes play key roles, with sometimes overlapping cytokine production (i.e., TNFα and IFN-γ) but also with unique mediators that influence the pathologic picture. Numerous studies have shown that in a sizable number of sarcoidosis patients, granulomas spontaneously resolve, usually within 3 years. Other sarcoidosis patients, however, may develop a chronic granulomatous disease which may subsequently lead to fibrosis. This chapter will outline our current understanding of inflammatory pathways in sarcoidosis which initiate and mediate granulomatous changes or onset of pulmonary fibrosis.

Keywords

Alveolar macrophages · Lymphocytes · Innate immunity · Granuloma mediators · Th17 · PPARγ

Abbreviations

ABCA1	ATP-binding cassette lipid transporter-A1
ABCG1	ATP-binding cassette lipid transporter-G1
CCL	C-C motif chemokine ligand
CCR	C-C chemokine receptor
CXCL	C-X-C motif ligand
DAMPs	danger-associated molecular patterns
GWAS	genome-wide association studies
HLA	human leukocyte antigen
IFN-γ	interferon gamma
IL	interleukin

B. P. Barna · M. J. Thomassen (✉)
Program in Lung Cell Biology and Translational Research, Division of Pulmonary and Critical Care Medicine, East Carolina University, Greenville, NC, USA
e-mail: thomassenm@ecu.edu

M. A. Judson
Division of Pulmonary and Critical Care Medicine, Albany Medical College, Albany, NY, USA

IL-23R   interleukin-23 receptor
MMP12   matrix metalloproteinase 12
mTOR   mammalian target of rapamycin
NLRs   NOD-like receptors
PAMPs   pathogen-associated molecular patterns
PBMC   peripheral blood mononuclear cells
PD-1   programmed cell death-1
PPARγ   peroxisome proliferator-activated
          receptor gamma
PRRs   pattern recognition receptors
RORγt   retinoic acid receptor-related orphan
          nuclear receptor γt
SAA   serum amyloid A
sIL-2R   soluble interleukin 2 receptor
STAT   signal transducer and activator of
          transcription
TGFβ   transforming growth factor beta
Th17   T-helper 17
TLRs   Toll-like receptors
TNFα   tumor necrosis factor alpha
Tregs   regulatory T cells
TSC2   tuberous sclerosis 2

3.1   Introduction

Concepts regarding etiology and pathophysiology of sarcoidosis have changed remarkably within the past 5 years [1]. Sarcoidosis is now viewed as a complex multi-causation disease related to a diverse collection of external environmental or infectious signals [2, 3]. It is generally accepted that the cause of sarcoidosis is unknown. Moreover, concepts of the inflammatory pathway have been modified by the realization that intrinsic genetic factors and innate immunity may modify adaptive immune responses to external triggers. With those potential regulatory pathways in mind, we will attempt to discuss the current understanding of the inflammatory response in sarcoidosis with emphasis on development of pulmonary granulomatous pathology. In that context, we will emphasize that both macrophages and T lymphocytes play key roles, with sometimes overlapping cytokine production (i.e., TNFα and IFN-γ) but also with unique mediators that influence the pathologic picture. Numerous

studies have shown that in a sizable number of sarcoidosis patients, granulomas spontaneously resolve, usually within 3 years [4]. Other sarcoidosis patients, however, may develop a chronic granulomatous disease which may subsequently lead to fibrosis. This chapter will outline our current understanding of inflammatory pathways in sarcoidosis which initiate and mediate granulomatous changes or onset of pulmonary fibrosis.

3.2   Induction of Inflammation

3.2.1   Environmental Factors

Inorganic Materials Pulmonary Inflammation associated with sarcoidosis has been linked to multiple environmental factors including inorganic and organic substances. Some of the putative organic substances are antigens of infectious agents although the disease itself is not infectious. Inorganic factors may include particulate matter such as dusts and silicates, as were observed in the "sarcoid-like" granulomatous pulmonary disease found in responders to the World Trade Center disaster [5]. Sarcoidosis cases have also been associated with occupational exposures to various inorganic agents such as encountered in firefighting, construction, and machining (reviewed in [3]).

Infectious Agents The possibility of an infectious origin for sarcoidosis has been suggested by its similarity to tuberculosis. Infectious organisms associated with sarcoidosis include mycobacteria and *Cutibacterium acnes* (formerly named *Propionibacterium acnes*). DNA residues from both agents have been reported in sarcoidosis granulomas, but no live bacteria have been found (reviewed in [2]). Moreover, in both cases, related proteins (mycobacterial catalase-peroxidase and *C. acnes* catalase) were found to elicit elevated T lymphocyte responses in sarcoidosis patients, suggesting that sarcoidosis might represent an immune-mediated disease to bacterial components [2]. The concept of sarcoidosis-associated immune reactivity stems from early studies in which intradermal injection of a pasteurized suspension of sarcoid lymphoid tissue was utilized by Kveim, a Norwegian pathologist, as a diagnostic test for sarcoidosis (reviewed in [6]). This "Kveim test" found that after 4–6 weeks, granulomatous responses were elicited in the skin from sarcoidosis patients but not from controls. Later proteonomic analyses of Kveim "antigen" reported that the presence of the mycobacterial catalase-peroxidase protein [7] has been shown to trigger elevated interferon gamma (IFN-γ) responses in sarcoidosis patients [8]. Currently, association of prior mycobacterial infection is present in many sarcoidosis cases, but there is no direct evidence for an infection-initiated etiology (reviewed in [9]).

3.2.2 Intrinsic Factors

Genetic Profiles and Autoimmunity Autoimmune mechanisms have been considered in sarcoidosis because of its complex diversity of symptoms. Proteonomic studies have identified numerous sarcoidosis-related proteins in Kveim preparations and have shown that one of them (vimentin) can elicit elevated IFN-γ responses in peripheral blood mononuclear cells (PBMC) of sarcoidosis patients but not healthy controls [6]. Vimentin is an intermediate filament protein secreted by alveolar macrophages and is important to cellular interactions (reviewed in [10]). Antibodies specific to vimentin have been detected in bronchoalveolar lavage fluid (BALF) from sarcoidosis patients [11]. Immune reactivity to vimentin is not specific to sarcoidosis, however, and antibodies to vimentin have been found in autoimmune conditions such as systemic lupus erythematosus (SLE) and rheumatoid arthritis (RA) [10]. In addition, low levels have been detected in BALF from healthy individuals [11]. In sarcoidosis, additional studies are needed to define the role of vimentin as it is unclear whether it is a cause or effect of inflammation.

Analysis of gene markers in sarcoidosis has focused on CD4+ T lymphocytes which accumu-

late in the lungs prior to granuloma formation (reviewed in [12]). Characterization of these cells indicates the presence of adaptive immune responses in sarcoidosis, with antigen-driven activation involving antigen-presenting cells such as macrophages. Several human leukocyte antigen (HLA) gene patterns have emerged from sarcoidosis analyses, with HLA-DRB1*01 and HLA-DRB1*04 appearing to be protective in Caucasian populations, while risk factors for sarcoidosis are represented by HLA-DRB1*03, HLA-DRB1*11, HLA-DRB1*12, HLA-DRB1*14, and HLA-DRB1*15 [12]. Interestingly, HLA-DRB1*03 is also present in patients with Lofgren's syndrome, a self-limiting form of sarcoidosis characterized by acute onset, fever, and clinical symptomology such as bilateral hilar lymphadenopathy [11] (reviewed in [2]). Clinical disease course in Lofgren's syndrome correlates with DRB1*03 presence: disease resolution occurs in 95% of DRB1*03 positives but in only 49% of DRB1*03 negatives [12]. With respect to vimentin antibodies in BALF, higher titers have been found in HLA-DRB1*03-positive sarcoidosis patients compared to HLA-DRB1*03 negatives [11]. The significance of these findings requires further study.

Genetic studies in sarcoidosis have also analyzed candidate genes associated with sarcoidosis such as the immune-related genes, tumor necrosis factor (TNF), and the interleukin-23 receptor (IL-23R) [12]. TNF meta-analysis studies revealed a significant association of the −308 polymorphism with sarcoidosis compared to controls, while other studies also found a −307 haplotype associated with good prognosis and a −857T containing one associated with persistent disease (reviewed in [12]). Studies of IL-23R variants have noted an association with Crohn's disease (CD) as well as sarcoidosis. The IL-23R gene codes for a subunit of the IL-23 receptor which is important to T-helper 17 (Th17)-mediated processes, and several alleles have been found with significant associations to sarcoidosis (reviewed in [12]). More recent approaches to the genetics of sarcoidosis have utilized genome-wide association studies

(GWAS) (reviewed in [13]). The major findings from these studies have confirmed the role of TH1 adaptive immune responses in sarcoidosis with emphasis on IFN-γ functions and signaling pathways. The authors concluded that future GWAS studies in sarcoidosis need to utilize more cutting-edge approaches to allow analyses of single-cell subpopulations as well as sequential monitoring of patients for changes in disease status [13].

Innate Immunity: Overview This section will consider several avenues of innate immune reactivity that have been reported in sarcoidosis. Innate immune responses are considered to be direct contributors to granulomatous inflammation in sarcoidosis (reviewed in [14]). Relevant innate immune factors to be discussed include the following: (a) pattern recognition receptors (PRRs); (b) the NOD-like receptor NLRP3 inflammasome network [15]; (c) the mTOR signaling pathway; (d) serum amyloid A (SAA), an acute-phase protein; and (e) chitotriosidase, an enzyme involved in pathogen defense [16] (Table 3.1).

Innate Immunity: Pattern Recognition Receptors (PRRs) Within the lung, PRRs represent an innate defense against infections within the alveolus and are expressed on alveolar macrophages, epithelial cells, endothelial cells, and other cell types (reviewed in [17]). PRRs sense microbial invaders by recognition of conserved microbial molecules classically defined as "pathogen-associated molecular patterns" (PAMPs). PRR encounter activates cellular production of inflammatory cytokines and chemokines which in turn recruit and activate macrophages and neutrophils. In the case of noninfectious injury elicited by large, inert particles such as silica crystals, some PRRs may also become active. In addition, release of endogenous intracellular molecules from injured cells, defined as "danger-associated molecular patterns" (DAMPs), also mediates further cellular inflammatory responses to noninfectious injuries [17].

Among the PRRs are Toll-like receptors (TLRs) and the cytosolic NOD-like receptors (NLRs) which include NLRP3. These molecules are expressed in alveolar macrophages, lung epithelial cells, and dendritic cells as well as on lymphocytes

Table 3.1 Mediators of innate immunity in sarcoidosis lung

Innate immune mediators	Role in pathophysiology	Pulmonary location
Pattern recognition receptors (PRRs) [17]	Host defense against pathogens; mediate inflammatory responses	Expressed on alveolar macrophages, epithelial cells, endothelial cells, and others
Pathogen-associated molecular patterns (PAMPs) [17]	Conserved microbial molecules recognized by PRRs; activate release of inflammatory cytokines	Responders to PAMPs include alveolar macrophages, epithelial cells, and endothelial cells
Danger-associated molecular patterns (DAMPs) [17]	Endogenous intracellular molecules from injured cells; mediate inflammatory responses	Responders to DAMPs include alveolar macrophages, epithelial cells, and endothelial cells
Toll-like receptors (TLRs) [14, 17]	Bind both microbial and endogenous ligands; mediate inflammatory responses	Expressed on lymphocytes, alveolar macrophages, epithelial cells, and dendritic cells
NLRP3 inflammasome network [15, 18, 19, 20]	Host defense against pathogens; initiates pyroptosis and release of pro-inflammatory cytokines	Granuloma tissues
MTOR signaling pathway [21, 22]	Regulates rapid responses of innate immune cells; involved in initiation and progression of granulomas	Granuloma tissues
Serum amyloid A (SAA) [23–25]	Acute-phase reactant; constituent of granulomas	Macrophages and giant cells within granuloma tissues
Chitotriosidase [16–26]	Host defense against chitin-containing fungi and protozoa	Produced by activated macrophages; detectable in sarcoidosis sera

(reviewed in [17]). There are ten members of the human TLR family located either at the cell surface (TLR2, TLR4–6, TLR10) or in lysosomal/endosomal membranes (TLR3, TLR7–9). Ligands include both microbial and endogenous factors which can induce production of antimicrobial peptides and pro-inflammatory mediators such as TNFα, IL-8, and proIL-1β. Further processing of proIL-1β is required by caspase activation via NLRP3 (discussed below). Mycobacterial ligands may recognize TLR2 and TLR9, and polymorphisms in both have been associated with susceptibility to mycobacterial infection and granulomatous pathobiology (reviewed in [14]). In sarcoidosis, enhanced TNFα responses to TLR2 stimulation have been found in cells from both blood and the lungs [14]. A positive feedback loop has been suggested by the capacities of TNFα and IFN-γ to enhance TLR2 expression on pulmonary epithelial cells [14].

Innate Immunity: The NLR Inflammasome Network NLRs comprise another innate immune receptor family of 22 members (in humans), but few of these have been functionally characterized. Five of these NLRs have been shown to assemble "inflammasomes" which are high-molecular-weight protein complexes present in the cytosol of stimulated immune cells (reviewed in [18]). These complexes have a critical role in host defense against pathogens and can be triggered by diverse stimuli. Basic functions of the NLRP3 inflammasome (which has been more intensely studied than other NLRs) are to initiate an inflammatory form of cell death (pyroptosis) and to release pro-inflammatory cytokines IL-1β and IL-18 (reviewed in [19]). In macrophages, NLRP3 activity begins with a priming signal which can be supplied by TLR4 agonists that induce expression of NLRP3 and proIL-1β. Next, an activation signal may be triggered by the above-described PAMPs and DAMPs, which promote inflammasome assembly, caspase-1-mediated IL-1β and IL-18 secretion, and pyroptosis. Interestingly, the priming step alone has been found to be sufficient for human monocytes to mediate caspase-1 activation and IL-1β release [19]. In sarcoidosis, the

NLRP3 inflammasome network has been shown to be constitutively activated and involved in granuloma formation [20]. Upregulation of NLRP3 components including cleaved caspase-1 and IL-1β has been found in sarcoid pulmonary granulomas, and NLRP3 mRNA was elevated in cell-sorted sarcoid alveolar macrophages compared to controls [20]. Of interest were findings in a mouse granuloma model which showed that NLRP3 KO mice exhibited decreased granuloma formation compared to wild type. Additional murine studies also showed that microRNA miR-223 acted as a downregulator of NLRP3, and in miR-223 KO mice, granulomas were increased in size [20]. In sarcoidosis alveolar macrophages, miR-223 levels were decreased in contrast to the elevation of NLRP3. These findings illustrate the importance of NLRP3 inflammasomes as an active component of innate immunity in sarcoidosis pathogenesis.

Innate Immunity: The mTOR Pathway The mammalian target of rapamycin (mTOR) is a serine/threonine kinase regulatory-associated protein which acts as a central regulator of cellular metabolism (reviewed in [21]). MTOR forms a part of two complexes: mTOR complex 1 (mTORC1) and mTOR complex 2 (mTORC2) which primarily shape rapid responses of innate immune cells represented by monocytes, macrophages, dendritic cells, neutrophils, mast cells, and innate-like NK cells. Rapamycin inhibits these mTOR complexes and has been used as an immunosuppressive drug to prevent kidney allograft rejection [21]. MTOR pathways become activated by various extracellular signals such as growth factors, TLR ligands, and cytokines. Activation controls a wide range of cellular functions including translation, protein synthesis, cell growth, metabolism, and anabolic processes. Studies in mice have shown that mTORC1 can also become activated by deletion of the gene encoding tuberous sclerosis 2 (TSC2) resulting in induced hypertrophy and proliferation culminating in excessive pulmonary granuloma formation in vivo [22]. Inhibition of mTORC1 induced apoptosis and completely resolved granulomas in

TSC2-deficient mice [22]. In studies of sarcoidosis patients, clinical disease progression was found to correlate with mTORC1 activation, macrophage proliferation, and glycolysis [22]. Findings support a role for the innate immune mTOR pathway in initiation and progression of granuloma pathophysiology.

Innate Immunity: Serum Amyloid A Serum amyloid A (SAA) is a highly conserved acute-phase reactant primarily synthesized by the liver. Circulating SAA levels can increase by as much as 1000-fold during inflammation (reviewed in [23]). In sarcoidosis, staining for SAA revealed high levels of expression in granulomatous tissues compared to tissues from patients with other granulomatous disorders [24, 25]. SAA is not specific to sarcoidosis, however, and statistical analyses indicate SAA staining is not sensitive enough for use in diagnostic testing [25]. SAA expression in sarcoid granulomas localizes to macrophages and giant cells and was found to correlate with numbers of CD3+ T cells within granulomas, suggesting an SAA linkage to local Th1 immune responses [24]. Findings from a murine granuloma model indicated that SAA regulated granuloma size, in part via TLR2 signaling, with production of IFN-γ, TNF, and IL-10 accompanying increased granuloma size [24]. Anti-TLR2 antibodies attenuated these effects [24]. Authors suggest that data points to SAA as a constituent and innate regulator of chronic granulomatous inflammation in sarcoidosis [24].

Innate Immunity: Chitotriosidase Chitotriosidase is the major active chitinase enzyme in humans and is produced mainly by activated macrophages (reviewed in [26]). The enzyme is an innate immune defense against chitin-containing pathogens such as fungi and protozoa. In several human diseases, including sarcoidosis, chitotriosidase is elevated in serum and is a marker of disease severity [26]. A recent study of 694 sarcoidosis patients and 101 healthy controls confirmed the presence of significantly elevated values for serum chitotriosidase in sarcoidosis [16]. Values were also found to be increased in patients with extrapulmonary involvement and in patients requiring increased steroid dosage [16]. The mechanisms by which chitotriosidase is implicated in sarcoidosis granulomatous pathophysiology, however, remain to be determined.

3.3 Inflammatory Granuloma Formation

Initiation of Granuloma Structure Granuloma structure is the product of coordinated responses from both innate and adaptive immunity to poorly degradable antigens (reviewed in [27]). Foreign antigenic materials deposited in the lungs are acquired by macrophages which release many pro-inflammatory cytokines (TNFα, IL-1β, IL-6) that persistently stimulate immune response pathways. Dendritic cells pick up the degraded antigens and migrate to regional lymph nodes where antigens are presented to naïve CD4+ T cells. Clonal T cell differentiation and proliferation occur in the nodes, and activated CD4+ T cells, via chemokines, migrate back to inflammatory sites within the lung. Both induction and maintenance of granuloma formation in sarcoidosis require CD4+ cell activation (reviewed in [28]). Once in the lungs, CD4+ T cells release cytokines (IFN-γ, TNFα, IL-12, and IL-18) which stimulate macrophages to organize into giant cells and granulomas. Evidence from sarcoidosis studies suggests that mediastinal lymph nodes constitute the initial site of granuloma formation prior to pulmonary granuloma development (reviewed in [29]). Unlike granulomas of infectious origin, sarcoid granulomas rarely have significant necrotic areas. A mature sarcoid granuloma is composed of both epithelioid and multinucleated giant cells in a tight configuration surrounded mainly by CD4+ T-helper cells. Historically, sarcoidosis granulomas have been considered to be driven by Th1 lymphocytes, but recent data suggest Th-17 cells have major roles in granuloma formation and persistence (reviewed in [2, 30]).

Granuloma Mediators Macrophages are one of the major producers of the enzyme matrix metalloproteinase 12 (MMP12) [31]. MMP12 is one of a family of proteolytic enzymes that degrades extracellular matrix elastin and enables infiltration of immune cells responsible for inflammation and granuloma formation. In sarcoidosis, MMP12 has been found to be one of the most highly expressed (>25-fold) enzymes in lung tissues, and MMP12 protein is elevated in BAL fluids [32]. Strikingly, MMP12 expression was highest near areas of active granulomatous inflammation, and MMP12 levels in BALF correlated with disease severity. Inhibition studies with a global MMP inhibitor (Marimastat) have reported reduced granuloma formation in lung tissue models of mycobacterial infection [33]. More recently, a comparative study of *mmp12* KO and wild-type mice was carried out in a carbon-nanotube-induced granuloma model [34]. Results indicated that granulomas formed in wild-type mice were detected as early as 10 days post instillation. These early granulomas were poorly formed, but by 60 days post instillation, granulomas appeared to be well defined. Surprisingly, no histological differences in granuloma formation were noted acutely in *mmp12* KO mice compared to wild type at day 10, but by 60 days, the granulomas were smaller and less well formed. These results suggest that MMP12 is required to maintain chronic granuloma pathology [34]. Currently, pursuit of novel MMP inhibitors continues with efforts to define MMP roles in specific disease-related immunological responses and inflammation [35].

Serum Markers of Inflammation in Sarcoidosis: Chemokines Inflammatory chemokines CXCL9, CXCL10, and CXCL11 are produced in the sarcoid granulomatous lung by several cell types, including macrophages, endothelial cells, and fibroblasts (reviewed in [36]). The primary induction signal for these chemokines is IFN-γ produced by activated CD4+ T-helper lymphocytes (reviewed in [37]). The major effect of these chemokines is to promote an influx of T cells into inflammatory tissues by binding and signaling through the CXCR3 receptor present on activated T cells [37]. Analyses of IFN-γ-inducible chemokines in sarcoidosis patients have shown that serum levels are higher than those of healthy controls and correlate with each other [36, 38]. Additional analyses in sarcoidosis have shown that CXCL9 levels correlate best with systemic organ involvement, and both CXCL10 and CXCL11 levels correlate with pulmonary function decline.

Serum Markers of Inflammation in Sarcoidosis: sIL-2R Quantitation of serum levels of the soluble interleukin 2 receptor (sIL-2R) represents a standardized clinical assay for assessment of T lymphocyte activation in various immune disorders [39]. T cells express the receptor for IL-2 and, during activation, begin to secrete the receptor in soluble form (reviewed in [40]). As in other immune-related disorders, elevated blood sIL-2R levels correlate with disease activity in sarcoidosis patients and have been shown to be predictive of spontaneous remission [41]. In a recent study of undiagnosed patients whose sIL-2R results were available before a definitive diagnosis had been made, sensitivity and specificity of serum sIL-2R for detection of sarcoidosis were 88% and 85%, respectively. Additional analyses revealed that the sIL-2R assay was superior to serum angiotensin-converting enzyme (ACE) measurement used previously in diagnosis of sarcoidosis cases (62% sensitivity and 76% specificity) [40].

Serum Markers of Inflammation in Sarcoidosis: Chitotriosidase As reviewed previously above, chitotriosidase, a component of innate immunity, has been shown to be elevated in sera from sarcoidosis patients with active disease [26]. In a study population of some 232 sarcoidosis patients, sensitivity and specificity of serum chitotriosidase for detection of sarcoidosis compared to healthy controls were calculated to be 88.6% and 92.8%, respectively [42].

Granuloma Components: Macrophages

Macrophages represent the basic building blocks of granulomas. Early events in granuloma construction involve aggregation of macrophages for transformation into epithelioid cells (reviewed in [27]). The influence of CD4-driven inflammatory cytokines (IFN-γ, TNFα, IL-12, IL-18) further promotes cell-cell fusion between macrophages and dendritic cells or monocytes, creating multinucleated giant cells which form a large portion of mature granuloma core structure [43]. Outer portions of granulomas contain large numbers of T lymphocytes and a few B lymphocytes.

Alveolar Macrophages: Lipid Homeostasis and Inflammation

Alveolar macrophages exhibit unique lipid metabolic properties compared to other tissue macrophages because of the complex lung microenvironment [43, 44]. The lung is constantly bombarded with foreign material and, further, is coated with a lipid-rich surfactant that serves to prevent pulmonary collapse [45]. The lung is the most active lipid-secreting organ because of surfactant production [46]. Alveolar macrophages represent an essential component of surfactant clearance and lipid homeostasis within the lung [47]. Surfactant contains phospholipids and neutral lipids, the bulk of which is cholesterol [48]. Alveolar macrophage ATP-binding cassette (ABC) lipid transporters ABCA1 and ABCG1 participate in clearance of cholesterol [49]. Deficiencies of lipid transporters result in increased intracellular cholesterol together with overproduction of pro-inflammatory cytokines and chemokines (reviewed in [50]). Overloading macrophages with cholesterol has been shown to activate inflammasome pathways and cytokine production (reviewed [51]).

The transcription factor PPARγ directly or indirectly regulates many genes involved in cholesterol metabolism and transport, including the ATP lipid transporters (reviewed in [43]). PPARγ also antagonizes transcription of many pro-inflammatory genes via mechanisms that may include direct association with coactivators or transrepression of transcription factors [43]. Healthy alveolar macrophages, unlike other tis-

sue macrophages, display high levels of PPARγ (reviewed in [44]). Alveolar macrophages from macrophage-specific PPARγ KO mice exhibit impaired surfactant lipid metabolism characterized by accumulation of intracellular neutral lipids, dysregulated lipid transporters, and elevated inflammatory cytokines [52, 53]. In addition, PPARγ deficiency also resulted in dysregulation of alveolar macrophage liver X receptor pathways which are critical to the promotion of cellular cholesterol export [54]. Dysregulation of PPARγ, ABCA1/ABCG1 lipid transporters, and pro-inflammatory cytokines are significant findings in alveolar macrophages from sarcoid patients [55].

In addition to cholesterol regulation in the lung, lipid transporters play key roles in host innate immunity; deficiencies lead to impaired immune cell homeostasis, further aggravating pulmonary inflammation [51].

Macrophage Profiles in the Lung

Studies of macrophage activation have led to a concept of classic and alternative activation termed M1 and M2, analogous to that of Th1 and Th2 cells (reviewed in [56]). More recently, however, it has become clear that whereas M1 and M2 phenotypes were derived from in vitro studies, tissue macrophages in vivo may display mixed responses [57].

Macrophage plasticity allows participation in both promotion and resolution of inflammation (reviewed in [58]). Generally, sarcoidosis data illustrate a model of persistent inflammation with macrophages exhibiting an M1 profile induced by pro-inflammatory cytokines such as IFN-γ [28]. Alveolar macrophages from pulmonary sarcoidosis patients have also been shown to produce IFN-γ which may provide further stimulation for granuloma formation [59]. The M1 or "classical activation" phenotype renders macrophages efficient killers of bacteria as well as transmitters of pro-inflammatory cytokines (IL-1β, IL-12, TNFα) (reviewed in [60]). An M2 or "alternative activation" phenotype, in contrast, allows macrophages to promote tissue repair which is a necessary function for the later resolution phase of inflammation (reviewed in [60]). Inducers of the

M2 macrophage phenotype include cytokines IL-4, IL-13, IL-10, and TGFβ [60]. The participation of M2 macrophages in fibrosis is still poorly understood, but M2 macrophages and giant cells have been noted within fibrotic areas of muscle specimens from sarcoidosis patients [61]. Additional studies are needed to more specifically define macrophage functions in sarcoidosis, particularly with respect to changes from a chronic granulomatous status to fibrosis.

Granuloma Components: Lymphocytes

Lymphocytes infiltrating sarcoid granulomas were once considered to be mostly Th1-polarized CD4+ cells, but current findings have established the presence of elevated Th17 phenotype lymphocytes in sarcoid granulomas and lymph nodes [62, 63] (Table 3.2). Th17 cells are a subgroup of CD4+ T lymphocytes that secrete IL-17A, a cytokine which can induce IFN-γ and TNFα production in macrophages (reviewed in [2, 30]). These Th17 products stimulate macrophages and promote both giant cell and granuloma formation. Th17 cells are not found in all granulomatous diseases; for example, they are not present in chronic beryllium disease [64]. IL-17A itself, however, can be produced in the lung by other cells such as NK cells and has an important role in mucosal immunity, including the respiratory tract [65]. Th17 cells are generated from CD4+ T cells exposed to TGFβ and IL-6 via a signaling cascade which culminates in tyrosine phosphorylation of STAT3 and STAT1 (reviewed in [66]). STAT3 induces Th17-related genes such as IL-17A and IL-23R as well as transcription factor retinoic acid receptor-related orphan nuclear receptor γt (RORγt) which is a negative regulator of alternative lineage phenotypes.

Th17 cells are remarkable for their plasticity in adopting pro-inflammatory or regulatory functions depending upon the microenvironment. Exposure to IL-1β, IL-12, and IL-23 transforms Th17 cells into a dual phenotype of Th1/Th17 which further drives inflammation with secretion of IFN-γ in addition to IL-17A. Further exposure of Th1/Th17 cells to Th1 cytokines IFN-γ and IL-12 can induce a Th17.1 phenotype which secretes mainly IFN-γ. Expression of the transcription factor T-bet has also been detected on some Th17 cells and thought to represent a transition state leading to the Th17.1 phenotype [38]. Chemokine receptors CCR6, CXCR3, and CCR4 have been helpful in differentiating among these CD4+ T-helper cells with (1) Th1 expressing CXCR3+, (2) Th17 expressing CCR6+ and CCR4 +, and (3) Th17.1 expressing CCR6+ and CXCR3+ [62] (Table 3.2). Studies indicate that CCR6-expressing Th17 cells are recruited to the lungs in pulmonary sarcoidosis by the macrophage chemokine CCL20. Newly diagnosed sarcoidosis patients with stage 2 pulmonary disease display elevated serum levels of IL-6, CCL20,

Table 3.2 T lymphocytes present in sarcoidosis lung

Initial phenotype	Inducing molecules	Induced phenotype	Secreted cytokines	Role in pathophysiology
CD4+	TGFβ, IL-6	**Th17** (CCR6+, CCR4+, CXCR3-, RORγt+)	IL-17A, IL-22, IL-23	Maintains mucosal surfaces [74]; stimulates granuloma-forming cells [2]
Th17	IL-1β, IL-12, IL-23	**Th1/Th17** (RORγt+, T-bet+)	IFN-γ, IL-17A	Promotes granuloma formation [30]
Th1/Th17	IFN-γ, IL-12	**Th17.1** (CCR6+, CCR4-, CXCR3+)	IFN-γ	Possible drivers of sarcoid immune response [64]
CD4+	Antigen affinity; ICOS[a] binding to macrophage ICOS[a] ligand [30]	**Regulatory T cells (Tregs)** (CD4+/CD25+, FoxP3+, CCR6+, CTLA-4+)	IL-10, TGFβ	Immunosuppressive activity [2]
CD4+	IFN-γ, IL-1, IL-12, CCL5, others [30, 61]	**Th1** (T-bet+, CCR6-, CCR4-, CXCR3+)	IFN-γ, IL-2	Th1% < Th17.1 in sarcoid lung [64]

[a]Inducible co-stimulator

IL-17A, and TGFβ [30, 67]. Ultimately, transformed Th17.1 cells appear in the lungs and do not proliferate, but have been considered to be end-stage cells which continue involvement in chronic pulmonary sarcoidosis [62].

Conventional regulatory T cells (Tregs) (CD4+/CD25+) are also present in sarcoid granulomas, and some studies suggest that sarcoidosis disease progression or regression is determined by the balance of Th17 and Treg cells (reviewed in [2, 30]). These Tregs also express FoxP3 and secrete both IL-10 and TGFβ. Data cited from newly diagnosed sarcoidosis patients in the study mentioned above indicated an imbalance of Th17/Tregs that coincided with the diagnosis of active disease [67]. Interestingly, murine studies have shown that Th17 cells may further differentiate via TGFβ and SMAD3 signaling pathways into IL-10-secreting, Foxp3+ Treg cells which no longer produce IL-17 [30]. These CD4+ Th17 Tregs appear to suppress inflammation by direct contact with pro-inflammatory cells. The immunosuppressive ability of conventional Tregs, however, appears to be reduced in sarcoidosis patients [2].

Inflammation and Fibrosis Less than 20% of sarcoidosis patients develop fibrosis which strongly associates with non-resolving pulmonary granulomatous inflammation (reviewed in [4, 28]). More research is needed to improve our understanding of how fibrosis can begin in a sarcoidosis pulmonary milieu characterized by elevated IFN-γ which is known to inhibit collagen expression [28]. Recent studies in human patients (including sarcoidosis) and murine models have cited evidence that expression of the programmed cell death-1 (PD-1) marker on CD4+ T cells promotes pulmonary fibrosis via STAT3-mediated secretion of IL-17A and TGFβ1 [68]. PD-1, a marker of cell exhaustion, is frequently displayed by chronically activated T effector cells and has been noted in patients with chronic sarcoidosis (reviewed in [69]). Blockade of PD-1 has been shown to restore CD4+ T cell proliferative function in sarcoidosis patients [70] and to significantly reduce fibrosis in murine models [68]. Currently, there are no clinically validated biomarkers for identifying sarcoidosis patients at risk of pulmonary fibrosis.

Summary Sarcoidosis remains a heterogeneous disease of unknown cause(s). The pathophysiology of sarcoidosis granuloma formation begins when alveolar macrophages contact inhaled particulate matter that is antigenic and capable of initiating immune activation of macrophages, dendritic cells, and lymphocytes. Initial studies in sarcoidosis focused on Th1-polarized lymphocytes, but current studies have shown the vigorous presence of Th17.1 phenotypic lymphocytes in granuloma stimulation and formation [62]. Moreover, several forms of innate immune mechanisms have also been found to participate in pulmonary inflammatory changes [14]. Both lymphocytes and alveolar macrophages generate numerous inflammatory signals (chemokines, cytokines, enzymes) that organize macrophages into giant cells and granuloma structures. Many of these products can be detected in sera and provide clinically relevant data with respect to sarcoidosis disease status [36, 40]. Figure 3.1 summarizes our current understanding of mechanisms of granuloma formation in sarcoidosis. In most patients, granulomas spontaneously resolve, but questions remain regarding markers and pathways leading to fibrosis in sarcoidosis.

New Directions The interplay between macrophages and lymphocytes in granuloma formation and resolution has not been fully defined. In the last decade, advances in flow cytometry, lineage tracing systems, and single-cell transcriptomics are making it possible to define the spectrum of macrophages phenotypes in different microdomains within healthy and diseased tissue [44, 71]. Questions regarding the role of resident macrophages and monocyte-derived macrophages have not been addressed in sarcoidosis, and application of these newer techniques may provide insight into factors which drive progressive versus resolving disease. Furthermore, the varied complexity and plasticity of T effector cells currently reported in sarcoidosis have

Mechanisms of Granuloma Formation in Sarcoidosis

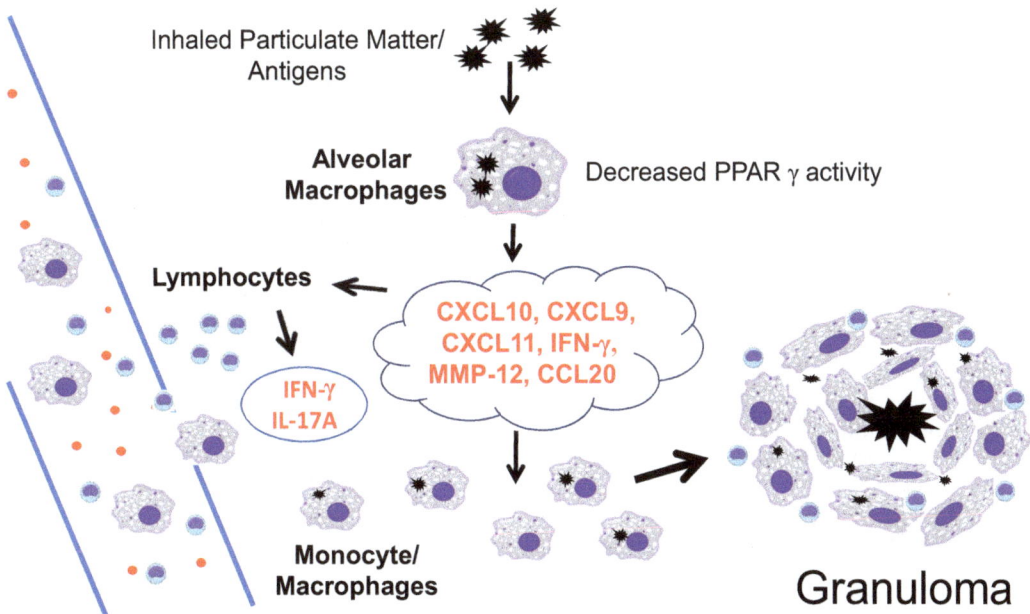

Fig. 3.1 The pathophysiology of sarcoidosis granuloma formation begins when alveolar macrophages contact inhaled particulate matter that is antigenic and capable of initiating immune activation of macrophages, dendritic cells, and lymphocytes. Lymphocytes and macrophages generate products (chemokines CXCL9, CSCL10, CSCL11, CCL20; cytokines IFN-γ, IL-17A; enzyme MMP12) that organize macrophages into giant cells and granuloma structures

opened additional questions regarding pathways of sarcoid disease progression or remission. The roles of Th1 and Th17 regulatory cells in sarcoidosis are unclear because of some confusing findings such as elevated Th17.1 in the self-limiting Lofgren's syndrome compared to non-Lofgren's patients (reviewed in [30]). It has been suggested that initial findings at diagnosis may provide clues to later sarcoidosis outcomes, and therefore, careful monitoring of patients is required in larger long-term studies. Unfortunately, current knowledge of cellular immune pathways in sarcoidosis is not yet sufficient to allow prediction of final sarcoid disease status from findings obtained at diagnosis (reviewed in [69]). Application of newer analytical technologies is needed to improve our understanding of mechanisms in disease progression or remission and, possibly, to provide better clinical treatment for sarcoidosis patients [72, 73].

References

1. Crouser ED, Maier LA, Wilson KC, Bonham CA, Morgenthau AS, Patterson KC, et al. Diagnosis and detection of sarcoidosis. An official American Thoracic Society clinical practice guideline. Am J Respir Crit Care Med. 2020;201(8):e26–51.
2. Bennett D, Bargagli E, Refini RM, Rottoli P. New concepts in the pathogenesis of sarcoidosis. Expert Rev Respir Med. 2019;13(10):981–91.
3. Judson MA. Environmental risk factors for sarcoidosis. Front Immunol. 2020;11:1340.
4. Culver DA, Judson MA. New advances in the management of pulmonary sarcoidosis. BMJ. 2019;367:l5553.
5. Crowley LE, Herbert R, Moline JM, Wallenstein S, Shukla G, Schechter C, et al. "Sarcoid like" granulomatous pulmonary disease in World Trade Center disaster responders. Am J Ind Med. 2011;54(3):175–84.
6. Eberhardt C, Thillai M, Parker R, Siddiqui N, Potiphar L, Goldin R, et al. Proteomic analysis of Kveim reagent identifies targets of cellular immunity in sarcoidosis. PLoS One. 2017;12(1):e0170285.

7. Song Z, Marzilli L, Greenlee BM, Chen ES, Silver RF, Askin FB, et al. Mycobacterial catalase-peroxidase is a tissue antigen and target of the adaptive immune response in systemic sarcoidosis. J Exp Med. 2005;201(5):755–67.

8. Chen ES, Wahlstrom J, Song Z, Willett MH, Wiken M, Yung RC, et al. T cell responses to mycobacterial catalase-peroxidase profile a pathogenic antigen in systemic sarcoidosis. J Immunol. 2008;181(12):8784–96.

9. Chen ES, Moller DR. Etiologic role of infectious agents. Semin Respir Crit Care Med. 2014;35(3):285–95.

10. Kaiser Y, Eklund A, Grunewald J. Moving target: shifting the focus to pulmonary sarcoidosis as an autoimmune spectrum disorder. Eur Respir J. 2019;54(1):1802153.

11. Kinloch AJ, Kaiser Y, Wolfgeher D, Ai J, Eklund A, Clark MR, et al. In situ humoral immunity to vimentin in HLA-DRB1*03(+) patients with pulmonary sarcoidosis. Front Immunol. 2018;9:1516.

12. Fischer A, Grunewald J, Spagnolo P, Nebel A, Schreiber S, Muller-Quernheim J. Genetics of sarcoidosis. Semin Respir Crit Care Med. 2014;35(3):296–306.

13. Schupp JC, Vukmirovic M, Kaminski N, Prasse A. Transcriptome profiles in sarcoidosis and their potential role in disease prediction. Curr Opin Pulm Med. 2017;23(5):487–92.

14. Chen ES. Innate immunity in sarcoidosis pathobiology. Curr Opin Pulm Med. 2016;22(5):469–75.

15. Riteau N, Bernaudin JF. In addition to mTOR and JAK/STAT, NLRP3 inflammasome is another key pathway activated in sarcoidosis. Eur Respir J. 2020;55(3):2000149.

16. Bennett D, Cameli P, Lanzarone N, Carobene L, Bianchi N, Fui A, et al. Chitotriosidase: a biomarker of activity and severity in patients with sarcoidosis. Respir Res. 2020;21(1):6.

17. Opitz B, van Laak V, Eitel J, Suttorp N. Innate immune recognition in infectious and noninfectious diseases of the lung. Am J Respir Crit Care Med. 2010;181(12):1294–309.

18. Broz P, Dixit VM. Inflammasomes: mechanism of assembly, regulation and signalling. Nat Rev Immunol. 2016;16(7):407–20.

19. Yang Y, Wang H, Kouadir M, Song H, Shi F. Recent advances in the mechanisms of NLRP3 inflammasome activation and its inhibitors. Cell Death Dis. 2019;10(2):128.

20. Huppertz C, Jager B, Wieczorek G, Engelhard P, Oliver SJ, Bauernfeind FG, et al. The NLRP3 inflammasome pathway is activated in sarcoidosis and involved in granuloma formation. Eur Respir J. 2020;55(3):1900119.

21. Weichhart T, Hengstschlager M, Linke M. Regulation of innate immune cell function by mTOR. Nat Rev Immunol. 2015;15(10):599–614.

22. Linke M, Pham HT, Katholnig K, Schnoller T, Miller A, Demel F, et al. Chronic signaling via the metabolic checkpoint kinase mTORC1 induces macrophage granuloma formation and marks sarcoidosis progression. Nat Immunol. 2017;18(3):293–302.

23. Uhlar CM, Whitehead AS. Serum amyloid A, the major vertebrate acute-phase reactant. Eur J Biochem. 1999;265(2):501–23.

24. Chen ES, Song Z, Willett MH, Heine S, Yung RC, Liu MC, et al. Serum amyloid A regulates granulomatous inflammation in sarcoidosis through Toll-like receptor-2. Am J Respir Crit Care Med. 2010;181(4):360–73.

25. Huho A, Foulke L, Jennings T, Koutroumpakis E, Dalvi S, Chaudhry H, et al. The role of serum amyloid A staining of granulomatous tissues for the diagnosis of sarcoidosis. Respir Med. 2017;126:1–8.

26. Elmonem MA, van den Heuvel LP, Levtchenko EN. Immunomodulatory effects of Chitotriosidase enzyme. Enzyme Res. 2016;2016:2682680.

27. Sakthivel P, Bruder D. Mechanism of granuloma formation in sarcoidosis. Curr Opin Hematol. 2017;24(1):59–65.

28. Patterson KC, Chen ES. The pathogenesis of pulmonary sarcoidosis and implications for treatment. Chest. 2018;153(6):1432–42.

29. Broos CE, van NM, Hoogsteden HC, Hendriks RW, Kool M, van den Blink B. Granuloma formation in pulmonary sarcoidosis. Front Immunol. 2013;4:437.

30. Crouser ED. Role of imbalance between Th17 and regulatory T-cells in sarcoidosis. Curr Opin Pulm Med. 2018;24(5):521–6.

31. Fingleton B. Matrix metalloproteinases as regulators of inflammatory processes. Biochim Biophys Acta, Mol Cell Res. 2017;1864(11 Pt A):2036–42.

32. Crouser ED, Culver DA, Knox KS, Julian MW, Shao G, Abraham S, et al. Gene expression profiling identifies MMP-12 and ADAMDEC1 as potential pathogenic mediators of pulmonary sarcoidosis. Am J Respir Crit Care Med. 2009;179(10):929–38.

33. Parasa VR, Muvva JR, Rose JF, Braian C, Brighenti S, Lerm M. Inhibition of tissue matrix metalloproteinases interferes with Mycobacterium tuberculosis-induced granuloma formation and reduces bacterial load in a human lung tissue model. Front Microbiol. 2017;8:2370.

34. Mohan AN, Malur A, Soliman E, McPeek M, Leffler N, Ogburn D, Tokarz, Knudson W, Gharib SA, Schnapp LM, Barna BP, Thomassen MJ. Matrix Metalloproteinase-12 is required for granuloma progression. Front Immunol. 2020;11:553949.

35. Fields GB. The rebirth of matrix metalloproteinase inhibitors: moving beyond the Dogma. Cell. 2019;8(9):984.

36. Arger NK, Ho M, Woodruff PG, Koth LL. Serum CXCL11 correlates with pulmonary outcomes and disease burden in sarcoidosis. Respir Med. 2019;152:89–96.

37. Su R, Nguyen ML, Agarwal MR, Kirby C, Nguyen CP, Ramstein J, et al. Interferon-inducible chemokines reflect severity and progression in sarcoidosis. Respir Res. 2013;14:121.

38. Arger NK, Machiraju S, Allen IE, Woodruff PG, Koth LL. T-bet expression in peripheral Th17.0 cells is associated with pulmonary function changes in sarcoidosis. Front Immunol. 2020;11:1129.
39. Rubin LA, Nelson DL. The soluble interleukin-2 receptor: biology, function, and clinical application. Ann Intern Med. 1990;113(8):619–27.
40. Eurelings LEM, Miedema JR, Dalm V, van Daele PLA, van Hagen PM, van Laar JAM, et al. Sensitivity and specificity of serum soluble interleukin-2 receptor for diagnosing sarcoidosis in a population of patients suspected of sarcoidosis. PLoS One. 2019;14(10):e0223897.
41. Zhou Y, Zhang Y, Zhao M, Li Q, Li H. sIL-2R levels predict the spontaneous remission in sarcoidosis. Respir Med. 2020;171:106115.
42. Bargagli E, Bennett D, Maggiorelli C, Di Sipio P, Margollicci M, Bianchi N, et al. Human chitotriosidase: a sensitive biomarker of sarcoidosis. J Clin Immunol. 2013;33(1):264–70.
43. Wilson JL, Mayr HK, Weichhart T. Metabolic programming of macrophages: implications in the pathogenesis of granulomatous disease. Front Immunol. 2019;10:2265.
44. Watanabe S, Alexander M, Misharin AV, Budinger GRS. The role of macrophages in the resolution of inflammation. J Clin Invest. 2019;129(7):2619–28.
45. Hawgood S, Poulain FR. The pulmonary collectins and surfactant metabolism. Annu Rev Physiol. 2001;63:495–519.
46. Tarling EJ, Edwards PA. Dancing with the sterols: critical roles for ABCG1, ABCA1, miRNAs, and nuclear and cell surface receptors in controlling cellular sterol homeostasis. Biochim Biophys Acta. 2012;1821(3):386–95.
47. Whitsett JA, Weaver TE. Hydrophobic surfactant proteins in lung function and disease. N Engl J Med. 2002;347(26):2141–8.
48. Veldhuizen R, Nag K, Orgeig S, Possmayer F. The role of lipids in pulmonary surfactant. Biochim Biophys Acta. 1998;1408(2-3):90–108.
49. Tarling EJ, de Aguiar Vallim TQ, Edwards PA. Role of ABC transporters in lipid transport and human disease. Trends Endocrinol Metab. 2013;24(7):342–50.
50. Fessler MB. A New Frontier in Immunometabolism. Cholesterol in lung health and disease. Ann Am Thorac Soc. 2017;14(Supplement_5):S399–405.
51. Gowdy KM, Fessler MB. Emerging roles for cholesterol and lipoproteins in lung disease. Pulm Pharmacol Ther. 2013;26(4):430–7.
52. Malur A, Mccoy AJ, Arce S, Barna BP, Kavuru MS, Malur AG, et al. Deletion of PPARγ in alveolar macrophages is associated with a Th-1 pulmonary inflammatory response. J Immunol. 2009;182:5816–22.
53. Baker AD, Malur A, Barna BP, Ghosh S, Kavuru MS, Malur AG, et al. Targeted PPAR{gamma} deficiency in alveolar macrophages disrupts surfactant catabolism. J Lipid Res. 2010;51(6):1325–31.
54. Baker AD, Malur A, Barna BP, Kavuru MS, Malur AG, Thomassen MJ. PPARgamma regulates the expression of cholesterol metabolism genes in alveolar macrophages. Biochem Biophys Res Commun. 2010;393(4):682–7.
55. Barna BP, McPeek M, Malur A, Fessler MB, Wingard CJ, Dobbs L, et al. Elevated MicroRNA-33 in sarcoidosis and a carbon nanotube model of chronic granulomatous disease. Am J Respir Cell Mol Biol. 2016;54(6):865–71.
56. Mosser DM, Edwards JP. Exploring the full spectrum of macrophage activation. Nat Rev Immunol. 2008;8(12):958–69.
57. Murray PJ, Allen JE, Biswas SK, Fisher EA, Gilroy DW, Goerdt S, et al. Macrophage activation and polarization: nomenclature and experimental guidelines. Immunity. 2014;41(1):14–20.
58. Smigiel KS, Parks WC. Macrophages, wound healing, and fibrosis: recent insights. Curr Rheumatol Rep. 2018;20(4):17.
59. Robinson BW, McLemore TL, Crystal RG. Gamma interferon is spontaneously released by alveolar macrophages and lung T lymphocytes in patients with pulmonary sarcoidosis. J Clin Invest. 1985;75(5):1488–95.
60. Mills CD, Ley K. M1 and M2 macrophages: the chicken and the egg of immunity. J Innate Immun. 2014;6(6):716–26.
61. Prokop S, Heppner FL, Goebel HH, Stenzel W. M2 polarized macrophages and giant cells contribute to myofibrosis in neuromuscular sarcoidosis. Am J Pathol. 2011;178(3):1279–86.
62. Ramstein J, Broos CE, Simpson LJ, Ansel KM, Sun SA, Ho ME, et al. IFN-gamma-producing T-helper 17.1 cells are increased in sarcoidosis and are more prevalent than T-helper type 1 cells. Am J Respir Crit Care Med. 2016;193(11):1281–91.
63. Broos CE, Koth LL, van NM, In 't Veen JCCM, Paulissen SMJ, van Hamburg JP, et al. Increased T-helper 17.1 cells in sarcoidosis mediastinal lymph nodes. Eur Respir J. 2018;51(3):1701124.
64. Greaves SA, Atif SM, Fontenot AP. Adaptive immunity in pulmonary sarcoidosis and chronic Beryllium disease. Front Immunol. 2020;11:474.
65. Iwanaga N, Kolls JK. Updates on T helper type 17 immunity in respiratory disease. Immunology. 2019;156(1):3–8.
66. Harbour SN, DiToro DF, Witte SJ, Zindl CL, Gao M, Schoeb TR, et al. TH17 cells require ongoing classic IL-6 receptor signaling to retain transcriptional and functional identity. Sci Immunol. 2020;5(49):eaaw2262.
67. Ding J, Dai J, Cai H, Gao Q, Wen Y. Extensively disturbance of regulatory T cells - Th17 cells balance in stage II pulmonary sarcoidosis. Int J Med Sci. 2017;14(11):1136–42.
68. Celada LJ, Kropski JA, Herazo-Maya JD, Luo W, Creecy A, Abad AT, et al. PD-1 up-regulation on CD4(+) T cells promotes pulmonary fibrosis through STAT3-mediated IL-17A and TGF-beta1 production. Sci Transl Med. 2018;10(460):eaar8356.

69. Bonham CA, Strek ME, Patterson KC. From granuloma to fibrosis: sarcoidosis associated pulmonary fibrosis. Curr Opin Pulm Med. 2016;22(5):484–91.

70. Braun NA, Celada LJ, Herazo-Maya JD, Abraham S, Shaginurova G, Sevin CM, et al. Blockade of the programmed death-1 pathway restores sarcoidosis CD4(+) T-cell proliferative capacity. Am J Respir Crit Care Med. 2014;190(5):560–71.

71. Alexander MJ, Budinger GRS, Reyfman PA. Breathing fresh air into respiratory research with single-cell RNA sequencing. Eur Respir Rev. 2020;29(156):200060.

72. Garman L, Montgomery CG, Rivera NV. Recent advances in sarcoidosis genomics: epigenetics, gene expression, and gene by environment (G x E) interaction studies. Curr Opin Pulm Med. 2020;26(5):544–53.

73. Bhargava M, Viken KJ, Barkes B, Griffin TJ, Gillespie M, Jagtap PD, et al. Novel protein pathways in development and progression of pulmonary sarcoidosis. Sci Rep. 2020;10(1):13282.

74. Ma Q. Polarization of immune cells in the pathologic response to inhaled particulates. Front Immunol. 2020;11:1060.

Innate Immune Responses and Pulmonary Diseases

4

Tao Liu, Siqi Liu, and Xiaobo Zhou

Abstract

Innate immunity is the first defense line of the host against various infectious pathogens, environmental insults, and other stimuli causing cell damages. Upon stimulation, pattern recognition receptors (PRRs) act as sensors to activate innate immune responses, containing NF-κB signaling, IFN response, and inflammasome activation. Toll-like receptors (TLRs), retinoic acid-inducible gene I-like receptors (RLRs), NOD-like receptors (NLRs), and other nucleic acid sensors are involved in innate immune responses. The activation of innate immune responses can facilitate the host to eliminate pathogens and maintain tissue homeostasis. However, the activity of innate immune responses needs to be tightly controlled to ensure the optimal intensity and duration of activation under various contexts. Uncontrolled innate immune responses can lead to various disorders associated with aberrant inflammatory response, including pulmonary diseases such as COPD, asthma, and COVID-19. In this chapter, we will have a broad overview of how innate immune responses function and the regulation and activation of innate immune response at molecular levels as well as their contribution to various pulmonary diseases. A better understanding of such association between innate immune responses and pulmonary diseases may provide potential therapeutic strategies.

Keywords

Pattern recognition receptors · NF-κB signaling · IFN response · Inflammasome · Pulmonary diseases

Abbreviations

AHR	Airway hyperresponsiveness
AIM2	Absent in melanoma 2
ALS	Amyotrophic lateral sclerosis
AnkRs	Ankyrin repeats
ASC	Apoptosis-associated speck-like protein containing a caspase recruitment domain or CARD
ATP	Adenosine triphosphate
BAF1	Barrier-to-autointegration factor 1
BAK	BCL2 antagonist/killer
BAX	BCL2-associated X
BHR	Bronchial hyperreactivity
C/EBPε	CCAAT enhancer-binding protein epsilon

T. Liu · S. Liu · X. Zhou (✉)
Channing Division of Network Medicine, Brigham and Women's Hospital and Harvard Medical School, Boston, MA, USA
e-mail: xiaobo.zhou@channing.harvard.edu

© The Author(s), under exclusive license to Springer Nature Switzerland AG 2021
Y. -X. Wang (ed.), *Lung Inflammation in Health and Disease, Volume II*, Advances in Experimental Medicine and Biology 1304, https://doi.org/10.1007/978-3-030-68748-9_4

CCDC50	Coiled-coil domain-containing protein 50
cGAMP	Cyclic GMP-AMP
cGAS	Cyclic GMP-AMP synthase
CHIKV	Chikungunya virus
COPD	Chronic obstructive pulmonary disease
COVID-19	Coronavirus disease 2019
CS	Cigarette smoke
CYLD	CYLD lysine 63 deubiquitinase
DAMPs	Danger-associated molecular patterns
DDX3	DEAD (Asp-Glu-Ala-Asp)-box helicase 3
DHX15	DEAH-box helicase 15
DHX9	DExH-box helicase 9
DRAIC	Downregulated RNA in cancer, inhibitor of cell invasion and migration
eATP	Extracellular ATP
FEV1	Forced expiratory volume in 1 s
FSTL-1	Follistatin-like 1
FVC	Forced vital capacity
HCV	Hepatitis C virus
HDAC6	Histone deacetylase 6
HIV	Human immunodeficiency virus
HOPS	Hepatocyte odd protein shuttling
IFI16	Interferon-γ (IFNγ)-inducible protein 16
IFN	Interferon
IFNAR	IFN-I receptor
IFN-α	Type I interferon-alpha
IFN-β	Type I interferon-beta
IKK	IκB kinase
IL-1	Interleukin-1
IPF	Idiopathic pulmonary fibrosis
IRAKs	IL-1R-associated kinases
IRF9	IFN-regulatory factor 9
ISGF3	IFN-stimulated gene factor 3
ISGs	IFN-stimulated genes
ISREs	IFN-stimulated response elements
IκB	Inhibitor of κB
JAK1	Tyrosine kinases Janus kinase 1
JNK	Jun N-terminal kinase
LDH	Lactate dehydrogenase
LF	Lethal factor
LGP2	Laboratory of genetics and physiology 2
LPS	Lipopolysaccharide
m(6)A	N(6)-Methyladenosine
MAD5	Melanoma differentiation-associated factor 5
MAVS	Mitochondrial antiviral signaling protein
Miz1	c-Myc-interacting zinc finger protein-1
MSU	Monosodium urate
mtDNA	Mitochondrial DNA
mtROS	Mitochondrial reactive oxygen species
MyD88	Myeloid differentiation primary response 88
MYSM1	Myb-like, SWIRM, and MPN domains 1
NADPH	Nicotinamide adenine dinucleotide phosphate hydrogen
NAIPs	NLR family, apoptosis inhibitory proteins
NF-κB	Nuclear factor-κB
NLRs	NOD-like receptors
NLS	Nuclear localization sequence
NOD	Nucleotide oligomerization domain
NSP6	Nonstructural protein 6
OGT	O-GlcNAc transferase
ORF6	Open reading frame 6
OTUB1	OTU deubiquitinase, ubiquitin aldehyde binding 1
PAH	Pulmonary arterial hypertension
PAMPs	Pathogen-associated molecular patterns
PBMCs	Peripheral blood mononuclear cells
PRRs	Pattern recognition receptors
RA	Rheumatoid arthritis
RHD	Rel homology domain
RIG-I	Retinoic acid-inducible gene I
RKIP	Raf kinase inhibitor protein
RLRs	Retinoic acid-inducible gene I (RIG-I)-like receptors
ROS	Reactive oxygen species
rRNA	Ribosomal RNA
SARS-CoV-2	Severe acute respiratory syndrome coronavirus 2

SPOP	Speckle-type POZ protein
STAT	Signal transducer and activator of transcription
STING	Stimulator of interferon genes
TBK1	TANK-binding kinase 1
TIRAP	TIR domain-containing adaptor protein
TLRs	Toll-like receptors
TRAF6	Tumor necrosis factor receptor-associated factor 6
TRIF	Toll/IL-1R domain-containing adaptor-inducing IFN-β
TRIKAs	TRAF6-regulated IKK activators
TRIM14	Tripartite-motif containing 14
TYK2	Tyrosine kinase 2
USP18	Ubiquitin-specific peptidase 18
USP19	Ubiquitin-specific protease 19
VEEV	Venezuelan equine encephalitis virus
WNV	West Nile virus
YFV	Yellow fever virus
YY1	Yin Yang 1
ZBP1	Z-DNA-binding protein 1
ZCCHC3	Zinc finger CCHC-type containing 3
ZNFX1	Zinc finger NFX1-type containing 1

4.1   Introduction

The innate immune system is crucial for the host to provide a protective response to infection or tissue injury. It utilizes distinct pattern recognition receptors (PRRs) to mediate diverse sets of pathogen-associated molecular pattern (PAMP) or danger-associated molecular pattern (DAMP) recognition, leading to infection removal and maintenance of tissue homeostasis. PRRs can be categorized based on their subcellular location, including Toll-like receptors (TLRs), retinoic acid-inducible gene I (RIG-I)-like receptors (RLRs), NOD-like receptors (NLRs), and several nucleic acid sensors that detect viral DNA or RNA. Upon stimuli recognition, PRRs activate a series of intracellular signaling molecules to ini-

tiate signal transduction pathways, including the nuclear factor-κB (NF-κB) signaling, interferon (IFN) response, and inflammasome activation.

4.2   TLRs

TLRs are the earliest discovered and the best characterized PRRs. Ten TLRs (TLR1–10) had been identified for recognizing distinct PAMPs and DAMPs in humans. TLR2 forms heterodimers with TLR1 or TLR6, sensing bacterial lipoproteins and lipopeptides [1]. TLR3, TLR7, TLR8, and TLR9 recognize viral RNA and DNA in the endosome [2, 3]. TLR4 functions as a lipopolysaccharide (LPS) sensor. TLR5 specifically detects flagellins and type IV secretion system components in various bacterial pathogens, including *Salmonella*, *Vibrio*, and *Helicobacter pylori* [4]. TLR7 recognizes the GUrich miR-Let7b, secreted from rheumatoid arthritis (RA) synovial fluid macrophages, resulting in synovitis [5]. Conversely, TLR10, the unique anti-inflammatory TLR, promotes HIV-1 infection and exerts anti-inflammatory effects [6, 7]. The mouse genome encodes 13 TLRs, although humans do not harbor the gene to encode functional TLR11, TLR12, and TLR13 [8]. TLR11 and TLR12 working as heterodimers directly bind to the profilin-like molecule from the protozoan parasite *Toxoplasma gondii* [9]. TLR13 recognizes a conserved 23S ribosomal RNA (rRNA) sequence, which is crucial for binding macrolide, lincosamide, and streptogramin group antibiotics in bacteria [10].

4.3   RLRs

RLRs are a family of RNA helicases and are described as cytoplasmic sensors responsible for viral RNA sensing. Three RLRs have been well defined including retinoic acid-inducible gene I (RIG-I), melanoma differentiation-associated factor 5 (MAD5), and laboratory of genetics and physiology 2 (LGP2). RIG-I recognizes short cytosol viral RNA derived from various virus species including influenza virus, hantavirus, reovirus, hepatitis, and rhinovirus [11, 12]. In comparison with RIG-I, MDA5 recognizes long strands of viral dsRNA following coronavirus, picornavirus, or influenza A virus infection [13, 14]. Negative regulator for this step includes LGP2, a homolog of RIG-I and MDA5, competing with RIG-I and MDA5 to interact with viral RNA, thereby inhibiting downstream signaling activation [15].

4.4 NLRs

The NLRs represent the largest and most diverse family. It is a group of evolutionarily conserved intracellular proteins that are responsible for the host against DAMPs or PAMPs. It harbors an N-terminal effector domain, a NOD domain that mediates ATP-dependent self-oligomerization, and a C-terminal LRR domain responsible for ligand recognition [16]. According to the characteristics of N-terminus, NLRs could be divided into two subgroups: the PYD domain-containing NLRP group and the CARD-containing NLRC group [17]. Most of the NLRPs, including NLRP1, NLRP2, NLRP3, NLRP6, NLRP7, and NLRP9, assemble inflammasome. NLRP1 is the first described receptor for inflammasome activation. It recognizes the stimulation of lethal factor (LF) protease secreted by *Bacillus anthracis* and is activated via proteasome-mediated degradation [18]. NLRP2 associates with the P2X7 receptor and the pannexin 1 channel to sense adenosine triphosphate (ATP) [19]. NLRP3 is activated by various stimuli, including monosodium urate (MSU), silica, asbestos, amyloid-β, alum, ATP, apolipoprotein E, nigericin, and viral RNA [12, 20–24]. NLRP6 and NLRP7 promote host defense against bacterial by detecting lipoteichoic acid and microbial acylated lipopeptides, respectively [25, 26]. NLRP9 recognizes short dsRNA from *Rotavirus* by concerting with the RNA sensor DExH-box helicase 9 (DHX9) [27]. Besides, some other NLRPs are involved in the inflammasome-independent pathway. NLRP4 inhibits double-stranded RNA or DNA-mediated type I interferon [28]. NLRP10 has significant effects on helper T-cell-driven immune responses

in response to adjuvants, including lipopolysaccharide, aluminum hydroxide, and complete Freund's adjuvant [29]. NLRP11 impairs LPS-induced NF-κB activation [30]. NLRP14 promotes fertilization by blockading cytosolic nucleic acid sensing [31]. NLRCs are involved in immune responses, and they consist of six members: nucleotide oligomerization domain 1 (NOD1), NOD2, NLRC3, NLRC4, NLRC5, and NLRX1 [32]. NOD1 and NOD2 recognize peptidoglycan (PGN) fragment produced by bacteria [33]. NLRC3 binds viral DNA and other nucleic acids through its LRR domain and licenses immune responses [34]. NLRC4 is an important gatekeeper against gram-negative bacteria including *Legionella pneumophila*, *Salmonella enterica* serovar *Typhimurium* (*Salmonella*), and *Shigella flexneri* [20, 35]. NLRC5 impairs gastric inflammation and mucosal lymphoid formation in response to *Helicobacter* infection [36]. Moreover, crystal analysis of the NLRX1 C-terminal fragment indicates a role for NLRX1 in intracellular viral RNA sensing in antiviral immunity [37].

4.5 Other Nucleic Acid Sensors

Notably, several other nucleic acid sensors have been identified recently. cGAS (cyclic GMP-AMP synthase) is known to be the most important DNA sensor that generates the second messenger cyclic GMP-AMP (cGAMP) for downstream cascade activation [38, 39]. Absent in melanoma 2 (AIM2) as well as interferon-γ (IFNγ)-inducible protein 16 (IFI16) are reported to recognize intracellular DNA. Additionally, Z-DNA-binding protein 1 (ZBP1; also known as DAI or DLM-1), DEAD (Asp-Glu-Ala-Asp)-box helicase 3 (DDX3), and zinc finger NFX1-type containing 1 (ZNFX1) are involved in RNA sensing and promoting innate immune responses [40–42]. These intracellular nucleic acid sensors are widely or ubiquitously expressed in almost all cell types and responsible for viral pathogen detection as well as endogenous nucleic acid recognition.

4.6 NF-κB Signaling

NF-κB is a collective name for a transcription factor family which consists of five different DNA-binding proteins (RelA, RelB, c-Rel, p105/p50, and p100/p52) [43]. Those five family members all contain an N-terminal Rel homology domain (RHD) responsible for dimerization and cognate DNA element binding [44]. Three of them (RelA, RelB, c-Rel) are synthesized as mature proteins and harbor C-terminal transactivation domains, which are essential for transcriptional activation [45]. The other two members (p105/p50 and p100/p52) are synthesized as large precursors (p105 and p100) and partially proteolyzed by the proteasome to yield active forms (p50 and p52) for DNA binding [46, 47]. The NF-κB family members can assemble into several homodimeric and heterodimeric dimers, and two paradigmatic dimers are p50:p65 and p52:RelB [48]. Different NF-κB dimers regulate various gene expressions, which are critical for immune responses, cell proliferation, migration, and apoptosis [49].

The activation of NF-κB dimers has sophisticated controls at multiple levels. In unstimulated cells, NF-κBs are inactive and retained in the cytoplasm by the binding of its specific inhibitors called "inhibitor of κB" (IκB) family [48]. The IκB proteins contain 5–7 tandem ankyrin repeats (AnkRs) that bind to the RHD of NF-κB, thus covering its nuclear localization sequence (NLS) [48]. Upon stimulation, IκB kinase (IKK) complex, including catalytic (IKKα and IKKβ) and regulatory (NEMO, also called IKKγ) subunits, was activated. The activated IKK complex catalyzes the phosphorylation and polyubiquitination of IκB family members, leading to degradation of IκB family members via proteasome and subsequent nuclear translocation of NF-κB family members [50]. Tumor necrosis factor receptor-associated factor 6 (TRAF6), a RING domain E3 ligase, together with two TRAF6-regulated IKK activators (TRIKAs) were identified as responsible for the IKK complex activation [51]. TRIKA1 is an E2 enzyme complex containing Ubc13 and Uev1A (or the functionally equivalent Mms2). Together with TRAF6, it mediates the K63-linked ubiquitination of NEMO and TRAF6

itself. TRIKA2 is a trimeric complex composed of the protein kinase TAK1 and two other proteins as TAB1 and TAB2 [52, 53]. TAK1 is a direct kinase in TRIKA2 to phosphorylate and activate IKK in a manner that depends on TRAF6 and Ubc13-Uev1A [51]. Of note, TAK1 also activates the Jun N-terminal kinase (JNK)-p38 kinase pathway by mediating MKK6 phosphorylation [51]. Additionally, the E3 ubiquitin-ligase TRAF2 (and/or TRAF5) and the kinase RIP1 are also reported to mediate the recruitment of the TRIKA2, contributing to the downstream cascade activation [54]. Adaptors, such as myeloid differentiation primary response 88 (MyD88), TIR domain-containing adaptor protein (TIRAP), and Toll/IL-1R domain-containing adaptor-inducing IFN-β (TRIF), are reported to engage and activate TRAFs by cytoplasmic intermediate IL-1R-associated kinases (IRAKs), such as the kinase IRAK1, IRAK2, and IRAK4 [55]. Importantly, IRAK4 acts upstream of IRAK1, and the kinase activity of IRAK4 might be required for IRAK1's modification [56]. Thus, upon stimulation, PRRs (TLR1, 2, 3, 4, 6, 7, 9) mediate PAMP or DAMP recognition and subsequently recruit adaptors for TRAF and TRIKA recruitment, leading to IKK complex activation, IκB degradation, and release of NF-κB for transcription. Those stimulations include viral and bacterial infections, necrotic cell products, DNA damage, oxidative stress, and pro-inflammatory cytokines (Fig. 4.1) [57].

The regulation of NF-κB signaling has been extensively studied. Additional regulators of NF-κB signaling include OTU deubiquitinase, ubiquitin aldehyde binding 1 (OTUB1), CYLD lysine 63 deubiquitinase (CYLD), and A20 that modulates the ubiquitination of various components [58–62]. Furthermore, phosphorylation, acetylation, methylation, and palmitoylation have also been reported to fine-tune the activity of the NF-κB signaling through multiple post-translational modifications on signal proteins. Besides, Speckle-type POZ protein (SPOP) is recruited to MyD88 to inhibit the aggregation of MyD88 and recruitment of the downstream signaling kinases IRAK4, IRAK1, and IRAK2 [63]. S100A10 interacts with TLR4 and inhibits its association with adaptor proteins including

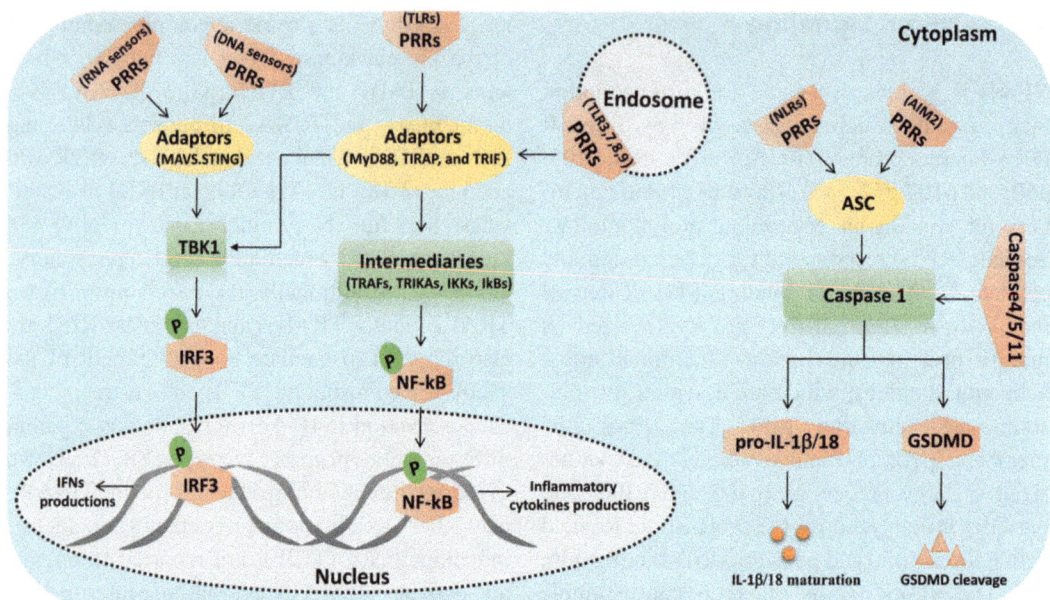

Fig. 4.1 Activation of innate immune responses. In response to distinct stimulation, different PRRs recruit various adaptors for downstream signaling cascades. In detail, cytosolic RNA or DNA sensors recruit MAVS or STING for TBK1 activation, respectively. Activated TBK1 mediates IRF3 phosphorylation, and the phosphorylated IRF3 translocates from the cytoplasm to the nucleus, promoting IFN production. TLRs on plasma or endosome membrane associate with distinct adaptors including MyD88, TIRAP, and TRIF, triggering interme-diate activation and subsequent NF-κB phosphorylation. Phosphorylated NF-κBs enter into the nucleus, inducing inflammatory cytokine production. NLRs and AIM2 bind to ASC and enhance the caspase-1 activity for cleaving pro-IL-1β and pro-IL-18, leading to IL-1β/18 maturation. On the other hand, activated caspase-1 mediates the cleavage of GSDMD, and the N-terminal of GSDMD mediates membrane pore formation and pyroptosis. Also, caspase 4/5/11 directly recognize LPS and bind to caspase-1 for downstream signaling activation

MyD88 and TRIF [64]. Downregulated RNA in cancer, inhibitor of cell invasion and migration (DRAIC) impairs IKK complex assembly and inhibits the phosphorylation of IκBα and the activity of NF-κB [65]. Lamtor5 and hepatocyte odd protein shuttling (HOPS) control TRAF6 and TLR4 stability for regulating NF-κB signaling, respectively [66, 67]. A well-controlled NF-κB signaling is crucial for the maintenance of tissue homeostasis, and the dysfunction of NF-κB signaling leads to many pathological conditions such as combined immunodeficiency, type 2 diabetes, and pulmonary diseases [43, 68–70].

4.7   IFN Response

Type I interferons have long been characterized as key players in antiviral responses, inhibiting viral replication and spread by sensing PAMPs, including viral DNA and RNA [71]. Upon virus infection, PRRs promote type I interferon expression, triggering pro-inflammatory cytokine and chemokine production, as well as the expression of innate immune genes to establish an intracellular antiviral state [72]. Fourteen subtypes of type I alpha IFNs (IFN-α) in mice and thirteen in humans, and one beta (IFN-β) IFNs are engaged in that signal through the same IFN-I receptor (IFNAR) [73]. IFNAR, which is composed of IFNAR1 and IFNAR2 subunits, employs the receptor-associated protein tyrosine kinases Janus kinase 1 (JAK1) and tyrosine kinase 2 (TYK2) to phosphorylate cytoplasmic transcription factors signal transducer and activator of transcription 1 (STAT1) and STAT2. Subsequently, phosphorylated STAT1 and STAT2 assemble heterodimers and translocate to the nucleus, together with IFN-regulatory factor 9 (IRF9), to form a transcriptionally active IFN-stimulated gene factor 3 (ISGF3) for directly activating the transcription of IFN-stimulated genes (ISGs) through binding IFN-stimulated response elements (ISREs; consensus sequence TTTCNNTTTC) [74, 75]. Several discovery-based screens demonstrate hundreds of ISGs for their ability to inhibit the replication of several important viruses including influenza A H1N1 virus, hepatitis C virus (HCV), yellow fever virus (YFV), West Nile virus (WNV), chikungunya virus (CHIKV), Venezuelan equine encephalitis virus (VEEV), and HIV-1 [76, 77].

Many types of PRRs can promote IFN-I production. These receptors mediate recognition of foreign and self-nucleic acids as well as a limited number of other non-nucleic acid PAMPs and recruit distinct adaptors for downstream TANK-binding kinase 1 (TBK1) phosphorylation. For example, RNA sensors including MDA5, RIG-I, and zinc finger NFX1-type containing 1 (ZNFX1) recruit mitochondrial antiviral signaling protein (MAVS) to activate and propagate antiviral response [42, 78–80]. Then, MAVS protein forms fibrils and behaves like prions to convert endogenous MAVS into functional aggregates to promote downstream signaling cascade [81]. Likely, DNA sensors including cyclic GMP-AMP synthase (cGAS) and IFI16 recruit stimulator of interferon genes (STING) for antiviral response. STING is an endoplasmic reticulum membrane protein. The cytoplasmic domain of STING undergoes a 180° rotation and unwinds around the crossover point between the proteins to form oligomers [82]. Oligomerized STING adopts a β-strand-like conformation and inserts into a groove between the kinase domain of one TBK1 through a conserved PLPLRT/SD motif within the C-terminal tail of STING [83, 84]. Activated TBK1 directly targets IRF3 for its phosphorylation and the phosphorylated IRF3 translocated from the cytosol to the nucleus for IFN production and subsequent ISG expression for the antiviral response [85]. Of note, MAVS and STING not only activate TBK1 but also recruit IRF3 to bind TBK1 to activate the IRF3 pathway [86]. In addition to MAVS and STING, TLR3 and TLR4 signaling activate TBK1 and IRF3 through the adaptor protein TRIF (Fig. 4.1) [87].

The dysfunction of IFNs results in multiple diseases. For example, activated variants in STING lead to a rare auto-inflammatory disease named STING-associated vasculopathy with onset in infancy via preventing the development of lymph nodes and Peyer's patches [88, 89]. Dysfunction of TDP-43- or C9orf72-induced STING activation causes amyotrophic lateral sclerosis (ALS) [90, 91]. Aberrant mitochondrial DNA (mtDNA)-induced cGAS-STING activation promotes lupus-like disease, acute kidney injury, renal Inflammation, and fibrosis [92–94]. Thus, the activation of the IFN response should be precisely controlled. Various regulators have been reported to modulate IFN signaling through distinct mechanisms. Myb-like, SWIRM, and MPN domains 1 (MYSM1), coiled-coil domain-containing protein 50 (CCDC50), USP15, MARCH8, OTUB1, and OTUB5 regulate IFN response through ubiquitination [95–100]. O-GlcNAc transferase (OGT), histone deacetylase 6 (HDAC6), and palmitoyltransferases modulate IFN production through O-GlcNAcylation, deacetylation, and palmitoylation, respectively [101–103]. N(6)-Methyladenosine (m(6)A) modification controls IFN response by dictating the fast turnover of IFNα and IFNβ mRNA [94]. G3BP1 and barrier-to-autointegration factor 1 (BAF) interfere DNA binding of cGAS for IFN regulation [104, 105]. Furthermore, zinc finger CCHC-type containing 3 (ZCCHC3) and DEAH-box helicase 15 (DHX15) are shown to facilitate RLR-mediated RNA recognition [106, 107].

4.8 Inflammasome Activation

Inflammasome is a molecular platform that mediates the processing of caspases, maturation, and secretion of interleukin-1 (IL-1) family members, and activation of inflammatory cell death called pyroptosis [20, 108]. It can be categorized into apoptosis-associated speck-like protein containing a caspase recruitment domain or CARD (ASC)-dependent or CARD (ASC)-independent inflammasome activation. Upon stimulation, NLRP3 and absent in melanoma 2 (AIM2) interact with ASC for inflammasome assembly. However, NLRC4 and NLRP1 could directly activate caspase-1 for downstream cascade activation without binding to ASC. Of note, ASC binding for NLRC4 or NLRP1 could enhance its inflammasome activity, although it is dispensable for NLRC4 or NLRP1 inflammasome activation [20]. Caspase-4/5/11 are directly activated by LPS sensing and cleave GSDMD for pyroptosis independent of ASC [109]. Intriguingly, those inflammatory caspases also target NLRP3-dependent caspase-1 activation in an ASC-dependent manner [110].

NLRP1, NLRC4, AIM2, and NLRP3 inflammasomes are most widely reported. Over the past decade, numerous mechanisms have been demonstrated in those inflammasome activations. During the *Bacillus anthracis* infection, bacterial secreted lethal factor (LF) protease was reported to mediate the degradation of amino-terminal domains of NLRP1B, leading to the release of a carboxyl-terminal fragment and subsequently caspase-1 activation [18, 111]. NLRC4 is responsible for bacterial detection. However, it is not the direct sensor for its activator. NAIP (NLR family, apoptosis inhibitory protein)-mediated ligand recognition is required for NLRC4 inflammasome activation. During bacterial infection, mouse NAIP1 and NAIP2 act as cytosolic innate immune sensors for bacterial T3SS needle and rod protein recognition, respectively [112]. In comparison to NAIP1 and NAIP2, NAIP5 and NAIP6 bind to the bacterial protein flagellin for NLRC4 inflammasome activation [113, 114]. AIM2 is a direct sensor, binding double-stranded DNA and utilizes ASC to form a caspase-1-activating inflammasome. The HIN domain of AIM2 is responsible for recognizing sugar-phosphate backbone of various double-stranded DNA, including bacterial DNA, viral DNA, and radiation-induced damaged DNA [115, 116]. NLRP3 inflammasome is the most extensively characterized inflammasome. The activation of NLRP3 inflammasome involves sophisticated regulations. NF-κB signaling activation acts as

the first step to mediate the priming process, including induction of both NLRP3 and pro-IL-1β. Subsequently, NLRP3 is activated. Three working models of NLRP3 activation have been proposed: (1) lysosomal rupture and release of the proteinase cathepsin B caused by crystal phagocytosis result in NLRP3 activation [117]; (2) mitochondrial reactive oxygen species (mtROS)-induced oxidized mtDNA conversion leads to NLRP3 activation [118]; and (3) ATP triggered the efflux of K+ contributing to NLRP3 activation [119]. Nonetheless, more details about NLRP3 activation need further investigation (Fig. 4.1).

Emerging evidence shows that sustained and uncontrolled inflammasome activation contributes to the development of many diseases, such as lung injury, vitiligo, very-early-onset inflammatory bowel disease, neutrophilic chronic rhinosinusitis with nasal polyps, adipose tissue inflammation, and auto-inflammatory diseases [12, 120–123]. Many regulators have been reported to regulate inflammasome activation through distinct mechanisms. Raf kinase inhibitor protein (RKIP) and synthetic vitamins K3 and K4 block inflammasome activation through interrupting inflammasome assembly [124, 125]. CCAAT enhancer-binding protein epsilon (C/EBPε), IRF4, and IRF8 modulate transcription level of inflammasome-associated genes for inflammasome regulation [126, 127]. Nicotinamide adenine dinucleotide phosphate hydrogen (NADPH) and ubiquitin-specific protease 19 (USP19) regulate inflammasome activation by reactive oxygen species (ROS) [128, 129]. Besides, various post-translational modifications were implicated in inflammasome regulation, including ubiquitination, phosphorylation, S-nitrosylation, prenylation, deglutathionylation, and ADP-ribosylation. Notably, several drugs have been developed for therapy via targeting to inflammasome activation, such as rasagiline, ticagrelor, kaempferol, and metformin [130–133]. Therefore, targeting inflammasome activation by the deployment of those drugs will shed light on inflammasome-related disease therapy.

4.9 Pulmonary Diseases

4.9.1 Role of Innate Immune Responses in COPD

NF-κB signaling and COPD Chronic obstructive pulmonary disease (COPD), a common chronic inflammatory disease of the airways, the alveoli, and the microvasculature, affects millions of people worldwide. The diagnosis of COPD is based on the reduced ratio of the post-bronchodilator forced expiratory volume in 1 s to the forced vital capacity (FEV1:FVC ratio) (<0.7) [134]. It is characterized by three pathological phenotypes including small airway obstruction due to remodeling, emphysema, and chronic bronchitis [134]. Cigarette smoking and indoor or outdoor air pollution are the most important risk factors and causes for COPD [135]. Emerging evidence indicates that innate immune responses are involved in COPD pathogenesis. The severity of COPD is reported to associate with an increased epithelial expression of NF-κB by analyzing bronchial biopsies from smokers with COPD, smokers with normal lung function, and nonsmokers with normal lung function [136]. Further analysis identified that IκB-α levels in lung tissue were significantly reduced and IKK complex activity in peripheral blood mononuclear cells (PBMCs) is dramatically enhanced in patients with COPD than in control subjects [137, 138]. Consistently, in the mouse model, cigarette smoke (CS) exposure regulates RelB by IKKa in B-lymphocytes, leading to inflammatory cytokine release [139]. Of note, loss of function of Miz1 (also known as c-Myc-interacting zinc finger protein-1 and Zbtb17) in the murine lung epithelium spontaneously develops a COPD-like phenotype via inducing sustained NF-κB signaling activation [70]. In addition, follistatin-like 1 (FSTL-1) hypomorphic mice develop spontaneous emphysema by promoting NF-κB p65 phosphorylation in a Nr4a1-dependent manner [140].

Inflammasome and COPD Except for NF-κB signaling, inflammasome activation also contrib-utes to the onset of COPD pathogenesis. The expression levels of IL-1β and IL-18, two hallmarks of inflammasome activation, are increased in COPD patients [141, 142]. Moreover, overexpression of IL-1β or IL-18 in the lungs of mice present chronic inflammatory changes similar to COPD, and lacking IL-1R or IL-18R in mice are protected against CS-induced lung inflammation [143, 144]. Likely, elevated caspase-1 activity is also observed in the lungs from both COPD patients and the CS-treated mice model [145]. Strikingly, in the mice model, acute smoke-mediated lung inflammation is blocked by z-VAD-fmk, a pan-caspase inhibitor, or z-WEHD-fmk, a caspase-1 inhibitor [146]. Notably, high levels of two inflammasome stimulators, extracellular ATP (eATP) and ROS, are observed in patients with COPD as well as in the genetic mouse models of COPD, indicating possible inflammasome activation in COPD pathogenesis [147, 148].

IFN response and COPD The role of IFN response in COPD pathogenesis needs more investigation. Deficient IFN-β expression in the lungs and reduced sputum expression of ISGs were detected in COPD patients [149, 150]. However, acute CS exposure leads to cGAS-STING-dependent IFN response by releasing self-DNA in mice model [151]. Thus, whether CS exposure induces COPD phenotype is IFN dependent or not needs to be further explored.

4.9.2 Role of Innate Immune Responses in Asthma

NF-κB signaling and asthma Asthma, one of the major chronic non-communicable diseases, affects as many as 334 million people in the world [152]. It is defined by mucus overproduction, bronchial hyperreactivity (BHR), airway wall remodeling, and airway narrowing [153]. The symptoms of asthma include repeated periods of shortness of breath, cough, wheezing, and chest tightness [154]. Genetic susceptibility and environmental exposures as well as aberrant immune responses contribute to the onset of disease [155]. Recent studies implicated NF-κB signaling activation as a key modulator in asthma pathogenesis. Increased activation of NF-κB was observed in asthma patients [156]. Furthermore, NF-κB activation in airway epithelial is sufficient to promote allergic sensitization to an inhaled antigen [157]. In contrast, repressed NF-κB signaling activation in airway epithelial impaired inflammation, led to decreased levels of chemokines and cytokines and circulating IgE, and ameliorated mucus cell metaplasia [158]. Notably, inhibition of NF-κB by a chimeric decoy oligodeoxynucleotide transfer prevents asthma exacerbation in a mouse model [159]. Besides, ex vivo farm dust or LPS stimulation restored anti-inflammatory TNFAIP3 gene and protein levels in asthmatic patients and shifted NF-κB signaling-associated gene expression toward an anti-inflammatory state [160]. Thus, targeting NF-κB signaling may provide a novel therapeutic approach to asthma.

Inflammasome and asthma Emerging evidence showed that inflammasome activation plays a crucial role in asthma pathogenesis. In neutrophilic asthma patients, the protein level of IL-1β was significantly higher, and sputum IL-1β protein level was associated with NLRP1, NLRP3, and NLRC4 expression [161]. Similar results with increased inflammasome components including Nlrp3, Nlrc4, caspase-1, and Il-1β were observed in eosinophilic, mixed, and neutrophilic experimental asthma in mice [162]. Lacking NLRP3 inflammasome activation in mice led to ameliorated allergic airway inflammation, reduced eosinophil infiltration, and dampened Th2 lymphocyte activation in the lung [163, 164]. Strikingly, treatment with an inhibitor of caspase-1 or NLRP3 suppresses airway hyperresponsiveness (AHR) in severe, steroid-resistant asthma [165]. Most importantly, uric acid, protein serum amyloid A, apolipoprotein E, and fatty acid exposure may contribute to inflammasome activation in allergic asthma [24, 166–168].

IFN response and asthma The role of IFN response in asthma pathogenesis is more complicated and warrants more investigation. On the one hand, increased expression of IFN-β, IFN-λ1/IL-29, OAS, and viperin in neutrophilic asthmatics and high IFN-α, IFN-β, and IFN-λ1 were detected in atopic asthmatic [169, 170]. Moreover, elevated ISG expression in epithelial in asthma is related to lung inflammation and FEV1 [171]. On the other hand, reduced IFN-α/β expression level in the bronchial epithelium in asthmatic cells was also reported [172]. Thus, how IFN response activated in asthma patients needs to be further explored.

4.9.3   Role of Innate Immune Responses in COVID-19

IFN response and COVID-19 Coronavirus disease 2019 (COVID-19), an ongoing pandemic of acute respiratory disease, affects millions of people in the world since late 2019. It is caused by a highly transmissible and pathogenic coronavirus, severe acute respiratory syndrome coronavirus 2 (SARS-CoV-2). A wide range of clinical features of COVID-19 patients were reported including fever, cough, myalgia or fatigue, sputum production, headache, hemoptysis, and diarrhea [173]. Bilateral diffused alveolar damage, hyaline membrane formation, desquamation of pneumocytes, and fibrin deposits are observed in the lungs of patients with severe COVID-19 via histopathology analyses [174]. Several hypotheses have been proposed for the mechanisms of COVID-19 including imbalanced innate immune responses promoting the pathogenesis of COVID-19 [175], in which aberrant IFN response is the key player driving the progression of COVID-19. Appropriate activation of IFN signaling controls SARS-CoV-2 infection [176]. However, over-activated IFN response amplifies inflammatory signals and induces inflammation in COVID-19 patients [177]. People genetically deficient in IFN response are more vulnerable to SARS-CoV-2 infection [178, 179]. Moreover, the mice model infected with SARS-CoV-2 demonstrates the activation of type I interferon signaling [180].

Thus, early interferon therapy is associated with reduced mortality and accelerated recovery [181, 182]. Of note, a truncated isoform of ACE2, the receptor for SARS-CoV-2, could be induced by interferon response activation [183]. On the other side, SARS-CoV-2 proteins, such as nonstructural protein 6 (nsp6), nsp13, and open reading frame 6 (ORF6), could antagonize cellular IFN response [184].

NF-κB signaling and COVID-19 IL-6, an inflammatory cytokine controlled by the activated NF-κB signaling, is commonly increased in COVID-19 patients [185, 186]. The maximal level of IL-6 and C-reactive protein level, lactate dehydrogenase (LDH) level, ferritin level, d-dimer level, neutrophil count, and neutrophil-to-lymphocyte ratio are highly predictive of the need for mechanical ventilation and mortality in COVID-19 patients [187, 188]. Strikingly, repurposing of anti-IL-6 therapeutics by tocilizumab reduces mortality and/or morbidity in severe COVID-19 from clinical trials [189, 190].

Inflammasome and COVID-19 Activation of the inflammasome was also found in COVID-19 lungs [191]. Fatal COVID-19 cases showed a higher number of ASC inflammasome specks [192, 193]. Thus, innate immune responses may represent a new target for COVID-19 therapy.

4.9.4 Role of Innate Immune Responses in Other Pulmonary Diseases

Dysfunctions of innate immune responses also lead to other pulmonary diseases, such as idiopathic pulmonary fibrosis (IPF) and pulmonary arterial hypertension (PAH). The SNPs in TOLLIP, an important regulator of innate immune responses mediated by the Toll-like receptor, are associated with IPF susceptibility [194]. Yin Yang 1 (YY1), a downstream gene of NF-κB signaling, regulates fibrogenesis by increasing α-SMA and collagen expression [195]. Statin, uric acid, and extracellular ATP enhance lung fibrosis through promoting NLRP3 inflammasome activation [196, 197]. In patients with PAH, serum IFN levels were elevated, and expression of TLR3 in lung tissue is reduced [198, 199]. IFNAR1-deficient mice were protected from PAH [198]. In contrast, Tlr3−/− mice showed a more severe PAH phenotype [199]. Besides, NLRP3 inflammasome activation may contribute to the pathogenesis of PAH, representing a possible target for PAH treatment [200].

4.10 Conclusion

In response to environmental stimulation signals, PRRs, including TLRs, NLRs, RLRs, and other nucleic acid sensors, trigger a variety of signaling pathways for defense to control and eventually eliminate such stimulation. However, aberrant immune responses lead to severe inflammatory diseases especially pulmonary diseases, such as COPD, asthma, COVID-19, IPF, and PAH. Thus, the optimal regulation and fine-tuning of innate immune responses are necessary. Distinct mechanisms have been revealed in immune response regulation. Various post-translational modifications control the intensity, duration, and timing of activated innate immune responses by manipulating protein stability, activity, and subcellular localization. Additional regulators control mRNA levels and stability to regulate innate immune responses. Discoveries of these and additional mechanisms modulating innate immune response will guide and illuminate current and future clinical trials for pulmonary diseases.

References

1. Jin MS, et al. Crystal structure of the TLR1-TLR2 heterodimer induced by binding of a tri-acylated lipopeptide. Cell. 2007;130:1071–82. https://doi.org/10.1016/j.cell.2007.09.008.
2. Zhou X, et al. The function and clinical application of extracellular vesicles in innate immune regulation. Cell Mol Immunol. 2020;17:323–34. https://doi.org/10.1038/s41423-020-0391-1.

3. Greulich W, et al. TLR8 is a sensor of RNase T2 degradation products. Cell. 2019;179:1264–1275 e1213. https://doi.org/10.1016/j.cell.2019.11.001.

4. Tegtmeyer N, et al. Toll-like receptor 5 activation by the CagY repeat domains of helicobacter pylori. Cell Rep. 2020;32:108159. https://doi.org/10.1016/j.celrep.2020.108159.

5. Umar S, et al. IRAK4 inhibition: a promising strategy for treating RA joint inflammation and bone erosion. Cell Mol Immunol. 2020; https://doi.org/10.1038/s41423-020-0433-8.

6. Fore F, Indriputri C, Mamutse J, Nugraha J. TLR10 and its unique anti-inflammatory properties and potential use as a target in therapeutics. Immune Netw. 2020;20:e21. https://doi.org/10.4110/in.2020.20.e21.

7. Henrick BM, et al. TLR10 senses HIV-1 proteins and significantly enhances HIV-1 infection. Front Immunol. 2019;10:482. https://doi.org/10.3389/fimmu.2019.00482.

8. Roach JC, et al. The evolution of vertebrate Toll-like receptors. Proc Natl Acad Sci U S A. 2005;102:9577–82. https://doi.org/10.1073/pnas.0502272102.

9. Andrade WA, et al. Combined action of nucleic acid-sensing Toll-like receptors and TLR11/TLR12 heterodimers imparts resistance to Toxoplasma gondii in mice. Cell Host Microbe. 2013;13:42–53. https://doi.org/10.1016/j.chom.2012.12.003.

10. Oldenburg M, et al. TLR13 recognizes bacterial 23S rRNA devoid of erythromycin resistance-forming modification. Science. 2012;337:1111–5. https://doi.org/10.1126/science.1220363.

11. Kell AM, Gale M Jr. RIG-I in RNA virus recognition. Virology. 2015;479-480:110–21. https://doi.org/10.1016/j.virol.2015.02.017.

12. Liu T, et al. NOD-like receptor family, pyrin domain containing 3 (NLRP3) contributes to inflammation, pyroptosis, and mucin production in human airway epithelium on rhinovirus infection. J Allergy Clin Immunol. 2019;144:777–787 e779. https://doi.org/10.1016/j.jaci.2019.05.006.

13. Zalinger ZB, Elliott R, Rose KM, Weiss SR. MDA5 is critical to host defense during infection with murine coronavirus. J Virol. 2015;89:12330–40. https://doi.org/10.1128/JVI.01470-15.

14. Saito T, Gale M Jr. Differential recognition of double-stranded RNA by RIG-I-like receptors in antiviral immunity. J Exp Med. 2008;205:1523–7. https://doi.org/10.1084/jem.20081210.

15. Li X, et al. The RIG-I-like receptor LGP2 recognizes the termini of double-stranded RNA. J Biol Chem. 2009;284:13881–91. https://doi.org/10.1074/jbc.M900818200.

16. Kanneganti TD, Lamkanfi M, Nunez G. Intracellular NOD-like receptors in host defense and disease. Immunity. 2007;27:549–59. https://doi.org/10.1016/j.immuni.2007.10.002.

17. Kufer TA, Sansonetti PJ. NLR functions beyond pathogen recognition. Nat Immunol. 2011;12:121–8. https://doi.org/10.1038/ni.1985.

18. Sandstrom A, et al. Functional degradation: a mechanism of NLRP1 inflammasome activation by diverse pathogen enzymes. Science. 2019;364 https://doi.org/10.1126/science.aau1330.

19. Minkiewicz J, de Rivero Vaccari JP, Keane RW. Human astrocytes express a novel NLRP2 inflammasome. Glia. 2013;61:1113–21. https://doi.org/10.1002/glia.22499.

20. Liu T. Regulation of inflammasome by autophagy. Adv Exp Med Biol. 2019;1209:109–23. https://doi.org/10.1007/978-981-15-0606-2_7.

21. Eisenbarth SC, Colegio OR, O'Connor W, Sutterwala FS, Flavell RA. Crucial role for the Nalp3 inflammasome in the immunostimulatory properties of aluminium adjuvants. Nature. 2008;453:1122–6. https://doi.org/10.1038/nature06939.

22. Dostert C, et al. Innate immune activation through Nalp3 inflammasome sensing of asbestos and silica. Science. 2008;320:674–7. https://doi.org/10.1126/science.1156995.

23. Ising C, et al. NLRP3 inflammasome activation drives tau pathology. Nature. 2019;575:669–73. https://doi.org/10.1038/s41586-019-1769-z.

24. Gordon EM, et al. Apolipoprotein E is a concentration-dependent pulmonary danger signal that activates the NLRP3 inflammasome and IL-1beta secretion by bronchoalveolar fluid macrophages from asthmatic subjects. J Allergy Clin Immunol. 2019;144:426–441 e423. https://doi.org/10.1016/j.jaci.2019.02.027.

25. Mukherjee S, et al. Deubiquitination of NLRP6 inflammasome by Cyld critically regulates intestinal inflammation. Nat Immunol. 2020;21:626–35. https://doi.org/10.1038/s41590-020-0681-x.

26. Hara H, et al. The NLRP6 inflammasome recognizes lipoteichoic acid and regulates gram-positive pathogen infection. Cell. 2018;175:1651–1664 e1614. https://doi.org/10.1016/j.cell.2018.09.047.

27. Zhu S, et al. Nlrp9b inflammasome restricts rotavirus infection in intestinal epithelial cells. Nature. 2017;546:667–70. https://doi.org/10.1038/nature22967.

28. Cui J, et al. NLRP4 negatively regulates type I interferon signaling by targeting the kinase TBK1 for degradation via the ubiquitin ligase DTX4. Nat Immunol. 2012;13:387–95. https://doi.org/10.1038/ni.2239.

29. Eisenbarth SC, et al. NLRP10 is a NOD-like receptor essential to initiate adaptive immunity by dendritic cells. Nature. 2012;484:510–3. https://doi.org/10.1038/nature11012.

30. Wu C, et al. NLRP11 attenuates Toll-like receptor signalling by targeting TRAF6 for degradation via the ubiquitin ligase RNF19A. Nat Commun. 2017;8:1977. https://doi.org/10.1038/s41467-017-02073-3.

31. Abe T, et al. Germ-cell-specific inflammasome component NLRP14 negatively regulates cytosolic nucleic acid sensing to promote

fertilization. Immunity. 2017;46:621–34. https://doi.org/10.1016/j.immuni.2017.03.020.

32. Velloso FJ, Trombetta-Lima M, Anschau V, Sogayar MC, Correa RG. NOD-like receptors: major players (and targets) in the interface between innate immunity and cancer. Biosci Rep. 2019;39 https://doi.org/10.1042/BSR20181709.

33. Nakamura N, et al. Endosomes are specialized platforms for bacterial sensing and NOD2 signalling. Nature. 2014;509:240–4. https://doi.org/10.1038/nature13133.

34. Li X, et al. Viral DNA binding to NLRC3, an inhibitory nucleic acid sensor, unleashes STING, a cyclic dinucleotide receptor that activates type I interferon. Immunity. 2019;50:591–599 e596. https://doi.org/10.1016/j.immuni.2019.02.009.

35. Franchi L, Nunez G. Immunology. Orchestrating inflammasomes. Science. 2012;337:1299–300. https://doi.org/10.1126/science.1229010.

36. Chonwerawong M, et al. Innate immune molecule NLRC5 protects mice from helicobacter-induced formation of gastric lymphoid tissue. Gastroenterology. 2020;159:169–182 e168. https://doi.org/10.1053/j.gastro.2020.03.009.

37. Hong M, Yoon SI, Wilson IA. Structure and functional characterization of the RNA-binding element of the NLRX1 innate immune modulator. Immunity. 2012;36:337–47. https://doi.org/10.1016/j.immuni.2011.12.018.

38. Sun L, Wu J, Du F, Chen X, Chen ZJ. Cyclic GMP-AMP synthase is a cytosolic DNA sensor that activates the type I interferon pathway. Science. 2013;339:786–91. https://doi.org/10.1126/science.1232458.

39. Gao P, et al. Cyclic [G(2′,5′)pA(3′,5′)p] is the metazoan second messenger produced by DNA-activated cyclic GMP-AMP synthase. Cell. 2013;153:1094–107. https://doi.org/10.1016/j.cell.2013.04.046.

40. Jiao H, et al. Z-nucleic-acid sensing triggers ZBP1-dependent necroptosis and inflammation. Nature. 2020;580:391–5. https://doi.org/10.1038/s41586-020-2129-8.

41. Gringhuis SI, et al. HIV-1 blocks the signaling adaptor MAVS to evade antiviral host defense after sensing of abortive HIV-1 RNA by the host helicase DDX3. Nat Immunol. 2017;18:225–35. https://doi.org/10.1038/ni.3647.

42. Wang Y, et al. Mitochondria-localised ZNFX1 functions as a dsRNA sensor to initiate antiviral responses through MAVS. Nat Cell Biol. 2019;21:1346–56. https://doi.org/10.1038/s41556-019-0416-0.

43. Kracht M, Muller-Ladner U, Schmitz ML. Mutual regulation of metabolic processes and proinflammatory NF-kappaB signaling. J Allergy Clin Immunol. 2020;146:694–705. https://doi.org/10.1016/j.jaci.2020.07.027.

44. Hoesel B, Schmid JA. The complexity of NF-kappaB signaling in inflammation and cancer. Mol Cancer. 2013;12:86. https://doi.org/10.1186/1476-4598-12-86.

45. Zhou J, Ching YQ, Chng WJ. Aberrant nuclear factor-kappa B activity in acute myeloid leukemia: from molecular pathogenesis to therapeutic target. Oncotarget. 2015;6:5490–500. https://doi.org/10.18632/oncotarget.3545.

46. Beinke S, Belich MP, Ley SC. The death domain of NF-kappa B1 p105 is essential for signal-induced p105 proteolysis. J Biol Chem. 2002;277:24162–8. https://doi.org/10.1074/jbc.M201576200.

47. Yilmaz ZB, et al. Quantitative dissection and modeling of the NF-kappaB p100-p105 module reveals interdependent precursor proteolysis. Cell Rep. 2014;9:1756–69. https://doi.org/10.1016/j.celrep.2014.11.014.

48. Zhang Q, Lenardo MJ, Baltimore D. 30 years of NF-kappaB: a blossoming of relevance to human pathobiology. Cell. 2017;168:37–57. https://doi.org/10.1016/j.cell.2016.12.012.

49. Smale ST. Dimer-specific regulatory mechanisms within the NF-kappaB family of transcription factors. Immunol Rev. 2012;246:193–204. https://doi.org/10.1111/j.1600-065X.2011.01091.x.

50. Sun SC. The non-canonical NF-kappaB pathway in immunity and inflammation. Nat Rev Immunol. 2017;17:545–58. https://doi.org/10.1038/nri.2017.52.

51. Wang C, et al. TAK1 is a ubiquitin-dependent kinase of MKK and IKK. Nature. 2001;412:346–51. https://doi.org/10.1038/35085597.

52. Chen ZJ. Ubiquitin signalling in the NF-kappaB pathway. Nat Cell Biol. 2005;7:758–65. https://doi.org/10.1038/ncb0805-758.

53. Sato S, et al. Essential function for the kinase TAK1 in innate and adaptive immune responses. Nat Immunol. 2005;6:1087–95. https://doi.org/10.1038/ni1255.

54. Israel A. The IKK complex, a central regulator of NF-kappaB activation. Cold Spring Harb Perspect Biol. 2010;2:a000158. https://doi.org/10.1101/cshperspect.a000158.

55. Walsh MC, Lee J, Choi Y. Tumor necrosis factor receptor- associated factor 6 (TRAF6) regulation of development, function, and homeostasis of the immune system. Immunol Rev. 2015;266:72–92. https://doi.org/10.1111/imr.12302.

56. Suzuki N, Suzuki S, Yeh WC. IRAK-4 as the central TIR signaling mediator in innate immunity. Trends Immunol. 2002;23:503–6. https://doi.org/10.1016/s1471-4906(02)02298-6.

57. Taniguchi K, Karin M. NF-kappaB, inflammation, immunity and cancer: coming of age. Nat Rev Immunol. 2018;18:309–24. https://doi.org/10.1038/nri.2017.142.

58. Yang Z, et al. USP18 negatively regulates NF-kappaB signaling by targeting TAK1 and NEMO for deubiquitination through distinct mechanisms. Sci Rep. 2015;5:12738. https://doi.org/10.1038/srep12738.

59. Mulas F, et al. The deubiquitinase OTUB1 augments NF-kappaB-dependent immune responses in dendritic cells in infection and inflammation by stabi-

lizing UBC13. Cell Mol Immunol. 2020; https://doi.org/10.1038/s41423-020-0362-6.

60. Sun SC. CYLD: a tumor suppressor deubiquitinase regulating NF-kappaB activation and diverse biological processes. Cell Death Differ. 2010;17:25–34. https://doi.org/10.1038/cdd.2009.43.

61. Sun SC, Chang JH, Jin J. Regulation of nuclear factor-kappaB in autoimmunity. Trends Immunol. 2013;34:282–9.   https://doi.org/10.1016/j.it.2013.01.004.

62. Boone DL, et al. The ubiquitin-modifying enzyme A20 is required for termination of Toll-like receptor responses. Nat Immunol. 2004;5:1052–60. https://doi.org/10.1038/ni1110.

63. Hu YH, et al. SPOP negatively regulates Toll-like receptor-induced inflammation by disrupting MyD88 self-association. Cell Mol Immunol. 2020; https://doi.org/10.1038/s41423-020-0411-1.

64. Lou Y, et al. Essential roles of S100A10 in Toll-like receptor signaling and immunity to infection. Cell Mol Immunol. 2020;17:1053–62. https://doi.org/10.1038/s41423-019-0278-1.

65. Saha S, et al. Long noncoding RNA DRAIC inhibits prostate cancer progression by interacting with IKK to inhibit NF-kappaB activation. Cancer Res. 2020;80:950–63.   https://doi.org/10.1158/0008-5472.CAN-19-3460.

66. Bellet MM, et al. HOPS/Tmub1 involvement in the NF-kB-mediated inflammatory response through the modulation of TRAF6. Cell Death Dis. 2020;11:865. https://doi.org/10.1038/s41419-020-03086-5.

67. Zhang W, et al. The metabolic regulator Lamtor5 suppresses inflammatory signaling via regulating mTOR-mediated TLR4 degradation. Cell Mol Immunol. 2020;17:1063–76. https://doi.org/10.1038/s41423-019-0281-6.

68. Mandola AB, et al. Combined immunodeficiency caused by a novel homozygous NFKB1 mutation. J Allergy Clin Immunol. 2020; https://doi.org/10.1016/j.jaci.2020.08.040.

69. Abbott J, et al. Heterozygous IKKbeta activation loop mutation results in a complex immunodeficiency syndrome. J Allergy Clin Immunol. 2020; https://doi.org/10.1016/j.jaci.2020.06.007.

70. Do-Umehara HC, et al. Epithelial cell-specific loss of function of Miz1 causes a spontaneous COPD-like phenotype and up-regulates Ace2 expression in mice. Sci Adv. 2020;6:eabb7238. https://doi.org/10.1126/sciadv.abb7238.

71. Kaur BP, Secord E. Innate immunity. Pediatr Clin N Am. 2019;66:905–11. https://doi.org/10.1016/j.pcl.2019.06.011.

72. Cui J, et al. USP3 inhibits type I interferon signaling by deubiquitinating RIG-I-like receptors. Cell Res. 2014;24:400–16. https://doi.org/10.1038/cr.2013.170.

73. Ng CT, Mendoza JL, Garcia KC, Oldstone MB. Alpha and Beta type 1 interferon signaling: passage for diverse biologic outcomes. Cell. 2016;164:349–52. https://doi.org/10.1016/j.cell.2015.12.027.

74. Ivashkiv LB, Donlin LT. Regulation of type I interferon responses. Nat Rev Immunol. 2014;14:36–49. https://doi.org/10.1038/nri3581.

75. Stark GR, Darnell JE Jr. The JAK-STAT pathway at twenty. Immunity. 2012;36:503–14. https://doi.org/10.1016/j.immuni.2012.03.013.

76. Schoggins JW, et al. A diverse range of gene products are effectors of the type I interferon antiviral response. Nature. 2011;472:481–5. https://doi.org/10.1038/nature09907.

77. Jiang D, et al. Identification of five interferon-induced cellular proteins that inhibit west nile virus and dengue virus infections. J Virol. 2010;84:8332–41. https://doi.org/10.1128/JVI.02199-09.

78. Seth RB, Sun L, Ea CK, Chen ZJ. Identification and characterization of MAVS, a mitochondrial antiviral signaling protein that activates NF-kappaB and IRF 3. Cell. 2005;122:669–82. https://doi.org/10.1016/j.cell.2005.08.012.

79. Yoneyama M, et al. The RNA helicase RIG-I has an essential function in double-stranded RNA-induced innate antiviral responses. Nat Immunol. 2004;5:730–7. https://doi.org/10.1038/ni1087.

80. Wu B, et al. Structural basis for dsRNA recognition, filament formation, and antiviral signal activation by MDA5. Cell. 2013;152:276–89. https://doi.org/10.1016/j.cell.2012.11.048.

81. Hou F, et al. MAVS forms functional prion-like aggregates to activate and propagate antiviral innate immune response. Cell. 2011;146:448–61. https://doi.org/10.1016/j.cell.2011.06.041.

82. Shang G, Zhang C, Chen ZJ, Bai XC, Zhang X. Cryo-EM structures of STING reveal its mechanism of activation by cyclic GMP-AMP. Nature. 2019;567:389–93.   https://doi.org/10.1038/s41586-019-0998-5.

83. Zhao B, et al. A conserved PLPLRT/SD motif of STING mediates the recruitment and activation of TBK1. Nature. 2019;569:718–22. https://doi.org/10.1038/s41586-019-1228-x.

84. Zhang C, et al. Structural basis of STING binding with and phosphorylation by TBK1. Nature. 2019;567:394–8.   https://doi.org/10.1038/s41586-019-1000-2.

85. Hopfner KP, Hornung V. Molecular mechanisms and cellular functions of cGAS-STING signalling. Nat Rev Mol Cell Biol. 2020;21:501–21. https://doi.org/10.1038/s41580-020-0244-x.

86. Liu S, et al. Phosphorylation of innate immune adaptor proteins MAVS, STING, and TRIF induces IRF3 activation. Science. 2015;347:aaa2630. https://doi.org/10.1126/science.aaa2630.

87. Akira S, Uematsu S, Takeuchi O. Pathogen recognition and innate immunity. Cell. 2006;124:783–801. https://doi.org/10.1016/j.cell.2006.02.015.

88. Lin B, et al. A novel STING1 variant causes a recessive form of STING-associated vasculopathy with onset in infancy (SAVI). J Allergy Clin Immunol. 2020; https://doi.org/10.1016/j.jaci.2020.06.032.

89. Bennion BG, et al. STING gain-of-function disrupts lymph node organogenesis and innate lymphoid cell development in mice. Cell Rep. 2020;31:107771. https://doi.org/10.1016/j.celrep.2020.107771.

90. Yu CH, et al. TDP-43 triggers mitochondrial DNA release via mPTP to activate cGAS/STING in ALS. Cell. 2020; https://doi.org/10.1016/j.cell.2020.09.020.

91. McCauley ME, et al. C9orf72 in myeloid cells suppresses STING-induced inflammation. Nature. 2020;585:96–101. https://doi.org/10.1038/s41586-020-2625-x.

92. Kim J, et al. VDAC oligomers form mitochondrial pores to release mtDNA fragments and promote lupus-like disease. Science. 2019;366:1531–6. https://doi.org/10.1126/science.aav4011.

93. Chung KW, et al. Mitochondrial damage and activation of the STING pathway lead to renal inflammation and fibrosis. Cell Metab. 2019;30:784–799 e785. https://doi.org/10.1016/j.cmet.2019.08.003.

94. Winkler R, et al. m(6)A modification controls the innate immune response to infection by targeting type I interferons. Nat Immunol. 2019;20:173–82. https://doi.org/10.1038/s41590-018-0275-z.

95. Tian M, et al. MYSM1 represses innate immunity and autoimmunity through suppressing the cGAS-STING pathway. Cell Rep. 2020;33:108297. https://doi.org/10.1016/j.celrep.2020.108297.

96. Hou P, et al. A novel selective autophagy receptor, CCDC50, delivers K63 polyubiquitination-activated RIG-I/MDA5 for degradation during viral infection. Cell Res. 2020; https://doi.org/10.1038/s41422-020-0362-1.

97. Huang L, et al. Ubiquitin-conjugating enzyme 2S enhances viral replication by inhibiting type I IFN production through recruiting USP15 to Deubiquitinate TBK1. Cell Rep. 2020;32:108044. https://doi.org/10.1016/j.celrep.2020.108044.

98. Jahan AS, et al. OTUB1 is a key regulator of RIG-I-dependent immune signaling and is targeted for proteasomal degradation by influenza A NS1. Cell Rep. 2020;30:1570–1584 e1576. https://doi.org/10.1016/j.celrep.2020.01.015.

99. Guo Y, et al. OTUD5 promotes innate antiviral and antitumor immunity through deubiquitinating and stabilizing STING. Cell Mol Immunol. 2020; https://doi.org/10.1038/s41423-020-00531-5.

100. Chen M, et al. TRIM14 inhibits cGAS degradation mediated by selective autophagy receptor p62 to promote innate immune responses. Mol Cell. 2016;64:105–19. https://doi.org/10.1016/j.molcel.2016.08.025.

101. Song N, et al. MAVS O-GlcNAcylation is essential for host antiviral immunity against lethal RNA viruses. Cell Rep. 2019;28:2386–2396 e2385. https://doi.org/10.1016/j.celrep.2019.07.085.

102. Choi SJ, et al. HDAC6 regulates cellular viral RNA sensing by deacetylation of RIG-I. EMBO J. 2016;35:429–42. https://doi.org/10.15252/embj.201592586.

103. Hansen AL, Mukai K, Schopfer FJ, Taguchi T, Holm CK. STING palmitoylation as a therapeutic target. Cell Mol Immunol. 2019;16:236–41. https://doi.org/10.1038/s41423-019-0205-5.

104. Liu ZS, et al. G3BP1 promotes DNA binding and activation of cGAS. Nat Immunol. 2019;20:18–28. https://doi.org/10.1038/s41590-018-0262-4.

105. Guey B, et al. BAF restricts cGAS on nuclear DNA to prevent innate immune activation. Science. 2020;369:823–8. https://doi.org/10.1126/science.aaw6421.

106. Lian H, et al. The zinc-finger protein ZCCHC3 binds RNA and facilitates viral RNA sensing and activation of the RIG-I-like receptors. Immunity. 2018;49:438–448 e435. https://doi.org/10.1016/j.immuni.2018.08.014.

107. Rehwinkel J, Gack MU. RIG-I-like receptors: their regulation and roles in RNA sensing. Nat Rev Immunol. 2020;20:537–51. https://doi.org/10.1038/s41577-020-0288-3.

108. Liu T, et al. TRIM11 suppresses AIM2 inflammasome by degrading AIM2 via p62-dependent selective autophagy. Cell Rep. 2016;16:1988–2002. https://doi.org/10.1016/j.celrep.2016.07.019.

109. Shi J, et al. Inflammatory caspases are innate immune receptors for intracellular LPS. Nature. 2014;514:187–92. https://doi.org/10.1038/nature13683.

110. Kayagaki N, et al. Caspase-11 cleaves gasdermin D for non-canonical inflammasome signalling. Nature. 2015;526:666–71. https://doi.org/10.1038/nature15541.

111. Chui AJ, et al. N-terminal degradation activates the NLRP1B inflammasome. Science. 2019;364:82–5. https://doi.org/10.1126/science.aau1208.

112. Rauch I, et al. NAIP proteins are required for cytosolic detection of specific bacterial ligands in vivo. J Exp Med. 2016;213:657–65. https://doi.org/10.1084/jem.20151809.

113. Tenthorey JL, et al. The structural basis of flagellin detection by NAIP5: a strategy to limit pathogen immune evasion. Science. 2017;358:888–93. https://doi.org/10.1126/science.aao1140.

114. Zhao Y, et al. The NLRC4 inflammasome receptors for bacterial flagellin and type III secretion apparatus. Nature. 2011;477:596–600. https://doi.org/10.1038/nature10510.

115. Man SM, et al. IRGB10 liberates bacterial ligands for sensing by the AIM2 and Caspase-11-NLRP3 inflammasomes. Cell. 2016;167:382–396 e317. https://doi.org/10.1016/j.cell.2016.09.012.

116. Hornung V, et al. AIM2 recognizes cytosolic dsDNA and forms a caspase-1-activating inflammasome with ASC. Nature. 2009;458:514–8. https://doi.org/10.1038/nature07725.

117. Hornung V, et al. Silica crystals and aluminum salts activate the NALP3 inflammasome through phagosomal destabilization. Nat Immunol. 2008;9:847–56. https://doi.org/10.1038/ni.1631.

118. Zhong Z, et al. New mitochondrial DNA synthesis enables NLRP3 inflammasome activation. Nature. 2018;560:198–203. https://doi.org/10.1038/s41586-018-0372-z.

119. Di A, et al. The TWIK2 potassium Efflux channel in macrophages mediates NLRP3 inflammasome-induced inflammation. Immunity. 2018;49:56–65 e54. https://doi.org/10.1016/j.immuni.2018.04.032.

120. Li S, et al. Activated NLR family pyrin domain containing 3 (NLRP3) inflammasome in keratinocytes promotes cutaneous T-cell response in patients with vitiligo. J Allergy Clin Immunol. 2020;145:632–45. https://doi.org/10.1016/j.jaci.2019.10.036.

121. Zhou L, et al. Excessive deubiquitination of NLRP3-R779C variant contributes to very-early-onset inflammatory bowel disease development. J Allergy Clin Immunol. 2020; https://doi.org/10.1016/j.jaci.2020.09.003.

122. Wei Y, et al. Activated pyrin domain containing 3 (NLRP3) inflammasome in neutrophilic chronic rhinosinusitis with nasal polyps (CRSwNP). J Allergy Clin Immunol. 2020;145:1002–1005 e1016. https://doi.org/10.1016/j.jaci.2020.01.009.

123. Zhang H, et al. AIM2 Inflammasome is critical for influenza-induced lung injury and mortality. J Immunol. 2017;198:4383–93. https://doi.org/10.4049/jimmunol.1600714.

124. Qin Q, et al. The inhibitor effect of RKIP on inflammasome activation and inflammasome-dependent diseases. Cell Mol Immunol. 2020; https://doi.org/10.1038/s41423-020-00525-3.

125. Zheng X, et al. Synthetic vitamin K analogs inhibit inflammation by targeting the NLRP3 inflammasome. Cell Mol Immunol. 2020; https://doi.org/10.1038/s41423-020-00545-z.

126. McDaniel MM, Kottyan LC, Singh H, Pasare C. Suppression of inflammasome activation by IRF8 and IRF4 in cDCs is critical for T cell priming. Cell Rep. 2020;31:107604. https://doi.org/10.1016/j.celrep.2020.107604.

127. Goos H, et al. Gain-of-function CEBPE mutation causes noncanonical autoinflammatory inflammasomopathy. J Allergy Clin Immunol. 2019;144:1364–76. https://doi.org/10.1016/j.jaci.2019.06.003.

128. Benyoucef A, Marchitto L, Touzot F. CRISPR gene-engineered CYBB(ko) THP-1 cell lines highlight the crucial role of NADPH-induced reactive oxygen species for regulating inflammasome activation. J Allergy Clin Immunol. 2020;145:1690–1693 e1695. https://doi.org/10.1016/j.jaci.2019.12.913.

129. Liu T, et al. USP19 suppresses inflammation and promotes M2-like macrophage polarization by manipulating NLRP3 function via autophagy. Cell Mol Immunol. 2020; https://doi.org/10.1038/s41423-020-00567-7.

130. Sanchez-Rodriguez R, et al. Targeting monoamine oxidase to dampen NLRP3 inflammasome activation in inflammation. Cell Mol Immunol. 2020; https://doi.org/10.1038/s41423-020-0441-8.

131. Huang B, et al. Ticagrelor inhibits the NLRP3 inflammasome to protect against inflammatory disease independent of the P2Y12 signaling pathway. Cell Mol Immunol. 2020; https://doi.org/10.1038/s41423-020-0444-5.

132. Han X, et al. Small molecule-driven NLRP3 inflammation inhibition via interplay between ubiquitination and autophagy: implications for Parkinson disease. Autophagy. 2019;15:1860–81. https://doi.org/10.1080/15548627.2019.1596481.

133. Yang F, et al. Metformin inhibits the NLRP3 inflammasome via AMPK/mTOR-dependent effects in diabetic cardiomyopathy. Int J Biol Sci. 2019;15:1010–9. https://doi.org/10.7150/ijbs.29680.

134. Postma DS, Bush A, van den Berge M. Risk factors and early origins of chronic obstructive pulmonary disease. Lancet. 2015;385:899–909. https://doi.org/10.1016/S0140-6736(14)60446-3.

135. Rennard SI, Drummond MB. Early chronic obstructive pulmonary disease: definition, assessment, and prevention. Lancet. 2015;385:1778–88. https://doi.org/10.1016/S0140-6736(15)60647-X.

136. Di Stefano A, et al. Increased expression of nuclear factor-kappaB in bronchial biopsies from smokers and patients with COPD. Eur Respir J. 2002;20:556–63. https://doi.org/10.1183/09031936.02.00272002.

137. Szulakowski P, et al. The effect of smoking on the transcriptional regulation of lung inflammation in patients with chronic obstructive pulmonary disease. Am J Respir Crit Care Med. 2006;174:41–50. https://doi.org/10.1164/rccm.200505-725OC.

138. Gagliardo R, et al. IkappaB kinase-driven nuclear factor-kappaB activation in patients with asthma and chronic obstructive pulmonary disease. J Allergy Clin Immunol. 2011;128:635–645 e631-632. https://doi.org/10.1016/j.jaci.2011.03.045.

139. Yang SR, et al. RelB is differentially regulated by IkappaB Kinase-alpha in B cells and mouse lung by cigarette smoke. Am J Respir Cell Mol Biol. 2009;40:147–58. https://doi.org/10.1165/rcmb.2008-0207OC.

140. Henkel M, et al. FSTL-1 attenuation causes spontaneous smoke-resistant pulmonary emphysema. Am J Respir Crit Care Med. 2020;201:934–45. https://doi.org/10.1164/rccm.201905-0973OC.

141. Chung KF. Cytokines in chronic obstructive pulmonary disease. Eur Respir J Suppl. 2001;34:50s–9s.

142. Petersen AM, et al. Elevated levels of IL-18 in plasma and skeletal muscle in chronic obstructive pulmonary disease. Lung. 2007;185:161–71. https://doi.org/10.1007/s00408-007-9000-7.

143. Botelho FM, et al. IL-1alpha/IL-1R1 expression in chronic obstructive pulmonary disease and mechanistic relevance to smoke-induced neutrophilia in mice. PLoS One. 2011;6:e28457. https://doi.org/10.1371/journal.pone.0028457.

144. Kang MJ, et al. IL-18 is induced and IL-18 receptor alpha plays a critical role in the pathogenesis of cigarette smoke-induced pulmonary emphysema

and inflammation. J Immunol. 2007;178:1948–59. https://doi.org/10.4049/jimmunol.178.3.1948.

145. Eltom S, et al. P2X7 receptor and caspase 1 activation are central to airway inflammation observed after exposure to tobacco smoke. PLoS One. 2011;6:e24097. https://doi.org/10.1371/journal.pone.0024097.

146. Churg A, Zhou S, Wang X, Wang R, Wright JL. The role of interleukin-1beta in murine cigarette smoke-induced emphysema and small airway remodeling. Am J Respir Cell Mol Biol. 2009;40:482–90. https://doi.org/10.1165/rcmb.2008-0038OC.

147. Wiegman CH, et al. Oxidative stress-induced mitochondrial dysfunction drives inflammation and airway smooth muscle remodeling in patients with chronic obstructive pulmonary disease. J Allergy Clin Immunol. 2015;136:769–80. https://doi.org/10.1016/j.jaci.2015.01.046.

148. Lao T, et al. Hhip haploinsufficiency sensitizes mice to age-related emphysema. Proc Natl Acad Sci USA. 2016;113:E4681–7. https://doi.org/10.1073/pnas.1602342113.

149. Garcia-Valero J, et al. Deficient pulmonary IFN-beta expression in COPD patients. PloS One. 2019;14:e0217803. https://doi.org/10.1371/journal.pone.0217803.

150. Hilzendeger C, et al. Reduced sputum expression of interferon-stimulated genes in severe COPD. Int J Chron Obstruct Pulmon Dis. 2016;11:1485–94. https://doi.org/10.2147/COPD.S105948.

151. Nascimento M, et al. Self-DNA release and STING-dependent sensing drives inflammation to cigarette smoke in mice. Sci Rep. 2019;9:14848. https://doi.org/10.1038/s41598-019-51427-y.

152. Vos T, et al. Years lived with disability (YLDs) for 1160 sequelae of 289 diseases and injuries 1990-2010: a systematic analysis for the Global Burden of Disease Study 2010. Lancet. 2012;380:2163–96. https://doi.org/10.1016/S0140-6736(12)61729-2.

153. Lambrecht BN, Hammad H. The immunology of asthma. Nat Immunol. 2015;16:45–56. https://doi.org/10.1038/ni.3049.

154. Papi A, Brightling C, Pedersen SE, Reddel HK. Asthma. Lancet. 2018;391:783–800. https://doi.org/10.1016/S0140-6736(17)33311-1.

155. von Mutius E, Smits HH. Primary prevention of asthma: from risk and protective factors to targeted strategies for prevention. Lancet. 2020;396:854–66. https://doi.org/10.1016/S0140-6736(20)31861-4.

156. Ogasawara N, et al. TNF induces production of type 2 cytokines in human group 2 innate lymphoid cells. J Allergy Clin Immunol. 2020;145:437–440 e438. https://doi.org/10.1016/j.jaci.2019.09.001.

157. Ather JL, Hodgkins SR, Janssen-Heininger YM, Poynter ME. Airway epithelial NF-kappaB activation promotes allergic sensitization to an innocuous inhaled antigen. Am J Respir Cell Mol Biol. 2011;44:631–8. https://doi.org/10.1165/rcmb.2010-0106OC.

158. Poynter ME, et al. NF-kappa B activation in airways modulates allergic inflammation but not hyperresponsiveness. J Immunol. 2004;173:7003–9. https://doi.org/10.4049/jimmunol.173.11.7003.

159. Miyake T, et al. Prevention of asthma exacerbation in a mouse model by simultaneous inhibition of NF-kappaB and STAT6 activation using a chimeric decoy strategy. Mol Ther Nucleic Acids. 2018;10:159–69. https://doi.org/10.1016/j.omtn.2017.12.005.

160. Krusche J, et al. TNF-alpha-induced protein 3 is a key player in childhood asthma development and environment-mediated protection. J Allergy Clin Immunol. 2019;144:1684–1696 e1612. https://doi.org/10.1016/j.jaci.2019.07.029.

161. Rossios C, et al. Sputum transcriptomics reveal upregulation of IL-1 receptor family members in patients with severe asthma. J Allergy Clin Immunol. 2018;141:560–70. https://doi.org/10.1016/j.jaci.2017.02.045.

162. Tan HT, et al. Tight junction, mucin, and inflammasome-related molecules are differentially expressed in eosinophilic, mixed, and neutrophilic experimental asthma in mice. Allergy. 2019;74:294–307. https://doi.org/10.1111/all.13619.

163. Ritter M, et al. Functional relevance of NLRP3 inflammasome-mediated interleukin (IL)-1beta during acute allergic airway inflammation. Clin Exp Immunol. 2014;178:212–23. https://doi.org/10.1111/cei.12400.

164. Besnard AG, et al. NLRP3 inflammasome is required in murine asthma in the absence of aluminum adjuvant. Allergy. 2011;66:1047–57. https://doi.org/10.1111/j.1398-9995.2011.02586.x.

165. Kim RY, et al. Role for NLRP3 inflammasome-mediated, IL-1beta-dependent responses in severe, steroid-resistant asthma. Am J Respir Crit Care Med. 2017;196:283–97. https://doi.org/10.1164/rccm.201609-1830OC.

166. Ather JL, et al. Serum amyloid A activates the NLRP3 inflammasome and promotes Th17 allergic asthma in mice. J Immunol. 2011;187:64–73. https://doi.org/10.4049/jimmunol.1100500.

167. Kool M, et al. An unexpected role for uric acid as an inducer of T helper 2 cell immunity to inhaled antigens and inflammatory mediator of allergic asthma. Immunity. 2011;34:527–40. https://doi.org/10.1016/j.immuni.2011.03.015.

168. Wood LG, et al. Saturated fatty acids, obesity, and the nucleotide oligomerization domain-like receptor protein 3 (NLRP3) inflammasome in asthmatic patients. J Allergy Clin Immunol. 2019;143:305–15. https://doi.org/10.1016/j.jaci.2018.04.037.

169. da Silva J, et al. Raised interferon-beta, type 3 interferon and interferon-stimulated genes - evidence of innate immune activation in neutrophilic asthma. Clin Exp Allergy: journal of the British Society for Allergy and Clinical Immunology. 2017;47:313–23. https://doi.org/10.1111/cea.12809.

170. Moskwa S, et al. Innate immune response to viral infections in primary bronchial epithelial cells is modified by the atopic status of asthmatic patients. Allergy, Asthma Immunol Res. 2018;10:144–54. https://doi.org/10.4168/aair.2018.10.2.144.

171. Ravi A, et al. Interferon-induced epithelial response to rhinovirus 16 in asthma relates to inflammation and FEV1. J Allergy Clin Immunol. 2019;143:442–447 e410. https://doi.org/10.1016/j.jaci.2018.09.016.

172. Zhu J, et al. Bronchial mucosal IFN-alpha/beta and pattern recognition receptor expression in patients with experimental rhinovirus-induced asthma exacerbations. J Allergy Clin Immunol. 2019;143:114–125 e114. https://doi.org/10.1016/j.jaci.2018.04.003.

173. Huang C, et al. Clinical features of patients infected with 2019 novel coronavirus in Wuhan, China. Lancet. 2020;395:497–506. https://doi.org/10.1016/S0140-6736(20)30183-5.

174. Hu B, Guo H, Zhou P, Shi ZL. Characteristics of SARS-CoV-2 and COVID-19. Nat Rev Microbiol. 2020; https://doi.org/10.1038/s41579-020-00459-7.

175. Blanco-Melo D, et al. Imbalanced host response to SARS-CoV-2 drives development of COVID-19. Cell. 2020;181:1036–1045 e1039. https://doi.org/10.1016/j.cell.2020.04.026.

176. Stanifer ML, et al. Critical role of type III interferon in controlling SARS-CoV-2 infection in human intestinal epithelial cells. Cell Rep. 2020;32:107863. https://doi.org/10.1016/j.celrep.2020.107863.

177. Zhou Z, et al. Heightened innate immune responses in the respiratory tract of COVID-19 patients. Cell Host Microbe. 2020;27:883–890 e882. https://doi.org/10.1016/j.chom.2020.04.017.

178. Zhang Q, et al. Inborn errors of type I IFN immunity in patients with life-threatening COVID-19. Science. 2020;370 https://doi.org/10.1126/science.abd4570.

179. Meyts I, et al. Coronavirus disease 2019 in patients with inborn errors of immunity: an international study. J Allergy Clin Immunol. 2020; https://doi.org/10.1016/j.jaci.2020.09.010.

180. Israelow B, et al. Mouse model of SARS-CoV-2 reveals inflammatory role of type I interferon signaling. J Exp Med. 2020;217 https://doi.org/10.1084/jem.20201241.

181. Wang N, et al. Retrospective Multicenter Cohort Study shows early interferon therapy is associated with favorable clinical responses in COVID-19 patients. Cell Host Microbe. 2020;28:455–464 e452. https://doi.org/10.1016/j.chom.2020.07.005.

182. Trouillet-Assant S, et al. Type I IFN immunoprofiling in COVID-19 patients. J Allergy Clin Immunol. 2020;146:206–208 e202. https://doi.org/10.1016/j.jaci.2020.04.029.

183. Onabajo OO, et al. Interferons and viruses induce a novel truncated ACE2 isoform and not the full-length SARS-CoV-2 receptor. Nat Genet. 2020; https://doi.org/10.1038/s41588-020-00731-9.

184. Xia H, et al. Evasion of type I interferon by SARS-CoV-2. Cell Rep. 2020;33:108234. https://doi.org/10.1016/j.celrep.2020.108234.

185. Huang L, et al. Sepsis-associated severe interleukin-6 storm in critical coronavirus disease 2019. Cell Mol Immunol. 2020;17:1092–4. https://doi.org/10.1038/s41423-020-00522-6.

186. Copaescu A, Smibert O, Gibson A, Phillips EJ, Trubiano JA. The role of IL-6 and other mediators in the cytokine storm associated with SARS-CoV-2 infection. J Allergy Clin Immunol. 2020;146:518–534 e511. https://doi.org/10.1016/j.jaci.2020.07.001.

187. Herold T, et al. Elevated levels of IL-6 and CRP predict the need for mechanical ventilation in COVID-19. J Allergy Clin Immunol. 2020;146:128–136 e124. https://doi.org/10.1016/j.jaci.2020.05.008.

188. Laguna-Goya R, et al. IL-6-based mortality risk model for hospitalized patients with COVID-19. J Allergy Clin Immunol. 2020;146:799–807 e799. https://doi.org/10.1016/j.jaci.2020.07.009.

189. Galvan-Roman JM, et al. IL-6 serum levels predict severity and response to Tocilizumab in COVID-19: an observational study. J Allergy Clin Immunol. 2020; https://doi.org/10.1016/j.jaci.2020.09.018.

190. Crisafulli S, Isgro V, La Corte L, Atzeni F, Trifiro G. Potential role of anti-interleukin (IL)-6 drugs in the treatment of COVID-19: rationale, clinical evidence and risks. BioDrugs: clinical immunotherapeutics, biopharmaceuticals and gene therapy. 2020;34:415–22. https://doi.org/10.1007/s40259-020-00430-1.

191. Saeedi-Boroujeni A, Mahmoudian-Sani MR, Bahadoram M, Alghasi A. COVID-19: a case for inhibiting NLRP3 inflammasome, suppression of inflammation with Curcumin? Basic Clin Pharmacol Toxicol. 2020; https://doi.org/10.1111/bcpt.13503.

192. Toldo S, et al. Inflammasome formation in the lungs of patients with fatal COVID-19. Inflamm Res. 2020; https://doi.org/10.1007/s00011-020-01413-2.

193. Ratajczak MZ, Kucia M. SARS-CoV-2 infection and overactivation of Nlrp3 inflammasome as a trigger of cytokine "storm" and risk factor for damage of hematopoietic stem cells. Leukemia. 2020;34:1726–9. https://doi.org/10.1038/s41375-020-0887-9.

194. Noth I, et al. Genetic variants associated with idiopathic pulmonary fibrosis susceptibility and mortality: a genome-wide association study. Lancet Respir Med. 2013;1:309–17. https://doi.org/10.1016/S2213-2600(13)70045-6.

195. Lin X, et al. Yin yang 1 is a novel regulator of pulmonary fibrosis. Am J Respir Crit Care Med. 2011;183:1689–97. https://doi.org/10.1164/rccm.201002-0232OC.

196. Xu JF, et al. Statins and pulmonary fibrosis: the potential role of NLRP3 inflammasome activation. Am J Respir Crit Care Med. 2012;185:547–56. https://doi.org/10.1164/rccm.201108-1574OC.

197. Gasse P, et al. Uric acid is a danger signal activating NALP3 inflammasome in lung injury inflammation and fibrosis. Am J Respir Crit Care Med. 2009;179:903–13. https://doi.org/10.1164/rccm.200808-1274OC.
198. George PM, et al. Evidence for the involvement of type I interferon in pulmonary arterial hypertension. Circ Res. 2014;114:677–88. https://doi.org/10.1161/CIRCRESAHA.114.302221.
199. Farkas D, et al. Toll-like receptor 3 is a therapeutic target for pulmonary hypertension. Am J Respir Crit Care Med. 2019;199:199–210. https://doi.org/10.1164/rccm.201707-1370OC.
200. Yin J, et al. Role of P2X7R in the development and progression of pulmonary hypertension. Respir Res. 2017;18:127. https://doi.org/10.1186/s12931-017-0603-0.

Interstitial Lung Disease Associated with Connective Tissue Diseases

5

Ruben A. Peredo, Vivek Mehta, and Scott Beegle

Abstract

Pulmonary manifestations of connective tissue diseases (CTD) carry high morbidity and potential mortality, and the most serious pulmonary type is interstitial lung disease (ILD). Identifying and promptly intervening CTD-ILD with immune suppressor therapy will change the natural course of the disease resulting in survival improvement. Compared to idiopathic pulmonary fibrosis, the most common presentation of idiopathic interstitial pneumonia (IIP), CTD-ILD carries a better prognosis due to the response to immune suppressor therapy. Nonspecific interstitial pneumonia (NSIP) is the most common type of CTD-ILD that is different from the fibrotic classical presentation of IPF, known as usual interstitial pneumonia (UIP). An exception is rheumatoid arthritis that presents more frequently with UIP type. Occasionally, IPF may not have typical radiographic features of UIP, and a full assessment to differentiate IPF from CTD-ILD is necessary, including the intervention of a multidisciplinary team and the histopathology. Interstitial pneumonia with autoimmune features (IPAF) shows promising advantages to identify patients with ILD who have some features of a CTD without a defined autoimmune disease and who may benefit from immune suppressors. A composition of clinical, serological, and morphologic features in patients presenting with ILD will fulfill criteria for IPAF. In summary, the early recognition and treatment of CTD-ILD, differentiation from IPF-UIP, and identification of patients with IPAF fulfill the assessment by the clinician for an optimal care.

Keywords

Interstitial lung disease · Connective tissue disease · Interstitial pneumonia with autoimmune features

R. A. Peredo (✉)
Division of Rheumatology, Department of Medicine, Albany Medical College, Albany, NY, USA
e-mail: peredor@amc.edu

V. Mehta
Rheumatology, Alaska Native Medical Center, Anchorage, AK, USA

S. Beegle
Division of Pulmonary & Critical Care Medicine, Albany Medical College, Albany, NY, USA
e-mail: beegles@amc.edu

5.1 Introduction

Connective tissue disease (CTD) is a loosely defined term which is now often replaced with "systemic rheumatic diseases," usually referring to disturbance in immune tissue resulting in widespread inflammatory tissue injury [1, 2]. Connective tissue diseases frequently target the respiratory system, affecting one or several of the pulmonary compartments, such as the airways, pleura, vasculature, and interstitium. When it comes to the latter, the lung involvement usually carries more morbidity and mortality [3–6]. Interstitial lung disease (ILD) refers to variable degrees of invasion and replacement of normal parenchymal space by inflammatory cell infiltrates and fibrosis or a mixture of both with different degrees. Depending on the magnitude of pulmonary inflammation, ILD may present with respiratory features ranging from asymptomatic to impending respiratory failure or even death. Early identification, classification, and intervention, subsequently, will define the treatment and outcome. Fibrosing lung disease may be associated with environmental exposure but also with other conditions, like sarcoidosis [7], chronic asbestosis and other occupational exposures [8], hobbies, drugs, and secondary to CTD (CTD-ILD) [9]. In the absence of any known possible causes of ILD, then the diagnosis of idiopathic interstitial pneumonia (IIP) is likely [10]. In addition, high-resolution chest tomography (HRCT) is required to narrow the differential diagnosis of type of ILD. Histopathology may be required to assist in confirming the diagnosis [11]. Idiopathic pulmonary fibrosis/usual interstitial pneumonia is a progressive fibrotic lung disease which carries a poor prognosis compared to other ILDs [12–17]. Idiopathic interstitial pneumonias (IIPs) in general have distinct clinical, histological, and radiographic features. Interstitial pulmonary fibrosis is the most common cause for morbidity (Table 5.1) [10].

Many clinical challenges complicate identifying a CTD to be the inflammatory component underlying the ILD process. A thorough and systematic evaluation will achieve the accurate CTD-ILD diagnosis and management [18]. Patients with CTD-ILD, classified as IIP, will have less chances to receive immune-modulating drug therapy, depriving them from all the benefits they could obtain. Conversely, classifying mistakenly a patient with CTD-ILD of a true case of

Table 5.1 Revised American Thoracic Society/European Respiratory Society classification of idiopathic interstitial pneumonias: multidisciplinary diagnoses [10]

Major idiopathic interstitial pneumonias
Idiopathic pulmonary fibrosis
Idiopathic nonspecific interstitial pneumonia
Respiratory bronchiolitis-interstitial lung disease
Desquamative interstitial pneumonia
Cryptogenic organizing pneumonia
Acute interstitial pneumonia
Rare idiopathic interstitial pneumonias
Idiopathic lymphoid interstitial pneumonia
Idiopathic pleuroparenchymal fibroelastosis
Unclassifiable idiopathic interstitial pneumonias*

Causes of unclassifiable idiopathic interstitial pneumonia include (1) inadequate clinical, radiological, or pathologic data and (2) major discordance between clinical, radiological, and pathologic findings that may occur in the following situations: (a) previous therapy resulting in substantial alteration of radiological or histological findings (e.g., biopsy of desquamative interstitial pneumonia after steroid therapy, which shows only residual nonspecific interstitial pneumonia; (b) new entity or unusual variant of recognized entity, not adequately characterized by the current American Thoracic Society/European Respiratory Society classification (e.g., variant of organizing pneumonia with supervening fibrosis)[2]; and (c) multiple high-resolution computed tomography and/or pathologic patterns that may be encountered in patients with idiopathic interstitial pneumonia. Information obtained from Travis et al. [10]

IIP may cause potential harm by exposing the patient with a drug that has no impact on IIP. Moreover, some immune-modulating drugs may cause more deleterious effects on the overall outcomes in patients with IIP [19]. The systematic and comprehensive evaluation of a patient with ILD will define presence of underlying CTDs, for which an integrated approach with contribution of different specialists, pulmonologists, radiologists, pathologists, and rheumatologists is necessary [20]. We will assess the key points for this methodology.

5.2 Evaluation of ILD Associated with CTD

Connective tissue diseases encompass a series of heterogeneous autoimmune diseases, each one of them differentiating from one another, based on the specific pathogenic mechanisms and a plethora of clinical and laboratory characteristics (Table 5.2). In addition, the disease expression in the lungs and the different compartments affected varies within each specific CTD, complicating the interpretation, classification, treatment modality, and prognosis. For example, rheumatoid arthritis (RA) and Sjogren's syndrome may involve the airway compartment with or without ILD (parenchyma) [21], which complicates the pulmonary function tests' (PFTs) interpretation consisting in obstruction/restriction or both. On other situations, like in lupus, pleural effusions or shrinking lung syndrome may cause restriction, leading to the same difficulties [22, 23]. In idiopathic inflammatory muscle disease, the presence of a restrictive pattern on the spirometry might simply (or additionally to the presence of ILD) be a diaphragmatic muscular weakness [24]. Esophageal dysmotility and recurrent aspiration will contribute to the same in systemic sclerosis, idiopathic inflammatory myopathies, and Sjogren's syndrome [25]. Furthermore, frequently, patients with CTD are on corticosteroids and other immune suppressor drugs, a fact that predisposes them a higher susceptibility for infections. *Pneumocystis jiroveci* pneumonia (PCP) may present with pulmonary changes mimicking ILD as an example and needs to be ruled out with a thorough clinical history and the bronchoalveolar lavage [26].

Many caveats will complicate determining the underlying CTD in presence of ILD. Interstitial lung disease may be the initial presenting feature of a CTD in an estimate of 15% [21, 27, 28], and the absence of the classical overt manifestations expected to be seen for the specific CTD may mistakenly classify the disease as IIP. Moreover, subclinical and subtle manifestations of a CTD may be unidentified or overlooked. For instance, subset of patients with Sjogren's syndrome may run with minimal sicca symptoms and positive anti-SSA/Ro antibodies and the disease to be confirmed, once the histopathology of minor salivary glands is explored [29, 30]. The latter procedure is available in only few specialized centers, limiting the access for a complete assessment, a fact to keep in mind at the time of evaluation of patients with ILD.

5.2.1 History and Physical Exam

The history may reveal symptoms such as dyspnea, tachypnea, and cough. In addition, nonspecific symptomatology like malaise, weight loss, and a decreased functional capacity may contribute poorly as a screening method. On the physical exam, the most common finding might relate to pulmonary basilar inspiratory Velcro crackles and, rarely, clubbing nails. These findings are nonspecific but easily identified. However, not every patient will disclose the respiratory symptoms, such as dyspnea in patients with active and symptomatic rheumatoid arthritis that may run inadvertently as they avoid any physical activity. Even patients with systemic sclerosis may have no respiratory symptoms, and to avoid missing the lung involvement, a baseline screening HRCT is recommended. In fact, a demographic screening study, using the HRCT, demonstrated that 90% of consecutive patients with systemic sclerosis had variable degrees of ILD [31].

Occasionally, patients with a CTD undergo procedures for other reasons that will uncover ILD, especially if the disease is mild, subclinical, or asymptomatic (e.g., pre-surgical evaluation, radiographic imaging for coronary artery calcification scoring, routine exams, lung cancer screening, and other circumstances) [24].

Each one of the CTDs will present with specific features with a plethora of manifestations (Table 5.2), allowing the clinician to identify the specific disease. Occasionally, the CTD may be inapparent at early stages, and/or the clinical features are not always apparent. We suggest looking for specificities on the history and exam for each disease (Table 5.2). Additionally, most of the CTDs may reveal capillary changes, visible on the periungual skin in the fingers. They are more notorious in systemic sclerosis and inflammatory myopathies, but also, the changes may occur in other diseases like systemic lupus erythematosus, Sjogren's syndrome, mixed connective tissue diseases, and rheumatoid arthritis [32]. The nailfold capillaroscopy will guide the clinician toward continuing searching for a CTD. If abnormal, it may associate with Raynaud's phenomenon.

Table 5.2 Clinical, serological, and pulmonary manifestations of connective tissue diseases [21]

Autoimmune condition	Common clinical presentations	Specific clues	Common associated serological markers	Common pulmonary manifestation
Rheumatoid arthritis	Inflammatory polyarthritis Rheumatoid nodules Inflammatory eye disease Rheumatoid vasculitis Pericarditis	Early tenosynovitis Extra-articular manifestations (e.g., scleritis, pericarditis)	Rheumatoid factor Anti-CCP antibody	RA-ILD Pleural effusion Obliterative bronchiolitis Follicular bronchiolitis Bronchiectasis Rheumatoid lung nodules
SLE	Inflammatory arthritis Cutaneous lupus Raynaud's phenomenon Hematologic abnormalities Neuropsychiatric lupus Lupus nephritis	Subtle CNS manifestations Microhematuria Photosensitivity Serositis Cytopenia in the three series	ANA Anti-ds DNA antibody Anti-Smith antibody Anti- RNP antibody SSA and SSB Anti-ribosomal P antibody Low C3 and C4	Pleural effusion Acute pneumonitis Pulmonary hemorrhage SLE-ILD Thromboembolic disease Pulmonary hypertension Shrinking lung syndrome
Systemic sclerosis	Sclerodactyly Raynaud's phenomenon GI dysmotility Calcinosis Telangiectasia Scleroderma Renal crisis	Abnormal nailfold capillaries Tendon friction rubs	ANA Centromere antibody Scl-70 antibody RNA polymerase III antibody	ILD Pulmonary hypertension Pleural effusion
Sjogren's syndrome	Sicca syndrome (the eyes, mouth, dyspareunia) Inflammatory arthritis Raynaud's phenomenon Neuropathy Renal tubular acidosis	Extra-glandular manifestations Neonatal lupus Fibromyalgia-like symptoms	ANA Anti-SSA/Ro and anti-SSB/La antibodies Sjogren's syndrome-specific antibodies	ILD Cystic lung disease Bronchiolitis
Myositis	Muscle weakness Raynaud's phenomenon Morbilliform Rash Inflammatory arthritis Dysphagia	Gottron's papules in the elbows, knees, and elsewhere Abnormal nailfold capillaries	ANA Myositis-specific antibodies Anti-SSA antibodies	ILD Organizing pneumonia Diffuse alveolar damage
Undifferentiated connective tissue disease	Arthralgia Arthritis Raynaud's phenomenon Rashes Dry mouth and/or eyes Morning stiffness Proximal muscle weakness	Not full enough criteria to define a specific CTD	ANA Rheumatoid factor Anti-SSA/Ro and anti-SSB/La antibodies	ILD with more prevalence on nonspecific interstitial pneumonia

Abbreviations: *ILD* interstitial lung disease, *CTD* connective tissue disease, *anti-CCP* antibodies: anti-citrullinated cyclic peptide antibodies, *ANA* antinuclear antibodies

5.2.1.1 Raynaud's Phenomenon and the Nailfold Capillaroscopy in the Assessment of CTD-ILD

Raynaud's phenomenon (also named as Raynaud) is the cutaneous change to the cold exposure and/or to situations with high level of anxiety (Fig. 5.1a). Digits (the hands or feet) and areas distant to the core body temperature are the most frequent target (e.g. the nose, ears). Raynaud associates with small-vessel and capillary dysfunction, given the underlying abnormal vasospasm and dysregulation of vessels, causing a slowed blood flow to the distal tissues. Raynaud has three phases, starting with a pale discoloration, turning bluish-purplish, and then transitions to erythema, once the environmental insult has ceased. The latter's color associates a compensatory distal blood flow, while the prior ones associate with ischemia. Raynaud can be primary or without a relationship to a systemic disease or secondary. Belonging to the latter group, Raynaud is part of almost all CTDs, with systemic sclerosis being the most common disease. The nailfold capillaroscopy (Fig. 5.1b) will be abnormal in association with CTDs, information that will give a hint along with other clinical features (Table 5.2), and positive abnormal serology to establish a CTD-ILD diagnosis. Under normal conditions, the capillaries at the periungual skin, and seen on a magnifying capillaroscopy, are uniformly aligned revealing tight loops lined up one next to the other. Loss of capillaries; dilated, misaligned loops; and leaking hemorrhage disclose evidence of a CTD. Based on the findings, experts have established the scleroderma pattern that can determine early, active, and late scleroderma [33]. The nailfold capillaroscopy plays a major role to identify this disease when systemic sclerosis has no apparent skin thickening as is scleroderma *sine scleroderma* [34]. Other authors have proposed a predictive role of the nailfold capillary patterns to determine internal organ involvement [35, 36]. Moreover, the giant loops seem to be predictive of pulmonary involvement across CTDs [37], but other additional patterns also have been described for lung involvement (e.g., pulmonary artery hypertension) [38, 39].

5.2.2 Serology

It is worth mentioning that the initial traditional screening resource for a CTD-ILD evaluation relies solely on serology testing (Table 5.3). However, serology may be absent in several CTDs [21, 29, 30]. Again, Sjogren's syndrome

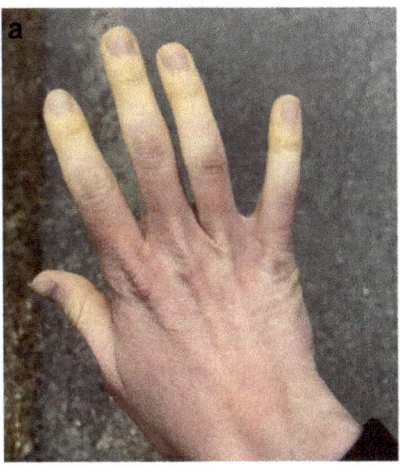
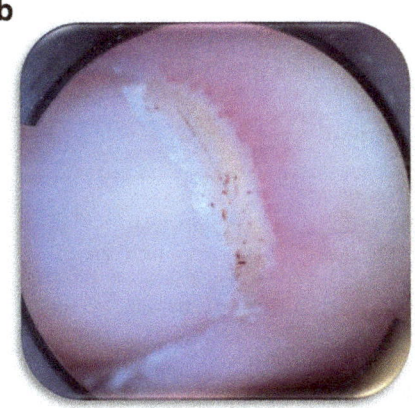

Fig. 5.1 Evaluation of Raynaud's phenomenon (RP). Evaluation of RP in a patient with clinical features of an underlying connective tissue disease: (**a**) finger blanching in a patient with cold exposure. RP in young women suggests the presence of an underlying connective tissue disease. (**b**) Abnormal nailfold capillaries support presence of RP as part of the underlying probable interstitial lung disease. Thickened capillaries (arrows) are visible

Table 5.3 Antibodies and clinical aspects associated with connective tissue diseases [77, 100–102]

Disease	Antibodies	Clinical features
Systemic sclerosis		
	Anti-topoisomerase (Scl-70)	Linked with diffuse systemic sclerosis
	Anti-Th/To	Linked with limited cutaneous systemic sclerosis
	Anti-U3 RNP	Associated with myositis and PH
	Anti-U11/U12 RNP	Selective for systemic sclerosis
Systemic sclerosis and myositis overlap		
	Anti-RuvBL1/2	Diffuse skin thickening and myositis
	Anti-EIF2B	Present with overlap syndrome
	Anti-PM-Scl	Inflammatory myositis and systemic sclerosis
	Anti-Ku	Associated with myositis and lupus overlap
	Anti-Trim21	High risk for ILD
Mixed connective tissue disease		
	Anti-U1 RNP	Features of systemic sclerosis, arthritis, and myositis
Sjogren's syndrome		
	Anti-SSA/Ro	Associated with myositis-specific antibodies and anti-synthetase syndrome
Systemic lupus erythematosus		
	ANA, speckled	Pleuritis and ILD
	Anti-RNP	Overlap; myositis; Sjogren's syndrome
Rheumatoid arthritis		
	Anti-CCP antibodies	High risk for ILD
	Rheumatoid factor	
Inflammatory myopathy		
Dermatomyositis		
	Anti-MDA-5	Associated with cancer; amyopathic dermatomyositis; rapidly progressive ILD (East Asians)
Anti-synthetase syndrome		Mechanic hands, Raynaud's phenomenon, arthritis, myositis, ILD, and fevers

(continued)

Table 5.3 (continued)

Disease	Antibodies	Clinical features
	Anti-Jo1 (histidyl)	
	Anti-PL12 (alanyl)	
	Anti-PL-7 (threonyl)	
	Anti-KS (asparaginyl)	
	Anti-OJ (isoleucyl)	
	Anti-EJ (glycyl)	
	Anti-Zo (phenylalanyl)	
	Anti-Ha (tyrosyl)	

Table adapted from Cotton CV et al. [100] and Betteridge ZE et al. [103]

Abbreviations: *PH* pulmonary hypertension, *ILD* interstitial lung disease, *MDA5* melanoma differentiation-associated gene 5

frequently presents with negative serology, so do idiopathic inflammatory muscle diseases. Rheumatoid arthritis presents with positive rheumatoid factor only in 85% of patients at the onset of the disease, and anti-CCP antibodies may be negative. Moreover, there are specific antibodies not tested regularly, as is in idiopathic inflammatory myopathy, including the ones pertinent to anti-synthetase antibody syndromes and systemic sclerosis (and the anti-Th/To ribonucleoprotein) [40]. Conversely and to complicate the serology screening method, low titers of rheumatoid factor or antinuclear antibody still fall under the IIP definition, based on the American Thoracic Society/European Respiratory Society/Japanese Respiratory Society 2011 statement [32]. In other situations and as a confounder, other autoimmune diseases may present with positive serology (e.g., autoimmune thyroiditis, celiac disease), unrelated with the infiltrative fibrotic pulmonary process that may probably lead to misinterpretations. The ANA pattern contributes to the differentials, being the nucleolar and centromere of most significance in association with a CTD and pulmonary involvement [33]. This assumption, however, does not exclude cases with a speckled or even a homogeneous pattern at high titers. In our experience, the ANA pattern and other results should be interpreted in the context of each patient and especially in association with demographic and clinical features. For instance, the age (young patients) and gender (females) suggest that the ANA test might associate with a CTD. We, again, advocate for a structured history and directed interview and specific physical exam aimed at detecting a possible underlying CTD in presence of ILD.

The thoracic radiographic images may reveal interstitial features, with reticulations and nodularity, and the pulmonary function tests may reflect various degrees of a restrictive pattern. These tests, although useful and accessible for most of the providers, may overlook subtle changes. The most sensitive test is the HRCT, which should ensure ILD confirmation and baseline data, useful as a predictor for CTD-ILD [41–47].

5.2.3 Pulmonary Function Tests

The restrictive ventilatory pattern and a reduced diffusion capacity of carbon monoxide (DLCO) are the usual findings in ILD. The reduced total lung capacity (TLC) and forced vital capacity (FVC) with normal or increased forced expiratory volume per second (FEV1) and a normal FEV1/FVC ratio are the usual features. However, the PFTs are not always sensitive enough to pinpoint an ongoing inflammatory/fibrotic parenchymal ILD [48]. The best and most sensitive additional exam is the complementary HRCT. Once adjusted to the reduced lung volumes, the DLCO may still remain reduced, suggestive of an overlap with pulmonary artery hypertension or emphysema overlap [49]. In addition, an obstructive physiology with reduced FEV1 and FEV1/FVC is common in CTDs affecting the airways (Table 5.4).

5.2.4 Radiology

The chest tomographic analysis has revolutionized the current approach on assessing patients with ILD. The technique of thin (1.5 mm or less) slices of coronal and sagittal volumetric reconstruction allows accurate description and fine interpretation of pulmonary findings (and their compartments) [50]. Based on the patterns seen on the HRCT, studies have been able to determine predictors for survival [51]. The different patterns are described according to the ATS consensus criteria used for IIPs [11, 55–57]. In CTD-ILD, as in IIP-ILD, there are similar radiographic presentations, like the nonspecific interstitial pneumonia (NSIP), usual interstitial pneumonia (UIP), lymphocytic interstitial pneumonia (LIP), organizing pneumonia (OP), acute interstitial pneumonia/diffuse alveolar damage (DAD), and a combination of them.

The radiological patterns often correlate with the histopathological findings, but this depends on the type. In IIP with a UIP variant, the concordance between the HRCP and the histopathology is excellent [52, 53]. In intermediate UIP definition or NSIP pattern obtained on the HRCT, this concordance with histopathology is not as optimal as desired, showing patterns of either NSIP or UIP in the biopsy [54]. These findings, in addition, have a predictive value. Patients with a radiological definitive or probable UIP pattern have a shorter survival than those with indeterminate UIP or definitive NSIP pattern. The biopsy of patients consistent with a UIP profile and with a non-UIP radiological pattern does better as those with histological UIP pattern and radiological of definite UIP but worse than those with a radiological NSIP pattern [53]. Even though this concept applies to IIPs, the similar findings seem to be extended to CTD-ILDs. In both, the correlations of definite UIP pattern on the HRCT predict a similar pattern on the histopathology, while in NSIP and other presentations, the HRCT is not an accurate predictor of an expected equivalent histopathology pattern [55].

Some patterns prevail over the others, and classical combinations are almost the signature for defined CTDs (Table 5.4). Organizing pneumonia is more frequently seen in combination of NSIP in inflammatory myositis (dermatomyositis, polymyositis) and RA; LIP is found in RA but

Table 5.4 Computed tomography imaging patterns of different connective tissue diseases

Disease	UIP	NSIP	OP	LIP	DAD	Airway[a]	Serositis[b]	Vascular	DAH
RA	+++	+++	++	+	+	+++	+++	+	–
SSc	+	+++	+	–	+	–	–	+++	+
PM/DM	+	+++	+++	–	++	–	–	+	–
SjS	+	++	–	++	+	+	+	++	–
SLE	+	++	+	++	++	+	+++	++	+++
MCTD	+	++	+	–	–	+	+	+	–

Table adapted from Mira-Avendano et al. [61], Bryson et al. [101], and Fischer et al. [25]
Legend: absence of findings: –; lowest: +; highest: +++
CTD connective tissue disease, *DAD* diffuse alveolar damage pattern, *LIP* lymphocytic interstitial pneumonia pattern, *MCTD* mixed connective tissue disease, *NSIP* nonspecific interstitial pneumonia pattern, *OP* organizing pneumonia pattern, *PM/DM* polymyositis/dermatomyositis, *RA* rheumatoid arthritis, *SjS* Sjogren's syndrome, *SLE* systemic lupus erythematosus, *SSc* systemic sclerosis, *UIP* usual interstitial pneumonia pattern
[a]Airway disease includes bronchiectasis, bronchial wall thickening, small centrilobular nodules (representing follicular bronchiolitis), and constrictive bronchiolitis
[b]Pleural or pericardial fluid or thickening

mainly in Sjogren's syndrome; DAD is seen in RA, systemic lupus erythematosus (SLE), inflammatory myositis, and undifferentiated connective tissue disease (UCTD). The overall prevalence of NSIP runs across all the CTDs, except for RA, which presents more frequently with a UIP form [15, 56, 57]. Presence of other compartment involvement is highly likely associated with an underlying CTD: pleural–pericardial effusions, pulmonary arterial hypertension (PAH) (either alone or associated with ILD), and the diaphragm pathology (as seen in inflammatory myopathies) or the skin (as seen in systemic sclerosis of diffuse type) and causing a restrictive pattern on the pulmonary function tests [24, 58] (Table 5.4). Even though about 25% of cases with ILD cannot classify into a defined CTD, they fall under the category of UCTD [28]. The differential of IIP with ILD-UCTD is crucial given that the prognosis is much better in the later [28, 59, 60].

We describe the most common types of HRCT.

5.2.4.1 Usual Interstitial Pneumonia (UIP)

This pattern is characterized by fibrosis and mostly localized in the periphery, basal, and subpleural areas, adjacent to the pleura. The hallmark is honeycombing, with reticulation and peripheral traction bronchiectasis or bronchiolectasis. The latter (traction bronchiectasis) represents the presence of architectural distortion, varicosity, coexisting with honeycombing and the absence of features

suggesting an alternative diagnosis [11] (Fig. 5.2). The latter includes profuse ground-glass attenuation, peribronchovascular predominance, perilymphatic distribution, discrete cysts, micronodules, centrilobular nodules, significant mosaic perfusion, and air trapping and consolidation [11, 61]. Probable UIP pattern presents on the HRCT in the absence of honeycombing but presence of reticular abnormalities with basal and subpleural distribution, traction bronchiectasis, or bronchiolectasis [11, 62]. Indeterminate form of UIP pattern refers to the atypical features of UIP but a histopathological pattern of UIP. In this category, the HRCT features show very limited subpleural ground-glass opacities or reticulation without clear features for fibrosis. The prone position will help in excluding subpleural atelectasis (Table 5.5). Additionally, other diagnoses should be investigated in images of pleural plaques, dilated esophagus, distal clavicular erosions, extensive lymph node enlargement, or pleural effusion or thickening (Table 5.5) [11].

5.2.4.2 Non-interstitial Pneumonia (NSIP)

This other pattern, the most frequent type present in CTD-ILD, displays bilateral basal-predominant ground-glass opacities with or without reticulation. The subpleural space is usually spared, opposed to the UIP pattern, being a hallmark for NSIP [63]. The spectrum ranging from inflammatory NSIP with cellular predominance

Fig. 5.2 Usual interstitial pneumonia in a patient with rheumatoid arthritis. Peripheral reticulation, traction bronchiectasis, and honeycombing in made lower lung predominant distribution. Subpleural nodularity without subpleural sparing or focal consolidation

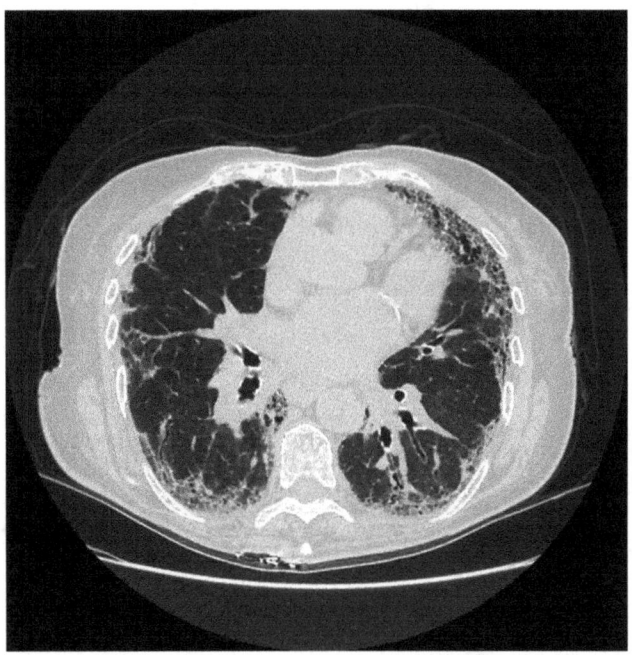

(Fig. 5.3) to the fibrotic pattern (Fig. 5.4) depends on the variable fibrotic tissue causing traction bronchiectasis and deformity in the latter. Moreover, the differences between the fibrotic NSIP and UIP patterns rely on the temporal uniformity of the pathologic features seen in NSIP as compared to the key findings described in UIP (honeycombing pattern, lack of subpleural sparing). In NSIP, the appearance is homogeneous of dense or loose interstitial fibrosis. Lower lobe volume loss and other signs of pulmonary fibrosis are another feature of NSIP [64].

5.2.4.3 Organizing Pneumonia (OP)

Infiltrates in a patchy, peripheral distribution may suggest OP. It prefers the subpleural spaces, but also, it is peribronchiolar. Other features are the perilobular distribution, nodules, and the reversed ground-glass halo sign (Fig. 5.5) [61].

5.2.4.4 Lymphocytic Interstitial Pneumonia (LIP)

Many presentations are considered, including the variable cystic lesions within the parenchyma. Lymphocytic interstitial pneumonia along with NSIP patterns is highly suspicious for an underlying CTD [65]. In addition, as in Sjogren's syndrome, the presence of mild thickening of interlobular septa, bronchovascular bundles and cystic lesions may represent the associated follicular bronchiolitis [66]. Additional frequent localizations are in the perilymphatic interstitium and the pleura. Areas of ground-glass attenuation and poorly defined centrilobular and subpleural nodules of varying sizes might be also present. Lymph node enlargement is also seen. Cysts are usually less than 3 cm and have variable wall thickness and shape. Occasional cysts may show a multiseptated appearance [67].

5.2.4.5 Airway Disease and Respiratory Bronchiolitis-Associated Interstitial Lung Disease

Various patterns may represent bronchial airway disease, like bronchiectasis, mosaic attenuation, bronchial wall thickening, air trapping, and centrilobular or branching nodules [66]. The airways may accompany with interstitial disease, like in respiratory bronchiolitis-associated interstitial lung disease. Representing features are bronchial wall thickening with ground-glass opacity (patchy) with a diffuse distribution and additionally centrilobular nodules.

Table 5.5 High-resolution computed tomography scanning patterns

UIP	Probable UIP	Indeterminate for UIP	Alternative diagnosis
Subpleural and basal predominant; distribution is often heterogeneous[a]	Subpleural and basal predominant; distribution is often heterogeneous	Subpleural and basal predominant	Findings suggestive of another diagnosis, including:
Honeycombing with or without peripheral traction bronchiectasis or bronchiolectasis[b]	Reticular pattern with peripheral traction bronchiectasis or bronchiolectasis	Subtle reticulation; may have mild GGO or distortion ("early UIP pattern")	CT features: Cysts Marked mosaic attenuation Predominant GGO Profuse micronodules Centrilobular nodules Nodules Consolidation
	May have mild GGO	CT features and/or distribution of lung fibrosis that does not suggest any specific etiology ("truly indeterminate for UIP")	Predominant distribution: Peribronchovascular Perilymphatic Upper or mid-lung
			Others: Pleural plaques (consider asbestosis) Dilated esophagus (consider CTD) Distal clavicular erosions (consider RA) Extensive lymph node enlargement (consider other etiologies) Pleural effusions and pleural thickening (consider CTD/drugs)

Definition of abbreviations: *CT* computed tomography, *CTD* connective tissue disease, *GGO* Ground-glass opacities, *RA* rheumatoid arthritis, *UIP* usual interstitial pneumonia
Information obtained from Raghu et al. [11]
[a]Variants of distribution: occasionally diffuse; may be asymmetrical
[b]Superimposed CT features: mild GGO, reticular pattern, and pulmonary ossification

5.2.4.6 Diffuse Alveolar Damage (DAD)

Overall, the pattern is of a diffuse ground-glass opacity pattern, with lower and peripheral zone and posterior preference. Details show ground-glass attenuation and reticular lines.

5.3 Pathogenesis

While the several mechanisms are implicated in the pathogenesis for each CTD, several hypotheses try to explain the inflammatory etiology in the lung parenchyma.

5.3.1 Sjogren's Syndrome Model

In Sjogren's syndrome, the concept of autoimmune epithelitis, or the invasion of inflammatory cells surrounding the epithelial cells of a specific organ, is present in different tissues as part of a systemic inflammation [68, 69]. The histopathological pattern is seen in interstitial nephritis, cholangitis, and lymphocytic clusters surrounding salivary ducts as part of the pathognomonic histopathology features of Sjogren's syndrome [70]. Epithelial cells seem to be the antigenic source to elicit the cellular

Fig. 5.3 Cellular nonspecific interstitial pneumonia in a patient with systemic lupus erythematosus de novo. Extensive bilateral ground-glass infiltrates

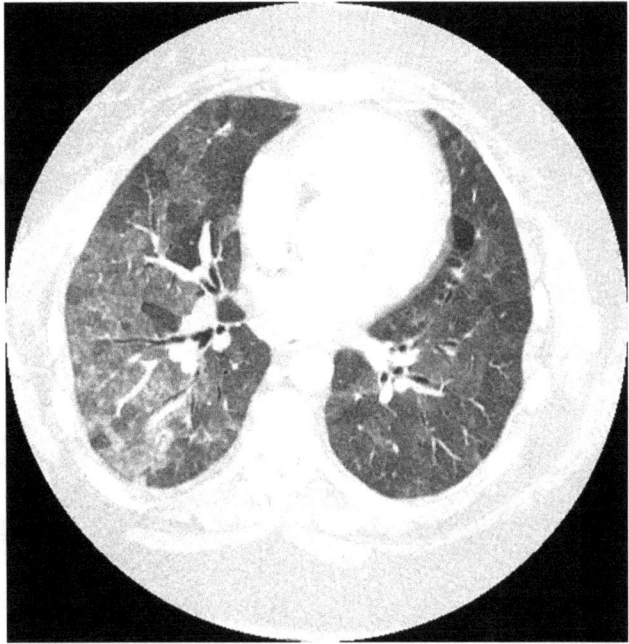

Fig. 5.4 Fibrotic non-interstitial pneumonia. Prominent diffuse bilateral subpleural interstitial thickening and subpleural preservation. Scattered traction bronchiectasis. Scattered ill-defined ground-glass infiltrates

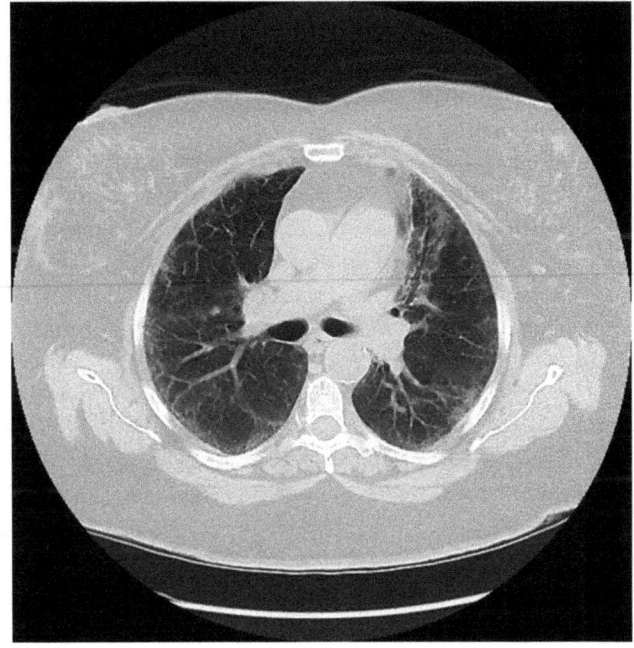

immune attack. In the terminal bronchioles, this phenomenon replicates the same pattern of clusters of inflammation surrounding the epithelial cells in bronchioles, termed as follicular bronchiolitis [71]. The latter is seen in the HRCT as enhancement of the bronchovascular bundles and air trapping [72]. The proposed inflammatory cell invasion into the lung parenchyma and as a continuum of the adjacent follicular bronchiolitis is described as nonspecific inflammatory pneumonitis (NSIP), a subtype of ILD.

Fig. 5.5 Organizing pneumonia. Patient with rheumatoid arthritis. Mild to diffuse bronchial wall thickening, with mild bronchiectasis. A diffuse mosaicism pattern throughout the lungs bilaterally is seen, with linear opacities in the right middle lobe and right lower lobe possibly associated with atelectasis vs. scaring

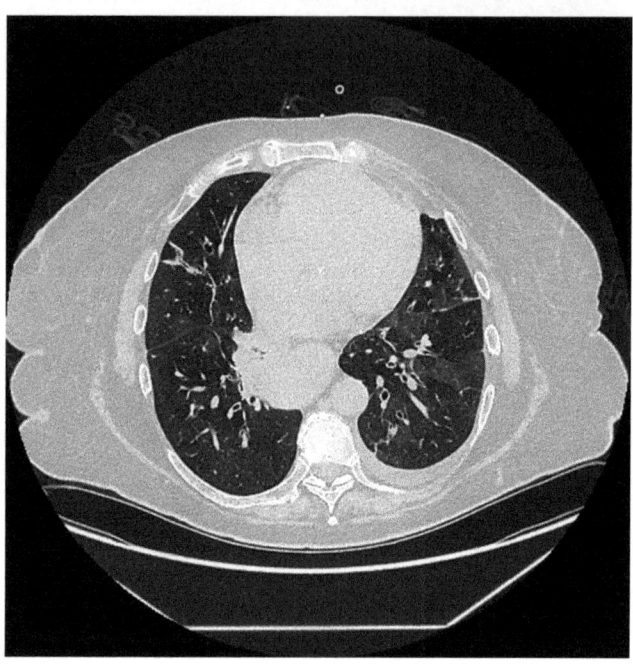

5.3.2 Rheumatoid Arthritis Model

In another hypothesis, the lungs may account for the initial tissue for the immune tolerance loss and possibly secondary to a local insult. In a recent epidemiological prospective study among patients with established asthma and chronic obstructive pulmonary disease (COPD) vs. controls and the incident rheumatoid arthritis, asthma was associated with the latter, so was COPD and more pronounced in ever-smokers older than 55 years [73]. Moreover, RA-related autoantibodies are detectable in sputum in preclinical RA and in early RA [74]. Not only this but also the generation of anti-citrullinated protein antibodies (ACPA) is present in bronchial tissue and bronchoalveolar lavage in patients with early untreated RA without clinical signs of lung involvement, strengthening the role of the lung compartment as an important player in ACPA-positive RA [75]. Moreover, this is seen in cases of ILD with positive serology and no findings of RA [76–78]. In time, it appears that the antibodies may play an active role in the pathogenesis of lung disease [79].

5.3.3 Pathology

Histopathology analysis is required under uncertain cases to define the type of ILD and the etiology. In other situations, the biopsy will help in identifying a different disorder (e.g., carcinoma, sarcoidosis). This is especially important in patients with an indeterminate UIP pattern on the HRCT, needed to differentiate between IIP from other causes with potential immune suppressor treatment efficacy, like hypersensitive pneumonitis or CTD-ILD. Finally, proceeding to the biopsy is ideal in younger patients with features of vasculitis or patients with clinical features of ILD without clear-cut findings on the HRCT. Different techniques will yield an optimal tissue sample, depending on the preferred localization and patient's clinical status. Available procedures are the video-assisted thoracoscopic surgery as the standard and most important technique. Other options are the transbronchial lung biopsy and thoracotomy. The transbronchial cryobiopsy technique is currently evolving.

5.3.3.1 Usual Interstitial Pneumonia (UIP)

The nature of ILD will help in defining management and will predict response to therapy. Therefore, it is key to differentiate the UIP from other non-UIP patterns, as the latter have more options for treatment. Currently, the concept of unresponsive therapy to fibrotic lung disease is challenged given the available contribution of new anti-fibrotic therapies (nintedanib, pirfenidone) that show promising favorable outcomes.

The hallmark for a UIP pattern shows dense fibrosis with distortion of lung architecture, subpleural and/or paraseptal fibrosis, patchy involvement with fibrotic lung alternating with areas of normal parenchyma, and honeycombing [11, 80]. In indeterminate UIP, the findings are consistent with fibrosis with or without architectural distortion or features of either UIP (fibrosis) or other pathologies. In the latter, there might be granulomas, hyaline membranes, prominent airway-centered changes, interstitial inflammation without fibrosis, chronic fibrous pleuritis, or organizing pneumonia [11].

5.3.3.2 Nonspecific Interstitial Pneumonia (NSIP)

Nonspecific interstitial pneumonia is an ILD type of different etiologies (e.g., HIV and hypersensitivity pneumonitis) and is the most frequent presentation of CTD-associated lung disease. It is important to differentiate NSIP from UIP as the former has shown to have better outcomes with a lower mortality rate of 18% among patients with idiopathic NSIP [81]. During the evaluation of new-onset idiopathic NSIP, 17% of them may associate with a CTD. This frequency increases at different time points of follow-up [82], while new features of a CTD ensue. A fraction of patients will classify as having interstitial lung disease with autoimmune features, a feature that is still under development (see ahead) [83]. The histopathological features in NSIP reveal varying degrees of interstitial inflammation and fibrosis, with uniform appearance. The lung architecture is usually preserved, unless traction bronchiectasis was present, a notable feature of the fibrotic NSIP type [84]. Some authors hypothesize a transition between NSIP and UIP, and this is represented by the excessive accumulation of collagen synthetized by an abnormal fibroblast-myofibroblasts and supported by the presence of both histopathological patterns in individual patients from biopsies of different pulmonary sites [52]. Furthermore, it is clearly difficult to differentiate NSIP from UIP in situations when these two changes coexist in an individual with biopsies from different locations [52, 85, 86].

5.3.3.3 Organizing Pneumonia (OP)

This type considered as ILD is characterized by the intra-alveolar accumulation of fibroblasts and loose connective matrix, potentially reversible with corticosteroid (CS) therapy. Depending on the alveolar infiltrate material, different variants may suggest the stages of OP. In the early stages, fibrin fills the alveoli, along with inflammatory cells, known as the fibrinous variant. It associates with extensive necrosis of alveolar epithelial type I cells. During the second stage, the formation of fibroinflammatory buds is gradually intermixed with newly invading fibroblasts and myofibroblasts that in turn will ensue abundant collagen deposits. In the third stage, mature fibrotic buds are composed of myofibroblasts, organized in concentric rings alternating with collagen bundles. This process clears out gradually if the basal laminae had remained intact during the inflammation. The granulation tissue may extend to the adjacent bronchiole, and the lumen might get obstructed. Inflammation in alveolar interstitium is present, and foamy alveolar macrophages may be localize in alveoli without granulation tissue [87].

5.3.3.4 Lymphocytic Interstitial Pneumonia (LIP)

It is characterized by the dense interstitial lymphocytic infiltrates, consisting of polymorphic lymphocytes, admixed with variable number of plasma cells, immunoblasts, macrophages, and rarely histiocytes. The extent of infiltrative cellular proliferation may expand the interlobular and alveolar septa [67]. A feature of CTD-associated LIP is the presence of B lymphocytes expressing CD20, usually from germinal centers localized in nodular lymphocytic aggregates. In contrast the CD3(+) T lymphocytes are predominant in the interstitial compartment. The infiltrates are present in areas surrounding the lymphatic channels, alveoli septa, peribronchovascular regions, and subpleural lung zones [67].

5.3.3.5 Diffuse Alveolar Damage (DAD)

This acute lung injury shows different phases, with the initial exudative (acute) phase with associated edema, hyaline membranes, and interstitial acute inflammation and then followed by a subacute phase. The latter consists of loosely organizing fibrosis localized in the alveolar septa with type II pneumocyte hyperplasia. These findings associate with thrombi in small- and medium-sized pulmonary arterioles [88]. The hallmark in this presentation is the presence of hyaline membranes. They consist of dense eosinophilic amorphous material of cellular debris, plasma proteins, and surfactant components and lined up along the alveolar septa [89]. Even DAD is considered the pathologic correlate of acute respiratory distress syndrome (ARDS); pathologic findings for DAD may be present in half of ARDS cases [90].

5.4   Differences Between CTD and IIP

The official ATS/ERS/JRS/ALAT Clinical Practice Guideline defines interstitial pulmonary fibrosis (IPF, a common form of IIP) as UIP pattern in the HRCT and the exclusion of other known causes of ILD (e.g., domestic, occupational, and environmental exposures; infections; CTDs; or drug toxicity). In addition, there should be either presence of the HRCT pattern of UIP or specific combinations of HRCT patterns and histopathology findings of UIP in patients who had the lung biopsy [11]. Usual interstitial pneumonia is an atypical feature linked with CTD-ILD, except for rheumatoid arthritis that has a predilection for the latter type.

Subsequently, the analog features between UIP in IPF and RA-UIP make the latter to be the most susceptible group for misclassification. In cases of uncertainty, the histopathology gives many clues favoring a CTD-ILD (e.g., RA-UIP): 1) in CTD-UIP, presence of fibroblast foci is scarce [17, 91] and shows smaller honeycomb spaces as compared to IPF-UIP [92]; 2) in RA-UIP presentations, there are more lymphoplasmacytic aggregates as in IPF-UIP; 3) germinal centers are a feature of CTD-ILD [92]; and 4) IPF, contrary to CTD-ILD, presents with heterogeneous ILD patterns in an individual patient from different pulmonary biopsy sites (lobes) (e.g., NISP in one sample and UIP in another sample), which can be considered as a hallmark for IPF [85, 91]. Furthermore, survival in RA- UIP is better compared to IPF (specifically IIP-UIP) [17]. This probably relates to the potentially treatable inflammatory niches in the histopathology (e.g., as described, lymphoplasmacytic aggregates, germinal centers), but not the honeycombing that implies the fibrotic tissue.

Given the better survival in CTDs-ILD, it is crucial to determine if the lung disease associates with any of them. Some authors suggested that all ILD patients with a NSIP pattern on the HRCT might have a potential underlying CTD [93]; however, these criteria are too loose, and if applied, many patients may falsely be classified as having some form of CTD. Refinements on this concept, however, showed that cumulative incident diagnosis of CTD-ILDs from previously labeled idiopathic NSIP patterns may increase in time, at one (3.6%), two (15.2%), and three years (20.0%). In this study, however, no predictor to determine CTD-ILD (serology, extrapulmonary findings) was found, but results imply re-evaluating patients prospectively [82]. An alternative to differentiate idiopathic vs. CTD-related ILD, as said, is the histopathology review in the context of clinical findings suspicious for a CTD. In a study assessing demographic, clinical, and laboratory associations with histopathological NSIP vs. IPF, patients who had a diagnosis of undifferentiated connective tissue disease (UCTD) were more frequently associated with NSIP (31% vs. 13%, respectively) [94]. In this same study, predictors identified for NSIP in the histopathology were the absence of HRCT typical for idiopathic pulmonary fibrosis (IPF), demographic features (women younger than 50 years), and features of UCTD, like Raynaud's phenomenon and positive serology for autoimmunity [94]. Compared to IPF, NSIP has the potential to respond to the immune suppressor therapy, which is the main reason to explore for the latter. The definition of UCTD is the clinical findings of a systemic CTD, without meeting the criteria for a particular disease [95]. Most of the patients may have a positive ANA and clinical features such as Raynaud's phenomenon, nonspecific rash, sicca symptoms, arthralgias, or arthritis. Over time, 34%–50% will develop a specific CTD [96–98], but others will remain as having UCTD. Definition for UCTD has changed several times, adding more difficulties to accurately define who would fall into the UCTD criteria. Despite these difficulties and in order to favor patients to have a justifiable immunosuppressor treatment (rather than anti-fibrotic therapy), many authors, in an attempt to classify patients within IPF (on HRCT) with some features of a CTD without fulfillment of a unique entity, have grouped clinical, radiological, histological, and laboratory findings to come up with a syndromic definition. They have generated several of such syndromic entities, named as "undifferentiated CTD," "autoimmune-featured ILD," and "lung-dominant CTD" (Table 5.6) [25]. Each one of them varies in sensitivity and specificity and has different criteria that for the medical community cause more confusion. To unify criteria, the one proposed by the ERS/ATS research statement known as interstitial pneumonia with autoimmune features appears to have great value.

5.4.1 Interstitial Pneumonia with Autoimmune Features (IPAF) [20]

This term is the one that is currently under universal approval. As seen, to use different criteria, trial designs and retrospective reviews

for comparison become cumbersome. The current definition for IPAF includes presence of an interstitial pneumonia by HRCT or surgical lung biopsy (SLB) and a full scrutiny to exclude alternative etiologies, and the patient should not meet criteria for a defined CTD. In addition, patient should have presence of at least one feature from at least two of the following domains (Table 5.6):

5.4.1.1 Clinical Domain
It includes all possible extra-thoracic presentations of a CTD without a full definition for a specific disease. They include distal digital fissuring (i.e., "mechanic hands"), distal digital tip ulceration, inflammatory arthritis or polyarticular morning joint stiffness ≥ 60 min, palmar telangiectasia, Raynaud's phenomenon, unexplained digital edema, and unexplained fixed rash on the digital extensor surfaces (Gottron's sign). In this domain, other nonspecific features were left aside as in previous definitions (e.g., arthralgia, myalgia, photosensitivity, and sicca symptoms).

5.4.1.2 Serological Domain
Conditions to include the ANA and rheumatoid factors are to have high titers to avoid false positives. In addition, regardless of the ANA titers, if they have a centromere or nucleolar patterns, they are incorporated as positive, as they are specific for possible lung involvement. They include (1) ANA $\geq 1:320$ titer, diffuse, speckled, and homogeneous patterns; (2) rheumatoid factor $\geq 2x$ upper limit of normal; (3) anti-CCP; (4) anti-ds-DNA; (5) anti-SSA/Ro; (6) anti-SSB/La; (7) anti-ribonucleoprotein; (8) anti-Smith; (9) anti-topoisomerase (Scl-70); (10) anti-tRNA synthetase (e.g., Jo-1, PL-7, PL-12, EJ, OJ, KS, Zo, tRS); (11) anti-PM/Scl; and (12) anti-MDA-5.

5.4.1.3 Morphologic Domain
It consists of three subdomains, radiographic, histopathological, and multicompartment (thoracic compartments supporting presence of CTD features). If any of these subdomains is present, then this domain has fulfilled the crite-

Table 5.6 Clinical, radiographic, histopathological, and serological features of syndromic autoimmune disorders linked with interstitial lung disease

Lung-dominant connective tissue disease [65]	Autoimmune-featured interstitial lung disease (ILD) [54]	Interstitial pneumonia with autoimmune features [20, 83]
(1) All of the following three clinical features: (a) NSIP, UIP, LIP, OP, or DAD (or desquamative interstitial pneumonia if no smoking history, defined by the lung biopsy or suggested by the HRCT) (b) Insufficient extra-thoracic features of a definite CTD (c) No identifiable alternative cause for interstitial pneumonia	This term includes as follows: (1) One or more of the following symptoms: dry eyes or dry mouth; gastro-esophageal reflux disease; weight loss; foot or leg swelling; joint pain or swelling; rash; photosensitivity; dysphagia; hand ulcers; mouth ulcers; Raynaud's phenomenon; morning stiffness; proximal muscle weakness	The current definition includes the following: (1) Presence of an interstitial pneumonia by HRCT or SLB (2) Exclusion of alternative etiologies (3) The patient does not meet criteria for a defined CTD (4) Presence of at least one feature from at least two of the following domains
(2) (a) One or more ANA ≥1:320; nucleolar or centromere patterns; rheumatoid factor >60 IU/mL; anti-SSA/Ro; anti-SSB/La; anti-Smith; anti-RNP; anti-ds-DNA; anti-topoisomerase I (Scl-70); anti-CCP; anti-tRNA synthetase; anti-PM/Scl	(2) One or more of the following positive or high laboratories: ANA ≥1:160; rheumatoid factor; anti-SSA/Ro, anti-SSB/La; anti-Smith; anti-ribonucleoprotein; anti-ds-DNA; anti-topoisomerase (Scl-70); anti-Jo-1; aldolase; creatinine phosphokinase	A. Clinical domain: (1) Distal digital fissuring (i.e., "mechanic hands") (2) Distal digital tip ulceration (3) Inflammatory arthritis or polyarticular morning joint stiffness ≥60 minutes (4) Palmar telangiectasia (5) Raynaud's phenomenon (6) Unexplained digital edema (7) Unexplained fixed rash on the digital extensor surfaces (Gottron's sign)
(b) Two or more of lymphoid aggregates with germinal centers; extensive pleuritis; prominent plasmacytic infiltration; dense perivascular collagen		B. Serological domain: (1) ANA ≥1:320 titer, diffuse, speckled, homogeneous patterns or a) ANA nucleolar pattern (any titer) or b) ANA centromere pattern (any titer) (2) Rheumatoid factor ≥2x upper limit of normal (3) Anti-CCP (4) Anti-ds-DNA (5) Anti-SSA/Ro (6) Anti-SSB/La (7) Anti-ribonucleoprotein (8) Anti-Smith (9) Anti-topoisomerase (Scl-70) (10) Anti-tRNA synthetase (e.g., Jo-1, PL-7, PL-12, EJ, OJ, KS, Zo, tRS) (11) Anti-PM/Scl (12) Anti-MDA-5

(continued)

Table 5.6 (continued)

Lung-dominant connective tissue disease [65]	Autoimmune-featured interstitial lung disease (ILD) [54]	Interstitial pneumonia with autoimmune features [20, 83]
		C. Morphologic domain: (1) Suggestive radiology patterns by HRCT: (a) NSIP; (b) OP; (c) NSIP with OP overlap; d) LIP (2) Histopathology patterns or features by surgical lung biopsy: (a) NSIP; (b) OP; (c) NSIP with OP overlap; (d) LIP; (e) interstitial lymphoid aggregates with germinal centers; (f) diffuse lymphoplasmacytic infiltration (with or without lymphoid follicles) (3) Multicompartment involvement (in addition to interstitial pneumonia): (a) unexplained pleural effusion or thickening, (b) unexplained pericardial effusion or thickening, (c) unexplained intrinsic airways disease (by PFT, imaging, or pathology, including airflow obstruction, bronchiolitis, or bronchiectasis), and (d) unexplained pulmonary vasculopathy

Source and adaptation: Fisher A et al. [65], Vij R et al. [54], and Graney BA et al. [20]
Abbreviations: *NSIP* nonspecific interstitial lung disease, *UIP* usual interstitial pneumonia, *LIP* lymphocytic interstitial pneumonia, *OP* organizing pneumonia, *DAD* diffuse alveolar damage, *HRCT* high-resolution chest computed tomography, *CTD* connective tissue disease, *ANA* antinuclear antibody, *anti-CCP* anti-citrullinated cyclic peptide, *SLB* surgical lung biopsy, *PFT* pulmonary function test

ria: (1) suggestive radiology patterns by HRCT: (a) NSIP, (b) OP, (c) NSIP with OP overlap, and (d) LIP; (2) histopathology patterns or features by surgical lung biopsy: (a) NSIP, (b) OP, (c) NSIP with OP overlap, (d) LIP, (e) interstitial lymphoid aggregates with germinal centers, and f) diffuse lymphoplasmacytic infiltration (with or without lymphoid follicles); (3) multicompartment involvement (in addition to interstitial pneumonia): (a) unexplained pleural effusion or thickening, (b) unexplained pericardial effusion or thickening, (c) unexplained intrinsic airways disease (by PFT, imaging, or pathology, including airflow obstruction, bronchiolitis, or bronchiectasis), and (d) unexplained pulmonary vasculopathy [83].

According to the domain definitions, UIP was not included as it has less likelihood to associate with features of CTD-ILD. A bias to this definition criteria is that patients with RA-UIP or any other CTD-UIP presentation would not be considered to have the same weight as other patterns in the HRCT. Despite this presentation, patients may classify as IPAF with a UIP pattern, if the other two domains (at least one item of each domain) fulfill criteria.

The utility of IPAF application has allowed authors to analyze predictive features for survival. Patients with IPAF criteria had slightly better outcomes in survival, when separated between those with the UIP and non-UIP patterns. Those with IPAF-UIP did worse and was comparable to IPF as compared to those patients with IPAF and non-UIP whose survival was similar to CTD-ILD [99]. Overall, however, the heterogeneity of patients fulfilling IPAF criteria within several cohorts makes difficult to identify unique features of all the studies performed that may predict better a survival other than when they present with UIP pattern [20]. Future diagnostic criteria will help unify IPAF with more precision, allowing researchers to avoid heterogeneity of inclusion criteria (especially in the morphologic domain) to achieve possible predictors of survival.

5.5   Conclusions

The current review is an approach focused on the integrated evaluation of patients with ILD. The main purpose is to establish a CTD in the initial assessment. A stepwise and thorough review is necessary to follow, including the history, physical exam, exam of key components for a connective tissue disease, laboratory review, and evaluation of the HRCT and histopathology. This methodology will avoid missing key components, including the evaluation of the nailfold capillaries and, if necessary, a biopsy of minor salivary glands from the lower lip. The knowledge of the new serology is important for the evaluation of specific syndromes among inflammatory myopathies, including the anti-synthetase syndrome.

A subset of patients will not fulfill a diagnosis of a specific CTD but may have several components favoring autoimmunity. In such situations, the three domains that are part of IPAF need to be reviewed to establish the diagnosis. This modality will help the clinician make appropriate and accurate management decisions. Except for RA-UIP, most of the CTD-ILD may respond to the immune suppressor therapy that ultimately will improve survival. Future therapies (anti-fibrotic) will improve survival in patients with a fibrotic pattern of ILD.

References

1. Klemperer P. The concept of connective-tissue disease. Circulation. 1962;25:869–71.
2. Klemperer P. The concept of collagen diseases in medicine. Am Rev Respir Dis. 1961;83:331–9.
3. Steen VD, Medsger TA. Changes in causes of death in systemic sclerosis, 1972-2002. Ann Rheum Dis. 2007;66(7):940–4.
4. Connors GR, Christopher-Stine L, Oddis CV, Danoff SK. Interstitial lung disease associated with the idiopathic inflammatory myopathies: what progress has been made in the past 35 years? Chest. 2010;138(6):1464–74.
5. Olson AL, Swigris JJ, Sprunger DB, Fischer A, Fernandez-Perez ER, Solomon J, et al. Rheumatoid arthritis-interstitial lung disease-associated mortality. Am J Respir Crit Care Med. 2011;183(3):372–8.
6. Turesson C, O'Fallon WM, Crowson CS, Gabriel SE, Matteson EL. Extra-articular disease manifestations in rheumatoid arthritis: incidence trends and risk factors over 46 years. Ann Rheum Dis. 2003;62(8):722–7.
7. Nobata K, Kasai T, Fujimura M, Mizuguchi M, Nishi K, Ishiura Y, et al. Pulmonary sarcoidosis with usual interstitial pneumonia distributed predominantly in the lower lung fields. Intern Med. 2006;45(6):359–62.
8. Yamamoto S. Histopathological features of pulmonary asbestosis with particular emphasis on the comparison with those of usual interstitial pneumonia. Osaka City Med J. 1997;43(2):225–42.
9. Kelly C, Iqbal K, Iman-Gutierrez L, Evans P, Manchegowda K. Lung involvement in inflammatory rheumatic diseases. Best Pract Res Clin Rheumatol. 2016;30(5):870–88.
10. Travis WD, Costabel U, Hansell DM, King TE Jr, Lynch DA, Nicholson AG, et al. An official American Thoracic Society/European Respiratory Society statement: update of the international multidisciplinary classification of the idiopathic interstitial pneumonias. Am J Respir Crit Care Med. 2013;188(6):733–48.
11. Raghu G, Remy-Jardin M, Myers JL, Richeldi L, Ryerson CJ, Lederer DJ, et al. Diagnosis of idiopathic pulmonary fibrosis. An official ATS/ERS/JRS/ALAT clinical practice guideline. Am J Respir Crit Care Med. 2018;198(5):e44–68.
12. Bouros D, Wells AU, Nicholson AG, Colby TV, Polychronopoulos V, Pantelidis P, et al. Histopathologic subsets of fibrosing alveolitis in patients with systemic sclerosis and their relationship to outcome. Am J Respir Crit Care Med. 2002;165(12):1581–6.
13. Park JH, Kim DS, Park IN, Jang SJ, Kitaichi M, Nicholson AG, et al. Prognosis of fibrotic interstitial pneumonia: idiopathic versus collagen vascular disease-related subtypes. Am J Respir Crit Care Med. 2007;175(7):705–11.
14. Fischer A, Swigris JJ, Groshong SD, Cool CD, Sahin H, Lynch DA, et al. Clinically significant interstitial lung disease in limited scleroderma: histopathology, clinical features, and survival. Chest. 2008;134(3):601–5.
15. Lee HK, Kim DS, Yoo B, Seo JB, Rho JY, Colby TV, et al. Histopathologic pattern and clinical features of rheumatoid arthritis-associated interstitial lung disease. Chest. 2005;127(6):2019–27.
16. Kim DS, Yoo B, Lee JS, Kim EK, Lim CM, Lee SD, et al. The major histopathologic pattern of pulmonary fibrosis in scleroderma is nonspecific interstitial pneumonia. Sarcoidosis Vasc Diffuse Lung Dis. 2002;19(2):121–7.
17. Flaherty KR, Colby TV, Travis WD, Toews GB, Mumford J, Murray S, et al. Fibroblastic foci in usual interstitial pneumonia: idiopathic versus collagen vascular disease. Am J Respir Crit Care Med. 2003;167(10):1410–5.

18. Hernandez-Gonzalez F, Prieto-Gonzalez S, Brito-Zeron P, Cuerpo S, Sanchez M, Ramirez J, et al. Impact of a systematic evaluation of connective tissue disease on diagnosis approach in patients with interstitial lung diseases. Medicine (Baltimore). 2020;99(4):e18589.

19. Idiopathic Pulmonary Fibrosis Clinical Research N, Raghu G, Anstrom KJ, King TE Jr, Lasky JA, Martinez FJ. Prednisone, azathioprine, and N-acetylcysteine for pulmonary fibrosis. N Engl J Med. 2012;366(21):1968–77.

20. Graney BA, Fischer A. Interstitial pneumonia with autoimmune features. Ann Am Thorac Soc. 2019;16(5):525–33.

21. Strange C, Highland KB. Interstitial lung disease in the patient who has connective tissue disease. Clin Chest Med. 2004;25(3):549–59, vii.

22. Duron L, Cohen-Aubart F, Diot E, Borie R, Abad S, Richez C, et al. Shrinking lung syndrome associated with systemic lupus erythematosus: a multicenter collaborative study of 15 new cases and a review of the 155 cases in the literature focusing on treatment response and long-term outcomes. Autoimmun Rev. 2016;15(10):994–1000.

23. Borrell H, Narvaez J, Alegre JJ, Castellvi I, Mitjavila F, Aparicio M, et al. Shrinking lung syndrome in systemic lupus erythematosus: a case series and review of the literature. Medicine (Baltimore). 2016;95(33):e4626.

24. Demoruelle MK, Mittoo S, Solomon JJ. Connective tissue disease-related interstitial lung disease. Best Pract Res Clin Rheumatol. 2016;30(1):39–52.

25. Fischer A, du Bois R. Interstitial lung disease in connective tissue disorders. Lancet. 2012;380(9842):689–98.

26. Pitcher RD, Zar HJ. Radiographic features of paediatric pneumocystis pneumonia -- a historical perspective. Clin Radiol. 2008;63(6):666–72.

27. Mittoo S, Gelber AC, Christopher-Stine L, Horton MR, Lechtzin N, Danoff SK. Ascertainment of collagen vascular disease in patients presenting with interstitial lung disease. Respir Med. 2009;103(8):1152–8.

28. Marigliano B, Soriano A, Margiotta D, Vadacca M, Afeltra A. Lung involvement in connective tissue diseases: a comprehensive review and a focus on rheumatoid arthritis. Autoimmun Rev. 2013;12(11):1076–84.

29. Fischer A, Swigris JJ, du Bois RM, Groshong SD, Cool CD, Sahin H, et al. Minor salivary gland biopsy to detect primary Sjogren syndrome in patients with interstitial lung disease. Chest. 2009;136(4):1072–8.

30. Shiboski CH, Shiboski SC, Seror R, Criswell LA, Labetoulle M, Lietman TM, et al. 2016 American College of Rheumatology/European League Against Rheumatism classification criteria for primary Sjogren's syndrome: a consensus and data-driven methodology involving three international patient cohorts. Ann Rheum Dis. 2017;76(1):9–16.

31. Schurawitzki H, Stiglbauer R, Graninger W, Herold C, Polzleitner D, Burghuber OC, et al. Interstitial lung disease in progressive systemic sclerosis: high-resolution CT versus radiography. Radiology. 1990;176(3):755–9.

32. Maricq HR, LeRoy EC. Patterns of finger capillary abnormalities in connective tissue disease by "wide-field" microscopy. Arthritis Rheum. 1973;16(5):619–28.

33. Cutolo M, Sulli A, Pizzorni C, Accardo S. Nailfold videocapillaroscopy assessment of microvascular damage in systemic sclerosis. J Rheumatol. 2000;27(1):155–60.

34. Fischer A, Meehan RT, Feghali-Bostwick CA, West SG, Brown KK. Unique characteristics of systemic sclerosis sine scleroderma-associated interstitial lung disease. Chest. 2006;130(4):976–81.

35. Smith V, Decuman S, Sulli A, Bonroy C, Piettte Y, Deschepper E, et al. Do worsening scleroderma capillaroscopic patterns predict future severe organ involvement? A pilot study. Ann Rheum Dis. 2012;71(10):1636–9.

36. Smith V, Riccieri V, Pizzorni C, Decuman S, Deschepper E, Bonroy C, et al. Nailfold capillaroscopy for prediction of novel future severe organ involvement in systemic sclerosis. J Rheumatol. 2013;40(12):2023–8.

37. van Roon AM, Huisman CC, van Roon AM, Zhang D, Stel AJ, Smit AJ, et al. Abnormal Nailfold Capillaroscopy is common in patients with connective tissue disease and associated with abnormal pulmonary function tests. J Rheumatol. 2019;46(9):1109–16.

38. Corrado A, Correale M, Mansueto N, Monaco I, Carriero A, Mele A, et al. Nailfold capillaroscopic changes in patients with idiopathic pulmonary arterial hypertension and systemic sclerosis-related pulmonary arterial hypertension. Microvasc Res. 2017;114:46–51.

39. Hofstee HM, Vonk Noordegraaf A, Voskuyl AE, Dijkmans BA, Postmus PE, Smulders YM, et al. Nailfold capillary density is associated with the presence and severity of pulmonary arterial hypertension in systemic sclerosis. Ann Rheum Dis. 2009;68(2):191–5.

40. Fischer A, Pfalzgraf FJ, Feghali-Bostwick CA, Wright TM, Curran-Everett D, West SG, et al. Anti-th/to-positivity in a cohort of patients with idiopathic pulmonary fibrosis. J Rheumatol. 2006;33(8):1600–5.

41. Wells AU, Cullinan P, Hansell DM, Rubens MB, Black CM, Newman-Taylor AJ, et al. Fibrosing alveolitis associated with systemic sclerosis has a better prognosis than lone cryptogenic fibrosing alveolitis. Am J Respir Crit Care Med. 1994;149(6):1583–90.

42. Wells AU, Margaritopoulos GA, Antoniou KM, Denton C. Interstitial lung disease in systemic sclerosis. Semin Respir Crit Care Med. 2014;35(2):213–21.

43. Moore OA, Goh N, Corte T, Rouse H, Hennessy O, Thakkar V, et al. Extent of disease on high-resolution computed tomography lung is a predictor of decline and mortality in systemic sclerosis-related interstitial lung disease. Rheumatology (Oxford). 2013;52(1):155–60.

44. Goh NS, Desai SR, Veeraraghavan S, Hansell DM, Copley SJ, Maher TM, et al. Interstitial lung disease in systemic sclerosis: a simple staging system. Am J Respir Crit Care Med. 2008;177(11):1248–54.

45. Hyldgaard C, Hilberg O, Pedersen AB, Ulrichsen SP, Lokke A, Bendstrup E, et al. A population-based cohort study of rheumatoid arthritis-associated interstitial lung disease: comorbidity and mortality. Ann Rheum Dis. 2017;76(10):1700–6.

46. Kelly CA, Saravanan V, Nisar M, Arthanari S, Woodhead FA, Price-Forbes AN, et al. Rheumatoid arthritis-related interstitial lung disease: associations, prognostic factors and physiological and radiological characteristics--a large multicentre UK study. Rheumatology (Oxford). 2014;53(9):1676–82.

47. Solomon JJ, Chung JH, Cosgrove GP, Demoruelle MK, Fernandez-Perez ER, Fischer A, et al. Predictors of mortality in rheumatoid arthritis-associated interstitial lung disease. Eur Respir J. 2016;47(2):588–96.

48. Kondoh Y, Taniguchi H, Ogura T, Johkoh T, Fujimoto K, Sumikawa H, et al. Disease progression in idiopathic pulmonary fibrosis without pulmonary function impairment. Respirology. 2013;18(5):820–6.

49. Cottin V, Nunes H, Brillet PY, Delaval P, Devouassoux G, Tillie-Leblond I, et al. Combined pulmonary fibrosis and emphysema: a distinct under-recognised entity. Eur Respir J. 2005;26(4):586–93.

50. Kim EA, Lee KS, Johkoh T, Kim TS, Suh GY, Kwon OJ, et al. Interstitial lung diseases associated with collagen vascular diseases: radiologic and histopathologic findings. Radiographics : a review publication of the Radiological Society of North America, Inc. 2002;22 Spec No:S151-65.

51. Walsh SL, Sverzellati N, Devaraj A, Keir GJ, Wells AU, Hansell DM. Connective tissue disease related fibrotic lung disease: high resolution computed tomographic and pulmonary function indices as prognostic determinants. Thorax. 2014;69(3):216–22.

52. Monaghan H, Wells AU, Colby TV, du Bois RM, Hansell DM, Nicholson AG. Prognostic implications of histologic patterns in multiple surgical lung biopsies from patients with idiopathic interstitial pneumonias. Chest. 2004;125(2):522–6.

53. Flaherty KR, Thwaite EL, Kazerooni EA, Gross BH, Toews GB, Colby TV, et al. Radiological versus histological diagnosis in UIP and NSIP: survival implications. Thorax. 2003;58(2):143–8.

54. Vij R, Noth I, Strek ME. Autoimmune-featured interstitial lung disease: a distinct entity. Chest. 2011;140(5):1292–9.

55. Flaherty KR, Khanna D. Idiopathic or connective tissue disease-associated interstitial lung disease: a case of HRCT mimicry. Thorax. 2014;69(3):205–6.

56. Kim EJ, Elicker BM, Maldonado F, Webb WR, Ryu JH, Van Uden JH, et al. Usual interstitial pneumonia in rheumatoid arthritis-associated interstitial lung disease. Eur Respir J. 2010;35(6):1322–8.

57. Zamora-Legoff JA, Krause ML, Crowson CS, Ryu JH, Matteson EL. Progressive decline of lung function in rheumatoid arthritis-associated interstitial lung disease. Arthritis Rheumatol. 2017;69(3):542–9.

58. Mira-Avendano IC, Abril A. Pulmonary manifestations of Sjogren syndrome, systemic lupus erythematosus, and mixed connective tissue disease. Rheum Dis Clin N Am. 2015;41(2):263–77.

59. Kinder BW, Shariat C, Collard HR, Koth LL, Wolters PJ, Golden JA, et al. Undifferentiated connective tissue disease-associated interstitial lung disease: changes in lung function. Lung. 2010;188(2):143–9.

60. Mosca M, Tani C, Neri C, Baldini C, Bombardieri S. Undifferentiated connective tissue diseases (UCTD). Autoimmun Rev. 2006;6(1):1–4.

61. Mira-Avendano I, Abril A, Burger CD, Dellaripa PF, Fischer A, Gotway MB, et al. Interstitial lung disease and other pulmonary manifestations in connective tissue diseases. Mayo Clin Proc. 2019;94(2):309–25.

62. Lynch DA, Sverzellati N, Travis WD, Brown KK, Colby TV, Galvin JR, et al. Diagnostic criteria for idiopathic pulmonary fibrosis: a Fleischner Society White Paper. Lancet Respir Med. 2018;6(2):138–53.

63. Capobianco J, Grimberg A, Thompson BM, Antunes VB, Jasinowodolinski D, Meirelles GS. Thoracic manifestations of collagen vascular diseases. Radiographics : a review publication of the Radiological Society of North America, Inc. 2012;32(1):33–50.

64. Kligerman SJ, Groshong S, Brown KK, Lynch DA. Nonspecific interstitial pneumonia: radiologic, clinical, and pathologic considerations. Radiographics : a review publication of the Radiological Society of North America, Inc. 2009;29(1):73–87.

65. Fischer A, West SG, Swigris JJ, Brown KK, du Bois RM. Connective tissue disease-associated interstitial lung disease: a call for clarification. Chest. 2010;138(2):251–6.

66. Flament T, Bigot A, Chaigne B, Henique H, Diot E, Marchand-Adam S. Pulmonary manifestations of Sjogren's syndrome. Eur Respir Rev. 2016;25(140):110–23.

67. Panchabhai TS, Farver C, Highland KB. Lymphocytic interstitial pneumonia. Clin Chest Med. 2016;37(3):463–74.

68. Argyropoulou OD, Chatzis LG, Rontogianni D, Tzioufas AG. Autoimmune epithelitis beyond the exocrine glands: an unusual case of anti-Ro/La and Scl-70 lymphocytic interstitial pneumonia. Clin Exp Rheumatol. 2019;37 Suppl 118(3):249–51.

69. Skopouli FN, Moutsopoulos HM. Autoimmune epi-theliitis: Sjogren's syndrome. Clin Exp Rheumatol. 1994;12(Suppl 11):S9–11.
70. Daniels TE, Cox D, Shiboski CH, Schiodt M, Wu A, Lanfranchi H, et al. Associations between salivary gland histopathologic diagnoses and phenotypic features of Sjogren's syndrome among 1,726 registry participants. Arthritis Rheum. 2011;63(7):2021–30.
71. Yousem SA, Colby TV, Carrington CB. Follicular bronchitis/bronchiolitis. Hum Pathol. 1985;16(7):700–6.
72. Ryu JH, Myers JL, Swensen SJ. Bronchiolar disorders. Am J Respir Crit Care Med. 2003;168(11):1277–92.
73. Ford JA, Liu X, Chu SH, Lu B, Cho MH, Silverman EK, et al. Asthma, chronic obstructive pulmonary disease, and subsequent risk for incident rheumatoid arthritis among women: a prospective cohort study. Arthritis Rheumatol. 2020;72(5):704–13.
74. Willis VC, Demoruelle MK, Derber LA, Chartier-Logan CJ, Parish MC, Pedraza IF, et al. Sputum autoantibodies in patients with established rheumatoid arthritis and subjects at risk of future clinically apparent disease. Arthritis Rheum. 2013;65(10):2545–54.
75. Reynisdottir G, Olsen H, Joshua V, Engstrom M, Forsslund H, Karimi R, et al. Signs of immune activation and local inflammation are present in the bronchial tissue of patients with untreated early rheumatoid arthritis. Ann Rheum Dis. 2016;75(9):1722–7.
76. Gizinski AM, Mascolo M, Loucks JL, Kervitsky A, Meehan RT, Brown KK, et al. Rheumatoid arthritis (RA)-specific autoantibodies in patients with interstitial lung disease and absence of clinically apparent articular RA. Clin Rheumatol. 2009;28(5):611–3.
77. Demoruelle MK, Solomon JJ, Fischer A, Deane KD. The lung may play a role in the pathogenesis of rheumatoid arthritis. Int J Clin Rheumatol. 2014;9(3):295–309.
78. Fischer A, Solomon JJ, du Bois RM, Deane KD, Olson AL, Fernandez-Perez ER, et al. Lung disease with anti-CCP antibodies but not rheumatoid arthritis or connective tissue disease. Respir Med. 2012;106(7):1040–7.
79. Rangel-Moreno J, Hartson L, Navarro C, Gaxiola M, Selman M, Randall TD. Inducible bronchus-associated lymphoid tissue (iBALT) in patients with pulmonary complications of rheumatoid arthritis. J Clin Invest. 2006;116(12):3183–94.
80. Raj R, Raparia K, Lynch DA, Brown KK. Surgical lung biopsy for interstitial lung diseases. Chest. 2017;151(5):1131–40.
81. Travis WD, Hunninghake G, King TE Jr, Lynch DA, Colby TV, Galvin JR, et al. Idiopathic nonspecific interstitial pneumonia: report of an American Thoracic Society project. Am J Respir Crit Care Med. 2008;177(12):1338–47.
82. Kono M, Nakamura Y, Yoshimura K, Enomoto Y, Oyama Y, Hozumi H, et al. Nonspecific interstitial pneumonia preceding diagnosis of collagen vascular disease. Respir Med. 2016;117:40–7.
83. Fischer A, Antoniou KM, Brown KK, Cadranel J, Corte TJ, du Bois RM, et al. An official European Respiratory Society/American Thoracic Society research statement: interstitial pneumonia with auto-immune features. Eur Respir J. 2015;46(4):976–87.
84. Enomoto Y, Takemura T, Hagiwara E, Iwasawa T, Fukuda Y, Yanagawa N, et al. Prognostic factors in interstitial lung disease associated with primary Sjogren's syndrome: a retrospective analysis of 33 pathologically-proven cases. PLoS One. 2013;8(9):e73774.
85. Flaherty KR, Travis WD, Colby TV, Toews GB, Kazerooni EA, Gross BH, et al. Histopathologic variability in usual and nonspecific interstitial pneumonias. Am J Respir Crit Care Med. 2001;164(9):1722–7.
86. Katzenstein AL, Fiorelli RF. Nonspecific interstitial pneumonia/fibrosis. Histologic features and clinical significance. Am J Surg Pathol. 1994;18(2):136–47.
87. Cottin V, Cordier JF. Cryptogenic organizing pneumonia. Semin Respir Crit Care Med. 2012;33(5):462–75.
88. Thille AW, Esteban A, Fernandez-Segoviano P, Rodriguez JM, Aramburu JA, Vargas-Errazuriz P, et al. Chronology of histological lesions in acute respiratory distress syndrome with diffuse alveolar damage: a prospective cohort study of clinical autopsies. Lancet Respir Med. 2013;1(5):395–401.
89. Cardinal-Fernandez P, Lorente JA, Ballen-Barragan A, Matute-Bello G. Acute respiratory distress syndrome and diffuse alveolar damage. New insights on a complex relationship. Ann Am Thorac Soc. 2017;14(6):844–50.
90. Cardinal-Fernandez P, Bajwa EK, Dominguez-Calvo A, Menendez JM, Papazian L, Thompson BT. The presence of diffuse alveolar damage on open lung biopsy is associated with mortality in patients with acute respiratory distress syndrome: a systematic review and meta-analysis. Chest. 2016;149(5):1155–64.
91. Cipriani NA, Strek M, Noth I, Gordon IO, Charbeneau J, Krishnan JA, et al. Pathologic quantification of connective tissue disease-associated versus idiopathic usual interstitial pneumonia. Arch Pathol Lab Med. 2012;136(10):1253–8.
92. Song JW, Do KH, Kim MY, Jang SJ, Colby TV, Kim DS. Pathologic and radiologic differences between idiopathic and collagen vascular disease-related usual interstitial pneumonia. Chest. 2009;136(1):23–30.
93. Kinder BW, Collard HR, Koth L, Daikh DI, Wolters PJ, Elicker B, et al. Idiopathic nonspecific interstitial pneumonia: lung manifestation of undifferentiated connective tissue disease? Am J Respir Crit Care Med. 2007;176(7):691–7.

94. Corte TJ, Copley SJ, Desai SR, Zappala CJ, Hansell DM, Nicholson AG, et al. Significance of connective tissue disease features in idiopathic interstitial pneumonia. Eur Respir J. 2012;39(3):661–8.
95. LeRoy EC, Maricq HR, Kahaleh MB. Undifferentiated connective tissue syndromes. Arthritis Rheum. 1980;23(3):341–3.
96. Bodolay E, Csiki Z, Szekanecz Z, Ben T, Kiss E, Zeher M, et al. Five-year follow-up of 665 Hungarian patients with undifferentiated connective tissue disease (UCTD). Clin Exp Rheumatol. 2003;21(3):313–20.
97. Conti V, Esposito A, Cagliuso M, Fantauzzi A, Pastori D, Mezzaroma I, et al. Undifferentiated connective tissue disease - an unsolved problem:revision of literature and case studies. Int J Immunopathol Pharmacol. 2010;23(1):271–8.
98. Vaz CC, Couto M, Medeiros D, Miranda L, Costa J, Nero P, et al. Undifferentiated connective tissue disease: a seven-center cross-sectional study of 184 patients. Clin Rheumatol. 2009;28(8):915–21.
99. Oldham JM, Adegunsoye A, Valenzi E, Lee C, Witt L, Chen L, et al. Characterisation of patients with interstitial pneumonia with autoimmune features. Eur Respir J. 2016;47(6):1767–75.
100. Cotton CV, Spencer LG, New RP, Cooper RG. The utility of comprehensive autoantibody testing to differentiate connective tissue disease associated and idiopathic interstitial lung disease subgroup cases. Rheumatology (Oxford). 2017;56(8):1264–71.
101. Bryson T, Sundaram B, Khanna D, Kazerooni EA. Connective tissue disease-associated interstitial pneumonia and idiopathic interstitial pneumonia: similarity and difference. Semin Ultrasound CT MR. 2014;35(1):29–38.
102. Hervier B, Uzunhan Y. Inflammatory myopathy-related interstitial lung disease: from pathophysiology to treatment. Front Med. 2019;6:326.
103. Betteridge ZE, Gunawardena H, McHugh NJ. Novel autoantibodies and clinical phenotypes in adult and juvenile myositis. Arthritis Res Ther. 2011;13(2):209.

Molecular Mechanisms of Vascular Damage During Lung Injury

6

Ramon Bossardi Ramos and Alejandro Pablo Adam

Abstract

A variety of pulmonary and systemic insults promote an inflammatory response causing increased vascular permeability, leading to the development of acute lung injury (ALI), a condition necessitating hospitalization and intensive care, or the more severe acute respiratory distress syndrome (ARDS), a disease with a high mortality rate. Further, COVID-19 pandemic-associated ARDS is now a major cause of mortality worldwide. The pathogenesis of ALI is explained by injury to both the vascular endothelium and the alveolar epithelium. The disruption of the lung endothelial and epithelial barriers occurs in response to both systemic and local production of pro-inflammatory cytokines. Studies that evaluate the association of genetic polymorphisms with disease risk did not yield many potential therapeutic targets to treat and revert lung injury. This failure is probably due in part to the phenotypic complexity of ALI/ARDS, and genetic predisposition may be obscured by the multiple environmental and behavioral risk factors. In the last decade, new research has uncovered novel epigenetic mechanisms that control ALI/ARDS pathogenesis, including histone modifications and DNA methylation. Enzyme inhibitors such as DNMTi and HDACi may offer new alternative strategies to prevent or reverse the vascular damage that occurs during lung injury. This review will focus on the latest findings on the molecular mechanisms of vascular damage in ALI/ARDS, the genetic factors that might contribute to the susceptibility for developing this disease, and the epigenetic changes observed in humans, as well as in experimental models of ALI/ADRS.

R. Bossardi Ramos (✉)
Department of Molecular and Cellular Physiology, Albany Medical College, Albany, NY, USA
e-mail: bossarr@amc.edu

A. P. Adam (✉)
Department of Molecular and Cellular Physiology, Albany Medical College, Albany, NY, USA

Department of Ophthalmology, Albany Medical College, Albany, NY, USA
e-mail: adama1@amc.edu

Keywords

Lung injury; Vascular damage; Epithelial barrier; Endothelial activation; Epigenetics; Histone acetylation, DNA methylation; Inflammation

Abbreviations

ALI	Acute lung injury
ANGPT2	Angiopoietin 2
APC	Activated protein C
ARDS	Acute respiratory distress syndrome
BMDM	Bone marrow-derived macrophage
COPD	Chronic obstructive pulmonary disease
COVID	Coronavirus disease
DNAm	DNA methylation
DNMT	DNA methyltransferase
DNMTi	DNA methyltransferase inhibitor
ECs	Endothelial cells
EPCR	Endothelial protein C receptor
EWAS	Epigenome-wide association study
GWAS	Genome-wide association studies
HAT	Histone acetyltransferase
HDAC	Histone deacetylases
HLMVEC	Human lung microvascular endothelial cells
HPAEC	Human pulmonary artery endothelial cells
IL	Interleukin
LPS	Lipopolysaccharides
MLE-12	Murine lung epithelial cell line
MODS	Multiorgan dysfunction syndrome
MYLK	Myosin light chain kinase
ncRNAs	Noncoding RNAs
PA	Plasminogen activators
PAI	Plasminogen activator inhibitor
PAN	Panobinostat
ROS	Reactive oxygen species
S1P	Sphingosine-1-phosphate
S1PRs	S1P receptors
SARS-CoV-2	Severe acute respiratory syndrome coronavirus 2
SB	Sodium butyrate
SELPLG	Selectin P ligand
SFTPB	Surfactant protein B
SNPs	Single-nucleotide polymorphisms
TF	Tissue factor
TM	Thrombomodulin
TNF-α	Tumor necrosis factor α
TSA	Trichostatin A
Tub A	Tubastatin A
VAP	Ventilator-associated pneumonia
VEGFR1	Vascular endothelial growth factor receptor 1
VILI	Ventilator-induced lung injury
VPA	Valproic acid
VWF	Von Willebrand factor

6.1   Introduction

Acute lung injury (ALI) and the more severe form, acute respiratory distress syndrome (ARDS), are described as clinical syndromes of acute respiratory failure with substantial morbidity and mortality [1]. ALI and ARDS develop most commonly in the setting of pneumonia, non-pulmonary sepsis (with sources that include the peritoneum, urinary tract, soft tissue, and skin), aspiration of gastric and/or oral and esophageal contents, and major trauma (such as blunt or penetrating injuries or burns) and may be part of a systemic failure consisting in multiorgan dysfunction syndrome (MODS) [2, 3]. Mortality rates, while improving, remain persistently high at near 30% [4]. ARDS is a major public health burden that causes substantial morbidity, mortality, and healthcare cost [5]. ALI/ARDS mortality is associated with the development of pulmonary edema, a major factor for hypoxemia [6], as well as fibrin microthrombi in the alveoli and pulmonary vasculature [7]. The clinical features of the patient with ALI/ARDS consist of acute hypoxemic respiratory failure with bilateral pulmonary infiltrates [8, 9]. The disease is characterized by an acute exudative phase involving pulmonary edema and inflammation, accompanying abnormal gas exchange and variable late-phase responses [10]. All these features can lead to severe hypoxemia often necessitating mechanical ventilation. Patients with multiple comorbidities, chronic alcohol abuse, or chronic lung disease have increased risk for lung injury [9]. Moreover, recent clinical data indicate that novel coronavirus disease (COVID)-19 most commonly manifests as viral pneumonia-induced ARDS mediated by the severe acute respiratory syndrome coronavirus 2 (SARS-CoV-2) [11]. COVID-19 pandemic-associated ARDS is now a major cause of mortality worldwide. This virus directly impacts the lung endothelium to promote leakage and microthrombi [11, 12]. However, the mechanisms that govern how the comorbidities affect these pathophysiological processes driving lung damage and our understanding of the mechanisms and heterogeneity of sepsis- and COVID-19-associated ARDS, and in particular the effects of SARS-CoV-2 in the endothelium, are still in an early stage [13]. It is becoming increasingly clear that endothelial dysfunction is a common denominator in all these processes. Critically, the mechanical ventilation that is required to restore blood oxygenation during ARDS can in itself promote local tissue damage through a mechanism called ventilator-induced lung injury (VILI). Many of these changes may be driven by intracellular signals in the endothelium, as reviewed recently [2]. For example, ROS generation through the multiple cell types within the lungs, including epithelial cells (EC) and immune cells, may mediate ARDS/VILI-induced damage, and previous studies showed that antioxidants can reduce the severity of ARDS/VILI [14, 15]. Other mechanical stress signals are mediated through changes in the cytoskeleton network, as well as transmembrane proteins mediating cell-cell adhesion [16, 17].

6.2 Molecular Mechanisms of Vascular Damage in ALI/ARDS

6.2.1 Endothelial Activation

The lung vascular endothelium forms a continuous uninterrupted layer of endothelial cells (ECs) that is located at the interface between the bloodstream and lung tissue and provides a semi-selective barrier. This endothelial layer is held together by two complex junctional structures: the adherens junctions and the tight junctions. These junctions are gatekeepers for the movement of fluid, proteins, and cells from the bloodstream to the lung interstitium (reviewed elsewhere [18–20]). Due to their localization, ECs play key roles not only in optimizing gas exchange but also in controlling barrier integrity and function [21].

A lack of understanding of the pathologic mechanisms and how the ECs are involved in ALI remains an obstacle to new and effective therapies that reduce vascular leakage [22, 23]. It is well accepted that the inflammatory process during ALI causes disruption of the lung endothelial and epithelial barriers, leading to respiratory epithelial dysfunction and surfactant depletion [1, 24, 25]. This disruption of the lung endothelial barriers is modulated by multiple signaling pathways, such as those involving reactive oxygen species (ROS), actomyosin contractility through action of Rho GTPases, and tyrosine phosphorylation of junctional proteins [20, 26–29]. The breakdown of endothelial and epithelial barriers in ALI occurs in response to intense inflammation and local production of pro-inflammatory cytokines (e.g., tumor necrosis factor α (TNF-α), interleukin (IL)-1β, IL-6), chemokines such as IL-8, and other edemagenic factors such as thrombin- or platelet-activating factor. Inflammatory cytokine signaling leads to a compromised EC barrier, elevated vascular permeability, and edema [30]. The activated endothelium then shows increased activation of pro-inflammatory transcription factors and release of inflammatory mediators, which in turn lead to increased oxidative stress, a pro-thrombotic phenotype, and miscommunication with adjacent vascular cell walls [31]. Loss of the endothelial barrier during ALI causes pulmonary edema by increasing alveolar-capillary permeability to fluid, proteins, neutrophils, and red blood cells [24]. Direct exposure to bacterial endotoxins, including lipopolysaccharides (LPS) and intercellular interactions with activated inflammatory cells, may also play an important role [32, 33]. Alveolar macrophages, neutrophils, and other immune effector cells, including mono-cytes and platelets, are critical components of a successful lung defense against external challenges and have key activities in acute lung injury (reviewed by Matthai et al. [2]). The endothelium is not only a major target of cytokine signaling but also a major source of danger signals. At least during an influenza virus-induced lung injury, the pulmonary endothelium is a key regulator of the cytokine storm, demonstrating its central role as a mediator of innate immunity cellular and cytokine responses [34].

Lung infiltration is intrinsically dependent on endothelial regulation. In response to cytokine stimuli, the EC promotes a large increase in the expression levels of multiple leukocyte adhesion molecules on the EC surface, including E-selectin, P-selectin, ICAM-1, and VCAM-1 [35]. These changes are accompanied by alterations in oxygen and nitrogen free radicals as the superoxide radical and NO, prostaglandins, leukotrienes, and growth factors, which mediate leukocyte adhesion and migration, as well as platelet accumulation [36].

A hypercoagulatory state is another hallmark of ARDS/ALI pathophysiology [5, 37]. The endothelium is a key regulator of the coagulation cascade through modulation of the thrombogenicity of its surface. During basal conditions, the endothelial surface is non-thrombogenic through the expression of inhibitors of the coagulation cascade and platelet activation and aggregation, including tissue factor (TF) inhibitor and thrombin inhibitors [38]. Vascular injury during ALI/ARDS initiates coagulation by promoting both activation of procoagulant factors and platelets and a reduction of anticoagulant components and fibrinolysis. Chiefly, the endothelial surface may express high levels of TF in response to TNF-α and IL-1β [39]. Further, the endothelium can control platelet adhesion and activation through the release of Von Willebrand factor (VWF), which promotes binding of platelets; plasminogen activators (t-PA and u-PA) that promote fibrinolysis; and electrical repulsion by negatively charged heparan sulfates, among others [38]. Conversely, the regulation of TF and the coagulation cascade is not only a critical effector of

inflammation but also a modulator of the inflammatory response [40]. In human studies, the intravenous administration of LPS leads to an increase in microparticles containing TF into the circulation, resulting in thrombin production [37, 41], thereby increasing the formation of fibrin, and resulting in the consumption of clotting factors. Similarly, a study with 45 patients with ALI/ARDS from septic and non-septic causes reported that the low activated protein C (APC) and high plasminogen activator inhibitor (PAI)-1 levels were associated with increased mortality [42]. Protein C is an endogenous anticoagulant that is activated by a complex consisting of thrombin, thrombomodulin (TM), and the endothelial protein C receptor (EPCR) within the endothelial lumen [43, 44]. APC inhibits coagulation via the suppression of fVa and fVIIIa and exerts anti-inflammatory effects as well, by suppressing cytokine production [45]. In addition, APC activates the fibrinolytic process, as it inhibits PAI-1 [37]. In a murine model, EPCR is a critical participant in APC-mediated protection of pulmonary vascular permeability [46]. Despite the important roles of APC in the regulation of coagulation and inflammation, recombinant activated protein C, the only approved pharmacological agent targeting the endothelial response, was withdrawn because of its lack of improvement of patient survival [47], emphasizing our lack of a full understanding of this complex response.

6.2.2 Genetic Factors

Studies designed to identify genetic factors that might contribute to the susceptibility for developing ALI/ARDS have been reported in the past decade. However, no specific loci with genome-wide significance for associations with the diseases have been identified, probably in part because of the phenotypic complexity of ALI/ARDS engendered by the different risk factors [48]. The challenges for finding a specific molecular marker can be explained by the heterogeneity of the disease, in which multiple different pathogenic processes contribute in varying ways in different patients, depending on the specific combination of clinical factors and genetics [49]. Moreover, the requirement for such a large environmental insult prevents the use of genetic linkage studies of family pedigrees to identify genetic influences [50]. One example of genetic background that affects the ALI is the identification of Nrf2 as a susceptibility gene in the murine model of ALI [51, 52]. Genomic approaches have also led to success in identifying VILI-associated genes, novel biomarkers (NAMPT/PBEF), and pro-inflammatory cytokines such as IL-6. IL-6 is produced by the endothelium, monocytes, and macrophages stimulated by the TLRs and is directly associated with the inflammatory response in ALI (reviewed in [53]). A previous study described two single-nucleotide polymorphisms (SNPs) consisting of T-1001G and C-1535 T transversions in the promoter region of the NAMPT gene (also known as pre-B-cell colony-enhancing factor, PBEF) that confer susceptibility to sepsis-induced ARDS (the GC haplotype conferred a 7.7-fold increased risk). Consistent with a lower risk, the -1543T variant showed an 1.8-fold reduced gene expression [54, 55]. In vitro, NAMPT gene downregulation led to lower IL-8 expression and reduced monolayer permeability of human pulmonary artery endothelial cells after IL-1β or thrombin challenges [56, 57]. Consistently, NAMPT overexpression led to increased IL-8 in these cells [55, 57].

Few reports exist regarding genome-wide association studies (GWAS) on ARDS. Data obtained from 600 trauma-induced ARDS patients and 2266 health controls of European descent found around 2.5 million SNP candidates across the genome [58]. Follow-up of the top 1% most significant SNPs with 212 trauma-induced ARDS patients and 283 at-risk controls showed an association between ARDS and variants from PPFIA1, a protein-associated focal adhesions, and the actin cytoskeleton [59], as well as other genes including IL-10, myosin light chain kinase (MYLK), angiopoietin 2 (ANGPT2), and Fas cell surface death receptor (FAS) [58]. A second and more recent GWAS study compared 232 African American patients with ARDS with 162 control subjects with an ARDS risk factor who never developed ARDS. The authors identified one SNP rs2228315 in intron 3 of the selectin P ligand (SELPLG) gene. The selectin pathway is critically involved in the transmigration of leukocytes from the vascular lumen to sites of inflammation [60]. Confirming a role in lung injury, SELPLG gene expression was significantly increased in VILI- and LPS-induced mouse models of lung injury. Moreover, antibody-mediated neutralization of PSGL-1, the SELPLG-encoded protein, attenuated experimental VILI- and LPS-induced lung, and SELPLG knockout mice are protected against LPS-induced lung injury [60]. Together, these findings suggest that variants in genes regulating leukocyte adhesion and migration may affect the risk of developing ARDS.

Importantly, other variants of genes regulating the vasculature may also be involved. By performing a prioritization of candidate genes integrating genomic data from a transcriptomic study in an animal model of ARDS and the data from the first published genome-wide association study described above, Hernandez-Pacheco et al. identified a variant in FLT1 (rs9513106) associated with ARDS, with on odds ratio of 0.75 (CI = 0.58–0.98) [61]. FLT1 encodes for the vascular endothelial growth factor receptor 1 (VEGFR1), a molecule critically involved in angiogenesis and endothelial barrier regulation [62].

6.2.3 Epigenetic Changes in Experimental Models of ALI/ADRS

Because it is generally considered that genetic variants contribute only a small proportion of ARDS/ALI risk [48], it is imperative to understand other mechanisms that can explain this disease. The increasing understanding about the many epigenetic mechanisms that strongly influence multifactorial diseases led to a growing interest in exploring how nongenetic variations, including epigenetic factors, influence the etiology of ALI and ARDS [63, 64].

Epigenetic processes play a critical role in transcriptional regulation by influencing genomic regulatory sequences in a cell-type-specific manner, in response to the extracellular environment [65]. The cell epigenome is extremely dynamic, being coordinated by complex mechanisms that interface genetics and environmental factors [66]. Different epigenetic mechanisms regulate gene expression in different ways in the different cell types and during development to orchestrate cellular responses to external stimuli [67]. Epigenetic information in mammals can be transmitted in a variety of ways, including DNA methylation (DNAm), posttranslational modifications of histones, and noncoding RNAs (ncRNAs) [68]. DNA methylation plays an important role in DNA repair, recombination, and replication, as well as regulation of gene activity, while the histones are highly conserved proteins that can become posttranslationally modified and play an important role in the organization of chromatin structure, particularly in the homeostasis between euchromatin, transcriptionally active, and heterochromatin, transcriptionally inactive (reviewed elsewhere [69, 70]). Multiple epigenetic modifications have been evaluated as markers of ALI. However, their role in driving the disease is yet not well understood.

Zhang et al. [71] analyzed microarray data of human blood samples from ARDS patients and healthy volunteers and found significant changes in 44,439 DNA methylation sites. The regions that showed differences in the DNA methylation comprised genes involved in the regulation of inflammation or immunity, endothelial function, epithelial function, and coagulation function, including the chemokine receptor CX3CR1 and the tyrosine kinase Fyn in both ARDS animal models and in vitro experiments [71]. Another study conducted an epigenome-wide association study (EWAS) between DNA methylation and 28-day survival time in 185 moderate-to-severe ARDS patients from intensive care units, and the results showed four CpG sites that were significantly associated with ARDS survival, with two sites related to inflammation and infection, on the genes encoding prostaglandin D2 receptor, and an integral membrane P4 ATPase [72], which has been suggested to act as a lipid flippase essential to inhibit the response to LPS [73]. Reduced endothelial histone acetylation during ALI can lead to an increased expression of adhesion molecules and regulate endothelial permeability [74].

Most of the human studies with epigenetic changes and lung disease are concentrated at chronic obstructive pulmonary disease (COPD). Recently, one study observed that a genetic variant in 19q13.2 region identified in genome-wide association studies of COPD is associated with differential DNA methylation in blood as well as gene expression in blood and the lungs [75]. Another study that evaluated the COPD in two family-based cohorts found 330 genes associated with 349 differentially methylated CpGs. Further analysis of these genes identified an enrichment in genes modulating the immune response, inflammatory pathways, and coagulation cascade [76]. However, another study aimed at identifying cell-free DNA methylation biomarkers to discriminate between overlapping and multifactorial lung diseases, such as cancer, ILD, and COPD, showed a moderate sensibility and specificity, suggesting a need for improved molecular techniques [77].

Specific therapies do not currently exist for the treatment of ALI or for the alleviation of the unremitting vascular leak, which serves as a defining feature of these illnesses. Given the possible links between DNA methylation and disease severity, a better understanding of the effect of DNA methyltransferase inhibitor (DNMTi) such as 5-Aza-2 deoxycytidine (Aza) or HDAC inhibitors such as trichostatin A (TSA) and sodium butyrate (SB) may be useful for studies in humans in the near future. Understanding these new mechanistic insights can lead to novel strategies, biomarkers, and therapies to reduce the morbidity and mortality of these acute and subacute inflammatory injuries [78]. Aza incorporates into DNA CpG sites opposite to a methylated CpG site, thereby inhibiting the DNA methyltransferase enzyme action in the DNA. This inhibition causes loss of DNA methylation in one daughter DNA strand because DNA methyltransferase is not available to remethylate the hemimethylated sites generated during the first round of DNA replication [79]. Also, the Aza and TSA have well-established biological activities and known safety and side-effect profiles [80, 81]. Evidence suggests that members of the HDAC family play a critical role in the development of ALI, and thus, manipulation of HDAC signaling pathways is a strategy that would provide novel therapeutic strategies for the protection of endothelial barrier function and for the intervention of the inflammatory profile in ALI [23].

Zhang et al. investigated the anti-inflammatory effects of HDAC inhibitors TSA and SB by assessing the degree of lung injury and the expression of inflammation-related genes after cecal ligation and puncture, a common animal model of sepsis. The results of the study showed that TSA and SB significantly alleviated septic lung injury, as evidenced by reduced lung wet to dry weight ratio, and attenuated pathohistological lesions. Moreover, treatment with TSA or SB significantly prolonged the survival time of CLP mice, providing evidence of the anti-inflammatory effect of HDAC inhibitors in sepsis [82]. It is important to highlight that TSA is a chemotherapeutic agent that induces differentiation and promotes apoptosis in transformed cells at higher concentrations [83] and also suppresses the production of inflammatory cytokines in animal models of autoimmune and inflammatory disease [84]. TSA also showed a reduction of the inflammatory cytokines MMP-9 and PAI-1, collagen gene expression, and total collagen production in mice exposed to bleomycin treatment, a process that required at least in part suppression of HDAC4 and Akt pathways [85]. Further support for these inhibitors was provided by studies evaluating the response of bone marrow-derived macrophage (BMDM) from endotoxin-treated mice treated with the same HDAC inhibitor, TSA in combination with Aza. This combination was more effective than either inhibitor alone to reduce pyroptosis and apoptosis [86]. Moreover, they observed significant changes in the mitochondrial membrane structure, lower levels of DNA fragmentation, and reduced expression of apoptotic and pyroptotic genes in LPS-induced BMDMs treated by a combination of Aza and TSA than in LPS-induced BMDMs treated with either drug alone. Consistently, inhibition of HDAC using panobinostat (Pan) and TSA showed

a protection against LPS-mediated endothelial barrier dysfunction in HLMVEC [87].

Thangavel et al. showed that a single dose of Aza+TSA in a mouse model of ALI prevented lung vascular hyperpermeability and inflammatory lung injury and promoted survival rate that compared with treatment with either Aza or TSA alone [88]. LPS-induced infiltration of neutrophils and generation of inflammatory cytokines in lung tissues were also significantly reduced in Aza+TSA-treated mice. The same treatment significantly reduced LPS-induced apoptosis of lung endothelial cells. Mechanistically, Aza+TSA treatment suppressed phosphorylation of MLC2 and eNOS and activated Cav1. The above effects are attributed to Aza+TSA-mediated epigenetic modulation of acetylated and methylated histone 3 protein of the VE-cadherin promoter [88]. Similarly, Aza+TSA treatment of endotoxemic mice induced a putative anti-inflammatory process, including lower levels of pro-inflammatory cytokines IL1β, TNFα, and IL-6 and higher levels of IL-10 that correlated with an increased M2/M1 macrophage ratio [89]. Mechanistic studies demonstrated that BMDM exposed to LPS showed a significant increase in phosphorylated p38MAPK and reduction in phosphorylated STAT3, which was ameliorated following treatment with Aza+TSA [89]. Multiple cell types may be mediating this response: Aza treatment promoted inflammation resolution in mouse models of lung injury, through a mechanism that involved an increase activity of regulatory T cells via an upregulation of FoxP3 [90].

Valproic acid (VPA) is another HDAC inhibitor [91] that may be effective in reducing lung injury. VPA treatment reduced the damage to the lungs in a rat hemorrhagic shock model [92]. Similarly, the lungs injured following an ischemia/reperfusion model displayed reduced histone H3 acetylation, and treatment with VPA significantly attenuated all the parameters of lung injury, oxidative stress, apoptosis, and inflammation [93].

Studies using intratracheal LPS instillation in mice as well as human lung microvascular ECs suggested a role for sphingosine-1-phosphate (S1P) and SphK1 signaling in histone acetylation and chromatin modification, since inhibition of S1P lyase attenuated LPS-induced histone acetylation and secretion of pro-inflammatory cytokines [78]. S1P, S1P receptors (S1PRs), and enzymes of S1P metabolism have been identified as key modulators of several human pathologies including pulmonary diseases [94]. A protective role of S1P in LPS-induced lung injury has been proposed [95, 96]. Intravenously administered S1P or S1P analogs reduced vascular leak and pulmonary edema in murine and canine models of sepsis-induced lung injury [97]. In vitro studies showed that LPS enhanced global acetylation of histones in human lung microvascular endothelial cells (HLMVEC). Moreover, treatment of HLMECs with a histone acetyltransferase (HAT) inhibitor reduced the LPS-induced release of IL-6 [78]. Consistently, histone modifications may mediate the inhibition of an angiogenic transcriptional profile observed in a murine model of ventilator-associated pneumonia (VAP) [98]. Interestingly, the changes on gene expression for genes encoding components of the Tie2/angiopoietin and VEGF/VEGFR systems were observed not only in the affected lungs but also in the kidneys and livers, suggesting that VAP leads to a systemic change in the epigenetic landscape of these genes [98].

Lung injury promotes important epigenetic changes in lung epithelium, but their effects are still not well understood. Using a murine lung epithelial cell line (MLE-12), Chen et al. demonstrated that changes in histone acetylation leads to altered cell proliferation [99]. Specifically, they found deacetylation of MORF4L1 at Lys-148, mediated by histone deacetylase (HDAC)-2. MORF4 produces massive cell death and senescence, but the biological function of MORF4L1 homodimers is not clear. One possibility is that MORF4L1 homodimers probably exist as an intermediate product for further assembly into larger functional complexes, such as a corepressor complex. MORF4L1 homodimerization augments MORF4L1 corepressor complex formation to repress cell proliferation [99]. These findings highlight a potential mechanism by which inflammatory factors can influence epigenetic

regulation and drive maladaptive changes in the endothelium.

Little effort has been devoted to inhibiting specific acetylases and deacetylases that can contribute to a reduction in pulmonary edema in the LPS-induced ALI. In vitro, pharmacological inhibition of HDAC6 by tubacin in HPAEC blocked the thrombin-induced endothelial barrier dysfunction through increased acetylation of α-tubulin and microtubule stabilization [100]. Consistently, the selective inhibition of HDAC6 by tubastatin A (Tub A) inhibited TNF-a-induced lung endothelial permeability and prevents endotoxin-induced pulmonary edema in HPAECs and HLMVECs [101]. However, HDAC6 is thought to regulate predominantly cytoplasmic proteins, not histones [102]. How this enzyme can modulate cytokine production is still unknown.

Research to unravel the interactions between genetic and epigenetic factors is critical to better understand lung disease. A well-designed study began to define the contribution of specific transcription factors, ARDS-associated SNPs, and promoter demethylation to NAMPT transcriptional regulation in response to mechanical stress. As described above, NAMPT has been found to be a novel biomarker of lung injury [54], and single-nucleotide polymorphisms in a GC sequence of NAMPT were associated with increased risk of ALI in a Caucasian population [54]. Further, pulmonary epithelial-cell-specific knockdown of NAMPT gene expression significantly attenuated LPS-induced mouse lung inflammation. Inhibition of NAMPT upregulated surfactant protein B (SFTPB) expression by enhancing histone acetylation to increase its transcription, whereas overexpression of NAMPT inhibited SFTPB expression in both H441 and A549 cells [103]. Follow-up research, using a mouse model of VILI, found NAMPT promoter demethylation on putative PAX5 binding sites. CHIP data from human pulmonary artery endothelial cells (HPAEC) demonstrated increased STAT5 binding to this promoter region in response to demethylation [104].

In summary, the studies using DNMTi and HDAC inhibitors strongly support the notion that modulation of epigenetic mechanisms is a promising therapeutic strategy to reduce the inflammatory response during ARDS and ALI. However, the mechanisms that mediate this response are not well understood, and more studies are necessary to understand how these drugs affect the inflammatory process and whether their actions directly impact the sustained vascular barrier breakdown as seen in lung injury.

6.3 Conclusion

Key aspects of the pathophysiology of ALI/ARDS involve endothelial dysfunction driven by a strong inflammatory milieu. These endothelial changes lead to increased edema, leukocyte infiltration, and thrombosis, inducing lung damage and further aggravating the local and systemic inflammatory response. Mechanical ventilation, usually required to sustain blood oxygenation levels, can induce further damage, leading to a self-feeding process that may lead to multiorgan dysfunction syndrome and death. Following acute inflammation, excessive reactive oxygen species (ROS) can induce damaged pulmonary epithelia to secrete pro-inflammatory and profibrotic cytokines that lead to imbalances between histone acetylation and deacetylation [85, 105].

Despite an ever-increasing understanding of the inflammatory mechanisms orchestrating this response, there are very few therapeutical advances to improve the survival odds. Here, we summarized the evidence accumulated over the last 15–20 years that strongly point to key roles for epigenetic mechanisms mediating vasculopathy. Use of enzyme inhibitors such as DNMTi and HDACi may offer new alternative strategies to prevent or reverse the vascular damage that occurs during lung injury. The use of HDAC inhibitors is associated with anti-inflammatory processes and activates an array of key anti-inflammatory signaling pathways not only in ALI but also in other inflammatory lung diseases including COPD and asthma [23, 106]. HDAC inhibition already showed a decrease in the severity of disease in several animal models of inflammation [107] as well as in in vitro models [87, 100]. Changes in histone acetylation modify chromatin structure and the activity of transcription factors that serve as an important mechanism for the regulation of gene expression [108], and with the advance of the knowledge of the molecular basis of the inflammation and pathways, new opportunities linked with the HDAC inhibitors surge as new treatment possibilities. HDAC inhibitors can act as effective anti-inflammatory or antifibrotic drugs by changing histone acetylation or suppressing the activity of multiple transcription factors [109–111]. Future work is required to fully understand how epigenetic mechanisms affect the vascular dysfunction in lung injury, and early-phase clinical trials will provide data to analyze the safety and efficacy of this strategy for the treatment of ALI and ARDS.

Acknowledgments Supported in part by a grant R01GM124133 from the NIH/NIGMS to APA.

References

1. Johnson ER, Matthay MA. Acute lung injury: epidemiology, pathogenesis, and treatment. J Aerosol Med Pulm Drug Deliv. 2010;23:243–52. https://doi.org/10.1089/jamp.2009.0775.
2. Matthay MA, et al. Acute respiratory distress syndrome. Nat Rev Dis Primers. 2019;5:18. https://doi.org/10.1038/s41572-019-0069-0.
3. Aziz M, Jacob A, Yang WL, Matsuda A, Wang P. Current trends in inflammatory and immunomodulatory mediators in sepsis. J Leukoc Biol. 2013;93:329–42. https://doi.org/10.1189/jlb.0912437.
4. Maca J, et al. Past and present ARDS mortality rates: a systematic review. Respir Care. 2017;62:113–22. https://doi.org/10.4187/respcare.04716.
5. Maniatis NA, Kotanidou A, Catravas JD, Orfanos SE. Endothelial pathomechanisms in acute lung injury. Vasc Pharmacol. 2008;49:119–33. https://doi.org/10.1016/j.vph.2008.06.009.
6. Herrero R, Sanchez G, Lorente JA. New insights into the mechanisms of pulmonary edema in acute lung injury. Ann Transl Med. 2018;6:32. https://doi.org/10.21037/atm.2017.12.18.
7. MacLaren R, Stringer KA. Emerging role of anticoagulants and fibrinolytics in the treatment of acute respiratory distress syndrome. Pharmacotherapy. 2007;27:860–73. https://doi.org/10.1592/phco.27.6.860.

8. Ashbaugh DG, Bigelow DB, Petty TL, Levine BE, Ashbaugh DG, Bigelow DB, Petty TL, Levine BE. Acute respiratory distress in adults. The Lancet, Saturday 12 August 1967. Crit Care Resusc. 2005;7:60–1.

9. Ware LB, Matthay MA. The acute respiratory distress syndrome. N Engl J Med. 2000;342:1334–49. https://doi.org/10.1056/NEJM200005043421806.

10. Hendrickson CM, Matthay MA. Endothelial biomarkers in human sepsis: pathogenesis and prognosis for ARDS. Pulm Circ. 2018;8:2045894018769876. https://doi.org/10.1177/2045894018769876.

11. Torres Acosta MA, Singer BD. Pathogenesis of COVID-19-induced ARDS: implications for an aging population. Eur Respir J. 2020. https://doi.org/10.1183/13993003.02049-2020.

12. Gupta A, et al. Extrapulmonary manifestations of COVID-19. Nat Med. 2020;26:1017–32. https://doi.org/10.1038/s41591-020-0968-3.

13. Fan E, et al. COVID-19-associated acute respiratory distress syndrome: is a different approach to management warranted? Lancet Respir Med. 2020;8:816–21. https://doi.org/10.1016/S2213-2600(20)30304-0.

14. Campos R, et al. N-acetylcysteine prevents pulmonary edema and acute kidney injury in rats with sepsis submitted to mechanical ventilation. Am J Physiol Lung Cell Mol Physiol. 2012;302:L640–50. https://doi.org/10.1152/ajplung.00097.2011.

15. Davidovich N, et al. Cyclic stretch-induced oxidative stress increases pulmonary alveolar epithelial permeability. Am J Respir Cell Mol Biol. 2013;49:156–64. https://doi.org/10.1165/rcmb.2012-0252OC.

16. Chatterjee S. Endothelial mechanotransduction, redox signaling and the regulation of vascular inflammatory pathways. Front Physiol. 2018;9:524. https://doi.org/10.3389/fphys.2018.00524.

17. Birukov K, Small G. GTPases in mechanosensitive regulation of endothelial barrier. Microvasc Res. 2009;77:46–52. https://doi.org/10.1016/j.mvr.2008.09.006.

18. Bazzoni G, Dejana E. Endothelial cell-to-cell junctions: molecular organization and role in vascular homeostasis. Physiol Rev. 2004;84:869–901. https://doi.org/10.1152/physrev.00035.2003.

19. Mehta D, Malik AB. Signaling mechanisms regulating endothelial permeability. Physiol Rev. 2006;86:279–367.

20. Adam AP. Regulation of endothelial adherens junctions by tyrosine phosphorylation. Mediat Inflamm. 2015;2015:272858. https://doi.org/10.1155/2015/272858.

21. Shapiro NI, et al. The association of endothelial cell signaling, severity of illness, and organ dysfunction in sepsis. Crit Care. 2010;14:R182. https://doi.org/10.1186/cc9290.

22. Huertas A, et al. Pulmonary vascular endothelium: the orchestra conductor in respiratory diseases. Eur Respir J. 2018;51:1700745. https://doi.org/10.1183/13993003.00745-2017.

23. Kovacs L, et al. Histone deacetylases in vascular permeability and remodeling associated with acute lung injury. Vessel Plus. 2018;2:15. https://doi.org/10.20517/2574-1209.2018.06.

24. Matthay MA, Zimmerman GA. Acute lung injury and the acute respiratory distress syndrome: four decades of inquiry into pathogenesis and rational management. Am J Respir Cell Mol Biol. 2005;33:319–27. https://doi.org/10.1165/rcmb.F305.

25. Parsons PE, et al. Lower tidal volume ventilation and plasma cytokine markers of inflammation in patients with acute lung injury. Crit Care Med. 2005;33:1–6. https://doi.org/10.1097/01.ccm.0000149854.61192.dc.

26. Vandenbroucke E, Mehta D, Minshall R, Malik AB. Regulation of endothelial junctional permeability. Ann N Y Acad Sci. 2008;1123:134–45.

27. Trani M, Dejana E. New insights in the control of vascular permeability: vascular endothelial-cadherin and other players. Curr Opin Hematol. 2015;22:267–72. https://doi.org/10.1097/MOH.0000000000000137.

28. Kuppers V, Vockel M, Nottebaum AF, Vestweber D. Phosphatases and kinases as regulators of the endothelial barrier function. Cell Tissue Res. 2014;355:577–86. https://doi.org/10.1007/s00441-014-1812-1.

29. Komarova YA, Kruse K, Mehta D, Malik AB. Protein interactions at endothelial junctions and signaling mechanisms regulating endothelial permeability. Circ Res. 2017;120:179–206. https://doi.org/10.1161/CIRCRESAHA.116.306534.

30. Kása A, Csortos C, Verin AD. Cytoskeletal mechanisms regulating vascular endothelial barrier function in response to acute lung injury. Tissue Barriers. 2015;3:e974448. https://doi.org/10.4161/21688370.2014.974448.

31. Huertas A, et al. Pulmonary vascular endothelium: the orchestra conductor in respiratory diseases: highlights from basic research to therapy. Eur Respir J. 2018;51:1700745. https://doi.org/10.1183/13993003.00745-2017.

32. Lucas R, Verin AD, Black SM, Catravas JD. Regulators of endothelial and epithelial barrier integrity and function in acute lung injury. Biochem Pharmacol. 2009;77:1763–72. https://doi.org/10.1016/j.bcp.2009.01.014.

33. Millar FR, Summers C, Griffiths MJ, Toshner MR, Proudfoot AG. The pulmonary endothelium in acute respiratory distress syndrome: insights and therapeutic opportunities. Thorax. 2016;71:462–73. https://doi.org/10.1136/thoraxjnl-2015-207461.

34. Teijaro JR, et al. Endothelial cells are central orchestrators of cytokine amplification during influenza virus infection. Cell. 2011;146:980–91. https://doi.org/10.1016/j.cell.2011.08.015.

35. Orfanos SE, Mavrommati I, Korovesi I, Roussos C. Pulmonary endothelium in acute lung injury: from basic science to the critically ill. Intensive Care

Med. 2004;30:1702–14. https://doi.org/10.1007/s00134-004-2370-x.

36. Kosmidou I, Karmpaliotis D, Kirtane AJ, Barron HV, Gibson CM. Vascular endothelial growth factors in pulmonary edema: an update. J Thromb Thrombolysis. 2008;25:259–64. https://doi.org/10.1007/s11239-007-0062-4.

37. Frantzeskaki F, Armaganidis A, Orfanos SE. Immunothrombosis in acute respiratory distress syndrome: cross talks between inflammation and coagulation. Respiration. 2017;93:212–25. https://doi.org/10.1159/000453002.

38. van Hinsbergh VW. Endothelium--role in regulation of coagulation and inflammation. Semin Immunopathol. 2012;34:93–106. https://doi.org/10.1007/s00281-011-0285-5.

39. Pober JS, Sessa WC. Evolving functions of endothelial cells in inflammation. Nat Rev Immunol. 2007;7:803–15. https://doi.org/10.1038/nri2171.

40. Levi M, van der Poll T, ten Cate H. Tissue factor in infection and severe inflammation. Semin Thromb Hemost. 2006;32:33–9. https://doi.org/10.1055/s-2006-933338.

41. Aras O, et al. Induction of microparticle- and cell-associated intravascular tissue factor in human endotoxemia. Blood. 2004;103:4545–53. https://doi.org/10.1182/blood-2003-03-0713.

42. Ware LB, Fang X, Matthay MA. Protein C and thrombomodulin in human acute lung injury. Am J Physiol Lung Cell Mol Physiol. 2003;285:L514–21. https://doi.org/10.1152/ajplung.00442.2002.

43. Isshiki T, et al. Recombinant human soluble thrombomodulin treatment for acute exacerbation of idiopathic pulmonary fibrosis: a retrospective study. Respiration. 2015;89:201–7. https://doi.org/10.1159/000369828.

44. Griffin JH, Fernandez JA, Gale AJ, Mosnier LO. Activated protein C. J Thromb Haemost. 2007;5(Suppl 1):73–80. https://doi.org/10.1111/j.1538-7836.2007.02491.x.

45. Stavenuiter F, Bouwens EA, Mosnier LO. Down-regulation of the clotting cascade by the protein C pathway. Hematol Educ. 2013;7:365–74.

46. Finigan JH, et al. Activated protein C protects against ventilator-induced pulmonary capillary leak. Am J Physiol Lung Cell Mol Physiol. 2009;296:L1002–11. https://doi.org/10.1152/ajplung.90555.2008.

47. Ranieri VM, et al. Drotrecogin alfa (activated) in adults with septic shock. N Engl J Med. 2012;366:2055–64. https://doi.org/10.1056/NEJMoa1202290.

48. Reilly JP, Christie JD, Meyer NJ. Fifty years of research in ARDS genomic contributions and opportunities. Am J Respir Crit Care Med. 2017;196:1113–21. https://doi.org/10.1164/rccm.201702-0405CP.

49. Reilly JP, et al. Heterogeneous phenotypes of acute respiratory distress syndrome after major trauma. Ann Am Thorac Soc. 2014;11:728–36. https://doi.org/10.1513/AnnalsATS.201308-280OC.

50. Calfee CS, et al. Distinct molecular phenotypes of direct vs indirect ARDS in single-center and multicenter studies. Chest. 2015;147:1539–48. https://doi.org/10.1378/chest.14-2454.

51. Cho HY, et al. Association of Nrf2 polymorphism haplotypes with acute lung injury phenotypes in inbred strains of mice. Antioxid Redox Signal. 2015;22:325–38. https://doi.org/10.1089/ars.2014.5942.

52. Cho HY, et al. Linkage analysis of susceptibility to hyperoxia – Nrf2 is a candidate gene. Am J Respir Cell Mol Biol. 2002;26:42–51. https://doi.org/10.1165/ajrcmb.26.1.4536.

53. Wang T, et al. Endothelial cell signaling and ventilator-induced lung injury: molecular mechanisms, genomic analyses, and therapeutic targets. Am J Physiol Lung Cell Mol Physiol. 2017;312:L452–76. https://doi.org/10.1152/ajplung.00231.2016.

54. Ye SQ, et al. Pre-B-cell colony-enhancing factor as a potential novel biomarker in acute lung injury. Am J Respir Crit Care Med. 2005;171:361–70. https://doi.org/10.1164/rccm.200404-563OC.

55. Zhang LQ, Heruth DP, Ye SQ. Nicotinamide phosphoribosyltransferase in human diseases. J Bioanal Biomed. 2011;3:13–25. https://doi.org/10.4172/1948-593x.1000038.

56. Ye SQ, et al. Pre-B-cell-colony-enhancing factor is critically involved in thrombin-induced lung endothelial cell barrier dysregulation. Microvasc Res. 2005;70:142–51. https://doi.org/10.1016/j.mvr.2005.08.003.

57. Liu P, et al. Critical role of PBEF expression in pulmonary cell inflammation and permeability. Cell Biol Int. 2009;33:19–30. https://doi.org/10.1016/j.cellbi.2008.10.015.

58. Christie JD, et al. Genome wide association identifies PPFIA1 as a candidate gene for acute lung injury risk following major trauma. PLoS One. 2012;7:e28268. https://doi.org/10.1371/journal.pone.0028268.

59. Serra-Pages C, et al. The LAR transmembrane protein tyrosine phosphatase and a coiled-coil LAR-interacting protein co-localize at focal adhesions. EMBO J. 1995;14:2827–38.

60. Bime C, et al. Genome-wide association study in African Americans with acute respiratory distress syndrome identifies the selectin P ligand gene as a risk factor. Am J Respir Crit Care Med. 2018;197:1421–32. https://doi.org/10.1164/rccm.201705-0961OC.

61. Hernandez-Pacheco N, et al. A vascular endothelial growth factor receptor gene variant is associated with susceptibility to acute respiratory distress syndrome. Intensive Care Med Exp. 2018;6:16. https://doi.org/10.1186/s40635-018-0181-6.

62. Apte RS, Chen DS, Ferrara N. VEGF in signaling and disease: beyond discovery and development. Cell. 2019;176:1248–64. https://doi.org/10.1016/j.cell.2019.01.021.

63. Petronis A. Epigenetics as a unifying principle in the aetiology of complex traits and diseases.

Nature. 2010;465:721–7. https://doi.org/10.1038/nature09230.

64. Feinberg AP, Irizarry RA. Stochastic epigenetic variation as a driving force of development, evolutionary adaptation, and disease. Proc Natl Acad Sci U S A. 2010;107:1757–64. https://doi.org/10.1073/pnas.0906183107.

65. Dawson MA, Kouzarides T. Cancer epigenetics: from mechanism to therapy. Cell. 2012;150:12–27. https://doi.org/10.1016/j.cell.2012.06.013.

66. Bernstein BE, Meissner A, Lander ES. The mammalian epigenome. Cell. 2007;128:669–81. https://doi.org/10.1016/j.cell.2007.01.033.

67. Bierne H, Hamon M, Cossart P. Epigenetics and bacterial infections. Cold Spring Harb Perspect Med. 2012;2:a010272. https://doi.org/10.1101/cshperspect.a010272.

68. Portela A, Esteller M. Epigenetic modifications and human disease. Nat Biotechnol. 2010;28:1057–68. https://doi.org/10.1038/nbt.1685.

69. Chervona Y, Costa M. Histone modifications and cancer: biomarkers of prognosis? Am J Cancer Res. 2012;2:589–97.

70. Buck-Koehntop BA, Defossez PA. On how mammalian transcription factors recognize methylated DNA. Epigenetics. 2013;8:131–7. https://doi.org/10.4161/epi.23632.

71. Zhang S, et al. DNA methylation exploration for ARDS: a multi-omics and multi-microarray interrelated analysis. J Transl Med. 2019;17:345. https://doi.org/10.1186/s12967-019-2090-1.

72. Guo Y, et al. Epigenome-wide association study for 28-day survival of acute respiratory distress syndrome. Intensive Care Med. 2018;44:1182–4. https://doi.org/10.1007/s00134-018-5100-5.

73. van der Mark VA, et al. Phospholipid flippases attenuate LPS-induced TLR4 signaling by mediating endocytic retrieval of toll-like receptor 4. Cell Mol Life Sci. 2017;74:715–30. https://doi.org/10.1007/s00018-016-2360-5.

74. Cross D, Drury R, Hill J, Pollard AJ. Epigenetics in sepsis: understanding its role in endothelial dysfunction, immunosuppression, and potential therapeutics. Front Immunol. 2019;10:1363. https://doi.org/10.3389/fimmu.2019.01363.

75. Nedeljkovic I, et al. COPD GWAS variant at 19q13.2 in relation with DNA methylation and gene expression. Hum Mol Genet. 2018;27:396–405. https://doi.org/10.1093/hmg/ddx390.

76. Qiu W, et al. Variable DNA methylation is associated with chronic obstructive pulmonary disease and lung function. Am J Respir Crit Care Med. 2012;185:373–81. https://doi.org/10.1164/rccm.201108-1382OC.

77. Wielscher M, et al. Diagnostic performance of plasma DNA methylation profiles in lung cancer, pulmonary fibrosis and COPD. EBioMedicine. 2015;2:929–36. https://doi.org/10.1016/j.ebiom.2015.06.025.

78. Ebenezer DL, Fu P, Suryadevara V, Zhao Y, Natarajan V. Epigenetic regulation of pro-inflammatory cytokine secretion by sphingosine 1-phosphate (S1P) in acute lung injury: role of S1P lyase. Adv Biol Regul. 2017;63:156–66. https://doi.org/10.1016/j.jbior.2016.09.007.

79. Christman JK. 5-Azacytidine and 5-aza-2′-deoxycytidine as inhibitors of DNA methylation: mechanistic studies and their implications for cancer therapy. Oncogene. 2002;21:5483–95. https://doi.org/10.1038/sj.onc.1205699.

80. Kaminskas E, Farrell AT, Wang YC, Sridhara R, Pazdur R. FDA drug approval summary: Azacitidine (5-azacytidine, Vidaza((TM))) for injectable suspension. Oncologist. 2005;10:176–82. https://doi.org/10.1634/theoncologist.10-3-176.

81. Avila AM, et al. Trichostatin a increases SMN expression and survival in a mouse model of spinal muscular atrophy. J Clin Investig. 2007;117:659–71. https://doi.org/10.1172/jci29562.

82. Zhang L, Jin SW, Wang CD, Jiang R, Wan JY. Histone deacetylase inhibitors attenuate acute lung injury during cecal ligation and puncture-induced polymicrobial sepsis. World J Surg. 2010;34:1676–83. https://doi.org/10.1007/s00268-010-0493-5.

83. Sassi FD, et al. Inhibitory activities of Trichostatin A in U87 glioblastoma cells and tumorsphere-derived cells. J Mol Neurosci. 2014;54:27–40. https://doi.org/10.1007/s12031-014-0241-7.

84. Blanchard F, Chipoy C. Histone deacetylase inhibitors: new drugs for the treatment of inflammatory diseases? Drug Discov Today. 2005;10:197–204. https://doi.org/10.1016/s1359-6446(04)03309-4.

85. Li LF, et al. Trichostatin A attenuates ventilation-augmented epithelial-mesenchymal transition in mice with bleomycin-induced acute lung injury by suppressing the Akt pathway. PLoS One. 2017;12:e0172571. https://doi.org/10.1371/journal.pone.0172571.

86. Samanta S, et al. DNMT and HDAC inhibitors together abrogate endotoxemia mediated macrophage death by STAT3-JMJD3 signaling. Int J Biochem Cell Biol. 2018;102:117–27. https://doi.org/10.1016/j.biocel.2018.07.002.

87. Joshi AD, et al. Histone deacetylase inhibitors prevent pulmonary endothelial hyperpermeability and acute lung injury by regulating heat shock protein 90 function. Am J Physiol Lung Cell Mol Physiol. 2015;309:L1410–9. https://doi.org/10.1152/ajplung.00180.2015.

88. Thangavel J, et al. Combinatorial therapy with acetylation and methylation modifiers attenuates lung vascular hyperpermeability in endotoxemia-induced mouse inflammatory lung injury. Am J Pathol. 2014;184:2237–49. https://doi.org/10.1016/j.ajpath.2014.05.008.

89. Thangavel J, et al. Epigenetic modifiers reduce inflammation and modulate macrophage phenotype during endotoxemia-induced acute lung injury. J Cell Sci. 2015;128:3094–105. https://doi.org/10.1242/jcs.170258.

90. Singer BD, et al. Regulatory T cell DNA methyltransferase inhibition accelerates resolution

of lung inflammation. Am J Respir Cell Mol Biol. 2015;52:641–52. https://doi.org/10.1165/rcmb.2014-0327OC.

91. Gottlicher M, et al. Valproic acid defines a novel class of HDAC inhibitors inducing differentiation of transformed cells. EMBO J. 2001;20:6969–78. https://doi.org/10.1093/emboj/20.24.6969.

92. Fukudome EY, et al. Pharmacologic resuscitation decreases circulating cytokine-induced neutrophil chemoattractant-1 levels and attenuates hemorrhage-induced acute lung injury. Surgery. 2012;152:254–61. https://doi.org/10.1016/j.surg.2012.03.013.

93. Wu SY, et al. Valproic acid attenuates acute lung injury induced by ischemia-reperfusion in rats. Anesthesiology. 2015;122:1327–37. https://doi.org/10.1097/ALN.0000000000000618.

94. Natarajan V, et al. Sphingosine-1-phosphate, FTY720, and sphingosine-1-phosphate receptors in the pathobiology of acute lung injury. Am J Respir Cell Mol Biol. 2013;49:6–17. https://doi.org/10.1165/rcmb.2012-0411TR.

95. Peng X, et al. Protective effects of sphingosine 1-phosphate in murine endotoxin-induced inflammatory lung injury. Am J Respir Crit Care Med. 2004;169:1245–51. https://doi.org/10.1164/rccm.200309-1258OC.

96. Dudek SM, et al. Pulmonary endothelial cell barrier enhancement by sphingosine 1-phosphate: roles for cortactin and myosin light chain kinase. J Biol Chem. 2004;279:24692–700. https://doi.org/10.1074/jbc.M313969200.

97. McVerry BJ, et al. Sphingosine 1-phosphate reduces vascular leak in murine and canine models of acute lung injury. Am J Respir Crit Care Med. 2004;170:987–93. https://doi.org/10.1164/rccm.200405-684oc.

98. Bomsztyk K, et al. Experimental acute lung injury induces multi-organ epigenetic modifications in key angiogenic genes implicated in sepsis-associated endothelial dysfunction. Crit Care. 2015;19:225. https://doi.org/10.1186/s13054-015-0943-4.

99. Chen Y, et al. Histone deacetylase 2 (HDAC2) protein-dependent deacetylation of mortality factor 4-like 1 (MORF4L1) protein enhances its homodimerization. J Biol Chem. 2014;289:7092–8. https://doi.org/10.1074/jbc.M113.527507.

100. Saito S, et al. Pharmacological inhibition of HDAC6 attenuates endothelial barrier dysfunction induced by thrombin. Biochem Biophys Res Commun. 2011;408:630–4. https://doi.org/10.1016/j.bbrc.2011.04.075.

101. Yu JY, Ma MS, Ma ZS, Fu J. HDAC6 inhibition prevents TNF-alpha-induced caspase 3 activation in lung endothelial cell and maintains cell-cell junctions. Oncotarget. 2016;7:54714–22. https://doi.org/10.18632/oncotarget.10591.

102. Moreno-Gonzalo O, Mayor F, Sanchez-Madrid F. HDAC6 at crossroads of infection and innate immunity. Trends Immunol. 2018;39:591–5. https://doi.org/10.1016/j.it.2018.05.004.

103. Bi G, et al. Up-regulation of SFTPB expression and attenuation of acute lung injury by pulmonary epithelial cell-specific NAMPT knockdown. FASEB J. 2018;32:3583–96. https://doi.org/10.1096/fj.201701059R.

104. Sun XG, et al. The NAMPT promoter is regulated by mechanical stress, signal transducer and activator of transcription 5, and acute respiratory distress syndrome-associated genetic variants. Am J Respir Cell Mol Biol. 2014;51:660–7. https://doi.org/10.1165/rcmb.2014-0117OC.

105. Rahman I, Marwick J, Kirkham P. Redox modulation of chromatin remodeling: impact on histone acetylation and deacetylation, NF-kappaB and pro-inflammatory gene expression. Biochem Pharmacol. 2004;68:1255–67. https://doi.org/10.1016/j.bcp.2004.05.042.

106. Adcock IM, Ito K, Barnes PJ. Histone deacetylation: an important mechanism in inflammatory lung diseases. COPD. 2005;2:445–55. https://doi.org/10.1080/15412550500346683.

107. Shakespear MR, Halili MA, Irvine KM, Fairlie DP, Sweet MJ. Histone deacetylases as regulators of inflammation and immunity. Trends Immunol. 2011;32:335–43. https://doi.org/10.1016/j.it.2011.04.001.

108. Jenuwein T, Allis CD. Translating the histone code. Science (New York, NY). 2001;293:1074–80. https://doi.org/10.1126/science.1063127.

109. Cetinkaya M, et al. Protective effects of valproic acid, a histone deacetylase inhibitor, against hyperoxic lung injury in a neonatal rat model. PLoS One. 2015;10:e0126028. https://doi.org/10.1371/journal.pone.0126028.

110. Sanders YY, et al. Histone deacetylase inhibition promotes fibroblast apoptosis and ameliorates pulmonary fibrosis in mice. Eur Respir J. 2014;43:1448–58. https://doi.org/10.1183/09031936.00095113.

111. Sweet MJ, Shakespear MR, Kamal NA, Fairlie DP. HDAC inhibitors: modulating leukocyte differentiation, survival, proliferation and inflammation. Immunol Cell Biol. 2012;90:14–22. https://doi.org/10.1038/icb.2011.88.

Neurotrophin Regulation and Signaling in Airway Smooth Muscle

7

Benjamin B. Roos, Jacob J. Teske,
Sangeeta Bhallamudi, Christina M. Pabelick,
Venkatachalem Sathish, and Y. S. Prakash

Abstract

Structural and functional aspects of bronchial airways are key throughout life and play critical roles in diseases such as asthma. Asthma involves functional changes such as airway irritability and hyperreactivity, as well as structural changes such as enhanced cellular proliferation of airway smooth muscle (ASM), epithelium, and fibroblasts, and altered extracellular matrix (ECM) and fibrosis, all modulated by factors such as inflammation. There is now increasing recognition that disease maintenance following initial triggers involves a prominent role for resident nonimmune airway cells that secrete growth factors with pleiotropic autocrine and paracrine effects. The family of neurotrophins may be particularly relevant in this regard. Long recognized in the nervous system, classical neurotrophins such as brain-derived neurotrophic factor (BDNF) and nonclassical ligands such as glial-derived neurotrophic factor (GDNF) are now known to be expressed and functional in non-neuronal systems including lung. However, the sources, targets, regulation, and downstream effects are still under investigation. In this chapter, we discuss current state of knowledge and future directions regarding BDNF and GDNF in airway physiology and on pathophysiological contributions in asthma.

Benjamin B. Roos and Jacob J. Teske contributed equally with all other contributors.

B. B. Roos · J. J. Teske
Department of Anesthesiology and Perioperative Medicine, Mayo Clinic, Rochester, MN, USA

S. Bhallamudi · V. Sathish
Department of Pharmaceutical Sciences, North Dakota State University, Fargo, ND, USA

C. M. Pabelick · Y. S. Prakash (✉)
Department of Anesthesiology and Perioperative Medicine, Mayo Clinic, Rochester, MN, USA

Department of Physiology and Biomedical Engineering, Mayo Clinic, Rochester, MN, USA
e-mail: prakash.ys@mayo.edu

Keywords

Brain-derived neurotrophic factor · Glial-derived neurotrophic factor · Calcium · Contractility · Remodeling · Asthma · Lung disease

7.1 Introduction

Throughout life, including fetal growth, bronchial airways are key structural and functional elements of the respiratory system. Diseases of the bronchial airway such as asthma or chronic obstructive pulmonary disease (COPD) result from both intrinsic factors such as developmental abnormalities and environmental exposures including allergens, microbes, pollutants, and tobacco smoke products. These detrimental effects lead to functional changes such as airway irritability and hyperreactivity, as well as to structural changes including enhanced cellular proliferation of airway smooth muscle (ASM), epithelium, and fibroblasts, and altered extracellular matrix (ECM) and airway fibrosis. While environmental and other exposures trigger the initial inflammatory response within the airway leading to initiation of disease, there is now increasing recognition that disease maintenance after initial triggers or inflammatory cells disappear involves resident nonimmune airway cells taking a prominent role. Here, beyond their expected roles in barrier maintenance and airway tone, cells such as epithelium, ASM, and fibroblasts can modulate the local inflammatory milieu, produce pro-fibrotic factors, and enhance cell-cell interactions, overall resulting in maintenance of cellular dysfunction that can then be exacerbated by intermittent stimuli such as additional allergens, infection, or inflammation. Thus, understanding the mechanisms that initiate vs. maintain airway disease becomes important for developing targeted therapeutics for asthma and COPD, particularly in the chronic state. Here, it is imperative to determine whether mechanisms common to different cell types or airway diseases can be identified, allowing for development of a wide range of novel therapeutic approaches.

An emerging aspect of resident airway cell function is expression and function of growth factors that can have autocrine or paracrine effects in airway structure and function in both health and disease. In this regard, the family of neurotrophins may be particularly relevant. Long recognized in the brain in the context of neuromodulation, neuronal plasticity, growth, and particularly regeneration, and in the pathophysiology of diseases such as depression and Alzheimer's [1–3], there has been increasing interest and recognition regarding neurotrophins in non-neuronal systems including lung [4–14]. In this regard, from both neuronal and non-neuronal work, it is clear that while neurotrophins are a broad concept, there is substantial heterogeneity in their

expression and function which is key towards understanding their roles in health and disease. While information is less, this is also true in the lung (or airways). Most of the work, including our own, has focused on specific neurotrophin family members such as nerve growth factor (NGF) and brain-derived neurotrophic factor (BDNF) [5, 6, 11, 15–18], and even here, the cell sources vs. targets, and importantly physiological or pathophysiological relevance, are still under exploration. There is now even more limited but emerging information regarding another aspect, the glial-derived neurotrophic factor (GDNF) family of neurotrophins [19–28]. In this chapter, we discuss neurotrophins in airway physiology and focus on pathophysiological contributions in asthma, expanding on BDNF and the emerging GDNF family.

7.2 Mechanisms of Neurotrophin Expression

Understandably, much of our knowledge regarding neurotrophin expression or signaling derives from exploration in the nervous system, starting with studies in the 1950s (e.g., Levi-Montalcini, Hamburger, Hempstead, etc.) on neural regeneration in the embryonic limb [29–31]. Here, it is important to delineate what qualifies as neurotrophin. While several factors can act as growth factors in influencing neuronal structure and function, there are classical and nonclassical neurotrophins. The classical neurotrophins are considered to consist of four polypeptides of comparable structure and function: NGF, the original, best-characterized member, BDNF, neurotrophin-3 (NT3), and neurotrophin-4 (NT4) [31–33]. Separately, the nonclassical GDNF family consists of GDNF, neurturin, artemin, and persephin [34–37]. While their signaling pathways differ substantially (described below), neurotrophins do not function in a vacuum, and it is now increasingly clear that their expression and signaling are intricately tied to several regulatory pathways including other neurotrophins, sex steroids, glucocorticoids, and particularly inflammation [38–40]. Of course, in the context of asthma, these pathways are quite relevant [14, 41–43]. Accordingly, understanding how neurotrophin expression and signaling occur is insightful towards recognizing their role in airway diseases.

7.2.1 BDNF Production

The BDNF gene has at least nine promoters allowing for multiple mRNA transcripts each with the full ORF for BDNF protein [44–46]. Via alternative promoters, splicing, and poly(A) sites, at least 22 transcripts can be generated. Such transcriptional complexity can result in complex regulation of BDNF generation and its intracellular localization, transport, and extracellular release. For example, cellular stimulation leading to intracellular calcium ($[Ca^{2+}]_i$) enhancement can induce BDNF in the context of activity-dependent regulation [46] involving Ca^{2+}-responsive elements in regulatory regions of several BDNF exons via cAMP response element binding protein (CREB) and protein kinase A (PKA), NFκB, and NFAT. In addition to transcriptional regulation, at least in neuronal cells, epigenetic mechanisms such as histone acetylation/methylation and DNA methylation also regulate BDNF [44, 46] in a complex, context-dependent fashion, affecting the different BDNF exons differentially.

The elements of BDNF regulation (Fig. 7.1) happen to be important in the airway as well, particularly in the context of inflammation. However, unlike in neurons, there is currently limited data on the mechanisms by which BDNF transcription occurs in the airway, with no data on epigenetic regulation. We previously explored BDNF production in human ASM in the context of inflammation (TNFα) effects, focusing on specific $[Ca^{2+}]_i$ regulation- and inflammation-related signaling cascades [16]. We found that TNFα enhances BDNF via the Ca^{2+} regulatory channels transient receptor potential channels TRPC3 and TRPC6 (but not TRPC1) and signaling intermediates such as ERK 1/2, PI3K, PLC, and PKC, Rho kinase, CREB, and NFAT. These elements, albeit to different extents, are increased in asthmatic ASM and with TNFα exposure and contribute to greater BDNF expression in diseased tissues [16]. Such local regulation and upregulation of neurotrophins can thus allow for potential downstream influences on parameters such as contractility and remodeling in the context of asthma.

BDNF protein synthesis has again been largely examined in neurons (Fig. 7.1) and to a limited extent in ASM. Classically endoplasmic reticulum synthesis of BDNF occurs as a precursor protein (pre-pro-BDNF; ~27 kDa) [47–49] which is then cleaved of its signal peptide to produce pro-BDNF that is transported to the Golgi to be sorted into constitutive or regulated secretory vesicles. It is vesicular pro-BDNF that is detected intracellularly, but it can be converted into mature BDNF by endoproteases (e.g., furin) or intragranule by convertases [50], again providing another level of regulation for BDNF expression. Vesicles can thus contain both pro-BDNF and mature BDNF, which are then released into the extracellular space. Regulated release involves SNARE protein complexes as in neurotransmission [51] and requires Ca^{2+} and cAMP. Upon extracellular release, pro-BDNF is cleaved by plasmin or importantly matrix metalloproteinases (MMPs) into mature BDNF.

7.2.2 GDNF Production

GDNF, neurturin, persephin, and artemin all have low amino acid sequence homology, and they all function as via the Ret receptor along with ligand-specific GDNF family receptor (GFR) isoforms (see below). Although only slightly similar in sequence, GDNF is related to TGF-β2 and is a distant member of the TGFβ superfamily [52]. Originally isolated from cultured rat glial cells, GDNF was recognized to enhance survival and differentiation of dopaminergic neurons [52]. Since then, GDNF has been detected in multiple mammalian cell types [53] and is widely distributed in both central and peripheral nervous systems. Several neuronal cells such as glia, astrocytes, oligodendrocytes, and motor neurons can synthesize and secrete GDNF.

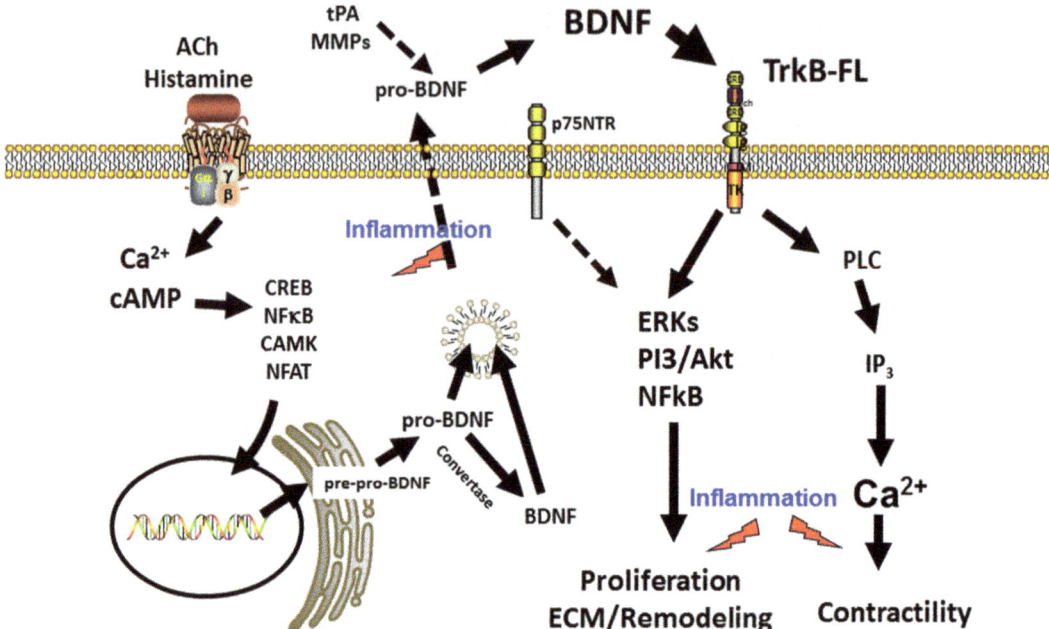

Fig. 7.1 Production and downstream effects of brain-derived neurotrophic factor (BDNF) in airway smooth muscle. Factors such as Ca²⁺ and signaling intermediates (relevant to asthma) stimulate production of pre-pro-BDNF that is cleaved into pro-BDNF that is packaged into vesicles and secreted. Pro-BDNF can also be intracellularly cleaved via convertases (e.g., furin) into mature BDNF that is also packaged into vesicles, and thus both pro- and mature BDNF are released into the extracellular space. Extracellular pro-BDNF can be cleaved to mature BDNF via proteases such as tissue plasminogen activator or matrix metalloproteinases. BDNF acts on plasma membrane high-affinity full-length tropomyosin-related kinase TrkB-FL receptors or low-affinity p75NTR receptors. In ASM, TrkB-FL is of functional importance. Downstream, TrkB increases [Ca²⁺]ᵢ via the PLC/IP₃ pathway and increases ASM proliferation and extracellular matrix (ECM) production via different signaling intermediates also relevant to inflammation. Thus, ASM-derived BDNF can interact with inflammatory signaling in the context of asthma

As a secretory protein (Fig. 7.2), GDNF is first formed as a 211-amino acid precursor that is trafficked to the endoplasmic reticulum for secretion when the protein folds with disulfide bonds and dimerizes, later being modified by N-linked glycosylation and finally proteolytic processing into a mature form of 134 amino acids [52, 54]. Thus, GDNF differs from BDNF in being first created as pro-form rather than a pre-pro form. However, as with BDNF, proteases such as furin and protein convertases are involved in cleavage of pro-GDNF to mature GDNF [55]. Additionally, GDNF can be secreted with or without proteases [56]. Nonetheless, glycosylation appears to be a key step for proper folding and processing of GDNF protein [54]. Studies have also found a shorter GDNF mRNA transcript in humans and rodents [57, 58], although it is unclear if the resultant protein is functional. With alternative splicing, two mRNAs, a full-length pre-(α)pro-GDNF and a shorter pre-(β)pro-GDNF, are created which can both be cleaved to mature GDNF [59, 60]. Interestingly, the two pro-forms of GDNF are secreted into the extracellular space using different regulatory pathways [55] where cellular depolarization promotes (β)pro-GDNF and mature GDNF secretion, while (α)pro-GDNF and its corresponding mature GDNF are mostly localized in the Golgi where they colocalize with secretory granules. In contrast, (β)pro-GDNF and its mature form are localized in other types of vesicles and are released much more rapidly via the secretory pathway [55, 58].

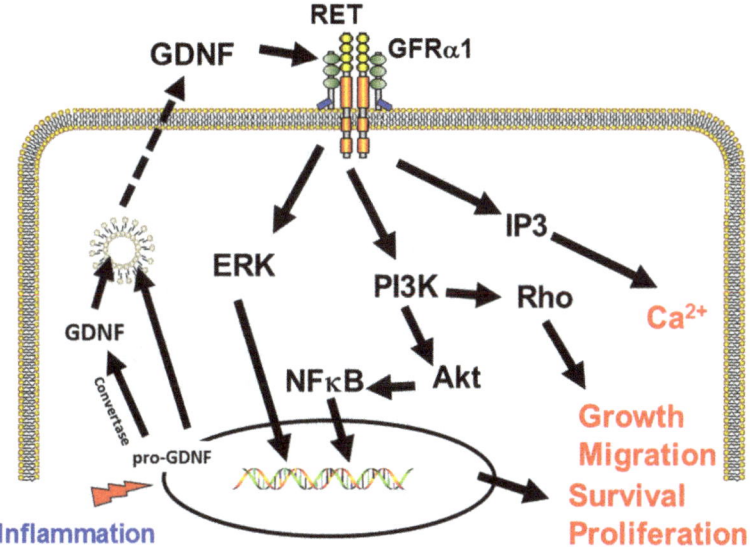

Fig. 7.2 Glial-derived neurotrophic factor (GDNF). While there is minimal data on GDNF in airways or ASM, information from other organ systems including brain and gut show that GDNF is first generated in a pro-form that is intracellularly cleaved into mature GDNF that can be packaged into vesicles and secreted. Extracellularly, pro-GDNF can be cleaved. GDNF acts via a heterodimer of the RET receptor and the GDNF family receptor alpha isoform 1 (GFRα1). Similar to BDNF, GDNF induces pathways that promote $[Ca^{2+}]_i$ as well as cell proliferation and fibrosis: elements that might be important in asthma

7.3 Mechanisms of Neurotrophin Signaling

There is substantial heterogeneity in neurotrophin signaling within the classical family as well as in the nonclassical GDNF family with the potential for interactions between neurotrophins, and cross-reactivity of ligands across different receptors, overall making for extremely complex systems that are likely cell type and context dependent. Nonetheless, again via work in neurons, the basics of neurotrophin signaling are now recognized.

7.3.1 BDNF Signaling

Interestingly, both pro- and mature BDNF are active ligands. They signal through two distinct, major receptor types: high-affinity tropomyosin-related kinase (Trk) receptors (~140 kDa), specifically TrkB which binds to BDNF and also to NT4 [61–63], and the low-affinity "pan-neurotrophin" receptor p75NTR that is a member of the TNF receptor superfamily. Other neurotrophins bind other Trks: NGF to TrkA and NT3 to TrkC [30, 31, 64].

The TrkB gene on chromosome 9q22 contains 24 exons [65, 66] which are responsible for four major isoforms, but 36 alternatively spliced isoforms. Among these, the full-length transmembrane TrkB receptor (TrkB-FL) is particularly important. Cellular responses to BDNF (Fig. 7.1) require expression of TrkB-FL which also contains an intracellular tyrosine kinase domain essential for downstream functionality [65–67]. Following ligand binding, TrkB-FL is autophosphorylated, engaging Shc, GRB2, ATP, and importantly PLCγ. Downstream, three tyrosine kinase pathways are activated [66–70]: PLCγ1/PKC important for the IP_3 pathway and thus $[Ca^{2+}]_i$ release, the ERK pathway, and PI3K/Akt, the later promoting cell survival and proliferation. There is also a truncated TrkB-T1 isoform, but since it lacks the tyrosine kinase domain, its role in downstream activation is not clear and thus may be more of a dominant negative receptor or serve to sequester BDNF limiting its signaling [71–74]. In terms of the low-affinity p75NTR receptor, upon binding either pro-BDNF or mature BDNF, a variety of intracellular pathways can be modulated including PI3KAkt, NFκB, MAPK, RhoA, and even PKA and HIF, all potentially relevant to airways in the context of asthma and inflammation [14, 41, 75–81]. Thus, in any cell type, the relative levels of pro- vs. mature BDNF and TrkB-FL vs. TrkB-T1 vs. p75NTR influence the overall effects of BDNF. These effects generally tend to be genomic in nature. However, there is also data that BDNF has rapid, "non-genomic" effects that occur in the timeframe of seconds to minutes via TrkB [82–85] which is important in the context of Ca^{2+} regulation and neurotransmission or neuromodulation [82–85].

7.3.2   GDNF Signaling

The Ret receptor is critical to signaling of all GDNF family member ligands. GDNF and other ligands use Ret for signaling, but these ligands cannot activate Ret alone. Ret activation requires a co-receptor glycosylphosphatidyli nositol-linked GDNF (GFRα). Similar to the classical neurotrophins such as BDNF, ligand-binding specificity for the GDNF family of ligands is dependent on four GFRα receptor proteins: GFRα1, GFRα2, GFRα3, and GFRα4. GDNF preferentially binds GFRα1 but can act via GFRα2 with lower affinity [37, 86, 87]. GDNF functionality involves a disulfide-stabilized homodimer that first binds to a glycosylphosphatidylinositol (GPI)–anchored GFRα1 (usually within lipid rafts) and forms a high-affinity complex. This complex recruits two Rets resulting in transphosphorylation of tyrosine residues and subsequent intracellular signaling [37, 87]. Downstream pathways (Fig. 7.2) involve MAPK, PI3K, ERK, and Akt that promote cell survival [37, 87, 88]. Further potentiation of GDNF effects can occur through membrane microdomains enriched with Src family kinases, making Src a major signaling molecule for GDNF [89]. GDNF signaling can also occur

without Ret, i.e., via GFRα alone [90], but involves transmembrane proteins such as NCAM [91] that activate Src-like kinase Fyn and focal adhesion kinase FAK, a non-receptor tyrosine kinase.

7.4   "Local" Neurotrophins in the Airway

7.4.1   Neurotrophins in Airway Nerves

Given that neurotrophins promote neuronal structure and function, it is reasonable to assume that innervated organs such as skin, skeletal muscles, epithelia, and smooth muscles express neurotrophins including BDNF [36, 92]. Indeed, neurotrophins have been found to contribute to neuronal plasticity in airways (recently reviewed in [93]). For example, in the context of asthma, there is a role for NGF, mediated through inflammatory cells, where NGF can inhibit eosinophil apoptosis and conversely NGF release from eosinophils increases [93]. In mouse models of asthma, anti-NGF antibody can reduce airway irritability while NGF overexpression is exacerbating [93, 94]. Overexpression of NGF results in increased density of tachykinin-expressing sensory fibers in the airway [95] resulting in increased irritability [96]. NGF can increase substance P in airway nodose neurons [97, 98], induce remodeling of airway parasympathetic ganglia, and enhance dendritic sprouting in parasympathetic neurons.

In spite of these data regarding NGF in airway neuronal plasticity, in terms of asthma pathogenesis, it may be BDNF that is of more interest [12, 99] given its expression in the airways (see below) [12]. BDNF treatment induces increased airway contractility in response to electrical field stimulation of airway nerves, whereas BDNF inhibition blunts ovalbumin effects [100]. There are some data that BDNF effects may be primarily via neuronal changes [100, 101], but several other groups (including ours) have reported BDNF effects on inflammation and ASM per se

[12, 13, 16–18, 102–104]. BDNF itself can induce a phenotypic shift of nodose ganglia neurons such that they express TRPV1 channels [23] and promote airway sensitivity.

Compared to BDNF, there is far less information on GDNF in the airways. GDNF is known to be important in development of airway innervation [105]. In the guinea pig, nodose tracheal Aδ fiber neurons express GFRα1 and are thus likely responsive to GDNF [23, 106]. Indeed, GDNF promotes allergen-induced TRPV1 in these neurons [23] and thus could contribute to AHR.

Locally production of BDNF or GDNF can result in autocrine/paracrine effects on non-neuronal cell types as long as the relevant receptors and signaling pathways are present. There is now data that Trks are widely distributed in non-neuronal tissues [5, 7, 11, 107]. BDNF, TrkB, and p75NTR are localized to different lung compartments including resident immune cells, bronchial and alveolar epithelium, smooth muscle, fibroblasts, and endothelium [11, 108, 109]. The relevance of BDNF to airways then lies in the finding that circulating BDNF and local (airway) receptor expression are both increased in asthma, while BDNF is increased in sputum and bronchoalveolar lavages in patients with asthma or those exposed to cigarette smoke [4, 110–113]. These studies highlight the concept of "local" BDNF as a source as well as target.

While one could expect a similar role for local GDNF in the airways, there is currently little to no data in this regard. GDNF is important for early lung innervation [105] with Ret signaling required for neurogenesis in trachea and primary bronchi [114]. Guinea pigs exposed to ovalbumin show elevated epithelial GDNF, while tracheal neurons show increased TRPV1 in response to GDNF, enhancing airway irritability [23]. However, these studies largely underline a neuronal aspect of GDNF, but there may be other airway cell types that are GDNF sources or targets, particularly in the context of AHR and remodeling. GDNF in bronchial biopsies of COPD patients is associated with mucus hypersecretion [21], and Ret fusion genes and Ret function are thought important in non-small cell lung cancer [115–117].

7.4.2 BDNF in ASM

The ASM layer occupies a major area and mass of the bronchial airways, and thus if any, BDNF produced by ASM has the potential to reach physiological levels. Such ASM-derived BDNF (or that from other resident cells of the airway) can act on ASM itself (i.e., autocrine effects) and have physiological effects. There is now evidence that ASM is a significant source of BDNF. Both the ASM within human lungs [109] and isolated human ASM cells [16] have constitutive ASM BDNF expression. BDNF is localized to ASM of larger airways as well as small airways that are involved in bronchial tone. Importantly, human ASM expresses TrkB [18, 103, 118–121] as well as p75NTR [118] and is thus responsive to BDNF, irrespective of the source of the ligand.

We and others have now demonstrated both mRNA and protein for BDNF in human ASM [18, 103, 120], further showing that asthmatic ASM expresses higher BDNF at baseline [16, 18] while pro-inflammatory cytokines such as TNFα and IL-13 can increase BDNF in non-asthmatic ASM [16], overall consistent with the idea of BDNF linking to asthma. In human ASM, elevation of $[Ca^{2+}]_i$ stimulates BDNF secretion [103]. The mechanisms that regulate $[Ca^{2+}]_i$ in ASM can also modulate BDNF secretion with particular roles for Ca^{2+} influx pathways such as TRPC and Orai1 channels [103, 122–124], which are also involved in exocytosis of other proteins [125]. Pathways that regulate BDNF expression such as PLCγ, TRPC6, Rho kinase, PI3K, ERK, and PKC can also modulate BDNF secretion in human ASM [16]. Thus, it is likely that agonist stimulation and resultant $[Ca^{2+}]_i$ increases in the context of bronchoconstriction enhance BDNF secretion in the airway, while factors such as inflammation increase BDNF. Indeed, smooth muscle-specific BDNF knockout mice that undergo a mixed allergen model of asthma show reduced AHR [17].

In ASM, BDNF can have both rapid, non-genomic as well as longer-term genomic effects. Application of BDNF can rapidly enhance $[Ca^{2+}]_i$ in human ASM cells and potentiate responses to agonists such as acetylcholine or histamine [118].

Such effects of BDNF on $[Ca^{2+}]_i$ mediated exclusively through TrkB (not involving p75) are enhanced by cytokines [119] and in asthmatics [16, 18]. As with neurons, rapid BDNF effects in ASM involve effects on Ca^{2+} influx [118], particularly via store-operated Ca^{2+} entry via TRPC3 and Orai1 [120]. Via PLCγ, BDNF also enhances Ca^{2+} release via IP_3 receptor channels of the sarcoplasmic reticulum. Prolonged BDNF exposure also increases expression of $[Ca^{2+}]_i$ regulatory proteins [120] that allow for increased $[Ca^{2+}]_i$ responses. Furthermore, with activation of pathways such as MAPK, PI3K/Akt, and NFκB, BDNF also increases ASM proliferation [15].

In addition to $[Ca^{2+}]_i$ increase and proliferation, BDNF has been shown to also affect airway fibrosis relevant to asthma [78, 126–129] by enhancing extracellular matrix (ECM) production [130]. In this regard, ECM and BDNF are linked in that matrix metalloproteinases (MMPs), particularly MMP-2 and MMP-9, are involved in extracellularly cleaving pro-BDNF [131–133], while MMPs also process ECM proteins (collagens and fibronectin) produced by ASM and are now shown to be modulated by BDNF. In human ASM, BDNF increases MMP-9 secretion [130, 134].

7.4.3 GDNF in ASM

There are currently no data on GDNF in the ASM beyond the understanding that GDNF is necessary for fetal airway development [105]. GDNF is known to be present in gut smooth muscle [135]. In intestinal smooth muscle, cytokines such as TNFα can upregulate GDNF [136] with downstream effects on enteric neurons that are exacerbated by MMP-9 [137]. The urinary bladder and intestine both express GFRα1 [138]. In ongoing studies (Roos, Teske, Bhallamudi, Sathish, Prakash, unpublished observations), we find that human ASM cells and tissues do express both GDNF and GFRα1 with some, albeit variable, expression of Ret. GDNF is released by human ASM indicating the presence of the machinery required for functional GDNF, especially given that the factors necessary for extracellular cleavage of pro-GDNF such as

MMP-9 are known to be expressed also by ASM. Inflammatory factors such as TNFα and TGFβ increase ASM GDNF, GFRα1, and Ret mRNA (albeit to different extents). Importantly, we also find that GDNF is increased in asthmatic ASM and TNFα or TGFβ enhance GDNF in asthmatic ASM, thus linking inflammation, GDNF, and asthma. From a functional perspective, ongoing studies suggest that similar to BDNF, GDNF can also increase $[Ca^{2+}]_i$ in human ASM and promote production of some ECM components such as fibronectin (unpublished observations). Thus, at least preliminarily, it appears GDNF may be important in the asthmatic airway.

7.5   Summary and Conclusions

While signaling via BDNF and to a lesser extent GDNF in the airway has been recognized, a more integrated model for neurotrophin expression and function in the airway, particularly in the context of disease, is relatively underdeveloped. A number of questions regarding neurotrophins in asthma become relevant for future research: (A) Other than ASM, what are the major sources of BDNF or GDNF in the airways, and do their contributions change with disease? (B) What are the primary targets for BDNF vs. GDNF, and by what mechanisms do these neurotrophins exert their effects? Here, given the complexity of signaling by either neurotrophin, understanding cell and context heterogeneity in terms of receptor expression becomes important. Furthermore, a related question becomes what functions BDNF or GDNF are intended to have in specific cell types, e.g., contractility of ASM, neural function, epithelial barrier function, etc. (C) Are there interactions between classical neurotrophins such as BDNF and nonclassical neurotrophins such as GDNF, i.e., can the two mutually enhance (or inhibit) their expression and function. (D) Do BDNF and GDNF play a role across the age spectrum, e.g., how do they contribute to lung development vs. changes in the airway (or in asthma) of the elderly? Ongoing research using in vitro and in vivo approaches ideally with human samples should help address these issues.

Acknowledgments This work was supported by grants from the US National Institutes of Health R01 HL088029 (Prakash), R01 HL142061 (Pabelick, Prakash), and R01 HL146705 (Sathish).

References

1. Allen SJ, Watson JJ, Shoemark DK, Barua NU, Patel NK. GDNF, NGF and BDNF as therapeutic options for neurodegeneration. Pharmacol Ther. 2013;138(2):155–75.
2. Miranda M, Morici JF, Zanoni MB, Bekinschtein P. Brain-derived neurotrophic factor: a key molecule for memory in the healthy and the pathological brain. Front Cell Neurosci. 2019;13:363.
3. Gudasheva TA, Povarnina P, Tarasiuk AV, Seredenin SB. The low molecular weight brain-derived neurotrophic factor mimetics with antidepressant-like activity. Curr Pharm Des. 2019;25(6):729–37.
4. Rochlitzer S, Nassenstein C, Braun A. The contribution of neurotrophins to the pathogenesis of allergic asthma. Biochem Soc Trans. 2006;34(Pt 4):594–9.
5. Lommatzsch M, Braun A, Renz H. Neurotrophins in allergic airway dysfunction: what the mouse model is teaching us. Ann N Y Acad Sci. 2003;992:241–9.
6. Yao Q, Zaidi SI, Haxhiu MA, Martin RJ. Neonatal lung and airway injury: a role for neurotrophins. Semin Perinatol. 2006;30(3):156–62.
7. Hoyle GW. Neurotrophins and lung disease. Cytokine Growth Factor Rev. 2003;14(6):551–8.
8. Jacoby DB. Pathophysiology of airway viral infections. Pulm Pharmacol Ther. 2004;17(6):333–6.
9. Piedimonte G. Contribution of neuroimmune mechanisms to airway inflammation and remodeling during and after respiratory syncytial virus infection. Pediatr Infect Dis J. 2003;22(2 Suppl):S66–74; discussion S-5.
10. Taylor-Clark T, Undem BJ. Transduction mechanisms in airway sensory nerves. J Appl Physiol. 2006;101(3):950–9.
11. Prakash Y, Thompson MA, Meuchel L, Pabelick CM, Mantilla CB, Zaidi S, et al. Neurotrophins in lung health and disease. Expert Rev Respir Med. 2010;4(3):395–411.
12. Prakash YS, Martin RJ. Brain-derived neurotrophic factor in the airways. Pharmacol Ther. 2014;143(1):74–86.
13. Thompson M, Britt RD Jr, Pabelick CM, Prakash YS. Hypoxia and local inflammation in pulmonary artery structure and function. Adv Exp Med Biol. 2017;967:325–34.
14. Prakash YS. Emerging concepts in smooth muscle contributions to airway structure and function: implications for health and disease. Am J Physiol Lung Cell Mol Physiol. 2016;311(6):L1113–L40.
15. Aravamudan B, Thompson M, Pabelick C, Prakash YS. Brain-derived neurotrophic factor induces pro-liferation of human airway smooth muscle cells. J Cell Mol Med. 2012;16(4):812–23.
16. Aravamudan B, Thompson MA, Pabelick CM, Prakash YS. Mechanisms of BDNF regulation in asthmatic airway smooth muscle. Am J Physiol Lung Cell Mol Physiol. 2016;311(2):L270–9.
17. Britt RD Jr, Thompson MA, Wicher SA, Manlove LJ, Roesler A, Fang YH, et al. Smooth muscle brain-derived neurotrophic factor contributes to airway hyperreactivity in a mouse model of allergic asthma. FASEB J. 2019;33(2):3024–34.
18. Freeman MR, Sathish V, Manlove L, Wang S, Britt RD Jr, Thompson MA, et al. Brain-derived neurotrophic factor and airway fibrosis in asthma. Am J Physiol Lung Cell Mol Physiol. 2017;313(2):L360–L70.
19. Kawai K, Takahashi M. Intracellular RET signaling pathways activated by GDNF. Cell Tissue Res. 2020;382(1):113–23.
20. Mulligan LM. GDNF and the RET receptor in cancer: new insights and therapeutic potential. Front Physiol. 2018;9:1873.
21. Dijkstra AE, Boezen HM, van den Berge M, Vonk JM, Hiemstra PS, Barr RG, et al. Dissecting the genetics of chronic mucus hypersecretion in smokers with and without COPD. Eur Respir J. 2015;45(1):60–75.
22. Liu XM, Aras-Lopez R, Martinez L, Tovar JA. Abnormal development of lung innervation in experimental esophageal atresia. Eur J Pediatr Surg. 2012;22(1):67–73.
23. Lieu TM, Myers AC, Meeker S, Undem BJ. TRPV1 induction in airway vagal low-threshold mechanosensory neurons by allergen challenge and neurotrophic factors. Am J Physiol Lung Cell Mol Physiol. 2012;302(9):L941–8.
24. Lieu T, Kollarik M, Myers AC, Undem BJ. Neurotrophin and GDNF family ligand receptor expression in vagal sensory nerve subtypes innervating the adult guinea pig respiratory tract. Am J Physiol Lung Cell Mol Physiol. 2011;300(5):L790–8.
25. Freem LJ, Escot S, Tannahill D, Druckenbrod NR, Thapar N, Burns AJ. The intrinsic innervation of the lung is derived from neural crest cells as shown by optical projection tomography in Wnt1-Cre;YFP reporter mice. J Anat. 2010;217(6):651–64.
26. Sparrow MP, Lamb JP. Ontogeny of airway smooth muscle: structure, innervation, myogenesis and function in the fetal lung. Respir Physiol Neurobiol. 2003;137(2–3):361–72.
27. Ibanez CF, Andressoo JO. Biology of GDNF and its receptors – relevance for disorders of the central nervous system. Neurobiol Dis. 2017;97(Pt B):80–9.
28. Ibanez CF, Paratcha G, Ledda F. RET-independent signaling by GDNF ligands and GFRalpha receptors. Cell Tissue Res. 2020;382(1):71–82.
29. Levi-Montalcini R, Skaper SD, Dal Toso R, Petrelli L, Leon A. Nerve growth factor: from neurotrophin to neurokine. Trends Neurosci. 1996;19(11):514–20.

30. Teng KK, Hempstead BL. Neurotrophins and their receptors: signaling trios in complex biological systems. Cell Mol Life Sci. 2004;61(1):35–48.

31. Reichardt LF. Neurotrophin-regulated signalling pathways. Philos Trans R Soc Lond Ser B Biol Sci. 2006;361(1473):1545–64.

32. Chao MV, Rajagopal R, Lee FS. Neurotrophin signalling in health and disease. Clin Sci (Lond). 2006;110(2):167–73.

33. Skaper SD. The biology of neurotrophins, signalling pathways, and functional peptide mimetics of neurotrophins and their receptors. CNS Neurol Disord Drug Targets. 2008;7(1):46–62.

34. Alberch J, Perez-Navarro E, Canals JM. Neuroprotection by neurotrophins and GDNF family members in the excitotoxic model of Huntington's disease. Brain Res Bull. 2002;57(6):817–22.

35. Saarma M, Sariola H. Other neurotrophic factors: glial cell line-derived neurotrophic factor (GDNF). Microsc Res Tech. 1999;45(4–5):292–302.

36. Sariola H. The neurotrophic factors in non-neuronal tissues. Cell Mol Life Sci. 2001;58(8):1061–6.

37. Sariola H, Saarma M. Novel functions and signalling pathways for GDNF. J Cell Sci. 2003;116(Pt 19):3855–62.

38. Pluchino N, Russo M, Santoro AN, Litta P, Cela V, Genazzani AR. Steroid hormones and BDNF. Neuroscience. 2013;239:271–9.

39. Tabakman R, Lecht S, Sephanova S, Arien-Zakay H, Lazarovici P. Interactions between the cells of the immune and nervous system: neurotrophins as neuroprotection mediators in CNS injury. Prog Brain Res. 2004;146:387–401.

40. Simpkins JW, Green PS, Gridley KE, Singh M, de Fiebre NC, Rajakumar G. Role of estrogen replacement therapy in memory enhancement and the prevention of neuronal loss associated with Alzheimer's disease. Am J Med. 1997;103(3A):19S–25S.

41. Panettieri RA Jr. Effects of corticosteroids on structural cells in asthma and chronic obstructive pulmonary disease. Proc Am Thorac Soc. 2004;1(3):231–4.

42. Sathish V, Martin YN, Prakash YS. Sex steroid signaling: implications for lung diseases. Pharmacol Ther. 2015;150:94–108.

43. Britt RD Jr, Faksh A, Vogel E, Martin RJ, Pabelick CM, Prakash YS. Perinatal factors in neonatal and pediatric lung diseases. Expert Rev Respir Med. 2013;7(5):515–31.

44. Boulle F, van den Hove DL, Jakob SB, Rutten BP, Hamon M, van Os J, et al. Epigenetic regulation of the BDNF gene: implications for psychiatric disorders. Mol Psychiatry. 2012;17(6):584–96.

45. Aid T, Kazantseva A, Piirsoo M, Palm K, Timmusk T. Mouse and rat BDNF gene structure and expression revisited. J Neurosci Res. 2007;85(3):525–35.

46. Zheng F, Zhou X, Moon C, Wang H. Regulation of brain-derived neurotrophic factor expression in neurons. Int J Physiol Pathophysiol Pharmacol. 2012;4(4):188–200.

47. McDonald NQ, Chao MV. Structural determinants of neurotrophin action. J Biol Chem. 1995;270(34):19669–72.

48. Lessmann V, Gottmann K, Malcangio M. Neurotrophin secretion: current facts and future prospects. Prog Neurobiol. 2003;69(5):341–74.

49. Lessmann V, Brigadski T. Mechanisms, locations, and kinetics of synaptic BDNF secretion: an update. Neurosci Res. 2009;65(1):11–22.

50. Mowla SJ, Farhadi HF, Pareek S, Atwal JK, Morris SJ, Seidah NG, et al. Biosynthesis and post-translational processing of the precursor to brain-derived neurotrophic factor. J Biol Chem. 2001;276(16):12660–6.

51. Shimojo M, Courchet J, Pierani S, Torabi-Rander N, Sando R 3rd, Polleux F, et al. SNAREs controlling vesicular release of BDNF and development of callosal axons. Cell Rep. 2015;11(7):1054–66.

52. Lin LF, Doherty DH, Lile JD, Bektesh S, Collins F. GDNF: a glial cell line-derived neurotrophic factor for midbrain dopaminergic neurons. Science. 1993;260(5111):1130–2.

53. Springer JE, Seeburger JL, He J, Gabrea A, Blankenhorn EP, Bergman LW. cDNA sequence and differential mRNA regulation of two forms of glial cell line-derived neurotrophic factor in Schwann cells and rat skeletal muscle. Exp Neurol. 1995;131(1):47–52.

54. Piccinini E, Kalkkinen N, Saarma M, Runeberg-Roos P. Glial cell line-derived neurotrophic factor: characterization of mammalian posttranslational modifications. Ann Med. 2013;45(1):66–73.

55. Lonka-Nevalaita L, Lume M, Leppanen S, Jokitalo E, Peranen J, Saarma M. Characterization of the intracellular localization, processing, and secretion of two glial cell line-derived neurotrophic factor splice isoforms. J Neurosci. 2010;30(34):11403–13.

56. Immonen T, Alakuijala A, Hytonen M, Sainio K, Poteryaev D, Saarma M, et al. A proGDNF-related peptide BEP increases synaptic excitation in rat hippocampus. Exp Neurol. 2008;210(2):793–6.

57. Cristina N, Chatellard-Causse C, Manier M, Feuerstein C. GDNF: existence of a second transcript in the brain. Brain Res Mol Brain Res. 1995;32(2):354–7.

58. Wang Y, Geng Z, Zhao L, Huang SH, Sheng AL, Chen ZY. GDNF isoform affects intracellular trafficking and secretion of GDNF in neuronal cells. Brain Res. 2008;1226:1–7.

59. Suter-Crazzolara C, Unsicker K. GDNF is expressed in two forms in many tissues outside the CNS. Neuroreport. 1994;5(18):2486–8.

60. Penttinen AM, Parkkinen I, Voutilainen MH, Koskela M, Back S, Their A, et al. Pre-alpha-pro-GDNF and pre-beta-pro-GDNF isoforms are neuroprotective in the 6-hydroxydopamine rat model of Parkinson's disease. Front Neurol. 2018;9:457.

61. Barker PA. p75NTR is positively promiscuous: novel partners and new insights. Neuron. 2004;42(4):529–33.
62. Blochl A, Blochl R. A cell-biological model of p75NTR signaling. J Neurochem. 2007;102(2):289–305.
63. Chen Y, Zeng J, Cen L, Chen Y, Wang X, Yao G, et al. Multiple roles of the p75 neurotrophin receptor in the nervous system. J Int Med Res. 2009;37(2):281–8.
64. Lu B, Pang PT, Woo NH. The yin and yang of neurotrophin action. Nat Rev Neurosci. 2005;6(8):603–14.
65. Barbacid M. The Trk family of neurotrophin receptors. J Neurobiol. 1994;25(11):1386–403.
66. Thiele CJ, Li Z, McKee AE. On Trk--the TrkB signal transduction pathway is an increasingly important target in cancer biology. Clin Cancer Res. 2009;15(19):5962–7.
67. Ohira K, Hayashi M. A new aspect of the TrkB signaling pathway in neural plasticity. Curr Neuropharmacol. 2009;7(4):276–85.
68. Barbacid M. Structural and functional properties of the TRK family of neurotrophin receptors. Ann N Y Acad Sci. 1995;766:442–58.
69. Chao MV, Hempstead BL. p75 and Trk: a two-receptor system. Trends Neurosci. 1995;18(7):321–6.
70. Pitts EV, Potluri S, Hess DM, Balice-Gordon RJ. Neurotrophin and Trk-mediated signaling in the neuromuscular system. Int Anesthesiol Clin. 2006;44(2):21–76.
71. Cheng A, Coksaygan T, Tang H, Khatri R, Balice-Gordon RJ, Rao MS, et al. Truncated tyrosine kinase B brain-derived neurotrophic factor receptor directs cortical neural stem cells to a glial cell fate by a novel signaling mechanism. J Neurochem. 2007;100(6):1515–30.
72. Hoo AF, Henschen M, Dezateux C, Costeloe K, Stocks J. Respiratory function among preterm infants whose mothers smoked during pregnancy. Am J Respir Crit Care Med. 1998;158(3):700–5.
73. Rose CR, Blum R, Pichler B, Lepier A, Kafitz KW, Konnerth A. Truncated TrkB-T1 mediates neurotrophin-evoked calcium signalling in glia cells. Nature. 2003;426(6962):74–8.
74. Fenner BM. Truncated TrkB: beyond a dominant negative receptor. Cytokine Growth Factor Rev. 2012;23(1–2):15–24.
75. Chiba Y, Misawa M. The role of RhoA-mediated Ca2+ sensitization of bronchial smooth muscle contraction in airway hyperresponsiveness. J Smooth Muscle Res. 2004;40(4–5):155–67.
76. Damera G, Tliba O, Panettieri RA Jr. Airway smooth muscle as an immunomodulatory cell. Pulm Pharmacol Ther. 2009;22(5):353–9.
77. Gosens R, Zaagsma J, Meurs H, Halayko AJ. Muscarinic receptor signaling in the pathophysiology of asthma and COPD. Respir Res. 2006;7(1):73.
78. Halayko AJ, Amrani Y. Mechanisms of inflammation-mediated airway smooth muscle plasticity and airways remodeling in asthma. Respir Physiol Neurobiol. 2003;137(2–3):209–22.
79. Halayko AJ, Solway J. Molecular mechanisms of phenotypic plasticity in smooth muscle cells. J Appl Physiol. 2001;90(1):358–68.
80. Hirota S, Helli PB, Catalli A, Chew A, Janssen LJ. Airway smooth muscle excitation-contraction coupling and airway hyperresponsiveness. Can J Physiol Pharmacol. 2005;83(8–9):725–32.
81. James A. Airway remodeling in asthma. Curr Opin Pulm Med 2005;11(1):1–6.
82. Blum R, Konnerth A. Neurotrophin-mediated rapid signaling in the central nervous system: mechanisms and functions. Physiology (Bethesda). 2005;20:70–8.
83. Kovalchuk Y, Holthoff K, Konnerth A. Neurotrophin action on a rapid timescale. Curr Opin Neurobiol. 2004;14(5):558–63.
84. Carvalho AL, Caldeira MV, Santos SD, Duarte CB. Role of the brain-derived neurotrophic factor at glutamatergic synapses. Br J Pharmacol. 2008;153(Suppl 1):S310–24.
85. Rose CR, Blum R, Kafitz KW, Kovalchuk Y, Konnerth A. From modulator to mediator: rapid effects of BDNF on ion channels. BioEssays. 2004;26(11):1185–94.
86. Cacalano G, Farinas I, Wang LC, Hagler K, Forgie A, Moore M, et al. GFRalpha1 is an essential receptor component for GDNF in the developing nervous system and kidney. Neuron. 1998;21(1):53–62.
87. Airaksinen MS, Holm L, Hatinen T. Evolution of the GDNF family ligands and receptors. Brain Behav Evol. 2006;68(3):181–90.
88. Kim M, Kim DJ. GFRA1: a novel molecular target for the prevention of osteosarcoma chemoresistance. Int J Mol Sci. 2018;19(4):1078.
89. Encinas M, Tansey MG, Tsui-Pierchala BA, Comella JX, Milbrandt J, Johnson EM Jr. c-Src is required for glial cell line-derived neurotrophic factor (GDNF) family ligand-mediated neuronal survival via a phosphatidylinositol-3 kinase (PI-3K)-dependent pathway. J Neurosci. 2001;21(5):1464–72.
90. Trupp M, Belluardo N, Funakoshi H, Ibanez CF. Complementary and overlapping expression of glial cell line-derived neurotrophic factor (GDNF), c-ret proto-oncogene, and GDNF receptor-alpha indicates multiple mechanisms of trophic actions in the adult rat CNS. J Neurosci. 1997;17(10):3554–67.
91. Paratcha G, Ledda F, Ibanez CF. The neural cell adhesion molecule NCAM is an alternative signaling receptor for GDNF family ligands. Cell. 2003;113(7):867–79.
92. Nockher WA, Renz H. Neurotrophins in clinical diagnostics: pathophysiology and laboratory investigation. Clin Chim Acta. 2005;352(1–2):49–74.

93. Kistemaker LEM, Prakash YS. Airway innervation and plasticity in asthma. Physiology (Bethesda). 2019;34(4):283–98.

94. Nassenstein C, Braun A, Erpenbeck VJ, Lommatzsch M, Schmidt S, Krug N, et al. The neurotrophins nerve growth factor, brain-derived neurotrophic factor, neurotrophin-3, and neurotrophin-4 are survival and activation factors for eosinophils in patients with allergic bronchial asthma. J Exp Med. 2003;198(3):455–67.

95. Hoyle GW, Graham RM, Finkelstein JB, Nguyen KP, Gozal D, Friedman M. Hyperinnervation of the airways in transgenic mice overexpressing nerve growth factor. Am J Respir Cell Mol Biol. 1998;18(2):149–57.

96. de Vries A, Dessing MC, Engels F, Henricks PA, Nijkamp FP. Nerve growth factor induces a neuro-kinin-1 receptor- mediated airway hyperresponsiveness in guinea pigs. Am J Respir Crit Care Med. 1999;159(5 Pt 1):1541–4.

97. Hunter DD, Myers AC, Undem BJ. Nerve growth factor-induced phenotypic switch in guinea pig airway sensory neurons. Am J Respir Crit Care Med. 2000;161(6):1985–90.

98. Undem BJ, Hunter DD, Liu M, Haak-Frendscho M, Oakragly A, Fischer A. Allergen-induced sensory neuroplasticity in airways. Int Arch Allergy Immunol. 1999;118(2–4):150–3.

99. Nockher WA, Renz H. Neurotrophins and asthma: novel insight into neuroimmune interaction. J Allergy Clin Immunol. 2006;117(1):67–71.

100. Braun A, Lommatzsch M, Neuhaus-Steinmetz U, Quarcoo D, Glaab T, McGregor GP, et al. Brain-derived neurotrophic factor (BDNF) contributes to neuronal dysfunction in a model of allergic airway inflammation. Br J Pharmacol. 2004;141(3):431–40.

101. Dragunas G, Woest ME, Nijboer S, Bos ST, van Asselt J, de Groot AP, et al. Cholinergic neuroplasticity in asthma driven by TrkB signaling. FASEB J. 2020;34(6):7703–17.

102. Wang SY, Freeman MR, Sathish V, Thompson MA, Pabelick CM, Prakash YS. Sex steroids influence brain-derived neurotropic factor secretion from human airway smooth muscle cells. J Cell Physiol. 2016;231(7):1586–92.

103. Vohra PK, Thompson MA, Sathish V, Kiel A, Jerde C, Pabelick CM, et al. TRPC3 regulates release of brain-derived neurotrophic factor from human airway smooth muscle. Biochim Biophys Acta. 2013;1833(12):2953–60.

104. Sathish V, Vanoosten SK, Miller BS, Aravamudan B, Thompson MA, Pabelick CM, et al. Brain-derived neurotrophic factor in cigarette smoke-induced airway hyperreactivity. Am J Respir Cell Mol Biol. 2013;48(4):431–8.

105. Tollet J, Everett AW, Sparrow MP. Development of neural tissue and airway smooth muscle in fetal mouse lung explants: a role for glial-derived neurotrophic factor in lung innervation. Am J Respir Cell Mol Biol. 2002;26(4):420–9.

106. Undem BJ, Nassenstein C. Airway nerves and dyspnea associated with inflammatory airway disease. Respir Physiol Neurobiol. 2009;167(1):36–44.

107. Nassenstein C, Kerzel S, Braun A. Neurotrophins and neurotrophin receptors in allergic asthma. Prog Brain Res. 2004;146:347–67.

108. Ricci A, Graziano P, Bronzetti E, Saltini C, Sciacchitano S, Cherubini E, et al. Increased pulmonary neurotrophin protein expression in idiopathic interstitial pneumonias. Sarcoidosis Vasc Diffuse Lung Dis. 2007;24(1):13–23.

109. Ricci A, Felici L, Mariotta S, Mannino F, Schmid G, Terzano C, et al. Neurotrophin and neurotrophin receptor protein expression in the human lung. Am J Respir Cell Mol Biol. 2004;30(1):12–9.

110. Andiappan AK, Parate PN, Anantharaman R, Suri BK, de Wang Y, Chew FT. Genetic variation in BDNF is associated with allergic asthma and allergic rhinitis in an ethnic Chinese population in Singapore. Cytokine. 2011;56(2):218–23.

111. Dagnell C, Grunewald J, Kramar M, Haugom-Olsen H, Elmberger GP, Eklund A, et al. Neurotrophins and neurotrophin receptors in pulmonary sarcoidosis – granulomas as a source of expression. Respir Res. 2010;11:156.

112. Ricci A, Mariotta S, Saltini C, Falasca C, Giovagnoli MR, Mannino F, et al. Neurotrophin system activation in bronchoalveolar lavage fluid immune cells in pulmonary sarcoidosis. Sarcoidosis Vasc Diffuse Lung Dis. 2005;22(3):186–94.

113. Tortorolo L, Langer A, Polidori G, Vento G, Stampachiacchere B, Aloe L, et al. Neurotrophin overexpression in lower airways of infants with respiratory syncytial virus infection. Am J Respir Crit Care Med. 2005;172(2):233–7.

114. Langsdorf A, Radzikinas K, Kroten A, Jain S, Ai X. Neural crest cell origin and signals for intrinsic neurogenesis in the mammalian respiratory tract. Am J Respir Cell Mol Biol. 2011;44(3):293–301.

115. Lin C, Wang S, Xie W, Chang J, Gan Y. The RET fusion gene and its correlation with demographic and clinicopathological features of non-small cell lung cancer: a meta-analysis. Cancer Biol Ther. 2015;16(7):1019–28.

116. Marsh DJ, Zheng Z, Arnold A, Andrew SD, Learoyd D, Frilling A, et al. Mutation analysis of glial cell line-derived neurotrophic factor, a ligand for an RET/coreceptor complex, in multiple endocrine neoplasia type 2 and sporadic neuroendocrine tumors. J Clin Endocrinol Metab. 1997;82(9):3025–8.

117. Roskoski R Jr, Sadeghi-Nejad A. Role of RET protein-tyrosine kinase inhibitors in the treatment RET-driven thyroid and lung cancers. Pharmacol Res. 2018;128:1–17.

118. Prakash YS, Iyanoye A, Ay B, Mantilla CB, Pabelick CM. Neurotrophin effects on intracellular Ca^{2+} and force in airway smooth muscle. Am J Physiol Lung Cell Mol Physiol. 2006;291(3):L447–56.

119. Prakash YS, Thompson MA, Pabelick CM. Brain-derived neurotrophic factor in TNF-alpha modula-

tion of Ca²⁺ in human airway smooth muscle. Am J Respir Cell Mol Biol. 2009;41(5):603–11.

120. Abcejo AJ, Sathish V, Smelter DF, Aravamudan B, Thompson MA, Hartman WR, et al. Brain-derived neurotrophic factor enhances calcium regulatory mechanisms in human airway smooth muscle. PLoS One. 2012;7(8):e44343.

121. Meuchel LW, Stewart A, Smelter DF, Abcejo AJ, Thompson MA, Zaidi SI, et al. Neurokinin-neurotrophin interactions in airway smooth muscle. Am J Physiol Lung Cell Mol Physiol. 2011;301(1):L91–8.

122. Amrani Y. TNF-alpha and calcium signaling in airway smooth muscle cells: a never-ending story with promising therapeutic relevance. Am J Respir Cell Mol Biol. 2007;36(3):387–8.

123. Sweeney M, McDaniel SS, Platoshyn O, Zhang S, Yu Y, Lapp BR, et al. Role of capacitative Ca²⁺ entry in bronchial contraction and remodeling. J Appl Physiol. 2002;92(4):1594–602.

124. White TA, Xue A, Chini EN, Thompson M, Sieck GC, Wylam ME. Role of TRPC3 in tumor necrosis factor-alpha enhanced calcium influx in human airway myocytes. Am J Respir Cell Mol Biol. 2006;35:243–51.

125. Bollimuntha S, Selvaraj S, Singh BB. Emerging roles of canonical TRP channels in neuronal function. Adv Exp Med Biol. 2011;704:573–93.

126. Bai TR. Evidence for airway remodeling in chronic asthma. Curr Opin Allergy Clin Immunol. 2010;10(1):82–6.

127. Holgate ST. Airway inflammation and remodeling in asthma: current concepts. Mol Biotechnol. 2002;22(2):179–89.

128. Joubert P, Hamid Q. Role of airway smooth muscle in airway remodeling. J Allergy Clin Immunol. 2005;116(3):713–6.

129. Lagente V, Boichot E. Role of matrix metalloproteinases in the inflammatory process of respiratory diseases. J Mol Cell Cardiol. 2010;48:440–4.

130. Freeman M, Manlove L, Wang S, Sathish V, Britt RD Jr, Thompson MA, et al. Brain derived neurotrophic factor and airway fibrosis in asthma. Am J Physiol Lung Cell Mol Physiol. 2017;313:L360–70.

131. Soleman S, Filippov MA, Dityatev A, Fawcett JW. Targeting the neural extracellular matrix in neurological disorders. Neuroscience. 2013;253:194–213.

132. Ethell IM, Ethell DW. Matrix metalloproteinases in brain development and remodeling: synaptic functions and targets. J Neurosci Res. 2007;85(13):2813–23.

133. Lee R, Kermani P, Teng KK, Hempstead BL. Regulation of cell survival by secreted proneurotrophins. Science. 2001;294(5548):1945–8.

134. Dagnell C, Kemi C, Klominek J, Eriksson P, Skold CM, Eklund A, et al. Effects of neurotrophins on human bronchial smooth muscle cell migration and matrix metalloproteinase-9 secretion. Transl Res. 2007;150(5):303–10.

135. Barrenschee M, Bottner M, Hellwig I, Harde J, Egberts JH, Becker T, et al. Site-specific gene expression and localization of growth factor ligand receptors RET, GFRalpha1 and GFRalpha2 in human adult colon. Cell Tissue Res. 2013;354(2):371–80.

136. Gougeon PY, Lourenssen S, Han TY, Nair DG, Ropeleski MJ, Blennerhassett MG. The pro-inflammatory cytokines IL-1beta and TNFalpha are neurotrophic for enteric neurons. J Neurosci. 2013;33(8):3339–51.

137. Zoumboulakis D, Cirella KR, Gougeon PY, Lourenssen SR, Blennerhassett MG. MMP-9 processing of intestinal smooth muscle-derived GDNF is required for neurotrophic action on enteric neurons. Neuroscience. 2020;443:8–18.

138. Dolatshad NF, Saffrey MJ. Differential expression of glial cell line-derived neurotrophic factor family receptor alpha-2 isoforms in rat urinary bladder and intestine. Neurosci Lett. 2007;415(3):215–8.

Novel Thoracic MRI Approaches for the Assessment of Pulmonary Physiology and Inflammation

8

Jonathan P. Brooke and Ian P. Hall

Abstract

Excessive pulmonary inflammation can lead to damage of lung tissue, airway remodelling and established structural lung disease. Novel therapeutics that specifically target inflammatory pathways are becoming increasingly common in clinical practice, but there is yet to be a similar stepwise change in pulmonary diagnostic tools. A variety of thoracic magnetic resonance imaging (MRI) tools are currently in development, which may soon fulfil this emerging clinical need for highly sensitive assessments of lung structure and function. Given conventional MRI techniques are poorly suited to lung imaging, alternate strategies have been developed, including the use of inhaled contrast agents, intravenous contrast and specialized lung MR sequences. In this chapter, we discuss technical challenges of performing MRI of the lungs and how they may be overcome. Key thoracic MRI modalities are reviewed, namely, hyperpolarized noble gas MRI, oxygen-enhanced MRI (OE-MRI), ultrashort echo time (UTE) MRI and dynamic contrast-enhanced (DCE) MRI. Finally, we consider potential clinical applications of these techniques including phenotyping of lung disease, evaluation of novel pulmonary therapeutic efficacy and longitudinal assessment of specific patient groups.

Keywords

Thoracic MRI · Hyperpolarized gas · Helium-3 · Xenon-129 · Oxygen-enhanced MRI · Ultrashort echo time · Dynamic contrast-enhanced MRI

Abbreviations

^{129}Xe	Xenon-129
^{3}He	Helium-3
ADC	Apparent diffusion coefficient
ASL	Arterial spin labelling
BOS	Bronchiolitis obliterans syndrome
CF	Cystic fibrosis
CFTR	Cystic fibrosis transmembrane receptor
COPD	Chronic obstructive pulmonary disease
CT	Computed tomography
CTPA	CT pulmonary angiography
DCE	Dynamic contrast enhancement
D_{LCO}	Diffusion capacity of the lung for carbon dioxide
DPD	Dynamic proton density

J. P. Brooke (✉) · I. P. Hall (✉)
Department of Respiratory Medicine, University of Nottingham, Queens Medical Centre, Nottingham, UK
e-mail: jonathan.brooke@nottingham.ac.uk; ian.hall@nottingham.ac.uk

DWI	Diffusion-weighted imaging
FEV$_1$	Forced expiratory volume in 1 second
GRE	Gradient recall echo
ILD	Interstitial lung disease
IPF	Idiopathic pulmonary fibrosis
LAM	Lymphangioleiomyomatosis
LCI	Lung clearance index
LVR	Lung volume reduction
MRA	Magnetic resonance angiography
MRI	Magnetic resonance imaging
OE-MRI	Oxygen-enhanced MRI
OTF	Oxygen transfer function
PE	Pulmonary embolism
PET	Positron emission tomography
PFT	Pulmonary function test
RBC	Red blood cell
RER	Relative enhancement ratio
RF	Radiofrequency
SS	Systemic sclerosis
SUV$_{max}$	Maximum standardized uptake value
T	Tesla
UTE	Ultrashort echo time
V/Q	Ventilation-perfusion
VDP	Ventilation defect percentage
VDV	Ventilation defect volume
ZTE	Zero echo time

8.1 Introduction

Inflammation dictates much of the discrete repertoire of responses the lungs employ against injury. Sophisticated filtration, removal and immune-mediated pulmonary defences work in parallel to limit entry of pathogens and environmental particulates. When initial defences are overwhelmed, a vigorous acute inflammatory response confines and destroys noxious agents and promotes recovery [1]. An extreme manifestation of this response is the acute respiratory distress syndrome, in which an exaggerated inflammatory cascade disrupts the alveolar-capillary membrane, leading to severe respiratory failure [2]. Persistence of acute inflammation may be followed by a chronic inflammatory response designed to clear necrotic tissue, isolate remaining infective organisms and repair damaged lung. Abnormal chronic inflammation is implicated in many lung diseases, including asthma, chronic obstructive pulmonary disease (COPD), cystic fibrosis (CF) and interstitial lung disease (ILD). In these conditions, chronic inflammation can lead to progressive tissue damage, airway remodelling or established structural lung disease [3–5]. The location and mechanism (including immune cell and cytokine expression) of this inflammation vary, which gives rise to distinct patterns of lung disease.

Novel pulmonary therapeutics such as biologics and cystic fibrosis transmembrane receptor (CFTR) modulators can modify abnormal and excessive inflammation, either by targeting an element of the inflammatory pathway or indirectly downregulating its activity. These therapies mark a paradigm shift in lung disease management towards precision medicine and personalized care, but there is yet to be a similar stepwise change in diagnostic tools.

Conventional respiratory diagnostics include blood tests, pulmonary function tests (PFTs) and a range of thoracic imaging. While the plain radiograph remains a key screening and diagnostic investigation, it is frequently complemented by several other modalities including ultrasound, computed tomography (CT), positron emission tomography (PET) and ventilation-perfusion (V/Q) imaging. CT remains the gold standard for assessment of lung structure and is tremendously versatile as a diagnostic tool for parenchymal lung disease, pulmonary vascular disorders, malignancy and pulmonary infection. The demand for CT imaging in healthcare is continually growing, and developments such as low-dose CT screening for lung cancer will likely see this increase further [6]. However, CT has limited scope for functional assessment, and while PFTs offer a global assessment of airflow, lung volumes and gas transfer, they are insensitive to regionally heterogeneous lung disease.

Although unlikely to replace CT, the novel thoracic magnetic resonance imaging (MRI) approaches discussed in this chapter may offer highly sensitive assessments of regionally heterogeneous lung disease that complement existing diagnostics. The absence of ionizing radiation also provides a new avenue for longitudinal imaging, which could be useful for the detection of early lung disease progression and evaluation of therapeutic interventions.

8.2 Basic Principles of MRI

MRI is widely used in clinical medicine, most notably for neuro-, cardiac, vascular, soft tissue and abdominal imaging. Standard images are acquired by exploiting the magnetic 'spin' property of protons in hydrogen atoms, applying a complex series of magnetic fields and then using radiofrequency (RF) pulses to localize and characterize those protons in a target tissue [7].

The MRI Scanner
The MR signals used to create an image are generated by a series of magnetic coils contained within the housing of the MRI scanner (Fig. 8.1). The main magnetic coil is a superconducting

magnet cooled to approximately −269 °C using cryogenic liquid helium. At this temperature, resistance to the flow of electric current is minimal. Thus, a high electric current flowing through the coil's loops of wire creates a high-strength magnetic field orientated in the z-axis. This strong magnetic field is referred to as B_0 and its field strength is measured in Tesla (T) units [8]. Clinical MRI scanners generally employ 1.5 or 3 T fields, but ultra-high-field scanners up to 11.7 T are currently in development for future research in human subjects [9].

Gradient coils create a secondary magnetic field that distorts B_0 in orthogonal directions. Most MR systems employ three sets of gradient coils, one for each orthogonal plane – x, y and z. The primary function of these coils is spatial encoding of MR signals. Shim coils also adjust B_0, but their purpose is to increase magnetic field homogeneity. This action minimizes local susceptibility effects (e.g. due to a person lying in the scanner) that would otherwise disrupt B_0 field homogeneity and degrade image quality.

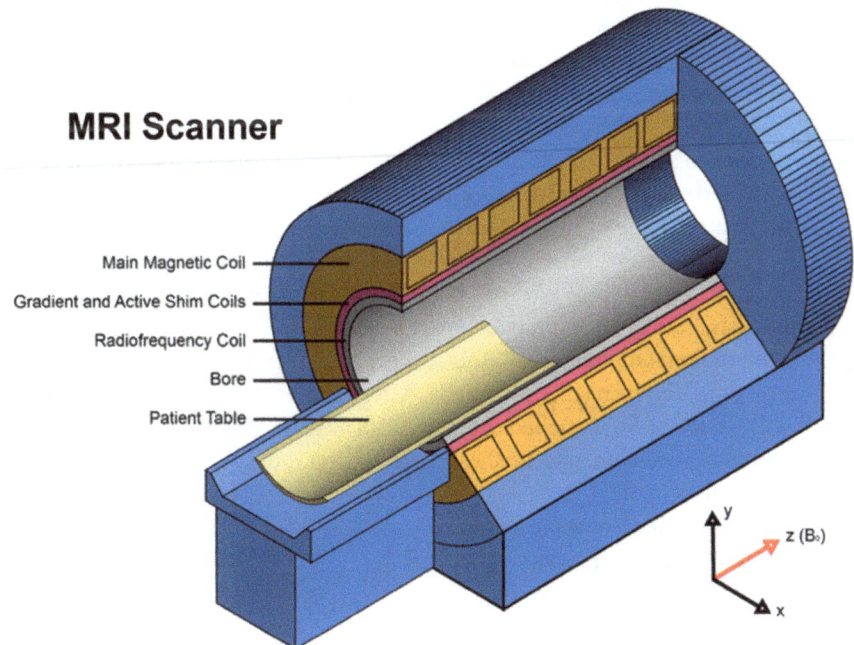

Fig. 8.1 Cross section of an MRI scanner. A closed bore MRI scanner is shown. A concentric series of magnet coils surrounds the bore of the scanner. The main magnetic field (B_0) is orientated in the z-axis, parallel to the scanner's bore. A sliding table allows a subject to be moved into and out of the scanner. (Adapted from Introduction to neuroimaging analysis [10])

RF coils transmit RF energy and detect proton RF signals. Some RF coils only transmit or receive, whereas others can do both. The transmit coil generates an RF pulse of electromagnetic energy that alters the spin of protons in B_0. When the RF pulse is switched off, protons create their own RF signal, which the receive coil can detect. The RF coil magnetic field is called B_1.

The sum of B_1, B_0 and gradient coil magnetic fields is a spatially localized output signal which is detected by the receive RF coil. Complex computer algorithms encode these output signals as volume elements (voxels) which are combined to create an MRI [8].

Basic MRI Physics

The following is an abbreviated explanation of MRI physics, which can be expanded upon with the literature referenced in this section.

The magnetic susceptibility of hydrogen and its abundance within the body are fundamental to conventional MRI. The atomic nucleus of hydrogen consists of a single positively charged proton that spins on its axis. This spinning proton generates a small magnetic field perpendicular to the direction of spin called a 'magnetic moment' [8]. When placed in the strong magnetic field of an MRI scanner, these magnetic moments become aligned to B_0. Most protons stay in a low energy state lining up parallel to B_0, but the remainder occupy a high energy state and instead line up anti-parallel to B_0. This results in protons having a net longitudinal magnetization parallel to B_0 [11, 12]. When in B_0, the spin axis of a proton is altered in a similar fashion to gravity acting upon a spinning top. The rotational motion of this spin axis is called precession. The resonance element of MRI occurs when an RF pulse is applied at the same frequency as the precession rate of these protons in a given magnetic field [13].

When an RF pulse is applied, some protons are flipped from a low to a high energy state, which reduces longitudinal magnetization. Proton spins are also pushed together so they precess in phase. The result is a net transverse magnetization vector relative to B_0 called a flip angle [7].

After the RF pulse is turned off, two types of proton relaxation effect can be measured. T1

relaxation occurs when high energy protons relax into the low energy state, release heat and restore longitudinal magnetization parallel to B_0. T2 relaxation occurs when the positively charged protons repel one another and no longer precess in phase [7]. Tissues in the human body exhibit characteristic differences in T1 and T2 relaxation dependent upon the quantity of hydrogen they contain and the molecular structure of hydrogen containing compounds. This allows for contrast between different tissues and anatomical structures to be detected on an MRI scan.

Greater contrast can be achieved by altering RF pulses or delivering multiple RF pulses in a so-called pulse sequence. Underpinning these concepts are two key MRI parameters: echo time and repetition time. Echo time is the time between delivery of an RF pulse and sampling of the resultant proton MR signal. Repetition time is the duration of one pulse sequence. Adjustment of these parameters can emphasize T1 and T2 relaxation effects resulting in T1- or T2-weighted images [14].

The eventual MRI is comprised of voxels, which represent the location and relative magnetic signal of protons in the scanned tissue. Voxels are conventionally displayed in grayscale, and their brightness varies based on MR signal strength: the stronger the signal, the brighter the voxel. MR signals are localized by gradient coils that apply a magnetic gradient to a slice of tissue and then encode the orthogonal position of protons by adjusting precession frequency (frequency encoding) and precession phase (phase encoding). A composite MR signal is then generated, which is initially stored as K-space data. After MRI acquisition, Fourier transformation is performed by computer software to convert the raw K-space data into an image [12, 15].

8.3   Technical Challenges of Thoracic MRI

MRI has advanced significantly since 1977 when the first in vivo human imaging of a finger and the thorax were performed [16, 17]. Despite this historical starting point, the lungs have been somewhat of an orphan organ amongst the otherwise extensive use of MRI in contemporary medical care. Instead, CT has been dominant in thoracic imaging given its affordability, speed and ease of interpretation [18]. Modern CT imaging rapidly delivers localized structural data and protocol improvements have significantly reduced ionizing radiation doses without compromising image quality [19]. As such, MRI is perhaps best thought of as a complement to, rather than a replacement for, current thoracic imaging methods. However, the difficulties of applying conventional MRI techniques to the lungs have been a hindrance to its mass adoption. The fundamental issues with performing MRI of the lungs are:

- Low tissue density
- Numerous air-tissue interfaces
- Physiological motion of the heart and thoracic cavity

MRI typically relies on protons in the target tissue to generate an MR signal. Lungs have a relatively low density, with a combined mass of approximately 1 kg in a thoracic volume of 4–5 L for a typical adult [20]. As such, lung proton density is significantly lower than that of other solid organs, which leads to reduced MR signal. Signal is further attenuated by magnetic susceptibility artefact due to the millions of air-tissue interfaces in the lower respiratory tract. These interfaces cause substantial local magnetic field inhomogeneity with rapid loss of proton MR signal during imaging [21]. The result is lung that appears uniformly black and featureless on standard MRI. Third, the physiological motion of the heart and thoracic cavity can produce ghost images and further artefact during MR acquisition, which also degrade image quality [22]. The key strategies used to tackle these issues are:

- Use of intravenous or inhaled contrast agents
- Implementation of specialized lung MR sequences
- Respiratory and cardiac gating

Intravenous gadolinium contrast is widely used for clinical MRI scans, notably in vascular and neuroimaging. Its ability to shorten T1 relaxation creates greater contrast between tissues, and thus pathology can be more easily identified [23]. While this is helpful for pulmonary vascular imaging, the approach does not appreciably improve lung parenchymal images. Instead, lung MRI has seen a predominant focus on inhaled contrast agents such as hyperpolarized noble gases, oxygen and more recently fluorinated hydrocarbons. These substances generate an inherently higher MR signal than the lung itself and facilitate visualization of the airways alongside functional information concerning flow, ventilation and gas exchange [24].

For structural imaging, specialized sequences have been designed to overcome the rapid signal decay that inhibits the use of conventional MRI in the lungs. These sequences minimize so-called echo time, which allows lung MR signal to be acquired before decay occurs. Ultrashort and zero echo time (UTE and ZTE) sequences employ this principle and can produce images of lung parenchyma comparable to CT [25].

Finally, thoracic motion can be addressed in a few ways. For short sequences, a single breath hold may suffice, but longer sequences typically require gating. Respiratory gating allows the subject to free breathe and can be achieved during imaging with an external sensor or belt worn by the subject set to trigger acquisition at the same point in each respiratory cycle [26]. Alternatively, respiratory navigated sequences that identify the diaphragm and automatically trigger acquisition when the diaphragm is in a specified imaging window can be used [27]. ECG gating may also be used to control for cardiac motion, but this significantly prolongs examination time and is probably a lesser priority than control of respiratory motion [28]. All of these methods also assume control of extra-thoracic motion, and as such, subjects must remain as still as possible during an imaging sequence.

8.4 Thoracic MRI Modalities

8.4.1 Overview of Hyperpolarized Gas MRI

As discussed, the lungs are an inherently challenging organ to image with conventional MRI techniques. An inhaled noble gas was one of the first solutions as demonstrated by Albert et al. in 1994, using hyperpolarized xenon-129 (^{129}Xe) to image ex vivo mouse lungs [29]. Compared with lung, the inherently greater signal of the hyperpolarized gas allowed the airways to be visualized, and this work was soon translated in vivo to human subjects in 1997 by Mugler et al. [30].

A close competitor for ^{129}Xe was another noble gas: helium-3 (^{3}He). The first hyperpolarized ^{3}He images were acquired in guinea pig lungs by Middleton et al., with an improved MR signal when compared to the ^{129}Xe images from 1 year before [31]. ^{3}He then became the favoured hyperpolarized gas in thoracic MRI for many years, as polarization methods and imaging quality were superior. However, scarcity and expense when compared to ^{129}Xe has seen ^{3}He use decline more recently.

A number of adjustments to conventional MRI are required for hyperpolarized gas imaging. First, a polarizer is needed to manufacture hyperpolarized gas by means of spin exchange optical pumping. The scanner itself must also be calibrated for resonance of the noble gases, given the gyromagnetic ratios, and hence Larmor frequencies of ^{129}Xe and ^{3}He differ from that of a proton. Finally, a dedicated radiofrequency (RF) coil, again attuned to the appropriate frequency, is necessary for an image to be generated [32]. While some of this equipment is commercially available, many research teams have developed their own bespoke polarizers and coils instead.

8.4.2 Hyperpolarized ^{3}He MRI

^{3}He is an inert gas, and unlike the more abundant ^{4}He isotope, the odd number of nucleons (two protons and one neutron) facilitates the magnetic spin required for MRI. Early adoption in thoracic MRI was facilitated by sophisticated hyperpolarization techniques, allowing levels of polarization of around 30%, compared with 1–2% for ^{129}Xe [33]. This equated to superior MR signal and image quality, which placed ^{3}He as the front-runner in hyperpolarized gas MRI research for many years. However, ^{3}He used in medical

research is derived from the radioactive decay of tritium (hydrogen-3), a substance historically used in the manufacture of nuclear weapons. In recent years, the supply of tritium has dwindled, which has made use of ^{3}He prohibitive and prompted a move back towards ^{129}Xe [34].

^{3}He imaging is typically performed during a breath hold, where hyperpolarized ^{3}He rapidly diffuses through the airways and remains confined in the lungs. Given its insolubility in lung tissue, side effects are uncommon (~6–7% of subjects) and often mild, for example a self-limiting cough or dry mouth [35]. The inhaled ^{3}He mixture is typically anoxic, and so modest oxygen desaturation (~4%) may be observed, but recovery is usually rapid upon free breathing of room air [36]. An anoxic mixture delays loss of polarity by limiting ^{3}He interaction with paramagnetic oxygen. However, once inhaled, ^{3}He undergoes rapid depolarization, both due to its interaction with oxygen in the airways and the application of RF pulses during imaging [37]. This rapid loss of polarity, alongside the constraint of a single breath hold, means images need to be obtained within approximately 20 seconds. Given the MRI scanner and RF coil are specifically attuned to ^{3}He, the resulting image isolates structures that contain the hyperpolarized gas – in this case, the airways – and excludes the surrounding tissues [38].

Static Ventilation ^{3}He Imaging

In static ventilation imaging, ventilation heterogeneity is localized and quantified during a ^{3}He breath hold. Areas of decreased ^{3}He signal imply a reduction in ventilation, referred to as ventilation defects. These are typically described as the ventilation defect volume (VDV) or ventilation defect percentage (VDP), which are regionally mapped and quantitative measures of pulmonary ventilation [39].

One of the earliest uses of hyperpolarized ^{3}He in human subjects by Kauczor et al. demonstrated ventilation defects in a spectrum of lung diseases. Defects were visible in patients with COPD, bronchiectasis, lung cancer and pleural effusion and correlated well with pathology visible on conventional imaging [40]. Mathew et al. showed defects in COPD patients varied little for same-day imaging, but changed more after a 1-week interval, despite stable spirometry [41]. As perhaps would be expected, similar studies in asthma have shown ventilation defects can be induced by exercise and methacholine challenge [42] and improve with inhaled beta-agonist administration [43]. More interestingly, longitudinal studies in asthma have shown defects may persist or recur in the same locations, which is thought to reflect areas of chronic inflammation and airway remodelling [44]. Small ventilation defects can also be seen in healthy patients with normal lung function, and these should not be mistaken as pathological during image analysis [45].

In CF, ventilation defects correlate with forced expiratory volume in 1 second (FEV_1) but are also seen in patients with normal range spirometry (Fig. 8.2) [46]. While studies with chest physiotherapy and nebulized DNase have shown little change in global measures of ^3He ventilation, regional differences before and after treatment can be detected, which may represent shifting airway secretions [46, 47]. Altes et al. demonstrated ventilation defects decrease in patients with G551D mutation taking ivacaftor, but return to baseline after washout of the drug. Improvements were even seen in patients with normal range spirometry and those with small changes in FEV_1, which supports the potential role of hyperpolarized gas MRI as a sensitive biomarker [48].

Diffusion-Weighted ^3He Imaging

Diffusion-weighted imaging (DWI) in conventional MRI assesses the net diffusion of water molecules in tissues [49]. This technique is commonly used in neuroimaging to identify restricted diffusion as seen in acute ischaemic stroke, malignancy and white matter disease [50]. DWI lung MRI instead measures the diffusion of hyperpolarized gas through the airways. ^3He DWI is performed during a breath hold, and diffusion is quantified as the apparent diffusion coefficient (ADC). ADC is a relative measure of diffusion restriction and can be represented visually as an ADC map to compare regions of interest [51].

ADC characteristically reduces as airway calibre decreases. This is well demonstrated in healthy lungs where ADC in the major airways is higher, and lower but homogeneous in peripheral airways [52]. In contrast, small airway destruction causes foci of increased ADC as seen in emphysematous lung (Fig. 8.3) [53]. Subclinical lung disease can also be detected, with even very mild emphysematous change demonstrating increased ADC values [54]. In COPD, mean ADC is well correlated with diffusion capacity of the lung for carbon monoxide (D_{LCO}), reinforcing the role of ADC as a surrogate marker of alveolar damage [55]. ADC also shows greater sensitivity than spirometry to detect deterioration in lung function of COPD patients over periods of up to 2 years [56].

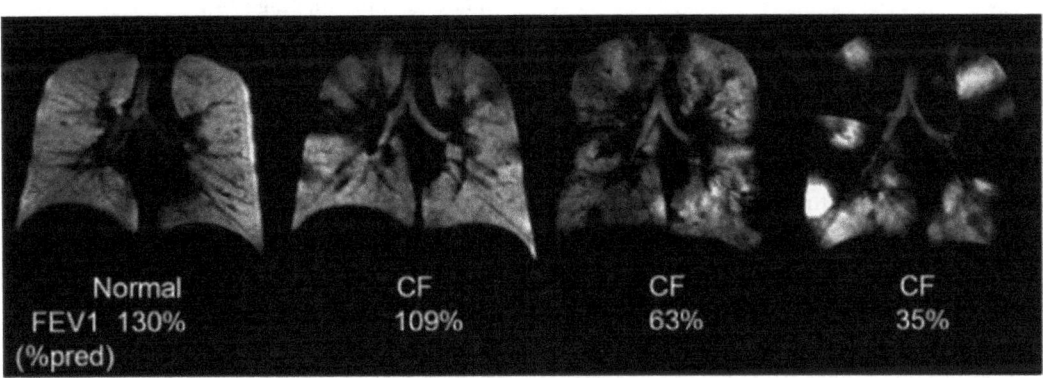

Fig. 8.2 ^3He MRI of static ventilation in CF. Comparison of a healthy subject and three adults with CF. An increasing burden of ventilation defects is demonstrated with deteriorating lung function [46]. (Image reproduced with permission of the rights holder. Elsevier – License number: 4858760327184)

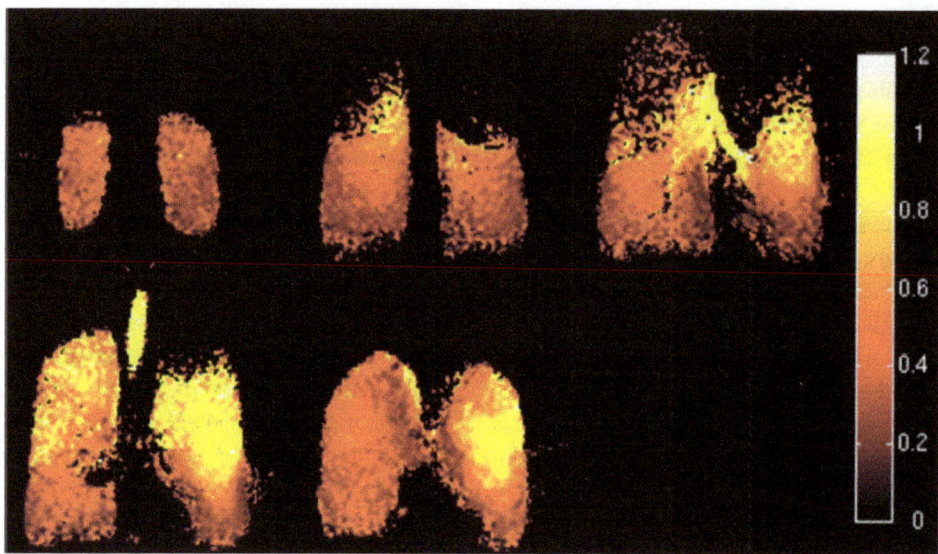

Fig. 8.3 [3]He MRI ADC map in severe COPD. Coronal views of a patient with severe COPD show increased ADC values (yellow pixels) in both upper lobes in keeping with emphysematous lung. Similar ADC values are also seen in large calibre airways (trachea and main bronchi). Coloured bar units are cm^2/second [53]. (Image reproduced with permission of the rights holder. John Wiley and Sons – License number: 4858820929850)

In idiopathic pulmonary fibrosis (IPF), mean ADC is also increased as damaged alveoli lose surface area and [3]He diffusion becomes less restricted [57]. Chan et al. showed mean ADC correlates with D_{LCO} and CT fibrosis scores and used an alternate DWI metric – the derived mean diffusive length scale – which estimates the mean alveolar dimension in a voxel [58] to demonstrate DWI is sensitive to small changes in IPF lung morphometry over a 1-year period [57].

8.4.3 Hyperpolarized [129]Xe MRI

In contrast to [3]He, xenon is lipophilic, making it readily soluble in pulmonary tissues as well as the bloodstream. This property is key to its anaesthetic effect but also the basis of dissolved phase MR imaging [30]. Approximately 2% of inhaled xenon dissolves, leaving ample gas for airspace MR signal [34]. The [129]Xe isotope used in hyperpolarized MRI is conveniently abundant, comprising roughly a quarter of naturally occurring xenon derived from the atmosphere [32]. [129]Xe imaging has demonstrated an excellent safety profile in both healthy volunteers and patients

with various lung diseases [59]. Common side effects include tingling, dizziness and euphoria, which are invariably mild and short-lived [60].

Many of the techniques developed for [3]He MRI have been applied to [129]Xe including ventilation and DWI. [129]Xe's solubility in pulmonary tissues has also led to great interest in dissolved phase imaging. As such, the use of [129]Xe MRI has grown considerably over the past decade as a virtue of favourable economics, better polarization technology and its scope for unique and sensitive functional lung imaging [34].

Static Ventilation [129]Xe Imaging

When the transition of hyperpolarized gas MRI from [3]He to [129]Xe began, various research groups sought to compare the two modalities. In 2010, Altes et al. showed the imaging quality of both was comparable, and similar ventilation defects could be detected in healthy volunteers and patients with COPD, asthma and CF [61]. However, subsequent work by other groups has highlighted important differences between [3]He and [129]Xe static ventilation imaging.

An intriguing finding was that some ventilation defects are seemingly missed or 'masked' with [3]He when compared side by side to [129]Xe imaging [62]. This was reflected by greater [129]Xe VDP in COPD [62] and asthma [63] compared to [3]He. Svenningsen et al. investigated asthmatic patients who had previously undergone [3]He imaging after methacholine challenge [64] with repeat [3]He and [129]Xe imaging 1 year later. They found that follow-up [129]Xe imaging without provocation testing revealed defects that had previously only been detectable with [3]He once methacholine had been given (Fig. 8.4) [63].

The rationale behind this may be the lower diffusion coefficient and higher density of [129]Xe, which results in its slower airway transit and delayed filling of poorly ventilated lung [62, 63]. Gas mixtures using [129]Xe are typically closer in density to air; hence, their diffusion coefficients are similar. This similarity allows clinically relevant ventilation defects to be visualized with [129]Xe, which may otherwise be undetectable with [3]He [62].

Static ventilation studies with [129]Xe have otherwise explored similar lung diseases to [3]He. In COPD, [129]Xe ventilation correlates with functional measures including spirometry [62] and V/Q imaging [65]. Ventilation defects improve after beta-agonist use in asthma [63] and may be a more sensitive measure than spirometry to detect lung function decline with advancing age [66]. Elevated VDP and ventilation heterogeneity have been shown in CF subjects with even mild lung disease (FEV$_1$ ≥ 100%) [67], and [129]Xe defects correlate with lung clearance index (LCI) in this group [68]. In lymphangioleiomyomatosis (LAM), co-registered [129]Xe and CT images have also shown significant ventilation heterogeneity between similarly sized cystic lung volumes, which highlights the merit of combining functional images alongside conventional structural assessments [69].

Diffusion-Weighted [129]Xe Imaging

The majority of [129]Xe DWI research has been conducted in COPD, where [3]He and [129]Xe ADC measurements have shown good correlation with one another [62]. When compared to [3]He, absolute [129]Xe ADC values are smaller [70], which likely reflects xenon's lower diffusion coefficient as described previously.

[129]Xe ADC correlates with emphysematous burden and alveolar destruction on both CT [71] and ex vivo histological samples [72]. Various groups have also shown strong correlations with spirometry and D$_{LCO}$ measurements. Research in other lung diseases has been more limited, but [129]Xe ADC measurements have been successfully performed in CF [73], LAM [69] and ex vivo IPF lungs [72].

Dissolved Phase [129]Xe Imaging

Dissolved phase [129]Xe MR takes advantage of xenon's solubility in pulmonary tissue to generate quantifiable and spatially localized gas exchange imaging. Xenon follows the same transfer pathway as oxygen, diffusing through the alveolar-capillary unit and then transiently

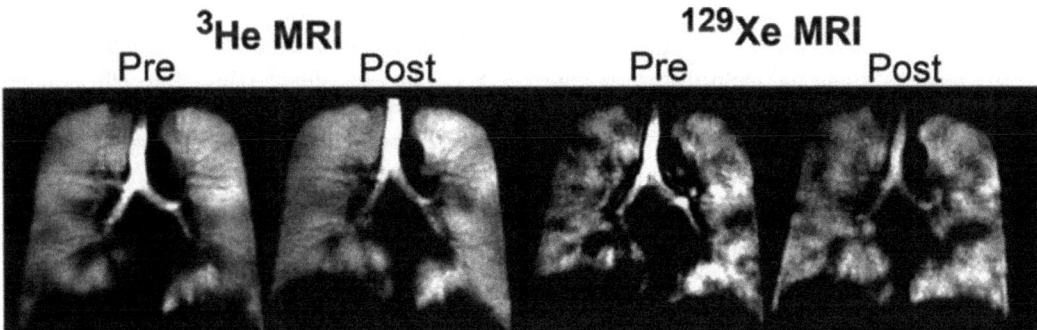

Fig. 8.4 Comparison of [3]He and [129]Xe static ventilation MRI in a subject with asthma. Coronal views of a patient with asthma taken pre- and post-bronchodilator. [129]Xe MRI reveals ventilation defects that are not detected with [3]He [63]. (Image reproduced with permission of the rights holder. John Wiley and Sons – License number: 4858821365032)

binding with haemoglobin once in the blood-stream [74]. A chemical shift relative to the gaseous phase occurs when ^{129}Xe diffuses into the pulmonary interstitium (referred to as 'barrier tissues' in dissolved phase imaging) and again upon diffusion into red blood cells (RBCs). Due to chemical shift, each compartment has a distinct ^{129}Xe MR signal – gas, barrier and RBC – that can be detected by selective excitation of the appropriate ^{129}Xe resonant frequency [75].

Early attempts to acquire a dissolved phase image were hampered by low ^{129}Xe polarization levels and poor MR signal [30]. In contrast to the gas phase, relatively little ^{129}Xe dissolves into barrier tissues (roughly 2%), and once there MR signal degrades rapidly [75]. These problems were overcome with frequency-selective RF pulses to isolate a combined barrier/RBC dissolved phase signal sufficient to create an image. These initial studies in healthy volunteers showed a dissolved phase gradient, where MR signal increases in the dependent lung. This is thought to represent increased perfusion and alveolar compression of those areas during imaging [75].

Later studies refined the technique further to generate separate barrier and RBC dissolved phase images [74]. This has allowed better discrimination of the gas exchange mechanics in lung pathology, with a particular focus on ILD. In IPF, Kaushik et al. showed mean RBC signal was decreased and barrier signal increased when compared with healthy controls. This results in a low RBC : barrier ratio which strongly correlates with D_{LCO} [74]. Low RBC/barrier ratio is also seen in COPD, again correlating with D_{LCO} as well as CT emphysema score [76] (Fig. 8.5).

Regional mapping has helped identify three patterns of impaired gas exchange in the dissolved phase imaging of IPF. These are thought to represent different levels of disease activity, which are [78]:

1. Diffusion block – high barrier and low RBC signal
2. End stage fibrosis – low/normal barrier and low RBC signal
3. Early/active disease – high barrier and normal RBC signal

While the 'end stage fibrosis' pattern overlaps with areas of severe fibrosis on CT, the 'early/active disease' pattern lacks such change on conventional imaging. Given a single MRI voxel contains upwards of 40,000 alveoli, these areas may represent early disease that could be used as a functional MR biomarker of inflammation [78]. Finally, recent work with higher field strength MRI has shown the sensitivity of dissolved phase ^{129}Xe can be augmented further to provide greater imaging resolution in IPF as well as other lung diseases [79].

8.4.4 Oxygen-Enhanced MRI (OE-MRI)

Oxygen can also be used as an MR contrast agent in oxygen-enhanced MRI (OE-MRI). However, unlike hyperpolarized gases, oxygen itself is not directly visualized. Instead, the weakly paramagnetic molecular oxygen (O_2) shortens T1 relaxation in lung tissue and the pulmonary circulation during MRI. When 100% oxygen is inhaled, T1 relaxation is shortened by approximately 9% [80], leading to a rise in T1-weighted signal.

Oxygen transport in the lungs is dependent upon ventilation, diffusion and perfusion, such that the T1-weighted signal in OE-MRI is representative of all three processes. It follows that disruption to any of these elements in lung pathology can then affect the oxygen-enhanced signal [81]. Maximal T1 signal is obtained when alveolar capillary blood is saturated with molecular oxygen, and Mai et al. demonstrated this could be achieved with an oxygen flow rate of 15 L/min [82]. Flow rates above this do not increase T1 signal further; hence, OE-MRI sequences typically employ sequential room air and 15 L/min acquisitions.

OE-MRI is typically performed during free breathing with acquisitions taking several minutes to complete. To improve image quality and reduce movement artefact, respiratory gating with navigator echoes that identify the diaphragm during imaging can be used [83]. High-flow oxygen and face masks are inexpensive and widely available, which means there are few barriers to performing

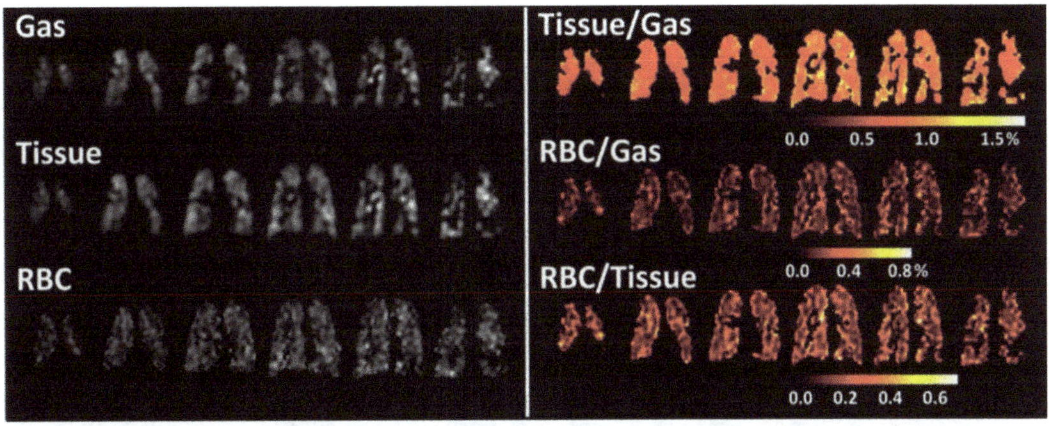

Fig. 8.5 ^{129}Xe dissolved phase MRI in severe COPD. Coronal views of the three ^{129}Xe phases and ratio maps in one subject. A large number of ventilation defects are seen in the gas phase. There is also marked heteroge- neity of the dissolved phase and ratio maps [77]. (Image reproduced with permission of the rights holder. John Wiley and Sons – License number: 4858830024322)

OE-MRI with standard clinical MRI scanners. However, MR signal is relatively small when compared to hyperpolarized gases and oxygen itself can alter pulmonary physiology [80]. Data from OE-MRI can be displayed and quantified in several ways, including T1 mapping and calculation of relative enhancement or oxygen transfer. In addition, OE-MRI may also be combined with intravenous gadolinium for V/Q imaging.

T1 Mapping in OE-MRI

T1 maps are one method of visualizing OE-MRI data and are typically colour-coded to represent different T1 relaxations (Fig. 8.6). Images performed with different fractions of inspired oxygen (FiO_2) can then be compared for qualitative evaluation of lung function. In healthy subjects, 100% oxygen causes widespread reduction of T1 relaxation with some heterogeneity [81]. The T1 maps in lung disease are generally even more heterogeneous and exhibit lower T1 relaxation values due to impaired oxygen transport [84].

Renne et al. demonstrated heterogeneity of T1 maps in lung transplant recipients with bronchiolitis obliterans syndrome (BOS), but median T1 relaxation values could not discriminate between stages of disease [86]. However, other studies have verified the regional sensitivity of lung T1 measurement. Jakob et al. identified focal lung disease in CF that exhibited both blunted T1 relaxation and matched perfusion abnormality [87]. Similar findings have also been shown in patients with CF receiving nebulized hypertonic saline [88].

Relative Enhancement and Oxygen Transfer Function

To help describe and visualize regional lung function in OE-MRI quantitatively, the terms relative enhancement ratio (RER) and oxygen transfer function (OTF) have been developed. First, RER describes the change in T1 signal intensity of spatially matched voxels between room air and 100% oxygen images [89]. It can be represented visually as a relative enhancement map or alternatively several regions of interest can be identified and the RER for each one calculated. Ohno et al. used these methods to show markedly lower RER in patients with lung cancer and emphysema compared with healthy volunteers. This study also showed that mean RER strongly correlated with FEV_1 and D_{LCO} [89]. The same research group has since corroborated these findings in a larger cohort of COPD patients [90] and found similar but weaker correlations in patients with asthma [91] and ILD [92].

OE-MRI VDP can also be derived from RER and has shown strong correlation of ^{3}He VDP

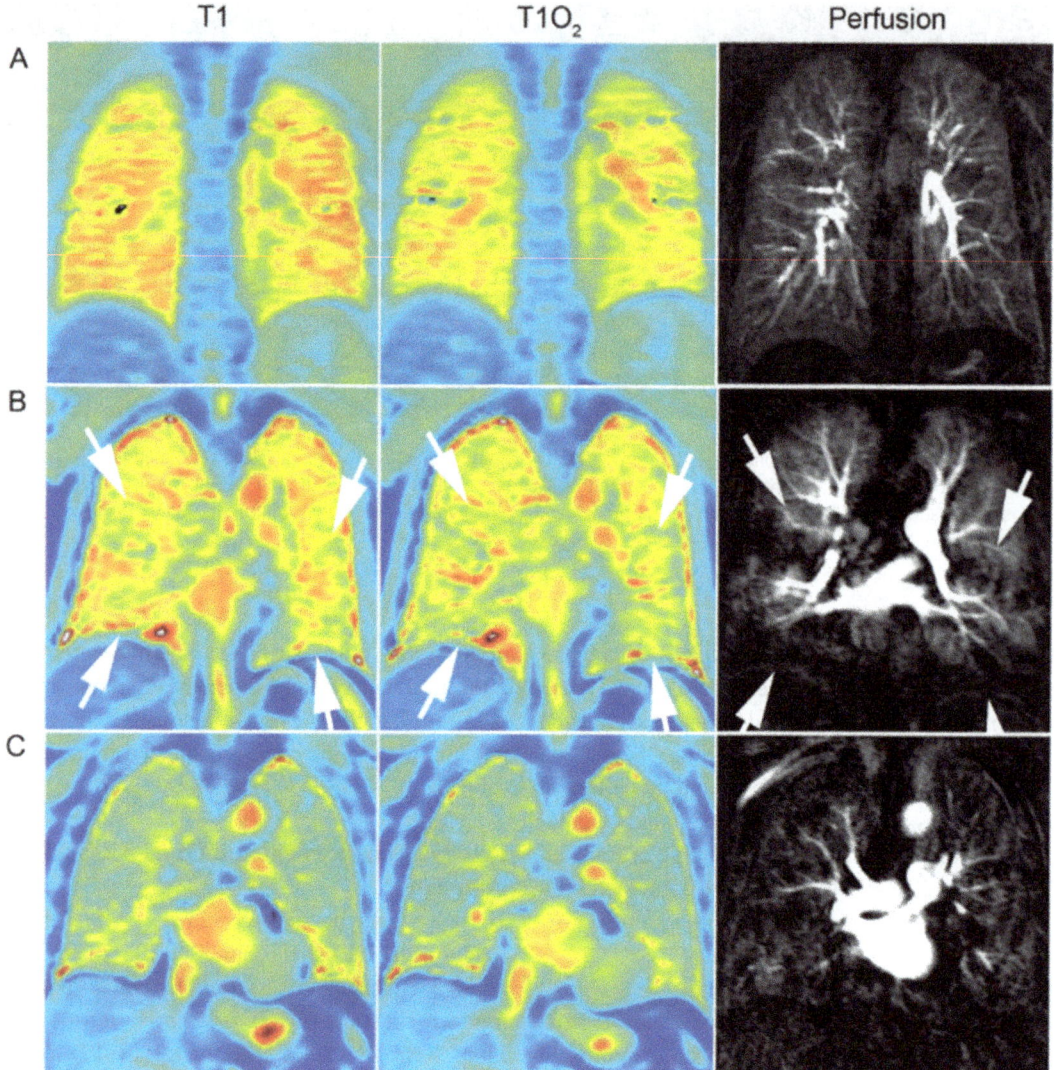

Fig. 8.6 OE-MRI T1 maps and DCE perfusion MRI. Coronal views of three subjects: (**a**) healthy subject, (**b**) and (**c**) COPD. Reduction in oxygen-enhanced signal with 100% oxygen (T1O₂). Signal is significantly diminished in COPD subjects. Minor ventilation and perfusion abnormalities are demonstrated in subject **b** (white arrows). More marked abnormalities are seen in subject **c** [85]. (Image reproduced under Creative Commons Attribution License)

measurements in adults with CF. However, OE-MRI VDP measures were approximately 5% lower than ³He VDP [93]. This likely reflects key differences in the performance of these techniques as OE-MRI has a longer wash-in time and lower spatial resolution and measures a combination of ventilation, diffusion and perfusion as previously discussed.

OTF describes the change in lung T1 relaxation rate (R1) for a given oxygen concentration and represents a combination of airflow, oxygen diffusion and lung perfusion [86, 87]. In the original paper, adults with CF showed significant OTF heterogeneity and areas of poor oxygen transfer also demonstrated MR perfusion abnormalities [87]. Renne et al. later examined the effect of endoscopic allergen testing on OTF and airway eosinophil count in asthma. Their study showed OTF decreased in airway segments exposed to allergen, with a corresponding airway eosinophilia. Follow-up MRI after 24 hours showed OTF then returned to baseline [94]. The same research group has also measured OTF in lung allograft recipients finding it is significantly lower in patients with evidence of BOS [86].

Ventilation/Perfusion OE-MRI
V/Q imaging was part of the inception of OE-MRI in 1996. Edelman et al. compared gadolinium-based perfusion imaging alongside oxygen-enhanced ventilation images to demonstrate V/Q mismatch [95]. In patients with COPD, oxygen enhancement is also markedly abnormal and strongly correlated with perfusion abnormalities [85].

Arterial spin labelling (ASL) has been used as alternative measure of perfusion for V/Q imaging in combination with OE-MRI. ASL precludes the need for an injected contrast agent and instead uses magnetically labelled arterial blood as a tracer [96]. In healthy volunteers, V/Q imaging has been successfully performed using this method, including the demonstration of a matched V/Q defect in a subject following left upper lobectomy [97].

8.4.5 Ultrashort Echo Time (UTE) MRI

Ultrashort echo time (UTE) is a proton-based MRI modality and was first described by Bergin et al. as a means of obtaining structural lung images [98]. This is achieved by minimizing the delay between application of an RF pulse and detection of the resulting MR signal. The resulting UTE images were comparable to CT of the time.

Given the rapid decay of lung MR signal during MRI, the echo time for UTE is by necessity ≤ 200 µs [21]. So-called zero echo time (ZTE) sequences reduce this further to as little as 5 µs, but the exact echo time achievable is limited by the software and hardware constraints of the MR platform used [99].

Typically, UTE sequences also employ specialized radial sampling approaches to acquire imaging data, which contrast with the Cartesian sampling methods used for conventional MRI [100]. This method helps to minimize motion artefact and facilitate free-breathing scans through respiratory gating [101]. Recent UTE techniques with self-gating discard any motion-corrupted data and use the remaining data to create the final images [102]. The result is a signal-averaged MRI scan with improved resolution by virtue of the pooled imaging data [103].

UTE MRI for Assessment of Structural Lung Disease
Neonatal intensive care patients are particularly prone to pulmonary morbidity due to prolonged periods of oxygen therapy and mechanical ventilation. Self-gated UTE can produce CT-like images in these patients by controlling for bulk movement and respiratory motion without the need for sedation [104]. In this way, end-inspiratory and -expiratory images can estimate tidal volume and identify structural lung change in bronchopulmonary dysplasia (BPD) [102, 105]. In BPD, preliminary studies have suggested UTE can be used to quantify hyperinflation to predict clinical outcomes, and could be used for longitudinal follow-up [104, 105].

In infants with CF, UTE has demonstrated good correlation with CT for bronchiectasis and bronchial wall thickening [106]. Similar structural imaging has also been demonstrated in adults with CF using free-breathing UTE acquisitions lasting 8–15 minutes (Fig. 8.7) [107].

In COPD, UTE shows short-term reproducibility over a 3-week period [108] and can produce structural information comparable to CT [109]. Chassagnon et al. compared a cohort of systemic sclerosis (SS) patients with and without evidence of ILD on CT. In this study, elastic registration of inspiratory and expiratory UTE showed increased lung stiffness in patients with SS-related ILD, corresponding with areas of fibrosis on HRCT [110].

UTE has also shown promising results in the detection and morphological characterization of lung nodules. Studies have shown UTE can detect

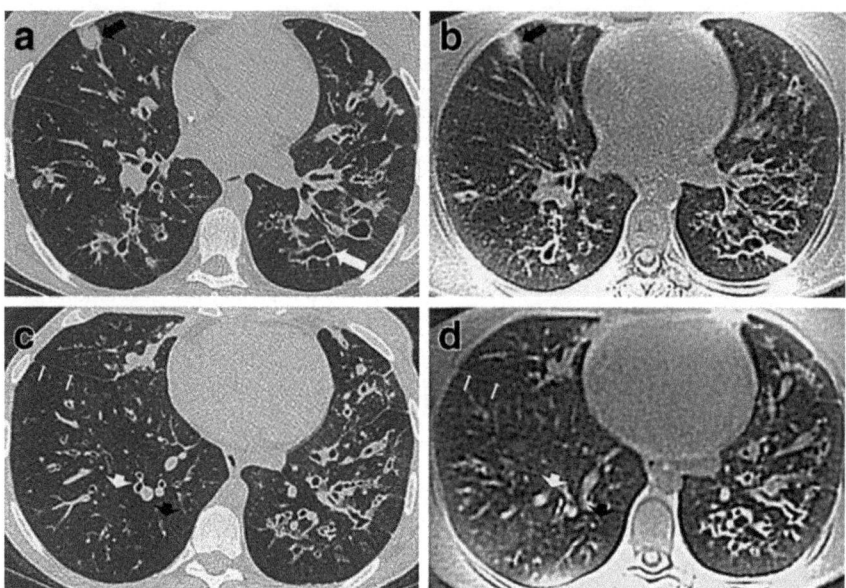

Fig. 8.7 Axial CT and UTE MRI in an adolescent with CF. CT (**a, c**) and UTE MRI (**b, d**) slices show evidence of nodular consolidation (black arrow in **a** and **b**) and wide- spread bronchiectatic change [107]. (Image reproduced with permission of the rights holder. Springer Nature – License number: 4858830473723)

between 73% and 86% of nodules when compared to CT, but is markedly less sensitive for identification of very small modules (i.e. those <4 mm) [111, 112]. It also tends to underestimate long- and short-axis nodule measurements by up to 1–2 mm [113].

UTE MRI for Assessment of Lung Function
UTE is an appealing foundation for co-registered structural and functional lung MRI as it is capable of reducing acquisition times and capturing motion-insensitive images during free breathing.

Sheikh et al. demonstrated a variation of UTE imaging – dynamic proton density (DPD) – in healthy volunteers and patients with asthma, comparing them to hyperpolarized [3]He MRI. Breath hold UTE images were acquired and used to generate DPD maps that reflect the difference in UTE signal intensity between full inspiration and expiration. This technique demonstrated ventilation heterogeneity and produced functional data comparable to [3]He imaging in patients given methacholine and salbutamol [114].

UTE can also be combined with inhaled contrast agents such as [129]Xe and oxygen to augment pulmonary assessment. For [129]Xe MRI, UTE sequences can improve image resolution and decrease breath hold times when compared to the conventionally used gradient recall echo (GRE) sequences [115]. Shortened T1 relaxation time in the presence of 100% oxygen also makes UTE desirable for OE-MRI and allows whole lung acquisitions with isotropic resolution in as little as 5 minutes alongside simultaneous structural imaging [116].

8.4.6 Dynamic Contrast-Enhanced MRI

Dynamic contrast-enhanced (DCE) MRI uses gadolinium-based intravenous contrast agents for pulmonary imaging. As discussed previously, gadolinium shortens T1 relaxation, which creates contrast between imaged tissues [23]. While this is less helpful for imaging lung parenchyma (given T1 relaxation values are already very short), gadolinium is well suited to pulmonary vascular imaging and can be used to evaluate abnormal lesions through measurement of changes in perfusion and vascular permeability.

Pulmonary Vascular Imaging with Gadolinium

MR angiography (MRA) with gadolinium can be used to evaluate acute and chronic pulmonary vascular disease. For the diagnosis of chronic thromboembolic pulmonary hypertension, the performance of MRA is comparable to CT angiography, but the extent of disease is still better characterized with CT [117]. The use of MRA in pulmonary embolism (PE) has historically been less successful; the multicentre PIOPED III study found a quarter of MR images were inadequate and overall 43% of PEs were missed [118]. A later study by Schiebler et al. significantly improved on this, with 97.4% of scans of diagnostic quality and a 1-year negative predictive value of 97% for PE [119]. In clinical practice, MRA may yet become a suitable alternative to CT pulmonary angiography (CTPA) when radiation exposure and iodinated contrast agents need to be avoided.

Assessment of Parenchymal Lung Disease with DCE MRI

DCE MRI may be used to evaluate inflammatory activity and characterize severity of ILD. Chin et al. measured time to peak contrast enhancement after gadolinium injection in ILD patients and correlated these findings with imaging-targeted lung biopsies [120]. Areas found to be inflammatory predominant on biopsy were characterized by early contrast enhancement, whereas fibrotic-predominant areas displayed late enhancement. More recently, Buzan et al. compared peak contrast enhancement with disease severity (assessed instead by CT) in a mixed group of ILD patients [121]. They found non-IPF patients had earlier peak enhancement than the IPF cohort, but only in the non-IPF group could severity of lung disease be discriminated by time to peak enhancement. These findings may represent different vascularity at the histological level, whereby less severe or actively inflamed lesions in ILD demonstrate greater vascularization and permeability to contrast. Conversely, severely fibrotic areas in IPF have low vascularity and hence delayed contrast enhancement. These inferences are similar to those encountered with

dissolved phase ^{129}Xe MRI [78], and so DCE MRI may present an alternate avenue for assessment of lung disease activity and response to pulmonary therapeutics in ILD.

In children with CF, early detection of lung disease may also be possible with DCE MRI. Amaxopoulou et al. demonstrated a spectrum of perfusion deficits despite normal appearances of lung parenchyma on structural imaging [122]. This is likely a reflection of small airways disease that would otherwise not be detectable with standard imaging. Kellenberger et al. also measured perfusion abnormalities in a mixed cohort of paediatric patients with congenital thoracic malformations [123]. At peak enhancement, perfusion deficits were evident in areas of hyperinflation, cystic malformation, bronchopulmonary sequestration and bronchogenic cysts. Both studies show DCE MRI can add highly sensitive functional data to proton-density imaging, which together provide a comprehensive assessment of complex thoracic abnormalities in the paediatric population.

Assessment of Pulmonary Nodules and Malignancy with DCE MRI

Complex pharmacokinetic parameters of gadolinium enhancement in pulmonary nodules and lung cancers have been explored by several groups as a means of differentiating benign and malignant aetiologies. Ohno et al. found DCE MRI measures of perfusion were comparable to the PET-CT maximum standardized uptake value (SUV_{max}) in the differentiation of solitary pulmonary nodules [124]. Other groups have also suggested a role for DCE MRI in the risk stratification of malignancy, notably for evaluation of radiologically indeterminate lesions [125, 126].

DCE MRI biomarkers have also been evaluated as potential predictors of anti-cancer therapy. Huang et al. found DCE MRI-derived markers of tumour perfusion and vascular permeability (K_{trans} and K_{ep}) were correlated with both SUV_{max} and decreased tumour size in lung cancer patients 6 weeks after radiotherapy [127]. Xu et al. similarly found that after 1 week of chemotherapy, tumour K_{trans} and K_{ep} decreased significantly in lung cancer patients defined later as treatment responders according to conventional CT criteria [128]. As such, these MR imaging-based biomarkers could offer sensitive early measures of treatment response to complement existing imaging techniques.

8.5 The Potential Role of Thoracic MRI in Respiratory Medicine

At present, the thoracic MRI modalities discussed in this chapter are predominantly used as research tools. Clinical translation of these techniques is a complex matter and requires technical hurdles to be addressed alongside consideration of application and health economics. For certain modalities, such as hyperpolarized gas MRI, their cost and technical requirements will likely be prohibitive for some institutions. Instead, delivery by specialist tertiary centres may be more feasible. It is then important to consider how these imaging techniques can answer key research questions and complement existing clinical methods of assessing lung structure and function.

One of the most often cited advantages of thoracic MRI is the absence of ionizing radiation. X-ray-based investigations can deliver substantial radiation doses, and so their use should be carefully rationalized by clinicians. The stewardship of ionizing radiation is particularly relevant in the following situations:

- For patients with long-term/lifelong conditions who may require serial imaging
- The investigation of thromboembolic disease in pregnancy, where radiation exposure to the breast tissue may increase future cancer risk [129]
- When minimizing radiation exposure is essential, such as in ataxia-telangiectasia where faulty DNA repair mechanisms greatly increase lifetime cancer risk [130]

However, potential limitations of thoracic MRI must also be considered as some techniques may be particularly challenging for certain patients:

- Modalities requiring prolonged scanning time may be uncomfortable for younger patients or patients who struggle to lie flat for extended periods due to breathlessness.
- Patients with significant breathlessness or cough may also find breath hold manoeuvres difficult during techniques like hyperpolarized gas MRI.
- Inhalation of anoxic contrast agents (viz. ^{3}He and ^{129}Xe) may be unsuitable in patients with severe lung disease due to risk of hypoxia.
- Some inhaled contrast agents can also cause sedation: ^{129}Xe is known to have anaesthetic effects [131], and high-flow oxygen can precipitate hypercapnic respiratory failure in susceptible patients [132].

Therefore, wider use of thoracic MRI requires evaluation of both its merits and its clinical practicality. As highlighted previously, there are some discrete situations where techniques such as UTE MRI or MRA could be the preferred imaging method for specific patient groups. However, the complete replacement of conventional studies that use ionizing radiation remains impractical given the current demands and financial models of healthcare [133].

Phenotyping Lung Disease with Thoracic MRI

Some thoracic MRI research has focused on the phenotyping of airways disease, namely in asthma and COPD. Increasingly, different asthma phenotypes are recognized, which has allowed treatments such as biologics to be tailored to the individual patient [134]. Phenotyping of asthma with MRI shows promise and highlights some of the difficulties encountered in the assessment of asthma severity. Hyperpolarized gas MRI can demonstrate regional ventilation defects, and while correlation with disease severity and spirometry is observed, some patients have marked ventilation defects in the presence of preserved spirometry [135]. When compared to other measurements such as CT and provocation testing, ventilation defects can also be indicative of regional airway remodelling [136]. Therefore, MRI may be particularly interesting when its findings are discordant with standard diagnostic tools like spirometry.

Historically, COPD has been divided into emphysematous and chronic bronchitis phenotypes, but broader classification based on anatomical, physiological or pathophysiological criteria has been suggested [137]. Some of these patterns of disease can be demonstrated using MRI, which could then be used to guide therapy. One example is lung volume reduction (LVR), given the increasing use of endobronchial valves. Patient selection for LVR is key, with favourable outcomes often linked to upper lobe predominant emphysema [138]. However, valve insertion in lower lobe predominant emphysema has also proven useful when collateral ventilation is minimal [139]. Hyperpolarized ^{3}He MRI has been used retrospectively to identify alternate targets for endobronchial valve insertion undetected on CT. In theory, this may increase the number of patients to whom LVR could be offered, but further research is required [140]. ^{3}He MRI has also been used to detect bronchodilator response [141] and identify patients with the 'frequent exacerbator' phenotype [142], but these applications would likely be harder to justify in clinical practice.

Assessing Response to Novel Therapeutics

It has been established that evidence of lung disease can be detected in patients with CF who have normal spirometry using MRI and lung clearance index (LCI) [143, 144]. As such, traditional measures like FEV_1 may lack the sensitivity required for future clinical practice and research trials. Instead, thoracic MRI using hyperpolarized gases or non-contrast functional measures could serve as imaging-based biomarkers to evaluate therapies like CFTR modulators and aid longitudinal monitoring of lung disease [48, 145].

As discussed previously, hyperpolarized ^{129}Xe and DCE MRI methods may become useful in assessing different levels of disease activity in ILD [78, 121]. In particular, functional imaging with hyperpolarized ^{129}Xe demonstrates longitudinal deterioration of gas exchange in IPF with a greater sensitivity than transfer factor (D_{LCO}) [146]. In this way, early detection of lung disease progression could be feasible and become an invaluable tool for the evaluation of novel therapeutics.

8.6 Conclusion

There have been tremendous advances in thoracic MRI since the pioneering hyperpolarized gas and structural imaging studies of the mid-1990s. However, widespread adoption is yet to be realized with CT remaining the workhorse of cross-sectional lung imaging in clinical practice.

Structural imaging without ionizing radiation already has clear clinical applications, but further refinements to improve spatial resolution are needed to bridge the gap between thoracic MRI and CT. Functional imaging shows tremendous promise, and we can expect this to become increasingly relevant in the delivery of cutting-edge medical therapies that require highly sensitive measures of lung function.

Substantial work is still required for clinical translation, but the potential of co-registered structural and functional imaging without ionizing radiation is an exciting prospect for the future of respiratory care.

References

1. Moldoveanu B, Otmishi P, Jani P, Walker J, Sarmiento X, Guardiola J, et al. Inflammatory mechanisms in the lung. J Inflamm Res. 2009;2:1–11.
2. Sweeney RM, McAuley DF. Acute respiratory distress syndrome. Lancet. 2016;388(10058):2416–30.
3. James AL, Wenzel S. Clinical relevance of airway remodelling in airway diseases. Eur Respir J. 2007;30(1):134–55.
4. Cottin V, Hirani NA, Hotchkin DL, Nambiar AM, Ogura T, Otaola M, et al. Presentation, diagnosis and clinical course of the spectrum of progressive-fibrosing interstitial lung diseases. Eur Respir Rev. 2018;27(150):180076.
5. Cantin AM, Hartl D, Konstan MW, Chmiel JF. Inflammation in cystic fibrosis lung disease: pathogenesis and therapy. J Cyst Fibros. 2015;14(4):419–30.
6. Oudkerk M, Devaraj A, Vliegenthart R, Henzler T, Prosch H, Heussel CP, et al. European position statement on lung cancer screening. Lancet Oncol. 2017;18(12):e754–e66.
7. Pooley RA. Fundamental physics of MR imaging. Radiographics. 2005;25(4):1087–99.
8. Currie S, Hoggard N, Craven IJ, Hadjivassiliou M, Wilkinson ID. Understanding MRI: basic MR physics for physicians. Postgrad Med J. 2013;89(1050):209–23.

9. Ertürk MA, Wu X, Eryaman Y, Van de Moortele P-F, Auerbach EJ, Lagore RL, et al. Toward imaging the body at 10.5 tesla. Magn Reson Med. 2017;77(1):434–43.

10. Jenkinson M, Chappell M. Introduction to neuroimaging analysis. New York: Oxford University Press; 2018.

11. Grover VPB, Tognarelli JM, Crossey MME, Cox IJ, Taylor-Robinson SD, McPhail MJW. Magnetic resonance imaging: principles and techniques: lessons for clinicians. J Clin Exp Hepatol. 2015;5(3):246–55.

12. Sands MJ, Levitin A. Basics of magnetic resonance imaging. Semin Vasc Surg. 2004;17(2):66–82.

13. Westbrook C. In: Westbrook C, Roth CK, Talbot JM, editors. MRI in practice. 4th ed. Oxford: Wiley-Blackwell; 2011.

14. Nitz WR, Reimer P. Contrast mechanisms in MR imaging. Eur Radiol. 1999;9(6):1032–46.

15. Vassiliou VS, Cameron D, Prasad SK, Gatehouse PD. Magnetic resonance imaging: physics basics for the cardiologist. JRSM Cardiovasc Dis. 2018;7:2048004018772237.

16. Mansfield P, Maudsley AA. Medical imaging by NMR. Br J Radiol. 1977;50(591):188–94.

17. Damadian R, Goldsmith M, Minkoff L. NMR in cancer: XVI. FONAR image of the live human body. Physiol Chem Phys. 1977;9(1):97–100, 108.

18. Iezzi R, Larici AR, Franchi P, Marano R, Magarelli N, Posa A, et al. Tailoring protocols for chest CT applications: when and how? Diagn Interv Radiol (Ankara, Turkey). 2017;23(6):420–7.

19. Thakur Y, McLaughlin PD, Mayo JR. Strategies for radiation dose optimization. Curr Radiol Rep. 2013;1(1):1–10.

20. Gustafsson T, Eriksson A, Wingren CJ. Multivariate linear regression modelling of lung weight in 24,056 Swedish medico-legal autopsy cases. J Forensic Legal Med. 2017;46:20–2.

21. Torres L, Kammerman J, Hahn AD, Zha W, Nagle SK, Johnson K, et al. Structure-function imaging of lung disease using ultrashort echo time MRI. Acad Radiol. 2019;26(3):431–41.

22. Kauczor HU, Kreitner KF. MRI of the pulmonary parenchyma. Eur Radiol. 1999;9(9):1755–64.

23. Lipton ML. In: Lipton ML, editor. Totally accessible MRI: a user's guide to principles, technology, and applications. New York: Springer; 2008.

24. Kruger SJ, Nagle SK, Couch MJ, Ohno Y, Albert M, Fain SB. Functional imaging of the lungs with gas agents. J Magn Reson Imaging: JMRI. 2016;43(2):295–315.

25. Wild JM, Marshall H, Bock M, Schad LR, Jakob PM, Puderbach M, et al. MRI of the lung (1/3): methods. Insights Imaging. 2012;3(4):345–53.

26. Johnson KM, Fain SB, Schiebler ML, Nagle S. Optimized 3D ultrashort echo time pulmonary MRI. Magn Reson Med. 2013;70(5):1241–50.

27. Oechsner M, Pracht ED, Staeb D, Arnold JFT, Köstler H, Hahn D, et al. Lung imaging under free-breathing conditions. Magn Reson Med. 2009;61(3):723–7.

28. Raptis CA, Ludwig DR, Hammer MM, Luna A, Broncano J, Henry TS, et al. Building blocks for thoracic MRI: challenges, sequences, and protocol design. J Magn Reson Imaging: JMRI. 2019;50:682–701.

29. Albert MS, Cates GD, Driehuys B, Happer W, Saam B, Springer CS, et al. Biological magnetic resonance imaging using laser-polarized 129Xe. Nature. 1994;370(6486):199–201.

30. Mugler JP III, Driehuys B, Brookeman JR, Cates GD, Berr SS, Bryant RG, et al. MR imaging and spectroscopy using hyperpolarized 129Xe gas: preliminary human results. Magn Reson Med. 1997;37(6):809–15.

31. Middleton H, Black RD, Saam B, Cates GD, Cofer GP, Guenther R, et al. MR imaging with hyperpolarized 3He gas. Magn Reson Med. 1995;33(2):271–5.

32. Mugler JP III, Altes TA. Hyperpolarized 129Xe MRI of the human lung. J Magn Reson Imaging. 2013;37(2):313–31.

33. Roos JE, McAdams HP, Kaushik SS, Driehuys B. Hyperpolarized gas MR imaging: technique and applications. Magn Reson Imaging Clin N Am. 2015;23(2):217–29.

34. Ebner L, Kammerman J, Driehuys B, Schiebler ML, Cadman RV, Fain SB. The role of hyperpolarized (129)xenon in MR imaging of pulmonary function. Eur J Radiol. 2017;86:343–52.

35. Altes T, Gersbach J, Mata J, Mugler III J, Brookeman J, de Lange E, editors. Evaluation of the safety of hyperpolarized helium-3 gas as an inhaled contrast agent for MRI. In: Proceedings of the Fifteenth Meeting of the International Society for Magnetic Resonance in Medicine Berkeley, CA: International Society for Magnetic Resonance in Medicine; 2007.

36. Lutey BA, Lefrak SS, Woods JC, Tanoli T, Quirk JD, Bashir A, et al. Hyperpolarized 3He MR imaging: physiologic monitoring observations and safety considerations in 100 consecutive subjects. Radiology. 2008;248(2):655–61.

37. Wild JM, Paley MNJ, Viallon M, Schreiber WG, van Beek EJR, Griffiths PD. k-Space filtering in 2D gradient-echo breath-hold hyperpolarized 3He MRI: spatial resolution and signal-to-noise ratio considerations. Magn Reson Med. 2002;47(4):687–95.

38. Altes TA, de Lange EE. Applications of hyperpolarized helium-3 gas magnetic resonance imaging in pediatric lung disease. Top Magn Reson Imaging: TMRI. 2003;14(3):231–6.

39. Fain S, Schiebler ML, McCormack DG, Parraga G. Imaging of lung function using hyperpolarized helium-3 magnetic resonance imaging: review of current and emerging translational methods and applications. J Magn Reson Imaging: JMRI. 2010;32(6):1398–408.

40. Kauczor HU, Ebert M, Kreitner KF, Nilgens H, Surkau R, Heil W, et al. Imaging of the lungs using 3He MRI: preliminary clinical experience in 18 patients with and without lung disease. J Magn Reson Imaging: JMRI. 1997;7(3):538–43.

41. Mathew L, Evans A, Ouriadov A, Etemad-Rezai R, Fogel R, Santyr G, et al. Hyperpolarized 3He magnetic resonance imaging of chronic obstructive pulmonary disease: reproducibility at 3.0 tesla. Acad Radiol. 2008;15(10):1298–311.

42. Samee S, Altes T, Powers P, de Lange EE, Knight-Scott J, Rakes G, et al. Imaging the lungs in asthmatic patients by using hyperpolarized helium-3 magnetic resonance: assessment of response to methacholine and exercise challenge. J Allergy Clin Immunol. 2003;111(6):1205–11.

43. Altes TA, Powers PL, Knight-Scott J, Rakes G, Platts-Mills TA, de Lange EE, et al. Hyperpolarized 3He MR lung ventilation imaging in asthmatics: preliminary findings. J Magn Reson Imaging: JMRI. 2001;13(3):378–84.

44. de Lange EE, Altes TA, Patrie JT, Battiston JJ, Juersivich AP, Mugler JP 3rd, et al. Changes in regional airflow obstruction over time in the lungs of patients with asthma: evaluation with 3He MR imaging. Radiology. 2009;250(2):567–75.

45. Lee EY, Sun Y, Zurakowski D, Hatabu H, Khatwa U, Albert MS. Hyperpolarized 3He MR imaging of the lung: normal range of ventilation defects and PFT correlation in young adults. J Thorac Imaging. 2009;24(2):110–4.

46. Mentore K, Froh DK, de Lange EE, Brookeman JR, Paget-Brown AO, Altes TA. Hyperpolarized HHe 3 MRI of the lung in cystic fibrosis: assessment at baseline and after bronchodilator and airway clearance treatment. Acad Radiol. 2005;12(11):1423–9.

47. Woodhouse N, Wild JM, van Beek EJ, Hoggard N, Barker N, Taylor CJ. Assessment of hyperpolarized 3He lung MRI for regional evaluation of interventional therapy: a pilot study in pediatric cystic fibrosis. J Magn Reson Imaging: JMRI. 2009;30(5):981–8.

48. Altes TA, Johnson M, Fidler M, Botfield M, Tustison NJ, Leiva-Salinas C, et al. Use of hyperpolarized helium-3 MRI to assess response to ivacaftor treatment in patients with cystic fibrosis. J Cyst Fibros. 2017;16(2):267–74.

49. Chilla GS, Tan CH, Xu C, Poh CL. Diffusion weighted magnetic resonance imaging and its recent trend-a survey. Quant Imaging Med Surg. 2015;5(3):407–22.

50. Baliyan V, Das CJ, Sharma R, Gupta AK. Diffusion weighted imaging: technique and applications. World J Radiol. 2016;8(9):785–98.

51. van Beek EJ, Wild JM, Kauczor HU, Schreiber W, Mugler JP III, de Lange EE. Functional MRI of the lung using hyperpolarized 3-helium gas. J Magn Reson Imaging. 2004;20(4):540–54.

52. Saam BT, Yablonskiy DA, Kodibagkar VD, Leawoods JC, Gierada DS, Cooper JD, et al. MR imaging of diffusion of 3He gas in healthy and diseased lungs. Magn Reson Med. 2000;44(2):174–9.

53. Salerno M, Altes TA, Brookeman JR, de Lange EE, Mugler JP 3rd. Rapid hyperpolarized 3He diffusion MRI of healthy and emphysematous human lungs using an optimized interleaved-spiral pulse sequence. J Magn Reson Imaging: JMRI. 2003;17(5):581–8.

54. Wang C, Miller GW, Altes TA, de Lange EE, Cates GD Jr, Mugler JP 3rd. Time dependence of 3He diffusion in the human lung: measurement in the long-time regime using stimulated echoes. Magn Reson Med. 2006;56(2):296–309.

55. Diaz S, Casselbrant I, Piitulainen E, Magnusson P, Peterson B, Wollmer P, et al. Validity of apparent diffusion coefficient hyperpolarized 3He-MRI using MSCT and pulmonary function tests as references. Eur J Radiol. 2009;71(2):257–63.

56. Kirby M, Mathew L, Wheatley A, Santyr GE, McCormack DG, Parraga G. Chronic obstructive pulmonary disease: longitudinal hyperpolarized (3) He MR imaging. Radiology. 2010;256(1):280–9.

57. Chan H-F, Weatherley N, Johns CS, Stewart N, Collier G, Bianchi SM, et al. Airway microstructure in idiopathic pulmonary fibrosis: assessment at hyperpolarized 3 He diffusion-weighted MRI. Radiology. 2019;291:181714.

58. Chan H-F, Stewart NJ, Parra-Robles J, Collier GJ, Wild JM. Whole lung morphometry with 3D multiple b-value hyperpolarized gas MRI and compressed sensing. Magn Reson Med. 2017;77(5):1916–25.

59. McCormack DG, Halko S, McKay S, Kirby M, Svenningsen S, Wheatley A, et al. Hyperpolarized 129Xe MRI feasibility, subject safety and tolerability: at the doorstep of clinical translation? In: A65 lung imaging: physiology and technology. American Thoracic Society; 2012. p. A2031-A.

60. Driehuys B, Martinez-Jimenez S, Cleveland ZI, Metz GM, Beaver DM, Nouls JC, et al. Chronic obstructive pulmonary disease: safety and tolerability of hyperpolarized 129Xe MR imaging in healthy volunteers and patients. Radiology. 2012;262(1):279–89.

61. Altes T, Mugler J, Dregely I, Ketel S, Ruset I, de Lange E, et al., editors. Hyperpolarized xenon-129 ventilation MRI: preliminary results in normal subjects and patients with lung disease. In: Proceedings 18th scientific meeting; 2010.

62. Kirby M, Svenningsen S, Owrangi A, Wheatley A, Farag A, Ouriadov A, et al. Hyperpolarized 3He and 129Xe MR imaging in healthy volunteers and patients with chronic obstructive pulmonary disease. Radiology. 2012;265(2):600–10.

63. Svenningsen S, Kirby M, Starr D, Leary D, Wheatley A, Maksym GN, et al. Hyperpolarized (3) He and (129) Xe MRI: differences in asthma before bronchodilation. J Magn Reson Imaging: JMRI. 2013;38(6):1521–30.

64. Costella S, Kirby M, Maksym GN, McCormack DG, Paterson NA, Parraga G. Regional pulmonary response to a methacholine challenge using

hyperpolarized (3)He magnetic resonance imaging. Respirology (Carlton, VIC). 2012;17(8):1237–46.

65. Doganay O, Matin T, Chen M, Kim M, McIntyre A, McGowan DR, et al. Time-series hyperpolarized xenon-129 MRI of lobar lung ventilation of COPD in comparison to V/Q-SPECT/CT and CT. Eur Radiol. 2018;29:4058–67.

66. He M, Driehuys B, Que LG, Huang YT. Using hyperpolarized (129)Xe MRI to quantify the pulmonary ventilation distribution. Acad Radiol. 2016;23(12):1521–31.

67. Thomen RP, Walkup LL, Roach DJ, Cleveland ZI, Clancy JP, Woods JC. Hyperpolarized (129)Xe for investigation of mild cystic fibrosis lung disease in pediatric patients. J Cyst Fibros. 2017;16(2):275–82.

68. Couch MJ, Thomen R, Kanhere N, Hu R, Ratjen F, Woods J, et al. A two-center analysis of hyperpolarized (129)Xe lung MRI in stable pediatric cystic fibrosis: potential as a biomarker for multi-site trials. J Cyst Fibros. 2019;18:728–33.

69. Walkup LL, Roach DJ, Hall CS, Gupta N, Thomen RP, Cleveland ZI, et al. Cyst ventilation heterogeneity and alveolar airspace dilation as early disease markers in lymphangioleiomyomatosis. Ann Am Thorac Soc. 2019;16(8):1008–16.

70. Kirby M, Svenningsen S, Kanhere N, Owrangi A, Wheatley A, Coxson HO, et al. Pulmonary ventilation visualized using hyperpolarized helium-3 and xenon-129 magnetic resonance imaging: differences in COPD and relationship to emphysema. J Appl Physiol (Bethesda, MD: 1985). 2013;114(6):707–15.

71. Matin TN, Rahman N, Nickol AH, Chen M, Xu X, Stewart NJ, et al. Chronic obstructive pulmonary disease: lobar analysis with hyperpolarized (129)Xe MR imaging. Radiology. 2017;282(3):857–68.

72. Thomen RP, Quirk JD, Roach D, Egan-Rojas T, Ruppert K, Yusen RD, et al. Direct comparison of (129) Xe diffusion measurements with quantitative histology in human lungs. Magn Reson Med. 2017;77(1):265–72.

73. Walkup LL, Thomen RP, Akinyi TG, Watters E, Ruppert K, Clancy JP, et al. Feasibility, tolerability and safety of pediatric hyperpolarized (129)Xe magnetic resonance imaging in healthy volunteers and children with cystic fibrosis. Pediatr Radiol. 2016;46(12):1651–62.

74. Kaushik SS, Freeman MS, Yoon SW, Liljeroth MG, Stiles JV, Roos JE, et al. Measuring diffusion limitation with a perfusion-limited gas--hyperpolarized 129Xe gas-transfer spectroscopy in patients with idiopathic pulmonary fibrosis. J Appl Physiol (Bethesda, MD: 1985). 2014;117(6):577–85.

75. Cleveland ZI, Cofer GP, Metz G, Beaver D, Nouls J, Kaushik SS, et al. Hyperpolarized Xe MR imaging of alveolar gas uptake in humans. PloS One. 2010;5(8):e12192-e.

76. Qing K, Shim Y, Tustison N, Altes T, Ruppert K, Mata J, et al. A19 getting polarized: MR imaging in obstructive lung disease: hyperpolarized xenon-129

MRI: a new tool to evaluate COPD. Am J Respir Crit Care Med. 2016;193:1.

77. Qing K, Ruppert K, Jiang Y, Mata JF, Miller GW, Shim YM, et al. Regional mapping of gas uptake by blood and tissue in the human lung using hyperpolarized xenon-129 MRI. J Magn Reson Imaging: JMRI. 2014;39(2):346–59.

78. Wang JM, Robertson SH, Wang Z, He M, Virgincar RS, Schrank GM, et al. Using hyperpolarized 129Xe MRI to quantify regional gas transfer in idiopathic pulmonary fibrosis. Thorax. 2018;73(1):21–8.

79. Wang Z, He M, Bier E, Rankine L, Schrank G, Rajagopal S, et al. Hyperpolarized (129) Xe gas transfer MRI: the transition from 1.5T to 3T. Magn Reson Med. 2018;80(6):2374–83.

80. Ohno Y, Hatabu H. Basics concepts and clinical applications of oxygen-enhanced MR imaging. Eur J Radiol. 2007;64(3):320–8.

81. Loffler R, Muller CJ, Peller M, Penzkofer H, Deimling M, Schwaiblmair M, et al. Optimization and evaluation of the signal intensity change in multisection oxygen-enhanced MR lung imaging. Magn Reson Med. 2000;43(6):860–6.

82. Mai VM, Liu B, Li W, Polzin J, Kurucay S, Chen Q, et al. Influence of oxygen flow rate on signal and T(1) changes in oxygen-enhanced ventilation imaging. J Magn Reson Imaging: JMRI. 2002;16(1):37–41.

83. Oechsner M, Pracht ED, Staeb D, Arnold JF, Kostler H, Hahn D, et al. Lung imaging under free-breathing conditions. Magn Reson Med. 2009;61(3):723–7.

84. Stadler A, Stiebellehner L, Jakob PM, Arnold JF, Eisenhuber E, von Katzler I, et al. Quantitative and o(2) enhanced MRI of the pathologic lung: findings in emphysema, fibrosis, and cystic fibrosis. Int J Biomed Imaging. 2007;2007:23624.

85. Jobst BJ, Triphan SM, Sedlaczek O, Anjorin A, Kauczor HU, Biederer J, et al. Functional lung MRI in chronic obstructive pulmonary disease: comparison of T1 mapping, oxygen-enhanced T1 mapping and dynamic contrast enhanced perfusion. PLoS One. 2015;10(3):e0121520.

86. Renne J, Lauermann P, Hinrichs JB, Schonfeld C, Sorrentino S, Gutberlet M, et al. Chronic lung allograft dysfunction: oxygen-enhanced T1-mapping MR imaging of the lung. Radiology. 2015;276(1):266–73.

87. Jakob PM, Wang T, Schultz G, Hebestreit H, Hebestreit A, Hahn D. Assessment of human pulmonary function using oxygen-enhanced T(1) imaging in patients with cystic fibrosis. Magn Reson Med. 2004;51(5):1009–16.

88. Kaireit TF, Sorrentino SA, Renne J, Schoenfeld C, Voskrebenzev A, Gutberlet M, et al. Functional lung MRI for regional monitoring of patients with cystic fibrosis. PLoS One. 2017;12(12):e0187483.

89. Ohno Y, Hatabu H, Takenaka D, Adachi S, Van Cauteren M, Sugimura K. Oxygen-enhanced MR ventilation imaging of the lung: preliminary clinical experience in 25 subjects. AJR Am J Roentgenol. 2001;177(1):185–94.

90. Ohno Y, Iwasawa T, Seo JB, Koyama H, Takahashi H, Oh Y-M, et al. Oxygen-enhanced magnetic resonance imaging versus computed tomography. Am J Respir Crit Care Med. 2008;177(10):1095–102.

91. Ohno Y, Koyama H, Matsumoto K, Onishi Y, Nogami M, Takenaka D, et al. Oxygen-enhanced MRI vs. quantitatively assessed thin-section CT: pulmonary functional loss assessment and clinical stage classification of asthmatics. Eur J Radiol. 2011;77(1):85–91.

92. Ohno Y, Nishio M, Koyama H, Yoshikawa T, Matsumoto S, Seki S, et al. Oxygen-enhanced MRI for patients with connective tissue diseases: comparison with thin-section CT of capability for pulmonary functional and disease severity assessment. Eur J Radiol. 2014;83(2):391–7.

93. Zha W, Nagle SK, Cadman RV, Schiebler ML, Fain SB. Three-dimensional isotropic functional imaging of cystic fibrosis using oxygen-enhanced MRI: comparison with hyperpolarized (3)He MRI. Radiology. 2019;290(1):229–37.

94. Renne J, Hinrichs J, Schonfeld C, Gutberlet M, Winkler C, Faulenbach C, et al. Noninvasive quantification of airway inflammation following segmental allergen challenge with functional MR imaging: a proof of concept study. Radiology. 2015;274(1):267–75.

95. Edelman RR, Hatabu H, Tadamura E, Li W, Prasad PV. Noninvasive assessment of regional ventilation in the human lung using oxygen–enhanced magnetic resonance imaging. Nat Med. 1996;2(11):1236–9.

96. Deibler AR, Pollock JM, Kraft RA, Tan H, Burdette JH, Maldjian JA. Arterial spin-labeling in routine clinical practice, part 1: technique and artifacts. AJNR Am J Neuroradiol. 2008;29(7):1228–34.

97. Mai VM, Bankier AA, Prasad PV, Li W, Storey P, Edelman RR, et al. MR ventilation-perfusion imaging of human lung using oxygen-enhanced and arterial spin labeling techniques. J Magn Reson Imaging: JMRI. 2001;14(5):574–9.

98. Bergin CJ, Pauly JM, Macovski A. Lung parenchyma: projection reconstruction MR imaging. Radiology. 1991;179(3):777–81.

99. Weiger M, Brunner DO, Dietrich BE, Muller CF, Pruessmann KP. ZTE imaging in humans. Magn Reson Med. 2013;70(2):328–32.

100. Miller GW, Mugler JP 3rd, Sá RC, Altes TA, Prisk GK, Hopkins SR. Advances in functional and structural imaging of the human lung using proton MRI. NMR Biomed. 2014;27(12):1542–56.

101. Tibiletti M, Paul J, Bianchi A, Wundrak S, Rottbauer W, Stiller D, et al. Multistage three-dimensional UTE lung imaging by image-based self-gating. Magn Reson Med. 2016;75(3):1324–32.

102. Higano NS, Hahn AD, Tkach JA, Cao X, Walkup LL, Thomen RP, et al. Retrospective respiratory self-gating and removal of bulk motion in pulmonary UTE MRI of neonates and adults. Magn Reson Med. 2017;77(3):1284–95.

103. Triphan SM, Breuer FA, Gensler D, Kauczor HU, Jakob PM. Oxygen enhanced lung MRI by simultaneous measurement of T1 and T2 * during free breathing using ultrashort TE. J Magn Reson Imaging: JMRI. 2015;41(6):1708–14.

104. Higano N, Walkup L, Hahn A, Thomen R, Merhar S, Kingma P, et al., editors. Quantification of neonatal lung parenchymal density via ultra-short echo-time (UTE) magnetic resonance imaging (MRI). American Journal of Respiratory and Critical Care Medicine. New York: American Thoracic Society; 2016.

105. Higano NS, Spielberg DR, Fleck RJ, Schapiro AH, Walkup LL, Hahn AD, et al. Neonatal pulmonary magnetic resonance imaging of bronchopulmonary dysplasia predicts short-term clinical outcomes. Am J Respir Crit Care Med. 2018;198(10):1302–11.

106. Roach DJ, Cremillieux Y, Fleck RJ, Brody AS, Serai SD, Szczesniak RD, et al. Ultrashort echo-time magnetic resonance imaging is a sensitive method for the evaluation of early cystic fibrosis lung disease. Ann Am Thorac Soc. 2016;13(11):1923–31.

107. Dournes G, Menut F, Macey J, Fayon M, Chateil JF, Salel M, et al. Lung morphology assessment of cystic fibrosis using MRI with ultra-short echo time at submillimeter spatial resolution. Eur Radiol. 2016;26(11):3811–20.

108. Ma W, Sheikh K, Svenningsen S, Pike D, Guo F, Etemad-Rezai R, et al. Ultra-short echo-time pulmonary MRI: evaluation and reproducibility in COPD subjects with and without bronchiectasis. J Magn Reson Imaging: JMRI. 2015;41(5):1465–74.

109. Roach DJ, Cremillieux Y, Serai SD, Thomen RP, Wang H, Zou Y, et al. Morphological and quantitative evaluation of emphysema in chronic obstructive pulmonary disease patients: a comparative study of MRI with CT. J Magn Reson Imaging: JMRI. 2016;44(6):1656–63.

110. Chassagnon G, Martin C, Marini R, Vakalopoulou M, Regent A, Mouthon L, et al. Use of elastic registration in pulmonary MRI for the assessment of pulmonary fibrosis in patients with systemic sclerosis. Radiology. 2019;291(2):487–92.

111. Burris NS, Johnson KM, Larson PE, Hope MD, Nagle SK, Behr SC, et al. Detection of small pulmonary nodules with ultrashort Echo time sequences in oncology patients by using a PET/MR system. Radiology. 2016;278(1):239–46.

112. Cha MJ, Park HJ, Paek MY, Stemmer A, Lee ES, Park SB, et al. Free-breathing ultrashort echo time lung magnetic resonance imaging using stack-of-spirals acquisition: a feasibility study in oncology patients. Magn Reson Imaging. 2018;51:137–43.

113. Wielputz MO, Lee HY, Koyama H, Yoshikawa T, Seki S, Kishida Y, et al. Morphologic characterization of pulmonary nodules with ultrashort TE MRI at 3T. AJR Am J Roentgenol. 2018;210(6):1216–25.

114. Sheikh K, Guo F, Capaldi DP, Ouriadov A, Eddy RL, Svenningsen S, et al. Ultrashort echo time MRI bio-

markers of asthma. J Magn Reson Imaging: JMRI. 2017;45(4):1204–15.

115. Willmering MM, Niedbalski PJ, Wang H, Walkup LL, Robison RK, Pipe JG, et al. Improved pulmonary (129) Xe ventilation imaging via 3D-spiral UTE MRI. Magn Reson Med. 2020;84(1):312–20.

116. Kruger SJ, Fain SB, Johnson KM, Cadman RV, Nagle SK. Oxygen-enhanced 3D radial ultrashort echo time magnetic resonance imaging in the healthy human lung. NMR Biomed. 2014;27(12):1535–41.

117. Ley S, Ley-Zaporozhan J, Pitton MB, Schneider J, Wirth GM, Mayer E, et al. Diagnostic performance of state-of-the-art imaging techniques for morphological assessment of vascular abnormalities in patients with chronic thromboembolic pulmonary hypertension (CTEPH). Eur Radiol. 2012;22(3):607–16.

118. Stein PD, Chenevert TL, Fowler SE, Goodman LR, Gottschalk A, Hales CA, et al. Gadolinium-enhanced magnetic resonance angiography for pulmonary embolism: a multicenter prospective study (PIOPED III). Ann Intern Med. 2010;152(7):434–43.

119. Schiebler ML, Nagle SK, François CJ, Repplinger MD, Hamedani AG, Vigen KK, et al. Effectiveness of MR angiography for the primary diagnosis of acute pulmonary embolism: clinical outcomes at 3 months and 1 year. J Magn Reson Imaging. 2013;38(4):914–25.

120. Yi CA, Lee KS, Han J, Chung MP, Chung MJ, Shin KM. 3-T MRI for differentiating inflammation- and fibrosis-predominant lesions of usual and nonspecific interstitial pneumonia: comparison study with pathologic correlation. AJR Am J Roentgenol. 2008;190(4):878–85.

121. Maria Ta B, Andreas W, Christopher MR, Michael K, Claus Peter H, Mark K, et al. Contrast agent accumulation patterns in chronic interstitial lung disease using 5D MRI. Br J Radiol. 2020;93:20190121.

122. Amaxopoulou C, Gnannt R, Higashigaito K, Jung A, Kellenberger CJ. Structural and perfusion magnetic resonance imaging of the lung in cystic fibrosis. Pediatr Radiol. 2018;48(2):165–75.

123. Kellenberger CJ, Amaxopoulou C, Moehrlen U, Bode PK, Jung A, Geiger J. Structural and perfusion magnetic resonance imaging of congenital lung malformations. Pediatr Radiol. 2020;50:1083–94.

124. Ohno Y, Fujisawa Y, Yui M, Takenaka D, Koyama H, Sugihara N, et al. Solitary pulmonary nodule: comparison of quantitative capability for differentiation and management among dynamic CE-perfusion MRI at 3 T system, dynamic CE-perfusion ADCT and FDG-PET/CT. Eur J Radiol. 2019;115:22–30.

125. Horn M, Oechsner M, Gardarsdottir M, Köstler H, Müller MF. Dynamic contrast-enhanced MR imaging for differentiation of rounded atelectasis from neoplasm. J Magn Reson Imaging. 2010;31(6):1364–70.

126. Wu W, Zhou S, Hippe DS, Liu H, Wang Y, Mayr NA, et al. Whole-lesion DCE-MRI intensity histogram analysis for diagnosis in patients with suspected lung cancer. Acad Radiol. 2020;28:e27–34.

127. Huang Y-S, Chen JL-Y, Hsu F-M, Huang J-Y, Ko W-C, Chen Y-C, et al. Response assessment of stereotactic body radiation therapy using dynamic contrast-enhanced integrated MR-PET in non-small cell lung cancer patients. J Magn Reson Imaging. 2018;47(1):191–9.

128. Xu J, Mei L, Liu L, Wang K, Zhou Z, Zheng J. Early assessment of response to chemotherapy in lung cancer using dynamic contrast-enhanced MRI: a proof-of-concept study. Clin Radiol. 2018;73(7):625–31.

129. Bourjeily G, Paidas M, Khalil H, Rosene-Montella K, Rodger M. Pulmonary embolism in pregnancy. Lancet. 2010;375(9713):500–12.

130. Rothblum-Oviatt C, Wright J, Lefton-Greif MA, McGrath-Morrow SA, Crawford TO, Lederman HM. Ataxia telangiectasia: a review. Orphanet J Rare Dis. 2016;11(1):159.

131. Cullen SC, Gross EG. The anesthetic properties of xenon in animals and human beings, with additional observations on krypton. Science (New York, NY). 1951;113(2942):580–2.

132. Abdo WF, Heunks LMA. Oxygen-induced hypercapnia in COPD: myths and facts. Crit Care. 2012;16(5):323.

133. Power SP, Moloney F, Twomey M, James K, O'Connor OJ, Maher MM. Computed tomography and patient risk: facts, perceptions and uncertainties. World J Radiol. 2016;8(12):902–15.

134. Papi A, Brightling C, Pedersen SE, Reddel HK. Asthma. Lancet. 2018;391(10122):783–800.

135. de Lange EE, Altes TA, Patrie JT, Gaare JD, Knake JJ, Mugler JP III, et al. Evaluation of asthma with hyperpolarized helium-3 MRI: correlation with clinical severity and spirometry. Chest. 2006;130(4):1055–62.

136. Svenningsen S, Kirby M, Starr D, Coxson HO, Paterson NAM, McCormack DG, et al. What are ventilation defects in asthma? Thorax. 2014;69(1):63–71.

137. Mirza S, Benzo R. Chronic obstructive pulmonary disease phenotypes: implications for care. Mayo Clin Proc. 2017;92(7):1104–12.

138. Davey C, Zoumot Z, Jordan S, McNulty WH, Carr DH, Hind MD, et al. Bronchoscopic lung volume reduction with endobronchial valves for patients with heterogeneous emphysema and intact interlobar fissures (the BeLieVeR-HIFi study): a randomised controlled trial. Lancet. 2015;386(9998):1066–73.

139. Eberhardt R, Herth FJF, Radhakrishnan S, Gompelmann D. Comparing clinical outcomes in upper versus lower lobe endobronchial valve treatment in severe emphysema. Respiration. 2015;90(4):314–20.

140. Adams CJ, Capaldi DPI, Di Cesare R, McCormack DG, Parraga G. On the potential role of MRI biomarkers of COPD to guide bronchoscopic lung volume reduction. Acad Radiol. 2018;25(2):159–68.

141. Kirby M, Mathew L, Heydarian M, Etemad-Rezai R, McCormack DG, Parraga G. Chronic obstructive pulmonary disease: quantification of bronchodilator

effects by using hyperpolarized (3)He MR imaging. Radiology. 2011;261(1):283–92.

142. Kirby M, Pike D, Coxson HO, McCormack DG, Parraga G. Hyperpolarized 3He ventilation defects used to predict pulmonary exacerbations in mild to moderate chronic obstructive pulmonary disease. Radiology. 2014;273(3):887–96.

143. Horsley AR, Gustafsson PM, Macleod KA, Saunders C, Greening AP, Porteous DJ, et al. Lung clearance index is a sensitive, repeatable and practical measure of airways disease in adults with cystic fibrosis. Thorax. 2008;63(2):135–40.

144. Smith L, Marshall H, Aldag I, Horn F, Collier G, Hughes D, et al. Longitudinal assessment of chil-dren with mild cystic fibrosis using hyperpolarized gas lung magnetic resonance imaging and lung clearance index. Am J Respir Crit Care Med. 2018;197(3):397–400.

145. Martini K, Gygax CM, Benden C, Morgan AR, Parker GJM, Frauenfelder T. Volumetric dynamic oxygen-enhanced MRI (OE-MRI): comparison with CT Brody score and lung function in cystic fibrosis patients. Eur Radiol. 2018;28(10):4037–47.

146. Weatherley ND, Stewart NJ, Chan H-F, Austin M, Smith LJ, Collier G, et al. Hyperpolarised xenon magnetic resonance spectroscopy for the longi-tudinal assessment of changes in gas diffusion in IPF. Thorax. 2019;74(5):500–2.

Overview on Interactive Role of Inflammation, Reactive Oxygen Species, and Calcium Signaling in Asthma, COPD, and Pulmonary Hypertension

9

Lillian Truong, Yun-Min Zheng, Sharath Kandhi, and Yong-Xiao Wang

Abstract

Inflammatory signaling is a major component in the development and progression of many lung diseases, including asthma, chronic obstructive pulmonary disorder (COPD), and pulmonary hypertension (PH). This chapter will provide a brief overview of asthma, COPD, and PH and how inflammation plays a vital role in these diseases. Specifically, we will discuss the role of reactive oxygen species (ROS) and Ca^{2+} signaling in inflammatory cellular responses and how these interactive signaling pathways mediate the development of asthma, COPD, and PH. We will also deliberate the key cellular responses of pulmonary arterial (PA) smooth muscle cells (SMCs) and airway SMCs (ASMCs) in these devastating lung diseases. The analysis of the importance of inflammation will shed light on the key questions remaining in this field and highlight molecular targets that are worth exploring. The crucial findings will not only demonstrate the novel roles of essential signaling molecules such as Rieske iron-sulfur protein and ryanodine receptor in the development and progress of asthma, COPD, and PH but also offer advanced insight for creating more effective and new therapeutic targets for these devastating inflammatory lung diseases.

Keywords

Inflammation · Ca^{2+} signaling · Reactive oxygen species · Rieske iron-sulfur protein · Nicotine · Contraction · Remodeling · Asthma · Chronic obstructive pulmonary disorder · Pulmonary hypertension

Abbreviations

ASK1	Apoptosis signal-regulating kinase 1
ASMC	Airway smooth muscle cell
ATM	Ataxia-telangiectasia mutated
BALF	Bronchoalveolar lavage fluid
CaM	Calmodulin
CDK	Cyclin-dependent kinase
CICR	Calcium-induced calcium release
CIRG	Calcium-induced ROS generation
COPD	Chronic obstructive pulmonary disease
CS	Cigarette smoke
DAG	Diacylglycerol

L. Truong · Y.-M. Zheng · S. Kandhi (✉) ·
Y.-X. Wang (✉)
Department of Molecular and Cellular Physiology,
Albany Medical College, Albany, NY, USA
e-mail: Sharath_kandhi@nymc.edu;
wangy@amc.edu

DAMP	Damage-associated molecular pattern
ECM	Extracellular matrix
ETC	Electron transport chain
GPCR	G protein-coupled receptor
Gpx	Glutathione peroxidase
GSH	Glutathione
HIF-1α	Hypoxia-inducible factor-1α
HPV	Hypoxic pulmonary vasoconstriction
IKK	IκB kinase
IP$_3$	Inositol triphosphate
IP$_3$R	Inositol-1,4,5-triphosphate receptor
Kv	Voltage-gated potassium
LTCC	L-type voltage-gated calcium channel
MAPK	Mitogen-activated protein kinase
MLCK	Myosin light chain kinase
MMP	Matrix metalloproteinases
nAChR	Nicotinic acetylcholine receptor
NADPH	Nicotinamide dinucleotide phosphate
NEMO	NF-κB essential modulator
NF-κB	Nuclear factor-κB
NIK	NF-κB-inducing kinase
NLR	NOD-like receptor
NOS	Nitric oxide synthase
PA	Pulmonary artery
PAMP	Pathogen-associated molecular pattern
PH	Pulmonary hypertension
PKC	Protein kinase C
PPA	Pulmonary arterial pressure
PRR	Pattern recognition receptor
PVR	Pulmonary vascular resistance
RHD	Rel homology domain
RICR	ROS-induced calcium release
RISP	Rieske iron-sulfur protein
ROS	Reactive oxygen species
RyR	Ryanodine receptor
SMC	Smooth muscle cells
SR	Sarcoplasmic reticulum
TAK1	TGF-β-activated kinase 1
TLR	Toll-like receptor
TNF-α	Tumor necrosis factor-α
TRP	Transient receptor potential
TRPC	Canonical TRP

9.1   Introduction

Asthma, COPD, and PH are considered common chronic lung inflammatory diseases [1–5]. These three diseases persist in the presence of specific triggers such as environmental irritants (e.g., airborne dust particles) or inhaled irritants such as cigarette smoke (CS) [1, 3, 5]. Exposure to these triggers leads to the activation of the inflammatory cells (i.e., macrophages, neutrophils, and T cells). Although these diseases are characterized as chronic inflammatory disorders, the underlying signaling mechanisms are shared and distinct to each disease.

9.1.1   Asthma

Asthma is a chronic respiratory condition characterized by narrowing airways due to inflammation and overproduction of mucus [3, 5]. While usually minor, asthma can be a significant problem that interferes with daily activities, worsened by strenuous physical activity, and may lead to a life-threatening asthma attack [5].

Asthmatic symptoms will vary from person to person, varying in frequency in asthma attacks and severity in attacks and other symptoms. Asthma symptoms often include shortness of breath, chest tightness, wheezing when exhaling, and trouble sleeping caused by shortness of breath, coughing, or wheezing [3]. Specific environments or situations can cause asthmatic flare-ups, which include exercise-induced asthma, occupational asthma (triggered by workplace irritants like fumes, gases, or dust), and allergy-induced asthma (triggered by an airborne substance like pollen, spores, or pet dander) [3, 5].

Complications that arise from asthma can include interference with sleep, work, other activities, permanent narrowing of the airways, hospitalizations for severe attacks, and side effects from long-term use of medications to stabilize severe asthma.

9.1.2 COPD

COPD is a chronic inflammatory disease characterized by the obstruction of airflow due to the narrowing of the airways [6, 7]. It is the third leading cause of death and affects approximately 16 million people in the United States [6]. According to a report from the Centers for Disease Control and Prevention, COPD is more likely to occur in people aged 65 and older, women, current and former smokers, and people with a history of asthma [4, 6, 8, 9].

COPD presents with pulmonary and extrapulmonary manifestations. As the disease progresses, respiratory symptoms include frequent coughing and/or wheezing, excess mucous production, shortness of breath, and tight chest. Approximately 10–30% of patients with moderate to severe COPD will develop pulmonary hypertension (PH), as defined by a PA pressure greater than 20 mmHg at rest [1, 10]. This physiological state of the pulmonary vasculature is correlated with increased mortality and morbidity.

Patients with COPD often show cardiac manifestations. Patients with COPD carry an increased risk of death due to cardiac arrhythmias, myocardial infarctions, coronary artery disease, or congestive heart failure, caused by increased sympathetic tone [11–13]. Two-thirds of these patients show evidence of right ventricular hypertrophy as indicated by right heart catheterization [12]. Additionally, these cardiac manifestations may also be linked to chronic or intermittent hypoxia-induced vascular remodeling and hyperresponsiveness. The systemic inflammation associated with COPD can also contribute to the development and progression of atherosclerosis. Pulmonary manifestations of COPD (i.e., hypoxemia, systemic inflammation, and arterial stiffness) often exacerbate cardiac manifestations, increasing the cardiovascular risk of disease and further worsening COPD symptoms.

9.1.3 PH

PH is characterized as an increased pulmonary arterial pressure (P_{PA}), $P_{PA} > 20$ mmHg at rest, and is a result of pulmonary vasoremodeling and vasoconstriction of the PA [1, 10, 14]. A high percentage of COPD patients develop pulmonary hypertension [1, 10]. PH is associated with increased exacerbation and decreased survival with COPD. Although this devastating disease is relatively common, the underlying mechanisms are still unknown, with treatments remaining ineffective.

9.1.4 Causes and Risk Factors

9.1.4.1 Asthma
Though asthma is prevalent, affecting 1 in every 13 people, the cause of asthma is not very clear. It is thought to be a combination of environmental and genetic factors. Exposure to irritants and allergens can trigger and exacerbate an asthma attack. These irritants and triggers include airborne allergens (i.e., pollen, dust mites, mold, pet dander), physical activity, cold air, and air pollutants (e.g., cigarette smoke) [3, 5, 15].

Several risk factors may increase the chances of developing asthma, such as having a familial history of asthma, allergies, being overweight, cigarette smoking habits, exposure to secondhand smoke, and exposure to occupational triggers [15].

9.1.4.2 COPD
Several factors can cause COPD; however, chronic obstructive pulmonary disease is ultimately due to long-term exposure to irritants that damage the lungs and airways, thereby promoting inflammation and remodeling of the vasculature and airways [2, 4, 6]. Smoking is recognized as the most important causative factor in developing COPD with nicotine as an active ingredient in cigarette smoke (CS). Studies have shown that approximately 50% of smokers will eventually develop COPD defined by the Global Initiative for Chronic Obstructive Lung Disease (GOLD) guidelines [16]. Although smoking remains the most causative agent in developing COPD, other risk factors such as air pollution, respiratory infections, poor nutrition, and chronic asthma can also promote the development of COPD [2, 4, 7, 17, 18].

In addition to chronic exposure to these irritants, numerous risk factors increase the chances of developing COPD. A hereditary deficiency in alpha-1 antitrypsin (AATD) is a genetic risk factor in COPD [2, 4, 7]. Age and gender also play a significant role in COPD, where those ages 65 or older and females are at an increased risk of developing COPD. People diagnosed with asthma are at a greater risk for COPD due to the pre-existing airflow limitation from airway hyperresponsiveness and remodeling. In addition to childhood asthma, a history of severe respiratory infections has been associated with reduced lung function and an increased risk for respiratory complications in adulthood [7].

9.1.4.3 PH

Although people of all ages, races, and sexes can be diagnosed with PH, certain risk factors make some more likely to get the disease. These risk factors include family history, obesity, gender, pregnancy, altitude, drug use, and cigarette smoking [1, 10, 14]. Obesity alone is not a risk factor; however, if combined with obstructive sleep apnea, it may result in mild PH. Idiopathic and heritable/familial PAH is more common in women with pregnant females, and females of childbearing age are more susceptible to develop PH. Living in high altitudes for years can influence the development of PH. Additionally, other diseases such as congenital heart disease, lung disease, and liver disease can lead to the development of PH [11, 13].

9.1.5 Treatments

There is currently no cure for asthma, COPD, or PH. In many ways, treatments for asthma, COPD, and COPD-associated PH are similar. Management of asthma- and COPD-exacerbating triggers and symptoms is the primary method of improving quality of life and slowing disease progression [2, 4, 7, 14]. For optimal management of the disease, both non-pharmacological and pharmacological treatments are required. In a non-pharmacological treatment, lifestyle changes such as smoking cessation (for smoking COPD patients) and avoiding airborne irritants such as secondhand CS and other air pollutants like dust and noxious fumes [2] are necessary to prevent exacerbating symptoms.

There are several pharmacological treatments for asthmatic and COPD patients. Often, treatments are combinations of different agents, individualized per patient based on the severity of symptoms, exacerbation triggers, and exposure to triggers. Comorbidities of patients must also be considered to avoid worsening conditions and complications.

Pharmacological treatments are aimed at (1) dilators of the airways to increase airflow (FEV_1) and (2) anti-inflammatory agents to reduce inflammation [8, 19, 20]. Bronchodilators are used to increase airflow, working as either β2 agonists or muscarinic receptor antagonists. β2 agonists, either short-acting or long-acting, relax airway smooth muscle by activating β-adrenergic receptors, thus dilating the airways. However, β2 agonists may have some adverse effects in older patients with disturbing cardiac rhythms. Muscarinic receptor antagonists, also known as antimuscarinic drugs, block bronchoconstriction caused by acetylcholine-induced activation of the M3 muscarinic receptors in the airway smooth muscle [6, 8, 19, 20].

Anti-inflammatory agents are used to treat the underlying systemic inflammation associated with asthma and COPD. The maintenance of asthma- and COPD-related inflammation is vital to reduce exacerbations and disease progression. Inhaled corticosteroids are often prescribed in combination with bronchodilators [20]. However, there are several risks in treating COPD with inhaled corticosteroids, which involve the risk of pneumonia, oral candidiasis, hoarseness, and withdrawal symptoms once treatment is ceased [6, 8, 20].

Although treatments have minimized complications and slowed disease progression in asthma and COPD patients, there is no cure for the disease. Further studies must be done to develop a specific treatment that targets the underlying pathophysiology that leads to the development and progression of asthma and COPD.

9.2 Cellular Responses in Asthma, COPD, and PH

9.2.1 Asthma

Airway smooth muscle cells (ASMCs) are seen as the primary effector cells of airway remodeling, contributing to the narrowing seen in asthma [21]. The dominant role of ASMCs may be due to airway hyperresponsiveness (AHR), defined as exaggerated airway narrowing due to nonspecific irritants and agonists, reversible by bronchodilators. ASM cellular processes that exacerbate AHR can have detrimental effects on downstream cellular responses and further worsen asthma symptoms.

9.2.1.1 ASM Hyperresponsiveness
Airway hyperresponsiveness (AHR) is a characteristic feature of asthma, which consists of an increased sensitivity of the airways to an inhaled stimuli that would produce little or no effect in healthy patients [5]. AHR causes excessive narrowing of the airways; thus, its measurement has provided insight into the underlying pathological mechanisms of the disease. AHR is increased in different risk-factor asthmatic groups, including the elderly, obese, and cigarette smoke patients. There are also varying associated pathophysiology with the increased AHR. For example, there are increased neutrophils and decreased elastic recoil in elderly asthmatic patients, whereas cigarette smoking patients display acutely reversible inflammation and structural changes.

9.2.1.2 ASM Remodeling
ASM remodeling is the main contributor to restricted airflow in asthmatic patients [5]. Structural changes correlate with airway wall thickening, reticular basement membrane thickness, and components of the extracellular matrix. The remodeling also includes changes like subepithelial fibrosis, ASM hypertrophy/hyperplasia, angiogenesis, and changes in the extracellular matrix composition [21, 22]. Hyperplasia or ASMC hyperproliferation is often influenced and stimulated by airway inflammation and is believed to be due to an upregulation of inflammatory mediators (i.e., cytokines and chemokines) [5, 23]. These pro-inflammatory cytokines that regulate ASMC proliferation include IL-1β and TNF-α.

9.2.1.3 ASM Cell Migration
ASMC migration is a contributing factor to airway remodeling and pathogenesis in asthma. Cell migration is influenced by chemotaxis or chemokines. It is possible that ASMC migration contributes to the increased smooth muscle mass in asthmatic airways [24]. Migration is composed of synchronized distinct steps of cell polarization, protrusion, adhesion, traction, and contraction [21, 24, 25]. Several molecular regulators control each step in cell migration, including Cdc42, Rac activation, PI3K, Rho family proteins, etc. [25]. Various pharmacological agents have been shown to have an inhibitory effect on ASMC migration and are beneficial in treating asthma [24, 25]. Additionally, traditional medications such as β-agonists and corticosteroids have been shown to reduce ASMC migration.

9.2.2 COPD and PH

Pulmonary arterial smooth muscle cells (PASMCs) are thought to be the main players in the pathogenesis of COPD and associated PH due to the overproduction of ROS and upregulated Ca^{2+} signaling. Regulation of these specific ROS and Ca^{2+} signaling pathways plays a crucial role in maintaining cellular homeostasis. Any disturbance to these processes can have detrimental effects on downstream cellular responses.

9.2.2.1 Vasoconstriction in COPD-Associated PH
COPD often leads to the development of PH, which accounts for the mortality of many COPD patients. Increased pulmonary vascular resistance (PVR) in PH is due to pulmonary artery constriction and remodeling. PASMCs serve as a critical player in vasoconstriction and remodeling through their hyperresponsiveness and increased proliferation, decreased or suppressed apoptosis, and migration.

Hypoxic pulmonary vasoconstriction is an adaptive mechanism to match perfusion and ventilation in response to alveolar hypoxia (a decrease in oxygen levels) in smaller, more resistant pulmonary arteries. Vasoconstriction is highly controlled by Ca^{2+} signaling in PASMCs. Ca^{2+} signaling may occur due to its release from the intracellular store (SR) and/or influx from the extracellular compartment through multiple ion channels. Intracellular Ca^{2+} release and/or extracellular Ca^{2+} influx lead to the increased cytosolic Ca^{2+}, which binds to calmodulin (CaM), activates myosin light chain kinase (MLCK), phosphorylates a 20 kDa light chain of myosin, initiates actin/myosin cross-bridge cycling, and initiates cell contraction.

9.2.2.2 Vascular Remodeling in COPD-Associated PH

Remodeling is characterized by structural changes to either the airway or the vasculature. Often, this involves hyperplasia (hyperproliferation) of airway (epithelial and smooth muscle) cells, thickening of the basement membrane, fibrosis, and collagen deposits [26]. Airway obstruction associated with COPD is a result of small airway remodeling. The vascular remodeling of pulmonary arteries involves hyperproliferation of PASMCs and contributes to COPD-associated PH [26].

Bronchoalveolar lavage fluid (BALF) collected from COPD patients has shown increased levels of matrix metalloproteases (MMPs), proteases that contribute to alveolar destruction and airway obstruction by degrading structural components of the extracellular matrix (ECM) [27]. MMPs are secreted by several inflammatory cell types, including neutrophils and macrophages. Thus, the observed increase in recruitment and infiltration of neutrophils and macrophages in COPD may account for the increased levels of MMPs found in COPD patient BALF [27, 28].

Several studies have shown that cigarette or tobacco smoke-induced ROS generation contributes to airway and pulmonary vascular remodeling in COPD. A study by Zhu et al. [29] has observed that tobacco smoke-induced ROS-mediated calpain activation is necessary for air-

way or pulmonary artery remodeling, whereas Churg et al. [27, 30] have shown CS induces activation of several growth factors and procollagen synthesis in airways.

9.3 Roles of Reactive Oxygen Species

Reactive oxygen species (ROS) are chemically reactive biomolecules that contain oxygen. Unregulated ROS generation can lead to detrimental effects, such as damage of nucleic acids, proteins, lipid peroxidation, etc. Oxidative stress on cells has been attributed to the development and progression of diseases such as asthma, PH, and other respiratory diseases [31, 32]. It is important to understand ROS signaling since the role of ROS in PASMCs has been shown to be involved in PA vasoconstriction and remodeling that lead to the development of PH.

9.3.1 Generation of ROS

Mitochondria have been noted as a primary and important source of ROS generation in many mammalian cells, including PASMCs [33, 34]. In the oxygen-rich environment of the mitochondria, one-electron reduction of oxygen to superoxide is the major reaction to produce ROS [34, 35]. It has also been shown that mitochondria can produce H_2O_2 [34, 35]. Complexes I and III of the mitochondrial electron transport chain (ETC) are the main contributors to mitochondrial ROS generation [36]. Conditions, such as the increase of electron movement and potential leakage, within these complexes, favor ROS generation [37]. Additionally, a high NADH/NAD+ ratio in the mitochondrial matrix has also been shown to contribute to the favorable ROS-generating environment [34].

Several studies have shown that within the mitochondrial ETC, the Rieske iron-sulfur protein (RISP), a catalytic subunit of complex III, is a significant molecule to ROS generation [38–41], which is evident by the fact that knockdown of RISP in PASMCs has shown to inhibit the increase in mitochondrial ROS generation associated with hypoxic exposure [39, 41].

Although mitochondria have been noted as a primary source of ROS, there are other sources such as nicotinamide dinucleotide phosphate (NADPH) oxidase (NOX), another well-studied source of ROS. NOX regulation is controlled through the interaction of cytoplasmic and membrane-associated proteins and in response to intracellular stimuli, such as Ca^{2+} signaling [42, 43]. For instance, protein kinase may phosphorylate and then modulate NOX activation [44]. Specifically, protein kinase C (PKC)-mediated phosphorylation can activate NOX [44, 45]. Other enzymes, such as xanthine oxidases, cyclooxygenases, and lipoxygenases, may also play a significant role in cellular responses in certain types of cells [46].

9.3.2 Upstream and Downstream Signaling of ROS

The unregulated overproduction of ROS plays a large role in many cellular processes. ROS can influence Ca^{2+} signaling through oxidizing key residues on several ion channels that allow for the Ca^{2+} influx [42, 47]. Conversely, it has also been shown that intracellular Ca^{2+} can modulate ROS generation as well as ROS scavengers [42, 47, 48]. Besides the mitochondria, Ca^{2+} regulates several enzymes that generate ROS, including NOX and nitric oxide synthase (NOS) [42, 49]. Antioxidant enzymes, such as catalase and GSH reductase, can be activated by Ca^{2+} [42, 50]. Ca^{2+} has also been shown to increase the level of superoxide dismutase (SOD), a key enzyme involved in the catalysis of superoxide into hydrogen peroxide [51]. Mitochondrial Ca^{2+} can activate several dehydrogenases of the citric acid cycle and ATP synthase to increase ROS generation [52, 53] subsequently. The ability of Ca^{2+} to induce ROS generation has been termed Ca^{2+}-induced ROS generation (CIRG). Additionally, the ability of ROS to stimulate the release of Ca^{2+} from the major intracellular Ca^{2+} store (sarcoplasmic reticulum (SR)) has been termed ROS-induced Ca^{2+} release (RICR).

ROS have also been shown to play a regulatory role in many cell signaling pathways, such as nuclear factor (NF)-κB inflammatory signaling, mitogen-activated protein kinase (MAPK) cascade signaling, PI3K-Akt signaling, and Keap1-Nrf2-ARE signaling [49]. ROS-induced oxidation of specific residues of kinases in the NF-κB signaling pathway has been shown to inhibit the activation of NF-κB and its subsequent translocation and transcriptional activity [52]. MAPK cascade signaling is involved in several cellular processes, which include cell growth, differentiation, survival, and death. ROS have been shown to prematurely activate the downstream signaling pathways of MAPK, such as JNK signaling and the ERK pathway [49]. Specifically, H_2O_2 induces the phosphorylation of phospholipase C, resulting in the production of diacylglycerol (DAG) and inositol triphosphate (IP$_3$) [49]. IP$_3$ has been shown to increase intracellular Ca^{2+} through the release of Ca^{2+} from the SR by activating its receptors [49, 53]. ROS-induced activation of protein kinase C (PKC) has been shown to increase Ca^{2+} as well as activation of NOX, influencing further ROS production via ROS-induced ROS generation (RIRG) [49, 54].

Keap1-Nrf-ARE signaling plays a critical role in maintaining cellular redox and inducing an adaptive response to oxidative stress. Increased levels of ROS have been shown to disrupt this Keap1-Nrf-ARE-dependent maintenance by influencing the dissociation of regulatory inhibitor Keap1 from Nrf. This dissociation is mediated by the oxidation of essential cysteine residues or activation of kinases such as PKC, MAPK, or PI3K that phosphorylate Nrf [49].

The unregulated generation of ROS plays a significant role in influencing many cellular processes. The oxidation of critical molecules in these pathways leads to unregulated and premature activation of the signaling pathways associated with the development and progression of diseases.

9.3.3 Role of ROS in Asthma

Oxidative stress plays a crucial role in the pathogenesis of asthma. Although accumulation and exposure to ROS can contribute to airway inflammation, it is also speculated that overproduction of ROS occurs after inflammation, post-asthmatic triggers, and attacks. Higher levels of key ROS signaling molecules were found in samples (e.g., breath condensates and sputum) taken from asthmatic patients compared with normal control subjects. In one study, scavenging of ROS prevented an immune response and inflammation in the airways, suggesting that ROS play an essential role in acting as the critical contributor in the initiation of allergic airway inflammation.

9.3.4 Role of ROS in COPD and PH

With a large percentage of COPD patients being smokers, chronic exposure to CS increases the production of ROS [31, 55]. This chronic exposure to cigarette smoke is detrimental for the cellular environment, including an injury to bronchiolar epithelial cells, infiltration of inflammatory cells in the lung, and an increase in the expression of oxidative stress markers and pro-inflammatory cytokines. The chrnoic exposure to CS (you can abbreviate cigarette smoke to CS) [56, 57].

Mitochondrial dysfunction, which can be linked to the increase in ROS generation, is seen in COPD patients compared to healthy patients [58]. Damaged mitochondria from H_2O_2 or CS have been shown to have a decreased mitochondrial membrane potential, impairment in oxidative phosphorylation and ATP production, and altered Ca^{2+} flux [58, 59].

The increase in ROS generation plays a large role in oxidative stress in patients with COPD. Importantly, studies have shown that COPD patients display a reduced anti-inflammatory defense, such as a reduction in glutathione (GSH) as seen in sputum or BAL collection from COPD patients [56], downregulation of SOD [56, 60], reduced catalase activity [56, 61], and reduced levels of GSH peroxidase (Gpx) [56, 61, 62]. Taken together, the reduced anti-inflammatory defense against ROS can contribute to the observed increase in oxidative stress in COPD patients.

Chronic inflammation is a crucial characteristic of COPD. Increased ROS generation has been shown to exacerbate the inflammation seen in COPD patients. ROS promote the recruitment and infiltration of inflammatory cells such as neutrophils and macrophages [7, 31]. ROS have been shown to inhibit the activation of NF-κB through oxidation of its specific residues; however, it has been demonstrated that in COPD patients, NF-κB activity is increased in sputum and BAL samples [63–65]. ROS derived from NOX4, the most abundant isoform of the NOX family in the pulmonary vasculature, has been shown to activate NF-κB and MAPK signaling pathways [66–68]. Thus, ROS can activate as well as repress NF-κB signaling.

COPD is also characterized by vascular remodeling and vasoconstriction. Studies have shown that both COPD-associated remodeling and vasoconstriction can be partly attributed to the increased ROS generation [10, 69]. Specifically, Zhu et al. [29] observed that remodeling is dependent on ERK phosphorylation with ERK activation dependent on ROS. Activation and phosphorylation of ERK influence collagen synthesis and cell proliferation in bronchial and pulmonary arterial SMCs [29]. Both collagen synthesis and cell proliferation of SMCs are critical characteristics of vascular and airway remodeling. TGF-β-induced NOX4-derived ROS generation has also been shown to play a key role in mediating pulmonary arteriolar remodeling, contributing to the development of COPD-associated pulmonary hypertension [70].

A key mechanism in the development of PH is highly associated with chronic hypoxia-induced cellular responses in PASMCs [71, 72]. Studies have shown that in COPD-associated PH, the mitochondria act as an oxygen sensor [31, 39, 73]. Although there is some controversy as to whether ROS generation increases or decreases in hypoxic conditions, the redox theory of hypoxic pulmonary vasoconstriction (HPV) suggests a hypoxia-induced change in ROS inhibits K^+ channels, which causes membrane depolarization and subsequent cell contraction due to the increase in cytosolic Ca^{2+} [74, 75]. Additionally, mitochondrial ETC-derived ROS have also been shown to modulate the mechanisms that mediate vasoconstriction [76, 77]. Sustained hypoxia has also been shown to activate Rho kinase, leading to Ca^{2+} sensitization and reinforcing vasoconstriction seen in PH [78–80].

Because the generation of ROS plays a large role in mediating several cellular processes associated with the development and progression of COPD, targeting the mechanisms that overproduce ROS will be critical to find effective therapies for COPD.

9.3.4.1 ROS, Nicotine, and Smoking

Nicotine, the main active ingredient in tobacco cigarettes, has been shown to enhance ROS generation in multiple cell types. Cigarette smoke, once inhaled, is divided into two phases: the tar (solid) and the gaseous phase [55]. Both phases contain high concentrations of free radicals, leading to oxidative stress within the cellular environment. Notably, cigarette tar can produce large amounts of H_2O_2 in the aqueous form [55]. Most importantly, CS and nicotine activate intracellular ROS generation systems, causing more ROS production within cells [81, 82].

CS/nicotine inhalation is a potent cellular stimulant that can cause DNA damage. Studies have shown that damage resulting from CS-induced oxidative stress is partially controlled by lack of ataxia-telangiectasia mutated (ATM) protein kinase activity, which in turn may be responsible for apoptosis [83, 84]. CS-induced DNA damage represses ATM protein kinase activity in endothelial cells, decreasing apoptosis. In support, inhibiting ATM protein kinase using pharmacological agent KU60019 increased PASMC proliferation and inhibited cell apoptosis [83].

Several studies suggest the expression of nicotinic receptors (nAChRs) on the mitochondria. Although 17 nAChR subunits have been identified in mammalian cells, mitochondria specifically express $\alpha 7$ nAChRs to regulate Ca^{2+} and cytochrome c release [85]. Apoptogens, such as high Ca^{2+} or oxidants (H_2O_2), stimulate this release of cytochrome c from the mitochondria [85]. Nicotine is detrimental to cells and can induce apoptosis by increasing ROS generation [86]. Additionally, it has been shown that CS activates hypoxia-inducible factor 1 (HIF-1) in a ROS-dependent manner [87].

Thus, studies have shown that CS/nicotine inhalation perpetuates the development of COPD and associated PH mechanistically by the overproduction of ROS.

9.4 Ca^{2+} Signaling

As the most common second messenger, Ca^{2+} signaling plays a significant role in many cellular processes and is critical in maintaining cellular homeostasis. Events that lead to the elevation of intracellular Ca^{2+}, such as chronic or intermittent hypoxia, contribute to PA or airway constriction, hyperproliferation of SMCs, and ultimately airway and pulmonary vascular remodeling and the development of asthma, COPD, and PH.

9.4.1 Important Ca^{2+} Channels

Multiple ion channels control and regulate Ca^{2+} signaling by causing intracellular Ca^{2+} release and extracellular Ca^{2+} influx. The release of Ca^{2+} from the intracellular store (SR) through RyR and/or IP_3R channel may initiate and promote the development of Ca^{2+} signaling [88–90]. All three subtypes of RyRs (RyR1, RyR2, and RyR3) have been shown to play a role in hypoxic cellular responses; however, RyR2 plays a more dominant role in PASMCs [54, 91, 92]. Three subtypes of IP_3Rs (IP_3R1, IP_3R2, and IP_3R3) are also found on the SR in PASMCs. Studies have shown that IP_3R1 plays a large role in mediating Ca^{2+} release in PASMCs, which is associated with PH [90].

L-type voltage-gated Ca^{2+} channels (LTCCs) have been shown to be the source for membrane depolarization-dependent Ca^{2+} influx in PASMCs [40]. Voltage-gated K^+ (K_V) channels are major regulators of the resting membrane potential in PASMCs. The inhibition of K_V channels can depolarize the cell membrane to activate LTCCs [40, 93]. Hypoxia, the main contributor to COPD-associated PH, may also inhibit K_V channels, thus influencing Ca^{2+} influx and vasoconstriction [94]. The increase in cytosolic Ca^{2+} stimulates cell proliferation and inhibits apoptosis, mediating vascular remodeling [91, 95].

The upregulation of transient receptor potential (TRP) channels, a family of nonselective ion channels, has been associated with increased Ca^{2+} influx and has been implicated in PH [96]. Canonical TRP (TRPC) channels include TRPC1–7, are a subfamily of TRP channels, and show mRNA and/or protein expression in pulmonary arteries [97]. Although all seven isoforms are expressed, TRPC1 and TRPC6 are the major players involved in Ca^{2+} responses in PASMCs [97]. Specifically, it has been shown that CS- and nicotine-induced vasoconstriction and remodeling in the PA and development of PH were highly influenced by the upregulation of TRPC1 and TRPC6 [98].

A recent study has shown that nicotine causes proliferation of cultured rat airway SMCs, and the role of nicotine is reliant on TRPC6-dependent Ca^{2+} influx through $\alpha7$ nAChR on the plasmalemma [99]. Consistent with a report by Wang et al. [98], Hong et al. [99] also showed that CS- and nicotine-exposure increase TRPC6 channel expression. Treatment of human airway SMCs with $\alpha7$ nAChR antagonists attenuated nicotine-induced proliferation; additionally, treatment of TRPC6 siRNA, with or without nicotine application, significantly affected cell proliferation as measured by EdU-positive staining [99]. Elevated intracellular Ca^{2+} levels have been shown to promote cell proliferation [100, 101]; thus, nicotine-induced increase in TRPC6 expression and Ca^{2+} influx via $\alpha7$ nAChR may be involved in hyperproliferation characteristic of vasoremodeling seen in COPD-associated PH. Therefore, although CS and its active agent nicotine have been shown to activate $\alpha7$-nAChR channels to induce Ca^{2+} influx and promote cell proliferation in airway SMCs [99], the role and underlying mechanisms of $\alpha7$ nAChR have not been well established in PASMCs.

9.4.2 Interaction Between Ca^{2+} and ROS Signaling

It has been shown that K_V channels are hypoxia sensitive, and its activity is mitochondrial ROS dependent in PASMCs [74]. It is interesting to point out that both hypoxic inhibition of K_V channels and activation of TRPC channels may occur due to RyR-mediated Ca^{2+} release [92, 102]. Additionally, the TRP channels are highly modulated by ROS, resulting in the regulation of several Ca^{2+}-mediated cellular processes in PASMCs, including contraction, migration, proliferation, and apoptosis, all of which are associated with COPD and COPD-associated PH [74, 103].

As previously mentioned, ROS have also been shown to influence intracellular Ca^{2+} release and extracellular Ca^{2+} influx. Thus, it is crucial to determine which residues on Ca^{2+} release and influx channels are oxidized by RISP-dependent ROS to play an important role in the development of COPD and associated PH.

The ability for Ca^{2+} to induce ROS generation (CIRG) and for ROS to induce Ca^{2+} release (RICR) in PASMCs, contributing to the development of COPD-associated PH, suggests targeting the underlying mechanisms that cause the upregulation of Ca^{2+} and ROS signaling is necessary.

9.5 Inflammatory Signaling

Three main signaling pathways mediate inflammatory responses, which include nuclear factor (NF)-κB signaling, Janus kinase/signaling transducer and activator of transcription (JAK/STAT) signaling pathway, and mitogen-activated protein kinase (MAPK) signaling pathway. Although all are important, we will focus on the significance of NF-κB signaling, as it has been shown to play a crucial role in lung diseases.

9.5.1 NF-κB Signaling

NF-κB is a family of inducible transcription factors involved in the regulation of inflammatory genes. There are five structurally related family subunits: p50, p51, RelA or p65, RelB, and c-Rel [104]. These subunits mediate target gene transcription through binding a specific DNA element, the κB enhancer [104]. The NF-κB proteins are sequestered in the cytoplasm by inhibitory proteins, which include the IκB family members [104, 105].

There are two major signaling pathways involved in the activation of NF-κB: the canonical and noncanonical pathways, both of which are important in regulating immune and inflammatory responses [65, 104]. In the canonical NF-κB pathway, stimuli such as ligands of various cytokine receptors, TNF receptors, and B-cell receptors induce the degradation of IκBα initiated by site-specific phosphorylation by multi-subunit IκB kinase (IKK) [65, 104, 105]. IKK has two catalytic subunits, IKKα and IKKβ, and a regulatory subunit, NF-κB essential modulator (NEMO) or IKKγ [65, 104, 105]. Once IKK is activated, the kinase phosphorylates IκBα at two N-terminal serine residues, which then triggers ubiquitin-dependent degradation of IκBα in the proteasome [65, 104, 105]. Nuclear translocation of canonical NF-κB members, mainly p50/p65 and p50/c-Rel dimers, results from the degradation of IκBα [65, 104, 105].

Innate immune cells, such as macrophages, dendritic cells, and neutrophils, serve important roles as the first line of defense in innate immunity and inflammation. These cells express pattern recognition receptors (PRRs) that detect microbial components, pathogen-associated molecular patterns (PAMPs), and damage-associated molecular patterns (DAMPs), specific molecules released by necrotic cells, and damaged tissues [104, 106]. Specifically, mammalian cells express five PRR families, including toll-like receptors (TLRs) and NOD-like receptors (NLRs) [104]. Each PRR family differs with distinct structural properties and distinct responses to different PAMPs and DAMPs. However, a common event of PRRs is the activation of the canonical NF-κB pathway [104, 106].

The activation of the canonical NF-κB pathway induces the transcription of pro-inflammatory cytokines, chemokines, and other inflammatory mediators such as anti-apoptotic factors, cell cycle regulators, and adhesion molecules [104, 106]. Transforming growth factor-β-activated kinase 1 (TAK1) is a signaling molecule that unites the different PRR pathways for the activation of NF-κB [104, 106]. Once activated, TAK1, composed of subunits TAB1 and TAB2, activates downstream kinase IKK, responsible for IκBα

phosphorylation and subsequent NF-κB activation [104, 106]. NF-κB acts as a key transcription factor of M1 macrophages and necessary for the induction of inflammatory genes, including genes for TNF-α, IL-1β, IL-6, and cyclooxygenase-2 [104, 106].

In the noncanonical NF-κB pathway, this pathway selectively responds to a set of stimuli, which includes a subset of TNF receptor ligands, and does not rely on the degradation of IκBα, but on the processing of the p65 precursor protein, p100 [65, 104, 105]. Briefly, NF-κB-inducing kinase (NIK) activates IKKα to mediate the phosphorylation of p100. Phosphorylated p100 induces subsequent ubiquitination and processing, which involved the degradation of its C-terminal IκB-like structure [65, 104, 105]. This degradation generates a mature NF-κB p52 and causes nuclear translocation of noncanonical p52/RelB [65, 104, 105].

9.5.2 Role of Reactive Oxygen Species in Inflammatory Signaling

It has been shown that ROS can affect inflammatory signaling pathways, such as NF-κB signaling [64, 65]. In response to toxic levels of ROS, it is critical for cells to prevent further oxidative damage to maintain cell survival. Based on the severity of oxidative stress, ROS can trigger apoptotic and necrotic cell death [32, 52]. The expression of NF-κB target genes, such as those encoding for antioxidant proteins, promotes cell survival by influencing cellular ROS levels [49, 52].

ROS has also been shown to play a regulatory role in NF-κB signaling. Cytoplasmic thioredoxin, an endogenous small redox protein that maintains protein thiol groups in the reduced state, has been shown to block the degradation of IκB [49, 52, 64]. It has also been shown that ROS can disturb the ubiquitination and degradation of IκB, thus inhibiting the activation of NF-κB [49]. Nucleic thioredoxin has been shown to influence NF-κB activity by enhancing DNA binding abilities [52, 64, 107]. NF-κB heterodimers can also be modified in increased oxidative stress conditions. Cysteine residue 62 in the Rel homology domain (RHD) of NF-κB p50 subunit is prone to oxidation, which subsequently decreases the ability to bind DNA [64]. However, the RHD has spatial redox regulation with oxidation occurring in the cytoplasm and the reduced form in the nucleus [64]. IKK is also a prime target for ROS, specifically S-glutathionylation of IKKβ on cysteine 179, thus inhibiting the catalytic activity of the IKK subunit and subsequently inhibiting NF-κB signaling [49].

9.5.3   Role of Ca²⁺ Signaling in NF-κB Signaling

Several studies have shown that Ca^{2+} mediates NF-κB signaling [108–112]. Specifically, increased intracellular Ca^{2+} from inherent increased resting intracellular Ca^{2+} [108] or depolarization of muscle cells [111] leads to an upregulation in NF-κB activity [108–111]. Ca^{2+}-induced NF-κB activation increases the transcriptional activity and p65 nuclear localization by mediating the phosphorylation of Ser536 [108, 110].

Intracellular Ca^{2+} can be regulated by several channels; however, only several ion channels are linked to Ca^{2+}-induced NF-κB activation. Release of Ca^{2+} from intracellular stores in the SR is mediated by either ryanodine receptor (RyR) or inositol-1,4,5-triphosphate receptor (IP_3R) channels [109, 111, 112]. Following the release of Ca^{2+} from the intracellular stores, NF-κB activity is increased. Further, the involvement of RyR channels has been confirmed by using ryanodine to inhibit Ca^{2+} release via RyR channels and the subsequent decrease in NF-κB activity [109, 111, 112].

There are several Ca^{2+} signaling pathways involved in the mediation of Ca^{2+}-induced NF-κB activation. Liu et al. showed that non-IκBα degradation-mediated signaling is regulated through Ca^{2+}-dependent PKCα-mediated phosphorylation of p65 [110]. Other studies have further supported the Ca^{2+}-dependent PKCα-mediated activation of NF-κB through pharmacological inhibition [109, 111].

Calmodulin and calmodulin kinases (i.e., CaMKII and CaMKIV) have also been shown to mediate NF-κB activity in a Ca^{2+}-dependent manner [109, 113]. It has been reported that calmodulin inhibitors W7 and calmidazolium decrease the basal activity of NF-κB in primary cultured neonatal cerebellar granule neurons and the role of CaMKII in mediating NF-κB signaling through the activation of IKK [109]. Additionally, it has been shown that CaMKII can induce cardiomyocyte hypertrophy through apoptosis signal-regulating kinase 1 (ASK1)-NF-κB signaling [114]. However, the role of calmodulin and associated kinase Ca^{2+}-dependent signaling is not largely known in PASMCs in the context of COPD and PH.

9.5.4   Role of NF-κB in Asthma and COPD

NF-κB signaling is responsible for mediating inflammatory responses seen in asthma and COPD.

Inflammation of the airways and vasculature is often caused by infiltration by inflammatory cells such as eosinophils, mast cells, monocytes, lymphocytes, and neutrophils. All these cell types contribute to the elevated levels of inflammatory mediators. Studies have shown evidence for upregulated activation of NF-κB and increased inflammatory cells in the bronchial biopsy of asthmatic and COPD patients, observed through bronchial biopsy [63, 115]. Specifically, IκBα levels are significantly lower in tissues and fluids collected from COPD patients than nonsmoking healthy patients [63, 116]. Upstream of upregulated IκBα degradation, IKK activity is increased in COPD patients and smokers [63, 116].

The collection of sputum and bronchoalveolar lavage fluid (BALF) is a common clinical assessment of inflammation in patients with respiratory diseases, such as asthma or COPD. Neutrophils isolated from COPD patient sputum and BALF show increased NF-κB activation following CS exposure, a common irritant and trigger of COPD inflammatory exacerbations [63, 117]. Macrophages isolated from BALF collected from patients with PH also displayed activated NF-κB [118]. This was further supported in an in vivo COPD model where CS increased NF-κB signaling activity and its subsequent nuclear recruitment to the promoters of several inflammatory genes [63, 117, 119].

Although NF-κB plays a large role in regulating inflammation seen in COPD, ubiquitously targeting this pathway is not an ideal therapeutic target in treating COPD since NF-κB is ubiquitously expressed and involved in other cellular processes. It is necessary to determine a specific and direct target within the NF-κB signaling pathway that contributes to COPD and associated PH.

9.5.4.1 NF-κB in Airway and Vascular Hyperresponsiveness and Remodeling in COPD and PH

Airway and vascular hyperresponsiveness and remodeling play a large role in the development of COPD and COPD-associated PH. Although it has not been well characterized in COPD and PH, NF-κB activity has been shown to be involved in airway hyperresponsiveness and remodeling in the context of asthma [63, 120]. Tully et al. [120] demonstrated that epithelial activation of NF-κB is required for airway remodeling and hyperresponsiveness in dust-mite-induced inflammation, whereas inhibition of NF-κB diminished neutrophil recruitment, remodeling, and hyperresponsiveness.

Additionally, little is known on the role of NF-κB in mediating these processes, specifically in PASMCs. Price et al. [121] showed that NF-κB was activated in PASMCs of the pulmonary vessel in idiopathic PH; however, its role in remodeling or vasoconstriction was not assessed. Another study by Hosokawa et al. [122] demonstrated that an inhibitor of NF-κB, IMD-0354, blocked the hyperproliferation of PASMCs, associated with PH. However, this study of PH was done using monocrotaline, a drug-induced PH model. The role of NF-κB in a hypoxia-induced model of PH in PASMCs was shown to mediate HIF-1α mRNA expression as a hypoxia-regulated transcription factor [123]. The accumulation of HIF-1 has been shown to influence hypoxia-induced apoptosis, which may contribute to vascular remodeling and PH [123, 124].

Further studies investigating the role of NF-κB in mediating cellular processes, such as airway or PA hyperresponsiveness and remodeling, that contribute to the development of COPD and associated PH may prove useful in finding therapeutic targets for the treatment of these devastating diseases.

9.6 Conclusions

We and other investigators have demonstrated that mitochondrial RISP is an essential molecule in ROS generation in PASMCs [38–41, 92], which plays a significant role in mediating Ca^{2+} signaling via RyR2-mediated SR Ca^{2+} release, serving as a significant contributor to PA vasoconstriction and remodeling in COPD-associated PH and possibly COPD as well [54, 91, 92]. In support, we have previously shown that RISP-mediated mitochondrial ROS can activate PKC-ε and then NOX, induce further ROS generation (RIRG), activate RyR2, and cause subsequent Ca^{2+} release from the SR in PASMCs [41, 54, 91, 125]. It is also known that ROS overproduction is a crucial characteristic of COPD, presumably leading to PA inflammation, vasoconstriction, and remodeling [46, 49, 50, 54, 125]. However, it is not known how RISP-mediated ROS generation affects inflammatory signaling pathways and nicotine-initiated signaling pathways.

NF-κB-dependent inflammatory signaling plays a crucial role in mediating multiple cellular responses in COPD [52, 63–65, 126]. Sputum and BALF collected from COPD patients have shown an upregulation in NF-κB activity such as decreased IκBα expression, a characteristic of upregulated IκBα degradation-dependent activation of NF-κB [63, 115, 116]. PASMCs have shown activated NF-κB following inflammatory and proliferative stimuli [121]. NF-κB is also activated in PASMCs of rats with pulmonary hypertension [127].

ROS are known to be the important mediators for NF-κB signaling. The increased ROS production during oxidative stress has shown to affect

NF-κB function. Inhibition of upstream kinases (e.g., IKK) or phosphorylation of specific residues on NF-κB signaling molecules (e.g., IκBα) has shown to either increase or decrease NF-κB signaling. Because inflammatory signaling can play a role in remodeling and vasoconstriction, an interesting question remains how ROS and NF-κB-dependent inflammatory signaling interact, specifically in PASMCs, to mediate PA vasoconstriction and remodeling, leading to the development of COPD and PH. It is also unclear how ROS are generated to activate NF-κB in PASMCs [34, 46, 49, 56, 57, 128]. As RISP is a primary molecule in ROS generation in PASMCs [39–41, 125], studies proposed to elucidate the role of RISP-mediated mitochondrial ROS in regulating NF-κB activity in PASMCs is of great significance as well.

Intracellular Ca²⁺, as the downstream signaling factor of ROS signaling in PASMCs, plays a crucial role in PA vasoconstriction and remodeling. Ca²⁺ release from the SR through RyRs and IP₃Rs increases the cytosolic Ca²⁺, which can activate actin-myosin complexes to induce PA vasoconstriction [88]. Additionally, Ca²⁺ may also mediate PA remodeling, a primary contributor to COPD-associated PH [42, 47, 89, 129]. Targeting readily oxidized residues on RyRs and IP₃Rs in PASMCs may help find more effective and specific therapies for these diseases.

Mitochondrial Ca²⁺ signaling has been attributed to the role of a nicotinamide acetylcholine receptor, α7 nAChR, on the mitochondrial membrane in isolated mitochondria from HEK cells [85]. It is unknown whether α7 nAChR has a role in mediating mitochondrial Ca²⁺ in PASMCs, which may play a role in RISP-mediated CIRG. It is unclear whether and how α7 nAChR contributes to CIRG and upstream calcium-induced calcium release calcium-induced calcium release (CICR) via RyRs and IP₃Rs in PASMCs, contributing to PA vasoconstriction and remodeling, COPD, and COPD-associated PH. Thus, the specific role of CIRG and CICR through mitochondrial ROS and important Ca²⁺ release channels RyRs and IP₃Rs should be the focus of future investigations.

9.6.1 Graphical Conclusions

Schematic of proposed interactions between ROS, Ca²⁺, and inflammatory signaling in respiratory diseases. Nicotine inhalation (including cigarette smoking) by inducing α7 nAChR-mediated mitochondrial Ca²⁺ release and hypoxia secondary to nicotine inhalation or cigarette smoking induce RISP-mediated ROS generation, oxidizes RyR2, which subsequently enhances its activity and increases Ca²⁺ release from the SR in PASMCs.

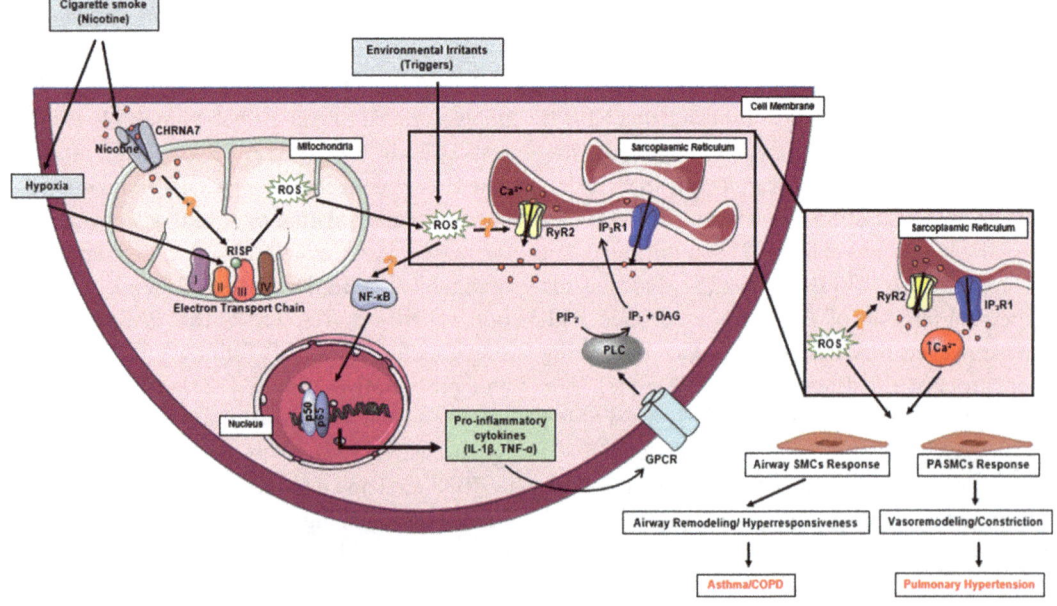

Additionally, environmental irritants, such as pollen and pet dander, can trigger ROS production, evident in asthmatic attacks and the development of asthma. Mitochondrial-generated ROS may increase NF-κB activity, increase pro-inflammatory factors, and trigger IP$_3$R1-mediated Ca^{2+} release from the SR. This Ca^{2+} release may further induce RyR2-mediated Ca^{2+} release, increasing the intracellular [Ca^{2+}], leading to PASM cellular responses (vasoconstriction and remodeling), contributing to the development of COPD-associated pulmonary hypertension. In ASMCs, this increase in intracellular [Ca^{2+}] can lead to ASM cellular responses (airway remodeling and airway hyperresponsiveness), seen in asthma.

References

1. Chaouat A, Naeije R, Weitzenblum E. Pulmonary hypertension in COPD. Eur Respir J. 2008;32(5):1371–85.
2. Devine JF. Chronic obstructive pulmonary disease: an overview. Am Health Drug Benefits. 2008;1(7):34–42.
3. Quirt J, et al. Asthma. Allergy Asthma Clin Immunol. 2018;14(Suppl 2):50.
4. Prevention, C.f.D.C.a. Chronic Obstructive Pulmonary Disease (COPD). 2018.
5. Bush A. Pathophysiological mechanisms of asthma. Front Pediatr. 2019;7:68.
6. NIH. COPD 2019.
7. Peter J, Barnes PGB, Silverman EK, Celli BR, Vestbo J, Wedzicha JA, Wouters EFM. Chronic obstructive pulmonary disease. Nature. 2015;1(15076)
8. Disease, G.I.f.C.O.L., Pocket guide to COPD diagnosis, management, and prevention, Hadfield R, editor. 2017.
9. Natalie Terzikhan KMCV, Hofman A, Stricker BH, Brusselle GG, Lahousse L. Prevalence and incidence of COPD in smokers and non-smokers: the Rotterdam study. Eur J Epidemiol. 2016;31(8):785–92.
10. Elwing J, Panos RJ. Pulmonary hypertension associated with COPD. Int J Chron Obstruct Pulmon Dis. 2008;3(1):55–70.
11. Sara Roversi LMF, Sin DD, Hawkins NM, Agusti A. Chronic obstructive pulmonary disease and cardiac diseases. Am J Respir Crit Care Med. 2016;194(11):1319–36.
12. Nathaniel M, Hawkins MCP, Jhund PS, Chalmers GW, Dunn FG, McMurray JJV. Heart failure and chronic obstructive pulmonary disease: diagnostic pitfalls and epidemiology. Eur J Heart Fail. 2009;11(2):130–9.
13. Jeremy A, Falk SK, Criner GJ, Scharf SM, Minaj OA, Diaz P. Cardiac disease in chronic obstructive pulmonary disease. Proc Am Thorac Soc. 2008;5(4):543–8.
14. Barberà JA, Peinado VI, Santos S. Pulmonary hypertension in chronic obstructive pulmonary disease. Eur Respir J. 2003;21(5):892.
15. Subbarao P, Mandhane PJ, Sears MR. Asthma: epidemiology, etiology and risk factors. CMAJ. 2009;181(9):E181–90.
16. Julio D, Antuni APJB. Evaluation of individuals at risk for COPD: beyond the scope of the global initiative for chronic obstructive lung disease. Chronic Obstr Pulm Dis. 2016;3(3):653–67.
17. Laniado-Laborin R. Smoking and chronic obstructive pulmonary disease (COPD). Parallel epidemics of the 21st century. Int J Environ Res Public Health. 2009;6(1):209–24.
18. Marsh S, Aldington S, Shirtcliffe P, Weatherall M, Beasley R. Smoking and COPD: what really are the risks? Eur Respir J. 2006;28:883–6.
19. Page C, Cazzola M. Bifunctional drugs for the treatment of asthma and chronic obstructive pulmonary disease. Eur Respir J. 2014;44(2):475.
20. Ernst P, Saad N, Suissa S. Inhaled corticosteroids in COPD: the clinical evidence. Eur Respir J. 2015;45:525–37.
21. Keglowich LF, Borger P. The three A's in asthma – airway smooth muscle, airway remodeling & angiogenesis. Open Respir Med J. 2015;9:70–80.
22. James AL, Wenzel S. Clinical relevance of airway remodelling in airway diseases. Eur Respir J. 2007;30(1):134.
23. Khan MA. Inflammation signals airway smooth muscle cell proliferation in asthma pathogenesis. Multidiscip Respir Med. 2013;8(1):11.
24. Salter B, et al. Regulation of human airway smooth muscle cell migration and relevance to asthma. Respir Res. 2017;18(1):156.
25. Goncharova EA, et al. Cyclic AMP-mobilizing agents and glucocorticoids modulate human smooth muscle cell migration. Am J Respir Cell Mol Biol. 2003;29(1):19–27.
26. Crosswhite P, Sun Z. Molecular mechanisms of pulmonary arterial remodeling. Mol Med (Cambridge, Mass). 2014;20(1):191–201.
27. Churg A, Zhou S, Wright JL. Matrix metalloproteinases in COPD. Eur Respir J. 2012;39(1):197.
28. Vandenbroucke RE, Dejonckheere E, Libert C. A therapeutic role for matrix metalloproteinase inhibitors in lung diseases? Eur Respir J. 2011;38(5):1200.
29. Zhu J, et al. Reactive oxygen species-dependent Calpain activation contributes to airway and pulmonary vascular remodeling in chronic obstructive pulmonary disease. Antioxid Redox Signal. 2019;
30. Churg A, et al. Cigarette smoke drives small airway remodeling by induction of growth factors in the Airway Wall. Am J Respir Crit Care Med. 2006;174(12):1327–34.

31. Boukhenouna S, et al. Reactive oxygen species in chronic obstructive pulmonary disease. Oxidative Med Cell Longev. 2018;2018:5730395.
32. Circu ML, Aw TY. Reactive oxygen species, cellular redox systems, and apoptosis. Free Radic Biol Med. 2010;48(6):749–62.
33. Smith KA, Waypa GB, Schumacker PT. Redox signaling during hypoxia in mammalian cells. Redox Biol. 2017;13:228–34.
34. Murphy MP. How mitochondria produce reactive oxygen species. Biochem J. 2009;417(1):1–13.
35. Zorov DB, Juhaszova M, Sollott SJ. Mitochondrial reactive oxygen species (ROS) and ROS-induced ROS release. Physiol Rev. 2014;94(3):909–50.
36. Dröse S, Brandt U. Molecular mechanisms of superoxide production by the mitochondrial respiratory chain. In: Kadenbach B, editor. Mitochondrial oxidative phosphorylation: nuclear-encoded genes, enzyme regulation, and pathophysiology. New York: Springer New York; 2012. p. 145–69.
37. Turrens JF. Mitochondrial formation of reactive oxygen species. J Physiol. 2003;552(Pt 2):335–44.
38. Diaz F, Enríquez JA, Moraes CT. Cells lacking Rieske iron-sulfur protein have a reactive oxygen species-associated decrease in respiratory complexes I and IV. Mol Cell Biol. 2012;32(2):415–29.
39. Korde AS, et al. Primary role of mitochondrial Rieske iron-sulfur protein in hypoxic ROS production in pulmonary artery myocytes. Free Radic Biol Med. 2011;50(8):945–52.
40. Truong L, Zheng Y-M, Wang Y-X. Mitochondrial Rieske iron-sulfur protein in pulmonary artery smooth muscle: a key primary signaling molecule in pulmonary hypertension. Arch Biochem Biophys. 2019;664:68–75.
41. Wang Y-X, et al. Rieske iron-sulfur protein in mitochondrial complex III is essential for the heterogeneity of hypoxic cellular responses in pulmonary and systemic (mesenteric) artery smooth muscle cells. FASEB J. 2015;29(1_supplement):859.1.
42. Görlach A, et al. Calcium and ROS: a mutual interplay. Redox Biol. 2015;6:260–71.
43. Meitzler JL, et al. NADPH oxidases: a perspective on reactive oxygen species production in tumor biology. Antioxid Redox Signal. 2014;20(17):2873–89.
44. Rastogi R, et al. NOX activation by subunit interaction and underlying mechanisms in disease. Front Cell Neurosci. 2017;10:301.
45. Cosentino-Gomes D, Rocco-Machado N, Meyer-Fernandes JR. Cell signaling through protein kinase C oxidation and activation. Int J Mol Sci. 2012;13(9):10697–721.
46. Di Meo S, et al. Role of ROS and RNS sources in physiological and pathological conditions. Oxidative Med Cell Longev. 2016;2016:-1245049.
47. Feno S, et al. Crosstalk between calcium and ROS in pathophysiological conditions. Oxidative Med Cell Longev. 2019;2019:9324018.
48. Hempel N, Trebak M. Crosstalk between calcium and reactive oxygen species signaling in cancer. Cell Calcium. 2017;63:70–96.
49. Zhang J, et al. ROS and ROS-mediated cellular signaling. Oxidative Med Cell Longev. 2016;2016:4350965.
50. Gordeeva AV, Zvyagilskaya RA, Labas YA. Crosstalk between reactive oxygen species and calcium in living cells. Biochem Mosc. 2003;68(10):1077–80.
51. Mondola P, et al. The cu, Zn superoxide dismutase: not only a dismutase enzyme. Front Physiol. 2016;7:594.
52. Morgan MJ, Liu Z-G. Crosstalk of reactive oxygen species and NF-κB signaling. Cell Res. 2011;21(1):103–15.
53. Banan A, et al. Phospholipase C-γ inhibition prevents EGF protection of intestinal cytoskeleton and barrier against oxidants. Am J Physiol Gastrointest Liver Physiol. 2001;281(2):G412–23.
54. Wang Y-X, Zheng Y-M. Role of ROS signaling in differential hypoxic Ca2+ and contractile responses in pulmonary and systemic vascular smooth muscle cells. Respir Physiol Neurobiol. 2010;174(3):192–200.
55. Valavanidis A, Vlachogianni T, Fiotakis K. Tobacco smoke: involvement of reactive oxygen species and stable free radicals in mechanisms of oxidative damage, carcinogenesis and synergistic effects with other respirable particles. Int J Environ Res Public Health. 2009;6(2):445–62.
56. McGuinness AJA, Sapey E. Oxidative stress in COPD: sources, markers, and potential mechanisms. J Clin Med. 2017;6(2):21.
57. Rahman I, Adcock IM. Oxidative stress and redox regulation of lung inflammation in COPD. Eur Respir J. 2006;28(1):219.
58. Liu X, Chen Z. The pathophysiological role of mitochondrial oxidative stress in lung diseases. J Transl Med. 2017;15(1):207.
59. Ježek J, Cooper KF, Strich R. Reactive oxygen species and mitochondrial dynamics: the yin and Yang of mitochondrial dysfunction and Cancer progression. Antioxidants (Basel, Switzerland). 2018;7(1):13.
60. Bernardo I, Bozinovski S, Vlahos R. Targeting oxidant-dependent mechanisms for the treatment of COPD and its comorbidities. Pharmacol Ther. 2015;155:60–79.
61. Marginean C, et al. Involvement of oxidative stress in COPD. Curr Health Sci J. 2018;44(1):48–55.
62. Domej W, Oettl K, Renner W. Oxidative stress and free radicals in COPD–implications and relevance for treatment. Int J Chron Obstruct Pulmon Dis. 2014;9:1207–24.
63. Schuliga M. NF-kappaB signaling in chronic inflammatory airway disease. Biomol Ther. 2015;5(3):1266–83.
64. Lingappan K. NF-κB in oxidative stress. Curr Opin Toxicol. 2018;7:81–6.
65. Lawrence T. The nuclear factor NF-kappaB pathway in inflammation. Cold Spring Harb Perspect Biol. 2009;1(6):-a001651.
66. Bedard K, Krause K-H. The NOX family of ROS-generating NADPH oxidases: physiology and pathophysiology. Physiol Rev. 2007;87(1):245–313.

67. Frazziano G, Champion HC, Pagano PJ. NADPH oxidase-derived ROS and the regulation of pulmonary vessel tone. Am J Physiol Heart Circ Physiol. 2012;302(11):H2166–77.
68. Santillo M, et al. NOX signaling in molecular cardiovascular mechanisms involved in the blood pressure homeostasis. Front Physiol. 2015;6:194.
69. Klaus F, Rabe JRH, Suissa S. Cardiovascular disease and COPD: dangerous liaisons? Eur Respir Rev. 2018;27:180057.
70. Guo X, et al. NOX4 expression and distal arteriolar remodeling correlate with pulmonary hypertension in COPD. BMC Pulm Med. 2018;18(1):111.
71. Pak O, et al. The effects of hypoxia on the cells of the pulmonary vasculature. Eur Respir J. 2007;30(2):364.
72. Veith C, et al. Molecular mechanisms of hypoxia-inducible factor-induced pulmonary arterial smooth muscle cell alterations in pulmonary hypertension. J Physiol. 2016;594(5):1167–77.
73. Waypa Gregory B, Chandel Navdeep S, Schumacker Paul T. Model for hypoxic pulmonary vasoconstriction involving mitochondrial oxygen sensing. Circ Res. 2001;88(12):1259–66.
74. Veit F, et al. Hypoxia-dependent reactive oxygen species signaling in the pulmonary circulation: focus on ion channels. Antioxid Redox Signal. 2015;22(6):537–52.
75. Olschewski A, Weir EK. Redox regulation of ion channels in the pulmonary circulation. Antioxid Redox Signal. 2015;22(6):465–85.
76. Sena LA, Chandel NS. Physiological roles of mitochondrial reactive oxygen species. Mol Cell. 2012;48(2):158–67.
77. Costa RM, et al. H2O2 generated from mitochondrial electron transport chain in thoracic perivascular adipose tissue is crucial for modulation of vascular smooth muscle contraction. Vasc Pharmacol. 2016;84:28–37.
78. Wang Z, et al. Rho-kinase activation is involved in hypoxia-induced pulmonary vasoconstriction. Am J Respir Cell Mol Biol. 2001;25(5):628–35.
79. Uehata M, et al. Calcium sensitization of smooth muscle mediated by a Rho-associated protein kinase in hypertension. Nature. 1997;389(6654):990–4.
80. Barman SA, Zhu S, White RE. RhoA/Rho-kinase signaling: a therapeutic target in pulmonary hypertension. Vasc Health Risk Manag. 2009;5:663–71.
81. Barr J, et al. Nicotine induces oxidative stress and activates nuclear transcription factor kappa B in rat mesencephalic cells. Mol Cell Biochem. 2007;297(1–2):93–9.
82. Rao P, et al. Effects of cigarette smoke condensate on oxidative stress, apoptotic cell death, and HIV replication in human Monocytic cells. PLoS One. 2016;11(5):e0155791.
83. Hu F, et al. Ataxia-telangiectasia mutated (ATM) protein signaling participates in development of pulmonary arterial hypertension in rats. Med Sci Monit. 2017;23:4391–400.
84. Aoshiba K, et al. DNA damage as a molecular link in the pathogenesis of COPD in smokers. Eur Respir J. 2012;39(6):1368.
85. Gergalova G, et al. Mitochondria express α7 nicotinic acetylcholine receptors to regulate Ca2+ accumulation and cytochrome c release: study on isolated mitochondria. PLoS One. 2012;7(2):–e31361.
86. Lan X, et al. Nicotine induces podocyte apoptosis through increasing oxidative stress. PLoS One. 2016;11(12):e0167071.
87. Daijo H, et al. Cigarette smoke reversibly activates hypoxia-inducible factor 1 in a reactive oxygen species-dependent manner. Sci Rep. 2016;6:34424.
88. Hill-Eubanks DC, et al. Calcium signaling in smooth muscle. Cold Spring Harb Perspect Biol. 2011;3(9):a004549.
89. Firth AL, Won JY, Park WS. Regulation of ca(2+) signaling in pulmonary hypertension. Korean J Physiol Pharmacol. 2013;17(1):1–8.
90. Yadav VR, et al. PLCγ1-PKCε-IP(3)R1 signaling plays an important role in hypoxia-induced calcium response in pulmonary artery smooth muscle cells. Am J Physiol Lung Cell Mol Physiol. 2018;314(5):L724–35.
91. Yadav VR, et al. Important role of PLC-γ1 in hypoxic increase in intracellular calcium in pulmonary arterial smooth muscle cells. Am J Physiol Lung Cell Mol Physiol. 2013;304(3):L143–51.
92. Yang Z, et al. Important role of sarcoplasmic reticulum Ca2+ release via ryanodine receptor-2 channel in hypoxia-induced Rieske iron-sulfur protein-mediated mitochondrial ROS generation in pulmonary artery smooth muscle cells. Antioxid Redox Signal. 2020;32(7):447–62.
93. Nieves-Cintrón, M., et al., Regulation of voltage-gated potassium channels in vascular smooth muscle during hypertension and metabolic disorders. Microcirculation (New York, NY : 1994), 2018. 25(1): https://doi.org/10.1111/micc.12423.
94. Pugliese SC, et al. The role of inflammation in hypoxic pulmonary hypertension: from cellular mechanisms to clinical phenotypes. Am J Physiol Lung Cell Mol Physiol. 2015;308(3):L229–52.
95. Perez-Zoghbi JF, et al. Ion channel regulation of intracellular calcium and airway smooth muscle function. Pulm Pharmacol Ther. 2009;22(5):388–97.
96. Ramirez GA, et al. Ion channels and transporters in inflammation: special focus on TRP channels and TRPC6. Cell. 2018;7(7):70.
97. Wang Y-X, Zheng Y-M. ROS-dependent signaling mechanisms for hypoxic Ca(2+) responses in pulmonary artery myocytes. Antioxid Redox Signal. 2010;12(5):611–23.
98. Wang J, et al. Effects of chronic exposure to cigarette smoke on canonical transient receptor potential expression in rat pulmonary arterial smooth muscle. Am J Physiol Cell Physiol. 2014;306(4):C364–73.
99. Hong W, et al. Nicotine-induced airway smooth muscle cell proliferation involves TRPC6-dependent

calcium influx via α7 nAChR. Cell Physiol Biochem. 2017;43(3):986–1002.

100. Lipskaia L, et al. Phosphatidylinositol 3-kinase and calcium-activated transcription pathways are required for VLDL-induced smooth muscle cell proliferation. Circ Res. 2003;92(10):1115–22.

101. Rokosova B, Peter Bentley J. Effect of calcium on cell proliferation and extracellular matrix synthesis in arterial smooth muscle cells and dermal fibroblasts. Exp Mol Pathol. 1986;44(3):307–17.

102. Lambert M, et al. Ion channels in pulmonary hypertension: a therapeutic interest? Int J Mol Sci. 2018;19(10):3162.

103. Song MY, Makino A, Yuan JXJ. Role of reactive oxygen species and redox in regulating the function of transient receptor potential channels. Antioxid Redox Signal. 2011;15(6):1549–65.

104. Liu T, Zhang L, Joo D, Sun S-C. NF-kB signaling in inflammation. Signal Transduct Target Ther. 2017;2:e17023)

105. Oeckinghaus A, Ghosh S. The NF-kappaB family of transcription factors and its regulation. Cold Spring Harb Perspect Biol. 2009;1(4):a000034.

106. Hayden MS, Ghosh S. NF-κB in immunobiology. Cell Res. 2011;21(2):223–44.

107. Matthews JR, et al. Thioredoxin regulates the DNA binding activity of NF-kappa B by reduction of a disulphide bond involving cysteine 62. Nucleic Acids Res. 1992;20(15):3821–30.

108. Altamirano F, et al. Increased resting intracellular calcium modulates NF-κB-dependent inducible nitric-oxide synthase gene expression in dystrophic mdx skeletal myotubes. J Biol Chem. 2012;287(25):20876–87.

109. Lilienbaum A, Israël A. From calcium to NF-kappa B signaling pathways in neurons. Mol Cell Biol. 2003;23(8):2680–98.

110. Liu X, et al. T cell receptor-induced NF-KB signaling and transcriptional activation are regulated by STIM1- and Orai1-mediated calcium entry. J Biol Chem. 2016;291:8440–52.

111. Valdes JA, et al. NF-KB activation by depolarization of skeletal muscle cells depends on ryanodine and IP3 receptor-mediated calcium signals. Am J Phys Cell Phys. 2007;292:C1960–70.

112. D R, et al. High-frequency field stimulation of primary neurons enhances ryanodine receptor-mediated Ca2+ release and generates hydrogen peroxide, which jointly stimulate NF-κB activity. Antioxid Redox Signal. 2011;14(7):1245–59.

113. Soon Bae J, et al. Phosphorylation of NF-κB by calmodulin-dependent kinase IV activates antiapoptotic gene expression. Biochem Biophys Res Commun. 2003;305(4):1094–8.

114. Kashiwase K, et al. CaMKII activates ASK1 and NF-κB to induce cardiomyocyte hypertrophy. Biochem Biophys Res Commun. 2005;327(1):136–42.

115. Stefano AD, et al. Increased expression of nuclear factor-kappaB in bronchial biopsies from smokers and patients with COPD. Eur Respir J. 2002;20:556–63.

116. Gagliardo R, et al. IkB kinase-driven nuclear factor-kB activation in patients with asthma and chronic obstructive pulmonary disease. J Allergy Clin Immunol. 2011;128(3):635–645.e2.

117. Brown V, et al. Dysregulated apoptosis and NFkappaB expression in COPD subjects. Respir Res. 2009;10(1):24.

118. Raychaudhuri B, et al. Nitric oxide blocks nuclear factor-κB activation in alveolar macrophages. Am J Respir Cell Mol Biol. 1999;21(3):311–6.

119. Yang S-R, et al. RelB is differentially regulated by IkappaB kinase-alpha in B cells and mouse lung by cigarette smoke. Am J Respir Cell Mol Biol. 2009;40(2):147–58.

120. Tully JE, et al. Epithelial NF-κB orchestrates house dust mite-induced airway inflammation, hyperresponsiveness, and fibrotic remodeling. J Immunol. 2013;191(12):5811–21.

121. Price LC, et al. Nuclear factor κ-B is activated in the pulmonary vessels of patients with end-stage idiopathic pulmonary arterial hypertension. PLoS One. 2013;8(10):e75415.

122. Hosokawa S, et al. Pathophysiological roles of nuclear factor kappaB (NF-kB) in pulmonary arterial hypertension: effects of synthetic selective NF-kB inhibitor IMD-0354. Cardiovasc Res. 2013;99(1):35–43.

123. Luo Y, et al. CD146-HIF-1α hypoxic reprogramming drives vascular remodeling and pulmonary arterial hypertension. Nat Commun. 2019;10(1):3551.

124. Greijer AE, van der Wall E. The role of hypoxia inducible factor 1 (HIF-1) in hypoxia induced apoptosis. J Clin Pathol. 2004;57(10):1009–14.

125. Song T, Zheng Y-M, Wang Y-X. Cross talk between mitochondrial reactive oxygen species and sarcoplasmic reticulum calcium in pulmonary arterial smooth muscle cells. In: Wang Y-X, editor. Pulmonary vasculature redox signaling in health and disease. Cham: Springer International Publishing; 2017. p. 289–98.

126. Hayden MS, West AP, Ghosh S. NF-kB and the immune response. Oncogene. 2006;25:6758–80.

127. Kimura S, et al. Nanoparticle-mediated delivery of nuclear factor κB decoy into lungs ameliorates Monocrotaline-induced pulmonary arterial hypertension. Hypertension. 2009;53(5):877–83.

128. Villegas L, Stidham T, Nozik-Grayck E. Oxidative stress and therapeutic development in lung diseases. J Pulmon Respirat Med. 2014;4(4):194.

129. Feissner RF, et al. Crosstalk signaling between mitochondrial Ca2+ and ROS. Front Biosci (Landmark edition). 2009;14:1197–218.

Protein S-Palmitoylation and Lung Diseases

10

Zeang Wu, Rubin Tan, Liping Zhu, Ping Yao, and Qinghua Hu

Abstract

S-palmitoylation of protein is a posttranslational, reversible lipid modification; it was catalyzed by a family of 23 mammalian palmitoyl acyltransferases in humans. S-palmitoylation can impact protein function by regulating protein sorting, secretion, trafficking, stability, and protein interaction. Thus, S-palmitoylation plays a crucial role in many human diseases including mental illness and cancers. In this chapter, we systematically reviewed the influence of S-palmitoylation on protein performance, the characteristics of S-palmitoylation regulating protein function, and the role of S-palmitoylation in pulmonary inflammation and pulmonary hypertension and summed up the treatment strategies of S-palmitoylation-related diseases and the research status of targeted S-palmitoylation agonists/inhibitors. In conclusion, we highlighted the potential role of S-palmitoylation and depalmitoylation in the treatment of human diseases.

Keywords

S-palmitoylation · Depalmitoylation · zDHHC · Protein modification

Zeang Wu and Rubin Tan contributed equally with all other contributors.

Z. Wu
School of Public Health, Tongji Medical College, Huazhong University of Science and Technology, Wuhan, China

First Affiliated Hospital, School of Medicine, Shihezi University, Shihezi, China

School of Basic Medicine, Tongji Medical College, Huazhong University of Science and Technology, Wuhan, China

R. Tan
School of Basic Medicine, Tongji Medical College, Huazhong University of Science and Technology, Wuhan, China

School of Basic Medicine, Xuzhou Medical University, Xuzhou, China

L. Zhu · Q. Hu (✉)
School of Basic Medicine, Tongji Medical College, Huazhong University of Science and Technology, Wuhan, China
e-mail: qinghuaa@mails.tjmu.edu.cn

P. Yao (✉)
School of Public Health, Tongji Medical College, Huazhong University of Science and Technology, Wuhan, China
e-mail: yaoping@mails.tjmu.edu.cn

© The Author(s), under exclusive license to Springer Nature Switzerland AG 2021
Y. -X. Wang (ed.), *Lung Inflammation in Health and Disease, Volume II*, Advances in Experimental Medicine and Biology 1304, https://doi.org/10.1007/978-3-030-68748-9_10

Abbreviations

ABHD17s	α/β hydrolase domain-containing 17 proteins
AML	Acute myeloid leukemia
AMPAR	α-Amino-3-hydroxy-5-methyl-4-isoxazolepropionic acid receptor
APTs	Acyl-protein thioesterases
BMP	Bone morphogenic protein
BMPR1a	BMP receptor 1a signaling
CCN3	Nephroblastoma-overexpressed protein
CCR5	CC chemokine receptor 5
CD-M6PR	Cation-dependent mannose-6-phosphate receptor
CHIKV	Chikungunya virus
CSP	Cysteine-string protein
D2R	D2 dopamine receptor
DR4	Death receptor 4
GPCR	G protein-coupled receptor
α1D AR	α-1 adrenergic receptor
ISO	Isoproterenol
5-HT1AR	Serotonin 1A receptor
IFITM3	Interferon-induced transmembrane protein 3
IRAK4	Interleukin-1 receptor-associated kinase 4
MC1R	Melanocortin-1 receptor
MYD88	Myeloid differentiation primary response protein
NPC	Nasopharyngeal carcinoma
NSPCs	Neural stem/progenitor cells
NTSR-1	Neurotensin receptor-1
NMNAT2	Nicotinamide mononucleotide adenylyltransferase 2
Ncdn	Neurochondrin
NCAM	Neural cell adhesion molecule
PD-L1	Programmed death-ligand 1
PA	Phospholipid acid
PRCD	Progressive rod-cone degeneration
PPHN	Persistent pulmonary hypertension of the newborn
PLCβ1	Phospholipase C β1
PLSCR1	Phospholipid scramblase 1
PPTs	Palmitoyl-protein thioesterases
Rab7α	Ras-related protein Rab-7α
SSTR5	Somatostatin receptor 5
SIRS	Systemic inflammatory response syndrome
TMD	Transmembrane domain
TSP1	Thrombospondin type-1
TLR	Toll-like receptor
TP	Thromboxane prostanoid
TPα	Thromboxane prostanoid α isoforms
β2AR	β2-adrenergic receptors
VSV-G	Vesicular stomatitis virus

10.1 Introduction of S-Palmitoylation

10.1.1 Concept

Palmitoylation is one kind of common posttranslational lipid modifications of proteins, which was first found about 40 years ago. Protein palmitoylation which generally occurs on cysteine residues is divided into three kinds (N-palmitoylation, O-palmitoylation, and S-palmitoylation) according to different connection modes. S-palmitoylation is a unique, reversible modification with a saturated 16-carbon long-chain fatty acid added to residues of particular cysteine(s); it regulates protein function by altering protein sorting, trafficking, localization, secretion, stability, and protein interaction [1]. A great number of proteins have been reported to undergo S-palmitoylation, including enzymes, receptors, viral glycoproteins, channels, and transporters. Many studies show that cycles of S-palmitoylation and depalmitoylation are pivotal to the occurrence, development, and treatment of diseases [2]. Hence, it is involved in many human diseases including DNA damage [3], neuropsychiatric diseases [4], virus infections [5], Crohn's disease [6], abnormal liver function [7], sepsis [8], alopecia [9], and osteoporosis [10] (Table 10.1).

S-palmitoylation is mainly mediated via a superfamily of 23 mammalian palmitoyl acyltransferases (zDHHCs) that contains a highly conserved domain of zinc finger (Asp-His-His-Cys) rich in cysteine. zDHHCs are membrane

Table 10.1 The characteristics, functions, target proteins, signaling pathways, and miRNAs of zDHHC proteins in related diseases

zDHHC	Intracellular localization	Tissue-specific distribution	Object	Function	Pathway	Disease	Target protein	Modified site	miRNA
zDHHC-1	ER	Brain, placenta, lung, uterus	Rat	Regulate the localization of Ncdn	/	/	Neurochondrin (Ncdn) [15]	Cys-3, Cys-4	/
zDHHC-2	ER/Golgi	Brain, kidney, testis, eye, lung, pancreas	HeLa, 293 T, Huh7.5.1 cells	Promote CHIKV replication	/	Chikungunya virus (CHIKV)	nsP1 [32]	Cys-417, Cys-418, Cys-419	/
			HeLa cells, MDCK cells	Essential for trafficking of CKAP4/p63 from the ER	/	Tumor	CKAP4/p63 [58]	Cys-100	/
			HEK-293, A431, MDA 231 cells	Promote physical associations between CD9 and CD151	/	/	CD9 and CD151 [59]	/	/
			CNE1, TW03, 293 T cells	Regulate cell migration	SOCS1, FOXO3	Nasopharyngeal carcinoma (NPC)	CKAP4/p63 [58]	Cys-100	MiR-155
zDHHC-3	Golgi	Liver, spleen, lung, brain, colon, eye, prostate, placenta	Neuroblastoma N2a cells	Stimulate the neurite outgrowth	/	/	Neural cell adhesion molecule (NCAM) [24]	Cys-18, Cys-295, Cys-297	/
			Rat	Regulate the localization of Ncdn	/	/	Neurochondrin (Ncdn) [15]	Cys-3, Cys-4	/
			C57BL/6	Reduce AMPAR trafficking at synapses	PI3K	Cognitive disorders	AMPA glutamate receptor subunit GluA1 [60]	Cys-585, Cys-811	/
			Human breast cancer	Tumor inhibitor	FAK, STAT3	Breast tumor	ERGIC3 [61]	/	/
			RS cell	Regulate UL20 localization and expression	/	HSV-1 infection	UL20 [42]	/	/
			hRPE1 cells	Regulate protein stability of PRCD	/	/	Progressive rod-cone degeneration (PRCD) [43]	Cys-322, Cys-323	/

(continued)

Table 10.1 (continued)

zDHHC	Intracellular localization	Tissue-specific distribution	Object	Function	Pathway	Disease	Target protein	Modified site	miRNA
zDHHC			MC38 cells	Stabilize PD-L1	ESCRT-MVB	Colon carcinoma	Programmed death-ligand 1 (PD-L1) [22]	Cys-272	/
			HEK-293 T	Regulate the stability of the D2R	/	/	D2 dopamine receptor [30]	Cys-443	/
			MDA-MB-231, HEK-293, AR230	Important for the homo-oligomerization and raft localization of DR4	/	Cell death	Death receptor DR4 [36]	C261-3S	/
zDHHC	Intracellular localization	Tissue-specific distribution	Object	Function	Pathway	Disease	Target protein	Modified site	miRNA
zDHHC-4	Golgi	Brain, lung, testis, prostate	HEK-293 T	Regulate the stability of the D2R	/	/	D2 dopamine receptor [30]	Cys-443	/
zDHHC-5	Plasma membrane	Prostate, testis, lung, colon, eye	RAW264.7, HCT116 cells	Essential for membrane recruitment and immune signaling	NF-kB, MAPK	Crohn's disease	(NOD1/2) [6]	Cys-557, Cys-567, Cys-952	/
			RPE-1, HeLa, HAP1 cells	Important for the cleavage of the anthrax toxin	/	Intoxication of the cell	Furin and PC7 [62]	Cys-711/ Cys-699, Cys-704	/
			Glioma tissue	Inhibitor of p53-mutated gliomas	/	Glioma	EZH2 [63]	Cys-571, Cys-576	/
			HEK293 cells	Regulate the stability of SSTR5	/	/	Somatostatin receptor 5 (SSTR5) [44]	Cys-319	/
			Neuronal stem cells	Regulate the neural stem cell differentiation	/	/	Flotillin-2 [64]	Cys-4, Cys-20	/
			HT-1080, A549, T98G, HEK 293 T	Regulate CIL56-induced cell death	/	/	Caspase-independent lethal 56 (CIL56) [21]	Cys-134	/
zDHHC-6	ER	Brain, colon, uterus, lung, kidney, thymus	C57BL/6 J	Regulate TLR-induced inflammation	NF-kB	Sepsis	Myeloid differentiation primary response protein (MYD88) [8]	Cys-113, Cys-274	/

	Intracellular localization	Tissue-specific distribution	Object	Function	Pathway	Disease	Target protein	Modified site	miRNA
zDHHC-7	Golgi	Lung, colon, brain, liver, skin, prostate, kidney	HEK-293T cells	Regulate calcium mobilization and TPα-mediated vasoconstrictor responses	/	PPHN	Gαq [47]	Cys-9, Cys-10	/
			HEK293A cells	Tumor suppression	PI3K, AKT	Cancers	Scribble (SCRIB) [65]	Cys-4, Cys-10	/
			HeLa cells	Affect transport of CCR5	/	HIV-1 infection	CCR5 [13]	Cys-3	/
			HEK-293T, A549, MEFs cells	Increase IFITM3 antiviral activity	/	Virus infections	Interferon-induced transmembrane protein 3[5]	Cys-71, Cys-72, Cys-105	/
			Breast cancer cells	Tumor inhibitor	FAK	Breast tumor	ERGIC3 [61]	/	/
zDHHC-8	Golgi	Brain, lung, kidney, ovary, pancreas, eye	C57BL/6	Promote the seizures	/	Epilepsy	AMPA receptor [66]	/	/
			Drosophila	Correlate with cancer survival	/	Cancer	Scribble and Ras64B [67]	Cys-46, Cys-120, Cys-147	/
			LgDel/+ mouse	Promote spine stabilization	/	Schizophrenia	cdc42 [68]	Cys-188, Cys-189	/
			HEK293 and HEK-nNOS cells	Important for neuronal connections, contribute to neurodevelopmental deficits	/	Neurodevelopmental deficits	PSD-95[4]	Cys-3, Cys-5	/
			HEK-293 T	Regulate the stability of the D2R	/	/	D2 dopamine receptor [30]	Cys-443	/

(continued)

Table 10.1 (continued)

zDHHC	Intracellular localization	Tissue-specific distribution	Object	Function	Pathway	Disease	Target protein	Modified site	miRNA
zDHHC-9	ER/Golgi	Brain, prostate, lung, kidney, thalamus	MDA-MB-231 and BT549	Regulate PD-L1 stability	/	Breast cancer	Programmed death-ligand 1 (PD-L1) [12]	Cys-272	/
			SD rats	Regulate membrane targeting of H-Ras	/	/	H-Ras [51]	Cys-181, Cys-184	miR-134
			C57BL/6	Regulate N-Ras plasma translocation	/	Leukemias	N-Ras [69]	/	/
			HEK293 cells	Stabilize the receptor	MAPK	/	β2-adrenergic receptors (β2AR) [28]	Cys-256, Cys-341	/
			HEK293 cells	Functional channel regulation	/	/	STREX [70]	Cys-12, Cys-13	/
zDHHC-11	ER	Testis, brain, lung, placenta, cerebellum	Rat	Regulate the localization of Ncdn	/	/	Neurochondrin (Ncdn) [15]	Cys-3, Cys-4	/
			HEK293, HeLa, THP1 cells	Mediate MITA-dependent innate immune responses against DNA viruses	MITA, STING	Viral infection	MITA [71]	/	/
			COS-7 cells	Regulate gp78 distribution	ERAD	/	Gp78 [72]	Cys-352	/
zDHHC	Intracellular localization	Tissue-specific distribution	Object	Function	Pathway	Disease	Target protein	Modified site	miRNA
zDHHC-12	ER/Golgi	Ascites, skin, lung, prostate, stomach, brain	C57BL/6	Modifies Aβ deposition	/	Alzheimer's disease	β-Amyloid peptides (Aβ) [73]	/	/

zDHHC	Intracellular localization	Tissue-specific distribution	Object	Function	Pathway	Disease	Target protein	Modified site	miRNA
zDHHC-13	ER	Uterus, brain, testis, stomach, placenta, colon	C57BL/6 J, HPMs, and B16 cells	Tumor inhibitor	/	Melanomagenesis	Melanocortin-1 receptor (MC1R) [54, 55]	Cys-315	/
			HEK-293 T, Hep1–6, MCF-7	Regulate mitochondrial activity	/	Abnormal liver function	MCAT and CTNND1 [7]	/	/
			C57BL/6NJ	Regulate anxiety-related behaviors	/	Behavioral abnormalities	Drp1 [74]	/	/
			HEK293 and MC3T3E1 cells	Regulator of postnatal skeletal development and bone mass acquisition	/	Osteoporosis	MT1-MMP [10]	Cys-574	/
			C57BL/6	Important for hair anchoring and skin barrier function	/	Alopecia and hyperkeratosis	Cornifelin [9]	Cys-95, Cys-101	/
			YAC128, mouse	Regulate the cell death-signaling pathways	/	Huntington disease	GluN2B [19]	Cys clusters I and II	/
zDHHC-14	ER	Brain, fetal eyes, uterus, prostate	HEK293 cells	Stabilize the receptor	MAPK	/	β2-adrenergic receptors (β2AR) [28]	Cys-256, Cys-341	/
			TMK-1 cells	Regulate GC cell migration and invasion	/	Gastric cancer	MMP-17[17]	/	/
			K562 cells	Regulate cellular differentiation	/	Acute biphenotypic leukemia	CD61 [18]	/	/
			Serum samples of CAD patients	Promote cell growth of vascular smooth muscle cells	TLR9	Coronary artery disease (CAD)	/	/	miR-574-5p [75]
zDHHC	Intracellular localization	Tissue-specific distribution	Object	Function	Pathway	Disease	Target protein	Modified site	miRNA

(continued)

Table 10.1 (continued)

zDHHC	Intracellular localization	Tissue-specific distribution	Object	Function	Pathway	Disease	Target protein	Modified site	miRNA
zDHHC-15	Golgi	Brain, trachea, cerebellum, ear, eye, kidney	HeLa cells	Affect the transport of CCR5	/	HIV-1 infection	CCR5 [13]	Cys-3	/
			HEK293 and HEK-nNOS cells	Important for neuronal connections, contribute to neurodevelopmental deficits	/	Neuro heteroplasia	PSD-95[4]	Cys-3, Cys-5	/
			PC12, HEK293	Promote CSP stabilization	/	/	Cysteine-string protein [14]	Cys-136	/
zDHHC-16	ER	Brain, placenta, lung, uterus, skin	C57Bl/6	Regulation of DNA damage responses	Atm	DNA damage	C-Abl [3]	/	/
			Zebrafish	Regulate NSPCs proliferation	FGF/ ERK	Neurological disorders	Neural stem/progenitor cells (NSPCs) [76]	/	/
zDHHC-17	Golgi	Brain, uterus, eye, lung, thalamus	Zebrafish	Affect TrkA binding to tubulin	ERK1/2	/	TrkA-tubulin [77]	/	/
			YAC128, mouse	Regulate the cell death-signaling pathways	/	Huntington disease	GluN2B [78]	Cys clusters I and II	/
			C57BL/6 JBabr mice	Influence NMNAT2 protein turnover and axon protective capacity	/	Axon degeneration and phenotypic disease	Nicotinamide mononucleotide adenylyltransferase 2 (NMNAT2) [79]	Cys-164, Cys-165	/
zDHHC-18	Golgi	Lung, testis, brain, placenta, carcinoid, kidney, prostate	HEK293 cells	Stabilize the receptor	MAPK	/	β2-adrenergic receptors (β2AR) [28]	Cys-256, Cys-341	/
			Human GBM specimens	Associated with the malignant development and progression of gliomas	/	Glioblastomas (GBM)	BMI1 [25]	/	/
zDHHC-19	ER	Testis, brain, medulla, placenta	HeLa, 293 T, Huh7.5.1 cells	Promote CHIKV replication	/	Chikungunya virus (CHIKV)	nsP1 [32]	Cys-417, Cys-418, Cys-419	/

zDHHC	Intracellular localization	Tissue-specific distribution	Object	Function	Pathway	Disease	Target protein	Modified site	miRNA
zDHHC-20	Plasma membrane	Placenta, uterus, brain, lung, testis	HEK-293T, A549, MEFs cells	Increase IFITM3 antiviral activity	/	Virus infections	Interferon-induced transmembrane protein 3[5]	Cys-71, Cys-72, Cys-105	/
			PC12, Jurkat cells	Impact the secretion of TAT	/	HIV-1 infections	HIV-1 Tat [37]	Cys-31	/
			HEK-293T, NIH 3 T3, and SW1573	Impact cancer cell survival	ERK/AKT	Cancers	EGFR [80]	Cys-1025, Cys-1122	/
			Primary mouse neural stem cells	Regulate the localization and trafficking of BMPR1a and alter BMP signaling	BMP	BMP	Bone morphogenic protein receptor 1a (BMPR1a) [35]	Cys-180	/
zDHHC-21	Plasma membrane	Brain, testis, uterus, eye, liver	C57BL/6 J	Regulate depression-like behavior	MAPK	Major depressive disorder (MDD)	Serotonin 1A receptor (5-HT1AR) [52]	Cys-417, Cys-420	miR-30e
			F233Δ mice	Affect vascular function	ERK1/2	/	α1D adrenoceptor [50]	Cys-120	/
			Rats	Regulate endothelial inflammation	GPCR	Endothelial dysfunction	PLCβ1 [49]	Cys-17	/
			Mouse	Regulate hair shaft differentiation	Wnt	Hair loss	Fyn [81]	/	/
			HUVEC, HEK 293 T, COS-7 cells	Regulate the levels of PECAM1 at the cell surface	/	/	PECAM1 [82]	Cys-595	/

(continued)

Table 10.1 (continued)

zDHHC	Intracellular localization	Tissue-specific distribution	Object	Function	Pathway	Disease	Target protein	Modified site	miRNA
zDHHC-22	ER/Golgi	Brain, eye, lung	Neuro2a, HEK293	Regulate CCN3 secretion	/	Cancer	NOV/CCN3 [38]	Cys-241	/
			HEK293 cells	Control BK channel cell surface expression	/	/	S0-S1 loop of BK channels [83]	Cys-53, Cys-54, Cys-56	/
zDHHC-23	ER	Testis, brain, colon, kidney	Human GBM specimens	Regulate the polyubiquitination and accumulation of BMI1		Glioblastomas (GBM)	BMI1 [25]	/	/
			HEK293 cells	Control BK channel cell surface expression	/	/	S0-S1 loop of BK channels [83]	Cys-53, Cys-54, Cys-56	/

"/": Not mention in the chapter

proteins with different subcellular localization, most zDHHCs are located to the secretory pathway such as endoplasmic reticulum and Golgi apparatus, while some are also identified in the cell membrane [11]. Many proteins have been identified as the substrates for zDHHCs, including programmed death-ligand 1 (PD-L1) [12], CC chemokine receptor 5 (CCR5) [13], cysteine-string protein (CSP) [14], and neurochondrin (Ncdn) [15]. When S-palmitoylation happens, zDHHCs first become transient acyl-intermediate by autoacylated, which are then transferred to the target protein. For some special proteins, physical contact between proteins and zDHHCs is essential for the rate and efficiency of S-palmitoylation. For instance, site mutations lead to physical isolation of the CSP from the zDHHCs, further inhibits the S-palmitoylation of CSP [14]. Not all the zDHHCs possess the same enzymatic activity, partly due to the sequence diversity at the C- and N-terminal cytoplasmic tails. Some zDHHCs have substrate specificity, while others do not. Someone attributes it to the different sequence and exterior region of the zDHHCs domain, as well as their subcellular localization, though more studies are needed to confirm this.

As the major enzyme regulating protein S-palmitoylation, zDHHCs are also potentially the targets for the treatment of S-palmitoylation-associated disorders, and the development of agonists/inhibitors of S-palmitoylation is also an alternative solution for disease treatment. In the following, we will review the influence of S-palmitoylation on protein performance, the characteristics of S-palmitoylation regulating protein function, and the potential role of targeted S-palmitoylation agonists/inhibitors in the treatment of related diseases.

10.2   Characteristics of S-Palmitoylation Regulating Protein

10.2.1   Same zDHHC Exerts Opposite Functions in Different Cancers

The same zDHHC has the opposite function in different cancer tissues, overexpress of zDHHC-

14 inhibits tumorigenesis [16] but promotes the invasion and migration of scirrhous gastric cancer [17], respectively. The different proteins regulated by zDHHC-14 in different tumors may be responsible for this phenomenon. For example, S-palmitoylation of MMP-17 is regulated by zDHHC-14 in scirrhous-type gastric cancer, and it can promote the growth, invasion, and metastasis of tumor cells, as well as the formation of tumor blood vessels [18]. Besides, S-palmitoylation of CD61 is also regulated by zDHHC-14 in leukemia and related to the inhibition of cellular differentiation [18]. So, if we try to use it as a target for the treatment of related diseases, we need to consider that it has the opposite effect in different tissues.

10.2.2   Same Protein Palmitoylated by Different zDHHCs Has Same/Different Function

One protein can be modified by several zDHHCs; for example, both zDHHC-17 and zDHHC-13 are responsible for the S-palmitoylation of huntingtin [19]. zDHHC-2, zDHHC-3, zDHHC-7, zDHHC-15, and zDHHC-17 are related to SNAP-25 S-palmitoylation [20]. The function of protein palmitoylated by different zDHHCs may be similar. For instance, knockout zDHHC-5 or mutate at its acylated Cys residues decrease the CIL56-induced cell death, overexpression of both zDHHC-5 and zDHHC-8 restore CIL56-induced cell death in 293 T zDHHC-5^{KO} cells [21]. Besides, PD-L1 palmitoylated by zDHHC-9 increases its protein level and cell surface distribution in breast cancer cells, thereby promoting the growth of RAS-activated tumors [12]. Meanwhile, zDHHC-3 can also palmitoylate PD-L1 and inhibit its mono-ubiquitination, thus block its trafficking to the multivesicular body. Consequently, this block increases the protein level of PD-L1 and inhibits T-cell cytotoxicity [22]. However, the same protein regulated by different zDHHCs may also play different roles in diseases. PSD-95 palmitoylated by zDHHC-8 can impact the nitrosylation of PSD-95, while zDHHC-15 can promote the S-palmitoylation of PSD-95 and increase the overall density of excitatory synapses [4].

10.2.3 zDHHC Itself Can Also Undergo S-Palmitoylation

zDHHC-6, as a palmitoyl acyltransferase regulating protein S-palmitoylation, is also palmitoylated by other enzymes. The zDHHC-16 can palmitoylate zDHHC-6, silence or knock out zDHHC-6 in HAP1 cells increases the mRNA level of zDHHC-16, indicating that they may interact physically and genetically [23]. What's more, phosphorylation of zDHHC affects auto S-palmitoylation and S-palmitoylation of target proteins. zDHHC-3 auto S-palmitoylation was increased by abolishing tyrosine phosphorylation, and it further enhanced interplay with neural cell adhesion molecule (NCAM) and promoted NCAM S-palmitoylation [24].

10.2.4 The Impact Between S-Palmitoylation and Other Protein Modifications

When the distance between the S-palmitoylation site and other protein modification sites is close, the two protein modifications may interact with each other. For example, PSD-95 was related to the neurodevelopmental and neurodegenerative diseases, and Cys-3 and Cys-5 are common sites of nitrosylation and S-palmitoylation. The nitrosylation of PSD-95 was influenced by its S-palmitoylation which was regulated by zDHHC-8. At the same time, the S-palmitoylation of PSD-95 was inhibited by NO produced in granule cells of the cerebellum, indicating a competitive cysteine modification between S-palmitoylation and nitrosylation of PSD-95 [4].

The interaction between S-palmitoylation and ubiquitination was observed in many proteins. The overexpression of zDHHC-18 reduced the amount of polyubiquitinated BMI1 and increased the tumor cell survival. Interestingly, the deletion of zDHHC-23 has the same effect as above [25]. It has been also demonstrated that PD-L1 ubiquitination is blocked by its S-palmitoylation on Cys-272, thereby inhibiting the degradation of PD-L1 caused by lysosomes. The ubiquitination of PD-L1 is increased by both 2-BP treatment and depletion of zDHHC-3 [22]. Proteins are more sensitive to ubiquitination when S-palmitoylation is disturbed. S-palmitoylation of lipoprotein receptor-related protein 6 (LRP6) is necessary for its export out of the endoplasmic reticulum and proper cell membrane trafficking. LRP6 deficient in palmitoylation retains in the endoplasmic reticulum, due to ubiquitination at Lys-1403, which locates nearby the palmitoylation sites (Cys-1394 and Cys-1399) [26].

S-palmitoylation of proteins with a weak membrane affinity (e.g., H- and N-Ras) can enhance the strength of their membrane interaction and trap proteins on a proper intracellular membrane. As a consequence, proteins bind to budding vesicles efficiently and do not separate from cell membranes during vesicle transport. S-palmitoylation may enhance protein membrane interaction due to cooperation with other modifications. Farnesylation at the C-terminal of Ras regulates its association with ER and Golgi membranes; however, farnesylation only provides Ras with a weak membrane affinity. When S-palmitoylation occurs, two lipid modifications in tandem can increase the membrane interaction of Ras [27].

Besides, the regulatory effect of S-palmitoylation on phosphorylation was also found in β_2AR. Ser-345/Ser-346 and Ser-261/Ser-262, the phosphorylation sites of β_2AR, are located next to its S-palmitoylation sites, Cys-341 and Cys-265. Studies show that phosphorylation by cAMP-dependent kinase (PKA) is essential for S-palmitoylation of β_2AR at Cys-265. Inhibition of S-palmitoylation by H-89 suppresses isoproterenol-caused, PKA-mediated phosphorylation at both Ser-261/Ser-262 and Ser-345/Ser-346. What's more, S-palmitoylation at Cys-341 restricts the accessibility of a PKA phosphorylation site in the carboxyl tail of the β_2AR and then regulates the phosphorylation state [28].

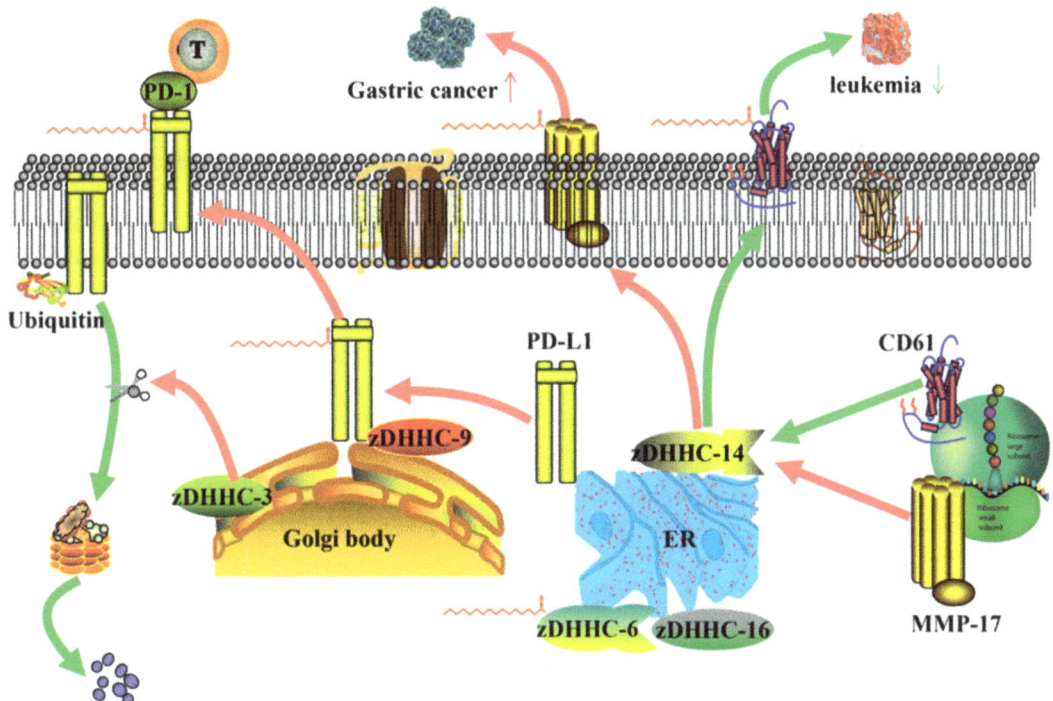

Fig. 10.1 Characteristics of S-palmitoylation regulating protein function

10.2.5 The Interaction Degree of zDHHCs on Different Proteins and S-Palmitoylation Modification Sites Are Various

Different proteins can be palmitoylated by an unequal number of zDHHCs due to their special spatial structure. For instance, interferon-induced transmembrane protein 3 (IFITM3) is a broad-spectrum virus inhibitor that can be palmitoylated by half of the zDHHCs [5]. However, β_2AR can be palmitoylated by zDHHC-9, zDHHC-14, and zDHHC-18 [28]. It is interesting to note that the nucleotide oligomerization domain-like receptors 1 and 2 are only palmitoylated by zDHHC-5 [6]. This phenomenon affects the therapeutic effect of targeting zDHHCs to some extent.

Further studies found that zDHHC has different effects on different S-palmitoylation modification sites. Cys-265 and Cys-341 are S-palmitoylation modification sites of β_2AR, knockdown zDHHC-9, zDHHC-14, zDHHC-18 alone or in combination reduce S-palmitoylation at Cys-265. However, the zDHHC-9/zDHHC-14/zDHHC-18 knockdown shows no obvious influence on S-palmitoylation at Cys-341, indicating the response of Cys-265 and Cys-341 to zDHHC-9/zDHHC-14/zDHHC-18 is different [28]. As a consequence, S-palmitoylation of the same protein at different sites may further exert distinct effects on physiological activity. AMPA receptor subunits (GluR1-GluR4) are palmitoylated at different sites (site 1 and site 2), which are located at various transmembrane domains (TMDs). S-palmitoylation of site 1 increases the accumulation of this receptor in Golgi apparatus and reduces expression levels on the cell surface, while S-palmitoylation of site 2 fails to regulate stable expression levels on the cell surface of the receptor [29] (Fig. 10.1).

10.3 S-Palmitoylation on Protein Function

10.3.1 Effect on Protein Sorting and Trafficking

Many membrane proteins are synthesized on ribosomes and then undergo a variety of processing and modification at the ER. Finally, these proteins with different markers are transported to the destinations via protein sorting and trafficking pathways. S-palmitoylation is certainly important in maintaining the stability of some proteins during sorting and trafficking. D2 dopamine receptor (D2R), a G protein-coupled receptor (GPCR), is indispensable in regulating mood [30]. Cys-443 represents the major amino acid of D2R S-palmitoylation by zDHHC-3 and zDHHC-4. S-palmitoylation of D2R can regulate the receptor proper trafficking to the cell membrane and maintain the receptor stability during the progress. C443 deletion or palmitoylation inhibition caused an obvious decline in D2R expression on the plasma membrane and a contingent increase in the Golgi.

The biophysical properties of many transmembrane and soluble proteins are altered by S-palmitoylation; these changes may affect the protein sorting and trafficking process including the transport rate of proteins across the Golgi. S-palmitoylation has been proved to be an anterograde signal for many proteins at the Golgi membrane interface; it leads to the concentration of curved regions along the Golgi edge by simple physicochemical action. For instance, the S-palmitoylation of G glycoprotein of the vesicular stomatitis virus (VSV-G) and transferrin receptor induced them to concentrate in the most curved regions of the Golgi membranes, thereby they are absorbed by highly curved tubular or vesicular carriers, finally promoted their efficient transport across the Golgi [31].

S-palmitoylation of protein can directly regulate its sorting and trafficking to affect subcellular localization. nsP1 which is one nonstructural protein in the open reading frames of the Chikungunya virus can locate in the plasma membrane and filopodial extensions after palmi-

toylated by zDHHC-2 and zDHHC-9. When S-palmitoylation is disturbed, nsP1 is degraded without proper localization, leading to reduced nsP1 plasma membrane levels and weakened viral replication [32]. Kim et al. find that S-palmitoylation of MCOLN3/TRPML3 regulates its trafficking between subcellular compartments to maintain its cellular function when autophagy is activated [33]. And some palmitoylated proteins can also regulate other proteins' sorting and trafficking. S-palmitoylation of PSD-95 at Cys-3 and Cys-5 is crucial for the synaptic trafficking of α-amino-3-hydroxy-5-methyl-4-isoxazolepropionic acid receptor (AMPAR) [34].

In addition to affect protein proper localization, S-palmitoylation further alters the signaling pathways and related functions involved in these proteins. As a major mediator of the bone morphogenic protein (BMP) signaling, BMP receptor 1a (BMPR1a) in embryonic stem cells and neural stem cells is also palmitoylated. BMPR1a S-palmitoylation at Cys-180 plays an essential role in the appropriate localization and intracellular trafficking of BMPR1a, further selectively alters ERK-dependent BMP signaling and increases the production of oligodendrocytes, ultimately affecting NSC activity and fate choices [35]. A similar phenomenon is also observed in death receptor 4 (DR4), which is palmitoylated by zDHHC-3. The S-palmitoylation of DR4 is required for its localization to lipid rafts, facilitating the efficient transmission of the cell death signal induced by TNF-related apoptosis-inducing ligand [36].

10.3.2 Effect on Protein Secretion

Abnormal protein secretion is related to several diseases and affected by S-palmitoylation. Tat is necessary for viral gene expression and HIV-1 virion production; S-palmitoylation of Tat prevents its secretion by enhancing its combination with the plasma membrane [37]. CCR5 is responsible for the dissemination and establishment of HIV-1 infection; site mutation will weaken the S-palmitoylation and secretion of CCR5. Single-site mutants slightly reduce the secretion of CCR5, and double- and triple-site mutants lead to a 50% drop in its secretion, which remarkably impairs HIV-1 infection of human macrophages [13]. A recent study shows that excessive secretion of nephroblastoma-overexpressed protein (CCN3) suppresses the neurons' axon growth, and thrombospondin type-1 (TSP1) domain of CCN3 is responsible for its secretion. S-palmitoylation of CCN3 at Cys-241 in its TSP1 domain is regulated by zDHHC-22, and mutation at Cys-241 suppresses the secretion of CCN3, indicating that S-palmitoylation is crucial for the secretion of CCN3 [38]. S-palmitoylation can directly or indirectly affect protein secretion. Wnts is a poorly secreted protein, which is important in the regulation of the growth of embryos and repairing the damaged tissues. There are studies show that Porcupine (PORCN) in combination with Wntless (WLS) regulates the secretion of Wnts, overexpression of PORCN promotes the S-palmitoylation of Wnts, and increase WLS's function of promoting Wnts secretion [39].

10.3.3 Effect on Protein Stability

S-palmitoylation can affect protein stability and alter degradation to cycling. Protein stability directly affects the amount of protein in the cell membrane and its function. As a posttranslational, reversible lipid modification, S-palmitoylation can indeed promote the association between proteins and lipid rafts, partly due to its ability to change the lipophilicity and hydrophobicity of proteins. S-palmitoylation of the D2R at C443 can affect the binding of PDZ-domain-containing proteins such as GIPC. GIPC has been shown to interact with the C-terminal cysteine of D2R and D3R (but not D4R) and prevent their lysosomal degradation. C443 deletion also resulted in decreased expression of D2R protein compared to wild-type D2R. ΔC443-D2R and wild-type receptor exhibited no detectable differences in mRNA expression, whereas the mutant protein showed an increased rate of degradation compared to wild-type D2R [30].

The lysosomal sorting receptor sortilin palmitoylated by zDHHC-15 is a prerequisite for the efficient retrograde trafficking of sortilin and recycling sortilin back to the Golgi. Sortilin fails to be recycled and is rapidly degraded in the absence of S-palmitoylation [40]. Recently study found that Ras is palmitoylated in the Golgi and then transported to the cell membrane, where it undergoes depalmitoylation to release Ras back into the cytosol, allowing Ras rebinds to the Golgi apparatus, and then begins the next cycle [41]. PD-L1 is a famous transmembrane protein that is pivotal for the immune escape of tumor cells. Studies show that S-palmitoylation of PD-L1 at Cys-272 by zDHHC-9 enhances the protein stability and cell surface distribution of PD-L1 and thus prevents the tumor from immune surveillance of T cells. Meanwhile, zDHHC-9 knockout or mutation of Cys-272 sharply abrogates PD-L1 S-palmitoylation, resulting in a lower protein level of PD-L1, and further inhibiting breast cancer cell growth with a better therapeutic efficacy [12]. The effect of S-palmitoylation on protein stability can also be observed in many proteins, including Ncdn [15], UL20 [42], progressive rod-cone degeneration (PRCD) [43], somatostatin receptor 5 (SSTR5) [44], CSP [14], and β2-adrenergic receptors (β_2AR) [28].

S-palmitoylation can gather or separate proteins under specific conditions by regulating the association between membrane proteins and lipid rafts. S-palmitoylation and glycosylation of neurotensin receptor-1 (NTSR-1) promote its trafficking to the structured membrane microdomains, interaction with $G\alpha_{q/11}$, and subsequent NTS-mediated ERK phosphorylation [45]. On the contrary, S-palmitoylation of toxin receptor TEM8 unexpectedly restrains its association with lipid rafts and segregates the receptor from its E3 ubiquitin ligase Cbl and consequently prevents its premature ubiquitination [46].

10.3.4 Effect on Protein Interactions

The biophysical properties of proteins are changed by S-palmitoylation, as well as their interaction with other proteins and the signal

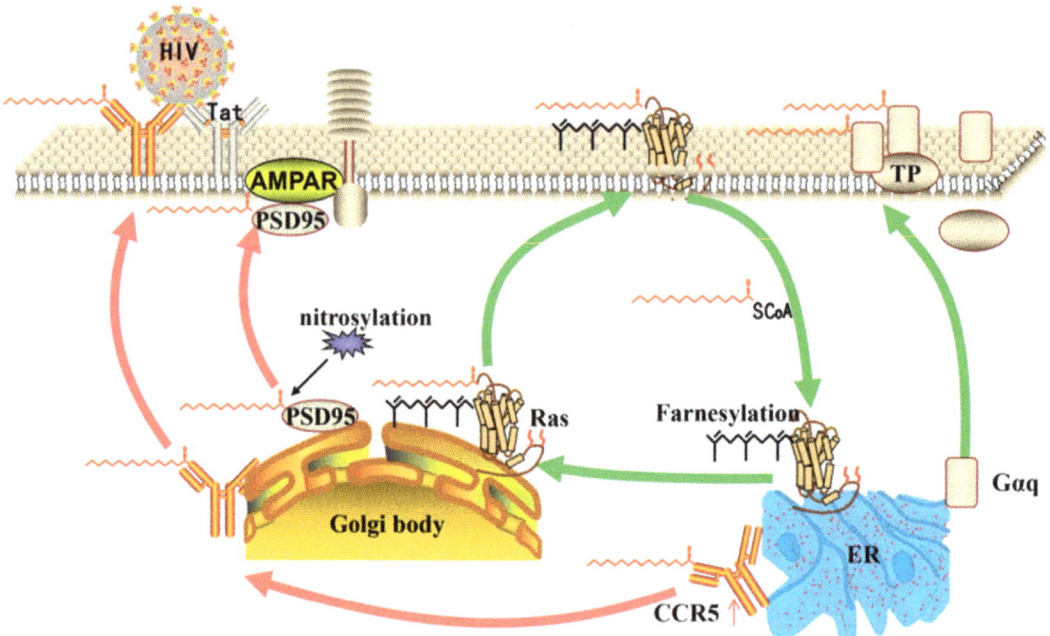

Fig. 10.2 Schematic illustration of S-palmitoylation regulating protein function

pathways they are involved in. A study about sepsis finds that myeloid differentiation primary response protein (MYD88) is responsible for the activation of Toll-like receptor (TLR) signaling by recruiting interleukin-1 receptor-associated kinase 4 (IRAK4). The S-palmitoylation sites of MYD88 (Cys-113 and Cys-274) may be involved in the MYD88-dependent TLR signaling activation. Mutant from cysteine to alanine at position 113 reduces the S-palmitoylation level of MYD88 and blocks the interaction between MYD88 and IRAK4, further inhibits TLR-caused inflammation, and facilitates the survival of mice with sepsis. However, all the phenomena above are rescuable by the CA113C reverse mutant, demonstrating that the S-palmitoylation of MYD88 at Cys-113 plays an important role in recruiting IRAK4 [8].

S-palmitoylation also regulates the interaction between membrane proteins and cytoplasmic proteins. Cation-dependent mannose-6-phosphate receptor (CD-M6PR) which is a type I transmembrane protein takes charge of arranging the new acid hydrolases. The CD-M6PR will return to the Golgi for recycling after it delivers acid hydrolases to endosomes from the trans-Golgi network. This step is under the control of the S-palmitoylation of CD-M6PR, which allows the interaction between cytosolic domain of CD-M6PR and retromer complex [40].

In addition, the degree of protein interaction depends on the level of S-palmitoylation. In pulmonary hypertension of the newborn (PPHN), hypoxia can increase the S-palmitoylation level of Gαq and promote the interaction of Gαq with thromboxane prostanoid (TP). This interaction depends on the S-palmitoylation degree of Gαq, the single-site mutant has little effect on the binding of Gαq and TPα, while double C9A/C10A mutants result in a significant reduction in the association of Gαq with TPα [47] (Fig. 10.2).

10.4 S-Palmitoylation on Pulmonary Inflammation and Disease

10.4.1 S-Palmitoylation and Lung Inflammation

In recent years, inflammatory lung diseases caused by pathogenic microorganisms, such as

SARS and COVID-19, have posed a great threat to human health. Studies have shown that S-palmitoylation is very important in the process of pathogenic microorganisms infecting the lungs. *Legionella pneumophila* was the bacterial pathogen of Legionnaires' disease, it could multiply in human alveolar macrophages and lung epithelial cells, and then cause a severe pneumonia. The virulence of *Legionella pneumophila* is closely related to the Dot/ICM IV type secretion system, which can manipulate the cell signal by carrying more than 300 kinds of effector protein to the host cell protein. This bacterium can build vacuoles (LCV) containing *Legionella* in the endoplasmic reticulum to avoid being decomposed by phagocytic enzymes. The process of localization of effector protein LpdA to LCV requires S-palmitoylation. S-palmitoylation-deficiency LpdA targets the plasma membrane and vesicle, where it hydrolyzes a range of substrates to produce phospholipid acid (PA). The disturbance of cell PA homeostasis will destroy the integrity of Golgi apparatus, thus increasing the virulence of *Legionella pneumophila* in the lung [48].

Endothelial dysfunction is an important sign of systemic inflammatory response syndrome (SIRS). 2-BP treatment can improve the vascular barrier damage in the lung by reducing leukocyte adhesion, microvascular leakage, and ICAM-1 expression. By studying the expression and function of DHHC-PAT in mouse lung microvascular endothelial cells, it was found that only the knockout of zDHHC-5 and zDHHC-21 significantly improved the barrier function. After IL-1b stimulation, the number of adherent white blood cells increased significantly, and inflammation can increase the S-palmitoylation level of phospholipase C β1 (PLCβ1). PLCβ1 is a major signal molecule downstream of GPCR, which mediates the inflammatory effects of thrombin and histamine. Thickening of alveolar capillary membranes and leukocyte infiltration have been noted in the zDHHC-21 group of mice. The overexpression of wild-type PLCβ1 in zDHHC-21[dep/des] endothelial cells failed to enhance the barrier dysfunction induced by thrombin, indicating that DHHC21-mediated S-palmitoylation of PLCβ1

is pivotal for the dysfunction of endothelial cells during inflammation, because it could activate a series of downstream events including calcium ion and phosphatase mobilization [49].

When the host's response to bacterial infection becomes unbalanced, a systemic inflammation called sepsis occurs. Under such conditions, the uncontrollable activation of inflammatory signals by TLR/MYD88 will damage the bactericidal activity of neutrophils and reduce the apoptosis and chemotaxis activity of neutrophils. zDHHC-6 palmitoylates MYD88 by using the endogenous fatty acids synthesized by FASN and CD36-mediated uptake of exogenous fatty acids. FASN inhibitor (C57) treatment or zDHHC-6 gene knockout can reduce the S-palmitoylation of MYD88 and the activation of TLR/MYD88 stimulated by lipopolysaccharide and further enhance the chemotactic activity of neutrophils and improve the survival of septic mice [8].

10.4.2 S-Palmitoylation and Pulmonary Hypertension

In addition to regulating lung inflammation, zDHHC-21 can also regulate the function of pulmonary blood vessels. zDHHC-21 is expressed in endothelial cells and is responsible for the S-palmitoylation of platelet endothelial cell adhesion molecule-1 and endothelial nitric oxide synthase, which are involved in pulmonary vasodilation and pulmonary vascular remodeling, angiogenesis, and transendothelial cell migration. α-1 adrenergic receptor (α1D AR), a key determinant of vascular tone, was also identified as new substrates for zDHHC-21 recently. zDHHC-21 has been reported to directly interact with α1D AR to form a complex, increasing the steady-state S-palmitoylation of α1D AR and the total expression of α1D AR, finally destroy α1AR-mediated vasoconstriction [50].

The S-palmitoylation of α1D AR is specific because zDHHC-21 cannot palmitoylate Gαq, which can regulate calcium mobilization by interacting with thromboxane prostanoid α isoforms (TPα). zDHHC-7 is responsible for the S-palmitoylation of Gαq in pulmonary hypertension of the newborn, and this lipid modification of Gαq plays an essential role in receptor-Gαq-phospholipase-C interactions. Hypoxic treatment of HEK-293T cells for 24 hours could promote the S-palmitoylation of Gαq and further increase the generation of inositol-1,4,5-trisphosphate mediated by TP. 2-BP treatment can significantly inhibit the S-palmitoylation of Gαq and reduce TP-mediated calcium mobilization. Cysteine mutations can change the S-palmitoylation and membrane localization of Gαq and have a dose-dependent effect on the calcium response induced by TP [47].

10.5 Treatment Strategies for S-Palmitoylation-Related Diseases

10.5.1 Targeting zDHHC-Related miRNAs

The S-palmitoylation of protein controls its activation and functions, since the most S-palmitoylation of proteins is regulated by zDHHC, it may be a workable plan to regulate protein S-palmitoylation by targeting zDHHC. Present studies show that zDHHC expression can be regulated by miRNAs. miR-134 has been proved to regulate the expression of zDHHC-9 and the S-palmitoylation of H-Ras mediated by zDHHC-9 and further change the subcellular localization of H-Ras [51]. miR-30e can downregulate the expression of zDHHC-21, which further reduce the S-palmitoylation of the serotonin 1A receptor (5-HT1AR) and 5-HT1AR-mediated signaling, eventually lead to depressive symptoms [52].

10.5.2 Regulate the Protein Depalmitoylation

The cycle rates of S-palmitate and depalmitate between different protein types may vary greatly. This dynamic cycle is essential for protein transport, stability, and function.

Depalmitoylation can change the location and function of palmitoylated proteins and then deliver them to lysosomes for degradation, indicating the importance of depalmitoylation in protein turnover. AMPA receptor is involved in the higher Ca^{2+} permeability that associates with motor neuron degeneration; PSD-95 can stabilize AMPA receptor at the postsynaptic membrane. GluR2 regulates the trafficking of AMPARs to the cell membrane from the endoplasmic reticulum, and depalmitoylation decreases GluR2 and PSD-95 expressions, leading to reduced mediated toxicity and calcium signaling in motor neurons [53].

Protein depalmitoylation is mainly regulated by the α/β hydrolase domain-containing 17 proteins (ABHD17s), the acyl-protein thioesterases (APTs), and the palmitoyl-protein thioesterases (PPTs). Some drugs targeting these enzymes may exert a positive effect on protein S-palmitoylation-related diseases. The S-palmitoylation of melanocortin-1 receptor (MC1R) regulated by zDHHC-13 was negatively related to the incidence of melanoma. APT2, as the primary MC1R depalmitoylase, can easily abolish these beneficial effects. ML349 and Palm-B are specific and nonspecific inhibitors of depalmitoylation, respectively. Both of them show a surprisingly powerful preventative effect on melanomagenesis by recovering MC1R S-palmitoylation and inhibiting the malignant transformation [54, 55].

Other drugs were also proved to be able to regulate protein depalmitoylation. Wogonoside can significantly induce the depalmitoylation of N-Ras and phospholipid scramblase 1 (PLSCR1), whose abnormal expressions are involved in the acute myeloid leukemia (AML). Depalmitoylation of PLSCR1 controls its trafficking to the cell membrane or the nucleus. Depalmitoylation of N-Ras decreases its ability to bind with phospholipid bilayers in the plasma membrane and thereby suppresses RAF1 phosphorylation and eventually leads to the inactivation of N-Ras/RAF1 pathway. Although the effects of wogonoside on the depalmitoylation of N-Ras and PLSCR1 are different, they all end up having anti-AML effects [56].

10.5.3 Exploit the Agonists of S-Palmitoylation

Since a high level of protein S-palmitoylation exerts a beneficial effect on certain diseases, some agonists are exploited to promote protein S-palmitoylation. Fluoxetine can significantly increase the S-palmitoylation level of GLUT1 and then promote the translocation of GLUT1 to membrane compartments, which increases in transporter activity and cellular glucose uptake [57]. Studies about cell proliferation show that both the S-palmitoylation level of both α-tubulin and Ras-related protein Rab-7α (Rab7α) are increased upon a stimulation of androgen. The regulatory effects of agonists on protein S-palmitoylation are often dose dependent; this phenomenon is observed in the experiment about isoproterenol (ISO) and β_2AR-C341A: 1 nM ISO enhances the S-palmitoylation of β_2AR-C341A to some extent and 10μM ISO greatly increases its S-palmitoylation level [28].

10.5.4 Exploit the Inhibitors of S-Palmitoylation

Until now, there are some nonspecific inhibitors including 2-BP, tunicamycin, and cerulenin that have been identified as the inhibitors of protein S-palmitoylation. However, nonspecific inhibitors may have some potential risks; 2-BP can inhibit protein S-palmitoylation and many other metabolic enzymes including glucose-6-phosphatase and fatty acid-CoA ligase at the same time. So it is a good choice to treat diseases by exploiting specific inhibitor of S-palmitoylation with small molecules. In the past, the intracellular routes of proteins' trafficking were considered too generic to be a therapeutic target. However, recent studies have revealed the diversity of the intracellular routes and tried to regulate protein trafficking with S-palmitoylation. Cadmium chloride and Zn pyrithione were identified as selective inhibitors of CCR5; they could inhibit the S-palmitoylation of CCR5 and further reduce the trafficking of CCR5 to the cell membrane [13].

10.6 Conclusions

S-palmitoylation is a common, reversible post-translational lipid modification of proteins; the balance between S-palmitoylation and depalmitoylation is involved in many human diseases. It provides us a potential way to treat diseases by regulating protein S-palmitoylation and depalmitoylation with selective agonists/inhibitors. However, the exploitation of S-palmitoylation selective inhibitors/agonists has been challenging, partly due to the large number of sequence and structural homology between isoforms, so few new inhibitors/agonists have been reported in recent years. It provides a glimmer of hope for the exploitation of S-palmitoylation selective inhibitors/agonists by targeting the structural heterogeneity in the C- and N-terminal domains of isoforms. However, more experiments are needed to achieve this goal. Even if some small molecules are found to function as potential inhibitors/agonists of S-palmitoylation, their hydrophobicity and poor solubility may become another major challenge in application. Fortunately, the development of using liposomes to transport hydrophobic drugs in recent years may be a potential way out.

It is worth noting that the exploitation of selective agonists/inhibitors of S-palmitoylation opens the door for treating diseases, though many obstacles are needed to conquer. What's more, with the development of cryo-electron microscopy and structural biology, the 3-D structure of many zDHHCs may be revealed in the future. It will be a great help for people to exploit selective inhibitors/agonists with a better physiological understanding of zDHHCs.

Acknowledgments This work was supported by grants from the National Natural Science Foundation of China (81770055, 81861128024, 81330001, 81922001, 81770052, 81700020, 31771275, 81800053, 31800980, 81700055, 81470252, and 31400990), the National Key Research and Development Program of China (2016YFC1304400, 2016YFC0903702), an Open Project of State Key Laboratory of Respiratory Disease (SKLRD-OP-201807), a Scientific Research Project of Hubei Province Health and Family Planning (WJ2019Q026), a grant of Jiangsu Provincial Natural Science Foundation (BK20160229), and an Integrated

Innovative Team for Major Human Diseases Program of Tongji Medical College, HUST.

References

1. Jiang H, Zhang X, Chen X, Aramsangtienchai P, Tong Z, Lin H. Protein Lipidation: occurrence, mechanisms, biological functions, and enabling technologies. Chem Rev. 2018;118(3):919–88.
2. Linder ME, Deschenes RJ. Palmitoylation: policing protein stability and traffic. Nat Rev Mol Cell Biol. 2007;8(1):74–84.
3. Cao N, Li JK, Rao YQ, Liu H, Wu J, Li B, et al. A potential role for protein palmitoylation and zDHHC16 in DNA damage response. BMC Mol Biol. 2016;17(1):12.
4. Ho GP, Selvakumar B, Mukai J, Hester LD, Wang Y, Gogos JA, et al. S-nitrosylation and S-palmitoylation reciprocally regulate synaptic targeting of PSD-95. Neuron. 2011;71(1):131–41.
5. McMichael TM, Zhang L, Chemudupati M, Hach JC, Kenney AD, Hang HC, et al. The palmitoyltransferase ZDHHC20 enhances interferon-induced transmembrane protein 3 (IFITM3) palmitoylation and antiviral activity. J Biol Chem. 2017;292(52):21517–26.
6. Lu Y, Zheng Y, Coyaud E, Zhang C, Selvabaskaran A, Yu Y, et al. Palmitoylation of NOD1 and NOD2 is required for bacterial sensing. Science. 2019;366(6464):460–7.
7. Shen LF, Chen YJ, Liu KM, Haddad A, Song IW, Roan HY, et al. Role of S-Palmitoylation by ZDHHC13 in mitochondrial function and metabolism in liver. Sci Rep. 2017;7(1):2182.
8. Kim YC, Lee SE, Kim SK, Jang HD, Hwang I, Jin S, et al. Toll-like receptor mediated inflammation requires FASN-dependent MYD88 palmitoylation. Nat Chem Biol. 2019;15(9):907–16.
9. Liu KM, Chen YJ, Shen LF, Haddad A, Song IW, Chen LY, et al. Cyclic alopecia and abnormal epidermal cornification in Zdhhc13-deficient mice reveal the importance of Palmitoylation in hair and skin differentiation. J Invest Dermatol. 2015;135(11):2603–10.
10. Song IW, Li WR, Chen LY, Shen LF, Liu KM, Yen JJ, et al. Palmitoyl acyltransferase, Zdhhc13, facilitates bone mass acquisition by regulating postnatal epiphyseal development and endochondral ossification: a mouse model. PLoS One. 2014;9(3):e92194.
11. Ohno Y, Kihara A, Sano T, Igarashi Y. Intracellular localization and tissue-specific distribution of human and yeast DHHC cysteine-rich domain-containing proteins. Biochim Biophys Acta. 2006;1761(4):474–83.
12. Yang Y, Hsu JM, Sun L, Chan LC, Li CW, Hsu JL, et al. Palmitoylation stabilizes PD-L1 to promote breast tumor growth. Cell Res. 2019;29(1):83–6.
13. Boncompain G, Herit F, Tessier S, Lescure A, Del NE, Gestraud P, et al. Targeting CCR5 trafficking to inhibit HIV-1 infection. Sci Adv. 2019;5(10):x821.
14. Greaves J, Salaun C, Fukata Y, Fukata M, Chamberlain LH. Palmitoylation and membrane interactions of the neuroprotective chaperone cysteine-string protein. J Biol Chem. 2008;283(36):25014–26.
15. Oku S, Takahashi N, Fukata Y, Fukata M. In silico screening for palmitoyl substrates reveals a role for DHHC1/3/10 (zDHHC1/3/11)-mediated neurochondrin palmitoylation in its targeting to Rab5-positive endosomes. J Biol Chem. 2013;288(27):19816–29.
16. Yeste-Velasco M, Mao X, Grose R, Kudahetti SC, Lin D, Marzec J, et al. Identification of ZDHHC14 as a novel human tumour suppressor gene. J Pathol. 2014;232(5):566–77.
17. Oo HZ, Sentani K, Sakamoto N, Anami K, Naito Y, Uraoka N, et al. Overexpression of ZDHHC14 promotes migration and invasion of scirrhous type gastric cancer. Oncol Rep. 2014;32(1):403–10.
18. Yu L, Reader JC, Chen C, Zhao XF, Ha JS, Lee C, et al. Activation of a novel palmitoyltransferase ZDHHC14 in acute biphenotypic leukemia and subsets of acute myeloid leukemia. Leukemia. 2011;25(2):367–71.
19. Sanders SS, Mui KK, Sutton LM, Hayden MR. Identification of binding sites in huntingtin for the huntingtin interacting proteins HIP14 and HIP14L. PLoS One. 2014;9(2):e90669.
20. Fukata Y, Iwanaga T, Fukata M. Systematic screening for palmitoyl transferase activity of the DHHC protein family in mammalian cells. Methods. 2006;40(2):177–82.
21. Ko PJ, Woodrow C, Dubreuil MM, Martin BR, Skouta R, Bassik MC, et al. A ZDHHC5-GOLGA7 protein acyltransferase complex promotes nonapoptotic cell death. Cell Chem Biol. 2019;26(12):1716–24.
22. Yao H, Lan J, Li C, Shi H, Brosseau JP, Wang H, et al. Inhibiting PD-L1 palmitoylation enhances T-cell immune responses against tumours. Nat Biomed Eng. 2019;3(4):306–17.
23. Abrami L, Dallavilla T, Sandoz PA, Demir M, Kunz B, Savoglidis G, et al. Identification and dynamics of the human ZDHHC16-ZDHHC6 palmitoylation cascade. elife. 2017;6:e27826.
24. Lievens PM, Kuznetsova T, Kochlamazashvili G, Cesca F, Gorinski N, Galil DA, et al. ZDHHC3 tyrosine phosphorylation regulates neural cell adhesion molecule Palmitoylation. Mol Cell Biol. 2016;36(17):2208–25.
25. Chen X, Hu L, Yang H, Ma H, Ye K, Zhao C, et al. DHHC protein family targets different subsets of glioma stem cells in specific niches. J Exp Clin Cancer Res. 2019;38(1):25.
26. Abrami L, Kunz B, Iacovache I, van der Goot FG. Palmitoylation and ubiquitination regulate exit of the Wnt signaling protein LRP6 from the endoplasmic reticulum. Proc Natl Acad Sci U S A. 2008;105(14):5384–9.
27. Lin D, Davis NG, Conibear E. Targeting the Ras palmitoylation/depalmitoylation cycle in cancer. Biochem Soc Trans. 2017;45(4):913–21.

28. Adachi N, Hess DT, McLaughlin P, Stamler JS. S-Palmitoylation of a novel site in the beta2-adrenergic receptor associated with a novel intracellular itinerary. J Biol Chem. 2016;291(38):20232–46.

29. Keller CA, Yuan X, Panzanelli P, Martin ML, Alldred M, Sassoe-Pognetto M, et al. The gamma2 subunit of GABA(A) receptors is a substrate for palmitoylation by GODZ. J Neurosci. 2004;24(26):5881–91.

30. Ebersole B, Petko J, Woll M, Murakami S, Sokolina K, Wong V, et al. Effect of C-terminal S-Palmitoylation on D2 dopamine receptor trafficking and stability. PLoS One. 2015;10(11):e140661.

31. Ernst AM, Syed SA, Zaki O, Bottanelli F, Zheng H, Hacke M, et al. S-Palmitoylation sorts membrane cargo for anterograde transport in the Golgi. Dev Cell. 2018;47(4):479–93.

32. Zhang N, Zhao H, Zhang L. Fatty acid synthase promotes the Palmitoylation of chikungunya virus nsP1. J Virol. 2019;93(3):e01747-18.

33. Kim SW, Kim DH, Park KS, Kim MK, Park YM, Muallem S, et al. Palmitoylation controls trafficking of the intracellular Ca(2+) channel MCOLN3/TRPML3 to regulate autophagy. Autophagy. 2019;15(2):327–40.

34. Sohn H, Park M. Palmitoylation-mediated synaptic regulation of AMPA receptor trafficking and function. Arch Pharm Res. 2019;42(5):426–35.

35. Wegleiter T, Buthey K, Gonzalez-Bohorquez D, Hruzova M, Bin IM, Abegg A, et al. Palmitoylation of BMPR1a regulates neural stem cell fate. Proc Natl Acad Sci U S A. 2019;116(51):25688–96.

36. Rossin A, Derouet M, Abdel-Sater F, Hueber AO. Palmitoylation of the TRAIL receptor DR4 confers an efficient TRAIL-induced cell death signalling. Biochem J. 2009;419(1):185–92. 2-192

37. Chopard C, Tong P, Toth P, Schatz M, Yezid H, Debaisieux S, et al. Cyclophilin A enables specific HIV-1 Tat palmitoylation and accumulation in uninfected cells. Nat Commun. 2018;9(1):2251.

38. Kim Y, Yang H, Min JK, Park YJ, Jeong SH, Jang SW, et al. CCN3 secretion is regulated by palmitoylation via ZDHHC22. Biochem Biophys Res Commun. 2018;495(4):2573–8.

39. Galli LM, Zebarjadi N, Li L, Lingappa VR, Burrus LW. Divergent effects of porcupine and Wntless on WNT1 trafficking, secretion, and signaling. Exp Cell Res. 2016;347(1):171–83.

40. McCormick PJ, Dumaresq-Doiron K, Pluviose AS, Pichette V, Tosato G, Lefrancois S. Palmitoylation controls recycling in lysosomal sorting and trafficking. Traffic. 2008;9(11):1984–97.

41. Pedro MP, Vilcaes AA, Gomez GA, Daniotti JL. Individual S-acylated cysteines differentially contribute to H-Ras endomembrane trafficking and acylation/deacylation cycles. Mol Biol Cell. 2017;28(7):962–74.

42. Wang S, Mott KR, Cilluffo M, Kilpatrick CL, Murakami S, Ljubimov AV, et al. The absence of DHHC3 affects primary and latent herpes simplex virus 1 infection. J Virol. 2018;92(4):e01599-17.

43. Murphy J, Kolandaivelu S. Palmitoylation of progressive rod-cone degeneration (PRCD) regulates protein stability and localization. J Biol Chem. 2016;291(44):23036–46.

44. Kokkola T, Kruse C, Roy-Pogodzik EM, Pekkinen J, Bauch C, Honck HH, et al. Somatostatin receptor 5 is palmitoylated by the interacting ZDHHC5 palmitoyltransferase. FEBS Lett. 2011;585(17):2665–70.

45. Heakal Y, Woll MP, Fox T, Seaton K, Levenson R, Kester M. Neurotensin receptor-1 inducible palmitoylation is required for efficient receptor-mediated mitogenic-signaling within structured membrane microdomains. Cancer Biol Ther. 2011;12(5):427–35.

46. Abrami L, Leppla SH, van der Goot FG. Receptor palmitoylation and ubiquitination regulate anthrax toxin endocytosis. J Cell Biol. 2006;172(2):309–20.

47. Sikarwar AS, Hinton M, Santhosh KT, Chelikani P, Dakshinamurti S. Palmitoylation of Galphaq determines its association with the thromboxane receptor in hypoxic pulmonary hypertension. Am J Respir Cell Mol Biol. 2014;50(1):135–43.

48. Schroeder GN, Aurass P, Oates CV, Tate EW, Hartland EL, Flieger A, et al. Legionella pneumophila effector LpdA is a Palmitoylated phospholipase D virulence factor. Infect Immun. 2015;83(10):3989–4002.

49. Beard RJ, Yang X, Meegan JE, Overstreet JW, Yang CG, Elliott JA, et al. Palmitoyl acyltransferase DHHC21 mediates endothelial dysfunction in systemic inflammatory response syndrome. Nat Commun. 2016;7:12823.

50. Marin EP, Jozsef L, Di Lorenzo A, Held KF, Luciano AK, Melendez J, et al. The protein acyl transferase ZDHHC21 modulates alpha1 adrenergic receptor function and regulates hemodynamics. Arterioscler Thromb Vasc Biol. 2016;36(2):370–9.

51. Chai S, Cambronne XA, Eichhorn SW, Goodman RH. MicroRNA-134 activity in somatostatin interneurons regulates H-Ras localization by repressing the palmitoylation enzyme, DHHC9. Proc Natl Acad Sci U S A. 2013;110(44):17898–903.

52. Gorinski N, Bijata M, Prasad S, Wirth A, Abdel GD, Zeug A, et al. Attenuated palmitoylation of serotonin receptor 5-HT1A affects receptor function and contributes to depression-like behaviors. Nat Commun. 2019;10(1):3924.

53. Krishnamurthy K, Mehta B, Singh M, Tewari BP, Joshi PG, Joshi NB. Depalmitoylation preferentially downregulates AMPA induced Ca2+ signaling and neurotoxicity in motor neurons. Brain Res. 2013;1529:143–53.

54. Cao J, Wan L, Hacker E, Dai X, Lenna S, Jimenez-Cervantes C, et al. MC1R is a potent regulator of PTEN after UV exposure in melanocytes. Mol Cell. 2013;51(4):409–22.

55. Chen S, Han C, Miao X, Li X, Yin C, Zou J, et al. Targeting MC1R depalmitoylation to prevent melanomagenesis in redheads. Nat Commun. 2019;10(1):877.

56. Li H, Yu X, Liu X, Hu P, Shen L, Zhou Y, et al. Wogonoside induces depalmitoylation and transloca-

tion of PLSCR1 and N-Ras in primary acute myeloid leukaemia cells. J Cell Mol Med. 2018;22(4):2117–30.

57. Stapel B, Gorinski N, Gmahl N, Rhein M, Preuss V, Hilfiker-Kleiner D, et al. Fluoxetine induces glucose uptake and modifies glucose transporter palmitoylation in human peripheral blood mononuclear cells. Expert Opin Ther Targets. 2019;23(10):883–91.

58. Zhang J, Planey SL, Ceballos C, Stevens SJ, Keay SK, Zacharias DA. Identification of CKAP4/p63 as a major substrate of the palmitoyl acyltransferase DHHC2, a putative tumor suppressor, using a novel proteomics method. Mol Cell Proteomics. 2008;7(7):1378–88.

59. Sharma C, Yang XH, Hemler ME. DHHC2 affects palmitoylation, stability, and functions of tetraspanins CD9 and CD151. Mol Biol Cell. 2008;19(8):3415–25.

60. Spinelli M, Fusco S, Mainardi M, Scala F, Natale F, Lapenta R, et al. Brain insulin resistance impairs hippocampal synaptic plasticity and memory by increasing GluA1 palmitoylation through FoxO3a. Nat Commun. 2017;8(1):2009.

61. Sharma C, Wang HX, Li Q, Knoblich K, Reisenbichler ES, Richardson AL, et al. Protein acyltransferase DHHC3 regulates breast tumor growth, oxidative stress, and senescence. Cancer Res. 2017;77(24):6880–90.

62. Sergeeva OA, van der Goot FG. Anthrax toxin requires ZDHHC5-mediated palmitoylation of its surface-processing host enzymes. Proc Natl Acad Sci U S A. 2019;116(4):1279–88.

63. Chen X, Ma H, Wang Z, Zhang S, Yang H, Fang Z. EZH2 Palmitoylation mediated by ZDHHC5 in p53-mutant glioma drives malignant development and progression. Cancer Res. 2017;77(18):4998–5010.

64. Li Y, Martin BR, Cravatt BF, Hofmann SL. DHHC5 protein palmitoylates flotillin-2 and is rapidly degraded on induction of neuronal differentiation in cultured cells. J Biol Chem. 2012;287(1):523–30.

65. Chen B, Zheng B, DeRan M, Jarugumilli GK, Fu J, Brooks YS, et al. ZDHHC7-mediated S-palmitoylation of Scribble regulates cell polarity. Nat Chem Biol. 2016;12(9):686–93.

66. Yang Q, Zheng F, Hu Y, Yang Y, Li Y, Chen G, et al. ZDHHC8 critically regulates seizure susceptibility in epilepsy. Cell Death Dis. 2018;9(8):795.

67. Strassburger K, Kang E, Teleman AA. Drosophila ZDHHC8 palmitoylates scribble and Ras64B and controls growth and viability. PLoS One. 2019;14(2):e198149.

68. Moutin E, Nikonenko I, Stefanelli T, Wirth A, Ponimaskin E, De Roo M, et al. Palmitoylation of cdc42 promotes spine stabilization and rescues spine density deficit in a mouse model of 22q11.2 deletion syndrome. Cereb Cortex. 2017;27(7):3618–29.

69. Liu P, Jiao B, Zhang R, Zhao H, Zhang C, Wu M, et al. Palmitoylacyltransferase Zdhhc9 inactivation mitigates leukemogenic potential of oncogenic N-Ras. Leukemia. 2016;30(5):1225–8.

70. Tian L, McClafferty H, Jeffries O, Shipston MJ. Multiple palmitoyltransferases are required for palmitoylation-dependent regulation of large conduc-

tance calcium- and voltage-activated potassium channels. J Biol Chem. 2010;285(31):23954–62.

71. Liu Y, Zhou Q, Zhong L, Lin H, Hu MM, Zhou Y, et al. ZDHHC11 modulates innate immune response to DNA virus by mediating MITA-IRF3 association. Cell Mol Immunol. 2018;15(10):907–16.

72. Fairbank M, Huang K, El-Husseini A, Nabi IR. RING finger palmitoylation of the endoplasmic reticulum Gp78 E3 ubiquitin ligase. FEBS Lett. 2012;586(16):2488–93.

73. Meckler X, Roseman J, Das P, Cheng H, Pei S, Keat M, et al. Reduced Alzheimer's disease ss-amyloid deposition in transgenic mice expressing S-palmitoylation-deficient APH1aL and nicastrin. J Neurosci. 2010;30(48):16160–9.

74. Napoli E, Song G, Liu S, Espejo A, Perez CJ, Benavides F, et al. Zdhhc13-dependent Drp1 S-palmitoylation impacts brain bioenergetics, anxiety, coordination and motor skills. Sci Rep. 2017;7(1):12796.

75. Zhongmeng L, Pengtao L, Xianfeng W, Jiansheng S, Ye C, Ying H, et al. MicroRNA-574-5p promotes cell growth of vascular smooth muscle cells in the progression of coronary artery disease. Biomed Pharmacother. 2018;97:162–7.

76. Wei S, Xueran C, Fen W, Ming G, Yang Y, Zhaoxia D, et al. ZDHHC16 modulates FGF/ERK dependent proliferation of neural stem/progenitor cells in the zebrafish telencephalon. Dev Neurobiol. 2016;76(9):1014–28.

77. Wei S, Fen W, Ming G, Yang Y, Zhaoxia D, Chen W, et al. ZDHHC17 promotes axon outgrowth by regulating TrkA–tubulin complex formation. Mol Cell Neurosci. 2015;68:194–202.

78. Kang R, Wang L, Sanders SS, Zuo K, Hayden MR, Raymond LA. Altered regulation of striatal neuronal N-methyl-D-aspartate receptor trafficking by Palmitoylation in Huntington disease mouse model. Front Synaptic Neurosci. 2019;11:3.

79. Milde S, Coleman MP. Identification of palmitoyl-transferase and thioesterase enzymes that control the subcellular localization of axon survival factor nicotinamide mononucleotide adenylyltransferase 2 (NMNAT2). J Biol Chem. 2014;289(47):32858–70.

80. Runkle KB, Kharbanda A, Stypulkowski E, Cao XJ, Wang W, Garcia BA, et al. Inhibition of DHHC20-mediated EGFR Palmitoylation creates a dependence on EGFR signaling. Mol Cell. 2016;62(3):385–96.

81. Mill P, Lee AW, Fukata Y, Tsutsumi R, Fukata M, Keighren M, et al. Palmitoylation regulates epidermal homeostasis and hair follicle differentiation. PLoS Genet. 2009;5(11):e1000748.

82. Marin EP, Derakhshan B, Lam TT, Davalos A, Sessa WC. Endothelial cell palmitoylproteomic identifies novel lipid-modified targets and potential substrates for protein acyl transferases. Circ Res. 2012;110(10):1336–44.

83. Tian L, McClafferty H, Knaus HG, Ruth P, Shipston MJ. Distinct acyl protein transferases and thioesterases control surface expression of calcium-activated potassium channels. J Biol Chem. 2012;287(18):14718–25.

Redox Role of ROS and Inflammation in Pulmonary Diseases

Li Zuo and Denethi Wijegunawardana

Abstract

Reactive oxygen species (ROS), either derived from exogenous sources or overproduced endogenously, can disrupt the body's antioxidant defenses leading to compromised redox homeostasis. The lungs are highly susceptible to ROS-mediated damage. Oxidative stress (OS) caused by this redox imbalance leads to the pathogenesis of multiple pulmonary diseases such as asthma, chronic obstructive pulmonary disease (COPD), and acute respiratory distress syndrome (ARDS). OS causes damage to important cellular components in terms of lipid peroxidation, protein oxidation, and DNA histone modification. Inflammation further enhances ROS production inducing changes in transcriptional factors which mediate cellular stress response pathways. This deviation from normal cell function contributes to the detrimental pathological characteristics often seen in pulmonary diseases.

L. Zuo (✉)
College of Arts and Sciences, Molecular Physiology and Biophysics Lab, University of Maine, Presque Isle Campus, Presque Isle, ME, USA

Interdisciplinary Biophysics Graduate Program, The Ohio State University, Columbus, OH, USA
e-mail: zuo.4@osu.edu

D. Wijegunawardana
Department of Pathology, Yale School of Medicine, New Haven, CT, USA

Although antioxidant therapies are feasible approaches in alleviating OS-related lung impairment, a comprehensive understanding of the updated role of ROS in pulmonary inflammation is vital for the development of optimal treatments. In this chapter, we review the major pulmonary diseases—including COPD, asthma, ARDS, COVID-19, and lung cancer—as well as their association with ROS.

Keywords

NF-κB · Hypoxia-inducible factor-1 · Leukocytes · Mitochondria

11.1 Introduction

Reactive oxygen species (ROS) are chemically active species containing oxygen. They can be grouped into two subtypes: (1) free radical ROS which have one or more unpaired electrons in the valence shell such as superoxide anion radicals ($O_2^{\cdot-}$) and hydroxyl radicals ($^{\cdot}OH$); iIn an attempt to pair up their own electrons, they extract electrons from a stable molecule and leave this original molecule in an unstable state initiating cellular damage); (2) non-radical ROS which lack the presence of unpaired electrons but are still active and have the ability to generate new radicals (i.e., hydrogen peroxide (H_2O_2)) [1, 2]. In biological systems, ROS are mostly generated through endogenous metabolic reactions such as phagocyte activation or mitochondrial electron transport during respiration (Fig. 11.1). ROS can also be derived from exogenous sources like cigarette smoke and contribute to the maintenance of cellular redox homeostasis under physiological conditions [3]. The lungs are exposed to elevated oxygen levels, which together with its large blood supply and surface area make it vulnerable to injury mediated by ROS [4]. Interestingly, ROS can be produced due to mitochondrial damage; thus, mitochondria are regarded as both source and target of these oxidants [5]. Moreover, oxidative stress (OS) results from an oxidant/antioxidant imbalance in favor of oxidants and causes oxidation of DNA, proteins, and lipids [6]. Heightened ROS production during inflammation also induces inhibition of apoptosis and activation of proto-oncogenes by initiation of signal transduction pathways. Therefore, it is conceivable that chronic inflammation-induced pulmonary ROS production may predispose individuals to lung diseases caused by OS-mediated pulmonary damage [7]. Epithelial cells, endothelial cells, and recruited inflammatory cells, such as eosinophils, neutrophils, lymphocytes, and monocytes, generate ROS in response to increased levels of secretagogue stimuli [8]. Activation of these inflammatory cells results in the formation of $O_2^{\cdot-}$, which is rapidly converted to H_2O_2 by superoxide dismutase (SOD). As a secondary reaction, $^{\cdot}OH$ is formed nonenzymatically in the presence of Fe^{2+}. Under pathological conditions when there are large concentrations of cellular ROS, permanent changes in signal transduction and gene expression pathways occur. Thus, OS mediated by ROS plays an important physiological role in inflammation and pathogenesis of various lung disorders such as asthma, acute lung injury (ALI), and lung cancer [9] and contributes to conditions such as hypertension [10].

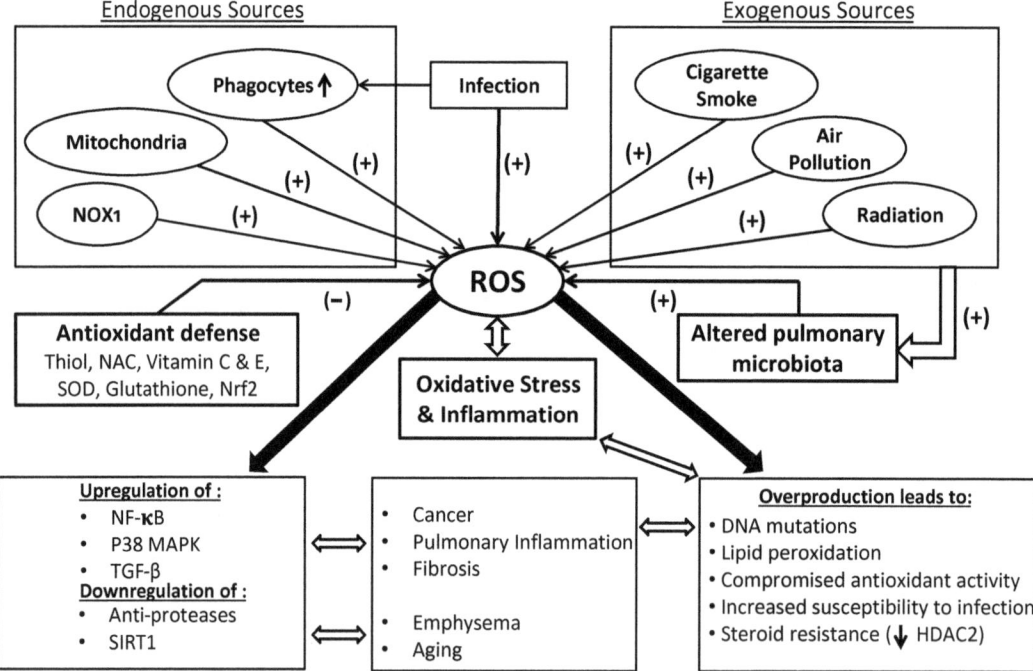

Fig. 11.1 Schematic illustrating key factors of ROS generation in biological and physiological pathways. The level of ROS is regulated by multiple antioxidant defense mechanisms (Thiol; NAC; Vitamin C & E; SOD; Glutathione; Nrf2). Enhanced levels of ROS lead to upregulation of NF-κB associated with cancer, p38 MAPK associated with pulmonary inflammation, TGF-β associated with fibrosis, downregulated anti-proteases with emphysema, and decreased SIRT1 with premature aging. Overproduction of ROS cause DNA mutations, lipid peroxidation and steroid resistance

11.2 Inflammation and Pulmonary Disease

Inflammation is part of the body's defense mechanism to recognize and remove foreign, harmful stimuli to begin the healing process [11]. Leukocytes and mast cells are recruited to the site of damage during inflammation. This causes a boost in the uptake of oxygen which leads to a "respiratory burst" causing an increase in the release and accumulation of ROS at the site of inflammation [12]. This increased inflammatory response exacerbates ROS production via phagocytosis (Fig. 11.1) [13]. Inflammatory cells produce mediators, such as metabolites of chemokines, cytokines, and arachidonic acid, which act by attracting more cells and producing more ROS. These key mediators have the ability to activate signal transduction cascades and induce changes in transcription factors such as nuclear factor-κB (NF-κB), activator protein-1 (AP-1), and hypoxia-inducible factor-1 (HIF-1), which all mediate immediate cellular stress responses. Aberrant expression of inflammatory cytokines such as interleukin-1 (IL-1, IL-6), chemokines (IL-8, CXC chemokine receptor 4 (CXCR4)), and tumor necrosis factor (TNF) have also been reported to play a role in inflammation induced by oxidative stress [14]. The sustained oxidative/inflammatory environment can damage healthy neighboring stromal and epithelial cells over prolonged periods leading to carcinogenesis and pulmonary disease [8, 15].

11.2.1 Nuclear Factor (NF)-κB

NF-κB is a protein transcription factor [16] and modulates gene expression in innate immunity, embryogenesis, cell proliferation, and apoptosis [17]. NF-κB is comprised of a heterodimer with one 50 kDa (p50) and one 65 kDa (p65) polypeptide [18]. NF-κB transcription factors are predominantly regulated by associating with inhibitor IκB proteins. Thus, in most cells, NF-κB is found in the cytoplasm bound to IκB in an inactive complex [19]. ROS molecules can medi-ate cellular toxicity by reacting with lipids (lipid peroxidation), proteins (especially cysteine residues), and nucleic acids (double strand breaks, histone modifications). Different models of OS have been studied to elucidate their effects on the NF-κB signaling pathway. The archetypal activators of the NF-κB pathway are comprised of TNF-α, lipopolysaccharide (LPS), and IL-1. When stimulated by chemokines such as TNF-α or other cellular stressors, TNF-α binds to TNF receptors. Such binding leads to an interaction with the IκB kinase (IκK) which then leads to the phosphorylation of IκB, subsequently resulting in IκB ubiquitination and degradation. Once degraded, the remaining NF-κB dimer (p65/p50 subunits) translocates to the nucleus, where it binds to various target genes. The selectivity of the NF-κB reaction pathway is based on several factors including cell type, timing, and dimer composition [18].

Endogenous redox regulators like thioredoxin play a role in NF-κB regulation in a model of OS involving UV irradiation. In the cytoplasm, it has been shown that thioredoxin blocks the degradation of IκB in the cytoplasm, while it enhances NF-κB activity in the nucleus by increasing its ability to bind DNA [20]. Antioxidants inhibit NF-κB activity by preventing ROS-induced translocation of this transcription factor into the nucleus. Activation of NF-κB is inhibited by preventing IκB degradation in response to various stimuli [21]. Taken together, these findings indicate that activation of NF-κB by ROS is a critical signaling mechanism for evoking inflammatory responses. NF-κB has also been shown to restrict inflammasome activation by the elimination of damaged mitochondria. Remarkably, in addition to NF-κB being an activator of inflammatory genes, it functions by limiting IL-1β production and NLRP3 inflammasome activation. The The induction of signaling adapter, p62 causes inflammasome inhibitory activity by NF-κB. It can control inflammation in macrophages by promoting p62-mediated removal of damaged mitochondria (mitophagy) after macrophages interact with different NLRP3 inflammasome activators [22].

11.2.2 Hypoxia-Inducible Factor-1

There are many instances, in both physiological and pathophysiological conditions, during which the lungs are subjected to localized or global hypoxia. Adaptation to hypoxia requires the coordinated regulation of a myriad of genes, and this collective response is mostly controlled at the level of transcription. Specifically, the hypoxia-inducible factors (HIFs) have been identified as key mediators of hypoxic conditions [23]. HIF-1α ubiquitination involves hydroxylation at two proline residues, Pro-402 and Pro-564 in human HIF-1α [24]. Under normoxic conditions, prolyl hydroxylase domain proteins (PHD) catalyze the hydroxylation of HIF-1α using molecular O_2 as a substrate. In hypoxic environments, HIF-1α hydroxylation at proline residues decreases due to the absence of sufficient molecular O_2. This causes the PHD activity to decline ensuing protein stabilization. Imminently, HIF-1α translocates into the nucleus where it binds HIF-1β and recruits coactivator proteins to the HIF binding site within the hypoxia response element, activating the transcription of several target genes [25].

HIF-1 is a heterodimeric helix-loop-helix-PAS domain transcription factor. HIF is known to activate hypoxia-responsive genes such as vascular endothelial growth factor (VEGF), an important biomarker of asthma [26]. In a hypoxic ischemia/reperfusion model, increased VEGF levels were associated with upregulation of HIF-1 protein levels resulting in augmented barrier disruption [27]. HIF-1 is composed of two subunits, HIF-1a and HIF-1b. HIF-1a, known to mediate gene expression by intracellular oxygen concentration, has also been found to be activated by nonhypoxic factors such as ROS in a redox-sensitive manner [28]. Furthermore, expression of HIF-1α and HIF-2α occurs in all solid tumors. HIFs represent an important signaling node in the switch to protumorigenic inflammatory responses. In hypoxic environments, antitumor immune responses are suppressed through altered immune cell effector functions and the recruitment of protumor immune cells. Tumor growth is promoted through ROS production, excessive growth-promoting cytokine production, and angiogenesis [29].

Hypoxia may occur as a consequence of acute lung injury (ALI), leading to aberrations in pulmonary function and repair. Initial events in ALI include damage of the alveolar lining layer, apoptosis of alveolar epithelial cells, and lung edema. At advanced phases, reactive hyperplasia of alveolar type II cells is prevalent, leading to fibrosis. During ALI, hypoxia can result in increased vascular permeability. Hypoxia has been reported to induce alveolar type II cell apoptosis via HIF-1α with respect to epithelial cell damage and subsequent fibrotic lung disease [30]. HIF-1α is activated in alveolar type II cells after lung injury and promotes proliferation during repair [31]. Moreover, inflammatory levels of nitric oxide ($^{\cdot}$NO) upregulate HIF-1 causing the suppression of epithelial cell wound repair [32]. This signifies that increased HIF-1 levels can render the injured, hypoxic lungs which are unable to support an appropriate healing response after epithelial injury. It is proposed that epithelial-mesenchymal transition (EMT) contributes to pulmonary fibrosis in patients with acute lung injuries [33]. This process is related to the hypoxia-induced increases in mitochondrial-derived ROS which stabilizes HIF-1α in multiple cell types, including alveolar epithelial cells [34]. Additionally, cancer cells were found to upregulate HIF-1 activity during metastatic colonization after extravasation in the lungs via ROS-dependent and hypoxia-independent manners. The administration of HIF-1 inhibitor, YC-1, repressed this reprogramming, amplified intratumoral ROS levels, and eventually inhibited the growth of metastatic tumors. These results denoted that HIF-1-mediated metabolic reprogramming is accountable for the survival of pulmonary metastatic cancers by reducing cytotoxic ROS levels [35].

11.2.3 Activator Protein-1

Activator protein-1 (AP-1) is a transcriptional activator composed mainly of the Jun and Fos family members. AP-1 is known to be involved in oxidant signaling, cell proliferation, and apoptosis [36]. TNF-α and transforming growth factor-β_1 (TGF-β_1) are peptides with multiple physiological functions that influence immunologic, neoplastic, and fibroproliferative diseases. TNF-α induces TGF-β_1 expression at the transcriptional level in lung fibroblasts via AP-1 activation [37]. It has been shown that many end-stage fibrotic diseases, including idiopathic pulmonary fibrosis, converge in the activation of the AP-1 transcription factor c-Jun in pathologic fibroblasts [38]. Another study analyzed tumor cells grown on an ex vivo 4D lung cancer models. The results show an amplification in components of AP-1, c-Fos, and c-Jun in circulating tumor cells (CTC). Administration of SR11302 (an AP-1 inhibitor) reduced metastatic lesion formation in 4D models [39]. Oxidants also induce AP-1 and AP-1-dependent gene expressions. Pyrrolidine dithiocarbamate (PDTC) inhibits NF-κB specifically. However, PDTC is shown to increase the binding of AP-1 and accumulation of ROS in cancer [40]. Aside from their detrimental effects, ROS function as messenger molecules during physiological processes. Glutamate treatment is also found to increase ROS production and activate AP-1 [41].

11.3 Oxidative Stress and Bronchial Asthma

Bronchial asthma is a chronic inflammatory disease and a serious worldwide health issue affecting 5–10% of people of all ages. Usually characterized by airway eosinophilia, bronchial hyperreactivity, chronic airway inflammation, and bronchoconstriction, its symptoms include recurrent episodes of acute shortness of breath, cough, wheezing, and a feeling of tightness in the chest [42]. According to the Asthma and Allergy Foundation of America (AAFA), over 300 million people suffered from this disorder globally since 2013. In the United States alone, asthma accounts for nine deaths per day and causes a yearly economic loss of $56 billion [43]. I Asthma affected 8.3% of American children in 2019 and is the most common chronic disease observed in childhood [44]. Emerging evidence also suggests that the elderly population is vulnerable to late-onset or even long-standing asthma. Commonly found in the elderly, nano-

topic asthma is believed to be related to OS in seniors and vastly affected by ageassociated increases in pathological ROS levels [45].

Several types of asthma have been defined by their phenotypes including allergic asthma, non-allergic asthma, and obesity-related asthma [46]. Most commonly found in childhood, allergic asthma is illustrated by eosinophilic airway inflammation, usually correlated with a familial history of allergic diseases. Allergic asthma is also accompanied by an increase in endogenous ROS formation, OS-induced damage, and mitigated antioxidant defenses [47]. Contrarily, non-allergic asthma is mostly seen in adults. The inflammatory cells present in these asthmatics include eosinophils and neutrophils. Some asthma patients with obesity exhibit significant respiratory symptoms while maintaining low levels of eosinophilic airway inflammation [48]. Excessive ROS can also result in direct oxidant damage and shedding of epithelial cells [49]. There is increasing evidence that inflammation, which is characteristic of asthma, results in increased oxidative stress in the pulmonary tract [50]. Alveolar macrophages from asthmatic subjects show an increased release of $O_2{}^{\bullet-}$, and other ROS compared with those of healthy controls [51]. The TGF-β-induced activation of the ROS-generating enzyme, NADPH oxidase 4 (Nox4), induces myofibroblast differentiation. This activity is involved in airway smooth muscle (ASM) cell proliferation and hypercontractility in asthma in addition to epithelial ciliary dysfunction in neutrophilic asthma [52].

11.3.1 Role of TLRs in Inflammatory Aspects of Asthma

Emerging data suggest that toll-like receptors (TLRs) may be responsible for the aberrant stimulation of immune responses, feasibly contributing to the long-lasting inflammation seen in asthma [53]. TLRs are a class of pattern recognition receptors (PRRs) that sense conserved molecular patterns for early immune recognition of a pathogen and initiate the innate immune response [54]. Serving as the first line of host defense, the pulmonary epithelia employ a variety of receptors including TLRs to detect antigens of numerous pathogens. The lung and respiratory tracts are particularly vulnerable to allergens and pathogens due to continuous contact with inhaled air. In addition, TLRs recognize exogenous pathogen-associated molecular patterns (PAMPs) and host-derived damage-associated molecular patterns (DAMPs) [55]. The activation of TLRs through PAMPs and DAMPs selectively induces cytokine release, inflammatory cell recruitment, and inflammation. Specifically, TLR-2 and TLR4 are identified to be the major TLRs responsible for sustaining the immune responses in both asthma and COPD. TLR4 detects gram-negative bacteria via their LPS, while TLR-2 recognizes gram-positive bacteria. Th2 cells, mast cells, and eosinophils are associated with the innate and adaptive immune responses in asthma.

The recognition of allergens activates TLR4 which consequently activates allergen-specific Th2 cells. TLR-2 stimulates Th2-biased immune responses, which may be correlated to the Th1/Th2 discrepancy in asthma. There are two major pathways in TLR signaling that is crucial to the innate immune response: myeloid differentiation factor 88 (MyD88) dependent and MyD88 independent. MyD88 and Toll/IL-1 receptor domain-containing adapter-inducing interferon-β (TRIF) bind autonomously to TLRs. This leads to the release of cytokines such as TNF-α, IL-1β, CXCL10, IL-6, and IFN-γ [56]. During acute asthmatic events, the cleavage products of proteinases, such as fibrinogen, bind to TLR4s resulting in allergic inflammation. Asthmatic patients who eventually died had increased expression of TLR-2, TLR3, and TLR4, suggesting their potential role in the advancement of severe or even fatal asthmatic exacerbations [57].

11.3.2 Clinical Aspects of Asthma

There is an increased oxidant production in patients with asthma compared to healthy subjects. Both EPO and MPO levels are increased, and many markers of OS such as glutathione

disulfide, malondialdehyde, and thiobarbituric acid are found in the sputum, peripheral blood, and bronchoalveolar lavage (BAL) fluid of patients with asthma. Levels of these markers correlate with the severity of asthma [58]. At the onset of an asthma attack, an increased production of ˙NO is paired with an increase in inducible NO synthase (NOS) activity. In the course of an allergic inflammatory response, ˙NO and O_2˙$^-$ form peroxynitrite, which exerts damaging effects in the respiratory tract. Reactive nitrogen species (RNS) directly heighten cytotoxicity and airway inflammation through nitrosative stress. Evidence suggests that examination of RNS in the pathobiology of asthma by monitoring the use of fractional exhaled nitric oxide (FE_{NO}) measurements can markedly contribute to asthma diagnosis. Peripheral airway inflammation correlates with an increase in alveolar FE_{NO} in patients with mild asthma. Hence, alveolar FE_{NO} can be used to evaluate the level of eosinophilic inflammation in the distal lungs. FE_{NO} measurement also presents a valuable clinical tool to gather information related to steroid responsiveness in patients with asthma [59]. In a recent study, an inverse correlation was observed between FE_{NO} and Asthma Control Test (ACT) score as well as between FE_{NO} and spirometry indicators of airway obstruction [60]. Therefore, FE_{NO} is a useful tool in asthma management.

11.3.3 Antioxidant Therapies for Asthma

Nutritional supplementation on chronic bronchial asthma plays an important role in the management of the disease. Antioxidant vitamins A, C, and E help counteract oxidants and reduce external attacks by bacteria, viruses, toxins, and xenobiotics in the lungs [61]. Thiol antioxidants that are metabolically converted to glutathione precursors, are also popular options for therapeutics. For example, N-acetylcysteine (NAC) is a common thiol precursor. Eosinophils, alveolar macrophages, and neutrophils from asthmatic patients produce more ROS than those from healthy specimen. In mice treated with NAC which was used as a ROS scavenger, there was a reduction in inflammatory cells in the lungs. Other antioxidants such as diphenyleneiodonium (DPI) and NAC can reduce eosinophil peroxidase (EPO), pro-inflammatory cytokines, NF-κB p65 immunocontent, and OS in the lungs, showing that these antioxidants are alternatives for reducing airway inflammation in asthma [62]. Other therapeutic strategies including inhibition of AP-1 and NF-κB have been developed in the signaling of airway inflammation. MOL 294 can be used to reduce airway hyperreactivity and inflammation in asthmatic mice by yet again inhibiting the activity of NF-κB and AP-1 [63]. The application of PNRI-299, a potent and specific inhibitor of AP-1, significantly attenuated both IL-4 release and airway eosinophil infiltration [63]. SOD and glutathione peroxidase mimetics are also effective antioxidants [64]. A study showed that ferroxidase enzymes such as ceruloplasmin are able to reduce active oxygen form (AOF) levels and improve immunity in patients with bronchial asthma [65]. Nontypeable *Haemophilus influenzae* (NTHi) is found in the upper respiratory tract of healthy individuals. It is also one of the most common strains found in the lower respiratory tracts of neutrophilic asthma patients. Recently, it was found that NTHi may cause the aggravation of neutrophilic asthma. Therefore, targeted treatment of NTHi with the use of antibiotics, anti-inflammatory drugs, or IL-17, has been shown to effectively reduce the symptoms of neutrophilic asthma [66].

11.4 Cigarette Smoke, Inhaled Oxidants, and COPD

Chronic obstructive pulmonary disease (COPD) is the most prevalent chronic respiratory disease in the world. According to the Centers for Disease Control and Prevention (CDC), in the United States, it is the third leading cause of death by disease after heart disease and cancer in 2020. Inhalation of volatile substances in cigarette smoke and air pollutants like O_3 and sulfur dioxide (SO_2) activates inflammatory responses and causes lung damage while increasing ROS levels

in the lungs (Fig. 11.1) [67]. Cigarette smoke is a complex mixture of thousands of chemical compounds, including two different populations of free radicals, one in the tar and one in the gas phase. The primary radical is a quinone/hydroquinone (Q/QH2) complex held in the tarry matrix capable of reducing molecular oxygen. This reduction process produces $O_2^{\cdot-}$, eventually forming H_2O_2 and $\cdot OH$. These gas-phase radicals are produced by the oxidation of $\cdot NO$ to NO_2, which reacts with active species in smoke such as isoprene [68]. Cigarette smoke induces damage to DNA in alveolar macrophages and suppresses the expression of important phagocytic activity antigen such as CD11b, TLR-2, and CD14 [69].

Emphysema and chronic bronchitis are the most common conditions for COPD. COPD causes irreversible damage to the lungs and is characterized by airflow blockage leading to difficulties in breathing. Its major features include chronic inflammation throughout the airways, parenchymal vasculature with increased numbers of macrophages, neutrophils, and T lymphocytes (especially CD8+). Markers of oxidative stress like $\cdot NO$ and H_2O_2 have been found in the urine, epithelial lining fluid, breath of patients with COPD, and cigarette smokers [70]. Disease development is associated with a protease/antiprotease imbalance [71] that leads to a lack of protection against elastolytic enzymes [72]. This imbalance creates an accumulation of endogenous ROS released by inflammatory cells like macrophages, neutrophils, and epithelial and endothelial cells [73].

Oxidants exacerbate inflammation by activating NF-κB, facilitating the expression of multiple inflammatory genes such as IL-8 and TNF-α which are thought to be important in COPD [74]. Furthermore, H_2O_2 narrows airway smooth muscle in vitro. Isoprostane F2a-III, formed by free radical peroxidation of arachidonic acid, is a potent constrictor of human airways and an important biomarker of pulmonary oxidative stress in vivo. There is an increased $O_2^{\cdot-}$ production and upregulation of adhesion molecules in circulating neutrophils from patients with COPD. Lipid peroxidation products such as F2-isoprostane, conjugated dienes of linoleic acid, and thiobarbituric acid reactive substances are notably higher in the plasma of healthy smokers and patients with acute exacerbations of chronic bronchitis compared with healthy nonsmokers. Thus, the products of lipid peroxidation formed by ROS also trigger signals that enhance pulmonary inflammation [75]. COPD is known to induce respiratory muscle dysfunction caused by constant resistive breathing. Respiratory muscle contractions increase significantly, resulting in ROS formation and OS. ROS activate molecules such as NF-κB and mitogen-activated protein kinases which stimulate cytokine release causing damage to the diaphragm and sarcomeric disruptions [76]. ROS play a vital role in vascular homeostasis; however, excessive ROS can impair lung function and alter pulmonary vasculature as implicated in COPD. Increased endothelial dysfunction and inflammatory cell infiltration contribute to the severity of the disease. Pulmonary hypertension caused by vascular remodeling reduces the long-term survival rate of patients [77]. COPD is commonly observed in the elderly population. Recently, it was proposed that a chronic state of inflammation associated with aging known as inflammaging is inflicted in COPD. A major aspect of inflammaging is the overabundance of ROS leading to OS, cellular damage, enhanced inflammation, and activation of apoptotic pathways [78].

11.4.1 COPD Therapeutics

COPD is a global health issue linked with high morbidity and mortality, especially in elderly patients. Several factors associated with aging include OS, shortened telomere length, and cellular senescence, closely related to chronic inflammatory responses in COPD [79]. Previous studies have shown that increased levels of biomarkers of OS (8oxodG, NT, F2-IsoPs, and AGEs) strongly correlate with the severity of airflow restriction in elderly patients with COPD [80]. A recent study showed that a combination of antioxidants, anti-inflammatory drugs, and mesenchymal stem cell treatments are a possible therapeutic strategy to treat elderly COPD

patients [81]. Preclinical studies and clinical trials have shown that small thiol molecules such as NACs [82], antioxidant enzymes such as glutathione peroxidases [83], activators of the Nrf2-regulated antioxidant defense system such as sulforaphane [84], and vitamins, for example, C, E, and D [85], can all reduce oxidative stress and boost the endogenous antioxidant system while slowing the progression of COPD [72].

11.5 Acute Lung Injury, Acute Respiratory Distress Syndrome

Acute respiratory distress syndrome (ARDS) is clinically characterized by the abrupt onset of severe hypoxemia and the presence of bilateral diffuse pulmonary infiltrates [86]. These infiltrates appear on a radiograph as pulmonary edema resulting from increased pulmonary vascular permeability. ARDS affects patients of all ages. The disproportionate generation of ROS under pathological conditions such as ALI and its most severe form, ARDS, leads to boosted endothelial permeability. Loss of junctional integrity in vascular microvessels and increased myosin contractions are hallmarks of ALI and ARDS. These stimulate the migration of polymorphonuclear leukocytes (PMNs) and the changeover of solutes/fluids in the alveolar lumen. Exacerbated ROS production by the injured endothelium/epithelium as well as recruited leukocytes play an important role in ARDS pathogenesis and lung damage. OS causes endothelial and epithelial barrier dysfunctions resulting in massive neutrophil penetration across barriers followed by secretion of cytotoxic agents. ROS upregulate the expression of proinflammatory cytokines and adhesion molecules amplifying tissue damage and pulmonary edema. A proper oxidant/antioxidant balance is critical for vasculature homeostasis [87]. Therefore, such biological generators for excessive ROS production such as leukocytes can be therapeutic target options in ARDS treatments [88]. Platelets are known to play an important role in the inflammatory cascade leading to the pathophysiology of

ARDS and are a vital component of regular homeostasis. Antiplatelet therapy (APT) functions by interfering with platelet activation that include adhesion, release, and aggregation [89]. New studies show that prehospital APT is associated with a reduced rate of ARDS [90].

Additionally, ROS contribute to cellular injury by various mechanisms including lipid peroxidation with formation of pro-inflammatory thromboxane molecules and direct damage to DNA resulting in strand breaks and point mutations. It also oxidizes proteins (predominantly at sulfhydryl groups) and alters their activity. Protein oxidation leads to the release of proteases and inactivation of antioxidant enzymes. In addition, ROS-mediated alteration of transcription factors such as AP-1 and NF-κB leads to enhanced expression of pro-inflammatory genes [91, 92]. To minimize the detrimental effects of ROS generated during cellular respiration, cells express a number of endogenous antioxidants such as catalase, superoxide dismutase, and glutathione peroxidase. Yet these antioxidants are rapidly overwhelmed during an acute inflammatory response [93]. In the context of ALI/ARDS, sources of ROS may include leukocytes, parenchymal cells, oxidant-generating enzymes, and inhaled gases with high concentrations of oxygen during mechanical ventilation. Patients with ARDS have increased levels of H_2O_2 in exhaled breath condensate. Moreover, BAL fluid from these patients usually contains an excess of oxidized proteins combined with a relative deficiency in antioxidant molecules such as glutathione. Thus, in ALI/ARDS, there is extensive overproduction of ROS to the extent that endogenous antioxidants are overwhelmed, leading to OS and oxidative cell damage [91, 94].

11.6 ROS-Induced Lipid Peroxidation and Protein Oxidation

Lipid is the main component of cellular, nuclear, and mitochondrial membranes in the lungs. Lipids, especially polyunsaturated fatty acids, are vulnerable to any damage mediated by ROS [95]. In polyunsaturated fatty acids, a hydrogen moiety of unsaturated carbon can be easily attacked by ROS to form water, leaving an unpaired electron which can be converted into a peroxyl radical [96]. These peroxyl radicals eventually produce malondialdehyde (MDA), 4-hydroxynonenal (4-HNE), and other toxic by-products [97]. It has been suggested that MDA is the major mutagenic and carcinogenic product of lipid peroxidation, while 4-HNE is less mutagenic but more toxic [98]. Peroxided lipid reacts with polyunsaturated fatty acids leading to further oxidation, ultimately disrupting plasma membranes [99]. It has been shown that ROS initiate protein oxidation directly as well as indirectly through lipid peroxidation and glycosylation [100]. Cytotoxic roles of lipid peroxidation include inhibition of gene expression and induced cell death in pulmonary tissue leading to lung cancer and hyperoxic lung injury [101]. Lipid peroxidation through TNF-induced ROS formation causes alterations in mitochondrial membrane properties such as permeabilization, acting as a key regulator in cell death [102]. Protein oxidation includes carbonyl group formation while crosslinking and fragmentating the proteins [103]. It is notable that surface-exposed cysteine and methionine residues of proteins are particularly sensitive to oxidation by ROS. Protein oxidation is known to affect cell survival via disrupting the active site of enzymes and subsequently interrupts both protein-protein and protein-DNA interactions [104]. Cellular injury mediated by radiation causes oxidative damage to DNA and proteins. While DNA damage can be fixed by highly efficient mechanisms, repair of oxidized proteins is limited. Oxidized proteins lead to endoplasmic reticulum stress and inflammation, eventually destined for programmed cell death [105]. Hence, cellular apoptosis and protein oxidation are associated with pulmonary diseases such as asthma, COPD, ARDS, and cystic fibrosis [106].

11.7 ROS-Induced DNA Oxidation

ROS are a persistent threat to DNA as they modify bases with the risk of inducing genome instability, as well as disrupting genome function and mutation. Such risks are primarily due to oxidative DNA damage and the repair process. The cell has to either repair the damaged base at a specific genomic site or leave it unrepaired. Persistent DNA damage can disrupt genomic function, but, conversely, it can also contribute to gene regulation by serving as an epigenetic marker. When such processes are out of balance, pathophysiological conditions get accelerated since oxidative DNA damage and resulting mutagenic processes are tightly linked to aging, inflammation, and the development of cancer [107]. It is known that ·OH can bind with DNA molecules leading to oxidation of both bases and the deoxyribose backbones [108]. The key product of DNA oxidation is 8-hydroxy-2 deoxyguanosine (8-OHdG), which results in transcriptional mutagenesis [109]. Remarkably, mitochondrial DNA (mtDNA) oxidation by ROS causes mtDNA abnormality and subsequently triggers the expression of aberrant mitochondrial proteins and mitochondrial dysfunction, collectively exacerbating ROS production. Therefore, there is a vicious cycle between mtDNA oxidation and increased ROS production, which ultimately leads to lung damage and ARDS pathogenesis [110].

11.8   COVID-19

The novel coronavirus disease (COVID-19) caused by severe acute respiratory syndrome coronavirus (SARS-CoV-2) spread throughout China in the year 2020 and received worldwide attention. The emergence of SARS-CoV-2 marks the third introduction of a highly pathogenic coronavirus into the human population in the twenty-first century. The first two included the Middle East respiratory syndrome coronavirus (MERS-CoV) in 2012 and SARS-CoV in 2002. According to the US Centers for Disease Control and Prevention, SARS-CoV-2 has infected more than 47 million people worldwide with an estimated 1.2 million deaths by November 2020. COVID-19 was declared a public health emergency of international concern on January 30, 2020, by the World Health Organization (WHO)

and a pandemic on March 11, 2020 [111]. SARS-CoV-2 is an enveloped, positive-sense, and single-stranded 29.9 kb RNA beta-coronavirus [112]. The first symptoms present as fever, dry cough, tachypnea, and shortness of breath [113], and latest studies report the occurrence of a cytokine storm [114]. Lung inflammation is the main cause of life-threatening respiratory disorders at the severe stages of a SARS-CoV-2 infection, characterized by the so-called cytokine release syndrome (CRS). Patients mostly die from acute respiratory distress syndrome, whereas in many cases the disease has a mild or even asymptomatic progression [115].

Moreover, a common factor in all conditions associated with COVID-19 appears to be the imbalanced redox homeostasis responsible for ROS accumulation. The inflammatory response can be traced back to the pathway of viral entry through its receptor, angiotensin-converting enzyme 2 (ACE2). ACE2 is a protease that takes part in the renin-angiotensin system (RAS) with its companion ACE. Located at the cell surface, they compete for the same substrates, angiotensin I and II (ANG). ACE2 counters the activity of ACE by reducing ANGII levels and increasing the amount of ANGI-7 peptide. The downstream effects of the two enzymes are antagonistic: ACE2 activity leads to vasodilatation, angiogenesis, anti-inflammatory, anti-oxidative, and anti-apoptotic effects, while ACE causes vasoconstriction, OS, inflammation, and apoptosis [116]. OS generated by ACE activity is due to the effects of its product, ANGII, which increases the production of ROS through the activation of NADPH oxidase and the generation of peroxynitrite anions. Contrarily, ACE2 synthesizes the ANG1-7 peptide leading to downregulation of pro-oxidant pathways, which attenuates cellular damage induced by OS. Each person has a distinctive balance between ACE and ACE2 and thus can be more prone to inflammation if ACE prevails. When this happens, an infection by SARS-CoV-2 further downregulates ACE2 prevalence on cell surfaces [117]. This results in the overaccumulation of toxic ANGII, exacerbated inflammation, and ARDS. One approach for the treatment of COVID-19 could be reducing OS secondary to the imbalance between ACE and ACE2. OS is a result of the failure of anti-oxidation defense systems to keep ROS and RNS in check. Hence, glutathione (GSH), a crucial endogenous antioxidant, is critical in extinguishing the exacerbated inflammation that triggers organ failure in such a disease [118].

Additionally, another aspect of COVID-19 involves induction of mitochondrial ROS functions in pulmonary host cells to promote viral replications [119] since cellular ROS levels are markedly increased in SARS-CoV 3CL^{pro}-expressing cells [120]. Mitochondrial ROS are an important factor for SARS-CoV 3a-induced NLRP3 inflammasome activation [121]. Underlying molecular mechanisms such as Nod-like receptor family and pyrin domain-containing 3 (NLRP3) have been reported to be activated by virus infection. This causes lung injury, dysregulation of inflammatory cytokines [122], and release of ROS from damaged mitochondria [123] leading to pathogenesis of SARS-CoV in the respiratory system. Activation of the NLRP3 inflammasome demonstrates significant dependency on ROS generation [124]. The mitophagy/autophagy blockade results in the buildup of damaged, ROS-generating mitochondria which activates the NLRP3 inflammasome. Thus, all known NLRP3 activators generate ROS, which results in the secretion of IL-1β in an NLRP3-ASC-caspase-1-dependent manner in THP-1 human macrophages [125]. Testing for COVID-19 includes nucleic acid amplification tests, direct viral antigen tests, and serological tests [126]. Fresh sputum contains plenty of virus-ridden lung epithelium [127]. Electrochemical diagnostic systems have also been developed to detect ROS levels in the sputum of candidates for COVID screening [128].

11.9 Lung Cancer

OS plays an essential role in the regulation of a variety of physiological processes, such as apoptosis, proliferative signaling pathways, and the pathogenesis of cancer. According to the 2018 global cancer statistics, lung cancer (LC) is con-sidered the most common cancer worldwide and is also the leading cause of death related to cancer [129]. There are two histologic types of lung cancer: small-cell lung cancer and non-small cell lung cancer (NSCLC). Particularly, NSCLC is further subdivided into three classifications: adenocarcinoma, squamous cell carcinoma, and large-cell carcinoma [130]. Cigarette smoking is a well-known environmental risk factor for LC, enhancing lung carcinogenesis by free radical-mediated pathways associated with OS [1]. It has been shown that the two main components of tobacco (nitrosamine 4-(methylnitrosamino)-1-(3-pyridyl)-1-butanone and polyaromatic hydrocarbons) are considered the predominant risk factors for LC [131]. Other inhaled carcinogens such as environmental pollutants and microorganisms can promote tumorigenesis through production of ROS and OS, leadsing to LC [132]. ROS induce DNA damage through oxidation of nucleobases. Repair of these altered bases can result in errors leading to mutagenesis. Consistently, radiation is one of the most well-known sources of ROS and has long been linked to tumor-initiating events [133].

ROS can also alter cellular processes through their effects on protein oxidation. Mild oxidation promotes cellular signaling and is usually reversible (e.g., conversion from disulfides to sulfenic or sulfinic acid and vice versa), allowing prompt changes in protein activity and signaling networks. On the contrary, excessive oxidation leads to terminal oxidation (formation of sulfonic acid) and a complete loss of protein function. Reversible modifications can be protective during stress, while irreversible cysteine modifications can be detrimental to protein function. Protein modifications play a key role in adaptation to OS by activating antioxidant (KEAP1) or metabolic (GAPDH, PKM2) pathways [134]. Other reversible modifications include CoAlation [135] and glutathionylation [136]. Several studies have reported that ROS act as a double-edged sword in cancers. Their role is dedicatedly governed by the amount, type, duration, and site of ROS production. Moderate amounts of ROS have been found to promote tumor survival, while excessive levels serve to suppress tumors and enhance tumor cell death [137]. Also, NOX-derived ROS in the cytoplasm in response to TNF-α promote tumor cell survival, while mitochondrial-derived ROS stimulate apoptosis [138]. In prostatic carcinoma, inhibition of ROS by antioxidants or NOX inhibitors is associated with an increase in apoptosis. This dual role of ROS in cancers provides a challenge for the development of different targeting therapeutic modalities for cancers [139].

11.9.1 Key Pathways of Lung Cancer Metastasis

Metastasis implicates the spread of cancerous cells from the primary tumor to neighboring tissues as well as distant organs. Metastasis of cancers is the primary cause of morbidity and mortality [140]. Tumor metastasis is not an independent program but a complex and multifaceted process. Metastasis mainly occurs due to the intrinsic mutational burden of cancerous cells and bidirectional interaction between nonmalignant and malignant cells [141]. It occurs due to the upregulation of several transcriptional factors such as NF-κB, ETS-1 (ETS proto-oncogene 1, transcription factor), Twist, AP-1, and Zeb (zinc finger E-box binding homeobox). ROS play a vital role in the migration and invasion of cancerous cells. Epithelial-mesenchymal transition (EMT) is the major cause of tumor metastasis, where epithelial cells lose their polarity and cell-cell adhesion and gain mobility [142]. Multiple studies show ROS to be a key source of EMT [143]. TGF-β1 regulates uPA (urokinase-type plasminogen activator) and MMP-9 to facilitate cell migration and invasion through ROS-dependent mechanisms. ROS also increases tumor migration by inducing hypoxia-mediated MMPs [144]. Mitochondrial dysfunction can lead to increased ROS production, which further upregulates C-X-C motif chemokine 14 (CXCL14) expression through the AP-1 signaling pathway and enhances cell mobility by elevating cytosolic Ca^{2+} levels [145]. ROS activate Nrf2 that then stimulates protein coding gene, Klf9 (Kruppel-like factor 9). This in turn acti-

vates ERK1/2 resulting in an increased ROS production in cancer cells. Hence, a premalignant growth can be suppressed by using topical antioxidants that target Klf9 [146].

Mitochondrial Ca^{2+} also plays an important role in cancer metastasis. MCUR1 (mitochondrial calcium uniporter regulator 1) is upregulated in hepatocellular carcinoma (HCC) which promotes EMT by activating ROS/Notch1/Nrf2 pathways. Accordingly, MCUR1 can be a potential target for the treatment of HCC [147]. NOX2 generates ROS, which influence metastasis by downgrading the function of natural killer (NK) cells (Fig. 11.1). NOX2 inhibition restores the NK cell-mediated clearance of myeloma cells [148]. A protein named vimentin plays a crucial role in cancer initiation and progression such as EMT and metastasis. OS caused by HIF-1 regulates vimentin gene transcription, aiding in the formation of invadopodia during cancer cell invasion and migration [149]. Suppression of vimentin expression by RNAi can reduce cell metastasis and hence decrease tumor volume [150]. ROS also induces epigenetic changes in the promoter region of E-cadherin and other tumor suppressor genes, resulting in tumor progression and metastasis. It causes hyper-methylation of the promoter gene by increasing the expression of transcriptional factor—Snail. This factor induces DNA methylation with the help of histone deacetylase 1 (HDAC1) and DNA methyltransferase 1 (DNMT1) [151]. ˙NO plays a role in angiogenesis and intravasation. Therefore, it has been clinically connected to a poor cancer prognosis. ˙NO donors inhibit cell proliferation and anti-apoptotic pathways, suggesting a novel therapy for various cancer treatments [152].

11.10 Conclusion

Over decades, studies have attempted to resolve the complex roles of ROS in pulmonary diseases. Physiological levels of ROS are produced by mitochondria as defenses to maintain pulmonary homeostasis. However, ROS overproduction causes inflammation and damage to the lungs. ROS play a critical role in inflammatory responses through the upregulation of redox-sensitive transcription factors. The resulting OS induce pro-inflammatory gene expression causing pulmonary diseases. antioxidants However, further studies are required to understand the molecular mechanisms of ROS-mediated pathophysiological pathways. ROS suppression by various antioxidants may aid in the design of novel therapies that target specific molecular pathways.

Conflict of Interest The authors confirm that this chapter content has no conflict of interest.

References

1. Filaire E, Dupuis C, Galvaing G, Aubreton S, Laurent H, Richard R, Filaire M. Lung cancer: what are the links with oxidative stress, physical activity and nutrition. Lung Cancer. 2013;82(3):383–9.
2. Leone A, Roca MS, Ciardiello C, Costantini S, Budillon A. Oxidative stress gene expression profile correlates with cancer patient poor prognosis: identification of crucial pathways might select novel therapeutic approaches. Oxidative Med Cell Longev. 2017;2017:2597581.
3. Kirkinezos IG, Moraes CT. Reactive oxygen species and mitochondrial diseases. Semin Cell Dev Biol. 2001;12(6):449–57.
4. Azad N, Rojanasakul Y, Vallyathan V. Inflammation and lung cancer: roles of reactive oxygen/nitrogen species. J Toxicol Environ Health B Crit Rev. 2008;11(1):1–15.
5. Gayathri L, Akbarsha MA, Ruckmani K. In vitro study on aspects of molecular mechanisms underlying invasive aspergillosis caused by gliotoxin and fumagillin, alone and in combination. Sci Rep. 2020;10(1):14473.
6. Sies H. Oxidative stress: a concept in redox biology and medicine. Redox Biol. 2015;4:180–3.
7. Rosanna DP, Salvatore C. Reactive oxygen species, inflammation, and lung diseases. Curr Pharm Des. 2012;18(26):3889–900; Agusti A, Hogg JC. Update on the pathogenesis of chronic obstructive pulmonary disease. N Engl J Med. 2019;381(13):1248–56.
8. Liu Z, Ren Z, Zhang J, Chuang CC, Kandaswamy E, Zhou T, Zuo L. Role of ROS and nutritional antioxidants in human diseases. Front Physiol. 2018;9:477.
9. Henricks PA, Nijkamp FP. Reactive oxygen species as mediators in asthma. Pulm Pharmacol Ther. 2001;14(6):409–20; Tong L, Chuang CC, Wu S, Zuo L. Reactive oxygen species in redox cancer therapy. Cancer Lett. 2015;367(1):18–25.
10. Zuo L, Rose BA, Roberts WJ, He F, Banes-Berceli AK. Molecular characterization of reactive oxygen species in systemic and pulmonary hypertension. Am J Hypertens. 2014;27(5):643–50.
11. Pahwa R, Goyal A, Bansal P, Jialal I. Chronic inflammation. In: StatPearls (Internet). Treasure Island: StatPearls Publishing; 2020.
12. Coussens LM, Werb Z. Inflammation and cancer. Nature. 2002;420(6917):860–7.
13. He F, Zuo L. Redox roles of reactive oxygen species in cardiovascular diseases. Int J Mol Sci. 2015;16(11):27770–80.
14. Hussain SP, Harris CC. Inflammation and cancer: an ancient link with novel potentials. Int J Cancer. 2007;121(11):2373–80; Song HK, Noh EM, Kim JM, You YO, Kwon KB, Lee YR. Reversine inhibits MMP-3, IL-6 and IL-8 expression through suppression of ROS and JNK/AP-1 activation in interleukin-1beta-stimulated human gingival fibroblasts. Arch Oral Biol. 2019;108:104530.
15. Federico A, Morgillo F, Tuccillo C, Ciardiello F, Loguercio C. Chronic inflammation and oxidative stress in human carcinogenesis. Int J Cancer. 2007;121(11):2381–6.
16. Salminen A, Huuskonen J, Ojala J, Kauppinen A, Kaarniranta K, Suuronen T. Activation of innate immunity system during aging: NF-kB signaling is the molecular culprit of inflamm-aging. Ageing Res Rev. 2008;7(2):83–105.
17. Baltimore D. Discovering NF-kappaB. Cold Spring Harb Perspect Biol. 2009;1(1):a000026.
18. Sen R, Smale ST. Selectivity of the NF-{kappa}B response. Cold Spring Harb Perspect Biol. 2010;2(4):a000257.
19. Baldwin AS Jr. The NF-kappa B and I kappa B proteins: new discoveries and insights. Annu Rev Immunol. 1996;14:649–83.
20. Muri J, Thut H, Feng Q, Kopf M. Thioredoxin-1 distinctly promotes NF-kappaB target DNA binding and NLRP3 inflammasome activation independently of Txnip. elife. 2020;9:e53627.
21. Gilmore TD, Herscovitch M. Inhibitors of NF-kappaB signaling: 785 and counting. Oncogene. 2006;25(51):6887–99.
22. Li F, Xu M, Wang M, Wang L, Wang H, Zhang H, Chen Y, Gong J, Zhang JJ, Adcock IM, Chung KF, Zhou X. Roles of mitochondrial ROS and NLRP3 inflammasome in multiple ozone-induced lung inflammation and emphysema. Respir Res. 2018;19(1):230.
23. Shimoda LA, Semenza GL. HIF and the lung: role of hypoxia-inducible factors in pulmonary development and disease. Am J Respir Crit Care Med. 2011;183(2):152–6.
24. Ivan M, Kondo K, Yang H, Kim W, Valiando J, Ohh M, Salic A, Asara JM, Lane WS, Kaelin WG Jr. HIFalpha targeted for VHL-mediated destruction by proline hydroxylation: implications for O2 sensing. Science. 2001;292(5516):464–8.
25. Appelhoff RJ, Tian YM, Raval RR, Turley H, Harris AL, Pugh CW, Ratcliffe PJ, Gleadle JM. Differential function of the prolyl hydroxylases PHD1, PHD2,

and PHD3 in the regulation of hypoxia-inducible factor. J Biol Chem. 2004;279(37):38458–65.

26. Lee HY, Min KH, Lee SM, Lee JE, Rhee CK. Clinical significance of serum vascular endothelial growth factor in young male asthma patients. Korean J Intern Med. 2017;32(2):295–301.

27. Becker PM, Alcasabas A, Yu AY, Semenza GL, Bunton TE. Oxygen-independent upregulation of vascular endothelial growth factor and vascular barrier dysfunction during ventilated pulmonary ischemia in isolated ferret lungs. Am J Respir Cell Mol Biol. 2000;22(3):272–9.

28. Bonello S, Zahringer C, BelAiba RS, Djordjevic T, Hess J, Michiels C, Kietzmann T, Gorlach A. Reactive oxygen species activate the HIF-1alpha promoter via a functional NFkappaB site. Arterioscler Thromb Vasc Biol. 2007;27(4):755–61.

29. Triner D, Shah YM. Hypoxia-inducible factors: a central link between inflammation and cancer. J Clin Invest. 2016;126(10):3689–98.

30. Krick S, Eul BG, Hanze J, Savai R, Grimminger F, Seeger W, Rose F. Role of hypoxia-inducible factor-1alpha in hypoxia-induced apoptosis of primary alveolar epithelial type II cells. Am J Respir Cell Mol Biol. 2005;32(5):395–403.

31. McClendon J, Jansing NL, Redente EF, Gandjeva A, Ito Y, Colgan SP, Ahmad A, Riches DWH, Chapman HA, Mason RJ, Tuder RM, Zemans RL. Hypoxia-inducible factor 1alpha signaling promotes repair of the alveolar epithelium after acute lung injury. Am J Pathol. 2017;187(8):1772–86.

32. Bove PF, Hristova M, Wesley UV, Olson N, Lounsbury KM, van der Vliet A. Inflammatory levels of nitric oxide inhibit airway epithelial cell migration by inhibition of the kinase ERK1/2 and activation of hypoxia-inducible factor-1 alpha. J Biol Chem. 2008;283(26):17919–28; Watts ER, Walmsley SR. Inflammation and hypoxia: HIF and PHD isoform selectivity. Trends Mol Med. 2019;25(1):33–46.

33. Kim KK, Kugler MC, Wolters PJ, Robillard L, Galvez MG, Brumwell AN, Sheppard D, Chapman HA. Alveolar epithelial cell mesenchymal transition develops in vivo during pulmonary fibrosis and is regulated by the extracellular matrix. Proc Natl Acad Sci U S A. 2006;103(35):13180–5.

34. Uzunhan Y, Bernard O, Marchant D, Dard N, Vanneaux V, Larghero J, Gille T, Clerici C, Valeyre D, Nunes H, Boncoeur E, Planes C. Mesenchymal stem cells protect from hypoxia-induced alveolar epithelial-mesenchymal transition. Am J Physiol Lung Cell Mol Physiol. 2016;310(5):L439–51.

35. Zhao T, Zhu Y, Morinibu A, Kobayashi M, Shinomiya K, Itasaka S, Yoshimura M, Guo G, Hiraoka M, Harada H. HIF-1-mediated metabolic reprogramming reduces ROS levels and facilitates the metastatic colonization of cancers in lungs. Sci Rep. 2014;4:3793; Wang M, Zhao X, Zhu D, Liu T, Liang X, Liu F, Zhang Y, Dong X, Sun B. HIF-1alpha promoted vasculogenic mimicry formation in hepatocellular carcinoma through LOXL2 up-regulation in hypoxic tumor microenvironment. J Exp Clin Cancer Res. 2017;36(1):60.

36. Garces de Los Fayos Alonso I, Liang HC, Turner SD, Lagger S, Merkel O, Kenner L. The role of activator protein-1 (AP-1) family members in CD30-positive lymphomas. Cancers (Basel). 2018;10(4):93.

37. Sullivan DE, Ferris M, Nguyen H, Abboud E, Brody AR. TNF-alpha induces TGF-beta1 expression in lung fibroblasts at the transcriptional level via AP-1 activation. J Cell Mol Med. 2009;13(8B):1866–76.

38. Wernig G, Chen SY, Cui L, Van Neste C, Tsai JM, Kambham N, Vogel H, Natkunam Y, Gilliland DG, Nolan G, Weissman IL. Unifying mechanism for different fibrotic diseases. Proc Natl Acad Sci U S A. 2017;114(18):4757–62.

39. Mishra DK, Kim MP. SR 11302, an AP-1 inhibitor, reduces metastatic lesion formation in ex vivo 4D lung cancer model. Cancer Microenviron. 2017;10(1–3):95–103.

40. Riera H, Afonso V, Collin P, Lomri A. A central role for JNK/AP-1 pathway in the pro-oxidant effect of pyrrolidine dithiocarbamate through superoxide dismutase 1 gene repression and reactive oxygen species generation in hematopoietic human cancer cell line U937. PLoS One. 2015;10(5):e0127571.

41. Aharoni-Simon M, Reifen R, Tirosh O. ROS-production-mediated activation of AP-1 but not NFkappaB inhibits glutamate-induced HT4 neuronal cell death. Antioxid Redox Signal. 2006;8(7–8):1339–49.

42. Huang Y, Liu H, Zuo L, Tao A. Key genes and co-expression modules involved in asthma pathogenesis. Peer J. 2020;8:e8456.

43. Zuo L, Otenbaker NP, Rose BA, Salisbury KS. Molecular mechanisms of reactive oxygen species-related pulmonary inflammation and asthma. Mol Immunol. 2013;56(1–2):57–63.

44. Trivedi M, Denton E. Asthma in children and adults-what are the differences and what can they tell us about asthma? Front Pediatr. 2019;7:256.

45. Zuo L, Pannell BK, Liu Z. Characterization and redox mechanism of asthma in the elderly. Oncotarget. 2016;7(18):25010–21.

46. Horak F, Doberer D, Eber E, Horak E, Pohl W, Riedler J, Szepfalusi Z, Wantke F, Zacharasiewicz A, Studnicka M. Diagnosis and management of asthma – statement on the 2015 GINA guidelines. Wien Klin Wochenschr. 2016;128(15–16):541–54.

47. Jiang L, Diaz PT, Best TM, Stimpfl JN, He F, Zuo L. Molecular characterization of redox mechanisms in allergic asthma. Ann Allergy Asthma Immunol. 2014;113(2):137–42.

48. Reddel HK, Bateman ED, Becker A, Boulet LP, Cruz AA, Drazen JM, Haahtela T, Hurd SS, Inoue H, de Jongste JC, Lemanske RF Jr, Levy ML, O'Byrne PM, Paggiaro P, Pedersen SE, Pizzichini E, Soto-Quiroz M, Szefler SJ, Wong GW, FitzGerald JM. A summary of the new GINA strategy: a roadmap to asthma control. Eur Respir J. 2015;46(3):622–39.

49. Brown DI, Griendling KK. Nox proteins in signal transduction. Free Radic Biol Med. 2009;47(9):1239–53.
50. Upton RL, Chen Y, Mumby S, Gutteridge JM, Anning PB, Nicholson AG, Evans TW, Quinlan GJ. Variable tissue expression of transferrin receptors: relevance to acute respiratory distress syndrome. Eur Respir J. 2003;22(2):335–41.
51. Riedl MA, Nel AE. Importance of oxidative stress in the pathogenesis and treatment of asthma. Curr Opin Allergy Clin Immunol. 2008;8(1):49–56.
52. Hough KP, Curtiss ML, Blain TJ, Liu RM, Trevor J, Deshane JS, Thannickal VJ. Airway remodeling in asthma. Front Med (Lausanne). 2020;7:191.
53. Phipps S, Lam CE, Foster PS, Matthaei KI. The contribution of toll-like receptors to the pathogenesis of asthma. Immunol Cell Biol. 2007;85(6):463–70.
54. Wallet SM, Puri V, Gibson FC. Linkage of infection to adverse systemic complications: periodontal disease, toll-like receptors, and other pattern recognition systems. Vaccines (Basel). 2018;6(2):21.
55. Lafferty EI, Qureshi ST, Schnare M. The role of toll-like receptors in acute and chronic lung inflammation. J Inflamm (Lond). 2010;7:57.
56. Piras V, Selvarajoo K. Beyond MyD88 and TRIF pathways in toll-like receptor signaling. Front Immunol. 2014;5:70.
57. Zuo L, Lucas K, Fortuna CA, Chuang CC, Best TM. Molecular regulation of toll-like receptors in asthma and COPD. Front Physiol. 2015;6:312.
58. Beier J, Beeh KM, Semmler D, Beike N, Buhl R. Increased concentrations of glutathione in induced sputum of patients with mild or moderate allergic asthma. Ann Allergy Asthma Immunol. 2004;92(4):459–63.
59. Zuo L, Koozechian MS, Chen LL. Characterization of reactive nitrogen species in allergic asthma. Ann Allergy Asthma Immunol. 2014;112(1):18–22.
60. Nguyen VN, Chavannes NH. Correlation between fractional exhaled nitric oxide and Asthma Control Test score and spirometry parameters in on-treatment-asthmatics in Ho Chi Minh City. J Thorac Dis. 2020;12(5):2197–209.
61. Riccioni G, Barbara M, Bucciarelli T, di Ilio C, D'Orazio N. Antioxidant vitamin supplementation in asthma. Ann Clin Lab Sci. 2007;37(1):96–101.
62. Silveira JS, Antunes GL, Kaiber DB, da Costa MS, Marques EP, Ferreira FS, Gassen RB, Breda RV, Wyse ATS, Pitrez P, da Cunha AA. Reactive oxygen species are involved in eosinophil extracellular traps release and in airway inflammation in asthma. J Cell Physiol. 2019;234(12):23633–46.
63. Comhair SA, Erzurum SC. Redox control of asthma: molecular mechanisms and therapeutic opportunities. Antioxid Redox Signal. 2010;12(1):93–124.
64. Sahiner UM, Birben E, Erzurum S, Sackesen C, Kalayci O. Oxidative stress in asthma. World Allergy Organ J. 2011;4(10):151–8.
65. Farkhutdinov UR, Farkhutdinov SU. Efficacy of ceruloplasmin in patients with asthma. Ter Arkh. 2012;84(12):45–8.
66. Zhang J, Zhu Z, Zuo X, Pan H, Gu Y, Yuan Y, Wang G, Wang S, Zheng R, Liu Z, Wang F, Zheng J. The role of NTHi colonization and infection in the pathogenesis of neutrophilic asthma. Respir Res. 2020;21(1):170.
67. Bezemer GF, Sagar S, van Bergenhenegouwen J, Georgiou NA, Garssen J, Kraneveld AD, Folkerts G. Dual role of toll-like receptors in asthma and chronic obstructive pulmonary disease. Pharmacol Rev. 2012;64(2):337–58.
68. Church DF, Pryor WA. Free-radical chemistry of cigarette smoke and its toxicological implications. Environ Health Perspect. 1985;64:111–26.
69. Zuo L, He F, Sergakis GG, Koozehchian MS, Stimpfl JN, Rong Y, Diaz PT, Best TM. Interrelated role of cigarette smoking, oxidative stress, and immune response in COPD and corresponding treatments. Am J Physiol Lung Cell Mol Physiol. 2014;307(3):L205–18.
70. Dekhuijzen PN, Aben KK, Dekker I, Aarts LP, Wielders PL, van Herwaarden CL, Bast A. Increased exhalation of hydrogen peroxide in patients with stable and unstable chronic obstructive pulmonary disease. Am J Respir Crit Care Med. 1996;154(3 Pt 1):813–6.
71. Fischer BM, Pavlisko E, Voynow JA. Pathogenic triad in COPD: oxidative stress, protease-antiprotease imbalance, and inflammation. Int J Chron Obstruct Pulmon Dis. 2011;6:413–21.
72. Boukhenouna S, Wilson MA, Bahmed K, Kosmider B. Reactive oxygen species in chronic obstructive pulmonary disease. Oxidative Med Cell Longev. 2018;2018:5730395.
73. Kirkham PA, Barnes PJ. Oxidative stress in COPD. Chest. 2013;144(1):266–73.
74. Zhang J, Bai C. The significance of serum interleukin-8 in acute exacerbations of chronic obstructive pulmonary disease. Tanaffos. 2018;17(1):13–21.
75. Park HS, Kim SR, Lee YC. Impact of oxidative stress on lung diseases. Respirology. 2009;14(1):27–38.
76. Li Zuo AHH, Marvin K, Yousif MT. Chien Oxidative stress, respiratory muscle dysfunction, and potential therapeutics in chronic obstructive pulmonary disease. Front Biol. 2012;7:506–13.
77. Zuo L, Chuang CC, Clark AD, Garrison DE, Kuhlman JL, Sypert DC. Reactive oxygen species in COPD-related vascular remodeling. Adv Exp Med Biol. 2017;967:399–411.
78. Zuo L, Prather ER, Stetskiv M, Garrison DE, Meade JR, Peace TI, Zhou T. Inflammaging and oxidative stress in human diseases: from molecular mechanisms to novel treatments. Int J Mol Sci. 2019;20(18):4472.
79. Liguori I, Russo G, Curcio F, Bulli G, Aran L, Della-Morte D, Gargiulo G, Testa G, Cacciatore F, Bonaduce D, Abete P. Oxidative stress, aging, and diseases. Clin Interv Aging. 2018;13:757–72.

80. Choudhury G, MacNee W. Role of inflammation and oxidative stress in the pathology of ageing in COPD: potential therapeutic interventions. COPD. 2017;14(1):122–35.

81. Xia S, Zhou C, Kalionis B, Shuang X, Ge H, Gao W. Combined antioxidant, anti-inflammaging and mesenchymal stem cell treatment: a possible therapeutic direction in elderly patients with chronic obstructive pulmonary disease. Aging Dis. 2020;11(1):129–40.

82. Messier EM, Day BJ, Bahmed K, Kleeberger SR, Tuder RM, Bowler RP, Chu HW, Mason RJ, Kosmider B. N-acetylcysteine protects murine alveolar type II cells from cigarette smoke injury in a nuclear erythroid 2-related factor-2-independent manner. Am J Respir Cell Mol Biol. 2013;48(5):559–67.

83. Vlahos R, Bozinovski S. Glutathione peroxidase-1 as a novel therapeutic target for COPD. Redox Rep. 2013;18(4):142–9.

84. Thimmulappa RK, Mai KH, Srisuma S, Kensler TW, Yamamoto M, Biswal S. Identification of Nrf2-regulated genes induced by the chemopreventive agent sulforaphane by oligonucleotide microarray. Cancer Res. 2002;62(18):5196–203.

85. Wu TC, Huang YC, Hsu SY, Wang YC, Yeh SL. Vitamin E and vitamin C supplementation in patients with chronic obstructive pulmonary disease. Int J Vitam Nutr Res. 2007;77(4):272–9.

86. Wheeler AP, Bernard GR. Acute lung injury and the acute respiratory distress syndrome: a clinical review. Lancet. 2007;369(9572):1553–64.

87. Zuo L, Zhou T, Pannell BK, Ziegler AC, Best TM. Biological and physiological role of reactive oxygen species--the good, the bad and the ugly. Acta Physiol (Oxf). 2015;214(3):329–48.

88. Kellner M, Noonepalle S, Lu Q, Srivastava A, Zemskov E, Black SM. ROS signaling in the pathogenesis of acute lung injury (ALI) and acute respiratory distress syndrome (ARDS). Adv Exp Med Biol. 2017;967:105–37.

89. Mohananey D, Sethi J, Villablanca PA, Ali MS, Kumar R, Baruah A, Bhatia N, Agrawal S, Hussain Z, Shamoun FE, Augoustides JT, Ramakrishna H. Effect of antiplatelet therapy on mortality and acute lung injury in critically ill patients: a systematic review and meta-analysis. Ann Card Anaesth. 2016;19(4):626–37.

90. Jin W, Chuang CC, Jin H, Ye J, Kandaswamy E, Wang L, Zuo L. Effects of pre-hospital antiplatelet therapy on the incidence of ARDS. Respir Care. 2020;65(7):1039–45.

91. Chow CW, Herrera Abreu MT, Suzuki T, Downey GP. Oxidative stress and acute lung injury. Am J Respir Cell Mol Biol. 2003;29(4):427–31.

92. Forrester SJ, Kikuchi DS, Hernandes MS, Xu Q, Griendling KK. Reactive oxygen species in metabolic and inflammatory signaling. Circ Res. 2018;122(6):877–902.

93. Fink MP. Role of reactive oxygen and nitrogen species in acute respiratory distress syndrome. Curr Opin Crit Care. 2002;8(1):6–11.

94. Puri G, Naura AS. Critical role of mitochondrial oxidative stress in acid aspiration induced ALI in mice. Toxicol Mech Methods. 2020;30(4):266–74.

95. Angelova PR, Horrocks MH, Klenerman D, Gandhi S, Abramov AY, Shchepinov MS. Lipid peroxidation is essential for alpha-synuclein-induced cell death. J Neurochem. 2015;133(4):582–9.

96. Patten DA, Germain M, Kelly MA, Slack RS. Reactive oxygen species: stuck in the middle of neurodegeneration. J Alzheimers Dis. 2010;20(Suppl 2):S357–67.

97. Zhong H, Yin H. Role of lipid peroxidation derived 4-hydroxynonenal (4-HNE) in cancer: focusing on mitochondria. Redox Biol. 2015;4:193–9.

98. McGrath LT, McGleenon BM, Brennan S, McColl D, McIlroy S, Passmore AP. Increased oxidative stress in Alzheimer's disease as assessed with 4-hydroxynonenal but not malondialdehyde. QJM. 2001;94(9):485–90.

99. Andersen JK. Oxidative stress in neurodegeneration: cause or consequence? Nat Med. 2004;10(Suppl):S18–25.

100. Yan LJ. Positive oxidative stress in aging and aging-related disease tolerance. Redox Biol. 2014;2:165–9.

101. Ayala A, Munoz MF, Arguelles S. Lipid peroxidation: production, metabolism, and signaling mechanisms of malondialdehyde and 4-hydroxy-2-nonenal. Oxidative Med Cell Longev. 2014;2014:360438.

102. Kotha SR. Mitochondrial lipid peroxidation in lung damage and disease. In: Parinandi NL, Natarajan V, editors. Mitochondrial function in lung health and disease. New York: Humana Press; 2014.

103. Mukherjee S, Kapp EA, Lothian A, Roberts AM, Vasil'ev YV, Boughton BA, Barnham KJ, Kok WM, Hutton CA, Masters CL, Bush AI, Beckman JS, Dey SG, Roberts BR. Characterization and identification of dityrosine cross-linked peptides using tandem mass spectrometry. Anal Chem. 2017;89(11):6136–45.

104. Jammes Y, Steinberg JG, Bregeon F, Delliaux S. The oxidative stress in response to routine incremental cycling exercise in healthy sedentary subjects. Respir Physiol Neurobiol. 2004;144(1):81–90.

105. Barshishat-Kupper M, McCart EA, Freedy JG, Tipton AJ, Nagy V, Kim SY, Landauer MR, Mueller GP, Day RM. Protein oxidation in the lungs of C57BL/6 J mice following X-irradiation. Proteomes. 2015;3(3):249–65.

106. Ciencewicki J, Trivedi S, Kleeberger SR. Oxidants and the pathogenesis of lung diseases. J Allergy Clin Immunol. 2008;122(3):456–68; quiz 469–70.

107. Poetsch AR. The genomics of oxidative DNA damage, repair, and resulting mutagenesis. Comput Struct Biotechnol J. 2020;18:207–19.

108. Alam ZI, Jenner A, Daniel SE, Lees AJ, Cairns N, Marsden CD, Jenner P, Halliwell B. Oxidative DNA damage in the parkinsonian brain: an apparent selec-

tive increase in 8-hydroxyguanine levels in substantia nigra. J Neurochem. 1997;69(3):1196–203.

109. Gmitterova K, Gawinecka J, Heinemann U, Valkovic P, Zerr I. DNA versus RNA oxidation in Parkinson's disease: Which is more important? Neurosci Lett. 2018;662:22–8.

110. Surmeier DJ, Guzman JN, Sanchez-Padilla J, Goldberg JA. The origins of oxidant stress in Parkinson's disease and therapeutic strategies. Antioxid Redox Signal. 2011;14(7):1289–301.

111. Guo YR, Cao QD, Hong ZS, Tan YY, Chen SD, Jin HJ, Tan KS, Wang DY, Yan Y. The origin, transmission and clinical therapies on coronavirus disease 2019 (COVID-19) outbreak – an update on the status. Mil Med Res. 2020;7(1):11.

112. Wu F, Zhao S, Yu B, Chen YM, Wang W, Song ZG, Hu Y, Tao ZW, Tian JH, Pei YY, Yuan ML, Zhang YL, Dai FH, Liu Y, Wang QM, Zheng JJ, Xu L, Holmes EC, Zhang YZ. Author correction: a new coronavirus associated with human respiratory disease in China. Nature. 2020;580(7803):E7.

113. Hui DS, Azhar EI, Madani TA, Ntoumi F, Kock R, Dar O, Ippolito G, Mchugh TD, Memish ZA, Drosten C, Zumla A. The continuing 2019-nCoV epidemic threat of novel coronaviruses to global health – The latest 2019 novel coronavirus outbreak in Wuhan, China. Int J Infect Dis. 2020;91:264–6.

114. Cascella M, Rajnik M, Cuomo A, Dulebohn SC, Di Napoli R. Features, evaluation, and treatment of coronavirus (COVID-19). In: StatPearls (Internet). Treasure Island: StatPearls Publishing; 2020.

115. Lotfi M, Hamblin MR, Rezaei N. COVID-19: transmission, prevention, and potential therapeutic opportunities. Clin Chim Acta. 2020;508:254–66.

116. Capettini LS, Montecucco F, Mach F, Stergiopulos N, Santos RA, da Silva RF. Role of renin-angiotensin system in inflammation, immunity and aging. Curr Pharm Des. 2012;18(7):963–70.

117. Kuba K, Imai Y, Rao S, Gao H, Guo F, Guan B, Huan Y, Yang P, Zhang Y, Deng W, Bao L, Zhang B, Liu G, Wang Z, Chappell M, Liu Y, Zheng D, Leibbrandt A, Wada T, Slutsky AS, Liu D, Qin C, Jiang C, Penninger JM. A crucial role of angiotensin converting enzyme 2 (ACE2) in SARS coronavirus-induced lung injury. Nat Med. 2005;11(8):875–9.

118. Forman HJ, Zhang H, Rinna A. Glutathione: overview of its protective roles, measurement, and biosynthesis. Mol Asp Med. 2009;30(1–2):1–12.

119. Cheng ML, Weng SF, Kuo CH, Ho HY. Enterovirus 71 induces mitochondrial reactive oxygen species generation that is required for efficient replication. PLoS One. 2014;9(11):e113234.

120. Lin CW, Lin KH, Hsieh TH, Shiu SY, Li JY. Severe acute respiratory syndrome coronavirus 3C-like protease-induced apoptosis. FEMS Immunol Med Microbiol. 2006;46(3):375–80.

121. Chen IY, Moriyama M, Chang MF, Ichinohe T. Severe acute respiratory syndrome coronavirus viroporin 3a activates the NLRP3 inflammasome. Front Microbiol. 2019;10:50.

122. Bauernfeind F, Ablasser A, Bartok E, Kim S, Schmid-Burgk J, Cavlar T, Hornung V. Inflammasomes: current understanding and open questions. Cell Mol Life Sci. 2011;68(5):765–83.

123. Nakahira K, Haspel JA, Rathinam VA, Lee SJ, Dolinay T, Lam HC, Englert JA, Rabinovitch M, Cernadas M, Kim HP, Fitzgerald KA, Ryter SW, Choi AM. Autophagy proteins regulate innate immune responses by inhibiting the release of mitochondrial DNA mediated by the NALP3 inflammasome. Nat Immunol. 2011;12(3):222–30.

124. Shimada K, Crother TR, Karlin J, Dagvadorj J, Chiba N, Chen S, Ramanujan VK, Wolf AJ, Vergnes L, Ojcius DM, Rentsendorj A, Vargas M, Guerrero C, Wang Y, Fitzgerald KA, Underhill DM, Town T, Arditi M. Oxidized mitochondrial DNA activates the NLRP3 inflammasome during apoptosis. Immunity. 2012;36(3):401–14.

125. Liu Q, Zhang D, Hu D, Zhou X, Zhou Y. The role of mitochondria in NLRP3 inflammasome activation. Mol Immunol. 2018;103:115–24.

126. La Marca A, Capuzzo M, Paglia T, Roli L, Trenti T, Nelson SM. Testing for SARS-CoV-2 (COVID-19): a systematic review and clinical guide to molecular and serological in-vitro diagnostic assays. Reprod Biomed Online. 2020;41(3):483–99.

127. Tse GM, Hui PK, Ma TK, Lo AW, To KF, Chan WY, Chow LT, Ng HK. Sputum cytology of patients with severe acute respiratory syndrome (SARS). J Clin Pathol. 2004;57(3):256–9.

128. Miripour ZS, Sarrami-Forooshani R, Sanati H, Makarem J, Taheri MS, Shojaeian F, Eskafi AH, Abbasvandi F, Namdar N, Ghafari H, Aghaee P, Zandi A, Faramarzpour M, Hoseinyazdi M, Tayebi M, Abdolahad M. Real-time diagnosis of reactive oxygen species (ROS) in fresh sputum by electrochemical tracing; correlation between COVID-19 and viral-induced ROS in lung/respiratory epithelium during this pandemic. Biosens Bioelectron. 2020;165:112435.

129. Bray F, Ferlay J, Soerjomataram I, Siegel RL, Torre LA, Jemal A. Global cancer statistics 2018: GLOBOCAN estimates of incidence and mortality worldwide for 36 cancers in 185 countries. CA Cancer J Clin. 2018;68(6):394–424.

130. Inamura K. Lung cancer: understanding its molecular pathology and the 2015 WHO classification. Front Oncol. 2017;7:193.

131. Shiels MS, Pfeiffer RM, Hildesheim A, Engels EA, Kemp TJ, Park JH, Katki HA, Koshiol J, Shelton G, Caporaso NE, Pinto LA, Chaturvedi AK. Circulating inflammation markers and prospective risk for lung cancer. J Natl Cancer Inst. 2013;105(24):1871–80.

132. Birben E, Sahiner UM, Sackesen C, Erzurum S, Kalayci O. Oxidative stress and antioxidant defense. World Allergy Organ J. 2012;5(1):9–19.

133. Keil AP, Richardson DB. Quantifying cancer risk from radiation. Risk Anal. 2018;38(7):1474–89.

134. Harris IS, DeNicola GM. The complex interplay between antioxidants and ROS in cancer. Trends Cell Biol. 2020;30(6):440–51.

135. Tsuchiya Y, Zhyvoloup A, Bakovic J, Thomas N, Yu BYK, Das S, Orengo C, Newell C, Ward J, Saladino G, Comitani F, Gervasio FL, Malanchuk OM, Khoruzhenko AI, Filonenko V, Peak-Chew SY, Skehel M, Gout I. Protein CoAlation and antioxidant function of coenzyme A in prokaryotic cells. Biochem J. 2018;475(11):1909–37.

136. Matsui R, Ferran B, Oh A, Croteau D, Shao D, Han J, Pimentel DR, Bachschmid MM. Redox regulation via glutaredoxin-1 and protein S-glutathionylation. Antioxid Redox Signal. 2020;32(10):677–700.

137. Gupta SC, Hevia D, Patchva S, Park B, Koh W, Aggarwal BB. Upsides and downsides of reactive oxygen species for cancer: the roles of reactive oxygen species in tumorigenesis, prevention, and therapy. Antioxid Redox Signal. 2012;16(11):1295–322.

138. Deshpande SS, Angkeow P, Huang J, Ozaki M, Irani K. Rac1 inhibits TNF-alpha-induced endothelial cell apoptosis: dual regulation by reactive oxygen species. FASEB J. 2000;14(12):1705–14.

139. Walton EL. The dual role of ROS, antioxidants and autophagy in cancer. Biom J. 2016;39(2):89–92.

140. Seyfried TN, Huysentruyt LC. On the origin of cancer metastasis. Crit Rev Oncog. 2013;18(1–2):43–73.

141. Brooks SA, Lomax-Browne HJ, Carter TM, Kinch CE, Hall DM. Molecular interactions in cancer cell metastasis. Acta Histochem. 2010;112(1):3–25.

142. Chitty JL, Filipe EC, Lucas MC, Herrmann D, Cox TR, Timpson P. Recent advances in understanding the complexities of metastasis. F1000Res. 2018;7.

143. Kamiya T, Goto A, Kurokawa E, Hara H, Adachi T. Cross talk mechanism among EMT, ROS, and histone acetylation in phorbol ester-treated human breast cancer MCF-7 cells. Oxidative Med Cell Longev. 2016;2016:1284372.

144. Liao Z, Chua D, Tan NS. Reactive oxygen species: a volatile driver of field cancerization and metastasis. Mol Cancer. 2019;18(1):65.

145. Pelicano H, Lu W, Zhou Y, Zhang W, Chen Z, Hu Y, Huang P. Mitochondrial dysfunction and reactive oxygen species imbalance promote breast cancer cell motility through a CXCL14-mediated mechanism. Cancer Res. 2009;69(6):2375–83.

146. Bagati A, Moparthy S, Fink EE, Bianchi-Smiraglia A, Yun DH, Kolesnikova M, Udartseva OO, Wolff DW, Roll MV, Lipchick BC, Han Z, Kozlova NI, Jowdy P, Berman AE, Box NF, Rodriguez C, Bshara W, Kandel ES, Soengas MS, Paragh G, Nikiforov MA. KLF9-dependent ROS regulate melanoma progression in stage-specific manner. Oncogene. 2019;38(19):3585–97.

147. Jin M, Wang J, Ji X, Cao H, Zhu J, Chen Y, Yang J, Zhao Z, Ren T, Xing J. MCUR1 facilitates epithelial-mesenchymal transition and metastasis via the mitochondrial calcium dependent ROS/Nrf2/ Notch pathway in hepatocellular carcinoma. J Exp Clin Cancer Res. 2019;38(1):136.

148. Aydin E, Johansson J, Nazir FH, Hellstrand K, Martner A. Role of NOX2-derived reactive oxygen species in NK cell-mediated control of murine melanoma metastasis. Cancer Immunol Res. 2017;5(9):804–11.

149. Kidd ME, Shumaker DK, Ridge KM. The role of vimentin intermediate filaments in the progression of lung cancer. Am J Respir Cell Mol Biol. 2014;50(1):1–6.

150. Paccione RJ, Miyazaki H, Patel V, Waseem A, Gutkind JS, Zehner ZE, Yeudall WA. Keratin down-regulation in vimentin-positive cancer cells is reversible by vimentin RNA interference, which inhibits growth and motility. Mol Cancer Ther. 2008;7(9):2894–903.

151. Aggarwal V, Tuli HS, Varol A, Thakral F, Yerer MB, Sak K, Varol M, Jain A, Khan MA, Sethi G. Role of reactive oxygen species in cancer progression: molecular mechanisms and recent advancements. Biomol Ther. 2019;9(11):735.

152. Cheng H, Wang L, Mollica M, Re AT, Wu S, Zuo L. Nitric oxide in cancer metastasis. Cancer Lett. 2014;353(1):1–7.

Semaphorin3E/plexinD1 Axis in Asthma: What We Know So Far!

12

Latifa Koussih and Abdelilah S. Gounni

Abstract

Semaphorin3E belongs to the large family of semaphorin proteins. Semaphorin3E was initially identified as axon guidance cues in the neural system. It is universally expressed beyond the nervous system and contributes to regulating essential cell functions such as cell migration, proliferation, and adhesion. Binding of semaphorin3E to its receptor, plexinD1, triggers diverse signaling pathways involved in the pathogenesis of various diseases from cancer to autoimmune and allergic disorders. Here, we highlight the novel findings on the role of semaphorin3E in airway biology. In particular, we highlight our recent findings on the function and potential mechanisms by which semaphorin3E and its receptor, plexinD1, impact airway inflammation, airway hyperresponsiveness, and remodeling in the context of asthma.

L. Koussih
Department of Immunology, Max Rady College of Medicine, Rady Faculty of Health Sciences, University of Manitoba, Winnipeg, MB, Canada

Department des sciences experimentales, Universite de Saint Boniface, Winnipeg, Manitoba, Canada

A. S. Gounni (✉)
Department of Immunology, Max Rady College of Medicine, Rady Faculty of Health Sciences, University of Manitoba, Winnipeg, MB, Canada
e-mail: abdel.gounni@umanitoba.ca

Keywords

Semaphorin3E · PlexinD1 · Asthma · Airway biology · Inflammation · Airway hyperresponsiveness · Tissue remodeling

12.1 Introduction

Asthma is one of the most common chronic diseases affecting 300 million people of all ages, with 250,000 annual deaths [1, 2]. Epidemiological studies have shown that asthma prevalence is more significant in high-income countries than low- and middle-income countries. In Western countries, asthma affects approximately 14% of children and 8% of adults [3–6]. The clinical symptoms of asthma are repeated wheezing episodes, shortness of breath, chest tightness, and nighttime or early morning coughing [7, 8].

From the pathological perspective, asthma is defined as a heterogeneous chronic disorder of the airways characterized by airflow obstruction, airway inflammation, airway hyperresponsiveness, and tissue remodeling [9–12].

Recent studies have implicated semaphorins and plexins in many airway diseases, including acute lung injury, allergic asthma, and pulmonary fibrosis [13]. This review summarized our recent work on the semaphorin3E (Sema3E)/plexinD1 axis on various aspects of asthma. Our studies' results using in vitro and in vivo approaches inform us that while critically important, the Sema3E/plexinD1 pathway is highly complex and touches upon all aspects of asthma. Further detailed understanding of the kinetics, cellular distribution, binding partners, and intracellular signaling networks of Sema3E/plexinD1 will help develop novel immunomodulatory strategies to target this pathway better and improve asthma outcomes.

12.2   Semaphorins

Semaphorins were initially discovered as axon guidance molecules in the nervous system [14, 15]. They are ubiquitously expressed in cardiovascular, endocrine, gastrointestinal, musculoskeletal, immune, and respiratory systems. Semaphorins play an essential role in regulating morphogenesis, angiogenesis, cell differentiation, adhesion, proliferation, and migration [16].

Semaphorins are phylogenetically categorized into eight classes. Classes 1 and 2 are found exclusively in invertebrates, whereas classes 3, 4, 6, and 7 are expressed in the vertebrates [17]. Class 5 semaphorins are found in both vertebrates and invertebrates, and class 8 is specific to viruses. Vertebrate-secreted semaphorins form class 3.

The semaphorins' structural hallmark is the N-terminal "Sema domain," which consists of ~500 amino acids with a seven-blade β-propeller fold conformation. Semaphorins function through binding to their receptors called plexins. Plexins are classified into four subfamilies based on the sequence similarity of their ectodomains [17]. In addition to plexins, neuropilins, tyrosine kinases, and integrins interact with semaphorins [18].

Binding of semaphorins to their receptors triggers diverse signaling pathways involved in the pathogenesis of various diseases [13]. These effects are mediated by shaping the immune system [22], regulating cell trafficking, and cell-to-cell interactions [19–21]. Semaphorins are involved in various phases of the immune response, including the activation of T cells [22] and dendritic cells [23], the regulation of T helper cell differentiation, and the navigation of immune cell trafficking [24].

12.3   Semaphorin3E and PlexinD1

12.3.1   Sema3E

Sema3E is a secreted protein originally known as a regulator of axonal growth of neurons [25]. Sema3E is biosynthesized as 85–90 kDa protein and can undergo cleavage by furin convertase giving rise to P61 kDa fragment [26]. Similar to other Sema3s, Sema3E comprises of Sema domain, PSI (plexin-semaphorin-integrin), Ig (immunoglobulin) domains, and a basic C-terminus tail. Sema3E exerts a significant role in immune responses [27], cell migration [28], and proliferation [29].

Recent studies demonstrated that Sema3E has a significant impact on macrophages and thymocytes. In the thymus, Sema3E is mainly expressed in the medulla and binds to plexinD1 receptors of CD69+ cells. Sema3E binding causes the inhibition of CD69+ cell migration toward the cortex via the repression of CCL25-CCR9 chemokine signaling. Thus, Sema3E plays a crucial role in thymocyte development by aiding the migration of CD69+ to thymic medulla [30].

Sema3E/plexinD1 axis is also involved in the process of inflammation and migration of macrophages. Sema3E is upregulated in macrophages when stimulated with oxidized low-density lipoprotein (LDL), LPS, and hypoxia. These stimulations convert macrophages toward M1 phenotype, which exerts proinflammatory action predominantly [31]. Sema3E also attracts monocytes via induction of p53 into adipose tissue, which eventually becomes pro-inflammatory macrophages [27]. In contrast, Sema3E also causes retention of M1 macrophages in atherosclerotic plaque by blocking CCL19 and CCL21 [31]. Also, Sema3E deficiency is associated with a reduction of macrophages and production of pro-inflammatory cytokines without any effect on vascularity in adipose tissue [27], thus highlighting the context-dependent manner of Sema3E function.

12.3.2 PlexinD1

PlexinD1 is considered the main binding partner of Sema3E. PlexinD1 is dynamically expressed in many embryonic tissues, and after development, in the endothelial cells, the podocytes, adrenal and mammary glands, osteoblastic cells and bone tissues, the lung mesenchyme, the smooth muscle of the small intestine, and immune cells [32]. Human *PLXND1* is located on chromosome 3 (3q22.1) which encodes a 1925 amino acid (212.07 kDa) protein. PlexinD1 expression is essential for cardiovascular development wherein the mice with genetic deletion of its gene will succumb 2 days postnatal because of cardiovascular defects [33]. On the other hand, *Sema3e*-deficient mice are viable after birth, and developmental cardiovascular defects are recapitulated suggestive of additional plexinD1 ligand(s) [34, 35]. Sema3E/plexinD1 axis has a significant contribution to many biological systems such as vascular and neuronal development [36, 37] and hormonal control [38]. It also exerts various functions in pathological conditions such as cancer metastasis [39], immune cell migration [28] and proliferation [29], insulin resistance, and release of pro-inflammatory cytokines [27].

Sema3E is recognized as the canonical ligand for plexinD1 [35], and it does not bind to other plexins. This situation is an exception to the typical pattern of class 3 semaphorin interaction with Nrp-Plexin complexes [40]. Binding of Sema3E to plexinD1 is an Nrp-independent process that leads to plexinD1-mediated endothelial cell repulsion [35]. In tumor models, Sema3E-plexinD1 signaling could be affected by ErbB2 [41, 42]. However, Nrp1 [34] and VEGFR2 [43] are the only known co-receptors that could be associated with plexinD1. Gating of Sema3E by these co-receptors is functionally crucial because it switches the repulsive effect of Sema3E-plexinD1 signaling into an attractive outcome. Nrp1 expression determines the functional pattern of Sema3E in neuronal cells in axon guidance. The absence of Nrp1 in the neurons ensures the repelling function of Sema3E, which is evident in the cortifugal and striatonigral tracts. In contrast, the subiculo-mammillary tract neurons express Nrp1, which leads to attraction by Sema3E [34]. Altogether, the precise mechanism underlying the Sema3E-plexinD1 function is not entirely understood, and the ultimate functional outcome more likely follows a cell- and context-dependent fashion. It should be defined in each cell/disease model.

In humans, expression of Sema3E was significantly suppressed in the airways of severe asthmatic patients [44] and an allergen-challenged mouse model of the disease. Interestingly, the surface expression of the Sema3E high-affinity receptor, plexinD1, was also reduced in ASM cells from asthmatic patients [29], suggesting that Sema3E/plexinD1 axis is dysregulated in allergic asthma.

12.4 Sema3E/plexinD1 Axis in Allergic Asthma

12.4.1 Airway Inflammation

Airway inflammation is a cardinal feature of many chronic airway diseases. Exposure to an allergen, a virus, or air pollutants, smoke, and cold air can induce inflammatory cell recruitment to the airway [12]. In asthma, chronic inflammation is associated with airway hyperresponsiveness that leads to recurrent episodes of wheezing, breathlessness, and coughing. These episodes are usually associated with widespread, but variable, airflow obstruction within the lung that is often reversible either spontaneously or with treatment [12].

Airway inflammation in asthma is mainly characterized by an increased number of eosinophils in most patients' peripheral blood and airways, which correlates with disease severity [44–46]. Eosinophils exert their functions by releasing preformed granular cytotoxic mediators, such as eosinophil peroxidase and major basic protein (MBP), oxygen radicals, and lipid mediators and cytokines and chemokines that aggravate allergen-induced airway inflammation and remodeling [47]. However, there are non-eosinophilic cases of asthma distinguishable from eosinophilic ones. The non-eosinophilic asthmatics are associated with clinical and pathological manifestations, for the most part, neutrophilia [48, 49]. Neutrophil number is increased in severe but not mild or moderate asthmatics [50–52]. In addition, neutrophilic inflammation is predominantly associated with asthma exacerbations [53], fatal asthma [53–55], occupational asthma [56], nocturnal asthma [57, 58], and childhood asthma [59–61].

Airway Neutrophilia and Eosinophilia One of our in vivo study's key findings is that *Sema3e*$^{-/-}$ mice displayed an enhanced pulmonary granulocytosis (i.e., eosinophils and neutrophils) at the steady state [22, 62]. This phenotype may account for the exacerbation of the allergic response upon HDM challenge. These data also suggest that Sema3E pathway may be a predisposing factor for human asthma and some asthmatics may have dysregulated Sema3E expression at the baseline that need further investigation. Furthermore, lung neutrophilia is a hallmark of subtypes of asthma [47]. In particular, severe asthmatics that are refractory to glucocorticoid therapies are characterized by massive recruitment of neutrophils to their lungs, where they accumulate due to higher migration or survival compared to normal conditions [63–65]. In acute and chronic HDM-induced models of allergic asthma, lung neutrophils were significantly higher in *Sema3e*$^{-/-}$ mice than those of wild-type littermates [62, 66]. These data suggest a regulatory role for Sema3E in pulmonary neutrophil extravasation. Besides, Sema3E intranasal treatment reduced HDM-induced recruitment of neutrophils into the lungs, associated with improved lung function, decreased IgE synthesis, and airway inflammation. It further suggests Sema3E as a potential treatment option for severe refractory asthma that deserves more mechanistic studies. Taken together, our data point out the importance of Sema3E endogenous defect as a predisposing factor for pulmonary neutrophilia and provide evidence of a previously unknown mechanism that regulates neutrophil migration in airway inflammatory disorders.

Dendritic Cell Function In allergic asthma, the most common form of the disease, inhalation and subsequent presentation of allergens by dendritic cells (DCs) induce Th2 immune deviation and recruitment of inflammatory cells into the lungs, which lead to tissue damage in atopic individuals [67]. Allergen-specific pulmonary Th2 cells producing interleukin (IL)-4, IL-5, and IL-13 play critical roles in IgE synthesis, eosinophil recruitment, mast cell growth, and AHR. Considering the essential role of DC in the induction of Th2 and Th17 responses [68], interference with their pro-allergic functions is regarded as a potential new therapeutic strategy in allergic asthma [69, 70].

In the lung, five lung DC subsets have been defined that include conventional DCs (cDCs), monocyte-derived DCs (Mo-DCs), and plasmacytoid DCs (pDCs) [71]. The cDCs could be further divided into CD11b$^+$ and CD11b$^-$ [72]. CD11b$^+$CD103$^-$ cDC (cDC2) are endowed with the ability to prime effector CD4 Th cells in both homeostatic and asthmatic conditions, whereas CD103$^+$CD11b$^-$ cDC (cDC1) play a crucial role in the development of tolerogenic protective response upon allergen inhalation [73]. cDCs and Mo-DCs contribute to HDM-induced airway inflammation, with lung CD11b$^+$ cDC2s being necessary and sufficient to induce allergic sensitization [74]. DC functional behavior is dictated by many local factors, including the presence of semaphorin 3 family members [75].

Sema3E plays an essential role in pulmonary DC composition in the HDM mouse model of allergic asthma. Sema3e-deficient mice revealed higher DC recruitment into the airways both at the baseline and after HDM sensitization. In particular, a higher frequency of pulmonary CD11b$^+$ DC observed in Sema3e-deficient mice is consistent with previous studies where CD11b$^+$ DC induced Th2 response upon HDM exposure [74], whereas CD103$^+$ DC induced pulmonary tolerance to inhaled allergens [73] and Th1 deviation [2]. The higher frequency of pulmonary CD11b$^+$ DC from *Sema3e$^{-/-}$* mice was also accompanied by privileged secretion of Th2/Th17 cytokines in our adoptive transfer model in vivo. This finding implies that Sema3E plays a role in regulating Th2/Th17 response in allergic airway inflammation. Also, adoptive transfer of CD11b$^+$ pulmonary DC from HDM-sensitized *Sema3e$^{-/-}$* into WT recipients induces higher airway inflammation compared to those of WT mice, suggesting an essential role of Sema3E in allergic asthma via modulating CD11b$^+$ pulmonary DC function.

12.4.2 Airway Hyperresponsiveness

AHR is a major clinical facet of allergic asthma [76]. Sema3E-Fc Ig, performed in prophylactic and therapeutic regimen, prevented HDM-induced airway resistance, tissue resistance, and elastance considered the characteristic parameters of AHR [66, 77]. Interestingly, in the chronic model of HDM challenge, Sema3E-Fc Ig effect was significant followed another week of HDM challenge that mimic life time scenario [66]. Conversely, we also showed that Sema3E-deficient mice have an exaggerated AHR upon HDM challenge [22]. A decreased level of IL-4 upon Sema3E-Fc Ig administration or after recall stimulation may explain, at least in part, the diminished AHR, since IL-4 and its signaling pathways such as STAT6 are required for the development of sustained AHR in mouse models of allergic asthma [78–80]. Downregulation of IL-4, as a key player for Ig class switching, by Sema3E treatment may further explain the reduction of pro-allergic antibody, IgE, considered a central driver of AHR [81, 82]. Also, a reduced TNF and IL-1β and possibly their direct effects on smooth muscle may be responsible for the decreased AHR upon Sema3E treatment which is in line with the important role of these pathways in AHR [12, 47]. In fact, TNF and IL-1β have been shown to induce impaired receptor-coupled airway relaxation in isolated segments of tracheal smooth muscle [83] and increased ASM contractility [84].

12.4.3 Airway Remodeling

In chronic asthma, airway remodeling involves alterations in structural cells in all of the layers of the airway wall. These include epithelial injury and repair, an increased number of goblet cells, deposition of extracellular matrix protein, increased development of myofibroblasts, neoangiogenesis, and thickness of the muscle bundles [47]. All these features seem to be enhanced in Sema3E-deficient mice subjected to challenge with HDM allergen and attenuated upon treatment with Sema3E-Fc prophylactically or in a therapeutic regimen [22, 66, 77].

Mechanistically, Sema3E significantly reduces growth factor-induced human ASM cell proliferation and migration that was associated with suppression of F-actin polymerization, Rac1 GTPase activity, ERK1/2, and Akt signaling [29]. In vivo, acute or chronic intranasal HDM exposure induces higher mucus overproduction and collagen deposition in the airways of Sema3E-deficient mice compared to the control littermates. Enhanced overexpression of *Col3a1* and *Muc5a genes* was observed in both naïve and upon HDM-challenged *Sema3e$^{-/-}$* mice compared to WT littermates [66], and exogenous treatment with Sema3E-Fc significantly reduced mucus overproduction and collagen deposition [77]. Altogether, these findings support the notion that Sema3E could modulate airway smooth muscle and mucus hyperplasia.

It is clearly known that Sema3E/plexinD1 interaction has a crucial role in vascular development [37]. The inhibitory effect of Sema3E/plexinD1 axis on endothelial cells culminates on inhibition of vessel growth and branching. This inhibition operates through various mechanisms that include the suppression of VEGF and Dll4/Notch signaling pathway [85–88] and the induction of soluble VEGFR-1 (sFlt-1) that suppresses VEGF-induced angiogenesis [89–92]. Consistent with these data, we recently showed that Sema3E has the ability to inhibit new blood vessel formation in the airways of HDM-challenged mice by shifting the ratio of VEGF/VEGFR2, enhancing the soluble VEGFR-1 production, and downregulating von Willebrand factor and CD31 expression [93], providing the first evidence that Sema3E modulates angiogenesis in allergic asthmatic airways via modulating pro- and antiangiogenic factors.

12.5 Conclusion

RNA-Seq data of normal human tissues has revealed that Sema3E is highly expressed in the lung and airway epithelial cells (AECs) are among the primary sources of Sema3E (www. lungmap.net). Furthermore, a genome-wide association study (GWAS) of African-American children, SAGE II (Study of African Americans, Asthma, Genes, and Environments; 812 asthma cases and 415 controls), revealed a *Sema3e* single-nucleotide polymorphism (SNP) genotype as a truly indicative of "direct" asthma associations [94]. AEC gene expression from 155 subjects with asthma and 26 healthy controls in the Severe Asthma Research Program (SARP) was analyzed by weighted gene co-expression network analysis to identify gene networks and profiles associated with severe asthma and its specific characteristics (i.e., pulmonary function tests, quality of life scores, urgent healthcare use, and steroid use). The authors found that gene modules linked to epithelial growth, repair, and neuronal function were markedly decreased in severe asthma. Further, *Sema3e* was among the top 25 genes correlated with asthma while adjusting for potential confounders, mainly age, sex, race, inhaled corticosteroids (ICS), and oral corticosteroids (OCS) [95]. Taken together, these data combined with our studies [22, 29, 62, 66, 77, 96] suggest that Sema3E functions as a critical factor in maintaining airway homeostasis and should be considered as therapeutic for the severe asthma.

Acknowledgments This work was supported by the Canadian Institutes of Health Research grant (MOP # 115115) and the Research Manitoba Chair to A. S. G.

References

1. World Health Organization: Global surveillance, prevention and control of chronic respiratory diseases: a comprehensive approach. 2007.
2. Khare A, Krishnamoorthy N, Oriss TB, Fei M, Ray P, Ray A. Cutting edge: inhaled antigen upregulates retinaldehyde dehydrogenase in lung CD103+ but not plasmacytoid dendritic cells to induce Foxp3 de novo in CD4+ T cells and promote airway tolerance. J Immunol. 2013;191(1):25–9.
3. Bateman ED, Hurd SS, Barnes PJ, Bousquet J, Drazen JM, FitzGerald M, Gibson P, Ohta K, O'Byrne P, Pedersen SE, et al. Global strategy for asthma management and prevention: GINA executive summary. Eur Respir J. 2008;31(1):143–78.
4. Holgate ST, Arshad HS, Roberts GC, Howarth PH, Thurner P, Davies DE. A new look at the pathogenesis of asthma. Clin Sci (Lond). 2010;118(7):439–50.
5. Romagnani S. The increased prevalence of allergy and the hygiene hypothesis: missing immune deviation, reduced immune suppression, or both? Immunology. 2004;112(3):352–63.
6. American Academy of Allergy Asthma and Immunology: Asthma Statistics. 2015.
7. Asthma Society of Canada: What is Asthma 2015.
8. What Are the Signs and Symptoms of Asthma. http://www.nhlbi.nih.gov/health/health-topics/topics/asthma/signs.
9. Barnes PJ. Immunology of asthma and chronic obstructive pulmonary disease. Nat Rev Immunol. 2008;8(3):183–92.
10. Hamid Q, Tulic MK, Liu MC, Moqbel R. Inflammatory cells in asthma: mechanisms and implications for therapy. J Allergy Clin Immunol. 2003;111(1 Suppl):S5–S12. discussion S12–17
11. Murdoch JR, Lloyd CM. Chronic inflammation and asthma. Mutat Res. 2010;690(1–2):24–39.
12. Maddox L, Schwartz DA. The pathophysiology of asthma. Annu Rev Med. 2002;53:477–98.
13. Movassagh H, Khadem F, Gounni AS. Semaphorins and their roles in airway biology: potential as therapeutic targets. Am J Respir Cell Mol Biol. 2018;58(1):21–7.
14. Kolodkin AL, Matthes DJ, O'Connor TP, Patel NH, Admon A, Bentley D, Goodman CS. Fasciclin IV: sequence, expression, and function during growth cone guidance in the grasshopper embryo. Neuron. 1992;9(5):831–45.
15. Luo Y, Raible D, Raper JA. Collapsin: a protein in brain that induces the collapse and paralysis of neuronal growth cones. Cell. 1993;75(2):217–27.
16. Roth L, Koncina E, Satkauskas S, Cremel G, Aunis D, Bagnard D. The many faces of semaphorins: from development to pathology. Cell Mol Life Sci. 2009;66(4):649–66.
17. Yazdani U, Terman JR. The semaphorins. Genome Biol. 2006;7(3):211.

18. Kumanogoh A, Kikutani H. Immunological functions of the neuropilins and plexins as receptors for semaphorins. Nat Rev Immunol. 2013;13(11):802–14.

19. Takamatsu H, Takegahara N, Nakagawa Y, Tomura M, Taniguchi M, Friedel RH, Rayburn H, Tessier-Lavigne M, Yoshida Y, Okuno T, et al. Semaphorins guide the entry of dendritic cells into the lymphatics by activating myosin II. Nat Immunol. 2010;11(7):594–600.

20. Morote-Garcia JC, Napiwotzky D, Kohler D, Rosenberger P. Endothelial Semaphorin 7A promotes neutrophil migration during hypoxia. Proc Natl Acad Sci U S A. 2012;109(35):14146–51.

21. Choi YI, Duke-Cohan JS, Chen W, Liu B, Rossy J, Tabarin T, Ju L, Gui J, Gaus K, Zhu C, et al. Dynamic control of beta1 integrin adhesion by the plexinD1-sema3E axis. Proc Natl Acad Sci U S A. 2014;111(1):379–84.

22. Movassagh H, Shan L, Mohammed A, Halayko AJ, Gounni AS. Semaphorin 3E deficiency exacerbates airway inflammation, Hyperresponsiveness, and remodeling in a mouse model of allergic asthma. J Immunol. 2017;198(5):1805–14.

23. Takegahara N, Takamatsu H, Toyofuku T, Tsujimura T, Okuno T, Yukawa K, Mizui M, Yamamoto M, Prasad DV, Suzuki K, et al. Plexin-A1 and its interaction with DAP12 in immune responses and bone homeostasis. Nat Cell Biol. 2006;8(6):615–22.

24. Takamatsu H, Okuno T, Kumanogoh A. Regulation of immune cell responses by semaphorins and their receptors. Cell Mol Immunol. 2010;7(2):83–8.

25. Chen H, Xie GH, Wang WW, Yuan XL, Xing WM, Liu HJ, Chen J, Dou M, Shen LS. Epigenetically downregulated Semaphorin 3E contributes to gastric cancer. Oncotarget. 2015;6(24):20449–65.

26. Klagsbrun M, Shimizu A. Semaphorin 3E, an exception to the rule. J Clin Invest. 2010;120(8):2658–60.

27. Shimizu I, Yoshida Y, Moriya J, Nojima A, Uemura A, Kobayashi Y, Minamino T. Semaphorin3E-induced inflammation contributes to insulin resistance in dietary obesity. Cell Metab. 2013;18(4):491–504.

28. Hughes A, Kleine-Albers J, Helfrich MH, Ralston SH, Rogers MJ. A class III semaphorin (Sema3e) inhibits mouse osteoblast migration and decreases osteoclast formation in vitro. Calcif Tissue Int. 2012;90(2):151–62.

29. Movassagh H, Shan L, Halayko AJ, Roth M, Tamm M, Chakir J, Gounni AS. Neuronal chemorepellent Semaphorin 3E inhibits human airway smooth muscle cell proliferation and migration. J Allergy Clin Immunol. 2014;133(2):560–7.

30. Choi YI, Duke-Cohan JS, Ahmed WB, Handley MA, Mann F, Epstein JA, Clayton LK, Reinherz EL. PlexinD1 glycoprotein controls migration of positively selected thymocytes into the medulla. Immunity. 2008;29(6):888–98.

31. Wanschel A, Seibert T, Hewing B, Ramkhelawon B, Ray TD, van Gils JM, Rayner KJ, Feig JE, O'Brien ER, Fisher EA, et al. Neuroimmune guidance cue Semaphorin 3E is expressed in atherosclerotic plaques

and regulates macrophage retention. Arterioscler Thromb Vasc Biol. 2013;33(5):886–93.

32. Gay CM, Zygmunt T, Torres-Vazquez J. Diverse functions for the semaphorin receptor PlexinD1 in development and disease. Dev Biol. 2011;349(1):1–19.

33. Zhang Y, Singh MK, Degenhardt KR, Lu MM, Bennett J, Yoshida Y, Epstein JA. Tie2Cre-mediated inactivation of plexinD1 results in congenital heart, vascular and skeletal defects. Dev Biol. 2009;325(1):82–93.

34. Chauvet S, Cohen S, Yoshida Y, Fekrane L, Livet J, Gayet O, Segu L, Buhot MC, Jessell TM, Henderson CE, et al. Gating of Sema3E/PlexinD1 signaling by neuropilin-1 switches axonal repulsion to attraction during brain development. Neuron. 2007;56(5):807–22.

35. Gu C, Yoshida Y, Livet J, Reimert DV, Mann F, Merte J, Henderson CE, Jessell TM, Kolodkin AL, Ginty DD. Semaphorin 3E and plexin-D1 control vascular pattern independently of neuropilins. Science. 2005;307(5707):265–8.

36. Mazzotta C, Romano E, Bruni C, Manetti M, Lepri G, Bellando-Randone S, Blagojevic J, Ibba-Manneschi L, Matucci-Cerinic M, Guiducci S. Plexin-D1/Semaphorin 3E pathway may contribute to dysregulation of vascular tone control and defective angiogenesis in systemic sclerosis. Arthritis Res Ther. 2015;17:221.

37. Oh WJ, Gu C. The role and mechanism-of-action of Sema3E and Plexin-D1 in vascular and neural development. Semin Cell Dev Biol. 2013;24(3):156–62.

38. Cariboni A, Andre V, Chauvet S, Cassatella D, Davidson K, Caramello A, Fantin A, Bouloux P, Mann F, Ruhrberg C. Dysfunctional SEMA3E signaling underlies gonadotropin-releasing hormone neuron deficiency in Kallmann syndrome. J Clin Invest. 2015;125(6):2413–28.

39. Christensen C, Ambartsumian N, Gilestro G, Thomsen B, Comoglio P, Tamagnone L, Guldberg P, Lukanidin E. Proteolytic processing converts the repelling signal Sema3E into an inducer of invasive growth and lung metastasis. Cancer Res. 2005;65(14):6167–77.

40. Kolodkin AL, Levengood DV, Rowe EG, Tai YT, Giger RJ, Ginty DD. Neuropilin is a semaphorin III receptor. Cell. 1997;90(4):753–62.

41. Casazza A, Finisguerra V, Capparuccia L, Camperi A, Swiercz JM, Rizzolio S, Rolny C, Christensen C, Bertotti A, Sarotto I, et al. Sema3E-Plexin D1 signaling drives human cancer cell invasiveness and metastatic spreading in mice. J Clin Invest. 2010;120(8):2684–98.

42. Casazza A, Kigel B, Maione F, Capparuccia L, Kessler O, Giraudo E, Mazzone M, Neufeld G, Tamagnone L. Tumour growth inhibition and anti-metastatic activity of a mutated furin-resistant Semaphorin 3E isoform. EMBO Mol Med. 2012;4(3):234–50.

43. Bellon A, Luchino J, Haigh K, Rougon G, Haigh J, Chauvet S, Mann F. VEGFR2 (KDR/Flk1) signaling mediates axon growth in response to semaphorin 3E in the developing brain. Neuron. 2010;66(2):205–19.

44. Bousquet J, Chanez P, Lacoste JY, Barneon G, Ghavanian N, Enander I, Venge P, Ahlstedt S, Simony-Lafontaine J, Godard P, et al. Eosinophilic inflammation in asthma. N Engl J Med. 1990;323(15):1033–9.

45. Louis R, Lau LC, Bron AO, Roldaan AC, Radermecker M, Djukanovic R. The relationship between airways inflammation and asthma severity. Am J Respir Crit Care Med. 2000;161(1):9–16.

46. Woodruff PG, Khashayar R, Lazarus SC, Janson S, Avila P, Boushey HA, Segal M, Fahy JV. Relationship between airway inflammation, hyperresponsiveness, and obstruction in asthma. J Allergy Clin Immunol. 2001;108(5):753–8.

47. Hamid Q, Tulic M. Immunobiology of asthma. Annu Rev Physiol. 2009;71:489–507.

48. Green RH, Brightling CE, McKenna S, Hargadon B, Parker D, Bradding P, Wardlaw AJ, Pavord ID. Asthma exacerbations and sputum eosinophil counts: a randomised controlled trial. Lancet. 2002;360(9347):1715–21.

49. Jayaram L, Pizzichini MM, Cook RJ, Boulet LP, Lemiere C, Pizzichini E, Cartier A, Hussack P, Goldsmith CH, Laviolette M, et al. Determining asthma treatment by monitoring sputum cell counts: effect on exacerbations. Eur Respir J. 2006;27(3):483–94.

50. Jatakanon A, Uasuf C, Maziak W, Lim S, Chung KF, Barnes PJ. Neutrophilic inflammation in severe persistent asthma. Am J Respir Crit Care Med. 1999;160(5 Pt 1):1532–9.

51. Shaw DE, Berry MA, Hargadon B, McKenna S, Shelley MJ, Green RH, Brightling CE, Wardlaw AJ, Pavord ID. Association between neutrophilic airway inflammation and airflow limitation in adults with asthma. Chest. 2007;132(6):1871–5.

52. Wenzel SE, Szefler SJ, Leung DY, Sloan SI, Rex MD, Martin RJ. Bronchoscopic evaluation of severe asthma. Persistent inflammation associated with high dose glucocorticoids. Am J Respir Crit Care Med. 1997;156(3 Pt 1):737–43.

53. Qiu Y, Zhu J, Bandi V, Guntupalli KK, Jeffery PK. Bronchial mucosal inflammation and upregulation of CXC chemoattractants and receptors in severe exacerbations of asthma. Thorax. 2007;62(6):475–82.

54. Choi JS, Jang AS, Park JS, Park SW, Paik SH, Park JS, Uh ST, Kim YH, Park CS. Role of neutrophils in persistent airway obstruction due to refractory asthma. Respirology. 2012;17(2):322–9.

55. Shiang C, Mauad T, Senhorini A, de Araujo BB, Ferreira DS, da Silva LF, Dolhnikoff M, Tsokos M, Rabe KF, Pabst R. Pulmonary periarterial inflammation in fatal asthma. Clin Exp Allergy. 2009;39(10):1499–507.

56. Anees W, Huggins V, Pavord ID, Robertson AS, Burge PS. Occupational asthma due to low molecular weight agents: eosinophilic and non-eosinophilic variants. Thorax. 2002;57(3):231–6.

57. Martin RJ, Cicutto LC, Smith HR, Ballard RD, Szefler SJ. Airways inflammation in nocturnal asthma. Am Rev Respir Dis. 1991;143(2):351–7.

58. Tan WC, Koh TH, Hay CS, Taylor E. The effect of inhaled budesonide on the diurnal variation in airway mechanics, airway responsiveness and serum neutrophil chemotactic activity in Asian patients with predominant nocturnal asthma. Respirology. 1998;3(1):13–20.

59. Gagliardo R, Chanez P, Gjomarkaj M, La Grutta S, Bonanno A, Montalbano AM, Di Sano C, Albano GD, Gras D, Anzalone G, et al. The role of transforming growth factor-beta1 in airway inflammation of childhood asthma. Int J Immunopathol Pharmacol. 2013;26(3):725–38.

60. Konradsen JR, James A, Nordlund B, Reinius LE, Soderhall C, Melen E, Wheelock AM, Lodrup Carlsen KC, Lidegran M, Verhoek M, et al. The chitinase-like protein YKL-40: a possible biomarker of inflammation and airway remodeling in severe pediatric asthma. J Allergy Clin Immunol. 2013;132(2):328–335.e325.

61. McDougall CM, Helms PJ. Neutrophil airway inflammation in childhood asthma. Thorax. 2006;61(9):739–41.

62. Movassagh H, Saati A, Nandagopal S, Mohammed A, Tatari N, Shan L, Duke-Cohan JS, Fowke KR, Lin F, Gounni AS. Chemorepellent Semaphorin 3E negatively regulates neutrophil migration in vitro and in vivo. J Immunol. 2017;198(3):1023–33.

63. Linden A. Role of interleukin-17 and the neutrophil in asthma. Int Arch Allergy Immunol. 2001;126(3):179–84.

64. Saffar AS, Dragon S, Ezzati P, Shan L, Gounni AS. Phosphatidylinositol 3-kinase and p38 mitogen-activated protein kinase regulate induction of Mcl-1 and survival in glucocorticoid-treated human neutrophils. J Allergy Clin Immunol. 2008;121(2):492–498.e410.

65. Trevor JL, Deshane JS. Refractory asthma: mechanisms, targets, and therapy. Allergy. 2014;69(7):817–27.

66. Movassagh H, Shan L, Duke-Cohan JS, Chakir J, Halayko AJ, Koussih L, Gounni AS. Downregulation of semaphorin 3E promotes hallmarks of experimental chronic allergic asthma. Oncotarget. 2017;8(58):98953–63.

67. Paul WE, Zhu J. How are T(H)2-type immune responses initiated and amplified? Nat Rev Immunol. 2010;10(4):225–35.

68. Kopf M, Schneider C, Nobs SP. The development and function of lung-resident macrophages and dendritic cells. Nat Immunol. 2015;16(1):36–44.

69. Lambrecht BN, Hammad H. Taking our breath away: dendritic cells in the pathogenesis of asthma. Nat Rev Immunol. 2003;3(12):994–1003.

70. Steinman RM, Banchereau J. Taking dendritic cells into medicine. Nature. 2007;449(7161):419–26.

71. van Helden MJ, Lambrecht BN. Dendritic cells in asthma. Curr Opin Immunol. 2013;25(6):745–54.

72. Guilliams M, Ginhoux F, Jakubzick C, Naik SH, Onai N, Schraml BU, Segura E, Tussiwand R, Yona S. Dendritic cells, monocytes and macrophages: a

unified nomenclature based on ontogeny. Nat Rev Immunol. 2014;14(8):571–8.

73. Furuhashi K, Suda T, Hasegawa H, Suzuki Y, Hashimoto D, Enomoto N, Fujisawa T, Nakamura Y, Inui N, Shibata K, et al. Mouse lung CD103+ and CD11bhigh dendritic cells preferentially induce distinct CD4+ T-cell responses. Am J Respir Cell Mol Biol. 2012;46(2):165–72.

74. Plantinga M, Guilliams M, Vanheerswynghels M, Deswarte K, Branco-Madeira F, Toussaint W, Vanhoutte L, Neyt K, Killeen N, Malissen B, et al. Conventional and monocyte-derived CD11b(+) dendritic cells initiate and maintain T helper 2 cell-mediated immunity to house dust mite allergen. Immunity. 2013;38(2):322–35.

75. Curreli S, Wong BS, Latinovic O, Konstantopoulos K, Stamatos NM. Class 3 semaphorins induce F-actin reorganization in human dendritic cells: role in cell migration. J Leukoc Biol. 2016;100(6):1323–34.

76. Busse WW. The relationship of airway hyperresponsiveness and airway inflammation: airway hyperresponsiveness in asthma: its measurement and clinical significance. Chest. 2010;138(2 Suppl):4S–10S.

77. Movassagh H, Shan L, Duke-Cohan JS, Halayko AJ, Uzonna JE, Gounni AS. Semaphorin 3E alleviates hallmarks of house dust mite-induced allergic airway disease. Am J Pathol. 2017;187(7):1566–76.

78. Chapoval SP, Dasgupta P, Smith EP, DeTolla LJ, Lipsky MM, Kelly-Welch AE, Keegan AD. STAT6 expression in multiple cell types mediates the cooperative development of allergic airway disease. J Immunol. 2011;186(4):2571–83.

79. Leigh R, Ellis R, Wattie JN, Hirota JA, Matthaei KI, Foster PS, O'Byrne PM, Inman MD. Type 2 cytokines in the pathogenesis of sustained airway dysfunction and airway remodeling in mice. Am J Respir Crit Care Med. 2004;169(7):860–7.

80. Sahoo A, Alekseev A, Obertas L, Nurieva R. Grail controls Th2 cell development by targeting STAT6 for degradation. Nat Commun. 2014;5:4732.

81. Erazo A, Kutchukhidze N, Leung M, Christ AP, Urban JF Jr, Curotto de Lafaille MA, Lafaille JJ. Unique maturation program of the IgE response in vivo. Immunity. 2007;26(2):191–203.

82. Wu LC, Zarrin AA. The production and regulation of IgE by the immune system. Nat Rev Immunol. 2014;14(4):247–59.

83. Hakonarson H, Herrick DJ, Serrano PG, Grunstein MM. Mechanism of cytokine-induced modulation of beta-adrenoceptor responsiveness in airway smooth muscle. J Clin Invest. 1996;97(11):2593–600.

84. Hakonarson H, Halapi E, Whelan R, Gulcher J, Stefansson K, Grunstein MM. Association between IL-1beta/TNF-alpha-induced glucocorticoid-sensitive changes in multiple gene expression and altered responsiveness in airway smooth muscle. Am J Respir Cell Mol Biol. 2001;25(6):761–71.

85. Kim J, Oh WJ, Gaiano N, Yoshida Y, Gu C. Semaphorin 3E-Plexin-D1 signaling regulates VEGF function in developmental angiogenesis via a feedback mechanism. Genes Dev. 2011;25(13):1399–411.

86. Suchting S, Freitas C, le Noble F, Benedito R, Breant C, Duarte A, Eichmann A. The Notch ligand Delta-like 4 negatively regulates endothelial tip cell formation and vessel branching. Proc Natl Acad Sci U S A. 2007;104(9):3225–30.

87. Fukushima Y, Okada M, Kataoka H, Hirashima M, Yoshida Y, Mann F, Gomi F, Nishida K, Nishikawa S, Uemura A. Sema3E-PlexinD1 signaling selectively suppresses disoriented angiogenesis in ischemic retinopathy in mice. J Clin Invest. 2011;121(5):1974–85.

88. Hellstrom M, Phng LK, Hofmann JJ, Wallgard E, Coultas L, Lindblom P, Alva J, Nilsson AK, Karlsson L, Gaiano N, et al. Dll4 signalling through Notch1 regulates formation of tip cells during angiogenesis. Nature. 2007;445(7129):776–80.

89. Zygmunt T, Gay CM, Blondelle J, Singh MK, Flaherty KM, Means PC, Herwig L, Krudewig A, Belting HG, Affolter M, et al. Semaphorin-PlexinD1 signaling limits angiogenic potential via the VEGF decoy receptor sFlt1. Dev Cell. 2011;21(2):301–14.

90. Torres-Vazquez J, Gitler AD, Fraser SD, Berk JD, Van NP, Fishman MC, Childs S, Epstein JA, Weinstein BM. Semaphorin-plexin signaling guides patterning of the developing vasculature. Dev Cell. 2004;7(1):117–23.

91. Rahimi N. VEGFR-1 and VEGFR-2: two nonidentical twins with a unique physiognomy. Fronti Biosci. 2006;11:818–29.

92. Krueger J, Liu D, Scholz K, Zimmer A, Shi Y, Klein C, Siekmann A, Schulte-Merker S, Cudmore M, Ahmed A, et al. Flt1 acts as a negative regulator of tip cell formation and branching morphogenesis in the zebrafish embryo. Development. 2011;138(10):2111–20.

93. Tatari N, Movassagh H, Shan L, Koussih L, Gounni AS. Semaphorin 3E inhibits house dust mite-induced angiogenesis in a mouse model of allergic asthma. Am J Pathol. 2019;189(4):762–72.

94. White MJ, Risse-Adams O, Goddard P, Contreras MG, Adams J, Hu D, Eng C, Oh SS, Davis A, Meade K, et al. Novel genetic risk factors for asthma in African American children: precision medicine and the SAGE II study. Immunogenetics. 2016;68(6–7):391–400.

95. Modena BD, Bleecker ER, Busse WW, Erzurum SC, Gaston BM, Jarjour NN, Meyers DA, Milosevic J, Tedrow JR, Wu W, et al. Gene expression correlated with severe asthma characteristics reveals heterogeneous mechanisms of severe disease. Am J Respir Crit Care Med. 2017;195(11):1449–63.

96. Movassagh H, Shan L, Chakir J, McConville JF, Halayko AJ, Koussih L, Gounni AS. Expression of semaphorin 3E is suppressed in severe asthma. J Allergy Clin Immunol. 2017;140(4):1176–9.

Serine Protease Inhibitors to Treat Lung Inflammatory Diseases

13

Chahrazade El Amri

Abstract

Lung is a vital organ that ensures breathing function. It provides the essential interface of air filtering providing oxygen to the whole body and eliminating carbon dioxide in the blood; because of its exposure to the external environment, it is fall prey to many exogenous elements, such as pathogens, especially viral infections or environmental toxins and chemicals. These exogenous actors in addition to intrinsic disorders lead to important inflammatory responses that compromise lung tissue and normal functioning. Serine proteases regulating inflammation responses are versatile enzymes, usually involved in pro-inflammatory cytokines or other molecular mediator's production and activation of immune cells. In this chapter, an overview on major serine proteases in airway inflammation as therapeutic targets and their clinically relevant inhibitors is provided. Recent updates on serine protease inhibitors in the context of the COVID-19 pandemic are summarized.

Keywords

Lung · Airway inflammation · Serine proteases · Clinical inhibitors · Respiratory viruses · Repurposing

Abbreviations

ALI	Acute lung injury
ARDS	Acute respiratory distress syndrome
COPD	Chronic obstructive pulmonary disease
COVID-19	Coronavirus disease 2019
HNE	Human neutrophil elastase
KLK	Kallikrein-related peptidase
PARs	Proteinase-activated receptors
TMPRSS2	Transmembrane protease, serine 2

13.1 Introduction: Overview on Serine Proteases

In humans, serine proteases exert their activities at various cellular levels and are responsible for the coordination of many physiological functions including digestion, cell cycle, coagulation, immunity, and reproduction [1, 2].

C. E. Amri (✉)
Sorbonne Université, Faculty of Sciences and Engineering, IBPS, UMR 8256 CNRS-UPMC, ERL INSERM U1164, Biological Adaptation and Ageing, Paris, France
e-mail: chahrazade.el_amri@sorbonne-universite.fr

Furthermore, serine proteases are key elements of the inflammation response due to their release from activated leukocytes and mast cells or generation through the coagulation cascade [3, 4]. Serine proteases which possess similar structures and catalytic properties are grouped into clans. The four main clans are clan PA represented by chymotrypsin, clan SB by subtilisin, and clans SC and SF which encompass various proteases [1, 5]. The specificity of cleavage of a given protease is based on the interaction of a protein or peptide substrate with the residues of the protease at the active site. The surface of the protease that is capable of hosting a side chain of a single substrate residue is referred to as the "subsite" [6–8]. The subsites are numbered from either side of the cleavage site: S_1–S_n to the N-terminus and S_1–$S_{n'}$ to the C-terminus of the substrate. The corresponding residues on the substrate are numbered P_1–P_n and P_1–$P_{n'}$, respectively [6, 9]. The structure of the active site of the protease thus determines the nature of the residues of the substrate which can bind specifically to the enzyme [9]. Depending on their specificity for cleaved residues (P_1), serine proteases can be classified as trypsin, chymotrypsin, or elastase analogs. This specificity is guided by the size and nature of the residue involved in the catalytic pocket of the enzyme. Trypsin or trypsin-like (TL) analogs cleave after a positively charged residue (lysine or arginine), their pocket S_1 being narrow, deep, and negatively charged. The S_1 pocket of "chymotrypsin-like (CTL)" is broad and hydrophobic in nature; this results in a specificity for the medium or large hydrophobic residues (tyrosine, phenylalanine, and leucine) [7, 8]. Elastase-like (EL) has a reduced pocket that receives short and uncharged residues [7]. For example, those which have a digestive role, such as chymotrypsin, generally have few requirements with respect to the substrate, and a large hydrophobic residue in position P_1 is sufficient. The activity of other proteases can be also based on the recognition of a complete peptide portion, such as enterokinase, which requires the binding of an Asp-Asp-Asp-Asp-Lys sequence in the S_1–S_5 subsites to operate properly [10].

Inflammation is a classical well-recognized essential step for the control of microbial invasion or tissue injury as well as for the maintenance of tissue homeostasis under various noxious conditions [11–13]. The causes of inflammation are numerous and varied: infectious agent, inert foreign substance, physical agent, post-traumatic cytotoxic injury, etc. Inflammation begins with a "recognition" reaction involving specific cells of the body (monocytes, macrophages, lymphocytes) or circulating proteins (antibodies, complement proteins, Hageman factor, etc.). The recognition phase follows the sequential activation of a whole set of cells and mediators whose order of intervention is complex and variable. Some mediators, such as prostaglandins and cytokines, are produced by different cell types, act on several cell types, and control sometimes their own production by retroactive regulation. Moreover, inflammation and coagulation constitute two host defense systems with complementary roles in eliminating invading pathogens, limiting tissue damage, and restoring homeostasis [12]. Infection leads to the production of pro-inflammatory cytokines that, in turn, stimulate the production of tissue factor. Conversely, activated coagulation proteases may affect specific receptors on inflammatory cells and endothelial cells and thereby modulate the inflammatory response [13]. Serine proteases are key actors of inflammation responses, both in physiological and pathophysiological contexts [2, 3]. In particular, in the microenvironment of inflammatory tissues, extracellular serine proteases (e.g., HNE or KLKs) can modulate cell signaling via the regulation of PARs. PARs (proteinase-activated receptors) (PAR-1, PAR-2, PAR-3, and PAR-4) belong to the superfamily of receptors coupled to G proteins and are involved in a number of pathways for physiological and pathological signaling in a wide variety of tissues [14]. These receptors are irreversibly activated by the action of proteases mainly from the class of serine proteases having a specificity of the trypsin-like substrate, cleaving following the arginine or lysine residues. A multitude of works suggest that the mechanism of this cleavage occurs within the extracellular domain of the receptor following an arginine or lysine residue, with the generation of a new N-terminus. This unmasked end (tethered ligand) binds as a ligand on the extracellular loops of the receptors causing allosteric changes followed by the coupling of receptors to heterotrimeric G proteins and signal transduction. Biased cleavage, namely, proteolytic processing outside this region, can occur and mediates various cellular responses especially in inflammatory context [15].

13.2 Lung Diseases and Serine Proteases: (Physio) pathological Implications and Overview

Lung inflammation is associated with a wide set of diseases and still represents a real challenge to public health. The airway epithelium is indeed the first site of contact with inhaled agents. Its epithelial cells secrete a variety of substances such as mucins, defensins, lysozyme, lactoferrin, and nitric oxide, which nonspecifically shield the respiratory tract from microbial attack [16]. For example, inflammation is an important feature of many pulmonary diseases such as pneumonia, ARDS (acute respiratory distress syndrome), asthma, and COPD (chronic obstructive pulmonary disease) [17]. COPD is a global epidemic, affecting nearly 300 million people worldwide and killing 3 million individuals each year. It is the only major cause of mortality that is increasing such that by 2030 the mortality rate will reach 7–8 million per annum. COPD which mainly affects cigarette smokers is characterized by lung inflammation, which intensifies with disease progression and is characterized by abnormal inflammatory response, ECM and age-related changes, structural changes in the small airways, and the role of sex-related differences [18]. Proteolytic enzymes also have a prominent role in particular in the emphysematous phenotype [19]. Varied and disparate strategies have been adopted to intervene in pulmonary inflammatory responses. In addition to looking at the cytokines, cytokine receptors, and cell-surface molecules, cellular signal transduction

and gene activation have been targeted for therapy [17]. Neutrophil proteases, especially HNE, have been early and extensively investigated as therapeutic target of prime importance in the above-cited lung diseases [20, 21]. Other selected serine proteases emblematic of lung inflammation processes are summarized in Table 13.1 and Fig. 13.1, namely, plasma kallikrein, kallikrein-related peptidases, and transmembrane serine protease 2, TMPRSS2, for which significant efforts have been recently put in the context of the COVID-19 pandemic.

13.3 Development of Serine Protease Inhibitors to Treat Inflammation in Lung Diseases

13.3.1 Neutrophil Elastase (HNE)

Neutrophils, key immune cells for protection against microbial infection, are also associated with a range of pathologies, including auto-inflammatory diseases, such as systemic lupus erythematosus (SLE) and psoriasis [22, 23]. Neutrophil infiltration is a common pathological feature in acute inflammatory diseases. Furthermore, neutrophils are a rich source of proteolytic enzymes, including serine proteases and their inhibitors involved in neutrophil programmed cell death pathways [24]. Four active serine proteases, neutrophil elastase (HNE), cathepsin G (CatG), proteinase 3 (PR3), and neutrophil serine protease 4 (NSP4), as well as azurocidin, an enzymatically inactive serine protease homolog, were characterized in neutrophils [25, 26]. Human neutrophil elastase (HNE) belongs to the chymotrypsin-like family of serine proteases stored in the azurophilic granules of the neutrophil cytoplasm. A very efficient HNE is necessary to many biological processes that necessitate structures breakdown, such as dynamics of extracellular matrix and tissue remodeling (elastin, collagens), host defense by the disabling of bacterial invasion (cell wall proteins) and in initiation steps of inflammation [27, 28]. Indeed, HNE is a key regulator by its inter-

Table 13.1 Serine proteases involved in lung inflammatory diseases selected in this chapter

Serine protease	Substrates in lung	Implication in lung inflammatory disorders	References
Neutrophil elastase	Elastin, collagens	COPD, ALI, ARDS	[22, 31, 39, 42, 92]
Plasma kallikrein	Kininogen, plasminogen	ARDS, asthma, COVID-19	[43, 93–95]
Tissue kallikreins	PARs, HA proteins	Asthma, influenza infections	[96]
TMPRSS2	AC2, spike SARS-CoV-2, hemagglutinins	Viral infections: influenza, SARS, MERS, SARS-CoV-2	[55, 64, 69, 97]

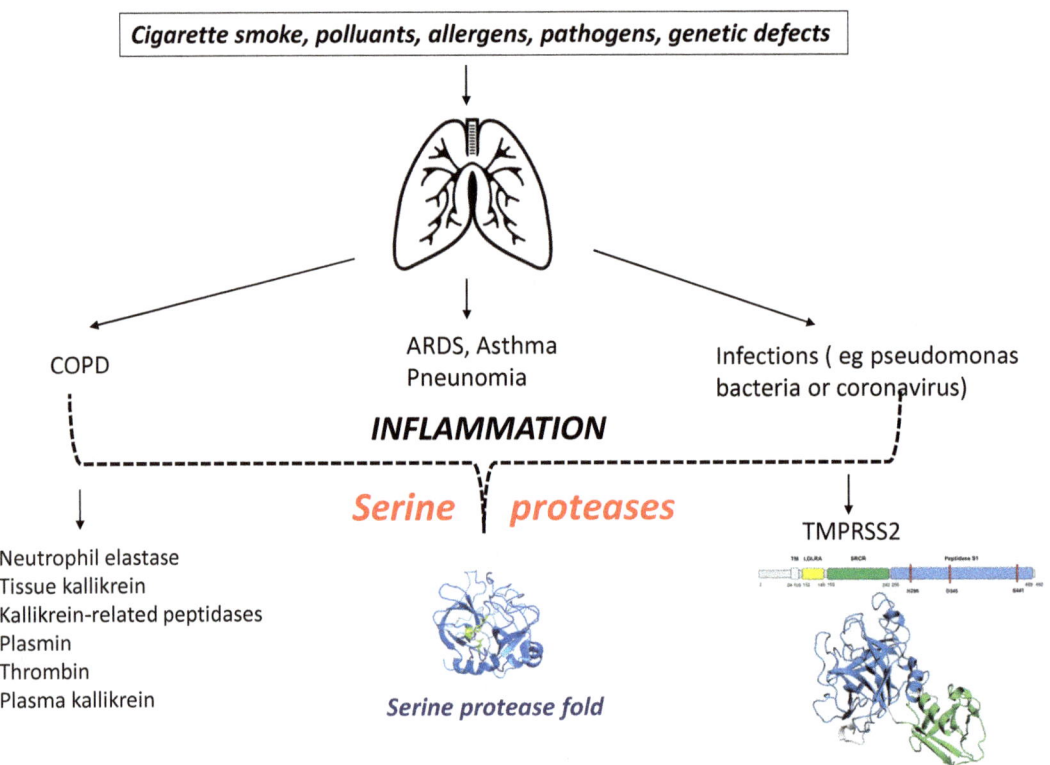

Fig. 13.1 General scheme illustrating lung inflammation in major diseases that imply serine proteases, constituting thus a key duo

vention in the activation of bioactive proteases like MMPs, release of growth factors, shedding of cell-surface-bound receptors, and degradation of endogenous protease inhibitors or virulence factors [20, 29]. The activity of HNE is tightly regulated by compartmentalization in storages granules and phagolysosomes, as well as by the intervention of endogenous serine protease inhibitors such as SERPINs (e.g., α-1 antitrypsin). Unopposed activity of HNE is implicated in the onset and progression of many inflammatory diseases especially of the cardiopulmonary system such as lung emphysema, chronic obstructive pulmonary disease (COPD), pulmonary arterial hypertension (PAH), and pulmonary fibrosis. Various rodent knockout models modulating either HNE or antiprotease expression revealed a significant decrease of phenotypic aspects of pulmonary diseases like emphysema. Elastase knockout mice are particularly sensitive to Gram-negative bacterial infection [30]. Thus, HNE constitutes an attractive target for drug discovery in the pharmaceutical industry by developing inhibitors [22, 31]. Despite the importance of HNE in the clinic (see cited pathologies), only few inhibitors reached the clinic. Only few chemical scaffolds showed profiles suitable for clinical development. The first potent HNE inhibitors to reach clinical investigations were biologicals such as elafin (tiprelestat) [32]. Among small-molecule inhibitors, bearing electrophilic properties like serine acylator sivelestat [33, 34] or transition state analog such as freselestat are very effective [35]. Therapeutic inhibition of HNE holds promise with powerful treatment effect in various preclinical models of lung, bowel, and skin inflammation and ischemia-reperfusion injury relevant to myocardial infarction, stroke, and transplant medicine [23]. Furthermore, sivelestat significantly attenuated LPS-induced acute lung injury (ALI) during recovery from neutropenia [36]. These findings suggest that HNE inhibition could be a promising way to decrease lung inflammation without increasing susceptibility to infection in ALI/ARDS during neutropenia recovery. Recently, phase II studies are in progress by the AstraZeneca firm on a reversible inhibitor AZD9668 (alvelestat) for patients with pulmonary diseases [37, 38]. Interestingly, HNE inhibitors demonstrated innovative strategies to cure lung inflammation [39].

Neutrophil elastase (HNE) plays also an important role in neutrophil extracellular traps (NETs) that contribute to the pathogenesis of acid aspiration-induced ALI/ARDS. NETs may contribute to ALI/ARDS by promoting tissue damage and systemic inflammation. Targeting NETs by inhibiting HNE using alvelestat was proposed as potential therapeutics [40].

The recent identification of BAY 85-8501, led to a very modern inhibitor which binds HNE via an induced fit with a frozen bioactive conformation leading to a significant increase in potency, selectivity, and stability [41]. This compound entered phase II clinical trials for patients with non-CF (cystic fibrosis) bronchiectasis [42]. Moreover, AZD9668, another HNE potent reversible inhibitor, with IC_{50} 12–50 nM, already

succeeded in preclinical studies and is now under phase II clinical trial for bronchiolitis obliterans syndrome (BOS) that is a complication for patients after hematopoietic stem cell transplant.

13.3.2 Plasma Kallikrein

Plasma kallikrein is a trypsin-like serine protease that proteolytically cleaves high molecular weight kininogen to generate the potent vasodilator and pro-inflammatory peptide, bradykinin [43, 44]. Unregulated plasma kallikrein activity is responsible for excessive and potentially fatal edema like hereditary angioedema with C1-inhibitor deficiency [43, 45]. Inhibitors of plasma kallikrein with an optimized selectivity profile were generated as therapeutics and diagnostic tools [46]. The potential of plasma kallikrein was recently explored in the context of the actual SARS-CoV-2 pandemic. SARS-CoV-2 enters cells employing angiotensin-converting enzyme 2 (ACE2) as a receptor, which is highly expressed in lung alveolar cells. ACE2 is one of the components of the cellular machinery that inactivates the potent inflammatory agent bradykinin, and SARS-CoV-2 infection could interfere with the catalytic activity of ACE2, leading to accumulation of bradykinin which induces vasodilation, lung injury, and inflammation [47].

In an open-label, randomized clinical trial, two pharmacological inhibitors of the kinin-kallikrein system that are currently approved for the treatment of hereditary angioedema were tested, icatibant and inhibitor of C1 esterase/kallikrein, in a group of 30 patients with severe COVID-19. Icatibant (Firazyr) is a synthetic peptidomimetic drug consisting of ten amino acids and acts as an effective and specific antagonist of bradykinin B2 receptors. Neither icatibant nor inhibitor of C1 esterase/kallikrein resulted in significant changes in disease mortality and time to clinical improvement. Icatibant may also improve oxygenation in patients with coronavirus disease 2019 (COVID-19) [48]. Both molecules rather promoted significant improvement of lung computed tomography scores and increased blood eosinophils, which has been reported as an indicator of disease recovery. Hence, in this small cohort, pharmacological inhibition of the kinin-kallikrein system seems to improve disease recovery [49].

13.3.3 Kallikrein-Related Peptidases

Multiple studies have revealed the crucial role of kallikrein-related peptidases in the pathophysiology of a number of chronic, infectious, and tumor lung diseases [50].

Among the relatively new serine proteases, "tissue" kallikreins or kallikrein-related peptidases – as distinct from "plasma" kallikrein – form a family of proteases present in at least six orders of mammals. In humans, tissue kallikreins (KLKs, hKLKs, or hKs for human kallikreins) are coded by 15 structurally similar genes (KLK) which co-locate in tandem on chromosome 19q13.4, thus representing the largest cluster of contiguous protease genes in the human genome [51–53]. They include human kallikrein KLK1 and the other 14 kallikrein-related peptidases (KLK2-KLK15). The first member of this family, KLK1, was characterized in the pancreas almost a century ago and named "kallikrein" in reference to this organ (καλλικρεας in Greek). With KLK3 or "PSA" (prostate-specific antigen), discovered in the 1960s, and KLK2, whose gene was isolated in the 1980s, KLK1 belongs to the subfamily of "classical kallikreins." These three proteases are more closely related to each other than the 12 "new kallikreins" (KLK4-15) whose progressive assignment to the KLK family only began in the late 1990s. Thus, the KLK family was best known for the role of KLK1 in the kallikrein-kinin system or for the use of KLK3 or "PSA" (and, to a lesser extent, KLK2) as a biomarker in screening for prostate cancer. However, during the last decade, great progress has been made in understanding the cellular and tissue localization and in vivo logical physio(patho)regulation of most KLKs [54].

KLK1, KLK3, and KLK14 are involved in asthma pathogenesis, and KLK1 could be also associated with the exacerbation of this inflammatory disease caused by rhinovirus. KLK5 was demonstrated as an influenza virus-activating protease in humans, and KLK1 and KLK12 could also be involved in the activation and spread of these viruses. KLK1 (tissue kallikrein 1) is a member of the tissue kallikrein family of serine proteases and is the primary kinin-generating enzyme in human airways. DX-2300 is a fully human antibody that inhibits KLK1 via a competitive inhibition mechanism (Ki = 0.13 nM). Proteolytic cleavage of the hemagglutinin (HA) of influenza virus by host trypsin-like proteases is required for viral infectivity [55]. Some serine proteases are capable of cleaving influenza virus HA, whereas some serine protease inhibitors (serpins) inhibit the HA cleavage in various cell types. Kallistatin, a serpin synthesized mainly in the liver and rapidly secreted into the circulation, forms complexes with KLK1 and inhibits its activity [56]. Kallistatin and other kallikrein inhibitors may be explored as antiviral agents against respiratory viruses [57].

13.3.4 TMPRSS2, a Key Trypsin-Like Protease in Viral Respiratory Diseases

The transmembrane protease serine type 2 (TMPRSS2), also known as epitheliasin, belongs to the hepsin/TMPRSS subfamily [58, 59] and was first largely investigated in the context of prostate cancer where its expression is upregulated by androgens [60, 61]. TMPRSS2 is a 492-amino acid protein organized in functional domains as follows: residues 1–84 constitute the cytoplasmic region; 85–105, the transmembrane region; and 106–492, the extracellular one; LDL receptor (LDLR) class A (residues 112–149), the scavenger receptor cysteine-rich domain 2 (SRCR-2) (residues 150–242), and the peptidase S1 C-terminal domain (residues 256–489), with the catalytic triad (H296, D345, and S441), two potential glycosylation sites (positions 213 and

249), as well as a cleavage site (residues 255–256) that may allow the shedding of the TMPRSS2 extracellular region [62]. TMPRSS2 was proven to be crucial for the activation of hemagglutinin of several human influenza viruses [55, 63]. In December 2019, a new coronavirus named SARS-CoV-2 was identified in the Hubei province of central China. This new coronavirus induces COVID-19, a severe respiratory disease with high death rate. Recently, validated *TMPRSS2* SNPs were proposed to be predictive biomarkers and to be incorporated in the CDC's current list of clinical biomarkers for COVID-19 disease severity [64]. TMPRSS2 protein has a key role in severe acute respiratory syndrome (SARS)-like coronavirus (SARS-CoV-2) infection. SARS-CoV-2 was reported to enter cells via binding to AC2 followed by priming of the virus's spike (S) protein by TMPRSS2 through trypsin-like activity [65–67] (Fig. 13.2). Hence, blocking its proteolytic activity appears as a valuable strategy. Recent systematic studies on the expression of AC2 and TMPRSS2 performed using different cell types of lung tissue from donors have shown that AC2 and TMPRSS2 co-expressed in bronchial and lung cells [68, 69] and underlined that both proteases were expressed in bronchial transient secretory cells [70]. In inflammatory common diseases like asthma, a study with patient cohorts showed that IL-13, a cytokine associated with type 2 asthma, suppresses *ACE2* expression and increases *TMPRSS2* expression in airway epithelial cells from participants with type 2 asthma and atopy; hence, AC2 and TMPRSS2 are modulated in type 2 inflammation in the upper and lower airways [71].

Moreover, naturally occurring genetic variations resulting in defective activity of TMPRSS2 may explain why some populations develop only weak symptoms [72, 73]. In murine models after coronavirus infection, it has been suggested that TMPRSS2 contributes to virus spread and immunopathology in the airways while its genetic inhibition reduces the severity of lung pathology after SARS-CoV or MERS-CoV infections [74]. All these literature data, although not exhaustive, underline that TMPRSS2 is a potential therapeutic target for the cure of respiratory viral infections. Since the beginning of the sanitary crisis, various research programs and clinical trials have been conducted to examine whether TMPRSS2 inhibitors may be useful therapeutics [75, 76]. Early researches have given rise to the design of the first synthetic subnanomolar inhibitors using classical medicinal chemistry to prevent influenza virus activation [77].

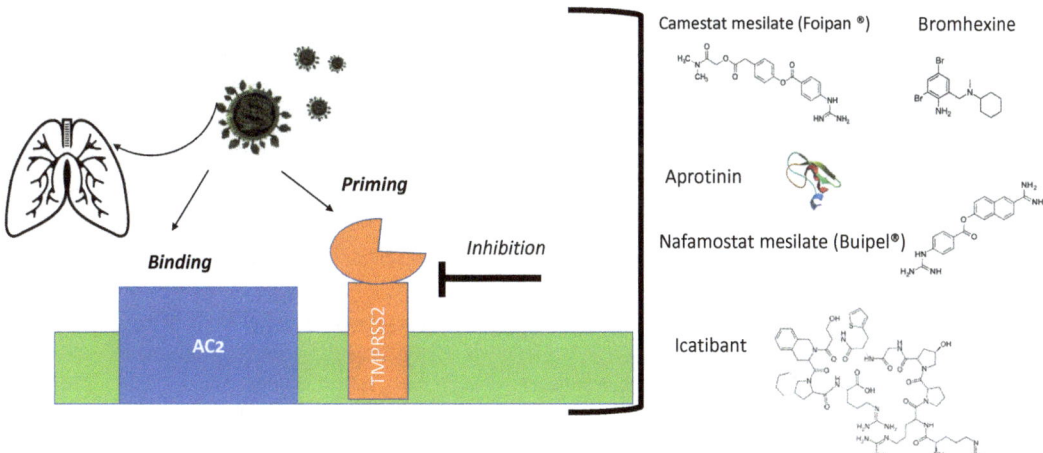

Fig. 13.2 Targeting TMPRSS2 to prevent viral infection and to attenuate COVID-19 symptoms. SARS-CoV-2 was reported to enter cells via binding to AC2 followed by priming of the virus's spike (S) protein by TMPRSS2 through trypsin-like activity [65–67]. Chemical structure of inhibitors under clinical consideration is given (see text for details)

Different strategies have been used to identify TMPRSS2 inhibitors including virtual screening on homology structural models of TMPRSS2 peptidase domain [78, 79], high-throughput screening, or drug repurposing [80].

Clinical serine protease inhibitors, namely, aprotinin, camostat, and nafamostat, are under consideration as TMPRSS2 inhibitors in several clinical trials: camostat (NCT04321096, NCT04338906, NCT043532284), nafamostat (NCT04352400), as well as bromhexine that is revealed as a TMPRSS2 inhibitor [81–83] (NCT04273763, NCT04340349, IRCT20200317046797N4); their chemical structure is given in Fig. 13.2 and their target in Table 13.2. Bromhexine is a bioflavonoid plant-derived mucolytic medication used to decrease viscosity of mucosal pulmonary secretions and was recently shown to inhibit TMPRSS2 in metastatic prostate cancer. The main active metabolite ambroxol of bromhexine was shown to have an anti-inflammatory effect through a decrease of the expression of various pro-inflammatory mediators (IL-1β, IL-6, IL-8, TNF-α) and antioxidative property [84, 85].

Aprotinin is a small protein bovine pancreatic trypsin inhibitor (BPTI), or basic trypsin inhibitor of bovine pancreas, that is part of antifibrinolytic arsenal [86]. Through its ability to inhibit kallikreins, thrombin, and plasmin, it contributes to attenuate inflammation, coagulation, and fibrinolytic pathways. Aprotinin aerosol were early shown to display a positive impact on influenza and paramyxovirus bronchopneumonia of mice [87]. In combination with furin inhibitors, aprotinin used as a TMPRSS2 inhibitor was shown to provide enhanced antiviral effect in human airway epithelial cells [88].

Camostat is an oral serine protease of plasmin, kallikreins, and thrombin used in the treatment of chronic pancreatitis. Hoffman et al. have shown in the recent report that camostat mesylate inhibits SARS-CoV-2 activation through TMPRSS2 and its metabolite GBPA exerts antiviral activity [89]. Nafamostat is another broad-spectrum synthetic serine protease inhibitor that displays anticoagulant and anti-inflammatory properties by inhibiting trypsin-1, kallikrein-related peptidase 1 (KLK1), coagulation factor XII, and coagulation factor X.

Nafamostat mesylate inhibits SARS-CoV-2 infection in the nanomolar range, making it an interesting candidate for clinical trials [90, 91].

13.4 Concluding Remarks

Lung inflammation is a physiological part of the complex biological response of tissues to counteract various harmful signals including pollutants and viruses. Respiratory diseases constitute an important and prioritary public health issue. Inflammatory processes involve diverse actors such as immune cells, blood vessels, and other molecular mediators to eliminate initial events of cell injury. Among them, serine proteases are key elements in both physiological and pathological inflammation; this was particularly emphasized by the recent and current COVID-19 epidemic. The active research programs have given the opportunity to repurpose well-established serine protease

Table 13.2 Summary of serine protease inhibitors that demonstrated potential therapeutic value or in clinical use

Target serine protease	Inhibitors	Targeted diseases	References
HNE	Sivelestat AZD9668	COPD, ARDS, ALI	[33]
Plasma kallikrein	Lanadelumab, icatibant	Lung inflammation, asthma, COVID-19, pulmonary edema	[47, 49]
Tissue kallikrein (kallikrein-related peptidase 1, KLK1)	Kallistatin	COPD, asthma, COVID-9, influenza	[57]
TMPRSS2	Camostat, nafamostat, gabexate, bromhexine, aprotinin	MERS, COVID-19, influenza, ARDS	[80, 89, 91]

inhibitors, illustrating common pathological pathways. Uncertainties around the development of a vaccine against SARS-CoV-2 make it essential to develop distinct and complementary therapeutic solutions. In this objective, serine proteases, namely, inhibitors of trypsin-like proteases, seem to be a valuable strategy to counteract at the same time virus entry, associated coagulopathy, and enhanced fibrinolysis. Several recent reports and ongoing clinical trials have shown that repurposing of already clinically used inhibitors constituted very appropriate strategies; however, definitive conclusions have not been yet established.

Declaration of Interest The author has no other relevant affiliations or financial involvement with any organization or entity with a financial interest in or financial conflict with the subject matter or materials discussed in the manuscript apart from those disclosed.

Funding The authors are grateful to Sorbonne Université (SU), Institut National pour la Recherche Médicale (INSERM), and Centre National de la Recherche Scientifique (CNRS) for research funding.

References

1. Di Cera E. Serine proteases. IUBMB Life. 2009;61:510–5.
2. Soualmia F, El Amri C. Serine protease inhibitors to treat inflammation: a patent review (2011–2016). Expert Opin Ther Pat. 2018;28:93–110.
3. Heutinck KM, ten Berge IJ, Hack CE, Hamann J, Rowshani AT. Serine proteases of the human immune system in health and disease. Mol Immunol. 2010;47:1943–55.
4. Sharony R, Yu PJ, Park J, Galloway AC, Mignatti P, Pintucci G. Protein targets of inflammatory serine proteases and cardiovascular disease. J Inflamm (Lond). 2010;7:45.
5. Rawlings ND, Barrett AJ, Bateman A. Using the MEROPS database for proteolytic enzymes and their inhibitors and substrates. Curr Protoc Bioinformatics. 2014;48:1.25.1–33.
6. Schechter I, Berger A. On the size of the active site in proteases. I. Papain. Biochem Biophys Res Commun. 1967;27:157–62.
7. Hedstrom L. Serine protease mechanism and specificity. Chem Rev. 2002;102:4501–24.
8. Hedstrom L. An overview of serine proteases. Curr Protoc Protein Sci. 2002;Chapter 21:Unit 21.10.
9. Page MJ, Di Cera E. Serine peptidases: classification, structure and function. Cell Mol Life Sci. 2008;65:1220–36.
10. Harris JL, Backes BJ, Leonetti F, Mahrus S, Ellman JA, Craik CS. Rapid and general profiling of protease specificity by using combinatorial fluorogenic substrate libraries. Proc Natl Acad Sci U S A. 2000;97:7754–9.
11. Medzhitov R. Inflammation 2010: new adventures of an old flame. Cell. 2010;140:771–6.
12. Scrivo R, Vasile M, Bartosiewicz I, Valesini G. Inflammation as "common soil" of the multifactorial diseases. Autoimmun Rev. 2011;10:369–74.
13. Mancek-Keber M. Inflammation-mediating proteases: structure, function in (patho) physiology and inhibition. Protein Pept Lett. 2014;21:1209–29.
14. Ramachandran R, Altier C, Oikonomopoulou K, Hollenberg MD. Proteinases, their extracellular targets, and inflammatory signaling. Pharmacol Rev. 2016;68:1110–42.
15. Hollenberg MD, Mihara K, Polley D, Suen JY, Han A, Fairlie DP, Ramachandran R. Biased signalling and proteinase-activated receptors (PARs): targeting inflammatory disease. Br J Pharmacol. 2014;171:1180–94.
16. West JB. Respiratory physiology: the essentials. Philadelphia: Lippincott Williams & Wilkins; 2012.
17. Moldoveanu B, Otmishi P, Jani P, Walker J, Sarmiento X, Guardiola J, Saad M, Yu J. Inflammatory mechanisms in the lung. J Inflamm Res. 2009;2:1–11.
18. Brandsma CA, Van den Berge M, Hackett TL, Brusselle G, Timens W. Recent advances in chronic obstructive pulmonary disease pathogenesis: from disease mechanisms to precision medicine. J Pathol. 2020;250:624–35.
19. Eapen MS, Myers S, Walters EH, Sohal SS. Airway inflammation in chronic obstructive pulmonary disease (COPD): a true paradox. Expert Rev Respir Med. 2017;11:827–39.
20. Kettritz R. Neutral serine proteases of neutrophils. Immunol Rev. 2016;273:232–48.
21. Dey T, Kalita J, Weldon S, Taggart CC. Proteases and their inhibitors in chronic obstructive pulmonary disease. J Clin Med. 2018;7:244.
22. von Nussbaum F, Li VM. Neutrophil elastase inhibitors for the treatment of (cardio)pulmonary diseases: into clinical testing with pre-adaptive pharmacophores. Bioorg Med Chem Lett. 2015;25:4370–81.
23. Henriksen PA. The potential of neutrophil elastase inhibitors as anti-inflammatory therapies. Curr Opin Hematol. 2014;21:23–8.
24. Benarafa C, Simon HU. Role of granule proteases in the life and death of neutrophils. Biochem Biophys Res Commun. 2017;482:473–81.
25. Pham CT. Neutrophil serine proteases: specific regulators of inflammation. Nat Rev Immunol. 2006;6:541–50.
26. Pham CT. Neutrophil serine proteases fine-tune the inflammatory response. Int J Biochem Cell Biol. 2008;40:1317–33.

27. Meyer-Hoffert U, Wiedow O. Neutrophil serine proteases: mediators of innate immune responses. Curr Opin Hematol. 2011;18:19–24.

28. Wiedow O, Meyer-Hoffert U. Neutrophil serine proteases: potential key regulators of cell signalling during inflammation. J Intern Med. 2005;257:319–28.

29. Ramachandran R, Mihara K, Chung H, Renaux B, Lau CS, Muruve DA, DeFea KA, Bouvier M, Hollenberg MD. Neutrophil elastase acts as a biased agonist for proteinase-activated receptor-2 (PAR2). J Biol Chem. 2011;286:24638–48.

30. Belaaouaj A, McCarthy R, Baumann M, Gao Z, Ley TJ, Abraham SN, Shapiro SD. Mice lacking neutrophil elastase reveal impaired host defense against gram negative bacterial sepsis. Nat Med. 1998;4:615–8.

31. Tsai YF, Hwang TL. Neutrophil elastase inhibitors: a patent review and potential applications for inflammatory lung diseases (2010–2014). Expert Opin Ther Pat. 2015;25:1145–58.

32. Twigg MS, Brockbank S, Lowry P, FitzGerald SP, Taggart C, Weldon S. The role of serine proteases and antiproteases in the cystic fibrosis lung. Mediat Inflamm. 2015;2015:293053.

33. Hagiwara S, Iwasaka H, Hidaka S, Hasegawa A, Noguchi T. Neutrophil elastase inhibitor (sivelestat) reduces the levels of inflammatory mediators by inhibiting NF-kB. Inflamm Res. 2009;58:198–203.

34. Hagiwara S, Iwasaka H, Togo K, Noguchi T. A neutrophil elastase inhibitor, sivelestat, reduces lung injury following endotoxin-induced shock in rats by inhibiting HMGB1. Inflammation. 2008;31:227–34.

35. Yoshimura Y, Hiramatsu Y, Sato Y, Homma S, Enomoto Y, Jikuya T, Sakakibara Y. ONO-6818, a novel, potent neutrophil elastase inhibitor, reduces inflammatory mediators during simulated extracorporeal circulation. Ann Thorac Surg. 2003;76:1234–9.

36. Lee JM, Yeo CD, Lee HY, Rhee CK, Kim IK, Lee DG, Lee SH, Kim JW. Inhibition of neutrophil elastase contributes to attenuation of lipopolysaccharide-induced acute lung injury during neutropenia recovery in mice. J Anesth. 2017;31:397–404.

37. Stevens T, Ekholm K, Granse M, Lindahl M, Kozma V, Jungar C, Ottosson T, Falk-Hakansson H, Churg A, Wright JL, et al. AZD9668: pharmacological characterization of a novel oral inhibitor of neutrophil elastase. J Pharmacol Exp Ther. 2011;339:313–20.

38. Kuna P, Jenkins M, O'Brien CD, Fahy WA. AZD9668, a neutrophil elastase inhibitor, plus ongoing budesonide/formoterol in patients with COPD. Respir Med. 2012;106:531–9.

39. Crocetti L, Quinn MT, Schepetkin IA, Giovannoni MP. A patenting perspective on human neutrophil elastase (HNE) inhibitors (2014-2018) and their therapeutic applications. Expert Opin Ther Pat. 2019;29:555–78.

40. Li H, Zhou X, Tan H, Hu Y, Zhang L, Liu S, Dai M, Li Y, Li Q, Mao Z, et al. Neutrophil extracellular traps contribute to the pathogenesis of acid-aspiration-induced ALI/ARDS. Oncotarget. 2018;9:1772–84.

41. von Nussbaum F, Li VM, Allerheiligen S, Anlauf S, Barfacker L, Bechem M, Delbeck M, Fitzgerald MF, Gerisch M, Gielen-Haertwig H, et al. Freezing the bioactive conformation to boost potency: the identification of BAY 85-8501, a selective and potent inhibitor of human neutrophil elastase for pulmonary diseases. ChemMedChem. 2015;10:1163–73.

42. Watz H, Nagelschmitz J, Kirsten A, Pedersen F, van der Mey D, Schwers S, Bandel TJ, Rabe KF. Safety and efficacy of the human neutrophil elastase inhibitor BAY 85-8501 for the treatment of non-cystic fibrosis bronchiectasis: a randomized controlled trial. Pulm Pharmacol Ther. 2019;56:86–93.

43. Bjorkqvist J, Jamsa A, Renne T. Plasma kallikrein: the bradykinin-producing enzyme. Thromb Haemost. 2013;110:399–407.

44. Bryant JW, Shariat-Madar Z. Human plasma kallikrein-kinin system: physiological and biochemical parameters. Cardiovasc Hematol Agents Med Chem. 2009;7:234–50.

45. Weidmann H, Heikaus L, Long AT, Naudin C, Schluter H, Renne T. The plasma contact system, a protease cascade at the nexus of inflammation, coagulation and immunity. Biochim Biophys Acta Mol Cell Res. 2017;1864(11 Pt B):2118–27.

46. Okada Y, Tsuda Y, Tada M, Wanaka K, Hijikata-Okunomiya A, Okamoto U, Okamoto S. Development of plasma kallikrein selective inhibitors. Biopolymers. 1999;51:41–50.

47. Xu Y, Liu S, Zhang Y, Zhi Y. Does hereditary angioedema make COVID-19 worse? World Allergy Organ J. 2020;13:100454.

48. van de Veerdonk FL, Kouijzer IJE, de Nooijer AH, van der Hoeven HG, Maas C, Netea MG, Bruggemann RJM. Outcomes associated with use of a kinin B2 receptor antagonist among patients with COVID-19. JAMA Netw Open. 2020;3:e2017708.

49. van de Veerdonk FL, Netea MG, van Deuren M, van der Meer JW, de Mast Q, Bruggemann RJ, van der Hoeven H. Kallikrein-kinin blockade in patients with COVID-19 to prevent acute respiratory distress syndrome. Elife. 2020;9:e57555.

50. Lenga Ma Bonda W, Iochmann S, Magnen M, Courty Y, Reverdiau P. Kallikrein-related peptidases in lung diseases. Biol Chem. 2018;399:959–71.

51. Clements J, Hooper J, Dong Y, Harvey T. The expanded human kallikrein (KLK) gene family: genomic organisation, tissue-specific expression and potential functions. Biol Chem. 2001;382:5–14.

52. Lundwall A, Band V, Blaber M, Clements JA, Courty Y, Diamandis EP, Fritz H, Lilja H, Malm J, Maltais LJ, et al. A comprehensive nomenclature for serine proteases with homology to tissue kallikreins. Biol Chem. 2006;387:637–41.

53. Paliouras M, Diamandis EP. The kallikrein world: an update on the human tissue kallikreins. Biol Chem. 2006;387:643–52.

54. Emami N, Diamandis EP. New insights into the functional mechanisms and clinical applications of

the kallikrein-related peptidase family. Mol Oncol. 2007;1:269–87.

55. Laporte M, Naesens L. Airway proteases: an emerging drug target for influenza and other respiratory virus infections. Curr Opin Virol. 2017;24:16–24.

56. Chao J, Chao L. Biochemistry, regulation and potential function of kallistatin. Biol Chem Hoppe Seyler. 1995;376:705–13.

57. Leu CH, Yang ML, Chung NH, Huang YJ, Su YC, Chen YC, Lin CC, Shieh GS, Chang MY, Wang SW, et al. Kallistatin ameliorates influenza virus pathogenesis by inhibition of kallikrein-related peptidase 1-mediated cleavage of viral hemagglutinin. Antimicrob Agents Chemother. 2015;59:5619–30.

58. Jacquinet E, Rao NV, Rao GV, Hoidal JR. Cloning, genomic organization, chromosomal assignment and expression of a novel mosaic serine proteinase: epitheliasin. FEBS Lett. 2000;468:93–100.

59. Jacquinet E, Rao NV, Rao GV, Zhengming W, Albertine KH, Hoidal JR. Cloning and characterization of the cDNA and gene for human epitheliasin. Eur J Biochem. 2001;268:2687–99.

60. Chen YW, Lee MS, Lucht A, Chou FP, Huang W, Havighurst TC, Kim K, Wang JK, Antalis TM, Johnson MD, et al. TMPRSS2, a serine protease expressed in the prostate on the apical surface of luminal epithelial cells and released into semen in prostasomes, is misregulated in prostate cancer cells. Am J Pathol. 2010;176:2986–96.

61. Mwamukonda K, Chen Y, Ravindranath L, Furusato B, Hu Y, Sterbis J, Osborn D, Rosner I, Sesterhenn IA, McLeod DG, et al. Quantitative expression of TMPRSS2 transcript in prostate tumor cells reflects TMPRSS2-ERG fusion status. Prostate Cancer Prostatic Dis. 2010;13:47–51.

62. Afar DE, Vivanco I, Hubert RS, Kuo J, Chen E, Saffran DC, Raitano AB, Jakobovits A. Catalytic cleavage of the androgen-regulated TMPRSS2 protease results in its secretion by prostate and prostate cancer epithelia. Cancer Res. 2001;61:1686–92.

63. Zmora P, Moldenhauer AS, Hofmann-Winkler H, Pohlmann S. TMPRSS2 isoform 1 activates respiratory viruses and is expressed in viral target cells. PLoS One. 2015;10:e0138380.

64. Strope JD, Pharm DC, Figg WD. TMPRSS2: potential biomarker for COVID-19 outcomes. J Clin Pharmacol. 2020;60:801–7.

65. Shang J, Wan Y, Luo C, Ye G, Geng Q, Auerbach A, Li F. Cell entry mechanisms of SARS-CoV-2. Proc Natl Acad Sci U S A. 2020;117:11727–34.

66. Shang J, Ye G, Shi K, Wan Y, Luo C, Aihara H, Geng Q, Auerbach A, Li F. Structural basis of receptor recognition by SARS-CoV-2. Nature. 2020;581:221–4.

67. Wan Y, Shang J, Graham R, Baric RS, Li F. Receptor recognition by the novel coronavirus from Wuhan: an analysis based on decade-long structural studies of SARS coronavirus. J Virol. 2020;94:e00127-20.

68. Zhou P, Yang XL, Wang XG, Hu B, Zhang L, Zhang W, Si HR, Zhu Y, Li B, Huang CL, et al. A pneumonia

outbreak associated with a new coronavirus of probable bat origin. Nature. 2020;579:270–3.

69. Hoffmann M, Kleine-Weber H, Schroeder S, Kruger N, Herrler T, Erichsen S, Schiergens TS, Herrler G, Wu NH, Nitsche A, et al. SARS-CoV-2 cell entry depends on ACE2 and TMPRSS2 and is blocked by a clinically proven protease inhibitor. Cell. 2020;181(271–280):e278.

70. Lukassen S, Chua RL, Trefzer T, Kahn NC, Schneider MA, Muley T, Winter H, Meister M, Veith C, Boots AW, et al. SARS-CoV-2 receptor ACE2 and TMPRSS2 are primarily expressed in bronchial transient secretory cells. EMBO J. 2020;39:e105114.

71. Kimura H, Francisco D, Conway M, Martinez FD, Vercelli D, Polverino F, Billheimer D, Kraft M. Type 2 inflammation modulates ACE2 and TMPRSS2 in airway epithelial cells. J Allergy Clin Immunol. 2020;146(80–88):e88.

72. Irham LM, Chou WH, Calkins MJ, Adikusuma W, Hsieh SL, Chang WC. Genetic variants that influence SARS-CoV-2 receptor TMPRSS2 expression among population cohorts from multiple continents. Biochem Biophys Res Commun. 2020;529:263–9.

73. Lee IH, Lee JW, Kong SW. A survey of genetic variants in SARS-CoV-2 interacting domains of ACE2, TMPRSS2 and TLR3/7/8 across populations. Infect Genet Evol. 2020;85:104507.

74. Iwata-Yoshikawa N, Okamura T, Shimizu Y, Hasegawa H, Takeda M, Nagata N. TMPRSS2 contributes to virus spread and immunopathology in the airways of murine models after coronavirus infection. J Virol. 2019;93:e01815–8.

75. Ragia G, Manolopoulos VG. Inhibition of SARS-CoV-2 entry through the ACE2/TMPRSS2 pathway: a promising approach for uncovering early COVID-19 drug therapies. Eur J Clin Pharmacol. 2020;76:1–8.

76. McKee DL, Sternberg A, Stange U, Laufer S, Naujokat C. Candidate drugs against SARS-CoV-2 and COVID-19. Pharmacol Res. 2020;157:104859.

77. Meyer D, Sielaff F, Hammami M, Bottcher-Friebertshauser E, Garten W, Steinmetzer T. Identification of the first synthetic inhibitors of the type II transmembrane serine protease TMPRSS2 suitable for inhibition of influenza virus activation. Biochem J. 2013;452:331–43.

78. Rahman N, Basharat Z, Yousuf M, Castaldo G, Rastrelli L, Khan H. Virtual screening of natural products against type II transmembrane serine protease (TMPRSS2), the priming agent of coronavirus 2 (SARS-CoV-2). Molecules. 2020;25:2271.

79. Singh N, Decroly E, Khatib AM, Villoutreix BO. Structure-based drug repositioning over the human TMPRSS2 protease domain: search for chemical probes able to repress SARS-CoV-2 spike protein cleavages. Eur J Pharm Sci. 2020;153:105495.

80. Shrimp JH, Kales SC, Sanderson PE, Simeonov A, Shen M, Hall MD. An enzymatic TMPRSS2 assay for assessment of clinical candidates and discovery of inhibitors as potential treatment of COVID-19. ACS Pharmacol Transl Sci. 2020;3:997–1007.

81. Maggio R, Corsini GU. Repurposing the mucolytic cough suppressant and TMPRSS2 protease inhibitor bromhexine for the prevention and management of SARS-CoV-2 infection. Pharmacol Res. 2020;157:104837.

82. Habtemariam SNS, Ghavami S, Cismaru CA, Berindan-Neagoe I, Nabavi SM. Possible use of the mucolytic drug, bromhexine hydrochloride, as a prophylactic agent against SARS-CoV-2 infection based on its action on the Transmembrane Serine Protease 2. Pharmacol Res. 2020;157:104853. https://doi.org/10.1016/j.phrs.2020.104853.

83. Li T, Sun L, Zhang W, Zheng C, Jiang C, Chen M, Dai Z, Chen D, Bao S, Shen X. Bromhexine hydrochloride tablets for the treatment of moderate COVID-19: an open-label randomized controlled pilot study. Clin Transl Sci. 2020;13:1096–102.

84. Olaleye OA, Kaur M, Onyenaka CC. Ambroxol hydrochloride inhibits the interaction between severe acute respiratory syndrome coronavirus 2 spike protein's receptor binding domain and recombinant human ACE2. bioRxiv [Preprint]. Sept 14:2020.09.13.295691. 2020; https://doi.org/10.1101/2020.09.13.295691.

85. Chikhale RV, Gupta VK, Eldesoky GE, Wabaidur SM, Patil SA, Islam MA. Identification of potential anti-TMPRSS2 natural products through homology modelling, virtual screening and molecular dynamics simulation studies. J Biomol Struct Dyn. 2020:1–16. https://doi.org/10.1080/07391102.2020.1798813.

86. Davis R, Whittington R. Aprotinin. A review of its pharmacology and therapeutic efficacy in reducing blood loss associated with cardiac surgery. Drugs. 1995;49:954–83.

87. Ovcharenko AV, Zhirnov OP. Aprotinin aerosol treatment of influenza and paramyxovirus bronchopneumonia of mice. Antivir Res. 1994;23:107–18.

88. Bestle D, Heindl MR, Limburg H, Van Lam van T, Pilgram O, Moulton H, Stein DA, Hardes K, Eickmann M, Dolnik O, Rohde C, Klenk HD, Garten W, Steinmetzer T, Böttcher-Friebertshäuser E. TMPRSS2 and furin are both essential for proteolytic activation of SARS-CoV-2 in human airway cells. Life Sci Alliance. 2020;3:e202000786.

89. Hoffmann M, Hofmann-Winkler H, Smith JC, Kruger N, Sorensen LK, Sogaard OS, Hasselstrom JB, Winkler M, Hempel T, Raich L, et al. Camostat mesylate inhibits SARS-CoV-2 activation by TMPRSS2-related proteases and its metabolite GBPA exerts antiviral activity. bioRxiv. 2020; https://doi.org/10.1101/2020.08.05.237651.

90. Yamamoto M, Kiso M, Sakai-Tagawa Y, Iwatsuki-Horimoto K, Imai M, Takeda M, Kinoshita N, Ohmagari N, Gohda J, Semba K, et al. The anticoagulant nafamostat potently inhibits SARS-CoV-2 S protein-mediated fusion in a cell fusion assay system and viral infection in vitro in a cell-type-dependent manner. Viruses. 2020;12:629.

91. Hoffmann M, Schroeder S, Kleine-Weber H, Muller MA, Drosten C, Pohlmann S. Nafamostat mesylate blocks activation of SARS-CoV-2: new treatment option for COVID-19. Antimicrob Agents Chemother. 2020;64:e00754-20.

92. Jugniot N, Voisin P, Bentaher A, Mellet P. Neutrophil elastase activity imaging: recent approaches in the design and applications of activity-based probes and substrate-based probes. Contrast Media Mol Imaging. 2019;2019:7417192.

93. Schranz J, Adelman B, Chyung Y (2017) Plasma kallikrein inhibitors and uses thereof for treating hereditary angioedema attack.

94. Li Z, Partridge J, Silva-Garcia A, Rademacher P, Betz A, Xu Q, Sham H, Hu Y, Shan Y, Liu B, et al. Structure-guided design of novel, potent, and selective macrocyclic plasma kallikrein inhibitors. ACS Med Chem Lett. 2017;8:185–90.

95. Kotian PL, Babu YS, Kumar VS, Zhang W, Vogeti L (2017) Human plasma kallikrein inhibitors.

96. Sexton DJ, Chen T, Martik D, Kuzmic P, Kuang G, Chen J, Nixon AE, Zuraw BL, Forteza RM, Abraham WM, et al. Specific inhibition of tissue kallikrein 1 with a human monoclonal antibody reveals a potential role in airway diseases. Biochem J. 2009;422:383–92.

97. Stopsack KH, Mucci LA, Antonarakis ES, Nelson PS, Kantoff PW. TMPRSS2 and COVID-19: serendipity or opportunity for intervention? Cancer Discov. 2020;10:779–82.

Sex and Gender Differences in Lung Disease

14

Patricia Silveyra, Nathalie Fuentes, and Daniel Enrique Rodriguez Bauza

Abstract

Sex differences in the anatomy and physiology of the respiratory system have been widely reported. These intrinsic sex differences have also been shown to modulate the pathophysiology, incidence, morbidity, and mortality of several lung diseases across the life span. In this chapter, we describe the epidemiology of sex differences in respiratory diseases including neonatal lung disease (respiratory distress syndrome, bronchopulmonary dysplasia) and pediatric and adult disease (including asthma, cystic fibrosis, idiopathic pulmonary fibrosis, chronic obstructive pulmonary disease, lung cancer, lymphangioleiomyomatosis, obstructive sleep apnea, pulmonary arterial hypertension, and respiratory viral infections such as respiratory syncytial virus, influenza, and SARS-CoV-2). We also discuss the current state of research on the mechanisms underlying the observed sex differences in lung disease susceptibility and severity and the importance of considering both sex and gender variables in research studies' design and analysis.

Keywords

Sex · Gender · Lung disease · Hormones · Chronic disease

P. Silveyra (✉)
Department of Environmental and Occupational Health, Indiana University Bloomington, Bloomington, IN, USA
e-mail: psilveyr@iu.edu

N. Fuentes
National Institute of Allergy, Asthma, and Infectious Diseases, Bethesda, MD, USA
e-mail: nathalie.fuentesortiz@nih.gov

D. E. Rodriguez Bauza
Clinical Simulation Center, The Pennsylvania State University College of Medicine, Hershey, PA, USA
e-mail: drrodriguezbauza@pennstatehealth.psu.edu

14.1 Introduction

Sex-related differences exist in many lung diseases throughout the life span [36, 277]. In neonates, the male disadvantage is a well-established clinical fact, especially in the preterm population [16] to the point that guidelines to predict outcomes from the *Eunice Kennedy Shriver* National Institute of Child Health and Human Development (NICHD) and Neonatal Research Network (NRN) for extremely preterm birth outcomes include sex as a critical biological variable [225]. In children and adults, some lung conditions are more commonly found in women and men, respectively, and can present with different degrees of severity and symptoms.

Overall, the literature shows that most lung diseases are more commonly found or present with higher degree of severity, exacerbation rate, hospitalizations, and mortality in women than in men [89]. These include asthma, chronic obstructive pulmonary disease (COPD), pulmonary hypertension, and some types of lung cancer such as adenocarcinoma [211, 228, 316]. Furthermore, some rare and less-understood lung conditions such as lymphangioleiomyomatosis (LAM) are almost exclusively found in women [309].

Although the terms sex and gender are commonly used interchangeably, they represent different concepts. According to the National Institutes of Health (NIH) Office of Research on Women's Health (ORWH), "sex" refers to the biological differences between females and males, including chromosomal, anatomical, hormonal, and other physiological and functional differences. "Gender," on the other hand, refers to the characteristics that a society or culture delineates as masculine or feminine, including social, environmental, cultural, and behavioral factors and choices that influence an individual's self-identity. As opposed to sex, gender is a social construct and not defined biologically. Importantly, an individual's gender does not necessarily need to be consistent with their biological sex given at birth nor be fixed or binary. However, because the health of men and women is influenced by both sex and gender, including these variables in research studies is crucial. In basic science, this means including both male and female cells and/or experimental animal models in study designs, as well as examining the influence of sex hormones down to the molecular level. For clinical, behavioral, and outcomes research, this means considering gender-specific social influences and their impact on health and disease. Only when we incorporate sex and gender factors in research studies, we will be able to understand the mechanisms underlying the numerous sex disparities observed in lung disease prevalence and severity and provide more efficient and personalized sex- and gender-specific medicine.

14.2 Sex and Gender Differences in Respiratory Disease

It is not possible to talk about sex differences in respiratory disease without discussing first sex differences in lung biology. From the 16th week of gestation to adult life, significant differences exist in the male and female lung. In addition, changes in sex hormone levels throughout development, puberty, and physiological events such as pregnancy and menopause also influence lung function and health. Early in life, while female sex hormones are beneficial, promoting lung development and maturation, androgens appear to exert the opposite effect [245]. After puberty, the opposite occurs in diseases such as severe asthma, where improvement is observed with increasing androgen levels, and fluctuations in female hormones throughout the menstrual cycle promote asthma exacerbations. Overall, the available body of research shows that the effect of sex hormones on lung health appears to depend on the timing of exposure and thus differentially affects disease prevalence and severity in males and females throughout the life span. Table 14.1 summarizes the information available on some of the most common lung diseases affecting men and women disproportionately. In the sections below, we describe the epidemiological information available as well as the status of the research aiming to understand the mechanisms behind the observed sex differences for each disease.

14.3 Neonatal Lung Disease

Infants born prematurely are at higher risk for cardiopulmonary and neurological comorbidities such as retinopathy, pulmonary hypoplasia, respiratory distress syndrome (RDS), bronchopulmonary dysplasia (BPD), intraventricular hemorrhage (IVH), and chronic neurocognitive developmental disorders [190]. Many of these comorbidities exhibit significant sex disparities that could be a consequence of differences in lung development and/or caused by a complex interaction between immunological, hormonal, and genetic factors earlier in life [189]. Overall,

Table 14.1 Sex differences in neonatal, pediatric, and adult lung disease prevalence

Disease	Population	Sex differences	References
Asthma	Children	Boys > girls	[3, 36, 185]
	Adults	Women > men	
Bronchopulmonary dysplasia (BPD)	Neonates	Boys > girls	[17, 85, 276]
Chronic bronchitis	Adults	Women > men	[11]
Chronic cough	Children	Boys > girls	[27]
Chronic obstructive pulmonary disease (COPD)	Adults	Women > men	[2, 111]
Cystic fibrosis	Children	Girls > boys[a]	[92, 289]
	Adults	Women > men	
Coronavirus disease 19 (COVID-19)	Adults	Men > women	[234]
Emphysema	Adults	Men > women	[160]
Idiopathic pulmonary fibrosis	Adults	Men > women	[213, 314]
Lung cancer	Adults	Women > men	[209, 252]
	Adults	Women > men	[309]
Pulmonary arterial hypertension	Adults	Women > men	[170]
Obstructive sleep apnea	Adults	Men > women	[144, 184]
Respiratory distress syndrome (RDS)	Neonates	Boys > girls	[17, 85, 276]
Respiratory syncytial virus infection (RSV)	Neonates/children	Boys > girls	[182]

[a]Infection rates and outcomes worse in girls than boys, but no sex differences in incidence

male infants are presumed to have an intrinsic disadvantage and to be more sensitive to adverse environmental exposures during development and after birth [181]. This sex-related disparity is particularly manifested during the neonatal period and is more pronounced in prematurely born infants.

14.3.1 Sex Differences in Lung Development

The development of the male and female lung is a highly regulated process controlled by genetic, epigenetic, hormonal, and environmental factors. This process is divided into five stages: embryonic, pseudoglandular, canalicular, saccular, and alveolar (Table 14.2). Each stage is characterized by specific cellular and structural events that are controlled by the expression of multiple developmental genes [296, 297]. Sex differences in structural, mechanical, and functional aspects of lung development, as well as in its control by sex hormones, have been widely documented [36, 37, 163]. These differences are thought to be associated with the sexual dimorphism observed not only in neonatal lung disease but also later in life

[245]. Respiratory diseases such as RDS and BPD contribute to a large proportion of the morbidity and mortality of prematurely born infants [276]. Importantly, even late preterm infants, born at gestational ages of 34–36 weeks, have been found to be greater risk for adverse respiratory morbidity and mortality than infants born at term [87].

During the fetal period, male lung maturation is usually delayed in comparison to female maturation. Pulmonary surfactant production initiates later in the male vs. the female lung [36, 235]. Consequently, male neonates are at increased risk of developing respiratory distress syndrome (RDS) and a higher risk of morbidity and mortality due to RDS compared with female neonates of similar gestational age [107, 269]. Furthermore, sex differences in overall neonatal survival and pulmonary outcomes have been described with a significantly higher incidence in males versus females [96]. One example is the high incidence observed in males for the development of bronchopulmonary dysplasia (BPD), a pulmonary pathology of the neonate for which RDS is not always an anterior event [207]. Differences in gene expression, particularly at the late developmen-

Table 14.2 Stages of human lung development

Developmental stage (gestational age)	Main events	Sex differences	References
Embryonic (3–6 weeks)	Lung buds emerge (foregut), and trachea and bronchial buds form	None reported	–
Pseudoglandular (6–16 weeks)	Bronchial development, airway branching	Fetal growth and breathing movements are detected earlier in the female fetus. AMH delays branching by promoting apoptosis in males	[41, 98]
Canalicular (16–26 weeks)	Subdivision of distal airways into canaliculi, vascularization Differentiation of type I and II cells, surfactant production	Surfactant secretion and phospholipid maturation are inhibited in males (androgens) and promoted in females (estrogens)	[207, 235, 274]
Saccular (26–36 weeks)	Cell differentiation, type II cell maturation, surfactant secretion Formation of sacs and primary septa	Surfactant production and phospholipid profile remain more advanced in females	[71]
Alveolar (36 weeks–adolescence)	Alveolar multiplication, enlargement, and maturation Lung growth continues and lung function increases with age and peaks in adolescence	Faster alveolarization in females. Higher flow rate per lung volume, but smaller lung size in girls. Better response to surfactant therapy in female newborns with RDS. FEV1 peaks earlier in females than males	[23, 30, 231, 271]

AMH Anti-Müllerian hormone, *FEV1* Forced expiratory volume in 1 minute

tal stages, have been shown to play significant roles in this sex disparity in lung health outcomes [6, 23, 85].

14.3.2 Respiratory Distress Syndrome

Respiratory distress syndrome is a condition of the premature born characterized by a deficiency in pulmonary surfactant [9]. Infants presenting with RDS show widespread alveolar atelectasis and a reduction in lung compliance, with secondary complications such as pneumothorax. Prior to the introduction of antenatal corticosteroids and postnatal surfactant replacement therapy, RDS was a major contributor to neonatal mortality, particularly in male newborns [194].

The main factors involved in the pathophysiology of RDS are surfactant deficiency and dysfunction in the immature lung, which occur at higher rates in males than females of the same gestational age [181]. Thus, the less developed or mature the lung, the higher the chance of disease manifestation after birth. As mentioned earlier,

pulmonary surfactant is produced earlier in females than males during gestation, and its production is stimulated by female sex hormones, and inhibited by male sex hormones [263, 278]. A meta-analysis of data from over 500,000 preterm newborn infants found that RDS was between 1.56 and 1.84 times higher to occur in newborn males than females [145]. This report indicated that males were also at higher risk for other diseases of the newborn, such as BPD, as well as lower respiratory tract infections, bronchiolitis, and pneumonia.

Preventative and treatment options for RDS include postnatal surfactant administration and antenatal corticosteroid therapy [276]. For a long time, corticosteroids (e.g., betamethasone) have been reported to have sex-specific effects on placental oxidative balance and microvascular blood flow [254, 255], as well as to improve the subsequent response of infants to surfactant administration [120], with more beneficial effects in females than males [194]. However, a more recent systematic review and meta-analysis on the topic did not find sex-specific differences, although the type of antenatal glucocorticoid used (betamethasone vs. dexamethasone) displayed a sex-specific effect [221].

14.3.3 Bronchopulmonary Dysplasia

Bronchopulmonary dysplasia is a lung disease of the prematurely newborn, characterized by an arrest in alveolarization and aberrant pulmonary vascular development [17, 113]. The disease diagnosis is performed by assessing the need for mechanical ventilation and oxygen respiratory support at 36 weeks' postmenstrual age [16] and displays a higher incidence in extremely low birth weight neonates [38].

The widespread use of antenatal corticosteroids, neonatal exogenous surfactant, and protective ventilation strategies has led to increased survival of more extremely preterm infants, with a consequent increase in BPD incidence in the past few decades [112, 113]. While mortality from the disease has declined significantly in the past several decades, children diagnosed with BPD still display long-term complications in lung health ranging from the need for tracheostomy and mechanical ventilation to pediatric pulmonary arterial hypertension and poor neurodevelopmental outcomes [55, 219]. Recent studies have also reported that adults who were born preterm display a higher incidence of airflow obstruction, gas trapping, and reduced gas exchange than those born term [310] and that worsening of lung function persists throughout childhood, particularly in males [93].

Multiple clinical studies have reported sex differences in BPD, including a higher incidence in males vs. females that persists after adjusting for other confounders. Moreover, males display higher death and oxygen dependency rates, as well as pulmonary hemorrhage and use of postnatal steroids [28, 112, 113]. Male sex is considered not only an independent major risk predictor of BPD but also to worsening of lung function during the neonatal and early childhood periods [268]. However, despite the well-established sexual dimorphism in the incidence of BPD, the mechanisms associated with these disparities are not completely understood. Recent studies in ani-

mal models have suggested a role for microRNAs (miRNAs) in mediating sex biases in BPD [138, 210, 317]. Others have related the sexual dimorphism of PBD to sex-specific differential activation of hypoxia-inducible factors and genes related to angiogenesis, supporting the pulmonary angiogenesis dysregulation in the pathobiology of BPD [49, 138].

14.4 Pediatric and Adult Lung Disease

Sex differences in lung and airway development persist throughout infancy and early childhood [270, 293]. While females display larger airways than males, the number of alveoli per unit area and the alveolar size do not differ between sexes. The age- and height-adjusted lung volume, however, is higher in boys than girls, which may result in a larger alveolar surface area and a higher diffusion capacity of carbon monoxide (DLCO) in males [20, 233, 270]. With age, differences in lung volumes, as well as in lung size and shape, become more evident [270, 275]. Together with differences in the distending forces of the lungs, these result in differences in the recoil pressure between males and females [51]. This sexual dimorphism in human lung morphometrics and function, together with physiological differences observed by pulmonary function testing, spirometry, and other techniques, has been used to partially explain the observed sex disparity in multiple pulmonary conditions.

Overall, while the majority of lung diseases presented below affect more adult women than men, several conditions are observed at higher rates in men than women and/or show opposite trends in childhood vs. adult life. A multitude of intrinsic factors, such as sex hormones, genetic and epigenetic factors, and comorbidities, along with other extrinsic factor,s have been suggested to contribute to these trends. In the next sections, we summarize the recent epidemiological data, as well as research aiming to understand the mechanisms behind sex disparities in lung disease throughout the life span.

14.4.1 Asthma

Asthma is a heterogeneous disease characterized by chronic airway inflammation. Some of its symptoms are wheezing, shortness of breath, chest tightness, cough, and airflow limitation [83]. Asthma is one of the most prevalent inflammatory diseases of the lung, affecting a significant portion of the world's population. The World Health Organization reported that more than 339 million people suffer from asthma, resulting in more than 400,000 deaths per year [42].

While asthma imposes a substantial public health burden in terms of impaired quality of life and mortality in men and women, clear sex differences exist in its risk, prevalence, and severity across life span [66, 89, 180, 237]. Depending on the sex and age of the patient, striking differences are observed in asthma incidence, prevalence, and severity [162]. An interesting fact is that asthma in children is more prevalent in boys than girls, and studies in adult populations frequently report more negative lung health outcomes for women than men, suggesting an involvement of sex hormones in mediating these effects [129, 228].

Epidemiological studies of childhood asthma have shown that prepubertal boys have more asthma than girls, especially at younger ages [75, 143]. Chronic cough in early childhood, whether from asthma or other causes, is also more common in boys than girls [27]. According to the Centers for Disease Control and Prevention, in the United States, it is estimated that 8.3% of boys and 6.7% of girls under 18 years old currently suffer from asthma. Interestingly, these patterns are reversed after puberty, where asthma prevalence rates for women are almost twice as those for men (5.5% vs. 9.8% for women and men over 18 years of age, respectively) [42]. These statistics have led investigators to hypothesize that hormonal changes starting in puberty contribute to asthma development. This notion is further supported by studies showing that girls who undergo menarche at an earlier age have a higher risk of developing asthma after puberty than girls in which menarche occurs later [226].

Studies showing variations in asthma symptoms and hospitalization rates throughout the menstrual cycle and a decline in asthma incidence in women after menopause also support this hypothesis [33, 204]. Also, women are more susceptible to asthma induced by air pollution and show worse adverse pulmonary health outcomes than men [141, 147]. In this regard, clinical studies and experimental evidence from mouse models have reported that female hormones such as estrogen can trigger lung inflammatory and allergic reactions, while male hormones such as androgens play the opposite role [76, 186, 307]. Interestingly, the severity of asthma in men increases later in life when androgen levels decrease [35]. Overall, more research is needed to elucidate the mechanisms underlying the observed sex differences in disease susceptibility and progression.

Sex differences in asthma have been linked to immunological factors, lung physiology and growth, and behavioral factors [74–130], as well as exposure to air pollutants [88, 86]. Human studies and in vivo models of asthma have shown that female hormones can trigger lung inflammatory and allergic reactions, and male hormones usually play the opposite role [77]. Interestingly, researchers have discovered that sex hormones can alter macrophage polarization and other immune-related cells such as the group 2 innate lymphocytes (ILC2s) and airway smooth muscle cells [22]. ILC2s that lack a killer cell lectin-like receptor G1 accumulate in the lungs of females after they have reached reproductive age but not in males [115]. Others have found that estrogen and testosterone increase and decrease Th2-mediated airway inflammation, respectively [78]. The authors of this study also concluded that females have augmented IL-17A-mediated airway inflammation compared to males [78].

Genetic associations with asthma have also been reported and found to be sex specific [105]. Two single nucleotide polymorphisms (SNPs) in the thymic stromal lymphopoietin (TSLP) gene (rs1837253 and rs2289276) have been associated with asthma in a sex-specific manner. Specifically, rs1837253 is associated with a lower risk for asthma in men, and rs2289276 is associated with a higher risk of asthma in women. While the underlying mechanisms for these sex-specific associations have not been elucidated, these genetic variants have been associated with changes in immunoglobulin E (IgE) levels, which in children are correlated with higher airway resistance and exacerbations triggered by dust, pollen, and pets [94].

14.4.2 Exercise-Induced Bronchoconstriction

Exercise-induced bronchospasm/bronchoconstriction (EIB) is a phenomenon of acute airway narrowing that occurs during or after exercise or physical exertion. As such, EIB can occur in the presence or absence of asthma. Traditionally, the terms exercise-induced asthma (EIA) and (EIB) have been used interchangeably. However, the current consensus is that EIB represents a more accurate reflection of the underlying pathophysiology of the condition, since exercise is not an independent risk factor for asthma but rather a trigger of bronchoconstriction in patients with underlying asthma [175, 196, 258].

As mentioned above, asthma prevalence is higher in boys than in girls; however, after puberty the prevalence is around 20% higher in women than men, indicating a potential contribution of hormones after puberty [195]. Moreover, sex and gender differences in response to exercise have clear implications for understanding gender-specific adaptations to exercise for athletic performance and overall health [188].

The estimated prevalence of EIB varies from approximately 5% to 20% in the general population to an estimated 30% to 70% in elite winter athletes and athletes who participate in summer endurance sports, and at least 90% in individuals with persistent asthma [298]. This condition has been reported in a range of sporting activities but is most common in participants of cold weather sports (e.g., Nordic skiing) and indoor sports (e.g., ice-skating and swimming) [224]. Shinohara et al. investigated whether sex differences influence the prevalence and severity of EIB in prepubertal children aged 5–6 years. They

found that the prevalence of EIB was higher in girls than in boys. In addition, the time to maximal bronchoconstriction was slower in girls than in boys, and the pattern of recovery after exercise was also faster in females than males [243]. Therefore, it is recommended that when evaluating the prevalence and severity of EIB in prepubertal children, the influence of sex is considered.

The pathogenesis of EIA is not fully elucidated. Minute ventilation, the volume of air inhaled or exhaled from a person's lungs per minute, rises with exercise. It is believed that EIB probably results from changes in airway physiology triggered by the large volume of relatively cool, dry air inhaled during vigorous activity [8, 167]. One of the major triggers for bronchoconstriction is water loss during periods of high ventilation. Strenuous exercise creates a hyperosmolar environment by introducing dry air into the airway with compensatory water loss, leading to transient osmotic changes in the airway surface. This hyperosmolar environment leads to mast cell degranulation and eosinophil activation with consequent release of inflammatory mediators, including leukotrienes. This process triggers bronchoconstriction and inflammation of the airway, as well as stimulation of sensory nerves and release of neurokinin and mucins [299]. All this is supported by several research findings concluding that it is not the type of exercise but the ventilation demand and humidity of the inspired air that are the main determinants of the occurrence and degree of bronchoconstriction [58, 119]. Therefore, the diagnosis of asthma in athletes should be confirmed by lung function test, usually with bronchial provocation testing [166] in association with a history consistent with EIB, because self-reported symptoms are not adequate. Varsity athletes show a high incidence of EIB when objectively diagnosed by a variety of pulmonary function criteria. The use of symptoms to diagnose EIB is not predictive of whether athletes have objectively documented EIB [197].

Management of EIB should be based on the understanding that EIB susceptibility varies widely among asthmatic patients, and it could also be present in individuals without underlying asthma. A study by Parsons et al. found that 36 out of 42 EIB-positive athletes (86%) had no prior history of EIB or asthma [197]. In patients with asthma, EIB can also be an indicator of poorly controlled disease, and underlying asthma should be treated prior to controlling EIB [299]. As mentioned above, asthma can deteriorate during the peri-menstrual period, a phenomenon known as peri-menstrual asthma (PMA) which is usually much more severe and troublesome than the reported periovulatory worsening of asthma [246]. In this context, Stanford et al. demonstrated for the first time that the menstrual cycle phase is an important determinant of the severity of EIB in female asthmatic athletes [253]. This study reported deterioration in the severity of EIB during the mid-luteal phase accompanied by worsening asthma symptoms and increased bronchodilator use [253]. Aiming that exercise is not avoided by patients with EIB, general measures and pharmacologic interventions can be assessed subjectively in terms of symptom control and exercise tolerance, considering the fact that sex hormones play an important role in lung inflammation. Thus, medical evaluation and medication adjustment would likely be based on the understanding of sex differences.

14.4.3 Chronic Obstructive Pulmonary Disease

Chronic obstructive pulmonary disease (COPD) affects an estimated 174 million people worldwide (104.7 million males and 69.7 million females) [53]. For many years, it was considered a disease of older men [211]. However, over the past 20 years, its prevalence and rates of hospitalization have increased among women, closing this prevalence gap [11, 56]. This phenomenon is due in part to increased rates of tobacco use – the single largest risk factor for the development of COPD – among women, together with recent evidence demonstrating that first- and secondhand tobacco smoke has more severe effects in women than men [26, 89]. Moreover, there is an increased recognition that the clinical presentation of COPD is different in women than men, which has

led to better and more accurate diagnosis in women in the past few decades [111, 137]. It has been shown that women with COPD have different disease burden, symptoms, and clinical trajectory than men [89, 199] and that women tend to develop COPD earlier in life and have more frequent respiratory exacerbations than men [199].

While asthma remains the most prevalent respiratory disease in the world, COPD is the fourth leading cause of death in the United States and the eighth leading cause of disability worldwide [248]. Recently, the World Health Organization has projected that COPD will be the third leading cause of death worldwide by 2030. Moreover, although the overall prevalence of COPD is increasing in both men and women [249], recent data from US Center for Disease Control's National Center for Health Statistics has shown that COPD prevalence in the United States not only is higher in women but also increasing at a higher rate among women than men [2]. Epidemiological data show that since the year 2000, the number of women in the United States dying from COPD has surpassed the number of men [10, 157]. Some studies have suggested that both asthma and the so-called asthma-COPD overlap syndrome (ACOS), which are more common among adult women than men, can predispose women to develop COPD [61, 272].

It is possible that the increased prevalence of COPD in women is not only due to increased tobacco use but also related to longer life expectancy for women in general, as well as changes in women's occupational exposures over the past few decades [10]. Historically, professions that predispose to lung disease were predominantly held by men. However, due to the reassignment of sex roles and more single-parent households in recent decades, a higher number of women are found in these jobs [11]. This may play a role in the increasing prevalence of the disease among women. It is estimated that 15% of COPD is work-related. In addition, it has long been theorized that indoor air pollution resulting from smoke from biomass fuel combustion for cooking and other purposes also contributes to the development of COPD in never smokers, with women being disproportionately exposed [284] and affected [202]. It is estimated that 50% of deaths from COPD are associated with indoor air pollution in developing countries, and about 75% of these are in women [227].

In recent years, there have been several clinical and experimental studies aiming to understand the contribution of sex to the biologic pathogenesis of COPD [18]. Levels of pro-inflammatory cytokines, including C-reactive protein, interleukin-6 (IL-6), tumor necrosis factor alpha (TNFα), matrix metalloproteinase 9 (MMP-9), pulmonary and activation-regulated chemokine (PARC), and vascular endothelial growth factor (VEGF), have been theorized to contribute to the development of COPD. VEGF helps regulate growth of new vessels and vascular leak and was found to be elevated in patients with COPD compared to healthy controls. In patients with COPD, statistically significant higher levels of VEGF and IL-6 have been found in men vs. women [10]. Additionally, studies in mouse models of chronic cigarette smoke have indicated that sex hormones may be contributing to the greater COPD susceptibility in females. Exposure to cigarette smoke in female mice results in higher peripheral airway obstruction and airway remodeling and less emphysema than male mice, an effect that is mediated by estrogens [267]. It was also found that in female mice, cigarette smoke was associated with activation of transforming growth factor-β (TGF-β), decreased expression of antioxidant genes and the transcription factor Nrf2 (nuclear factor erythroid-derived 2-like 2), as well as increased oxidative stress [267]. Overall, more research is needed to better understand the mechanisms behind sex differences in COPD susceptibility, as well as in the response of men and women to COPD available therapies.

14.4.4 Cystic Fibrosis

Cystic fibrosis (CF), an autosomal recessive multiorgan disease caused by mutations in the cystic fibrosis transmembrane conductance regulator (CFTR) gene, also displays sex differences. Epidemiological studies have reported a sex-based disparity in CF outcomes, where females experience higher rates of pulmonary exacerbations and a shortened life expectancy than males [289]. While the etiology of this disparity is not fully elucidated, it appears to be multifactorial. Studies have associated the sexual dimorphism in CF outcomes to bias in diagnosis [136], anatomical differences between males and females [60], socioenvironmental factors [218], medication adherence [239], physical activity level [229], and actions of male and female sex hormones [1, 264]. A combination of poor perception of disease prognosis, withdrawal, anxiety, decreased adherence to therapies, and decrease in physical activity after puberty has been associated with increased morbidity and mortality in CF females [236]. Moreover, despite reported earlier referral to lung transplantation in females than males, survival time after transplantation does not show sex differences. Not only females with CF experience higher rates of infection and exacerbations than males, they also require more intensified treatment regarding antibiotics, macrolides, steroids, and days of hospitalization than their male counterparts [171, 193]. However, despite earlier referral to lung transplantation in females than males, survival time after transplantation does not show sex differences.

As with other lung diseases described earlier, the sexual dimorphism in CF outcomes is also age dependent. In females, the predisposition to worse outcomes in CF has been found at a young age, where girls are more susceptible to bacterial infection than boys [50]. Females not only show higher lung bacterial colonization with *Pseudomonas aeruginosa*, *Burkholderia cepacia*, and *methicillin-resistant Staphylococcus aureus* than males [54, 103, 161, 217] but also earlier colonization in life, which is a predictor of negative outcomes and decline in survival for females [59]. Females have also been found to acquire *methicillin-sensitive Staphylococcus aureus*, *methicillin-resistant Staphylococcus aureus*, *Haemophilus influenzae*, *Achromobacter xylosoxidans*, *Aspergillus species*, and nontuberculous mycobacteria at earlier ages than males and often even prior to puberty [92].

During puberty and reproductive years, the predisposition to infection is enhanced in females, as well as an increased risk for pulmonary exacerbations and extrapulmonary complications [161]. Females also show a steeper decline in lung function, one of the key predictors of long-term health in CF patients, than males [50]. Because of the reduced life expectancy of patients with CF, little is known about the influence of menopause in the course of the disease [187]. A study in long-term survivors (older than 40 years old) showed that females with CF are also less likely to live to the age of 40 than males [187].

While the mechanisms underlying these sex-disparities have not been fully elucidated, a role of sex hormones in mediating inflammatory processes [101], and types of pathogens colonizing the lung has been suggested [280]. A study by Chotirmall et al. showed that the female hormones 17β-estradiol and estriol can induce conversion of *Pseudomonas aeruginosa* from a non-mucoid to mucoid phenotype in females with CF. The same study suggested that high levels of 17β-estradiol in females result in higher capture of more mucoid strains of *Pseudomonas aeruginosa* and subsequent pulmonary exacerbations [48]. In addition, not only postpubertal increases in pulmonary exacerbations are reported in females [261], but also women display cyclical symptoms in relation to their menstrual cycle, with higher lung function measures during the luteal phase than other cycle phases [114]. Studies in bronchial epithelial cells also showed that 17β-estradiol reduces expression of proinflammatory cytokines via upregulation of the secretory leucoprotease inhibitor (SLPI), which could contribute to the higher infection rate observed in females vs. males [48]. In mouse models of CF, 17β-estradiol stimulates expression of toll-like receptor 2, IL-23, and IL-17A and results in higher lung inflammatory infiltrates and mucin [295]. Abid et al. showed that female mice inoculated with *Pseudomonas aeruginosa* died earlier and showed slower bacterial clearance than male mice [1]. This effect was reversed by treatment with the estrogen receptor (ER)

antagonist ICI 182,780 and ovariectomy and recapitulated in ovariectomized females treated with exogenous 17β-estradiol [1].

Very few studies have evaluated the role of progesterone and testosterone in mediating sex differences in CF infection rates and outcomes. One study in human tracheal epithelial cells showed that exposure to progesterone results in decreased cilia beat frequency, an effect that was attenuated with the addition of 17β-estradiol [109]. While women with CF are able to carry on pregnancies, the role of progesterone in lung function and CF outcomes has not been studied in detail [191]. With regard to androgens, a few reports have indicated that adolescent and adult males with CF have lower salivary and serum levels of testosterone than healthy controls [29, 139], as well as higher rates of male infertility [312]. In rodent studies, testosterone was found to enhance expression and functional activity of epithelial sodium channels [174]. Overall, the direct impact of sex hormones on disease progression in patients with CF remains unknown.

14.4.5 Idiopathic Pulmonary Fibrosis

Idiopathic pulmonary fibrosis (IPF) is a specific form of chronic, progressive fibrosing interstitial pneumonia of unknown cause, occurring primarily in older adults, and limited to the lungs [212]. With a median survival of 3–5 years following diagnosis, IPF is characterized by a progressive worsening of dyspnea and decline in lung function and quality of life in most patients [68]. Sex discrepancies in this disorder have been suggested for some time. The incidence and prevalence of disease have been reported in multiple studies to be higher in males than in females, with ratios ranging from ~1.6:1 to 2:1. Prior reports have also suggested that female sex is associated with better survival [90].

Although our current understanding of the pathogenesis of IPF is incomplete, recent advances have delineated specific clinical and pathologic features. Epithelial cell dysfunction and aberrant epithelial-mesenchymal signaling lead to the activation of fibroblasts and extracellular matrix deposition and remodeling. This chronic activation appears to lead to profibrotic, pathologic changes in IPF fibroblasts. The myofibroblast is the classic pathologic fibroblast phenotype described in IPF lungs. Several mediators, including TGF-β, can elicit the differentiation of fibroblasts to myofibroblasts. Compared with resident lung fibroblasts, myofibroblasts secrete excessive amounts of matrix, including type I collagen. This excess matrix deposition may lead to pathologic lung fibrosis and remodeling [303]. Although these mechanisms have provided significant advances in our understanding of the disease, there is limited information on the molecular basis underlying the observed sex disparities in IPF. A study by Smith et al. suggested that estrogen may modulate the expression of genes involved in chromatin remodeling pathways, as well as the expression of genes in extracellular matrix turnover [247]. However, results from animal studies have provided mixed results. Genome-wide association studies have pointed to genetic influences mediating sex differences, including SNP polymorphisms in mucin 5B, near A-kinase anchoring protein 13, and desmoplakin genes [7].

Sex differences in IPF have been studied in the clinic. Han et al. studied whether the rate of increase in desaturation during serial 6-min walk testing, as well as survival, displayed sex differences. They noted several important observations: (1) males with IPF demonstrate more rapid deterioration in exertional desaturation over time when compared with females; (2) survival was worse in males than females; and (3) better survival for females persisted after additional adjustment for relative change in exertional desaturation and forced vital capacity (FVC) [90].

Among the clinical conditions that have been associated with a worse IPF prognosis is the presence of comorbidities and complications such as emphysema, pulmonary hypertension, cardiovascular diseases, and bronchogenic carcinoma [68]. As mentioned in other sections, some of these conditions also present a sexual dimorphism, which could potentially influence the progression and outcomes of IPF. Finally, prompt treatment of IPF is critical to preserving the patients' lung

function, reducing the risk of acute exacerbations, and improving outcomes [149]. Currently, two drugs are approved for the treatment of IPF in the United States and Europe: nintedanib and pirfenidone. In vitro studies have shown that by inhibiting signaling mediated via tyrosine kinases, nintedanib inhibits fundamental processes of fibrosis, such as the recruitment, proliferation, and differentiation of fibroblasts and fibrocytes and the deposition of extracellular matrix. Data from animal models of fibrosis suggest that nintedanib may also act to normalize the distorted microvascular architecture in the lungs. The mechanism of action of pirfenidone is less well defined, as its target remains unknown, but nonclinical studies suggest that it inhibits profibrotic behaviors in fibroblasts and fibrocytes. Both drugs have been shown to slow the disease progression but not significantly impact mortality [149]. However, studies have not addressed sex differences in the effectiveness of these and other therapies for IPF. Current efforts are directed at identifying key biomarkers that may direct more customized patient-centered healthcare to improve outcomes for these patients in the future, and it is essential that they address sex-specific mechanisms [19].

14.4.6 Lung Cancer

Lung cancer is a major public health problem worldwide and is the world's leading cause of cancer death [21]. Approximately 95% of all lung cancers are classified as either small cell lung cancer (SCLC) or non-small cell lung cancer (NSCLC) [242, 260]. This distinction is essential for staging, treatment, and prognosis. Lung cancer is relatively rare before the fifth decade of life, and risk increases with age thereafter. Over the past decade, the cancer incidence rate (2005–2014) has been found stable in women and declined by approximately 2% annually in men, while the overall cancer death rate (2006–2015) declined by about 1.5% annually in both men and women [165]. Lung adenocarcinoma is also more common among women than men [315].

Environmental risk factors for lung cancer include smoking cigarettes and other tobacco products, as well as exposure to secondhand tobacco smoke, occupational lung carcinogens, radiation, and indoor and outdoor air pollution [4, 132, 201]. However, lung cancer incidence patterns reflect trends in sex behaviors associated with cigarette smoking [220]. Generally speaking, any form of smoking exposure increases the lung cancer risk [4, 311]. A recent US report indicated that lung cancer incidence and death rates among women have increased in 18 states. Interestingly, the states with higher prevalence of smoking among adult women had the highest rates of lung cancer. This report showed that only one state had decreasing lung cancer incidence and death rates in women [110]. Currently, lung cancer incidence rates are declining about twice as fast in men as in women, reflecting historical differences in tobacco uptake and cessation, as well as upturns in female smoking prevalence in some birth cohorts [64]. In addition, the implementation of widespread lung cancer screening holds promise for the future.

Zang et al. found that the odds of developing major lung cancer types are consistently higher for women than for men at every level of exposure to cigarette smoke [315]. This sex difference, however, cannot be explained by differences in baseline exposure, smoking history, or body size, but it is likely due to the higher susceptibility to tobacco carcinogens in women [198, 26]. In this regard, higher levels of polycyclic aromatic hydrocarbon-derived DNA adducts have been reported in female smokers vs. male smokers [177]. A potential mechanism associated with these outcomes is related to the fact that estrogen synergizes with some tobacco compounds through the induction of CYP1B1, an enzyme responsible for estrogenic metabolism, which leads to enhanced reactive oxygen species formation and carcinogenesis [102]. Moreover, Kure et al. found a higher frequency of G:C-->T:A mutations and a higher average hydrophobic DNA adduct level in female patients than males, even though the level of exposure to carcinogens from cigarette smoke

was lower among females than males [131]. These findings lend support to epidemiological evidence that women are at greater risk than men of contracting tobacco-induced lung cancer.

As mentioned earlier, there is considerable evidence indicating that sex hormones can influence respiratory function throughout life [24, 277]. As with other lung diseases, sex hormones have also been implicated in lung cancer [176]. For example, estrogen, known to be a risk factor for the development of adenocarcinoma of the breast, ovary, and endometrium, has been postulated to contribute to lung cancer development and progression [244, 315]. Furthermore, estrogen has also been implicated in lung cancer therapy [13]. Women with advanced NSCLC survive longer than men after adjustment for other prognostic factors in the modern chemotherapy era, suggesting that estrogen levels may interact with the efficacy of current chemotherapy prescriptions or other as yet undefined factors. This finding, if validated, could be potentially exploited in designing new therapies [230, 294].

Regarding estrogen receptors (ERs), Kadota et al. reported that stage I lung adenocarcinoma cells express higher levels of ERα in females than males (19% vs. 14%) and that ERα expression correlates with smaller tumor size. The authors concluded that nuclear ERα expression is an independent predictor of recurrence in pT1a stage lung adenocarcinoma (i.e., tumor size of 2 cm or less) and correlates with poor prognostic immune microenvironments [118]. In addition, non-small cell lung cancer lines (both squamous cell and adenocarcinoma) have been found to express estrogen receptors [192].

Hormone replacement therapy (HRT) is a common treatment used in postmenopausal women. To date, there are several controversies in the relationship between the HRT and lung cancer. The Women's Health Initiative trial concluded that treatment with estrogen plus progestin in postmenopausal women did not increase the incidence of lung cancer. However, HRT was found to increase the number of deaths from lung cancer, in particular deaths from non-small cell lung cancer [47]. These findings should be incorporated into risk-benefit discussions with women considering combined hormone therapy, especially those with a high risk of lung cancer.

In summary, there is accumulating evidence to support the notion that the risk of development of lung cancer is different among women than among men. As expressed earlier, women may be more susceptible to the effects of carcinogens in tobacco and tobacco smoke as a result of hormonal, genetic, and metabolic differences between the sexes. Thus, the significance of sex as a separate contributing factor shall be considered in prognosis and therapeutic management.

14.4.7 Lymphangioleiomyomatosis

Lymphangioleiomyomatosis (LAM) is a rare progressive lung disease that occurs almost exclusively in women [309]. The incidence of LAM is estimated to range between 1 and 8 per million women, and the disease mostly affects women of childbearing age [91]. The average age of symptom onset among LAM patients in the United States and Europe ranges between 34 and 37 years of age [52]. LAM is characterized by infiltration of specific dysregulated smooth muscle-like cells (LAM cells) in various organs and tissues, including lymph nodes, kidneys, and the lungs. As a result, LAM patients experience a progressive decline in lung function due to parenchymal destruction and development of cysts in lung tissue.

The mechanisms underlying LAM development, and the marked sex disparity in its incidence, have not been fully elucidated [89]. However, the neoplastic phenotype of LAM cells is known to occur as a consequence of constitutive activation of the mechanistic target of rapamycin (mTOR) due to loss of heterozygosity in the tuberous sclerosis genes (TSC1 or TSC2) [97]. Advances in the understanding of TSC biology have provided critical clues to LAM pathogenesis and treatment and led to the use of the mTOR inhibitor sirolimus (i.e., rapamycin) as an effective suppressive therapy. Alternatively, lung transplantation is also an established option for women with severe pulmonary impairment due to LAM.

The striking sex disparity observed in LAM has led multiple investigators to consider a role of sex hormones, and specifically estrogen, in the development, progression, and severity of LAM disease [63]. LAM clinical presentation occurs after puberty, accelerated progression is frequently observed during pregnancy, and menopause is associated with attenuated progression [154]. Animal models and in vitro studies have also shown that estrogen increases cell proliferation and migration [313]. Moreover, LAM cells are known to express both estrogen and progesterone receptors. However, definitive evidence is lacking regarding manipulating sex hormones as a potential therapeutic approach, and additional efforts are needed to develop strategies for disease prevention and treatment.

14.4.8 Obstructive Sleep Apnea

The prevalence of OSA is similar between the sexes before puberty but becomes more common in boys than girls after puberty [216]. This sexual dimorphism persists throughout adulthood, where both the rate and severity of OSA are higher in men compared to women. These differences have been attributed to anatomical differences in the upper airway and increased accumulation in the neck of fluid displaced from the legs during recumbency while sleeping [122, 155]. Other risk factors include craniofacial abnormalities, genetic conditions, and neuromuscular disorders. Studies in hypogonadal men and obese adolescents with low testosterone levels have suggested a role of male sex hormones in the observed sex differences in OSA [164, 172]. In females, progesterone has been found to increase the tone of upper airway muscles and stimulate ventilation via chemoreceptor responses to hypoxia and hypercapnia [203, 238]. These sex hormone-mediated mechanisms have been proposed to contribute to the lower risk and severity of OSA observed in girls and women after puberty.

14.4.9 Pulmonary Arterial Hypertension

Pulmonary arterial hypertension (PAH) is a progressive and devastating disease of the pulmonary vasculature characterized by extreme elevation of pulmonary arterial pressure and subsequent right ventricular failure [72]. PAH is also characterized by progressive obstruction of the pulmonary arterial circulation due to formation of vaso-occlusive lesions arising from vigorous proliferation and migration of endothelial cells [214]. As a result of the increased pulmonary vascular resistance, higher right ventricular (RV) afterload causes adaptive RV hypertrophy that often progresses to maladaptive RV hypertrophy and fibrosis, leading to eventual premature death from RV failure. Despite improvements in the diagnosis and management of PAH, the disease continues to have a poor prognosis. A recent analysis showed that the 5-year survival for PAH is approximately 60% [169].

Accumulating evidence shows that more females than men are diagnosed with PAH; however, epidemiological data show that survival among females is better than among males, especially in older patients [12, 104, 108, 135, 159]. Interestingly, the survival benefit for females appears to decline with age [240] and correlate with declines in estradiol levels [285]. This discrepancy in incidence and disease outcomes in men and women is commonly referred to as the PAH "estrogen paradox" and has prompted research into the sex-based differences and hormonal regulation mechanisms in PAH. While the mechanisms behind the sex disparity are far from being understood, they likely involve contributions of genetics, as well as sex hormones and their metabolites.

A few studies have suggested a genetic component in PAH. Mutations in the gene encoding for the bone morphogenetic protein (BMP) receptor type 2 (BMPR2) have been shown to increase PAH penetrance and severity in mouse models [69]. Moreover, mutations in the *BMPR2* gene are the most common genetic cause of familial PAH [65, 148]. By using the "four core

genotypes" mouse model [57], it was found that the Y chromosome, and specifically upregulation of Y chromosomal genes in the lung, was protective against pulmonary vascular remodeling [281] irrespective of gonadal sex. More recent studies have investigated whether genes encoding for enzymes that mediate estrogen metabolism, such as CYP1B1, are associated with PAH in males and females. West et al. found that CYP1B1 expression was markedly downregulated in female but not male patients with PAH due to BMPR2 mutations [301].

Regarding sex hormones, there are multiple studies demonstrating that estrogens exert complex and context-dependent effects on the pulmonary vasculature [62, 133, 168, 208, 308]. Microarray analyses in animal models identified diverse set of pathways regulated by estradiol, including steroid metabolism, immune response, cytoskeletal function, extracellular matrix composition, bone morphogenetic protein (BMP), Notch, Wnt, and calcium signaling [73]. Studies in vascular cells have shown that estradiol affects proliferation [151, 152, 302]. In a rescue approach experimental animal model, it was shown that estradiol treatment reversed pulmonary vascular remodeling, fibrosis, and inflammatory signaling [282]. Overall, while some studies in animals have demonstrated that both exogenous and endogenous estradiol can be protective against PAH, others have suggested a more causative role [123, 134, 151]. Collectively, these studies demonstrate that both endogenous and exogenous estradiol can act as potent regulators of pulmonary vascular homeostasis and greatly impact the progression or resolution of vascular injury. However, these models do not display sex differences nor point to a female predisposition, indicating that more research is needed to fully understand the roles played by hormones in PAH in men and women. Accumulating evidence indicates that estrogen metabolites can also modulate PAH pathogenesis [273]. Thus, it is important to consider the role of metabolites when investigating the effects of estrogen in PAH. Interestingly, low levels of dehydroepiandrosterone (DHEA), a precursor for estrogens that can bind estrogen receptors, are associated with PA development in men [286]. A recent study in postmenopausal women also showed that women with PAH had lower levels of DHEA and higher levels of estradiol than those without cardiopulmonary disease [14]. In patients with PAH, low DHEA and high estradiol were also associated with worse prognosis and increased risk for death [14], as well as fluctuations in pulmonary function throughout the menstrual cycle [15]. Whether DHEA is a marker or mediator of PAH remains under investigation. Overall, more research is needed to understand the mechanisms mediating sex differences in PAH, in order to develop sex-specific therapies to prevent and treat this devastating disease.

14.4.10 Respiratory Infection

Respiratory infection remains a leading cause of morbidity and mortality across all age groups. While, overall, males are disadvantaged in the occurrence and severity of lower respiratory tract infections, females appear to be more susceptible to upper respiratory infections [67]. Multiple studies have suggested that a complex interplay of genetics, sex hormones, host immunity, anatomical and physiological differences, as well as sociocultural and behavioral is likely to underlie the observed sex differences in infection rates and severity [25, 45, 106, 117]. In the following sections, we describe the epidemiology and current knowledge on respiratory diseases that present with a sexual dimorphism.

14.4.10.1 Respiratory Syncytial Virus
During infancy and early childhood, infection with respiratory syncytial virus (RSV) occurs more frequently in boys than girls, especially those born prematurely [150]. Resulting from RSV infection, bronchiolitis is also more frequent and severe in male infants and young children and is often associated with higher risk of wheezing and childhood asthma, as well as higher risk of hospitalization [95, 183].

Sex differences in RSV infection and bronchiolitis have been attributed to anatomical and immunological factors, including smaller airway diameter in males than females [40], and sex differences in the Th2/Th17 response to viral infection [128, 124]. Animal mouse models of RSV infection show that infected female mice display better viral control than males, via mechanisms involving interferon-β expression. In addition, male mice show persistent immune alterations in Th2/Th17 cells, dendritic cells, and ILC2 responses that result in delayed control of viremia [156]. Similar studies have indicated that sex hormones and their receptors can also mediate these mechanisms, although their contributions to infant and pediatric infectious disease remain unclear [116].

14.4.10.2 Influenza
Influenza is an acute respiratory infectious disease caused by several types of influenza viruses. According to the World Health Organization, there are 3 to 5 million cases annually of severe illness and about 290,000 to 650,000 respiratory deaths [305]. The severity and mortality of influenza disease are worse for young children, the elderly, individuals with chronic and immunocompromised medical conditions, and pregnant women [179].

Researchers have reported sex differences in influenza severity, mortality, vaccine tolerance, responses, and outcomes [127]. Interestingly, males are more susceptible to infection than females, and females have greater immune responses but experience more adverse reactions to influenza vaccines than males [128, 290]. In addition, females of reproductive age have the worst outcome during pandemic influenza [304]. However, the causes and mechanisms for these discrepancies in susceptibility are not well-known. Research groups have reported that immunity to viruses can vary with changes in hormone concentrations caused by fluctuations over the menstrual cycle, contraception use, pregnancy, and menopause [32].

Most experiments using murine models have shown that young adult females develop greater respiratory inflammatory responses and have a more severe outcome from influenza infection than males, despite the sexes having similar virus titers [99, 222, 223]. For instance, proinflammatory cytokines (e.g., TNFα, IFNγ, IL6, and IL12) and chemokines (e.g., CCL2, CCL5, and CCL12) are higher in the lungs of influenza-infected females when compared to males [99]. It was also discovered that increased levels of testosterone and amphiregulin, which is an epidermal growth factor that mediates lung tissue repair, improve repair and recovery of lung damage in males [288]. Moreover, infection of female mice of reproductive age with influenza decreases ovarian function and levels of sex hormones suggesting that inhibition of sex hormones may contribute to severe outcomes in female mice [223, 287]. Independent research groups discovered that female mice with influenza that were treated with estrogen showed a decrease in the inflammatory response (e.g., CCL2, IFNγ, TNFα) and an increase in antibody response to influenza vaccine [43, 291]. Importantly, the expression of toll-like receptor-7 is higher in B cells from vaccinated females than males, and its deletion decreased sex differences in vaccine-induced antibody responses and protection [70]. Future research should focus on the molecular mechanisms that regulate how hormones and genes affect immunity to influenza and vaccines in males vs. females.

14.4.10.3 Coronavirus Disease 2019

The coronavirus disease 2019 (COVID-19) is a public health crisis caused by the novel severe acute respiratory syndrome coronavirus 2 (SARS-CoV-2). As of this writing, there have been over 112 million confirmed COVID-19 cases and 2.5 million deaths worldwide. Importantly, demographic and clinical data gathered by multiple health agencies around the globe have demonstrated profound sex differences in COVID-19 outcomes [206]. While the rate of SARS-CoV-2 infection is similar between males and females, male patients infected with the SARS-CoV-2 virus have a significantly higher risk of developing severe COVID-19, being

admitted to an intensive care unit (ICU), and dying when compared to female-infected patients [126].

As mentioned earlier, sex-specific immune responses to a diverse array of viral pathogens have been reported [31, 81, 200, 232, 290]. In addition, there are also prominent sex differences in the immune responses mounted by individuals receiving viral vaccines [127, 158]. In the case of infection with coronaviruses, there have been reports of sex differences during prior outbreaks, including the 2003 severe acute respiratory syndrome (SARS) and Middle East respiratory syndrome (MERS) epidemics, which had a higher case fatality rate and number of deaths in males than females [5, 121, 140].

While not all countries provide sex-disaggregated data, the Sex, Gender and COVID-19 Project [84] has combined efforts from agencies located in several continents to increase reporting of data by sex for confirmed cases, testing, hospitalizations, ICU admissions, confirmed cases among healthcare workers, and deaths. In almost all countries, a significant male predominance in COVID-19 morbidity and mortality has been reported, suggesting a biological mechanism involved [234]. In the United States, most of the states have made public sex-disaggregated data on COVID-19 morbidity and mortality. In an article published in June of 2020, Klein et al. reported that in states providing sex-disaggregated information, data shows that men are twice as likely to die from COVID-19 than women [126]. Moreover, sex differences in the immune response to SARS-CoV-2 have also been reported, where males with mild disease had higher plasma levels of pro-inflammatory cytokines and chemokines than females, but females had higher CD4 and CD8 T-cell activation than males [265]. A study comparing responses to convalescent plasma also showed higher microneutralization and IgG responses to SARS-CoV-2 in males than to females, which correlated with worse COVID-19 outcomes [125].

Some of these sex effects have been attributed to chromosomal differences, since the X chromosome has been shown to express a large number of immune-related genes, including some involved in cytokine and toll-like receptor (TLR) signaling, NF-kB signaling, and MAPK signaling [251]. In addition, the gene encoding the human angiotensin-converting enzyme 2 (ACE2), which serves as the receptor for the spike (S) protein of SARS-CoV-2 [46] is also expressed in the X chromosome and can escape X inactivation and be expressed from both the active and inactive X chromosome [39]. This has been shown to lead to sex differences in ACE2 gene expression [80, 142, 279], which has potential consequences for the vulnerability to SARS-CoV-2.

As with other inflammatory lung diseases and infectious processes, a role of sex hormones has been postulated in mediating sex differences in COVID-19 [250, 259, 266]. Estrogen can regulate the expression of SARS-CoV-2 viral entry receptors, including ACE2 and the transmembrane protease, serine 2 (TMPRSS2) [100, 153]. In this context, a recent report showed that post-pubertal females have lower levels of serum ACE2 when compared to age-matched males [262]. Furthermore, the serum activity of ACE2 is higher in postmenopausal women when compared to younger women, suggesting a regulation by sex hormones like estrogen [79]. Interestingly, Stelzig et al. recently showed that estrogen can downregulate the expression of ACE2 in normal human bronchial epithelial (NHBE) cells but had no effect on TMPRSS2 [256]. This correlates with prior work conducted in the four core genotypes model indicating that sex differences in enzymatic activity of ACE2 in mice are estrogen-dependent and sex chromosome-independent [146].

Regarding male sex hormones, it is unclear whether androgen levels contribute to SARS-CoV-2 or COVID-19 outcomes [178]. A recent report showed that in males with SARS-CoV-2 pneumonia, low testosterone levels were associated with higher rates of ICU admission and death [215]. This correlates with prior studies showing that testosterone can upregulate IL-1 and downregulate IL-1β, IL-6, and TNF-α leading to a suppression of inflammation [173, 205].

Future studies investigating the effects of androgen levels on COVID-19 should consider the timing of the androgen measurement in the course of the SARS-CoV-2 infection [259]. Testosterone can also regulate the expression of TMPRSS2 [82, 257], thus contributing to viral infection and disease outcomes. Interestingly, TMPRSS2 is also highly expressed in urogenital organs, such as the prostate [44].

Finally, it has been hypothesized that gender factors, i.e., smoking habits, handwashing, caregiver gender roles, etc., can influence the outcome of SARS-CoV-2 infections [34, 79, 283, 300]. There are also significant sex and gender differences in comorbidities that have been associated with COVID-19 progression and outcomes [241]. In general, these comorbidities tend to be more prevalent in men than women [306]. Thus, several structural gender health disparities will need to be addressed in order to effectively mitigate the negative effects of the COVID-19 pandemic [250].

14.5 Conclusion

Gender and sex differences in the prevalence, severity, and susceptibility to a variety of lung diseases have been reported across the life span (Fig. 14.1). While the causes of these disparities have not been fully elucidated, a lot has been accomplished in the past few decades. These investigations have revealed associations of biological factors (sex) such as airway anatomy and physiology, chromosomal contributions, genetics and epigenetics, and sex hormones with lung disease onset and outcomes in men and women. Others have shown that sociocultural and environmental factors (gender) can also influence differential outcomes in lung disease. Understanding the contributions of sex and gender, as well as their complex interplay in the context of respiratory health, represents a fundamental step toward precision medicine and the future development of more effective options to prevent and treat lung disease.

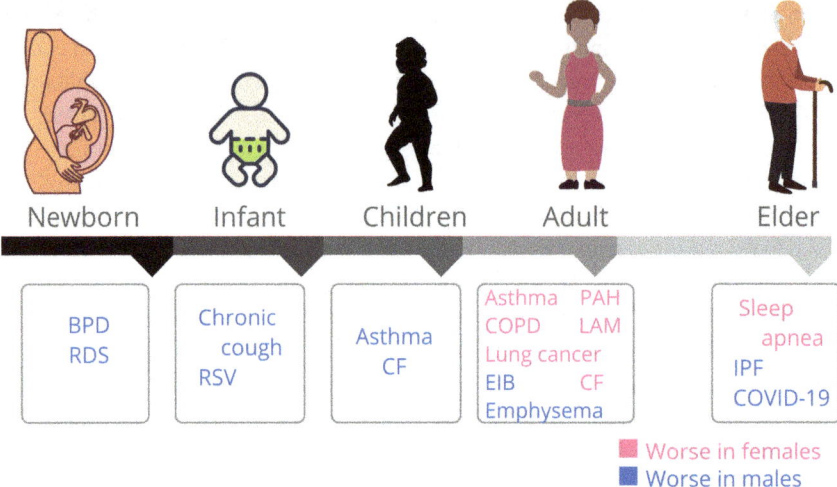

Fig. 14.1 Sex differences in lung disease progression across life span. There are sex differences in the prevalence of several lung diseases across the life span. In pink are lung diseases that are more prevalent in females than males (blue). (Abbreviations: BPD Bronchopulmonary dysplasia, RDS Respiratory distress syndrome, RSV Respiratory syncytial virus, CF Cystic fibrosis, PAH Pulmonary arterial hypertension, COPD Chronic obstructive pulmonary disease, LAM Lymphangioleiomyomatosis, EIB Exercise-induced bronchoconstriction, IPF Idiopathic pulmonary fibrosis, COVID-19 Coronavirus disease 2019)

Acknowledgments This research was supported by the National Heart, Lung, and Blood Institute of the National Institutes of Health under awards number HL133520 and HL141618 (PS).

References

1. Abid S, Xie S, Bose M, Shaul PW, Terada LS, Brody SL, Thomas PJ, Katzenellenbogen JA, Kim SH, Greenberg DE, Jain R. 17β-estradiol dysregulates innate immune responses to Pseudomonas aeruginosa respiratory infection and is modulated by estrogen receptor antagonism. Infect Immun. 2017; https://doi.org/10.1128/IAI.00422-17.
2. Akinbami LJ, Liu X. Chronic obstructive pulmonary disease among adults aged 18 and over in the United States, 1998-2009. NCHS Data Brief. 2011;1(63):1–8.
3. Akinbami LJ, Moorman JE, Bailey C, Zahran HS, King M, Johnson CA, Liu X. Trends in asthma prevalence, health care use, and mortality in the United States, 2001-2010. NCHS Data Brief. 2012;1(94):1–8.
4. Alberg AJ, Brock MV, Ford JG, Samet JM, Spivack SD. Epidemiology of lung cancer: Diagnosis and management of lung cancer, 3rd ed: American College of Chest Physicians evidence-based clinical practice guidelines. Chest. 2013;143(5 Suppl):e1S–e29S. https://doi.org/10.1378/chest.12-2345.
5. Alghamdi IG, Hussain II, Almalki SS, Alghamdi MS, Alghamdi MM, El-Sheemy MA. The pattern of Middle East respiratory syndrome coronavirus in Saudi Arabia: a descriptive epidemiological analysis of data from the Saudi Ministry of Health. Int J Gen Med. 2014;7:417–23. https://doi.org/10.2147/IJGM.S67061.
6. Ali K, Greenough A. Long-term respiratory outcome of babies born prematurely. Ther Adv Respir Dis. 2012;6(2):115–20. https://doi.org/10.1177/1753465812436803.
7. Allen RJ, Porte J, Braybrooke R, Flores C, Fingerlin TE, Oldham JM, Guillen-Guio B, Ma SF, Okamoto T, John AE, Obeidat M, Yang IV, Henry A, Hubbard RB, Navaratnam V, Saini G, Thompson N, Booth HL, Hart SP, Hill MR, Hirani N, Maher TM, McAnulty RJ, Millar AB, Molyneaux PL, Parfrey H, Rassl DM, Whyte MKB, Fahy WA, Marshall RP, Oballa E, Bosse Y, Nickle DC, Sin DD, Timens W, Shrine N, Sayers I, Hall IP, Noth I, Schwartz DA, Tobin MD, Wain LV, Jenkins RG. Genetic variants associated with susceptibility to idiopathic pulmonary fibrosis in people of European ancestry: a genome-wide association study. Lancet Respir Med. 2017;5(11):869–80. https://doi.org/10.1016/s2213-2600(17)30387-9.
8. Anderson SD, Schoeffel RE, Black JL, Daviskas E. Airway cooling as the stimulus to exercise-induced asthma--a re-evaluation. Eur J Respir Dis. 1985;67(1):20–30.
9. Angus DC, Linde-Zwirble WT, Clermont G, Griffin MF, Clark RH. Epidemiology of neonatal respiratory failure in the United States: projections from California and New York. Am J Respir Crit Care Med. 2001;164(7):1154–60. https://doi.org/10.1164/ajrccm.164.7.2012126.

10. Aryal S, Diaz-Guzman E, Mannino DM. COPD and gender differences: an update. Transl Res. 2013;162(4):208–18. https://doi.org/10.1016/j.trsl.2013.04.003.

11. Aryal S, Diaz-Guzman E, Mannino DM. Influence of sex on chronic obstructive pulmonary disease risk and treatment outcomes. Int J Chron Obstruct Pulmon Dis. 2014;9:1145–54. https://doi.org/10.2147/COPD.S54476.

12. Badesch DB, Raskob GE, Elliott CG, Krichman AM, Farber HW, Frost AE, Barst RJ, Benza RL, Liou TG, Turner M, Giles S, Feldkircher K, Miller DP, McGoon MD. Pulmonary arterial hypertension: baseline characteristics from the REVEAL Registry. Chest. 2010;137(2):376–87. https://doi.org/10.1378/chest.09-1140.

13. Baik CS, Eaton KD. Estrogen signaling in lung cancer: an opportunity for novel therapy. Cancers (Basel). 2012;4(4):969–88. https://doi.org/10.3390/cancers4040969.

14. Baird GL, Archer-Chicko C, Barr RG, Bluemke DA, Foderaro AE, Fritz JS, Hill NS, Kawut SM, Klinger JR, Lima JAC, Mullin CJ, Ouyang P, Palevsky HI, Palmisicano AJ, Pinder D, Preston IR, Roberts KE, Smith KA, Walsh T, Whittenhall M, Ventetuolo CE. Lower DHEA-S levels predict disease and worse outcomes in post-menopausal women with idiopathic, connective tissue disease- and congenital heart disease-associated pulmonary arterial hypertension. Eur Respir J. 2018 Jun 28;51(6):1800467. https://doi.org/10.1183/13993003.00467-2018. PMID: 29954925; PMCID: PMC6469347.

15. Baird GL, Walsh T, Aliotta J, Allahua M, Andrew R, Bourjeily G, Brodsky AS, Denver N, Dooner M, Harrington EO, Klinger JR, MacLean MR, Mullin CJ, Pereira M, Poppas A, Whittenhall M, Ventetuolo CE. Insights from the Menstrual Cycle in Pulmonary Arterial Hypertension. Ann Am Thorac Soc. 2021 Feb;18(2):218–228. https://doi.org/10.1513/AnnalsATS.202006-671OC. PMID: 32885987; PMCID: PMC7869782.

16. Bancalari E, Jain D. Bronchopulmonary dysplasia: 50 years after the original description. Neonatology. 2019;115(4):384–91. https://doi.org/10.1159/000497422.

17. Bancalari EH, Jobe AH. The respiratory course of extremely preterm infants: a dilemma for diagnosis and terminology. J Pediatr. 2012;161(4):585–8. https://doi.org/10.1016/j.jpeds.2012.05.054.

18. Barnes PJ. Sex differences in chronic obstructive pulmonary disease mechanisms. Am J Respir Crit Care Med. 2016;193(8):813–4. https://doi.org/10.1164/rccm.201512-2379ED.

19. Barratt SL, Creamer A, Hayton C, Chaudhuri N. Idiopathic Pulmonary Fibrosis (IPF): an overview. J Clin Med. 2018;7(8) https://doi.org/10.3390/jcm7080201.

20. Barre SF, Haberthur D, Cremona TP, Stampanoni M, Schittny JC. The total number of acini remains constant throughout postnatal rat lung devel-

opment. Am J Physiol Lung Cell Mol Physiol. 2016;311(6):L1082–9. https://doi.org/10.1152/ajplung.00325.2016.

21. Barta JA, Powell CA, Wisnivesky JP. Global epidemiology of lung cancer. Ann Glob Health. 2019;85(1) https://doi.org/10.5334/aogh.2419.

22. Becerra-Díaz M, Strickland AB, Keselman A, Heller NM. Androgen and androgen receptor as enhancers of M2 macrophage polarization in allergic lung inflammation. J Immunol. 2018; https://doi.org/10.4049/jimmunol.1800352.

23. Becklake MR, Kauffmann F. Gender differences in airway behaviour over the human life span. Thorax. 1999;54(12):1119–38.

24. Behan M, Wenninger JM. Sex steroidal hormones and respiratory control. Respir Physiol Neurobiol. 2008;164(1-2):213–21. https://doi.org/10.1016/j.resp.2008.06.006.

25. Ben-Shmuel A, Sheiner E, Wainstock T, Landau D, Vaknin F, Walfisch A. The association between gender and pediatric respiratory morbidity. Pediatr Pulmonol. 2018;53(9):1225–30. https://doi.org/10.1002/ppul.24083.

26. Ben-Zaken Cohen S, Paré PD, Man SF, Sin DD. The growing burden of chronic obstructive pulmonary disease and lung cancer in women: examining sex differences in cigarette smoke metabolism. Am J Respir Crit Care Med. 2007;176(2):113–120. doi:200611-1655PP. https://doi.org/10.1164/rccm.200611-1655PP.

27. Benscoter DT. Bronchiectasis, chronic suppurative lung disease and protracted bacterial bronchitis. Curr Probl Pediatr Adolesc Health Care. 2018;48(4):119–23. https://doi.org/10.1016/j.cppeds.2018.03.003.

28. Binet ME, Bujold E, Lefebvre F, Tremblay Y, Piedboeuf B, Canadian Neonatal N. Role of gender in morbidity and mortality of extremely premature neonates. Am J Perinatol. 2012;29(3):159–66. https://doi.org/10.1055/s-0031-1284225.

29. Boas SR, Cleary DA, Lee PA, Orenstein DM. Salivary testosterone levels in male adolescents with cystic fibrosis. Pediatrics. 1996;97(3):361–3.

30. Boezen HM, Jansen DF, Postma DS. Sex and gender differences in lung development and their clinical significance. Clin Chest Med. 2004;25(2):237–45. https://doi.org/10.1016/j.ccm.2004.01.012.

31. Bongen E, Lucian H, Khatri A, Fragiadakis GK, Bjornson ZB, Nolan GP, Utz PJ, Khatri P. Sex differences in the blood transcriptome identify robust changes in immune cell proportions with aging and influenza infection. Cell Rep. 2019;29(7):1961–1973 e1964. https://doi.org/10.1016/j.celrep.2019.10.019.

32. Brabin L. Interactions of the female hormonal environment, susceptibility to viral infections, and disease progression. AIDS Patient Care STDS. 2002;16(5):211–21. https://doi.org/10.1089/10872910252972267.

33. Brenner BE, Holmes TM, Mazal B, Camargo CA. Relation between phase of the menstrual cycle and asthma presentations in the emergency depart-

ment. Thorax. 2005;60(10):806–9. https://doi.
org/10.1136/thx.2004.033928.

34. Cai H. Sex difference and smoking predisposi-
tion in patients with COVID-19. Lancet Respir
Med. 2020;8(4):e20. https://doi.org/10.1016/
S2213-2600(20)30117-X.

35. Canguven O, Albayrak S. Do low testosterone
levels contribute to the pathogenesis of asthma?
Med Hypotheses. 2011;76(4):585–8. https://doi.
org/10.1016/j.mehy.2011.01.006.

36. Carey MA, Card JW, Voltz JW, Arbes SJ, Germolec
DR, Korach KS, Zeldin DC. It's all about sex:
gender, lung development and lung disease.
Trends Endocrinol Metab. 2007a;18(8):308–
313. doi:S1043-2760(07)00132-4. https://doi.
org/10.1016/j.tem.2007.08.003.

37. Carey MA, Card JW, Voltz JW, Germolec DR,
Korach KS, Zeldin DC. The impact of sex and sex
hormones on lung physiology and disease: lessons
from animal studies. Am J Physiol Lung Cell Mol
Physiol. 2007b;293(2):L272–278. doi:00174.2007.
https://doi.org/10.1152/ajplung.00174.2007.

38. Carlo WA, Stark AR, Wright LL, Tyson JE, Papile
LA, Shankaran S, Donovan EF, Oh W, Bauer
CR, Saha S, Poole WK, Stoll B. Minimal ventilation to prevent bronchopulmonary dysplasia in
extremely-low-birth-weight infants. J Pediatr.
2002;141(3):370–4.

39. Carrel L, Willard HF. X-inactivation profile reveals
extensive variability in X-linked gene expression in
females. Nature. 2005;434(7031):400–4. https://doi.
org/10.1038/nature03479.

40. Carvajal JJ, Avellaneda AM, Salazar-Ardiles C,
Maya JE, Kalergis AM, Lay MK. Host components
contributing to respiratory syncytial virus pathogenesis. Front Immunol. 2019;10:2152. https://doi.
org/10.3389/fimmu.2019.02152.

41. Catlin EA, Powell SM, Manganaro TF, Hudson PL,
Ragin RC, Epstein J, Donahoe PK. Sex-specific
fetal lung development and müllerian inhibiting substance. Am Rev Respir Dis. 1990;141(2):466–70.
https://doi.org/10.1164/ajrccm/141.2.466.

42. Centers for Disease Control and Prevention. Most
recent asthma data. 2020. https://www.cdc.gov/
asthma/most_recent_data.htm.

43. Celestino I, Checconi P, Amatore D, De Angelis
M, Coluccio P, Dattilo R, Alunni Fegatelli D,
Clemente AM, Matarrese P, Torcia MG, Mancinelli
R, Mammola CL, Garaci E, Vestri AR, Malorni W,
Palamara AT, Nencioni L. Differential redox state
contributes to sex disparities in the response to influ-
enza virus infection in male and female mice. Front
Immunol. 2018;9:1747. https://doi.org/10.3389/
fimmu.2018.01747.

44. Chakravarty D, Nair SS, Hammouda N, Ratnani
P, Gharib Y, Wagaskar V, Mohamed N, Lundon D,
Dovey Z, Kyprianou N, Tewari AK. Sex differences
in SARS-CoV-2 infection rates and the potential link
to prostate cancer. Commun Biol. 2020;3(1):374.
https://doi.org/10.1038/s42003-020-1088-9.

45. Chamekh M, Deny M, Romano M, Lefèvre N,
Corazza F, Duchateau J, Casimir G. Differential susceptibility to infectious respiratory diseases between
males and females linked to sex-specific innate
immune inflammatory response. Front Immunol.
2017;8:1806.

46. Chen Y, Guo Y, Pan Y, Zhao ZJ. Structure analysis
of the receptor binding of 2019-nCoV. Biochem
Biophys Res Commun. 2020; https://doi.
org/10.1016/j.bbrc.2020.02.071.

47. Chlebowski RT, Schwartz AG, Wakelee H, Anderson
GL, Stefanick ML, Manson JE, Rodabough RJ,
Chien JW, Wactawski-Wende J, Gass M, Kotchen
JM, Johnson KC, O'Sullivan MJ, Ockene JK, Chen
C, Hubbell FA, Investigators WsHI. Oestrogen
plus progestin and lung cancer in postmenopausal
women (Women's Health Initiative trial): a post-hoc
analysis of a randomised controlled trial. Lancet.
2009;374(9697):1243–51. https://doi.org/10.1016/
S0140-6736(09)61526-9.

48. Chotirmall SH, Greene CM, McElvaney
NG. Immune, inflammatory and infectious consequences of estrogen in women with cystic fibrosis.
Expert Rev Respir Med. 2012;6(6):573–5. https://
doi.org/10.1586/ers.12.59.

49. Coarfa C, Zhang Y, Maity S, Perera DN, Jiang W,
Wang L, Couroucli X, Moorthy B, Lingappan
K. Sexual dimorphism of the pulmonary transcriptome in neonatal hyperoxic lung injury: identification of angiogenesis as a key pathway. Am J Physiol
Lung Cell Mol Physiol. 2017;313(6):L991–L1005.
https://doi.org/10.1152/ajplung.00230.2017.

50. Cogen J, Emerson J, Sanders DB, Ren C, Schechter
MS, Gibson RL, Morgan W, Rosenfeld M, Group
ES. Risk factors for lung function decline in a
large cohort of young cystic fibrosis patients.
Pediatr Pulmonol. 2015;50(8):763–70. https://doi.
org/10.1002/ppul.23217.

51. Colebatch HJ, Greaves IA, Ng CK. Exponential
analysis of elastic recoil and aging in healthy
males and females. J Appl Physiol Respir Environ
Exerc Physiol. 1979;47(4):683–91. https://doi.
org/10.1152/jappl.1979.47.4.683.

52. Collaborators GCRD. Global, regional, and
national deaths, prevalence, disability-adjusted
life years, and years lived with disability for
chronic obstructive pulmonary disease and asthma,
1990-2015: a systematic analysis for the Global
Burden of Disease Study 2015. Lancet Respir
Med. 2017;5(9):691–706. https://doi.org/10.1016/
S2213-2600(17)30293-X.

53. Collaborators GCRD. Prevalence and attribut-
able health burden of chronic respiratory diseases,
1990-2017: a systematic analysis for the Global
Burden of Disease Study 2017. Lancet Respir
Med. 2020;8(6):585–96. https://doi.org/10.1016/
S2213-2600(20)30105-3.

54. Courtney JM, Bradley J, Mccaughan J, O'Connor
TM, Shortt C, Bredin CP, Bradbury I, Elborn
JS. Predictors of mortality in adults with cystic fibro-

sis. Pediatr Pulmonol. 2007;42(6):525–32. https://doi.org/10.1002/ppul.20619.

55. Davidson LM, Berkelhamer SK. Bronchopulmonary dysplasia: chronic lung disease of infancy and long-term pulmonary outcomes. J Clin Med. 2017;6(1) https://doi.org/10.3390/jcm6010004.

56. de Torres JP, Cote CG, Lopez MV, Casanova C, Diaz O, Marin JM, Pinto-Plata V, de Oca MM, Nekach H, Dordelly LJ, Aguirre-Jaime A, Celli BR. Sex differences in mortality in patients with COPD. Eur Respir J. 2009;33(3):528–35. https://doi.org/10.1183/09031936.00096108.

57. De Vries GJ, Rissman EF, Simerly RB, Yang L-Y, Scordalakes EM, Auger CJ, Swain A, Lovell-Badge R, Burgoyne PS, Arnold AP. A model system for study of sex chromosome effects on sexually dimorphic neural and behavioral traits. J Neurosci. 2002;22(20):9005–14. https://doi.org/10.1523/JNEUROSCI.22-20-09005.2002.

58. Deal EC, McFadden ER, Ingram RH, Strauss RH, Jaeger JJ. Role of respiratory heat exchange in production of exercise-induced asthma. J Appl Physiol Respir Environ Exerc Physiol. 1979;46(3):467–75. https://doi.org/10.1152/jappl.1979.46.3.467.

59. Demko CA, Byard PJ, Davis PB. Gender differences in cystic fibrosis: Pseudomonas aeruginosa infection. J Clin Epidemiol. 1995;48(8):1041–9. https://doi.org/10.1016/0895-4356(94)00230-n.

60. Diab Cáceres L, Girón Moreno RM, García Castillo E, Pastor Sanz MT, Olveira C, García Clemente M, Nieto Royo R, Prados Sánchez C, Caballero Sánchez P, Olivera Serrano MJ, Padilla Galo A, Nava Tomas E, Esteban Peris A, Fernández Velilla M, Torres MI, Ancochea Bermúdez J. Effect of sex differences on computed tomography findings in adults with cystic fibrosis: a multicenter study. Arch Bronconeumol. 2020; https://doi.org/10.1016/j.arbres.2019.12.028.

61. Dodd KE, Wood J, Mazurek JM. Mortality among persons with both asthma and chronic obstructive pulmonary disease aged ≥25 years, by industry and occupation - United States, 1999-2016. MMWR Morb Mortal Wkly Rep. 2020;69(22):670–9. https://doi.org/10.15585/mmwr.mm6922a3.

62. Earley S, Resta TC. Estradiol attenuates hypoxia-induced pulmonary endothelin-1 gene expression. Am J Physiol Lung Cell Mol Physiol. 2002;283(1):L86–93. https://doi.org/10.1152/ajplung.00476.2001.

63. El-Chemaly S, Henske EP. The next breakthrough in LAM clinical trials may be their design: challenges in design and execution of future LAM clinical trials. Expert Rev Respir Med. 2015;9(2):195–204. https://doi.org/10.1586/17476348.2015.1024663.

64. Escobedo LG, Peddicord JP. Smoking prevalence in US birth cohorts: the influence of gender and education. Am J Public Health. 1996;86(2):231–6. https://doi.org/10.2105/ajph.86.2.231.

65. Evans JD, Girerd B, Montani D, Wang XJ, Galiè N, Austin ED, Elliott G, Asano K, Grünig E, Yan Y, Jing ZC, Manes A, Palazzini M, Wheeler LA, Nakayama I, Satoh T, Eichstaedt C, Hinderhofer K, Wolf M, Rosenzweig EB, Chung WK, Soubrier F, Simonneau G, Sitbon O, Gräf S, Kaptoge S, Di Angelantonio E, Humbert M, Morrell NW. BMPR2 mutations and survival in pulmonary arterial hypertension: an individual participant data meta-analysis. Lancet Respir Med. 2016;4(2):129–37. https://doi.org/10.1016/s2213-2600(15)00544-5.

66. Falagas ME, Mourtzoukou EG, Vardakas KZ. Sex differences in the incidence and severity of respiratory tract infections. Respir Med. 2007a;101(9):1845–63.

67. Falagas ME, Mourtzoukou EG, Vardakas KZ. Sex differences in the incidence and severity of respiratory tract infections. Respir Med. 2007b;101(9):1845–63. https://doi.org/10.1016/j.rmed.2007.04.011.

68. Fernández Fabrellas E, Peris Sánchez R, Sabater Abad C, Juan Samper G. Prognosis and follow-up of idiopathic pulmonary fibrosis. Med Sci (Basel). 2018;6(2) https://doi.org/10.3390/medsci6020051.

69. Fessel JP, Chen X, Frump A, Gladson S, Blackwell T, Kang C, Johnson J, Loyd JE, Hemnes A, Austin E, West J. Interaction between bone morphogenetic protein receptor type 2 and estrogenic compounds in pulmonary arterial hypertension. Pulm Circ. 2013;3(3):564–77. https://doi.org/10.1086/674312.

70. Fink AL, Engle K, Ursin RL, Tang WY, Klein SL. Biological sex affects vaccine efficacy and protection against influenza in mice. Proc Natl Acad Sci U S A. 2018;115(49):12477–82. https://doi.org/10.1073/pnas.1805268115.

71. Fleisher B, Kulovich MV, Hallman M, Gluck L. Lung profile: sex differences in normal pregnancy. Obstet Gynecol. 1985;66(3):327–30.

72. Foderaro A, Ventetuolo CE. Pulmonary arterial hypertension and the sex hormone paradox. Curr Hypertens Rep. 2016;18(11):84. https://doi.org/10.1007/s11906-016-0689-7.

73. Frump AL, Albrecht ME, McClintick JN, Lahm T. Estrogen receptor-dependent attenuation of hypoxia-induced changes in the lung genome of pulmonary hypertension rats. Pulm Circ. 2017;7(1):232–43. https://doi.org/10.1177/2045893217702055.

74. Fuchs O, Genuneit J, Latzin P, Büchele G, Horak E, Loss G, Sozanska B, Weber J, Boznanski A, Heederik D, Braun-Fahrländer C, Frey U, von Mutius E; GABRIELA Study Group. Farming environments and childhood atopy, wheeze, lung function, and exhaled nitric oxide. J Allergy Clin Immunol. 2012 Aug;130(2):382–8.e6. https://doi.org/10.1016/j.jaci.2012.04.049. Epub 2012 Jun 28. PMID: 22748700.

75. Fuchs O, Bahmer T, Weckmann M, Dittrich AM, Schaub B, Rösler B, Happle C, Brinkmann F, Ricklefs I, König IR, Watz H, Rabe KF, Kopp MV, Hansen G, von Mutius E, (DZL) ASGapotGCfLR. The all age asthma cohort (ALLIANCE) - from early beginnings to chronic disease: a longitudinal cohort study. BMC Pulm Med. 2018;18(1):140. https://doi.org/10.1186/s12890-018-0705-6.

76. Fuentes N, Nicoleau M, Cabello N, Montes D, Zomorodi N, Chroneos ZC, Silveyra P. 17b-estradiol affects lung function and inflammation following ozone exposure in a sex-specific manner. Am J Physiol Lung Cell Mol Physiol. 2019; https://doi.org/10.1152/ajplung.00176.2019.

77. Fuentes N, Silveyra P. Endocrine regulation of lung disease and inflammation. Exp Biol Med (Maywood). 2018;1535370218816653 https://doi.org/10.1177/1535370218816653.

78. Fuseini H, Newcomb DC. Mechanisms driving gender differences in asthma. Curr Allergy Asthma Rep. 2017;17(3):19. https://doi.org/10.1007/s11882-017-0686-1.

79. Gebhard C, Regitz-Zagrosek V, Neuhauser HK, Morgan R, Klein SL. Impact of sex and gender on COVID-19 outcomes in Europe. Biol Sex Differ. 2020;11(1):29. https://doi.org/10.1186/s13293-020-00304-9.

80. Gemmati D, Bramanti B, Serino ML, Secchiero P, Zauli G, Tisato V. COVID-19 and individual genetic susceptibility/receptivity: role of ACE1/ACE2 genes, immunity, inflammation and coagulation. Might the double X-chromosome in females be protective against SARS-CoV-2 compared to the single X-chromosome in males? Int J Mol Sci. 2020;21(10) https://doi.org/10.3390/ijms21103474.

81. Ghosh S, Klein RS. Sex drives dimorphic immune responses to viral infections. J Immunol. 2017;198(5):1782–90. https://doi.org/10.4049/jimmunol.1601166.

82. Giagulli VA, Guastamacchia E, Magrone T, Jirillo E, Lisco G, De Pergola G, Triggiani V. Worse progression of COVID-19 in men: is testosterone a key factor? Andrology. 2020; https://doi.org/10.1111/andr.12836.

83. GINA. 2019.. Retrieved from: https://ginasthma.org/wp-content/uploads/2019/06/GINA-2019-main-report-June-2019-wms.pdf. Accessed 24 Sept 2020.

84. Global Health 50/50. The COVID-19 sex-disaggregated data tracker. 2020. https://globalhealth5050.org/the-sex-gender-and-covid-19-project/. Accessed 25 Sept 2020

85. Gortner L, Shen J, Tutdibi E. Sexual dimorphism of neonatal lung development. Klin Padiatr. 2013;225(2):64–9. https://doi.org/10.1055/s-0033-1333758.

86. Guarnieri M, Balmes JR. Outdoor air pollution and asthma. Lancet. 2014 May 3;383(9928):1581–92. https://doi.org/10.1016/S0140-6736(14)60617-6. PMID: 24792855; PMCID: PMC4465283.

87. Gyamfi-Bannerman C, Thom EA, Blackwell SC, Tita AT, Reddy UM, Saade GR, Rouse DJ, McKenna DS, Clark EA, Thorp JM Jr, Chien EK, Peaceman AM, Gibbs RS, Swamy GK, Norton ME, Casey BM, Caritis SN, Tolosa JE, Sorokin Y, VanDorsten JP, Jain L, Network NM-FMU. Antenatal Betamethasone for women at risk for late preterm delivery. N Engl J Med. 2016;374(14):1311–20. https://doi.org/10.1056/NEJMoa1516783.

88. Hafkamp-de Groen E, Lingsma HF, Caudri D, Levie D, Wijga A, Koppelman GH, Duijts L, Jaddoe VW, Smit HA, Kerkhof M, Moll HA, Hofman A, Steyerberg EW, de Jongste JC, Raat H. Predicting asthma in preschool children with asthma-like symptoms: validating and updating the PIAMA risk score. J Allergy Clin Immunol. 2013 Dec;132(6):1303–10. https://doi.org/10.1016/j.jaci.2013.07.007. Epub 2013 Aug 26. PMID: 23987795.

89. Han MK, Arteaga-Solis E, Blenis J, Bourjeily G, Clegg DJ, DeMeo D, Duffy J, Gaston B, Heller NM, Hemnes A, Henske EP, Jain R, Lahm T, Lancaster LH, Lee J, Legato MJ, McKee S, Mehra R, Morris A, Prakash YS, Stampfli MR, Gopal-Srivastava R, Laposky AD, Punturieri A, Reineck L, Tigno X, Clayton J. Female sex and gender in lung/sleep health and disease. Increased understanding of basic biological, pathophysiological, and behavioral mechanisms leading to better health for female patients with lung disease. Am J Respir Crit Care Med. 2018;198(7):850–8. https://doi.org/10.1164/rccm.201801-0168WS.

90. Han MK, Murray S, Fell CD, Flaherty KR, Toews GB, Myers J, Colby TV, Travis WD, Kazerooni EA, Gross BH, Martinez FJ. Sex differences in physiological progression of idiopathic pulmonary fibrosis. Eur Respir J. 2008;31(6):1183–8. https://doi.org/10.1183/09031936.00165207.

91. Harknett EC, Chang WY, Byrnes S, Johnson J, Lazor R, Cohen MM, Gray B, Geiling S, Telford H, Tattersfield AE, Hubbard RB, Johnson SR. Use of variability in national and regional data to estimate the prevalence of lymphangioleiomyomatosis. QJM. 2011;104(11):971–9. https://doi.org/10.1093/qjmed/hcr116.

92. Harness-Brumley CL, Elliott AC, Rosenbluth DB, Raghavan D, Jain R. Gender differences in outcomes of patients with cystic fibrosis. J Womens Health (Larchmt). 2014;23(12):1012–20. https://doi.org/10.1089/jwh.2014.4985.

93. Harris C, Zivanovic S, Lunt A, Calvert S, Bisquera A, Marlow N, Peacock JL, Greenough A. Lung function and respiratory outcomes in teenage boys and girls born very prematurely. Pediatr Pulmonol. 2020;55(3):682–9. https://doi.org/10.1002/ppul.24631.

94. Haselkorn T, Szefler SJ, Simons FE, Zeiger RS, Mink DR, Chipps BE, Borish L, Wong DA; TENOR Study Group. Allergy, total serum immunoglobulin E, and airflow in children and adolescents in TENOR. Pediatr Allergy Immunol. 2010 Dec;21(8):1157–65. https://doi.org/10.1111/j.1399-3038.2010.01065.x. PMID: 20444153.

95. Henderson J, Hilliard TN, Sherriff A, Stalker D, Al Shammari N, Thomas HM. Hospitalization for RSV bronchiolitis before 12 months of age and subsequent asthma, atopy and wheeze: a longitudinal birth cohort study. Pediatr Allergy Immunol. 2005;16(5):386–92. https://doi.org/10.1111/j.1399-3038.2005.00298.x.

96. Henderson-Smart DJ, Hutchinson JL, Donoghue DA, Evans NJ, Simpson JM, Wright I, Network

AaNZN. Prenatal predictors of chronic lung disease in very preterm infants. Arch Dis Child Fetal Neonatal Ed. 2006;91(1):F40–5. https://doi.org/10.1136/adc.2005.072264.

97. Henske EP, McCormack FX. Lymphangioleiomyomatosis - a wolf in sheep's clothing. J Clin Invest. 2012;122(11):3807–16. https://doi.org/10.1172/JCI58709.

98. Hepper PG, Shannon EA, Dornan JC. Sex differences in fetal mouth movements. Lancet. 1997;350(9094):1820. https://doi.org/10.1016/S0140-6736(05)63635-5.

99. Hoffmann J, Otte A, Thiele S, Lotter H, Shu Y, Gabriel G. Sex differences in H7N9 influenza A virus pathogenesis. Vaccine. 2015;33(49):6949–54. https://doi.org/10.1016/j.vaccine.2015.08.044.

100. Hoffmann M, Kleine-Weber H, Schroeder S, Kruger N, Herrler T, Erichsen S, Schiergens TS, Herrler G, Wu NH, Nitsche A, Muller MA, Drosten C, Pohlmann S. SARS-CoV-2 cell entry depends on ACE2 and TMPRSS2 and is blocked by a clinically proven protease inhibitor. Cell. 2020; https://doi.org/10.1016/j.cell.2020.02.052.

101. Holtrop M, Heltshe S, Shabanova V, Keller A, Schumacher L, Fernandez L, Jain R. A Prospective Study of the Effects of Sex Hormones on Lung Function and Inflammation in Women with Cystic Fibrosis. Ann Am Thorac Soc. 2021 Feb 5. https://doi.org/10.1513/AnnalsATS.202008-1064OC. Epub ahead of print. PMID: 33544657.

102. Hsu LH, Chu NM, Kao SH. Estrogen, Estrogen Receptor and Lung Cancer. Int J Mol Sci. 2017;18(8):1713. Published 2017 Aug 5. https://doi.org/10.3390/ijms18081713

103. Hubert D, Réglier-Poupet H, Sermet-Gaudelus I, Ferroni A, Le Bourgeois M, Burgel PR, Serreau R, Dusser D, Poyart C, Coste J. Association between Staphylococcus aureus alone or combined with Pseudomonas aeruginosa and the clinical condition of patients with cystic fibrosis. J Cyst Fibros. 2013;12(5):497–503. https://doi.org/10.1016/j.jcf.2012.12.003.

104. Humbert M, Sitbon O, Chaouat A, Bertocchi M, Habib G, Gressin V, Yaici A, Weitzenblum E, Cordier JF, Chabot F, Dromer C, Pison C, Reynaud-Gaubert M, Haloun A, Laurent M, Hachulla E, Simonneau G. Pulmonary arterial hypertension in France: results from a national registry. Am J Respir Crit Care Med. 2006;173(9):1023–30. https://doi.org/10.1164/rccm.200510-1668OC.

105. Hunninghake GM, Soto-Quirós ME, Avila L, Kim HP, Lasky-Su J, Rafaels N, Ruczinski I, Beaty TH, Mathias RA, Barnes KC, Wilk JB, O'Connor GT, Gauderman WJ, Vora H, Baurley JW, Gilliland F, Liang C, Sylvia JS, Klanderman BJ, Sharma SS, Himes BE, Bossley CJ, Israel E, Raby BA, Bush A, Choi AM, Weiss ST, Celedón JC. TSLP polymorphisms are associated with asthma in a sex-specific fashion. Allergy. 2010 Dec;65(12):1566–75. https://doi.org/10.1111/j.1398-9995.2010.02415.x. PMID: 20560908; PMCID: PMC2970693.

106. Ingersoll MA. Sex differences shape the response to infectious diseases. PLoS Pathog. 2017;13(12):e1006688. https://doi.org/10.1371/journal.ppat.1006688.

107. Ishak N, Sozo F, Harding R, De Matteo R. Does lung development differ in male and female fetuses? Exp Lung Res. 2014;40(1):30–9. https://doi.org/10.3109/01902148.2013.858197.

108. Jacobs W, van de Veerdonk MC, Trip P, de Man F, Heymans MW, Marcus JT, Kawut SM, Bogaard HJ, Boonstra A, Vonk Noordegraaf A. The right ventricle explains sex differences in survival in idiopathic pulmonary arterial hypertension. Chest. 2014;145(6):1230–6. https://doi.org/10.1378/chest.13-1291.

109. Jain R, Ray JM, Pan JH, Brody SL. Sex hormone-dependent regulation of cilia beat frequency in airway epithelium. Am J Respir Cell Mol Biol. 2012;46(4):446–53. https://doi.org/10.1165/rcmb.2011-0107OC.

110. Jemal A, Thun MJ, Ries LA, Howe HL, Weir HK, Center MM, Ward E, Wu XC, Eheman C, Anderson R, Ajani UA, Kohler B, Edwards BK. Annual report to the nation on the status of cancer, 1975-2005, featuring trends in lung cancer, tobacco use, and tobacco control. J Natl Cancer Inst. 2008;100(23):1672–94. https://doi.org/10.1093/jnci/djn389.

111. Jenkins CR, Chapman KR, Donohue JF, Roche N, Tsiligianni I, Han MK. Improving the management of COPD in women. Chest. 2017;151(3):686–96. https://doi.org/10.1016/j.chest.2016.10.031.

112. Jensen EA, Schmidt B. Epidemiology of bronchopulmonary dysplasia. Birth Defects Res A Clin Mol Teratol. 2014;100(3):145–57. https://doi.org/10.1002/bdra.23235.

113. Jobe AH, Bancalari E. Bronchopulmonary dysplasia. Am J Respir Crit Care Med. 2001;163(7):1723–9. https://doi.org/10.1164/ajrccm.163.7.2011060.

114. Johannesson M, Lúdvíksdóttir D, Janson C. Lung function changes in relation to menstrual cycle in females with cystic fibrosis. Respir Med. 2000;94(11):1043–6. https://doi.org/10.1053/rmed.2000.0891.

115. Kadel S, Ainsua-Enrich E, Hatipoglu I, Turner S, Singh S, Khan S, Kovats S. A major population of functional KLRG1. Immunohorizons. 2018;2(2):74–86. https://doi.org/10.4049/immunohorizons.1800008.

116. Kadel S, Kovats S. Sex hormones regulate innate immune cells and promote sex differences in respiratory virus infection. Front Immunol. 2018;9:1653. https://doi.org/10.3389/fimmu.2018.01653.

117. Kadioglu A, Cuppone AM, Trappetti C, List T, Spreafico A, Pozzi G, Andrew PW, Oggioni MR. Sex-based differences in susceptibility to respiratory and systemic pneumococcal disease in mice. J Infect Dis. 2011;204(12):1971–9. https://doi.org/10.1093/infdis/jir657.

118. Kadota K, Eguchi T, Villena-Vargas J, Woo KM, Sima CS, Jones DR, Travis WD, Adusumilli PS. Nuclear estrogen receptor-α expression is an

independent predictor of recurrence in male patients with pT1aN0 lung adenocarcinomas, and correlates with regulatory T-cell infiltration. Oncotarget. 2015;6(29):27505–18. https://doi.org/10.18632/oncotarget.4752.

119. Kallings LV, Emtner M, Bäcklund L. Exercise-induced bronchoconstriction in adults with asthma--comparison between running and cycling and between cycling at different air conditions. Ups J Med Sci. 1999;104(3):191–8. https://doi.org/10.3109/03009739909178962.

120. Kari MA, Hallman M, Eronen M, Teramo K, Virtanen M, Koivisto M, Ikonen RS. Prenatal dexamethasone treatment in conjunction with rescue therapy of human surfactant: a randomized placebo-controlled multicenter study. Pediatrics. 1994;93(5):730–6.

121. Karlberg J, Chong DS, Lai WY. Do men have a higher case fatality rate of severe acute respiratory syndrome than women do? Am J Epidemiol. 2004;159(3):229–31. https://doi.org/10.1093/aje/kwh056.

122. Kasai T, Motwani SS, Elias RM, Gabriel JM, Taranto Montemurro L, Yanagisawa N, Spiller N, Paul N, Bradley TD. Influence of rostral fluid shift on upper airway size and mucosal water content. J Clin Sleep Med. 2014;10(10):1069–74. https://doi.org/10.5664/jcsm.4102.

123. Kawut SM, Archer-Chicko CL, DeMichele A, Fritz JS, Klinger JR, Ky B, Palevsky HI, Palmisciano AJ, Patel M, Pinder D, Propert KJ, Smith KA, Stanczyk F, Tracy R, Vaidya A, Whittenhall ME, Ventetuolo CE. Anastrozole in pulmonary arterial hypertension. A randomized, double-blind, Placebo-controlled trial. Am J Respir Crit Care Med. 2017;195(3):360–8. https://doi.org/10.1164/rccm.201605-1024OC.

124. Klein SL. Sex influences immune responses to viruses, and efficacy of prophylaxis and treatments for viral diseases. Bioessays. 2012;34(12):1050-1059. https://doi.org/10.1002/bies.201200099

125. Klein S, Pekosz A, Park HS, Ursin R, Shapiro J, Benner S, Littlefield K, Kumar S, Naik HM, Betenbaugh M, Shrestha R, Wu A, Hughes R, Burgess I, Caturegli P, Laeyendecker O, Quinn T, Sullivan D, Shoham S, Redd A, Bloch E, Casadevall A, Tobian A. Sex, age, and hospitalization drive antibody responses in a COVID-19 convalescent plasma donor population. medRxiv. 2020a; https://doi.org/10.1101/2020.06.26.20139063.

126. Klein SL, Dhakal S, Ursin RL, Deshpande S, Sandberg K, Mauvais-Jarvis F. Biological sex impacts COVID-19 outcomes. PLoS Pathog. 2020b;16(6):e1008570. https://doi.org/10.1371/journal.ppat.1008570.

127. Klein SL, Flanagan KL. Sex differences in immune responses. Nat Rev Immunol. 2016;16(10):626–38. https://doi.org/10.1038/nri.2016.90.

128. Klein SL, Hodgson A, Robinson DP. Mechanisms of sex disparities in influenza pathogenesis. J Leukoc Biol. 2012;92(1):67–73. https://doi.org/10.1189/jlb.0811427.

129. Koper I, Hufnagl K, Ehmann R. Gender aspects and influence of hormones on bronchial asthma - Secondary publication and update. World Allergy Organ J. 2017;10(1):46. https://doi.org/10.1186/s40413-017-0177-9.

130. Kosti RI, Priftis KN, Anthracopoulos MB, Papadimitriou A, Grigoropoulou D, Lentzas Y, Yfanti K, Panagiotakos DB. The association between leisure-time physical activities and asthma symptoms among 10- to 12-year-old children: the effect of living environment in the PANACEA study. J Asthma. 2012 May;49(4):342–8. https://doi.org/10.3109/02770903.2011.652328. Epub 2012 Feb 2. PMID: 22300140.

131. Kure EH, Ryberg D, Hewer A, Phillips DH, Skaug V, Baera R, Haugen A. p53 mutations in lung tumours: relationship to gender and lung DNA adduct levels. Carcinogenesis. 1996 Oct;17(10):2201–5. https://doi.org/10.1093/carcin/17.10.2201. PMID: 8895489.

132. Kurt OK, Zhang J, Pinkerton KE. Pulmonary health effects of air pollution. Curr Opin Pulm Med. 2016;22(2):138–43. https://doi.org/10.1097/MCP.0000000000000248.

133. Lahm T, Albrecht M, Fisher AJ, Selej M, Patel NG, Brown JA, Justice MJ, Brown MB, Van Demark M, Trulock KM, Dieudonne D, Reddy JG, Presson RG, Petrache I. 17β-Estradiol attenuates hypoxic pulmonary hypertension via estrogen receptor-mediated effects. Am J Respir Crit Care Med. 2012;185(9):965–80. https://doi.org/10.1164/rccm.201107-1293OC.

134. Lahm T, Frump AL. Toward harnessing sex steroid signaling as a therapeutic target in pulmonary arterial hypertension. Am J Respir Crit Care Med. 2017;195(3):284–6. https://doi.org/10.1164/rccm.201609-1906ED.

135. Lahm T, Tuder RM, Petrache I. Progress in solving the sex hormone paradox in pulmonary hypertension. Am J Physiol Lung Cell Mol Physiol. 2014;307(1):L7–26. https://doi.org/10.1152/ajplung.00337.2013.

136. Lai HC, Kosorok MR, Laxova A, Makholm LM, Farrell PM. Delayed diagnosis of US females with cystic fibrosis. Am J Epidemiol. 2002;156(2):165–73. https://doi.org/10.1093/aje/kwf014.

137. Lamprecht B, Soriano JB, Studnicka M, Kaiser B, Vanfleteren LE, Gnatiuc L, Burney P, Miravitlles M, García-Rio F, Akbari K, Ancochea J, Menezes AM, Perez-Padilla R, Montes de Oca M, Torres-Duque CA, Caballero A, González-García M, Buist S, BOLD Collaborative Research Group tE-ST, the PLATINO Team, and the PREPOCOL Study Group. Determinants of underdiagnosis of COPD in National and International Surveys. Chest. 2015;148(4):971–85. https://doi.org/10.1378/chest.14-2535.

138. Leary S, Das P, Ponnalagu D, Singh H, Bhandari V. Genetic strain and sex differences in a hyperoxia-induced mouse model of varying sever-

ity of bronchopulmonary dysplasia. Am J Pathol. 2019;189(5):999–1014. https://doi.org/10.1016/j.ajpath.2019.01.014.

139. Leifke E, Friemert M, Heilmann M, Puvogel N, Smaczny C, von zur Muhlen A, Brabant G. Sex steroids and body composition in men with cystic fibrosis. Eur J Endocrinol. 2003;148(5):551–7. https://doi.org/10.1530/eje.0.1480551.

140. Leong HN, Earnest A, Lim HH, Chin CF, Tan C, Puhaindran ME, Tan A, Chen MI, Leo YS. SARS in Singapore--predictors of disease severity. Ann Acad Med Singapore. 2006;35(5):326–31.

141. Li X, Chen Q, Zheng X, Li Y, Han M, Liu T, Xiao J, Guo L, Zeng W, Zhang J, Ma W. Effects of ambient ozone concentrations with different averaging times on asthma exacerbations: a meta-analysis. Sci Total Environ. 2019;691:549–61. https://doi.org/10.1016/j.scitotenv.2019.06.382.

142. Li Y, Jerkic M, Slutsky AS, Zhang H. Molecular mechanisms of sex bias differences in COVID-19 mortality. Crit Care. 2020;24(1):405. https://doi.org/10.1186/s13054-020-03118-8.

143. Licari A, Castagnoli R, Brambilla I, Marseglia A, Tosca MA, Marseglia GL, Ciprandi G. Asthma endotyping and biomarkers in childhood asthma. Pediatr Allergy Immunol Pulmonol. 2018;31(2):44–55. https://doi.org/10.1089/ped.2018.0886.

144. Lin CM, Davidson TM, Ancoli-Israel S. Gender differences in obstructive sleep apnea and treatment implications. Sleep Med Rev. 2008;12(6):481–96. https://doi.org/10.1016/j.smrv.2007.11.003.

145. Liptzin DR, Landau LI, Taussig LM. Sex and the lung: observations, hypotheses, and future directions. Pediatr Pulmonol. 2015;50(12):1159–69. https://doi.org/10.1002/ppul.23178.

146. Liu J, Ji H, Zheng W, Wu X, Zhu JJ, Arnold AP, Sandberg K. Sex differences in renal angiotensin converting enzyme 2 (ACE2) activity are 17beta-oestradiol-dependent and sex chromosome-independent. Biol Sex Differ. 2010;1(1):6. https://doi.org/10.1186/2042-6410-1-6.

147. Liu Y, Pan J, Zhang H, Shi C, Li G, Peng Z, Ma J, Zhou Y, Zhang L. Short-term exposure to ambient air pollution and asthma mortality. Am J Respir Crit Care Med. 2019;200(1):24–32. https://doi.org/10.1164/rccm.201810-1823OC.

148. Machado RD, Eickelberg O, Elliott CG, Geraci MW, Hanaoka M, Loyd JE, Newman JH, Phillips JA, Soubrier F, Trembath RC, Chung WK. Genetics and genomics of pulmonary arterial hypertension. J Am Coll Cardiol. 2009;54(1 Supplement):S32. https://doi.org/10.1016/j.jacc.2009.04.015.

149. Maher TM, Strek ME. Antifibrotic therapy for idiopathic pulmonary fibrosis: time to treat. Respir Res. 2019;20(1):205. https://doi.org/10.1186/s12931-019-1161-4.

150. Mahmoud O, Granell R, Tilling K, Minelli C, Garcia-Aymerich J, Holloway JW, Custovic A, Jarvis D, Sterne J, Henderson J. Association of height growth in puberty with lung function: a lon-

gitudinal study. Am J Respir Crit Care Med. 2018; https://doi.org/10.1164/rccm.201802-0274OC.

151. Mair KM, Wright AF, Duggan N, Rowlands DJ, Hussey MJ, Roberts S, Fullerton J, Nilsen M, Loughlin L, Thomas M, MacLean MR. Sex-dependent influence of endogenous estrogen in pulmonary hypertension. Am J Respir Crit Care Med. 2014;190(4):456–67. https://doi.org/10.1164/rccm.201403-0483OC.

152. Mair KM, Yang XD, Long L, White K, Wallace E, Ewart M-A, Docherty CK, Morrell NW, MacLean MR. Sex affects bone morphogenetic protein type II receptor signaling in pulmonary artery smooth muscle cells. Am J Respir Crit Care Med. 2015;191(6):693–703. https://doi.org/10.1164/rccm.201410-1802OC.

153. Majdic G. Could sex/gender differences in ACE2 Expression in the lungs contribute to the large gender disparity in the morbidity and mortality of patients infected with the SARS-CoV-2 virus? Front Cell Infect Microbiol. 2020;10:327. https://doi.org/10.3389/fcimb.2020.00327.

154. Makrigiannis AP, Anderson SK. The murine Ly49 family: form and function. Arch Immunol Ther Exp (Warsz). 2001;49(1):47–50.

155. Malhotra A, Huang Y, Fogel RB, Pillar G, Edwards JK, Kikinis R, Loring SH, White DP. The male predisposition to pharyngeal collapse: importance of airway length. Am J Respir Crit Care Med. 2002;166(10):1388–95. https://doi.org/10.1164/rccm.2112072.

156. Malinczak CA, Fonseca W, Rasky AJ, Ptaschinski C, Morris S, Ziegler SF, Lukacs NW. Sex-associated TSLP-induced immune alterations following early-life RSV infection leads to enhanced allergic disease. Mucosal Immunol. 2019;12(4):969–79. https://doi.org/10.1038/s41385-019-0171-3.

157. Mannino DM, Buist AS. Global burden of COPD: risk factors, prevalence, and future trends. Lancet. 2007;370(9589):765–73. https://doi.org/10.1016/S0140-6736(07)61380-4.

158. Markle JG, Fish EN. SeXX matters in immunity. Trends Immunol. 2014;35(3):97–104. https://doi.org/10.1016/j.it.2013.10.006.

159. Marra AM, Benjamin N, Eichstaedt C, Salzano A, Arcopinto M, Gargani L, M DA, Argiento P, Falsetti L, Di Giosia P, Isidori AM, Ferrara F, Bossone E, Cittadini A, Grünig E. Gender-related differences in pulmonary arterial hypertension targeted drugs administration. Pharmacol Res. 2016;114:103–9. https://doi.org/10.1016/j.phrs.2016.10.018.

160. Martinez FJ, Curtis JL, Sciurba F, Mumford J, Giardino ND, Weinmann G, Kazerooni E, Murray S, Criner GJ, Sin DD, Hogg J, Ries AL, Han M, Fishman AP, Make B, Hoffman EA, Mohsenifar Z, Wise R, Group NETTR. Sex differences in severe pulmonary emphysema. Am J Respir Crit Care Med. 2007;176(3):243–52. https://doi.org/10.1164/rccm.200606-828OC.

161. Maselli JH, Sontag MK, Norris JM, MacKenzie T, Wagener JS, Accurso FJ. Risk factors for initial acquisition of Pseudomonas aeruginosa in children with cystic fibrosis identified by newborn screening. Pediatr Pulmonol. 2003;35(4):257–62. https://doi.org/10.1002/ppul.10230.

162. Masoli M, Fabian D, Holt S, Beasley R; Global Initiative for Asthma (GINA) Program. The global burden of asthma: executive summary of the GINA Dissemination Committee report. Allergy. 2004;59(5):469–78. https://doi.org/10.1111/j.1398-9995.2004.00526.x. PMID: 15080825.

163. Massaro GD, Mortola JP, Massaro D. Sexual dimorphism in the architecture of the lung's gas-exchange region. Proc Natl Acad Sci U S A. 1995;92(4):1105–7.

164. Matsumoto AM, Sandblom RE, Schoene RB, Lee KA, Giblin EC, Pierson DJ, Bremner WJ. Testosterone replacement in hypogonadal men: effects on obstructive sleep apnoea, respiratory drives, and sleep. Clin Endocrinol (Oxf). 1985;22(6):713–21. https://doi.org/10.1111/j.1365-2265.1985.tb00161.x.

165. Mattiuzzi C, Lippi G. Current cancer epidemiology. J Epidemiol Glob Health. 2019;9(4):217–22. https://doi.org/10.2991/jegh.k.191008.001.

166. McFadden ER. Hypothesis: exercise-induced asthma as a vascular phenomenon. Lancet. 1990;335(8694):880–3. https://doi.org/10.1016/0140-6736(90)90478-n.

167. McFadden ER, Ingram RH. Exercise-induced asthma: observations on the initiating stimulus. N Engl J Med. 1979;301(14):763–9. https://doi.org/10.1056/NEJM197910043011406.

168. McMurtry IF, Frith CH, Will DH. Cardiopulmonary responses of male and female swine to simulated high altitude. J Appl Physiol. 1973;35(4):459–62. https://doi.org/10.1152/jappl.1973.35.4.459.

169. Medrek S, Sahay S, Zhao C, Selej M, Frost A. Impact of race on survival in pulmonary arterial hypertension: results from the REVEAL registry. J Heart Lung Transplant. 2020;39(4):321–30. https://doi.org/10.1016/j.healun.2019.11.024.

170. Mehari A, Valle O, Gillum RF. Trends in pulmonary hypertension mortality and morbidity. Pulm Med. 2014;2014:105864. https://doi.org/10.1155/2014/105864.

171. Mehta G, Macek M, Mehta A, Group ERW. Cystic fibrosis across Europe: EuroCareCF analysis of demographic data from 35 countries. J Cyst Fibros. 2010;9(Suppl 2):S5–S21. https://doi.org/10.1016/j.jcf.2010.08.002.

172. Mogri M, Dhindsa S, Quattrin T, Ghanim H, Dandona P. Testosterone concentrations in young pubertal and post-pubertal obese males. Clin Endocrinol (Oxf). 2013;78(4):593–9. https://doi.org/10.1111/cen.12018.

173. Mohamad NV, Wong SK, Wan Hasan WN, Jolly JJ, Nur-Farhana MF, Ima-Nirwana S, Chin KY. The relationship between circulating testosterone and inflammatory cytokines in men. Aging Male. 2019;22(2):129–40. https://doi.org/10.1080/13685538.2018.1482487.

174. Mokhtar HM, Giribabu N, Muniandy S, Salleh N. Testosterone decreases the expression of endometrial pinopode and L-selectin ligand (MECA-79) in adult female rats during uterine receptivity period. Int J Clin Exp Pathol. 2014;7(5):1967–76.

175. Molis MA, Molis WE. Exercise-induced bronchospasm. Sports Health. 2010;2(4):311–7. https://doi.org/10.1177/1941738110373735.

176. Mollerup S, Jørgensen K, Berge G, Haugen A. Expression of estrogen receptors alpha and beta in human lung tissue and cell lines. Lung Cancer. 2002;37(2):153–9.

177. Mollerup S, Berge G, Baera R, Skaug V, Hewer A, Phillips DH, Stangeland L, Haugen A. Sex differences in risk of lung cancer: Expression of genes in the PAH bioactivation pathway in relation to smoking and bulky DNA adducts. Int J Cancer. 2006 Aug 15;119(4):741–4. https://doi.org/10.1002/ijc.21891. PMID: 16557573.

178. Moradi F, Enjezab B, Ghadiri-Anari A. The role of androgens in COVID-19. Diabetes Metab Syndr. 2020;14(6):2003–2006. https://doi.org/10.1016/j.dsx.2020.10.014

179. Morgan R, Klein SL. The intersection of sex and gender in the treatment of influenza. Curr Opin Virol. 2019;35:35–41. https://doi.org/10.1016/j.coviro.2019.02.009.

180. Naeem A, Silveyra P. Sex differences in paediatric and adult asthma. Eur Med J (Chelmsf). 2019;4(2):27–35.

181. Naeye RL, Burt LS, Wright DL, Blanc WA, Tatter D. Neonatal mortality, the male disadvantage. Pediatrics. 1971;48(6):902–6.

182. Nair H, Nokes DJ, Gessner BD, Dherani M, Madhi SA, Singleton RJ, O'Brien KL, Roca A, Wright PF, Bruce N, Chandran A, Theodoratou E, Sutanto A, Sedyaningsih ER, Ngama M, Munywoki PK, Kartasasmita C, Simões EA, Rudan I, Weber MW, Campbell H. Global burden of acute lower respiratory infections due to respiratory syncytial virus in young children: a systematic review and meta-analysis. Lancet. 2010;375(9725):1545–55. https://doi.org/10.1016/S0140-6736(10)60206-1.

183. Nair H, Simoes EA, Rudan I, Gessner BD, Azziz-Baumgartner E, Zhang JSF, Feikin DR, Mackenzie GA, Moiisi JC, Roca A, Baggett HC, Zaman SM, Singleton RJ, Lucero MG, Chandran A, Gentile A, Cohen C, Krishnan A, Bhutta ZA, Arguedas A, Clara AW, Andrade AL, Ope M, Ruvinsky RO, Hortal M, McCracken JP, Madhi SA, Bruce N, Qazi SA, Morris SS, El Arifeen S, Weber MW, Scott JAG, Brooks WA, Breiman RF, Campbell H, Severe Acute Lower Respiratory Infections Working G. Global and regional burden of hospital admissions for severe acute lower respiratory infections in young children in 2010: a systematic analysis. Lancet. 2013;381(9875):1380–90. https://doi.org/10.1016/S0140-6736(12)61901-1.

184. Nevšímalová S. Sleep and sleep-related disorders in women. Cas Lek Cesk. 2019;158(7-8):321–2.
185. Newcomb D. Chapter 4. Sex, gender, and asthma. In: Hemnes AH, editor. Gender, sex hormones and respiratory disease. A comprehensive guide; 2016. p. 87–103. https://doi.org/10.1007/978-3-319-23998-9.
186. Newcomb DC, Peebles RS Jr. Th17-mediated inflammation in asthma. Curr Opin Immunol. 2013;25(6):755–60. https://doi.org/10.1016/j.coi.2013.08.002.
187. Nick JA, Chacon CS, Brayshaw SJ, Jones MC, Barboa CM, St Clair CG, Young RL, Nichols DP, Janssen JS, Huitt GA, Iseman MD, Daley CL, Taylor-Cousar JL, Accurso FJ, Saavedra MT, Sontag MK. Effects of gender and age at diagnosis on disease progression in long-term survivors of cystic fibrosis. Am J Respir Crit Care Med. 2010;182(5):614–26. https://doi.org/10.1164/rccm.201001-0092OC.
188. Northoff H, Symons S, Zieker D, Schaible EV, Schäfer K, Thoma S, Löffler M, Abbasi A, Simon P, Niess AM, Fehrenbach E. Gender- and menstrual phase dependent regulation of inflammatory gene expression in response to aerobic exercise. Exerc Immunol Rev. 2008;14:86–103.
189. O'Driscoll DN, Greene CM, Molloy EJ. Immune function? A missing link in the gender disparity in preterm neonatal outcomes. Expert Rev Clin Immunol. 2017;13(11):1061–71. https://doi.org/10.1080/1744666X.2017.1386555.
190. O'Driscoll DN, McGovern M, Greene CM, Molloy EJ. Gender disparities in preterm neonatal outcomes. Acta Paediatr. 2018; https://doi.org/10.1111/apa.14390.
191. Ødegaard I, Stray-Pedersen B, Hallberg K, Haanaes OC, Storrøsten OT, Johannesson M. Prevalence and outcome of pregnancies in Norwegian and Swedish women with cystic fibrosis. Acta Obstet Gynecol Scand. 2002;81(8):693–7.
192. Olak J, Colson Y. Gender differences in lung cancer: have we really come a long way, baby? J Thorac Cardiovasc Surg. 2004;128(3):346–51. https://doi.org/10.1016/j.jtcvs.2004.05.025.
193. Olesen HV, Pressler T, Hjelte L, Mared L, Lindblad A, Knudsen PK, Laerum BN, Johannesson M, Consortium SCFS. Gender differences in the Scandinavian cystic fibrosis population. Pediatr Pulmonol. 2010;45(10):959–65. https://doi.org/10.1002/ppul.21265.
194. Papageorgiou AN, Colle E, Farri-Kostopoulos E, Gelfand MM. Incidence of respiratory distress syndrome following antenatal betamethasone: role of sex, type of delivery, and prolonged rupture of membranes. Pediatrics. 1981;67(5):614–7.
195. Papi A, Brightling C, Pedersen SE, Reddel HK. Asthma. Lancet. 2018;391(10122):783–800. https://doi.org/10.1016/S0140-6736(17)33311-1.
196. Parsons JP, Hallstrand TS, Mastronarde JG, Kaminsky DA, Rundell KW, Hull JH, Storms WW, Weiler JM, Cheek FM, Wilson KC, Anderson SD, Bronchoconstriction ATSSoE-i. An official American Thoracic Society clinical practice guideline: exercise-induced bronchoconstriction. Am J Respir Crit Care Med. 2013;187(9):1016–27. https://doi.org/10.1164/rccm.201303-0437ST.
197. Parsons JP, Kaeding C, Phillips G, Jarjoura D, Wadley G, Mastronarde JG. Prevalence of exercise-induced bronchospasm in a cohort of varsity college athletes. Med Sci Sports Exerc. 2007;39(9):1487–92. https://doi.org/10.1249/mss.0b013e3180986e45.
198. Peng J, Meireles SI, Xu X, Smith WE, Slifker MJ, Riel SL, Zhai S, Zhang G, Ma X, Kurzer MS, Ma GX, Clapper ML. Estrogen metabolism in the human lung: impact of tumorigenesis, smoke, sex and race/ethnicity. Oncotarget. 2017 Nov 1;8(63):106778-106789. https://doi.org/10.18632/oncotarget.22269. PMID: 29290988; PMCID: PMC5739773.
199. Perez TA, Castillo EG, Ancochea J, Pastor Sanz MT, Almagro P, Martínez-Camblor P, Miravitlles M, Rodríguez-Carballeira M, Navarro A, Lamprecht B, Ramírez-García Luna AS, Kaiser B, Alfageme I, Casanova C, Esteban C, Soler-Cataluña JJ, De-Torres JP, Celli BR, Marin JM, Lopez-Campos JL, Riet GT, Sobradillo P, Lange P, Garcia-Aymerich J, Anto JM, Turner AM, Han MK, Langhammer A, Sternberg A, Leivseth L, Bakke P, Johannessen A, Oga T, Cosío B, Echazarreta A, Roche N, Burgel PR, Sin DD, Puhan MA, Soriano JB. Sex differences between women and men with COPD: a new analysis of the 3CIA study. Respir Med. 2020;171:106105. https://doi.org/10.1016/j.rmed.2020.106105.
200. Piasecka B, Duffy D, Urrutia A, Quach H, Patin E, Posseme C, Bergstedt J, Charbit B, Rouilly V, MacPherson CR, Hasan M, Albaud B, Gentien D, Fellay J, Albert ML, Quintana-Murci L, Milieu Interieur C. Distinctive roles of age, sex, and genetics in shaping transcriptional variation of human immune responses to microbial challenges. Proc Natl Acad Sci U S A. 2018;115(3):E488–97. https://doi.org/10.1073/pnas.1714765115.
201. Pope CA, Burnett RT, Thun MJ, Calle EE, Krewski D, Ito K, Thurston GD. Lung cancer, cardiopulmonary mortality, and long-term exposure to fine particulate air pollution. JAMA. 2002;287(9):1132–41. https://doi.org/10.1001/jama.287.9.1132.
202. Pope D, Diaz E, Smith-Sivertsen T, Lie RT, Bakke P, Balmes JR, Smith KR, Bruce NG. Exposure to household air pollution from wood combustion and association with respiratory symptoms and lung function in nonsmoking women: results from the RESPIRE trial, Guatemala. Environ Health Perspect. 2015;123(4):285–92. https://doi.org/10.1289/ehp.1408200.
203. Popovic RM, White DP. Upper airway muscle activity in normal women: influence of hormonal status. J Appl Physiol (1985). 1998;84(3):1055–62. https://doi.org/10.1152/jappl.1998.84.3.1055.
204. Postma DS. Gender differences in asthma development and progression. Gend Med. 2007;4(Suppl B):S133–46.

205. Pozzilli P, Lenzi A. Commentary: testosterone, a key hormone in the context of COVID-19 pandemic. Metabolism. 2020;108:154252. https://doi.org/10.1016/j.metabol.2020.154252.

206. Pradhan A, Olsson PE. Sex differences in severity and mortality from COVID-19: are males more vulnerable? Biol Sex Differ. 2020 Sep 18;11(1):53. https://doi.org/10.1186/s13293-020-00330-7. PMID: 32948238; PMCID: PMC7498997.

207. Provost PR, Simard M, Tremblay Y. A link between lung androgen metabolism and the emergence of mature epithelial type II cells. Am J Respir Crit Care Med. 2004;170(3):296–305. https://doi.org/10.1164/rccm.200312-1680OC.

208. Rabinovitch M, Gamble WJ, Miettinen OS, Reid L. Age and sex influence on pulmonary hypertension of chronic hypoxia and on recovery. Am J Physiol. 1981;240(1):H62–72. https://doi.org/10.1152/ajpheart.1981.240.1.H62.

209. Radkiewicz C, Dickman PW, Johansson ALV, Wagenius G, Edgren G, Lambe M. Sex and survival in non-small cell lung cancer: a nationwide cohort study. PLoS One. 2019;14(6):e0219206. https://doi.org/10.1371/journal.pone.0219206.

210. Raffay TM, Dylag AM, Di Fiore JM, Smith LA, Einisman HJ, Li Y, Lakner MM, Khalil AM, MacFarlane PM, Martin RJ, Gaston B. S-Nitrosoglutathione attenuates airway hyperresponsiveness in murine bronchopulmonary dysplasia. Mol Pharmacol. 2016;90(4):418–26. https://doi.org/10.1124/mol.116.104125.

211. Raghavan D, Jain R. Increasing awareness of sex differences in airway diseases. Respirology. 2016;21(3):449–59. https://doi.org/10.1111/resp.12702.

212. Raghu G, Collard HR, Egan JJ, Martinez FJ, Behr J, Brown KK, Colby TV, Cordier JF, Flaherty KR, Lasky JA, Lynch DA, Ryu JH, Swigris JJ, Wells AU, Ancochea J, Bouros D, Carvalho C, Costabel U, Ebina M, Hansell DM, Johkoh T, Kim DS, King TE, Kondoh Y, Myers J, Müller NL, Nicholson AG, Richeldi L, Selman M, Dudden RF, Griss BS, Protzko SL, Schünemann HJ, Fibrosis AEJACoIP. An official ATS/ERS/JRS/ALAT statement: idiopathic pulmonary fibrosis: evidence-based guidelines for diagnosis and management. Am J Respir Crit Care Med. 2011;183(6):788–824. https://doi.org/10.1164/rccm.2009-040GL.

213. Raghu G, Weycker D, Edelsberg J, Bradford WZ, Oster G. Incidence and prevalence of idiopathic pulmonary fibrosis. Am J Respir Crit Care Med. 2006;174(7):810–6. https://doi.org/10.1164/rccm.200602-163OC.

214. Ranchoux B, Antigny F, Rucker-Martin C, Hautefort A, Péchoux C, Bogaard HJ, Dorfmüller P, Remy S, Lecerf F, Planté S, Chat S, Fadel E, Houssaini A, Anegon I, Adnot S, Simonneau G, Humbert M, Cohen-Kaminsky S, Perros F. Endothelial-to-mesenchymal transition in pulmonary hypertension. Circulation. 2015;131(11):1006–18. https://doi.org/10.1161/circulationaha.114.008750.

215. Rastrelli G, Di Stasi V, Inglese F, Beccaria M, Garuti M, Di Costanzo D, Spreafico F, Greco GF, Cervi G, Pecoriello A, Magini A, Todisco T, Cipriani S, Maseroli E, Corona G, Salonia A, Lenzi A, Maggi M, De Donno G, Vignozzi L. Low testosterone levels predict clinical adverse outcomes in SARS-CoV-2 pneumonia patients. Andrology. 2020; https://doi.org/10.1111/andr.12821.

216. Redline S, Storfer-Isser A, Rosen CL, Johnson NL, Kirchner HL, Emancipator J, Kibler AM. Association between metabolic syndrome and sleep-disordered breathing in adolescents. Am J Respir Crit Care Med. 2007;176(4):401–8. https://doi.org/10.1164/rccm.200703-375OC.

217. Ren CL, Morgan WJ, Konstan MW, Schechter MS, Wagener JS, Fisher KA, Regelmann WE, Fibrosis IaCotESoC. Presence of methicillin resistant Staphylococcus aureus in respiratory cultures from cystic fibrosis patients is associated with lower lung function. Pediatr Pulmonol. 2007;42(6):513–8. https://doi.org/10.1002/ppul.20604.

218. Rho J, Ahn C, Gao A, Sawicki GS, Keller A, Jain R. Disparities in mortality of hispanic patients with cystic fibrosis in the United States. A National and Regional Cohort Study. Am J Respir Crit Care Med. 2018;198(8):1055–63. https://doi.org/10.1164/rccm.201711-2357OC.

219. Rivera L, Siddaiah R, Oji-Mmuo C, Silveyra GR, Silveyra P. Biomarkers for bronchopulmonary dysplasia in the preterm infant. Front Pediatr. 2016;4:33. https://doi.org/10.3389/fped.2016.00033.

220. Rivera MP. Lung cancer in women: differences in epidemiology, biology, histology, and treatment outcomes. Semin Respir Crit Care Med. 2013;34(6):792–801. https://doi.org/10.1055/s-0033-1358550.

221. Roberge S, Lacasse Y, Tapp S, Tremblay Y, Kari A, Liu J, Fekih M, Qublan HS, Amorim MM, Bujold E. Role of fetal sex in the outcome of antenatal glucocorticoid treatment to prevent respiratory distress syndrome: systematic review and meta-analysis. J Obstet Gynaecol Can. 2011;33(3):216–26. https://doi.org/10.1016/s1701-2163(16)34822-8.

222. Robinson DP, Huber SA, Moussawi M, Roberts B, Teuscher C, Watkins R, Arnold AP, Klein SL. Sex chromosome complement contributes to sex differences in coxsackievirus B3 but not influenza A virus pathogenesis. Biol Sex Differ. 2011a;2:8. https://doi.org/10.1186/2042-6410-2-8.

223. Robinson DP, Lorenzo ME, Jian W, Klein SL. Elevated 17β-estradiol protects females from influenza A virus pathogenesis by suppressing inflammatory responses. PLoS Pathog. 2011b;7(7):e1002149. https://doi.org/10.1371/journal.ppat.1002149.

224. Rodriguez Bauza DE, Silveyra P. Sex Differences in Exercise-Induced Bronchoconstriction in Athletes: A Systematic Review and Meta-Analysis. Int J Environ Res Public Health. 2020 Oct 5;17(19):7270. https://doi.org/10.3390/ijerph17197270. PMID: 33027929; PMCID: PMC7579110.

225. Rysavy MA, Horbar JD, Bell EF, Li L, Greenberg LT, Tyson JE, Patel RM, Carlo WA, Younge NE, Green CE, Edwards EM, Hintz SR, Walsh MC, Buzas JS, Das A, Higgins RD, Network EKSNIoCHaHDNRNaVO. Assessment of an Updated Neonatal Research Network extremely preterm birth outcome model in the Vermont Oxford Network. JAMA Pediatr. 2020;174(5):e196294. https://doi.org/10.1001/jamapediatrics.2019.6294.

226. Salam MT, Wenten M, Gilliland FD. Endogenous and exogenous sex steroid hormones and asthma and wheeze in young women. J Allergy Clin Immunol. 2006;117(5):1001–7.   https://doi.org/10.1016/j.jaci.2006.02.004.

227. Salvi SS, Barnes PJ. Chronic obstructive pulmonary disease in non-smokers. Lancet. 2009;374(9691):733–43.   https://doi.org/10.1016/S0140-6736(09)61303-9.

228. Sathish V, Martin YN, Prakash YS. Sex steroid signaling: implications for lung diseases. Pharmacol Ther. 2015;150:94–108. https://doi.org/10.1016/j.pharmthera.2015.01.007.

229. Savi D, Simmonds N, Di Paolo M, Quattrucci S, Palange P, Banya W, Hopkinson NS, Bilton D. Relationship between pulmonary exacerbations and daily physical activity in adults with cystic fibrosis. BMC Pulm Med. 2015;15:151. https://doi.org/10.1186/s12890-015-0151-7.

230. Schiller JH, Harrington D, Belani CP, Langer C, Sandler A, Krook J, Zhu J, Johnson DH, Group ECO. Comparison of four chemotherapy regimens for advanced non-small-cell lung cancer. N Engl J Med. 2002;346(2):92–8. https://doi.org/10.1056/NEJMoa011954.

231. Schrader PC, Quanjer PH, Olievier IC. Respiratory muscle force and ventilatory function in adolescents. Eur Respir J. 1988;1(4):368–75.

232. Schurz H, Salie M, Tromp G, Hoal EG, Kinnear CJ, Moller M. The X chromosome and sex-specific effects in infectious disease susceptibility. Hum Genomics. 2019;13(1):2. https://doi.org/10.1186/s40246-018-0185-z.

233. Schwartz J, Katz SA, Fegley RW, Tockman MS. Sex and race differences in the development of lung function. Am Rev Respir Dis. 1988;138(6):1415–21. https://doi.org/10.1164/ajrccm/138.6.1415.

234. Scully EP, Haverfield J, Ursin RL, Tannenbaum C, Klein SL. Considering how biological sex impacts immune responses and COVID-19 outcomes. Nat Rev Immunol. 2020;20(7):442–7. https://doi.org/10.1038/s41577-020-0348-8.

235. Seaborn T, Simard M, Provost PR, Piedboeuf B, Tremblay Y. Sex hormone metabolism in lung development and maturation. Trends Endocrinol Metab. 2010;21(12):729–38. https://doi.org/10.1016/j.tem.2010.09.001.

236. Selvadurai HC, Blimkie CJ, Cooper PJ, Mellis CM, Van Asperen PP. Gender differences in habitual activity in children with cystic fibrosis. Arch Dis Child. 2004;89(10):928–33. https://doi.org/10.1136/adc.2003.034249.

237. Shah R, Newcomb DC. Sex bias in asthma prevalence and pathogenesis. Front Immunol. 2018;9:2997. https://doi.org/10.3389/fimmu.2018.02997.

238. Shahar E, Redline S, Young T, Boland LL, Baldwin CM, Nieto FJ, O'Connor GT, Rapoport DM, Robbins JA. Hormone replacement therapy and sleep-disordered breathing. Am J Respir Crit Care Med. 2003;167(9):1186–92. https://doi.org/10.1164/rccm.200210-1238OC.

239. Shakkottai A, Kidwell KM, Townsend M, Nasr SZ. A five-year retrospective analysis of adherence in cystic fibrosis. Pediatr Pulmonol. 2015;50(12):1224–9. https://doi.org/10.1002/ppul.23307.

240. Shapiro S, Traiger GL, Turner M, McGoon MD, Wason P, Barst RJ. Sex differences in the diagnosis, treatment, and outcome of patients with pulmonary arterial hypertension enrolled in the registry to evaluate early and long-term pulmonary arterial hypertension disease management. Chest. 2012;141(2):363–73.   https://doi.org/10.1378/chest.10-3114.

241. Sharma G, Volgman AS, Michos ED. Sex differences in mortality from COVID-19 pandemic: are men vulnerable and women protected? JACC Case Rep. 2020; https://doi.org/10.1016/j.jaccas.2020.04.027.

242. Sher T, Dy GK, Adjei AA. Small cell lung cancer. Mayo Clin Proc. 2008;83(3):355–67. https://doi.org/10.4065/83.3.355.

243. Shinohara M, Ogawa S, Nakaya T, Niino R, Ito M, Haro K, Ishii E. Sex differences in the prevalence and severity of exercise-induced bronchoconstriction in Kindergarteners in Japan. J Gen Fam Med. 2019;20(6):221–9. https://doi.org/10.1002/jgf2.270.

244. Siegfried JM. Women and lung cancer: does oestrogen play a role? Lancet Oncol. 2001;2(8):506–13. https://doi.org/10.1016/S1470-2045(01)00457-0.

245. Silveyra P. Chapter 9: Developmental lung disease. In: Hemnes AR, editor. Gender, sex hormones and respiratory disease. A comprehensive guide; 2016. p. 243. https://doi.org/10.1007/978-3-319-23998-9.

246. Skoczyński S, Semik-Orzech A, Szanecki W, Majewski M, Kołodziejczyk K, Sozańska E, Witek A, Pierzchała W. Perimenstrual asthma as a gynecological and pulmonological clinical problem. Adv Clin Exp Med. 2014;23(4):665–8. https://doi.org/10.17219/acem/37250.

247. Smith LC, Moreno S, Robertson L, Robinson S, Gant K, Bryant AJ, Sabo-Attwood T. Transforming growth factor beta1 targets estrogen receptor signaling in bronchial epithelial cells. Respir Res.   2018;19(1):160.   https://doi.org/10.1186/s12931-018-0861-5.

248. Soriano JB. An epidemiological overview of chronic obstructive pulmonary disease: what can real-life data tell us about disease management? COPD. 2017;14(sup1):S3–7. https://doi.org/10.1080/15412555.2017.1286165.

249. Soriano JB, Alfageme I, Miravitlles M, de Lucas P, Soler-Cataluña JJ, García-Río F, Casanova C, Rodríguez González-Moro JM, Cosío BG,

Sánchez G, Ancochea J. Prevalence and determinants of COPD in Spain: EPISCAN II. Arch Bronconeumol. 2020; https://doi.org/10.1016/j.arbres.2020.07.024.

250. Spagnolo PA, Manson JE, Joffe H. Sex and gender differences in health: what the COVID-19 pandemic can teach us. Ann Intern Med. 2020; https://doi.org/10.7326/M20-1941.

251. Spolarics Z, Pena G, Qin Y, Donnelly RJ, Livingston DH. Inherent X-linked genetic variability and cellular mosaicism unique to females contribute to sex-related differences in the innate immune response. Front Immunol. 2017;8:1455. https://doi.org/10.3389/fimmu.2017.01455.

252. Stabile LP, Siegfried JM. Sex and gender differences in lung cancer. J Gend Specif Med. 2003;6(1):37–48.

253. Stanford KI, Mickleborough TD, Ray S, Lindley MR, Koceja DM, Stager JM. Influence of menstrual cycle phase on pulmonary function in asthmatic athletes. Eur J Appl Physiol. 2006;96(6):703–10. https://doi.org/10.1007/s00421-005-0067-7.

254. Stark MJ, Hodyl NA, Wright IM, Clifton V. The influence of sex and antenatal betamethasone exposure on vasoconstrictors and the preterm microvasculature. J Matern Fetal Neonatal Med. 2011a;24(10):1215–20. https://doi.org/10.3109/14767058.2011.569618.

255. Stark MJ, Hodyl NA, Wright IM, Clifton VL. Influence of sex and glucocorticoid exposure on preterm placental pro-oxidant-antioxidant balance. Placenta. 2011b;32(11):865–70. https://doi.org/10.1016/j.placenta.2011.08.010.

256. Stelzig KE, Canepa-Escaro F, Schiliro M, Berdnikovs S, Prakash YS, Chiarella SE. Estrogen regulates the expression of SARS-CoV-2 receptor ACE2 in differentiated airway epithelial cells. Am J Physiol Lung Cell Mol Physiol. 2020;318(6):L1280–1. https://doi.org/10.1152/ajplung.00153.2020.

257. Stopsack KH, Mucci LA, Antonarakis ES, Nelson PS, Kantoff PW. TMPRSS2 and COVID-19: serendipity or opportunity for intervention? Cancer Discov. 2020;10(6):779–82. https://doi.org/10.1158/2159-8290.CD-20-0451.

258. Storms WW. Asthma associated with exercise. Immunol Allergy Clin North Am. 2005;25(1):31–43. https://doi.org/10.1016/j.iac.2004.09.007.

259. Strope JD, Chau CH, Figg WD. Are sex discordant outcomes in COVID-19 related to sex hormones? Semin Oncol. 2020; https://doi.org/10.1053/j.seminoncol.2020.06.002.

260. Suster DI, Mino-Kenudson M. Molecular pathology of primary non-small cell lung cancer. Arch Med Res. 2020; https://doi.org/10.1016/j.arcmed.2020.08.004.

261. Sutton S, Rosenbluth D, Raghavan D, Zheng J, Jain R. Effects of puberty on cystic fibrosis related pulmonary exacerbations in women versus men. Pediatr Pulmonol. 2014;49(1):28–35. https://doi.org/10.1002/ppul.22767.

262. Sward P, Edsfeldt A, Reepalu A, Jehpsson L, Rosengren BE, Karlsson MK. Age and sex differences in soluble ACE2 may give insights for

COVID-19. Crit Care. 2020;24(1):221. https://doi.org/10.1186/s13054-020-02942-2.

263. Sweezey NB, Ghibu F, Gagnon S, Schotman E, Hamid Q. Glucocorticoid receptor mRNA and protein in fetal rat lung in vivo: modulation by glucocorticoid and androgen. Am J Physiol. 1998;275(1):L103–9. https://doi.org/10.1152/ajplung.1998.275.1.L103.

264. Sweezey NB, Ratjen F. The cystic fibrosis gender gap: potential roles of estrogen. Pediatr Pulmonol. 2014;49(4):309–17. https://doi.org/10.1002/ppul.22967.

265. Takahashi T, Ellingson MK, Wong P, Israelow B, Lucas C, Klein J, Silva J, Mao T, Oh JE, Tokuyama M, Lu P, Venkataraman A, Park A, Liu F, Meir A, Sun J, Wang EY, Casanovas-Massana A, Wyllie AL, Vogels CBF, Earnest R, Lapidus S, Ott IM, Moore AJ, Irt Y, Shaw A, Fournier JB, Odio CD, Farhadian S, Dela Cruz C, Grubaugh ND, Schulz WL, Ring AM, Ko AI, Omer SB, Iwasaki A. Sex differences in immune responses that underlie COVID-19 disease outcomes. Nature. 2020; https://doi.org/10.1038/s41586-020-2700-3.

266. Takahashi, T., Ellingson, M.K., Wong, P. et al. Sex differences in immune responses that underlie COVID-19 disease outcomes. Nature 588, 315–320 (2020). https://doi.org/10.1038/s41586-020-2700-3

267. Tam A, Morrish D, Wadsworth S, Dorscheid D, Man SF, Sin DD. The role of female hormones on lung function in chronic lung diseases. BMC Womens Health. 2011;11:24. https://doi.org/10.1186/1472-6874-11-24.

268. Thomas MR, Marston L, Rafferty GF, Calvert S, Marlow N, Peacock JL, Greenough A. Respiratory function of very prematurely born infants at follow up: influence of sex. Arch Dis Child Fetal Neonatal Ed. 2006;91(3):F197–201. https://doi.org/10.1136/adc.2005.081927.

269. Thurlbeck WM. Lung growth and alveolar multiplication. Pathobiol Annu. 1975;5:1–34.

270. Thurlbeck WM. Postnatal human lung growth. Thorax. 1982;37(8):564–71. https://doi.org/10.1136/thx.37.8.564.

271. Thurlbeck WM. Pathology of chronic airflow obstruction. Chest. 1990;97(2 Suppl):6S–10S.

272. To T, Zhu J, Gray N, Feldman LY, Villeneuve PJ, Licskai C, Gershon A, Miller AB. Asthma and chronic obstructive pulmonary disease overlap in women. Incidence and risk factors. Ann Am Thorac Soc. 2018;15(11):1304–10. https://doi.org/10.1513/AnnalsATS.201802-078OC.

273. Tofovic SP, Jones TJ, Bilan VP, Jackson EK, Petrusevska G. Synergistic therapeutic effects of 2-methoxyestradiol with either sildenafil or bosentan on amelioration of monocrotaline-induced pulmonary hypertension and vascular remodeling. J Cardiovasc Pharmacol. 2010;56(5):475–83. https://doi.org/10.1097/FJC.0b013e3181f215e7.

274. Torday JS, Nielsen HC. The sex difference in fetal lung surfactant production. Exp Lung Res. 1987;12(1):1–19.

275. Torres-Tamayo N, Garcia-Martinez D, Lois Zlolniski S, Torres-Sanchez I, Garcia-Rio F, Bastir M. 3D analysis of sexual dimorphism in size, shape and breathing kinematics of human lungs. J Anat. 2018;232(2):227–37. https://doi.org/10.1111/joa.12743.

276. Townsel CD, Emmer SF, Campbell WA, Hussain N. Gender differences in respiratory morbidity and mortality of preterm neonates. Front Pediatr. 2017;5:6. https://doi.org/10.3389/fped.2017.00006.

277. Townsend EA, Miller VM, Prakash YS. Sex differences and sex steroids in lung health and disease. Endocr Rev. 2012;33(1):1–47. https://doi.org/10.1210/er.2010-0031.

278. Trotter A, Kipp M, Schrader RM, Beyer C. Combined application of 17beta-estradiol and progesterone enhance vascular endothelial growth factor and surfactant protein expression in cultured embryonic lung cells of mice. Int J Pediatr. 2009;2009:170491. https://doi.org/10.1155/2009/170491.

279. Tukiainen T, Villani AC, Yen A, Rivas MA, Marshall JL, Satija R, Aguirre M, Gauthier L, Fleharty M, Kirby A, Cummings BB, Castel SE, Karczewski KJ, Aguet F, Byrnes A, Consortium GT, Laboratory DA, Coordinating Center -Analysis Working G, Statistical Methods groups-Analysis Working G, Enhancing Gg, Fund NIHC, Nih/Nci, Nih/Nhgri, Nih/Nimh, Nih/Nida, Biospecimen Collection Source Site N, Biospecimen Collection Source Site R, Biospecimen Core Resource V, Brain Bank Repository-University of Miami Brain Endowment B, Leidos Biomedical-Project M, Study E, Genome Browser Data I, Visualization EBI, Genome Browser Data I, Visualization-Ucsc Genomics Institute UoCSC, Lappalainen T, Regev A, Ardlie KG, Hacohen N, DG MA. Landscape of X chromosome inactivation across human tissues. Nature. 2017;550(7675):244–8. https://doi.org/10.1038/nature24265.

280. Tyrrell J, Harvey BJ. Sexual dimorphism in the microbiology of the CF 'Gender Gap': Estrogen modulation of Pseudomonas aeruginosa virulence. Steroids. 2020;156:108575. https://doi.org/10.1016/j.steroids.2019.108575.

281. Umar S, Cunningham CM, Itoh Y, Moazeni S, Vaillancourt M, Sarji S, Centala A, Arnold AP, Eghbali M. The Y chromosome plays a protective role in experimental hypoxic pulmonary hypertension. Am J Respir Crit Care Med. 2018;197(7):952–5. https://doi.org/10.1164/rccm.201707-1345LE.

282. Umar S, Iorga A, Matori H, Nadadur RD, Li J, Maltese F, van der Laarse A, Eghbali M. Estrogen rescues preexisting severe pulmonary hypertension in rats. Am J Respir Crit Care Med. 2011;184(6):715–23. https://doi.org/10.1164/rccm.201101-0078OC.

283. Vardavas CI, Nikitara K. COVID-19 and smoking: a systematic review of the evidence. Tob Induc Dis. 2020;18:20. https://doi.org/10.18332/tid/119324.

284. Varkey AB. Chronic obstructive pulmonary disease in women: exploring gender differences. Curr Opin Pulm Med. 2004;10(2):98–103. https://doi.org/10.1097/00063198-200403000-00003.

285. Ventetuolo CE, Praestgaard A, Palevsky HI, Klinger JR, Halpern SD, Kawut SM. Sex and haemodynamics in pulmonary arterial hypertension. Eur Respir J. 2014;43(2):523–30. https://doi.org/10.1183/09031936.00027613.

286. Ventetuolo CE, Baird GL, Barr RG, Bluemke DA, Fritz JS, Hill NS, Klinger JR, Lima JA, Ouyang P, Palevsky HI, Palmisciano AJ, Krishnan I, Pinder D, Preston IR, Roberts KE, Kawut SM. Higher Estradiol and Lower Dehydroepiandrosterone-Sulfate Levels Are Associated with Pulmonary Arterial Hypertension in Men. Am J Respir Crit Care Med. 2016 May 15;193(10):1168–75. https://doi.org/10.1164/rccm.201509-1785OC. PMID: 26651504; PMCID: PMC4872665.

287. Vermillion MS, Ursin RL, Attreed SE, Klein SL. Estriol reduces pulmonary immune cell recruitment and inflammation to protect female mice from severe influenza. Endocrinology. 2018a;159(9):3306–20. https://doi.org/10.1210/en.2018-00486.

288. Vermillion MS, Ursin RL, Kuok DIT, Vom Steeg LG, Wohlgemuth N, Hall OJ, Fink AL, Sasse E, Nelson A, Ndeh R, McGrath-Morrow S, Mitzner W, Chan MCW, Pekosz A, Klein SL. Production of amphiregulin and recovery from influenza is greater in males than females. Biol Sex Differ. 2018b;9(1):24. https://doi.org/10.1186/s13293-018-0184-8.

289. Vidaillac C, Yong VFL, Jaggi TK, Soh MM, Chotirmall SH. Gender differences in bronchiectasis: a real issue? Breathe (Sheff). 2018;14(2):108–21. https://doi.org/10.1183/20734735.000218.

290. vom Steeg LG, Klein SL. SeXX matters in infectious disease pathogenesis. PLoS Pathog. 2016;12(2):e1005374. https://doi.org/10.1371/journal.ppat.1005374.

291. Vom Steeg LG, Vermillion MS, Hall OJ, Alam O, McFarland R, Chen H, Zirkin B, Klein SL. Age and testosterone mediate influenza pathogenesis in male mice. Am J Physiol Lung Cell Mol Physiol. 2016;311(6):L1234–44. https://doi.org/10.1152/ajplung.00352.2016.

292. von Mutius E, Vercelli D. Farm living: effects on childhood asthma and allergy. Nat Rev Immunol. 2010 Dec;10(12):861–8. https://doi.org/10.1038/nri2871. Epub 2010 Nov 9. Erratum in: Nat Rev Immunol. 2019 Sep;19(9):594. PMID: 21060319.

293. Wailoo MP, Emery JL. Normal growth and development of the trachea. Thorax. 1982;37(8):584–7. https://doi.org/10.1136/thx.37.8.584.

294. Wakelee HA, Wang W, Schiller JH, Langer CJ, Sandler AB, Belani CP, Johnson DH, Group ECO. Survival differences by sex for patients with advanced non-small cell lung cancer on Eastern Cooperative Oncology Group trial 1594. J Thorac Oncol. 2006;1(5):441–6.

295. Wang Y, Cela E, Gagnon S, Sweezey NB. Estrogen aggravates inflammation in Pseudomonas aeruginosa pneumonia in cystic fibrosis

mice. Respir Res. 2010;11:166. https://doi.org/10.1186/1465-9921-11-166.

296. Warburton D, Bellusci S. The molecular genetics of lung morphogenesis and injury repair. Paediatr Respir Rev. 2004;5 Suppl A:S283–7.

297. Warburton D, Zhao J, Berberich MA, Bernfield M. Molecular embryology of the lung: then, now, and in the future. Am J Physiol. 1999;276(5 Pt 1):L697–704.

298. Weiler JM, Bonini S, Coifman R, Craig T, Delgado L, Capão-Filipe M, Passali D, Randolph C, Storms W, Ad Hoc Committee of Sports Medicine Committee of American Academy of Allergy AtI. American Academy of Allergy, Asthma & Immunology Work Group report: exercise-induced asthma. J Allergy Clin Immunol. 2007;119(6):1349–58. https://doi.org/10.1016/j.jaci.2007.02.041.

299. Weiler JM, Brannan JD, Randolph CC, Hallstrand TS, Parsons J, Silvers W, Storms W, Zeiger J, Bernstein DI, Blessing-Moore J, Greenhawt M, Khan D, Lang D, Nicklas RA, Oppenheimer J, Portnoy JM, Schuller DE, Tilles SA, Wallace D. Exercise-induced bronchoconstriction update-2016. J Allergy Clin Immunol. 2016;138(5):1292–1295.e1236. https://doi.org/10.1016/j.jaci.2016.05.029.

300. Wenham C, Smith J, Morgan R, Gender, Group C-W. COVID-19: the gendered impacts of the outbreak. Lancet. 2020;395(10227):846–8. https://doi.org/10.1016/S0140-6736(20)30526-2.

301. West J, Cogan J, Geraci M, Robinson L, Newman J, Phillips JA, Lane K, Meyrick B, Loyd J. Gene expression in BMPR2 mutation carriers with and without evidence of pulmonary arterial hypertension suggests pathways relevant to disease penetrance. BMC Med Genomics. 2008;1:45. https://doi.org/10.1186/1755-8794-1-45.

302. White K, Johansen Anne K, Nilsen M, Ciuclan L, Wallace E, Paton L, Campbell A, Morecroft I, Loughlin L, McClure John D, Thomas M, Mair Kirsty M, MacLean Margaret R. Activity of the estrogen-metabolizing enzyme cytochrome P450 1B1 influences the development of pulmonary arterial hypertension. Circulation. 2012;126(9):1087–98. https://doi.org/10.1161/CIRCULATIONAHA.111.062927.

303. Wolters PJ, Collard HR, Jones KD. Pathogenesis of idiopathic pulmonary fibrosis. Annu Rev Pathol. 2014;9:157–79. https://doi.org/10.1146/annurev-pathol-012513-104706.

304. World Health Organization. Sex, gender and influenza. 2010. http://apps.who.int/iris/bitstream/10665/44401/1/9789241500111_eng.pdf.

305. World Health Organization. Influenza (Seasonal). 2020. https://www.who.int/news-room/fact-sheets/detail/influenza-(seasonal). Accessed 26 Sept 2020

306. Wu Z, McGoogan JM. Characteristics of and important lessons from the coronavirus disease 2019 (COVID-19) outbreak in China: summary of a report of 72314 cases from the Chinese Center for Disease Control and Prevention. JAMA. 2020; https://doi.org/10.1001/jama.2020.2648.

307. Wulfsohn NL, Politzer WM, Henrico JS. Testosterone therapy in bronchial asthma. S Afr Med J. 1964;38:170–2.

308. Xu D-Q, Luo Y, Liu Y, Wang J, Zhang B, Xu M, Wang Y-X, Dong H-Y, Dong M-Q, Zhao P-T, Niu W, Liu M-L, Gao Y-Q, Li Z-C. Beta-estradiol attenuates hypoxic pulmonary hypertension by stabilizing the expression of p27kip1 in rats. Respir Res. 2010;11(1):182. https://doi.org/10.1186/1465-9921-11-182.

309. Xu KF, Xu W, Liu S, Yu J, Tian X, Yang Y, Wang ST, Zhang W, Feng R, Zhang T. Lymphangioleiomyomatosis. Semin Respir Crit Care Med. 2020;41(2):256–68. https://doi.org/10.1055/s-0040-1702195.

310. Yang J, Kingsford RA, Horwood J, Epton MJ, Swanney MP, Stanton J, Darlow BA. Lung function of adults born at very low birth weight. Pediatrics. 2020;145(2) https://doi.org/10.1542/peds.2019-2359.

311. Yang P, Wang Y, Wampfler JA, Xie D, Stoddard SM, She J, Midthun DE. Trends in subpopulations at high risk for lung cancer. J Thorac Oncol. 2016;11(2):194–202. https://doi.org/10.1016/j.jtho.2015.10.016.

312. Yoon JC, Casella JL, Litvin M, Dobs AS. Male reproductive health in cystic fibrosis. J Cyst Fibros. 2019;18(Suppl 2):S105–10. https://doi.org/10.1016/j.jcf.2019.08.007.

313. Yu J, Astrinidis A, Howard S, Henske EP. Estradiol and tamoxifen stimulate LAM-associated angiomyolipoma cell growth and activate both genomic and nongenomic signaling pathways. Am J Physiol Lung Cell Mol Physiol. 2004;286(4):L694–700. https://doi.org/10.1152/ajplung.00204.2003.

314. Zaman T, Moua T, Vittinghoff E, Ryu JH, Collard HR, Lee JS. Differences in clinical characteristics and outcomes between men and women with idiopathic pulmonary fibrosis: a multicenter retrospective cohort study. Chest. 2020;158(1):245–51. https://doi.org/10.1016/j.chest.2020.02.009.

315. Zang EA, Wynder EL. Differences in lung cancer risk between men and women: examination of the evidence. J Natl Cancer Inst. 1996;88(3-4):183–92. https://doi.org/10.1093/jnci/88.3-4.183.

316. Zein JG, Erzurum SC. Asthma is different in women. Curr Allergy Asthma Rep. 2015;15(6):28. https://doi.org/10.1007/s11882-015-0528-y.

317. Zhang Y, Coarfa C, Dong X, Jiang W, Hayward-Piatkovskyi B, Gleghorn JP, Lingappan K. MicroRNA-30a as a candidate underlying sex-specific differences in neonatal hyperoxic lung injury: implications for BPD. Am J Physiol Lung Cell Mol Physiol. 2019;316(1):L144–56. https://doi.org/10.1152/ajplung.00372.2018.

Sex Hormones and Lung Inflammation

15

Jorge Reyes-García, Luis M. Montaño,
Abril Carbajal-García, and Yong-Xiao Wang

Abstract

Inflammation is a characteristic marker in numerous lung disorders. Several immune cells, such as macrophages, dendritic cells, eosinophils, as well as T and B lymphocytes, synthetize and release cytokines involved in the inflammatory process. Gender differences in the incidence and severity of inflammatory lung ailments including asthma, chronic obstructive pulmonary disease (COPD), pulmonary fibrosis (PF), lung cancer (LC), and infectious related illnesses have been reported. Moreover, the effects of sex hormones on both androgens and estrogens, such as testosterone (TES) and 17β-estradiol (E2), driving characteristic inflammatory patterns in those lung inflammatory diseases have been investigated. In general, androgens seem to display anti-inflammatory actions, whereas estrogens produce pro-inflammatory effects. For instance, androgens regulate negatively inflammation in asthma by targeting type 2 innate lymphoid cells (ILC2s) and T-helper (Th)-2 cells to attenuate interleukin (IL)-17A-mediated responses and leukotriene (LT) biosynthesis pathway. Estrogens may promote neutrophilic inflammation in subjects with asthma and COPD. Moreover, the activation of estrogen receptors might induce tumorigenesis. In this chapter, we summarize the most recent advances in the functional roles and associated signaling pathways of inflammatory cellular responses in asthma, COPD, PF, LC, and newly occurring COVID-19 disease. We also meticulously deliberate the influence of sex steroids on the development and progress of these common and severe lung diseases.

J. Reyes-García
Departamento de Farmacología, Facultad de Medicina, Universidad Nacional Autónoma de México, CDMX, Mexico City, Mexico

Department of Molecular and Cellular Physiology, Albany Medical College, Albany, NY, USA

L. M. Montaño · A. Carbajal-García
Departamento de Farmacología, Facultad de Medicina, Universidad Nacional Autónoma de México, CDMX, Mexico City, Mexico

Y.-X. Wang (✉)
Department of Molecular and Cellular Physiology, Albany Medical College, Albany, NY, USA
e-mail: wangy@amc.edu

Keywords

Testosterone · 17β-Estradiol · Inflammation · Asthma · COPD · Lung cancer · Pulmonary fibrosis

Abbreviations

AA	Arachidonic acid
AC	Adenylate cyclase
AECs	Airway epithelial cells
AHR	Airway hyperresponsiveness
AP-1	Activator protein 1
AR	Androgen receptor
ASM	Airway smooth muscle
ASMCs	Airway smooth muscle cells
BALF	Bronchoalveolar lavage fluid
cAMP	Cyclic adenosine monophosphate
CCL C-C	chemokine ligand
CCR C-C	chemokine receptor
CD	Cluster of differentiation
cGMP	Cyclic guanosine monophosphate
COPD	Chronic obstructive pulmonary disease
CYP11A1	P450 side chain cleavage enzyme
CYP17A1	P450 17α-hydroxylase
DAMPs	Danger-associated molecular patterns
DCs	Dendritic cells
DHEA	Dehydroepiandrosterone
E2	17β-Estradiol
ECM	Extracellular matrix
EGF	Epidermal growth factor
EMT	Epithelial-mesenchymal transition
eNOS	Endothelial nitric oxide synthase
ERK	Extracellular signal-regulated kinase
ERα	Estrogen receptor alpha
ERβ	Estrogen receptor beta
ET	Endothelin
FEV1	Forced expiratory volume in 1 second
FVC	Forced vital capacity
G-CSF	Granulocyte colony-stimulating factor
GM-CSF	Granulocyte-macrophage colony-stimulating factor
HDM	House dust mite
IFN-γ	Interferon gamma
Ig	Immunoglobulin
IL	Interleukin
ILC2	Type 2 innate lymphoid cell
IPF	Idiopathic pulmonary fibrosis
JNK	Jun N-terminal kinase
LC	Lung cancer
LTs	Leukotrienes
MAPK	Mitogen-activated protein kinase
MCP-1	Monocyte chemoattractant protein-1
MMP	Matrix metalloproteinase
NF-κB	Nuclear factor kappa B
NO	Nitric oxide
OC	Oral contraceptives
OVA	Ovalbumin
P4	Progesterone
PAMPs	Pathogen-associated molecular patterns
PBMCs	Peripheral blood mononuclear cells
PEFR	Peak expiratory flow rate
PF	Pulmonary fibrosis
PI3K	Phosphoinositide 3-kinase
PMA	Perimenstrual asthma
PR	Progesterone receptor
PRRs	Pattern recognition receptors
ROS	Reactive oxygen species
STAR	Steroidogenic acute regulatory protein
TAM	Tumor-associated macrophages
TES	Testosterone
TGF-β1	Transforming growth factor beta 1
Th	T-helper cell
TLR	Toll-like receptor
TNF-α	Tumor necrosis factor alpha
TSPO	Translocator protein

15.1 Introduction

15.1.1 Lung Inflammation

Inflammation is a complex biological response to harmful stimuli (i.e., bacterial and viral infections, irritants or environmental pollutants, and damaged cells), which is orchestrated by the immune system [1–4]. Two phases of inflammation can be distinguished: acute and chronic inflammation. In the acute phase (hours to days), host cells recognize danger-associated molecular patterns (DAMPs) or pathogen-associated molecular patterns (PAMPs) through the action of pattern recognition receptors (PRRs), expressed predominantly in monocytes, macrophages, neutrophils, and dendritic cells [1, 2, 5, 6]. These cells migrate to the injured site along a chemotactic gradient mediated by specific cytokines. Cellular stimulation leads to the inflammatory process through the activation of transcription factors like nuclear factor kappa B (NF-κB) and the release of pro-inflammatory cytokines such as tumor necrosis factor alpha (TNF-α), interleukin (IL)-1, IL-6, IL-8, IL-12, IL-17 and interferon (IFN)-I and IFN-II [3, 7–11]. IL-8 acts as a neutrophil chemotactic agent, and TNF-α augments the expression of adhesion molecules in the endothelial cells of lung capillaries. Then, antigen-presenting cells (APCs) present the T lymphocytes with the foreign antigen (virial/bacterial or damaged cell components) and evoke either a type 1 T helper (Th)1 lymphocyte- or Th2 lymphocyte-mediated response [10–13]. The persistence of inflammation due to a long-time exposure to inflammatory stimuli, and a failed or incomplete acute response resolution leads to the chronic phase of inflammation that may last for weeks or months and in some circumstances for years. In this phase, the inflammatory response is amplified, and tissue damage may occur. In most cases of chronic lung inflammation, profibrotic and immunoregulatory Th2 cytokines govern [1, 2, 10]. Acute and chronic inflammation are typical markers in numerous lung disorders, including infectious [14, 15], immunological [16, 17], genetic [18, 19], neoplastic [20, 21], and environmental [22, 23] ailment-related.

Commonly, cellular mechanisms of lung inflammation include the expression of adhesion molecules, the release of systemic inflammatory mediators, and the recruitment of distinct leukocytes into the lung vasculature [10, 13, 24]. Neutrophils are the first type of immune cells to be recruited, followed by the resident macrophages, including both alveolar and interstitial and, in some cases, pulmonary intravascular macrophages. The displayed profile of immune cells and cytokines will depend on the developed type of lung disease or injury [10, 16].

15.1.2 Sex Differences in Lung Inflammatory Diseases

Lung disease is a major health issue. According to the Centers for Disease Control and Prevention and to the National Center for Health Statistics in the USA, the number of deaths from chronic lower respiratory diseases reaches ~154,000 every year. Furthermore, it is estimated that this kind of illnesses affects more than 500 million people across the world [25]. It is well known that sex hormones play diverse regulatory effects on the human lung development and physiology [26–30]. Moreover, sex differences are essential predictors in a lot of common diseases, used in diagnosis, prognosis, and treatment recommendations [31, 32].

The influence of sex hormones in the incidence and severity of the inflammatory response in lung disease has been widely studied and recognized for years. For instance, one of the most prevalent lung inflammatory illnesses, chronic obstructive pulmonary disease (COPD), affects both men and women; nevertheless, recent studies point out that females are at a higher risk of developing COPD with lower exposures to tobacco smoke [31, 33–37]. During childhood, asthma symptoms seem to be more prevalent in boys than in girls, and this trend reverts during puberty; however, the incidence in asthma symptoms increases in older men when testosterone (TES) levels decrease, suggesting a potential protective role of the androgens in this ailment [25, 28, 31, 38–42]. Contrariwise, female sex hormones have been related to negative outcomes in asthma [25, 42–44]. Pulmonary fibrosis (PF), an interstitial lung illness, affects more men than women with a higher mortality rate in males [45–47]. Moreover, murine models of bleomycin-induced PF have shown the influence of sex hormones (both males and females) in the decrease of lung function [31, 48, 49]. In cystic fibrosis (CF), a genetic disease, women have been described to display more severe consequences than men, especially in response to bacterial respiratory infections [31, 50, 51]. Pulmonary arterial hypertension (PAH) is another pulmonary disease influenced by gender showing a female predominance [52–55] that has been related to the estrogen receptor alpha (ERα) [56, 57]. Lung cancer (LC) is also an ailment in which the evidence suggests that its incidence and progression are affected by sex hormones differences, particularly by the action of estrogens and their receptors [58, 59]. Furthermore, gender differences have been described as well in lung ill-nesses caused or enhanced by infectious agents [52, 60, 61].

In spite of the marked gender differences in several lung ailments and the great amount of studies related to them, the role of sex hormones in the mechanisms associated with these illnesses has not been fully elucidated. This chapter summarizes the advances in basic and clinical studies of sex hormones (mainly testosterone and estradiol) as modulators of the inflammatory response in lung disease with particular emphasis on asthma and COPD. The information obtained from sex-specific research on lung physiology and pathology would potentially help in the development of sex-specific therapeutics for inflammatory lung diseases regarding the hormonal status of the patient.

15.2   Sex Hormone Steroidogenesis

15.2.1   Classification of Sex Hormones

Sex hormones are involved in processes such as growth, development, reproduction, and systemic homeostasis [62, 63]. This group of hormones possesses a common structure of cyclopentane-perhydro-phenanthrene, a complex of 17-carbon atoms forming a 4-ring system. According to the number of carbon atoms, sex steroids can be divided into three major groups: progestins (21-carbon atoms), androgens (19-carbon atoms), and estrogens (18-carbon atoms) [64]. The same synthesis pathway of steroid hormone production is carried out by different organs in men and women, i.e., testis, adrenal cortex, ovary, brain, placenta, etc.; however, the amount and the main type of synthetized hormone molecule rely on the expression and activity specific to each tissue [64–66].

Cholesterol is the precursor of sex hormones, both male and female, which are synthetized in specialized cells [62, 65, 67]. This precursor is an indispensable element of the cellular plasma membrane contributing to the fluidity, permeability, and regulation of transmembrane signaling pathways [68]. 3-Hydroxy-3-methyl-glutaryl-coenzyme A reductase (HMG-CoA or HMGCR) is the enzyme responsible for the metabolic pathway that produces cholesterol from acetyl-CoA [69–71]. Additionally, cholesterol can be taken from low-density lipoprotein (LDL, through the LDL receptor pathway) and high-density lipoprotein (HDL), via the scavenger receptor class B type (SR-BI) pathway or lipid droplets [72–74].

15.2.2   Androgen Biosynthetic Pathway in Leydig Cells

In men, the synthesis of TES (the main androgen) is carried out in a major proportion (95%) by Leydig cells from the adult testis through the action of cytochrome P450 enzymes. The synthesis and secretion of this androgen are tightly regulated by luteinizing hormone (LH) [67]. In addition, smaller amounts of TES are produced in the adrenal cortex [65]. Gonadotropin-releasing hormone (GnRH) is secreted from the hypothalamus and stimulates gonadotropic cells in the anterior pituitary gland to release LH. This hormone stimulates its Gs protein-coupled receptor, resulting in the activation of adenylyl cyclase (AC). AC induces the formation of cyclic adenosine monophosphate (cAMP), which stimulates the mobilization of cholesterol to the mitochondria by activating protein kinase A (PKA) signaling [75]. The importation of cholesterol into mitochondria is carried out by the transduceosome, a protein complex conformed mainly by the steroidogenic acute regulatory protein (STAR) and the translocator protein (TSPO) [73, 74]. After the synthesis of cholesterol, this precursor is converted into pregnenolone by the P450 side chain cleavage enzyme (P450scc/CYP11A1) located in the mitochondrial membrane [63, 65, 74, 76]. Pregnenolone can either be converted to progesterone via 3β-hydroxysteroid dehydrogenase type 2 (3β-HSD2) or be hydroxylated to 17α-hydroxypregnenolone and then transformed to dehydroepiandrosterone (DHEA, an androgen) by cytochrome P450 17α-hydroxylase (P450c17/CYP17A1/C17-C20 lyase, an enzyme with hydroxylase and lyase activity) [63, 65, 77, 78]. DHEA is further reduced to androstenediol via 17β-hydroxysteroid dehydrogenase type 3 (17β-HSD3) [79, 80] or converted to androstenedione by 3β-HSD2 [78, 81]. Androstenediol and androstenedione are finally biotransformed to TES by 3β-HSD2 and 17β-HSD3, respectively [65, 78–81]. Furthermore, TES is reduced to 5α-dihydrotestosterone (5α-DHT) by 5α-reductase and to 5β-dihydrotestosterone (5β-DHT) by 5β-reductase [76, 82, 83]. In addition, TES can be converted to 17β-estradiol (E2, an estrogen) through the P450 aromatase (P450aro/CYP19A1 aromatase) action [63] (Fig. 15.1).

Although TES is necessary for estrogen production, men have much higher plasmatic levels of TES than women. The synthesis of TES by Leydig cells in men is seven to eight times greater than that produced in women ovaries. In men, TES plasma concentration reaches values between 6 and 50 nM depending on the person's age. On the other hand, women display stable TES values between 0.6 and 2.4 nM that are maintained along the different life stages, except during pregnancy when TES concentrations increment (3.5–5 nM) [31].

15.2.3 Estrogen Biosynthetic Pathway in Theca and Granulosa Cells

E2 and progesterone (P4) are considered the main female sex hormones. The former is a type of estrogen and the latter is a type of progestogen, both essentially produced in ovaries [36, 64, 84]. In addition to E2, two more estrogen molecules naturally occur in women: estrone (E1) and estriol (E3). Estriol is the predominant estrogen during pregnancy, while estradiol is the prevalent form in non-pregnant premenopausal females. In menopause, estrone is the predominant type of estrogen [62, 85]. Ovaries are the vastest source of estrogens before menopause. Nevertheless, in postmenopausal women and in men, these female hormones are locally produced from circulating

testosterone and adrenal cortex steroids in non-reproductive and reproductive tissues [86–88]. Ovarian steroids are synthetized through the interaction between theca (TCs) and granulosa cells (GCs), a process regulated by LH and follicle-stimulating hormone (FSH) [36, 62, 84, 85, 89, 90]. GnRH secreted from the hypothalamus stimulates LH and FSH. LH acts on both TCs and GCs, and FSH exerts its effects mainly on GCs. These hormones stimulate AC activity and cAMP formation. This cyclic nucleotide triggers PKA activation and the further expression of steroidogenic enzymes [91–93] (Fig. 15.2).

P4 is predominantly produced in luteal cells through a system of three cholesterol-modifying enzymes: STAR, P450scc, and 3β-HSD. STAR catalyzes cholesterol transfer within the mitochondria, which is considered the rate-limit step in the production of all steroids [84]. This regulatory protein is mostly expressed in luteal cells; however, STAR can be found in theca and granulosa cells during follicle development or during luteinizing phase, respectively, conferring to these cells the ability to synthetize progesterone [84, 85]. The first step in female steroidogenesis is the initial conversion of cholesterol into pregnenolone by the action of STAR and P450scc [36, 90, 94]. Subsequently, pregnenolone is converted to progesterone by 3β-HSD2 in both theca and granulosa cells [90, 94]. In theca cells, P450c17 hydroxylates pregnenolone to produce 17α-hydroxypregnenolone and subsequently removes the acetyl group in order to form DHEA. This last product can be either converted into androstenedione via 3β-HSD2 or metabolized into androstenediol by 17β-HSD1 [85, 94, 95]. These androgens are further biotransformed to TES via 17β-HSD5 and 3β-HSD2, respectively. TES and androstenedione diffuse across the follicle membrane into the follicular fluid, where they are taken up by granulosa cells [36, 85, 90, 95]. The endoplasmic reticulum of granulosa cells expresses P450aro, which converts TES to E2 and androstenedione to E1 by the addition of an aromatic A ring. Additionally, in granulosa cells, 17β-HSDs regulate the interconversion between E1 and E2. 17β-HSD1 catalyzes the formation of E2 from E1, and

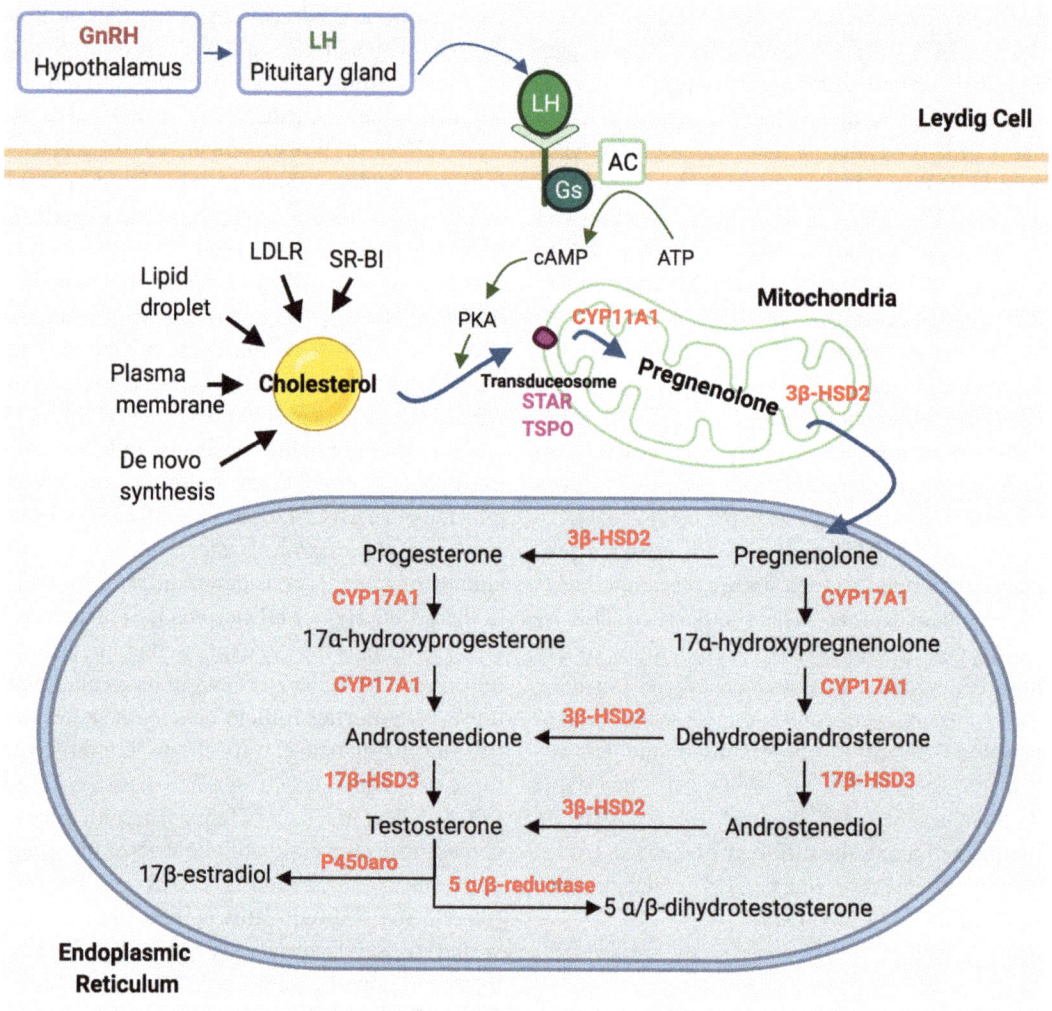

Fig. 15.1 Synthesis of androgens from cholesterol in Leydig cells. Gonadotropin-releasing hormone (GnRH) is secreted from the hypothalamus and stimulates gonadotropic cells in the anterior pituitary gland to release luteinizing hormone (LH), which tightly regulates the synthesis and secretion of androgens in Leydig cells. LH binds to its Gs protein-coupled receptor, resulting in the activation of adenylyl cyclase (AC) and increased intracellular cyclic adenosine monophosphate (cAMP) formation. cAMP stimulates the mobilization of cholesterol to the mitochondria by activating protein kinase A (PKA) signaling. Cholesterol is the precursor of all sex steroids. This lipid precursor can be produced de novo or taken from low-density lipoprotein (LDL) and high-density lipoprotein (HDL) via the scavenger receptor class B type (SR-BI) pathway, plasma membrane, or lipid droplets. Once synthetized, cholesterol is imported into mitochondria through the transduceosome (a protein complex), composed of the steroidogenic acute regulatory protein (STAR), the translocator protein (TSPO), and other proteins. The first step is the conversion of cholesterol to pregnenolone by the C27 cholesterol side chain cleavage cytochrome P450 enzyme (CYP11A1) located on the matrix side of the inner mitochondrial membrane. Then, pregnenolone is converted into testosterone by 3β-hydroxysteroid dehydrogenase (3β-HSD2, located at the mitochondria), 17α-hydroxylase/17,20 lyase (CYP17A1), and type 3 17β-hydroxysteroid dehydrogenase (17b-HSD3) in the endoplasmic reticulum. Furthermore, testosterone can be reduced to 5α- or 5β-dihydrotestosterone by 5α-/β-reductase. Finally, P450 aromatase (P450aro) can convert testosterone to 17β-estradiol in Leydig cells of the adult testis

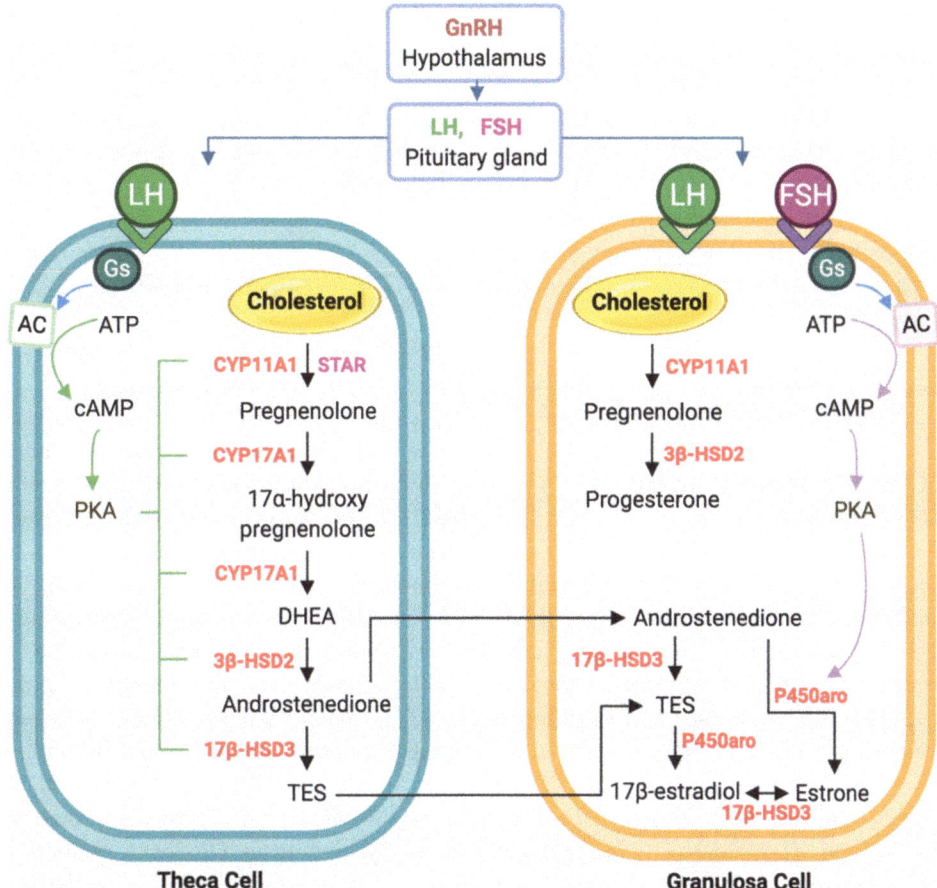

Fig. 15.2 Steroid hormone biosynthesis pathways in the ovary by theca and granulosa cells. Ovarian steroids are synthetized through the interaction between theca and granulosa cells, a process regulated by gonadotropin-releasing hormone (GnRH) secreted from the hypothalamus. GnRH stimulates luteinizing hormone (LH) and follicle-stimulating hormone (FSH). LH acts on both theca and granulosa cells; FSH acts only on granulosa cells. These hormones stimulate adenylyl cyclase (AC) via Gs protein-coupled receptors. The cyclic adenosine monophosphate (cAMP) generated from adenosine triphosphate (ATP) activates protein kinase A (PKA) to stimulate the expression of steroidogenic enzymes in theca and granulosa cells. The first step in female steroidogenesis is the initial conversion of cholesterol to pregnenolone by the action of steroidogenic acute regulatory protein (STAR) and C27 cholesterol side chain cleav- age cytochrome P450 enzyme (CYP11A1). Later, pregnenolone is converted to progesterone by 3β-hydroxysteroid dehydrogenase (3β-HSD2). In theca cells, 17α-hydroxylase/17,20 lyase (CYP17A1) hydroxyl- ates pregnenolone to produce 17α-hydroxypregnenolone and subsequently removes the acetyl group in order to form dehydroepiandrosterone (DHEA). This last product can be converted into androstenedione via 3β-HSD2 and further biotransformed into testosterone (TES) via 17β-hydroxysteroid dehydrogenase (17β-HSD3). TES and androstenedione diffuse across the follicle membrane into the follicular fluid where they are taken up by granu- losa cells. The endoplasmic reticulum of granulosa cells expresses P450 aromatase (P450aro) that converts TES to 17β-estradiol and androstenedione to estrone. Additionally, in granulosa cells, 17β-HSD3 catalyzes the formation of 17β-estradiol from estrone

17β-HSD2 catalyzes the oxidation of E2 to E1 [89, 90, 94] (Fig. 15.2). Moreover, estrogens can be produced in the Leydig cells, Sertoli cells, and mature spermatocytes from the male gonads [96].

Variation of estrogens levels during life is based on different factors such as age, menstrual cycle phase, and pregnancy. Interestingly, andro- gen levels in women are higher than estrogen lev- els most of the time. An exception occurs during the preovulatory and midluteal phases of the menstrual cycle [31, 85, 97]. In non-pregnant women, E2 serum levels fluctuate between 80 and 800 pM. These levels may increase up to 150 nM during pregnancy and highly decrease, oscillating between 40 and 120 pM in menopausal period. Progesterone plasmatic levels also vary between 1 and 60 nM in non-pregnant women and reach values of 1000 nM during pregnancy [31].

15.3 Sex Hormone Receptors

15.3.1 Sex Hormone Binding Globulin

The distribution of sex hormones to different tissues, including the lung, is regulated by sex hormone binding globulin (SHBG), a key steroid hormone binding protein in human plasma. Plasma SHBG is a homodimeric protein largely produced in hepatocytes [36, 98, 99]. Each monomer possesses a steroid binding pocket and a Ca^{2+}-binding site [99, 100]. This globulin binds steroids such as TES, 5α-DHT, and E2 with nanomolar affinities [36, 99–102]. Normally, in humans, between 40% and 65% of circulating TES and between 20% and 40% of circulating E2 are bound to SHBG [101, 103]. Literature suggests that sex hormones bound to SHBG do not display biological activity. Moreover, sex steroid dissociation from this globulin in the circulatory system allow them to bind and activate male or female sex hormone receptors, triggering gene transcription and protein synthesis [103, 104]. Interestingly, SHBG binds TES with a higher affinity than it does for E2, acting as an estrogen amplifier [36, 98, 103, 104].

Sex steroids exert their physiological effects mostly through the binding to their own receptors, e.g., the androgen receptor (AR), estrogen receptors (ERs), and progesterone receptors (PRs) [25, 31, 76, 105–107]. Sex steroid actions comprise genomic and non-genomic effects. Genomic effects occur from hours to days and involve a direct modulation of gene transcription and protein synthesis via the binding and activa-tion of nuclear hormone receptor complexes. Non-genomic effects are mediated by plasma membrane receptors or ion channels that trigger intracellular signaling pathways that mayresult in transcriptional regulation [25, 31, 36, 76, 106, 108].

15.3.2 Androgen Receptor

The AR modulates the activity and effects of male sex hormones and their influence on lung development. Testes from fetuses produce TES after sex differentiation, and this androgen retards the production of surfactant during gestation [109]. Also, it has been suggested that branching morphogenesis of the human lung is regulated by androgens' effects [110]. The AR is found in several lung tissues and immune cells, including airway smooth muscle (ASM) [111], lung parenchyma, bronchial epithelium [112, 113], dendritic cells (DCs) [114], macrophages, neutrophils, and T and B lymphocytes [115–118]. The AR, also known as NR3C4, is a cytoplasmatic/nuclear receptor that is activated by binding to TES or to its more active reduced metabolite, 5α-DHT, in the cytosol. Male sex steroids binding to the AR promote their dissociation from its chaperon proteins, forming a complex that is translocated into the nucleus. The formed complex acts as a DNA-binding transcription factor that modulates gene transcription and protein synthesis [25, 76, 95, 119, 120]. 5β-DHT, the other reduced metabolite of TES, possesses less androgenic activity than TES and 5α-DHT, because it has a lower binding affinity for the AR [121]. In addition, TES is capable of activating plasma membrane receptors such as ZIP9, a zinc transporter from the ZIP family [122–125], and GPRC6A (class C, group 6, subtype A, G protein-coupled receptor family member) [124, 125]. The binding of TES to ZIP9 or GPRC6A triggers signaling pathways that implicate the activation of G proteins, including Gs, Gi, and Gq11, and the stimulation of extracellular signal-regulated kinases 1 and 2 (ERK1/2) [122–128].

15.3.3 Estrogen Receptors

Estrogen and progesterone receptors contribute to sexual and lung development [30, 37, 95]. Two classes of ERs have been described: nuclear ERs, ERα (ESR1) and ERβ (ESR2), and the plasma membrane ER (mER), G protein-coupled receptor 30 (GPER/GPR30) [25, 129–132]. Estrogen receptors exert their physiological actions through either genomic pathways leading to gene transcription or non-genomic signaling pathways (rapid effects that involve phosphorylation processes) [133–137]. ERα and ERβ function as transcription factors. The stimulation of ERs in the cytoplasm causes dimerization, nuclear translocation, and binding to estrogen response elements (EREs) in the promoter region of target genes. Moreover, ERs are able to indirectly modulate gene transcription by forming complexes with proteins such as c-fos and c-jun, essential components of activator protein 1 (AP-1) [136, 137]. Also, the activation of ERs induces rapid effects mediated by protein members of the mitogen-activated protein kinase (MAPK) family, e.g., ERK1/2, p38 MAPK, and the c-Jun N-terminal kinase (JNK) [138–140]. ERα and ERβ occur in numerous lung cells and tissues, e.g., ASM, bronchial epithelial cells, macrophages, DCs, and T and B lymphocytes [25, 141–146]. Some studies have demonstrated the critical role of both ERα and ERβ in fetal lung development. The expression level of these receptors is significantly elevated in fetal lungs compared to postnatal and adult lungs from mice [147]. Furthermore, ERα and ERβ participate in alveolar formation [148, 149].

In addition, two shorter or truncated splice variants of the human ERα (hERα-66/ER66 due to its molecular weight) have been identified as mERs: 46 kDa ER (hERα-46/ER46) and 36 kDa ER (hERα-36/ER36) [150, 151]. It has been demonstrated that ER66 can translocate to the plasma membrane via the interaction with the scaffolding protein of caveolae (caveolin-1), a process dependent on palmitoylation [152–155]. ER66-induced transcription is mediated by two activation domains: the ligand-independent activation function (AF)-1, which is located in the N-terminal domain, and the ligand-dependent AF-2, situated in the C-terminal domain [156, 157]. The trun-

cated ER46 isoform lacks the AF-1 domain; however, this splice variant maintains the corresponding domains for caveolin-1 association and palmitoylation [150, 152]. The absence of the AF-1 domain in the ER46 supposes null influence in its ability to evoke non-genomic actions. In this context, it has been shown that ER46 variant stimulates the phosphorylation of the endothelial nitric oxide synthase (eNOS) in a higher degree than ER66 [158]. ER36, the other truncated isoform of ER66, lacks the AF-1 and AF-2 domains [151]. Non-genomic actions mediated by ER36 have been described. The activation of ER36 by E2 elicits the mobilization of intracellular Ca^{2+} in different breast cancer cell lines. Also, ER36 triggers MAPK pathways leading to cell growth and proliferation in HEK293 cells [159]. In HEK293 cells as well, saturation binding assays show that the equilibrium dissociation constant (K_d) for the binding of E2 to ER66 and ER46 corresponds with the serum levels of this estrogen found in women (68.8 pM and 60.7 pM, respectively), while ER36 exhibits no saturable specific binding [160]. Furthermore, the evidence about the molecular characterization of ERβ splice variants is still unclear. The presence of membranal estrogen receptor insinuates an additional regulation mechanism for estrogens' actions; however, little is known regarding the function of ER-truncated isoforms in the lung [133, 161].

GPR30 was identified as a functional membrane receptor, different from the ERα-truncated splice variants, involved in rapid E2 signaling pathways [131, 132, 162, 163]. GPR30 binds E2 with high affinity (K_d = 6 nM) [164, 165] and triggers numerous intracellular signal transduction pathways such as cyclic adenosine monophosphate (cAMP) production, Ca^{2+} mobilization, and the activation of phosphatidylinositol 3-kinase (PI3K) and ERK1/2 [129, 131, 132]. Moreover, the stimulation of GPR30 has been involved in the activation of the epidermal growth factor receptor (EGFR)-mediated signaling in breast and lung cancer [132, 162, 165, 166]. The expression and function of GPR30 in the lung have not been fully elucidated yet. In this context, Townsend et al. did not find a significant expression of this receptor in human ASM [167].

15.3.4 Progesterone Receptors

Progesterone (P4) plays a crucial role in the maintenance of pregnancy. This hormone modulates the transition of the endometrium from a proliferative stage to a secretory phase and promotes the implantation of the blastocyst [64]. The lipophilic nature of P4 allows it to cross the cell membrane and binding to the two types of progesterone receptors (PRs) identified in mammals: PR-A and PR-B. Once PRs are activated, enter the nucleus, and promote DNA modulation, gene transcription and protein synthesis [64, 168–171]. Both progesterone receptors are encoded by the same gene but display different patterns on progesterone response elements, i.e., distinct transcriptional activity [172, 173]. PR-B is considered a strong promoter of gene transcription, whereas PR-A is associated with repressor responses [172]. The stimulation of the PRs triggers the activation of transcriptional regulatory proteins known as steroid receptor coactivators 1, 2, and 3 (SRC-1, SRC-2, and SRC-3). SRCs contribute to the regulation of DNA transcription by assisting nuclear receptors [174, 175]. Moreover, PR-B activation allows the association with the N-terminal domain of the ER. The association of the PR-B with the ER causes the proto-oncogene tyrosine-protein kinase (Src)/p21-Ras/ERK pathway [176]. Progesterone receptors have been identified in ASM [111], cilia from the airway epithelium [177], and endothelial cells from vascular beds [178, 179]. The expression of PRs in immune cells is still under debate; however, some reports have suggested the occurrence of this receptors in DCs [179, 180].

Non-genomic effects of P4 are mediated by membrane receptors. The existence of non-classical membrane PRs (mPRs) has been pointed out and reviewed in mammalian tissues, including reproductive and non-reproductive systems [181, 182]. The non-classical PRs were first identified in fish where they modulate gamete physiology [183]. mPRs are classified into the progesterone and adipoQ receptor (PAQR) family, which belongs to the superfamily of GPCRs. Until now five different subtypes of mPRs have been characterized: mPRα (PAQR7), mPRβ (PAQR8), mPRγ (PAQR5), mPRδ (PAQR6), and mPRε (PAQR9) [181, 182, 184, 185]. These receptors have been described in immune cells, such as peripheral blood mononuclear cells (PBMCs) and T cells [181, 182, 186–190]. The stimulation of mPRs evokes intracellular signal pathways implicating Ca^{2+} mobilization, the increase in cAMP levels, and the activation of p38MAPK and JNK [181, 182, 190].

15.4 Immune Cells

15.4.1 Neutrophils

Neutrophils are essential granulocytes of the innate immune system. These cells are the most abundant circulating leukocytes, comprising around 70% of all white blood cells in healthy humans [191]. Neutrophils are commonly the first cells responding to damage and migrating to the injured tissues, including the lung. The activation of neutrophils and the release of chemokines induce the recruitment of monocytes into the inflamed tissues [10, 191, 192]. The evidence of sex steroid actions on the development and physiology of neutrophils is present. A study shows that the genetic ablation of the AR in mice drastically diminishes (90%) the proliferative activity of neutrophil precursors and retards neutrophil maturation. Also, neutropenia is exhibited in mice with testicular feminization (Tfm). The same study points out that androgens induce the production of neutrophils via the modulation of granulocyte colony-stimulating factor (G-CSF) [193]. Additionally, TES suppresses the production of superoxide and the anti-microbial capacity of human neutrophils [194]. Furthermore, in an ozone-induced lung inflammation mice model, the expression of neutrophil-attracting chemokines (*Ccl20*, *Cxcl5*, and *Cxcl2*) and the number of neutrophils are significantly higher in females compared with males [195].

15.4.2 Macrophages

In humans, the respiratory tract harbors mononuclear phagocytic cells to provide one of the first lines of defense against inhaled allergens, pathogens, particles, and gases. Most of these phagocytic cells are macrophages distributed along the lung, airways, and alveoli [196, 197]. Besides the resident macrophages, two more distinct populations are found in the lung: alveolar and interstitial macrophages [198, 199]. Alveolar macrophages are considered the main leukocyte subtype in the lung. Cell counting analysis of bronchoalveolar lavage fluid (BALF) from healthy adults reveals 72–96% of macrophages, 2–26% of lymphocytes, 0–4% of neutrophils, and 0–1% of eosinophils, basophils, and mast cells [200]. After an injury or damage is produced, monocytes derived from the bone marrow are recruited to the lung and differentiated into alveolar macrophages. The differentiation process and the following nesting into the alveoli are dependent on the production of granulocyte-macrophage colony-stimulating factor (GM-CSF) by alveolar type 2 cells. Signals triggered by alveolar type 1 and 2 cells, such as exposure to surfactant-rich fluid and elevated oxygen tension, may modulate different functional phenotypes of alveolar macrophages [196, 201]. In this context, classically activated macrophages (CAM/M1) and alternatively activated macrophages (AAM/M2) have been described. Macrophages may undergo M1 polarization in response to cytokines produced by Th1 lymphocytes, e.g., interferon gamma (IFN-γ), and by Toll-like receptor (TLR) ligands found in bacterial and viral products. The activation of M2 macrophages is provoked by Th2 cytokines such as IL-4 and IL-13 [202].

Regarding sex hormones' influence in these immune cells, it has been shown that TES decreases the expression of the TLR4 in macrophages from mice [203]. Also, TES reduces the production of IL-1β, IL-6, and TNF-α in human macrophages and human monocytes and increases the expression of IL-10 in macrophages from humans, as well as in a murine macrophage cell line [204, 205]. Moreover, TES is capable of diminishing nitric oxide (NO) production induced by the stimulus of lipopolysaccharide (LPS) [205]. Female sex hormones also exert a regulation in the activity of macrophages. Ovalbumin (OVA)-induced asthmatic mice display increased numbers of M2 macrophages in the lung compared with male mice [206]. Estrogens facilitate the resolution phase of inflammation toward a M2 phenotype dependent on IL-10, contributing to tissue remodeling and shortening the pro-inflammatory state [207]. Progesterone inhibits NO production and the release of microparticles (MPs) with pro-inflammatory and prothrombotic properties. MPs are discharged by macrophages stimulated by ligands of TLRs in a NO-dependent process [208].

15.4.3 Eosinophils

Eosinophils are key modulators in allergic inflammation as occurring in asthma [209]. They are developed from granulocytic precursor populations in the bone marrow and are activated through the action of the IL-5 secreted by Th2 cells. The maturation of eosinophils is also mediated by GM-CSF. Degranulation of activated eosinophils releases pro-inflammatory cytokines, leukotrienes (LTs), platelet-activating factor (PAF), reactive oxygen species (ROS), and the cationic proteins, e.g., basic protein and eosinophil cationic protein [209, 210]. LTs contribute to bronchospasm in asthma [211]. While the expression of sex steroid receptors in eosinophils is unclear, the impact of sex hormones on these immune cells has been investigated. In gonadectomized male mice, the peripheral and bone marrow eosinophils are augmented, and TES abolishes peripheral and bone marrow eosinophil responses at the early phase of infection by *Brugia pahangi* in females [212]. Moreover, TES decreases human eosinophil viability and adhesion properties in vitro [213]. Conversely, female mice display more severe eosinophilia than males, a phenomenon that is reversed after gonadectomy [214], and progesterone treatment enhances the recruitment of eosinophils and induces airway hyperresponsiveness (AHR) in a murine model of allergic asthma [215].

15.4.4 Mast Cells and Dendritic Cells

Mast cell development occurs in the bone marrow, and their maturation is mediated by the interaction between the stem cell factor (SCF) with its own receptor c-kit and by the influence of IL-3, IL-4, IL-9, and IL-10. Activated mast cells participate in the acute and chronic phases of inflammation. For instance, these cells contribute to the infiltration of leukocytes, tissue remodeling, and fibrosis [216, 217]. The expression of the AR has been reported in skin mast cells; nevertheless, the numbers and distribution of these cells seem not to be affected by androgens [218, 219]. Instead, TES interferes with the production of IL-6 [220] and induces the expression of IL-33 [221] by mast cells. IL-33 is known to regulate the formation of innate lymphoid cells (ILCs) and the production of Th2 cytokines [222]. Regarding female sex hormone action on mast cells, serum levels of immunoglobulin E (IgE, an antibody that induces the degranulation of mast cells) seem to fluctuate depending on the menstrual cycle phase [223]. Moreover, estrogen enhances IgE-induced mast cell degranulation and the release of histamine [224–226]. Meanwhile, progesterone diminishes the migration of mast cells and histamine secretion [227, 228].

Dendritic cells are originated from a CD34+ hematopoietic precursor that gives rise to myeloid (MP) and lymphoid (LP) progenitors [229]. The maturation of these cells is associated with the recognition of PAMPs and/or DAMPs. The process of maturation also implicates metabolic changes and gene transcription that lead to the migration of dendritic cells from peripheral tissues to secondary lymphoid organs. Mature dendritic cells express large amounts of C-C chemokine receptor type 7 (CCR7) and secrete IL-12 and IL-23, which promotes T-cell differentiation [230–232]. The effect of sex steroids on dendritic cells and their influence on lung disease have not been well explored. The acute exposure of bone marrow-derived dendritic cells (BMDCs) to 5α-DHT during antigen priming decreases the stimulation of T-cell cytokine production (IL-4, IL-10, and IL-13) in vitro [233]. On the other hand, estrogen promotes the differentiation of DCs from bone marrow precursors [234] and enhances their T-cell stimulatory capacity [233, 235]. Progesterone treatment decreases the production of the pro-inflammatory cytokines TNF-α and IL-1β by BMDCs [114].

15.4.5 T and B Lymphocytes

T and B lymphocytes or T and B cells are critical elements in the adaptive immune response. B cells carry out the antibody response, whereas T cells play a critical role in the cell-mediated response. In the antibody response, B cells secrete specialized proteins called immunoglobulins (Igs) that circulate across the bloodstream where they bind to and inactivate foreign antigens. Cell-mediated response employs T cells that react directly against foreign antigens, which are presented to them and further neutralize the cells that have been infected with distinct pathogens (bacteria or virus) [236–238]. The development of T cells occurs through phenotypic changes that involve the expression of essential membrane markers such as CD3, CD4, and CD8 [239]. Th1, Th2, Th17, and regulatory T (reg) cells conform the different lineages of CD4+ effector T cells [240].

T- and B-cell development is negatively regulated by androgen actions. For instance, castrated male mice show increased numbers of double positive CD4+ CD8+ T cells in the thymus and B cells in the spleen [241]. Furthermore, castration not only increases the number of B cells in the spleen but in the bone marrow, and this phenomenon is not altered by preceding thymectomy [242]. The negative regulation of B cells exerted by androgens has shown to be dependent on the AR activity. A murine model of general AR knockout (ARKO), and B-cell-specific ARKO, displays enhanced B-cell lymphopoiesis defined by increased numbers of B cells in the blood and bone marrow. Interestingly, the B-cell-specific ARKO group shows a lesser effect on B-cell lymphopoiesis compared with the general ARKO group, pointing out that the negative effect of androgens comprises a regulation of B cells and the stroma [243]. Contrariwise, estrogens restrain the B-cell lymphopoiesis by reducing the precursors, pro-B cell, pre-B cell, and mature B cells of the bone marrow [244]. Also, estrogens enhance the activity and antibody production of mature B cells [245].

Likewise, androgens negatively modulate T lymphocytes and monocytes in the blood. In this context, Yao et al. demonstrate in Sprague-Dawley rats that TES reduces the numbers of monocytes. However, the lymphocyte subpopulations show an increase in CD8+ T cells, while the numbers of CD3+, CD4+, and double positive CD4+ CD8+ T cells remain unaffected after TES treatment. Therefore, the immunosuppressive role of TES may be due to a decline in the number of monocytes, the change in CD4+/CD8+ ratio, and the increase of CD8+ T cells [246]. Moreover, TES causes a shift in the balance of Th1/Th2 cytokines through a reduction in TNF-α secretion in T-cell lines [247] and stimulates the production of IL-10 (an anti-inflammatory cytokine) in CD4+ T cells TES [248]. In addition, in a model of androgen deficiency, rats show decreased levels of IL-2, IL-6, IL-10, IL-12, and IL-13, whereas TES supplementation restores those levels [249].

Furthermore, it has been reported that estrogens modulate the differentiation and function of distinct T-cell phenotypes. In this context, E2 signaling through the ERα impedes the differentiation of Th1 and Th17 cells, conferring a protective mechanism against inflammation in experimental encephalomyelitis [250]. Besides, physiological concentrations of E2 promote the proliferation of T lymphocytes and the production of IFN-γ in vitro [251]. In Th2 cells, the stimulation of the ERα by E2 increases the expression of IL-4 [252].

The role of sex hormones on the function of immune cells in lung ailments has been established (Fig. 15.3). For instance, TES negatively regulates type 2 immune response seen in asthma evoked by Th2 cells [253]. E2 enhances the severity of pneumonia in adult mice with cystic fibrosis by improving the inflammation mediated by Th17 cells [254]. The involvement of immune cells, e.g., neutrophils, macrophages, eosino-phils, and T and B lymphocytes, in the pathogenesis of lung diseases, such as asthma and COPD, and the influence of sex hormones on these cells and their mediators are discussed in the next sections.

15.5 Sex Hormones in Inflammatory Lung Pathologies

15.5.1 Asthma and Chronic Obstructive Pulmonary Disease

Although asthma and COPD differ regarding pathogenesis, progression, prognosis, and treatment options, both ailments exhibit similar symptoms and inflammatory mechanisms [255–258]. Asthma is a chronic airway inflammatory disease that affects around 339 million people worldwide, as informed by the Global Asthma Report 2018. This airway disease is characterized by episodes of variable airflow obstruction (limitation of expiratory airflow), hyperresponsiveness, inflammation, and mucus production [41]. The etiology of this ailment has been related to heritability, environmental exposures, and sensitization to inhalant allergens [259–261]. Airflow obstruction is defined by spirometry parameters indicating a reduced forced vital capacity (FVC) and a reduced forced expiratory volume in 1 second (FEV1, less than 80%). Moreover, FEV1/FVC ratio is found minor than 0.7 in asthmatic patients [262]. Inhaled bronchodilators and corticosteroids are the main treatment in order to reduce inflammation and mitigate the symptoms of this illness [263]. Gender differences in the incidence and the severity of the asthma symptoms have been described and associated with hormonal changes through different life stages. During childhood, boys are more susceptible than girls to develop this ailment; however, in puberty, this trend reverses, and women display more severe symptoms [31, 264–266]. Fluctuations in progesterone and estrogens levels, such as those occurring during the menstrual cycle, have been correlated with the aggravation

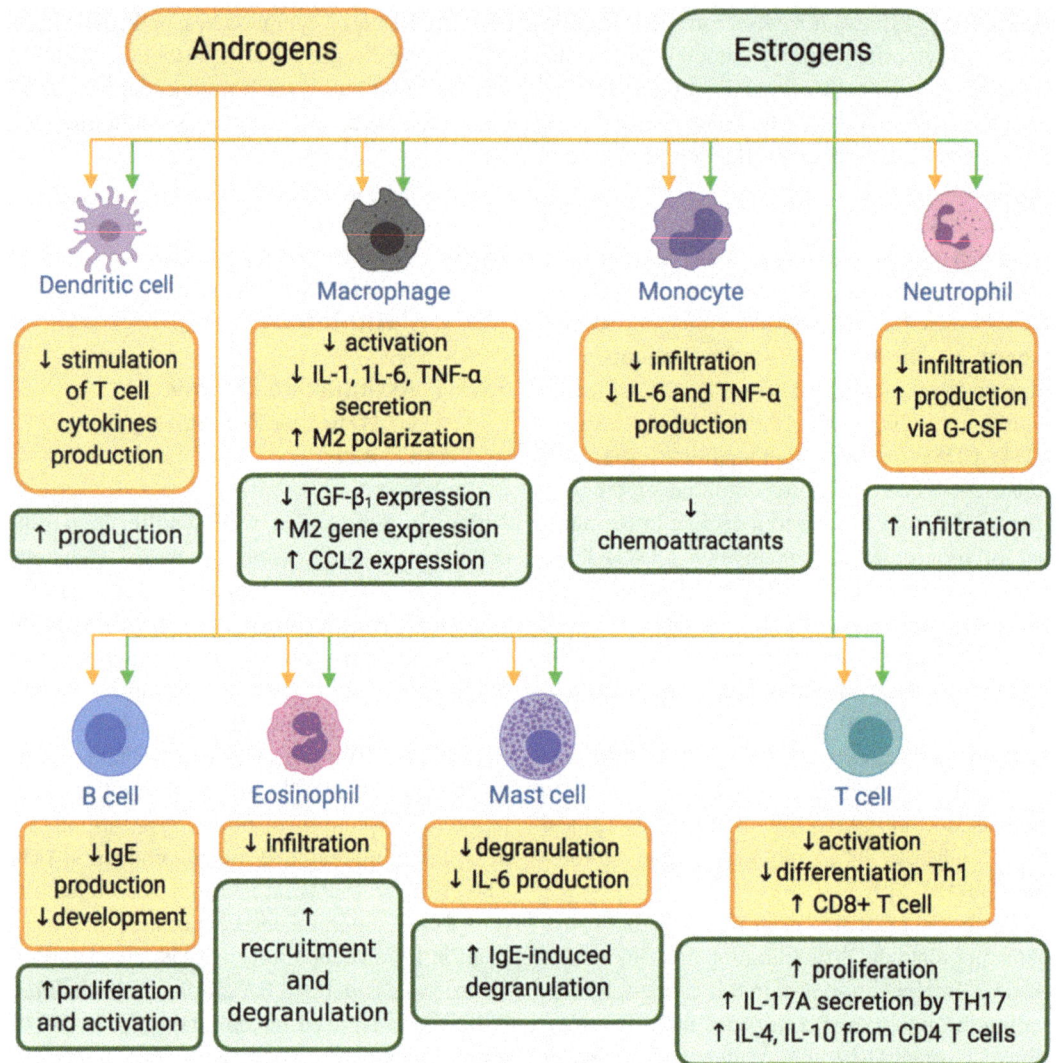

Fig. 15.3 Summary of androgen and estrogen effects on individual inflammatory cell types. Lung disease encompasses a substantial inflammatory component by the action of immune cells playing an essential role in the initial response such as dendritic cells, macrophages, monocytes, and T and B lymphocytes involved in the adaptive immune response. This figure summarizes current knowledge of androgen and estrogen actions on specific types of immune cells that are particularly important in lung ailments. Symbols' meaning and abbreviatures: ↓, inhibits; ↑, enhances. Tumor necrosis factor alpha (TNF-α); interleukin (IL)-1, 4, 6, 10, 17-A; monocyte chemoattractant protein, MCP-1 (CCL2); transforming growth factor beta1 (TGF-β1); granulocyte colony-stimulating factor (G-CSF); immunoglobulin E (IgE). For details, see the *Immune Cells* section

of this disease [25, 41, 267, 268]. Several studies have shown that 20–40% of premenopausal women experience pre-menstrual or perimenstrual asthma (PMA) [269–276], suffering from an exacerbation of the symptoms with increased bronchial inflammation in the week preceding menstruation [273, 276]. PMA is defined as a cyclical worsening of asthma during the luteal phase and/or during the first days of menstruation [274, 277]. PMA has also been associated with less atopy and poor lung function [273]. Furthermore, the use of oral contraceptives (OC) alleviates perimenstrual exacerbations in women with mild to moderate asthma [278, 279]. In addition, between 7% and 10% of all pregnant and childbearing-aged women have been reported to present asthma [280–282], and the use of OC does not improve the symptoms in women diagnosed with severe disease [283]. In women from 50 years of age, menopause may correspond with the beginning of asthma or be associated with exacerbations of a preexisting asthma condition, conforming a new phenotype described as menopausal-onset asthma [284, 285]. Around 18% of the total female asthma population suffers from menopausal-onset asthma, which is distinguished by the absence of atopy, aspirin sensitivity, persistent sinusitis, and frequent rate of hospitalizations [285]. In postmenopausal women with asthma, symptoms are commonly severe, and E2 levels have been found higher compared to those in non-asthmatic women [285–287]. Moreover, menopausal hormone therapy (MHT), and particularly the use of estrogen, increases the risk of developing asthma [288, 289]. In addition, progesterone in plasma reaches a peak about 24-fold higher than follicular phases. Interestingly, circulating progesterone is positively correlated with the peak expiratory flow rate (PEFR) during the luteal phase of the menstrual cycle [25].

On the other hand, androgens seem to reduce asthma exacerbations [76]. Higher plasma levels of TES in men, compared with those found in women, are thought to be useful favoring a bigger airway caliber and lung capacity [31, 290, 291]. It has been shown that, following the age of 11 years, the provocative concentration of methacholine necessary to produce a 20% decrement in FEV1 (PC20) increases in teenage boys but not in girls, pointing out an androgen-related improvement in airway responsiveness during puberty [292]. In this regard, non-genomic and genomic effects of androgens on airway smooth muscle [76, 293–297] and the inflammatory response in the asthmatic condition have been extensively explored [38, 42, 253, 298, 299]. Interestingly, the protective role of androgens against asthma in male patients declines after the fifth decade of life, and the symptoms of this ailment appear again. This could be explained by the decrease in plasmatic TES levels during this

life stage [31, 300, 301]. Moreover, Mileva and Maleeva [302] reported that male patients with moderate to severe asthma have lower levels of TES compared with those with mild symptoms. In addition, it has been noticed that asthmatic women carrying female fetuses are more susceptible to present symptoms of this illness compared to those with male fetuses [303]. The effects of androgens and estrogens (mainly E2) on the pathogenesis of asthma are discussed below with a special focus in inflammatory cells and their mediators.

COPD is a lung disease characterized by the presence of obstruction ventilator trouble (OVT), i.e., a persistent limitation of airflow that is not fully reversible [304]. This illness is usually caused by exposure to harmful particles or gases that induces emphysematous lung destruction and airway narrowing [305, 306]. Clinical symptoms include cough, sputum production, and progressive dyspnea that is unresponsive to steroids and bronchodilator therapy [307]. Similar to asthma, the diagnosis of COPD relies on clinical evidence and spirometry data, including FEV, FVC, and the ratio FEV1/FVC. A FEV1/FVC ratio less than 0.7, and the lack of full reversibility after the administration of salbutamol (400μg), indicate OVT [304]. It has been estimated that more than 12% of the world population suffers from this illness [308, 309]. The prevalence of COPD is more common in men than in women [308, 310]. This could be explained by the fact that men present higher occupational risks and/or higher rates of smokers [311, 312]. However, data suggests that women have a greater predisposition to this disease. The number of tobacco smoker women with a diagnosis of COPD has notably increased in the last years [33, 313]. The rate of lung function declines faster in female tobacco users than in male smokers. Also, the majority of COPD cases involving non-smokers or never smokers are women [314–317]. Typically, men with COPD have more emphysematous deterioration of the lung, while women tend to have more reactive airways and more pronounced airway narrowing [318, 319]. The apparent increased female susceptibility points out a sex hormone modulation of this illness [37, 320].

Asthma and COPD are characterized to produce chronic inflammation of the airways. Cells and mediators of inflammation implicated in these ailments are even targets for medical treatment [321, 322]. The inflammatory response in both lung diseases involves innate and adaptive immunity. Most asthmatic patients' course with eosinophilic inflammation that is driven by Th2 lymphocytes. This type of inflammation occurs due to exposure to allergens such as pollen, house dust mite (HDM), viruses, cockroach antigens, etc. [323–325]. Dendritic cells present allergenic peptides to uncommitted T lymphocytes and stimulate the production of allergen-specific T cells [326]. Then, epithelial alarmins IL-25, IL-33, and thymic stromal lymphopoietin (TSLP) recruit CD4$^+$ Th2 cells and group 2 innate lymphoid cells (ILC2s) that secrete IL-5 and other interleukins [327]. This interleukin regulates the generation of eosinophils in the bone marrow [325]. Eosinophil migration and recruitment into the lung is regulated by IL-5 as well and by the epithelial-secreted chemokines and C-C motif chemokines (CCL) 11 and 5 [328]. Moreover, type 2 inflammation is also associated with the production of other cytokines, including IL-4 and IL-13, which promote the synthesis of IgE by B lymphocytes [324, 329]. IgE regulates mast cell activity and degranulation. Significantly, mast cell infiltration in the airways has been associated with airway hyperresponsiveness (AHR) [330, 331] mediated by the release of bronchoconstrictor substances such as histamine, LTs, and prostaglandin (PG)-D$_2$. Mast cells and Th2 cells, additionally secrete IL-4, IL-5, IL-9, IL-13, and TNF-α [332, 333]. TNF-α, a well-known pro-inflammatory cytokine, is elevated in patients with asthma and is involved in the induction of AHR [334]. Experimental evidence has shown that TNF-α potentiates the agonist-induced increase in intracellular Ca^{2+} concentration ($[Ca^{2+}]_i$) [335] and contractile responses [334–339]. In addition to be involved in switching the isotype of B cells, IL-4 plays a key role in the differentiation of Th2 cells from uncommitted Th0 cells and in the initial sensitization to allergens [333]. IL-5 is deeply associated with the differentiation of eosinophils from their precursor

cells in the bone marrow as well as in eosinophil survival [340]. The use of glucocorticoids in asthma treatment decreases airway eosinophilia by inducing eosinophil apoptosis and inhibiting the response to IL-5 [341, 342]. IL-9 participates in mast cell proliferation and activation [343, 344]. IL-13 is another representative Th2 cytokine that not only participates in mucus production, airway inflammation, and remodeling but has been suggested to modulate steroid insensitivity [345–347]. In this regard, it has been shown that the administration of the anti-IL-13 antibody improves airflow obstruction in asthma patients when inhaled corticosteroid treatment seems not to work [348].

Nevertheless, not all asthmatic patients show this inflammatory pattern but develop an IL-17-mediated neutrophil inflammatory response that is mostly described in subjects with the most severe symptoms [42, 105]. Th17 cells are recognized as a distinct population of CD4+ T cells secreting IL-17A, IL-17F, and IL-22 [349]. The differentiation of this group of cells occurs after the stimulation of naive T cells by IL-1β, IL-6, IL-21, IL-22, IL-23, and transforming growth factor beta1 (TGF-β1) produced by macrophages and epithelial and dendritic cells [38, 42, 350, 351]. IL17-A is a key inductor of neutrophilic inflammation by evoking granulopoiesis and neutrophil chemotaxis [352, 353]. Furthermore, the inflammatory response mediated by Th17 cells is known to be highly associated with the resistance to corticosteroid treatment [42, 354].

COPD patients suffer from an accelerated decrease of lung function related to progressive airway obstruction caused by mucus hypersecretion and ciliary dysfunction [355, 356]. The inflammation developed in COPD subjects, located predominantly in the peripheral airways and lung parenchyma [357], is characterized by the presence of alveolar macrophages, neutrophils, T lymphocytes, dendritic cells, and B lymphocytes [358]. Most severe cases of COPD are commonly associated with higher numbers of B lymphocytes and neutrophils [304, 358]. T lymphocytes implicated in this disease are mostly CD8+ T-cytotoxic cells (Tc), but CD4+ Th1 cells are also increased [359]. In the blood of COPD patients, the proportions of IFN-γ- and TNF-α-producing CD8+ T cells are augmented compared with healthy ones [360]. Moreover, increased numbers of Th17 cells have been reported in COPD patients. IL-17A and IL-22 secreted by these cells modulate the recruitment of neutrophils and cause inflammation [361, 362]. Also, the number of macrophages is increased in the lungs from patients with COPD, and this increment corresponds to the severity of the disease [363]. Several inflammatory mediators are involved in the development of this ailment; for instance, alveolar macrophages and epithelial cells from COPD patients release more chemical mediators than in normal subjects [364, 365], including IL-1β, IL-6, TNF-α, CXCL1, CXCL8 (IL8), CCL2 (monocyte chemoattractant protein, MCP-1), reactive oxygen species (ROS), and LTB$_4$. Notably, some of the former cytokines act as chemotactic factors that promote neutrophil migration [366]. Cigarette smoke stimulates granulocyte production through granulocyte-macrophage colony-stimulating factor (GM-CSF), granulocyte colony-stimulating factor (G-CSF), and TGF-β1 released from airway epithelial cells and lung macrophages. TGF-β1 participates in the activation of myofibroblasts and airway smooth muscle cells, causing proliferation and fibrosis [358, 367]. Furthermore, granulocyte production leads to persistent neutrophilic inflammation present in most COPD patients [367]. Additionally, epithelial cells and alveolar macrophages release chemokines such as CXCL9 (induced by IFN-γ), CXCL10 (IFN-inducible protein 10), and CXCL11 (IFN-inducible T-cell alpha chemoattractant) [358]. The upregulation of CXCL10 and its receptor (CXCR3) contributes to the accumulation of CD4+ and CD8+ T cells in patients with this obstructive disease [368]. CD8+ T cells are cytotoxic entities that release perforins, granzyme B, and TNF-α, contributing to alveolar cell apoptosis in emphysema [369]. Matrix metalloproteinases (MMPs), e.g., MMP-9 and MMP-12, similarly contribute to proteolytic attack on the alveolar wall matrix [364, 370]. Furthermore, airway eosinophilia may occur when COPD exacerbations are mediated by viral infections [371].

IL-33 produced by epithelial cells has been suggested to be the key regulators in Th2- and ILC2-mediated eosinophilia in this illness [372]. Moreover, oral corticosteroid therapy has been shown to be effective in reducing eosinophilic inflammation in patients with COPD [373].

15.5.2 Androgens' Effects on Inflammation in Asthma and COPD

Airway smooth muscle (ASM) is one of the major structural elements of the airways. This smooth muscle layer controls airway caliber and tone [95, 295, 374]. Hyperresponsiveness to physical or chemical stimuli is a characteristic feature of asthma. AHR is described as an exaggerated airway narrowing [375] that is usually reversed with the use of bronchodilators [376]. Given the gender differences in the incidence and the outcomes of this illness, sex steroids' (e.g., androgen) non-genomic and genomic effects on the airways have been widely studied. Primordial evidence of non-genomic actions of androgens on ASM was found in 2006 by Kouloumenta et al. They observed that TES is capable of relaxing rabbit tracheal preparations pre-contracted with cholinergic agonists [377]. Further studies showed that TES and their metabolites 5α-DHT and 5β-DHT induce the relaxation of bovine and guinea pig ASM by blocking L-type voltage-dependent Ca^{2+} channels (L-VDCCs) [378]. In this context, published data from Dr. Montaño's research group confirmed that TES, 5α-DHT, 5β-DHT [379], and DHEA [294] relax the KCl and carbachol pre-contracted guinea pig tracheal tissues. Additional studies from the same research group pointed out that TES not only blocks L-VDCCs but interferes with store-operated Ca^{2+} channels (SOCCs) and inositol 1,4,5-trisphosphate (IP$_3$) receptor (ITPR), and induces the production of PGE$_2$ [296, 297]. Most recent published data shows that the chronic exposure to a physiological TES concentration increases the expression of β$_2$-adrenoceptor (β$_2$-AR) and upregulates delayed rectifier voltage-dependent K$^+$ channels (K$_V$) and high conductance

Ca^{2+}-activated K$^+$ channels (BK$_{Ca}$), enhancing the relaxing responses to salbutamol in guinea pig ASM [380].

Androgen effects on immune cells and their inflammatory mediators in healthy subjects and asthmatic patients have also gained relevance. For instance, a study on human peripheral blood mononuclear cells (PBMCs) showed that the number of cells secreting IFN-γ is correlated with the serum levels of DHEA-3-sulfate (DHEA-S, a metabolite of DHEA mainly produced in the suprarenal cortex that possesses weak androgen activity) in premenopausal women and men [381]. Moreover, it has been observed that patients with an asthmatic condition have reduced serum DHEA and DHEA-S concentrations compared with healthy subjects [382–384]. Also, DHEA significantly diminishes both Th1 and Th2 responses in cultured PMBCs from patients with asthma [385], and DHEA-S attenuates chemotaxis and migration of peripheral human neutrophils and inhibits chemokinesis of human ASM [386]. These observations point out that these androgens interfere with cell migration and inflammation in airways, maybe leading to the decrease of asthma symptoms. In this regard, the administration of nebulized DHEA-S to patients with poorly controlled moderate to severe asthma diminished the symptoms and improved the control of this disease [387]. Furthermore, asthmatic women with low serum levels of DHEA-S who received a supplementation with DHEA showed an upgraded lung function in asthma outcomes [388]. In allergic asthma, ILC2s seem to play a pivotal role in initiating and mounting the typical Th2 inflammatory response [343], which induces eosinophilic inflammation, AHR, and remodeling of the airways [42, 389]. In fact, ILC2s are known for producing higher quantities of IL-5 and IL-13 compared with Th2 cells [390–392]. Several reports have shown increased numbers and activation status of ILC2s in blood and sputum samples from adult and pediatric asthma patients [390, 393–397]. Likewise, genetic polymorphisms related to asthma susceptibility have also been shown localized to gene regulatory elements in ILC2s [398]. Animal models have served to prove that andro-gens exert a downregulation on ILC2s and Th2 inflammatory patterns [106, 253, 299, 399]. Two of these studies illustrate that DHEA reduces house dust mite-induced allergic inflammation by decreasing blood eosinophilia, IL-4, IL-5, and IFN-γ levels [399] and suppresses eosinophil infiltration and AHR through the modulation of Th2 cytokines in ovalbumin (OVA)-sensitized mice [400]. Male mice have low numbers of eosinophils and lymphocytes in bronchoalveolar fluid and lower IL-4 mRNA expression levels in splenic cells than castrated males and females [401]. Moreover, male mice show reduced numbers of ILC2 progenitors (ILC2Ps) and less severe IL-33-driven lung allergic inflammation. Furthermore, IL-5 and IL-13 levels, after IL-33-induced activation of ILC2s, are diminished in male mice compared to females [392]. Likewise, ILC2s from male mice display higher expression levels of killer cell lectin-like receptor subfamily G member 1 (KLRG1) and IL-33 receptor (ST2) [299]. Importantly, 5α-DHT through AR signaling limits the differentiation of ILC2Ps into mature ILC2s in the bone marrow [106, 299] (Fig. 15.4). These insights suggest marked differences between ILC2 development from male and female mice. Additionally, TES decreases house dust mite-induced airway eosinophilic and neutrophilic inflammation, IgE production, and AHR in castrated mice [253]. It is well known that mast cells infiltrate the bronchial epithelium and release bronchoconstrictor mediators in allergic asthma. In this regard, the systemic administration of 5α-DHT inhibits mast cell activation and degranulation. Also, this androgen reduces airway hyperplasia and mucus production in a murine model of OVA-sensitized females [402]. Mast cell degranulation leads to the release of leukotrienes, highly potent lipidic mediators also involved in the allergic response, specifically inducing ILC2 activation and lung inflammation [403]. Macrophages, eosinophils, and basophils are capable of releasing these lipidic molecules as well [404, 405]. Two classes of LTs have been described to modulate different targets. While LTB$_4$ acts as a potent leukocyte chemoattractant and enhances the macrophages' bacterial killing capability, the cysteinyl leukotrienes (Cys-LTs,

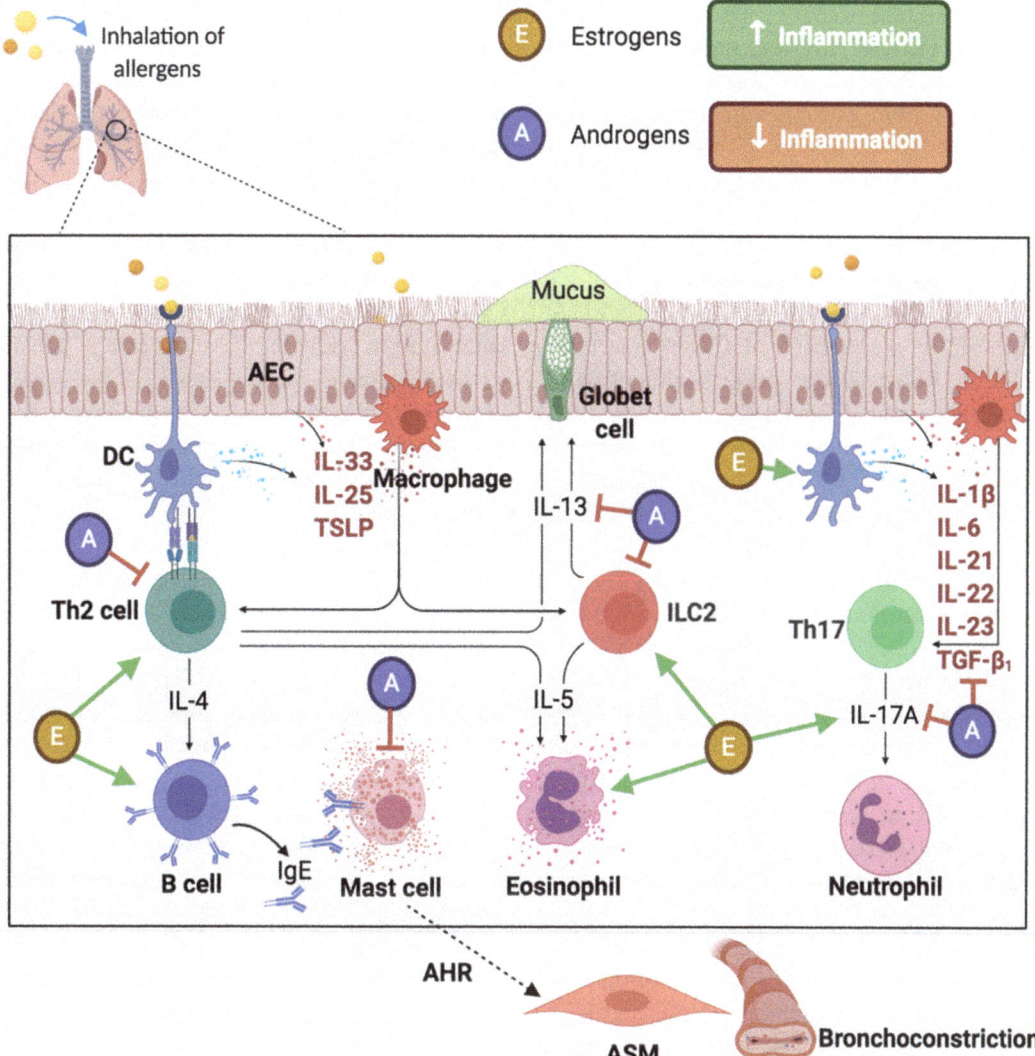

Fig. 15.4 Androgen and estrogen effects on the inflammation in asthma. Asthmatic disease involves an inflammatory-driven response of the airway by the exposure to allergens (pollen, house dust mite, cockroach, fungal antigens, etc.). Dendritic cells (DCs) present allergenic peptides to uncommitted T lymphocytes and stimulate the production of allergen-specific T cell. DCs, airway epithelial cells (AECs), and macrophages secrete epithelial alarmins IL-25, IL-33, and thymic stromal lymphopoietin (TSLP), which recruit Th2 cells and group 2 innate lymphoid cells (ILC2s) that secrete IL-4, IL-5, and other cytokines. Eosinophil migration is induced by IL-5, whereas IL-4 and IL-13 mainly favor immunoglobulin E (IgE) production by B cells with the consequent activation of mast cells. The infiltration of mast cells in the airways has been associated with airway hyperresponsiveness (AHR) mediated by the release of bronchoconstrictor substances such as histamine, leukotrienes, and prostaglandin D_2. These mediators activate their own receptors in the airway smooth muscle (ASM) provoking bronchoconstriction. Moreover, exposure to allergens also initiates an immune response through IL-1β, IL-6, IL-21, IL-22, IL-23, and TGF-β1 produced by DCs, AECs, and macrophages favoring naive Th differentiation to Th17. Th17 cells synthesize IL-17A, a key inductor of neutrophilic inflammation by evoking granulopoiesis and neutrophil chemotaxis. The effects of estrogens (E) or androgens (A) on the illustrated immune cells can differ depending on the hormone concentration and the timing and duration of the stimulus. Androgens diminish Th2 and ILC2 cell population, consequently limiting eosinophilic inflammation and IgE production. They also decrease IL-17A synthesis by Th17, lowering neutrophilic inflammation. Estrogens increase Th2 and ILC2 cell differentiation and induce IL-17A production from Th17 cells. Furthermore, estrogens augment total serum IgE, IL-5 production, and eosinophilia. In conclusion, estrogens may display detrimental effects on airway function by enhancing inflammation, and androgens exert opposite effects

LTC$_4$, LTD$_4$, and LTE$_4$) evoke bronchoconstriction and increase vascular permeability [406, 407]. For LT biosynthesis, arachidonic acid (AA) is excised from phospholipids in the plasma membrane by the action of phospholipase A$_2$ (PLA$_2$). Once AA is released, the nuclear membrane-bound 5-LOX-activating protein (FLAP) delivers it to 5-lipoxygenase (5-LOX) that converts AA into the different types of LTs [406, 408, 409]. In this regard, 5α-DHT and TES diminish the biosynthesis of LTs by interfering with 5-LOX localization via activation of type 2 extracellular signal-regulated kinase 2 (ERK 2) [407, 410] and by blocking the assembly of 5-LOX/FLAP [411]. The regulation of ERK2 and 5-LO trafficking exerted by androgens may explain the gender differences observed in the anti-leukotriene therapy, where young girls show better outcomes than boys [412]. Moreover, increase in [Ca^{2+}]$_i$ in ASM cells (ASMCs) is a primordial mechanism that triggers the exacerbated bronchoconstriction seen in asthma. In this context, a study in primary human ASMCs made by Kalidhindi et al. indicates that basal expression of AR is greater in males compared to females but increases with asthma or with an inflammatory condition in both genders. More interestingly, ASMCs from asthmatic females display a greater AR expression than males; however, androgen receptor may take minor functionality in females. In addition, TNF-α and IL-13 enhance histamine-induced increase in [Ca^{2+}]$_i$ in ASMCs; nevertheless, TES and 5α-DHT decrease the enhancement through AR signaling. AR effects on [Ca^{2+}]$_i$ increments are explained by the downregulation of stromal interaction molecule 1 (STIM1) and Orai1, key machineries in store-operated Ca^{2+} entry (SOCCE), and the increasing of SARAF (formerly known as TMEM66, a negative regulator of SOCCE) [413].

Not all patients with asthma course with type 2 inflammation but display an IL-17A-mediated neutrophil inflammatory pattern that is resistant to corticosteroid treatment. IL-17A is secreted by CD4+ Th17 cells and is associated with more severe asthma phenotypes [42, 105]. Moreover, macrophages and DCs express receptors for IL-17A and favor the synthesis of IL-6 and TNF-α [414]. Interestingly, the stimulation of the AR by TES decreases IL-17A protein expression and IL-23 receptor (IL-23R) mRNA expression from Th17 cells, reducing neutrophilic airway inflammation [253]. Also, IL-17A induces glucocorticoid receptor β (GR-β) expression and favors corticosteroid therapy resistance in patients with severe asthma [415]. α and β subtypes have been described as the two known glucocorticoid receptors. GR-β acts as an inhibitor of GR-α, and this latter isoform decreases the expression of inflammatory mediator genes [416, 417]. These findings point out that TES (by downregulating IL-17A) may decrease the expression of GR-β, promoting the anti-inflammatory effect of corticosteroids.

Persistent asthma (as seen in a IL-17A-mediated inflammatory response) is characterized by airway remodeling that involves epithelial-mesenchymal transition (EMT), cell mucus hypertrophy, subepithelial fibrosis, deposition of extracellular matrix proteins, and smooth muscle hypertrophy and hyperplasia [418, 419]. These cell modifications lead to pronounced airflow obstruction and epithelial damage that predispose to AHR. TGF-β1 induces EMT, and its increased expression levels correlate with severe asthma phenotypes [420, 421]. In this context, Xu et al. demonstrated that DHEA (via a genomic effect) inhibits the bronchial TGF-β1-induced EMT and preserves the epithelial morphology [422].

The influence of androgens on COPD patients and the inflammatory response in animal models have not been broadly investigated as occurring in asthma. However, since circulating TES levels have been positively correlated with cardiorespiratory function and muscle growth and strength [423, 424], low levels of this androgen may be associated with worse outcomes in patients with COPD [425]. In this context, numerous studies show that men with COPD have medical relevantly lower levels of TES and DHEA-S compared with healthy men [425–431]. Moreover, a recent research suggests that testosterone replacement therapy (TRT) decreases the rate of COPD patient hospitalizations and slows the progression of the disease [432]. COPD patients' usually display neutrophil-mediated inflammation, mainly located in the lung parenchyma, influenced by the action of IL-17A [433]. Macrophage activity also plays a crucial role in the lung of smokers and is even higher in COPD patients. Macrophages expressing IL-17A receptors have been observed in mice with myocarditis, and the genetic suppression of these receptors interferes with macrophage recruitment [434]. Similar to neutrophils, macrophages contribute to lung damage by synthetizing and releasing pro-inflammatory mediators, e.g., cytokines, chemokines, and ROS [363]. Moreover, cigarette smoke stimulates the activation and recruitment of macrophages, causing emphysema, involving the action of MMP-9, MMP-12, and CCL2 [197, 363, 435, 436]. In COPD, the role of the two distinct described phenotypes of macrophages (M1 and M2) is unclear, showing no sign of predominance for a determined subtype [197, 363]. The classical activation of macrophages (M1) is associated with Th1 immune response producing pro-inflammatory cytokines such as TNF-α, IL-1β, IL-6, and IL-12 [197, 202]. However, Th2 cytokines (IL-4 and IL-13) can alternatively activate macrophages as occurring in allergic asthma [437]. M2 macrophages commonly produce anti-inflammatory cytokines, including TGF-β, IL-10, CCL18, and CCL22 [197, 437]. Interestingly, evidence suggests that alveolar macrophages may be involved in the pathogenesis of COPD in a non-inflammatory manner; i.e., smoking induces a polarization pattern toward the down-regulation of M1-related inflammatory genes (CXCL9, CXCL10, CXCL11, and CCL5) and the induction of genes associated with the M2 polarization mechanisms (MMP-2, MMP-7, and the adenosine A3 receptor) [438]. Nevertheless, M2 alveolar macrophages contain inflammatory mediators that may cause an increase in cell recruitment, mucus secretion, and airway remodeling if they are immoderately discharged, as occurs in humans and mice asthmatic lungs [439, 440]. In this context and contrary to the anti-inflammatory androgen effects, it was revealed that, although 5-αDHT reduces lung inflammation, the same androgen enhances IL-4-stimulated M2 macrophage polarization in OVA-induced

allergic mice. Also, the genetic ablation of AR diminishes eosinophil recruitment and lung inflammation due to the compromised M2 polarization [440]. Therefore, further research is required in order to elucidate the role of androgens on alveolar M2 macrophages and in the inflammatory outcomes of diseases such as asthma and COPD.

The evidence indicates that male sex steroids have beneficial anti-inflammatory properties in asthma patients by attenuating innate lymphoid cells Type 2, Th2 cells, IL-17A-mediated response, and the leukotriene biosynthesis pathway, through different mechanisms (Fig. 15.4). Also, androgens might reduce the neutrophilic inflammation in COPD, but additional studies are required. Moreover, even though the use of androgens in order to increase muscle mass and improve cardiorespiratory functions is promising, the effect of these hormones in men with COPD seems to be modest, and more research is indispensable to resolve whether TRT could be an option in COPD treatment. Regarding the use of androgens as an anti-inflammatory treatment in asthma and COPD, it should be considered the metabolic pathway of DHEA and TES, leading to the production of estrogens. Furthermore, 5α-DHT, a reduced metabolite of TES with important androgenic actions, has been associated with prostate cancer [441]. In this context, 5β-DHT, the other reduced metabolite of TES with minor androgenic activity and without estrogenic effects, might be taken into account as a potential therapeutic choice [76], although clinical studies are imperative to support this notion.

15.5.3 Estrogens' Effects on Inflammation in Asthma and COPD

It has been established that the incidence of asthma is more common in young boys and adult women, and that the severity of the symptoms may increase during pregnancy [26, 31, 105]. The transition from childhood to adulthood is distinguished by a higher probability of persistence of wheezing in females [300, 442].

Therefore, the influence of female hormones (mainly estradiol) on the airway biology has been widely explored [95, 141, 142, 145, 443–446]. Interestingly, a contradicting role of estrogen suggests either the induction of AHR and inflammation [214, 447–449] or an improvement of asthma symptoms by downregulating inflammation and favoring ASM relaxation [214, 450–453]. In this regard, several bronchodilator mechanisms have been shown to be affected through the rapid actions of female sex steroids. In 1983, a group of researchers reported that E2 in supraphysiological concentrations enhances the bronchodilator response to adrenaline and noradrenaline in pig bronchus and the increment in the potency of catecholamine-induced bronchodilation may be mediated by an inhibition of catecholamine metabolism or uptake [454]. Later on, it was exhibited that E2 causes relaxation of isolated trachea muscle strips (pre-contracted with acetylcholine or KCl), independently of the adrenergic system, but possibly involving prostaglandin synthesis and cyclic guanosine monophosphate (cGMP) modulation [455]. It is well known that prostaglandins modulate cyclic adenosine monophosphate (cAMP) and cGMP formation. Cyclic nucleotides stimulate protein kinases that influence the inhibition of Ca^{2+} influx channels, which favors bronchodilation. In this context, E2-induced increase in cAMP has been observed in porcine coronary arteries [456]. Also, physiological concentrations of E2 potentiate the relaxation evoked by isoproterenol via cAMP production in human and guinea pig ASM. The cAMP pathway triggered by E2 in a non-genomic way promotes the blockade of Ca^{2+} influx (mainly through L-VDCCs) and the further decrease in $[Ca^{2+}]_i$ [453]. In addition, as demonstrated by Dimitropoulou et al. [457], the increase in cyclic nucleotides leads to the phosphorylation and the opening of K^+ channels. They observed that a physiological E2 concentration inhibits AHR in asthmatic mice via an ER signaling pathway that implicates the activation of protein kinase G (PKG) and the opening of BK_{Ca} channels. Nevertheless, a recent study demonstrates that 17β-estradiol has a little effect on the formation of cAMP in primary cultured human ASMCs.

Interestingly, the same study shows that progesterone is capable of promoting the formation of cAMP after 3 minutes of stimulation [458]. In the last years, some studies have further demonstrated that estrogens diminish Ca^{2+} levels in ASMCs from different species via rapid (non-genomic) effects through ERs or by directly blocking membranal Ca^{2+} channels. Supraphysiological concentrations of E2 lower ASM basal tone through the obstruction of L-VDCCs in guinea pig ASM [295]. Moreover, the acute exposure of physiological concentrations of E2 decreases histamine-evoked Ca^{2+} influx via the inhibition of L-VDCCs and SOCCE in ASMCs from women [167]. It has been shown that asthmatic airways are less responsive to nitric oxide (NO) donors; however, E2 through ERβ reverses this phenomenon favoring ASM relaxation [451]. Also, E2 (in a physiological range) rapidly increases NO production in human bronchial epithelium from women and produces relaxation of bronchial rings pre-contracted with acetylcholine (Ach) [459]. In addition, a study has demonstrated that P4 and 5β-pregnanolone prevent histamine- or carbachol-induced contraction in guinea pig ASM [460].

Chronic effects of estrogens, or the lack of them, on Ca^{2+} handling and AHR have also been investigated. For instance, the genetic ablation of ERα induces AHR among other lung function anomalies and interferes with airway smooth muscle and nerve physiology, probably involving the dysregulation of M2 muscarinic receptors [461]. Matsubara et al. reported that endogenous estrogens downregulate AHR in OVA-induced asthmatic female mice. Correspondingly, 17β-estradiol suppresses AHR in male mice challenged for 10 days with OVA [462]. Dimitropoulou et al. observed that estrogen replacement therapy prevents the development of AHR and inflammation (important markers in asthma) in ovariectomized asthmatic mice. Also, they found a reduction in TGF-β1 levels from bronchoalveolar lavage fluid (BALF) of estrogen-treated mice [463]. Recently, Bhallamudi et al. [464] found in non-asthmatic and asthmatic human ASMCs that long-term exposure (24 h) to propylpyrazoletriol (an ERα agonist) enhances the Ca^{2+} response to histamine. Differently, ERβ stimulation with the agonist WAY-200070 (WAY) evokes a decrease in histamine-induced $[Ca^{2+}]_i$ increase. Besides, TNF-α and IL-13 improve $Ca^{2+}]_i$ responses evoked by histamine, Ach, and bradykinin; and ERβ activation abolishes this phenomenon. Interestingly, E2, a non-selective ER agonist, does not show significant changes in $[Ca^{2+}]_i$ when ASM cells are stimulated only by histamine. However, E2 induces a decrease in $[Ca^{2+}]_i$ increase elicited by histamine bradykinin and Ach when ASMCs are pre-treated with TNF-α or IL-13. ERβ effects on agonist-induced $[Ca^{2+}]_i$ increases seem to be mediated by an augment on sarco(endo)plasmic reticulum Ca^{2+} ATPase 2 (SERCA2) function and by the inhibition of L-VDCC [464].

Estrogen actions on the inflammatory response and their cellular mechanisms in asthma have been explored as well. Most studies point out that estrogens, mainly through ERα signaling, increase allergic airway inflammation and allergen-induced AHR [141, 142, 145, 461]. Th2 inflammation in asthmatic patients influence B-cell activation that favors increased levels of IgE [465]. Women have been shown to have higher serum IgE levels during puberty, which are associated with more severe asthma symptoms provoked by histamine, IL-4, and IL-13 released from mast cells [442, 465, 466]. Likewise, OVA-induced female asthmatic mice with increased IgE levels (compared with those observed in males) show minor sensitivity to budesonide treatment against IL-5 production and the development of AHR [447]. The exposure to environmental tobacco smoke (ETS) increases Th2 cytokine response after allergic (OVA) sensitization. Female mice exposed to ETS display more IgE-positive cells in the lungs and augmented levels of IL-4, IL-5, IL-10, and IL-13 than males [449]. Additionally, acute exposure of physiological concentrations of E2 elicits the increase in $[Ca^{2+}]_i$ via ERα activation and the subsequent IgE-induced degranulation and LTC_4 production in a rat basophilic leukemia cell line (RBL-2H3M) and in a human mast cell line (HMC-1) [226]. These insights indicate that females are more susceptible to develop a more

damaging inflammatory response after the exposure to tobacco smoke as occurring in COPD. Furthermore, it has been proposed that female sex steroids are responsible for inducing eosinophilia in allergic asthma. A study performed by Riffo-Vázquez et al. shows that gonadectomy decreases total serum IgE, IL-5 production, and pulmonary eosinophilia in female mice sensitized to OVA [214]. In addition, it has been demonstrated in a murine model of allergic asthma that progesterone promotes airway eosinophilia and provokes AHR [215].

Dendritic cells and macrophages are also modulated by estrogen actions. These cells present allergenic peptides to uncommitted T lymphocytes and favor type 2-mediated airway inflammation [51]. In this context, it has been observed that E2 triggers the expression of DC-derived cytokines, such as IL-6, IL-8, and MCP-1 [467], and enhances the production of human DC population, promoting the proliferation and differentiation of Th cells into Th2 cells [468]. Asthmatic patients present significantly increased numbers of M2 macrophages in BALFs compared to normal subjects [469, 470]. OVA-sensitized female mice also have more M2 alveolar macrophages compared to males [471, 472]. In addition, alveolar macrophages from female mice exhibit greater expression of M2 genes, IL-4 receptor (IL-4R)-α, and ERα after OVA challenge. Furthermore, IL-4 induces M2 gene expression in female mice macrophages, and exogenous E2 potentiates this polarization [473].

Female sex hormones have also been associated with the IL-17-mediated inflammatory pattern displayed in some patients with asthma. This phenotype is distinguished by manifesting neutrophilic inflammation as occurring in COPD [42, 304]. In this context, Newcomb et al. found that women with severe asthma have increased numbers of Th2 cells and greater production of IL-17A compared with asthmatic men [474]. Given these insights, the authors claimed that female sex steroids are responsible for increasing Th17 cell differentiation and IL-17A production from these cells. However, they did not observe a correlation between E2 and P4 plasma levels (at the time of obtaining blood samples) with IL-17A protein expression in Th17 cells from women and further suggested that the increase in the production of IL-17A in women compared to men is due to the exposure of T cells to female sex hormones during the development of Th17 cells in the body. To further determine the mechanism by which sex steroids modulate the production of IL-17A, they performed an experiment where it was observed that the administration of E2 and P4 augments the expression of IL-17A and IL-23 receptor (IL-23R) in Th17 cells from ovariectomized mice [474]. In this regard, it has been established that IL-17A requires IL-23R signaling to maintain and stabilize the Th17 cell phenotype [475, 476]. Moreover, the same authors found an increased expression of IL-23R in Th17 cells from women compared to men [474]. Another study shows that E2 chronic stimulation enhances the expression of IL-6 and IL-17 in peripheral blood mononuclear cells from asthmatic patients, both females and males [477]. Interestingly, it has been demonstrated that the estrogen deficiency occurring in postmenopausal women is associated with increased serum levels of IL-17A [478]. Anti-inflammatory effects of estrogen in asthma have been reported as well. E2 inhibits cell recruitment into the lungs during the allergic inflammatory process, preventing the cell adhesion through the downregulation of E-selectin. Finally, estradiol treatment of ovariectomized OVA-induced allergic rats diminishes the release of LTB_4, IL-10, and TNF-α [267].

More severe asthma symptoms reported in women appear to be associated with hormonal changes occurring during the different life stages, i.e., menstruation, pregnancy, and menopause [38, 276]. During pregnancy, asthma can change its manifestations, and women with severe disease may present worsening of symptoms that do not improve with the use of OC [283]. It has been described that in the third trimester of pregnancy, one third of women with asthma show an improvement, one third show no change, and another third get worse [479, 480]. In this regard, highest estrogen levels during the third trimester of pregnancy may explain the different manifestations of the disease [265]. Cytokines with an anti-inflammatory role such as IL-10 and IL-4

have been shown augmented after the stimulation with E2 (in concentrations occurring during pregnancy) in CD4 T cells from humans and mice [252, 481, 482]. Furthermore, higher physiological concentrations of estrogen reduce the T-cell production of TNF-α [483]. In the menopausal period, asthmatic patients tend to experience more pronounced respiratory complications [484]. It has been proposed that E2 serum levels may function as an appropriate biomarker for asthma severity in postmenopausal women [285]. Height reduction of the thoracic spine due to osteoporosis related to estrogen deficit in menopausal women might be implicated in the decrease of lung function [485]. Menopause hormone therapy based on estrogen administration is used to alleviate climacteric symptoms, including osteoporosis; vasomotor disorders; skin, urogenital, and weight changes; etc. [486]. Nonetheless, risks as developing thromboembolic venous diseases via alterations in blood coagulation must be taken into account when high doses of MHT are used [486–488]. Aminoestrol, butolame, and pentolame, types of 17β-aminoestrogens, have been suggested as an alternative of MHT, considering their weak estrogenic and antithrombotic activity [489–491]. In this regard, Flores-Soto et al. observed that butolame and pentolame but not aminoestrol, cause hyperresponsiveness to carbachol, histamine, and KCl in guinea pig ASM through the activation of L-VDCCs [492]. This finding points out that aminoestrol is a good alternative for MHT, but further studies are required to critically asses the role of aminoestrogens in postmenopausal asthma exacerbations, particularly in the inflammatory response.

Several recent works have shown a differential expression and activation of both ERα and ERβ in non-asthmatic and asthmatic ASMCs that may explain the distinct outcomes regarding airway inflammation, remodeling, and responsiveness in females [141, 142, 145, 161, 443]. Aravamudan et al. quantified the expression of ERα and ERβ in ASMCs from asthmatic and non-asthmatic subjects. They observed that ERβ expression is greater in asthmatic ASMCs and in ASMCs exposed to TNF-α or IL-13 as well. The expression of the ER isoforms is regulated by inflammatory signaling pathways such as p42/pp44 mitogen-activated protein kinase (p42/44MAPK), phosphatidylinositol 3-kinase (PI3K), and NF-κB, proteins implicated in airway remodeling and asthma development [161]. Furthermore, the dual effects of 17β-estradiol on smooth muscle have been reported. Some works demonstrate that this estrogen promotes a mitogenic effect in female and male rabbit ASM [493] and induces rat vascular smooth muscle proliferation via ERs and MAPK signaling cascade [494], while other studies show that E2 inhibits the proliferation and migration of vascular smooth muscle cells [495, 496]. Recent studies by Ambhore et al. suggest that ERβ plays a protective role against airway remodeling and hyperresponsiveness. One of these studies confirmed that ERβ but not ERα activation inhibits platelet-derived growth factor (PDGF)-induced proliferation in ASMCs from asthmatic and non-asthmatic males and females by interfering with cell cycle mechanisms and suppressing proliferative proteins [443]. Also, the administration of a selective ERβ agonist (WAY) decreases airway remodeling and AHR in a murine model of allergen-induced asthma; specifically, ERβ activation abolishes the increase in vimentin, fibronectin, and collagen I caused by allergic sensitization [141]. Fibronectin and collagen are molecules highly involved in extracellular matrix (ECM) deposition that leads to airway remodeling and hyperresponsiveness [497–499]. In this regard, another study shows that ERβ activation diminishes TNF-α-induced increased protein expression of ECM proteins such as collagen I, collagen III, and fibronectin in human ASMCs from asthmatic and non-asthmatic subjects. Also, ERβ signaling reduces the activity of MMP-2 and the expression of MMP-2 and MMP-9 in both asthmatic and non-asthmatic ASMCs [142]. These proteolytic enzymes are key regulators in the ECM degradation and the progression of asthmatic airway remodeling [500–502]. Interestingly, the inhibitory effect of ERβ on MMP expression and activity seems to be mediated by the NF-kB signaling pathway [142]. The most recent study about the protective role was performed by Kalidhindi et al., using an allergen-induced asthma model in ERα and ERβ

knockout (KO) mice [145]. Initially, they observed that female mice exposed to the allergen show a more pronounced decay in lung function (in terms of airway resistance and compliance) compared to male mice. Correspondingly, KO animals display a deteriorate lung function compared to ERα KO and WT KO animals. In addition, the genetic ablation of ERβ results in a more prominent airway remodeling and responsiveness and increased expression of fibronectin, vimentin, and alpha smooth muscle actin (α-SMA), consistent with the observations made by Ambhore et al. [141, 142]. These explorations help to clarify the role of estrogen receptors both α and β, pointing out a protective role for the latter one, and contribute to explain the gender differences observed in asthma prevalence after puberty, when women present more severe symptoms.

Otherwise, the influence of estrogens in COPD has been poorly explored. Data suggest that women are more likely to have COPD and present a faster decline of lung function than men. Also, epidemiological studies have shown that hormone replacement therapy (HRT) containing estrogens exacerbates COPD [503]. Unlike the reported for asthma, the expression of the ERs seems not to play an essential role in the development of COPD, since the three major subtypes, ERα, ERβ, and GPR30, are expressed in the same grade in lung tissues from COPD patients and normal subjects, including both men and women. However, the intracellular pathways that lead to the production of female sex steroids are upregulated. Aromatase and 17β-HSD1, two of the main enzymes responsible for the generation of estrogens, are increased in alveolar macrophages of COPD patients compared with controls [504]. Moreover, once smoke cigarette is inhaled, chemicals are metabolized in two separate phases. Cytochrome P450 enzymes accomplish the phase I. These enzymes are a large family of proteins with the critical function of metabolizing cigarette smoke and other environmental irritants, turning them into intermediate compounds. Phase II enzymes conjugate and secrete the metabolites produced in the former clearance steps. A downregulation in either expression or function of phase II enzymes might suppose the accumulation of metabolites produced by CYP in the lung, causing oxidant injuries to the tissue [36, 505]. Interestingly, it has been suggested that E2 upregulates CYP enzymes. For instance, female mice show a more pronounced sensitivity to naphthalene (an important component of cigarette smoke) toxicity and a more prominent pattern of airway epithelial injury than male mice [506]. Also, female mice have more significant expression of CYP enzymes and augmented accumulation of naphthalene metabolites [507]. Furthermore, the stimulation of the ERα by E2 increases the basal and the smoke-induced expression of CYP1A1 and CYP1B1 (two members of CYP family) in human bronchial epithelial cells [508]. Moreover, a CYP1A1 differential metabolic activity and distinct outcomes of the produced metabolites have been claimed. It has been demonstrated that CYP1A1 shows high metabolic activity for E2 2-hydroxylation, followed by 15α-, 6α-, 4-, and 7α-hydroxylation [509]. 4- and 16α-hydroxylated estrogens may contribute to cancer development [510, 511], while 2-hydroxylated estrogens are thought to have anti-cancerogenic activity and a higher rate of excretion the in premenopausal female smokers [512–514]. The augment in the expression of CYP is associated with increased levels of estradiol and oxidants, mediated by the enhancement in the metabolism of cigarette smoke [36, 514]. These insights point out that COPD female patients may have increased production of E2 and a higher production of cigarette smoke metabolites leading to more severe lung damage.

These basic and clinical reports indicate contradictory outcomes regarding anti-inflammatory and pro-inflammatory properties of estrogen. Although several studies show that E2 is capable of relaxing ASM and reducing AHR, this sex steroid may lead to disadvantageous results in asthma patients by enhancing allergic and IL-17A-mediated inflammation (Fig. 15.5). Also, dual hormonal effects exerted by E2 and the differential expression and activation of estrogen receptors may limit the use of female hormones in asthma treatment. Moreover, the findings suggest that women under menopause hormone therapy and experiencing PMA might consider the risk of worsening the lung function. Furthermore, estrogen may promote neutrophilic inflammation in subjects with COPD, but further studies are necessary in order to corroborate this appreciation. Additionally, the use of aromatase inhibitors in a pulmonary illness such as lung cancer has been validated in preclinical assays [515, 516]. In this regard, the possibility of 17β-HSD1 and aromatase serve as potential therapeutic targets in COPD patients should be explored.

15.5.3.1   Pulmonary Fibrosis

Pulmonary fibrosis (PF) is a chronic interstitial lung disease characterized by progressive remodeling of the lung parenchyma with extracellular matrix deposition that leads to an abnormal tissue repair (lung scarring) [517, 518]. This disease affects approximately three million persons worldwide and is more common in men than in women [519–521]. The etiology of this disease has been related to infections [522–524], environmental and occupational pollutants [525, 526], cigarette smoke [527, 528], obstructive sleep apnea, and others [529, 530]. However, it can also manifest without any associated pathology, known as idiopathic pulmonary fibrosis (IPF). The PF symptoms, such as dyspnea and dry cough, progressively worsen, resulting in a median survival time of 3 to 5 years after diagnosis [518, 531]. Spirometry reveals a restrictive respiratory pattern, attributed to the accumulation of parenchymal scar tissue and the subsequent alteration of normal lung architecture [532, 533].

Fibrogenesis in the lung involves the fibroblast hyperproliferation at alveolar injuries and the excessive deposition of extracellular matrix (ECM) components, along with signaling pathways that degrade the ECM [534–536]. The alveolar endothelial injury caused by infections, tobacco smoke, gastroesophageal reflux contents, and environmental pollutants favors an impaired epithelial regrowth above an irregular matrix. During PF development, epithelial-mesenchymal transition (EMT) occurs. In this process, pro-fibrotic cytokines, including TGF-β1, confer to

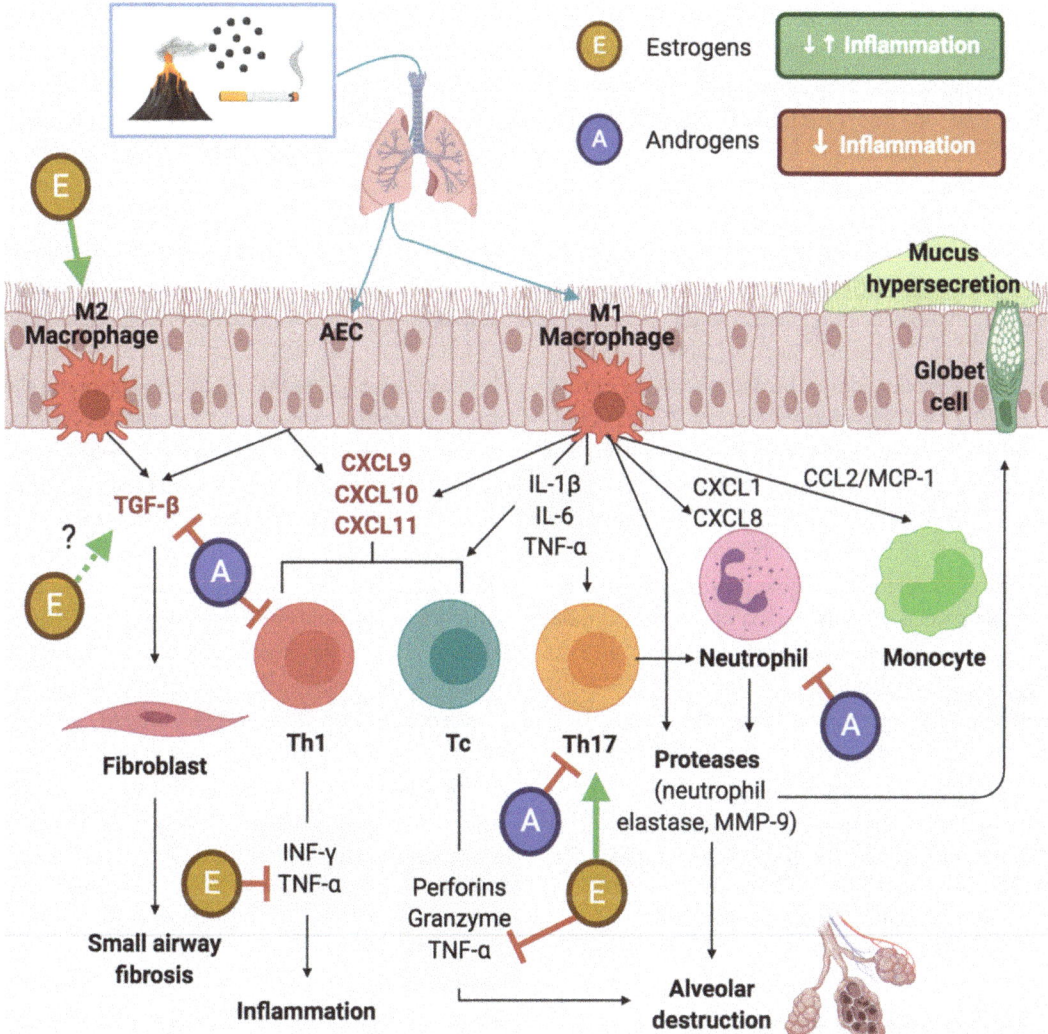

Fig. 15.5 Androgen and estrogen actions on inflammation in chronic obstructive pulmonary disease (COPD). Inhaled cigarette smoke and environmental and occupational exposures activate airway epithelial cells (AECs) and macrophages to release several chemotactic factors that attract T helper 1 (Th1) cells and Th1 CD8+ T cells (Tc) to the lungs, including CXC-chemokine ligand 9 (CXCL9), CXCL10, and CXCL11. Th1 cells release cytokines that induce inflammation, and Tc release perforins, granzyme B, and tumor necrosis factor α (TNF-α), contributing to alveolar cell apoptosis. Cytokines (IL-1β, IL-6, TNF-α) released from M1 macrophages (proinflammatory cells) activate Th17 cells, which modulate the neutrophilic inflammation. The attraction of neutrophils and monocytes is mediated by CXCL1, CXCL8, and CC-chemokine ligand 2 (CCL2), respectively. Inflammatory cells secrete matrix metalloproteinase 9 (MMP-9), which causes elastin degradation and emphysema. Neutrophil elastase evokes mucus hypersecretion. AECs and M2 macrophages (anti-inflammatory cells) release transforming growth factor β1 (TGF-β1), which stimulates fibroblast proliferation, provoking fibrosis of the small airways. The effects of estrogens (E) or androgens (A) on these immune cells can change depending on the hormone concentration and the timing and duration of the stimulus. Androgens diminish Th1 responses and inhibit TGF-β1-induced epithelial-to-mesenchymal transition (EMT) and preserve the epithelial morphology. In addition, androgens decrease IL-17A synthesis by Th17 cells, lowering neutrophilic inflammation. On the other hand, estrogens potentiate the M2 polarization of macrophages, favoring an anti-inflammatory condition, and might enhance the secretion of TGF-β1. Moreover, high physiological concentrations of estrogen reduce the T-cell production of TNF-α. Estrogens may present dual effects on airway inflammation in COPD, and androgens may reduce this response

epithelial cells, properties of mesenchymal cells to produce collagen [537–539]. This cytokine also promotes the accumulation of fibronectin in the ECM and the differentiation of fibroblasts into myofibroblasts [540, 541].

Moreover, T cells modulate lung fibrosis development. Th1 cytokines attenuate fibrosis, whereas Th2 cytokines promote fibrogenesis [542, 543]. In this regard, several studies have shown increased IL-13 expression in patients with IPF and in animal models [544–547]. IL-13 is mainly produced by Th2 lymphocytes, epithelial cells, ILC2s, and M2 macrophages [548–550]. This interleukin stimulates the proliferation of fibroblasts and induces pro-fibrotic cytokines, e.g., TGF-β1 and PDGF [551, 552]. In animals, exposure to bleomycin is the standard model of injury-associated PF since it promotes lung damage associated with apoptosis of alveolar epithelium, loss of the epithelial function, and an increased inflammatory response [553–555]. Interestingly, bleomycin also stimulates IL-13 production and myofibroblast differentiation [556]. Additionally, IL-17 has been linked to pro-fibrotic effects through interactions with TGF-β signaling [555, 557, 558]. It has been proved that, by blocking IL-17 production, the progression of PF is delayed in different murine models of lung fibrosis [557, 559, 560].

15.5.3.2 Androgens' Effect on Inflammation in Pulmonary Fibrosis

Gender is an important factor in determining the risk and prognosis for PF. The prevalence of PF is greater in men, who display faster progression and less survival rates compared to women [561, 562]. Moreover, the incidence of this disease increases with age, appearing between the fifth and seventh decades of life [563, 564]. Studies suggest that androgens may contribute to increase lung injury and fibrosis [48, 49]. However, the sex differences in PF have been studied mostly in animal models. In this context, Voltz et al. found in a murine model of bleomycin-induced PF that males show a decreased lung function and an increased fibrosis compared with females. Furthermore, mice castration restores lung func-

tion, and 5α-DHT replacement therapy aggravates it [49]. Another research group using the same model observed that aged male mice exhibit augmented collagen deposition and neutrophilic alveolitis, and an elevated mortality, compared to female mice or young mice [565]. Also, young and old mice exposed to bleomycin show exacerbatedlevels of TGF-β, and aged male mice present increments in neutrophil chemoattractants such as IL-17A, chemokine (C-X-C motif) ligand 1 (CXCL1), and CXCL2 [565, 566]. These observations suggest that androgens promote the fibroproliferative response associated with fibrocyte recruitment and favor the susceptibility to present lung fibrosis. In another study, since decreased levels of DHEA have been related to immunosenescence [567], the levels of DHEA and DHEA-S in BALF and plasma from patients with IPF were analyzed. Interestingly, DHEA/DHEA-S ratio is significantly decreased in plasma from males with IPF. Furthermore, the same study shows that DHEA (100μM) reduces human lung fibroblast proliferation and differentiation into myofibroblast induced by TGF-β1 [568]. Recently, Cephus et al. suggested that androgens negatively regulate ILC2 and ILC2-derived IL-13 production. They demonstrated that chronic administration of 5α-DHT to mice reduce IL-13 production from lung ILC2s. Moreover, IL-13+ ILC2s are also significantly decreased in sham-operated male mice compared to gonadectomized male mice and sham-operated female mice [298]. In conclusion, studies show discrepancies, indicating that androgens may play a promoting or protective role in the development of PF, and more information is needed related to androgen effects on the inflammatory response in lung fibrosis.

15.5.3.3 Estrogen Effects on Inflammation in Pulmonary Fibrosis

The development PF has also been thought to be influenced by estrogen actions. Distinct reports indicate a protective role for estrogen; however, animal models have shown mixed results. In this context, Gharaee-Kermani et al. observed that female Fisher rats with bleomycin-induced PF have higher mortality rates and more severe fibrosis compared to male rats. Furthermore, the authors found that ovariectomized animals presented less fibrosis and estradiol replacement therapy (ERP, 1 and 10 nM) increases procollagen 1, IL-4, and TGF-β1 mRNA expression in fibroblast from bleomycin-treated rats [48]. In contrast, a protective role of estrogen on lung fibrosis has also been proposed. A study conducted in ovariectomized female relaxin gene KO (Rln1−/−) mice revealed an increment in collagen concentration and deposition in the lung. Relaxin is a hormone capable of diminishing fibrosis in several organs [569]. Other research group demonstrated in Sprague-Dawley (SD) rats that ovariectomy exacerbates bleomycin-induced PF and pulmonary hypertension, and 2-methoxyestradiol attenuates this condition [570]. The discrepancy of results in which estrogen may promote or inhibit lung fibrosis can be explained by the difference among the species of rodents used in the former studies since ovariectomized Fisher rats develop a less severe inflammatory response to pneumotoxins and have twice higher plasma levels of E2 than SD rats. Moreover, elevated levels of E2 potentiate the activity of NF-κB, which, along with E2, contribute to a pro-inflammatory state [570–572].

Monocytes and their derivatives participate in the immune response developed in lung fibrosis. Studies confirmed that estrogen might inhibit the expression of monocyte chemoattractants [573–575]. Alveolar macrophages are a vast source of pro-fibrotic molecules such as TGF-β1, PDGF, and MMPs [566, 576]. PDFG is a potent fibrogenic molecule that promotes PF through fibroblast activation [577]. In the lungs of IPF patients, PDGF expression is increased in epithelial cells and macrophages [578, 579]. It has been reported that this growth factor induces Ca^{2+} waves through IP_3 receptors and modulates gene expression of ECM proteins in human pulmonary fibroblasts [580]. In this context, Ambhore et al. postulated that the activation of the ERβ by a selective ERβ agonist suppresses PDGF-elicited proliferation of human ASMCs [443]. Furthermore, estrogen regulation on the main pro-fibrotic cytokine has also been reported. In a rat model of congenital diaphragmatic hernia (CDH) induced by nitrofen, the administration of E2 (0.2 mg/kg) significantly reduces the TGF-β1 expression in lung tissue [581]. Correspondingly, tamoxifen (an anti-estrogen drug used in breast cancer treatment) seems to promote TGF-β1 expression, leading to lung fibrosis [582]. Recently, Smith et al. claimed that E2 might modulate TGF-β1-induced mesenchymal transition in bronchial epithelial cells; however, TGF-β1-induced EMT was not significantly affected by E2 [583]. According to the authors, this fact may be due to the decrease in mRNA expression of ERs (ERα, ERβ, and GPR30) and the reduction in protein expression of ERα elicited by TGF-β1. The authors also found that E2 downregulates the expression of chloride intracellular channel protein 3 (CLIC3) and retinol binding protein 7 (RBP7), proteins associated with pathogenic mechanisms of PF. Recently, Elliot et al. demonstrated that ERα is augmented in lung human tissues and myofibroblasts from IPF patients and in bleomycin-treated mice. Moreover, a decrease in the expression of ERα in human myofibroblasts lessens fibrosis-associated pathways [584]. The evidence so far indicates that E2 mostly downregulates the inflammatory response in PF, conferring a protective status in women. However, the mixed results observed in lung tissues in vitro and in animal models require further exploration.

15.5.3.4 Lung Cancer

Lung cancer (LC) is characterized by morphological cellular transformations and alterations of key pathways in cellular homeostasis. According to the World Health Organization, LC is one of the most frequent causes of cancer-related death in men and women worldwide. Interestingly, about 80–90% of these cases are produced by tobacco smoking; however, only ~15% of smokers develop lung cancer, suggesting a genetic susceptibility [585–587]. The estimated number of new cases and deaths in men continues to exceed those values in women [588]. However, trends suggest that the number of deaths from LC in women will exceed those in men in the future [589, 590]. The augmented incidence in women is related to an increase in smoking habits. However, non-smokers diagnosed with LC are more likely to be female, pointing out an additional hormonal component [591–593]. Non-smoker women with this pathology diagnosed at early ages have better survival rates than men [594–596].

Lung cancer is categorized in small cell lung carcinoma (SCLC) and non-small cell lung carcinoma (NSCLC, the most predominant type). NSCLS subtypes are squamous cell carcinoma, large cell carcinoma, and adenocarcinoma [597]. In LC, tumor microenvironment, made up of tumor cells, fibroblasts, vascular and lymphatic endothelial cells, growth factors, and others, favors a pro-inflammatory state. Fibroblast exposure to cigarette chemicals or another pro-carcinogenic factor leads to an increased inflammatory response by secreting PGE_2 and IL-8 and promoting the activity of ERK1/2 [598, 599]. Macrophages and neutrophils play a critical role in the inflammatory condition of LC. In this regard, two distinct types of macrophages are involved in the tumoral condition: resident macrophages which play a cytotoxic role against tumor development and tumor-associated macrophages (TAMs) with a pro-tumoral function [600]. TAMs are recruited at the tumor site by monocyte chemoattractant protein-1 (MCP-1/CCL2) [601, 602]. M1 macrophages (activated via IFN-γ) display pro-immunogenic characteristics, and M2 macrophages promote tumor growth, angiogenesis, invasion, and metastasis [603–606]. In addition, neutrophil infiltration into the tumor microenvironment, as described in the adenocarcinoma LC subtype, is associated with lower survival. In this context, neutrophils promote the EMT and potentiate the migration activity of tumor cells, and notably, neutrophil infiltration is associated with hemoptysis (coughing up of blood) [607]. Moreover, these immune cells release pro-inflammatory cytokines, proteases, and ROS, which damage DNA and activate oncogenes [608, 609]. Contrariwise, CD8+ and CD4+ T cells appear to have a protective role in LC, as they improve the survival rate by reducing the progression of the disease [610–612].

Several cytokines that regulate the tumor microenvironment have been studied in LC as well. For instance, TNF-α favors the survival of tumor cells by inducing genes encoding NF-κB-dependent anti-apoptotic molecules and inflammatory cytokines, including IL-1β, TNF-α, TGF-β1, IL-6, and IL-8 [613–615]. TGF-β is overexpressed in LC and displays pleiotropic effects such as cell growth, proliferation, differentiation, and apoptosis [616–619]. IL-10 has been shown to exert dual effects on the tumor microenvironment. In this context, it has been demonstrated that IL-10 possesses immunosuppressive functions by promoting T-cell apoptosis and anti-angiogenic properties [620–622]. Also, higher levels of IL-10 have been shown to be associated with metastasis [623, 624].

15.5.3.5 Androgens' Effects on Inflammation in Lung Cancer

The androgen pathway and its relevance to lung cancer cells have been studied. The AR (a member of the nuclear receptor superfamily) is found in normal and in lung cancer cells [625–627]. The expression of nuclear receptors has been suggested as a prognostic biomarker for survival and relapse of LC patients [628, 629]. In 2012, Jeong et al. proved that the exposure of SCLC and NSCLC cell lines to 5α-DHT (at physiological concentrations) increases the mRNA expression of the AR and stimulates cellular growth [630]. TES is also capable of stimulating the growth of SCLC cell lines expressing the AR [627]. An epidemiological study shows that high concentrations of total serum TES are associated with the presence of LC [631]. Furthermore, the use of 5α reductase inhibitors in patients with LC is associated with long better survival [632]. These insights suggest that the suppression of the androgen pathway may have a direct effect on lung cancer.

The correlation between androgens and inflammation on lung carcinogenesis has been scantly investigated. Regarding the survival of patients, T cells have been related to favorable prognosis in LC [610, 612]. In this context, it has been observed that androgen deprivation positively regulates the infiltration of T cells in the lung tissue [633]. Furthermore, Wu et al. found that androgen deprivation by castration increases the radiation-induced inflammatory response in mice [634]. They demonstrated that mRNA levels of TNF-α, IL-6, IL-1, and TGF-β are increased after castration. According to the authors, androgens could downregulate the actions of NF-κB and, consequently, inhibit the activation of genes encoding inflammatory cytokines. These data point out that TES might restrict the inflammatory response in LC by limiting the reactivity of T cells and interfering with the NF-κB signaling.

The influence of androgens on macrophages is well known; nonetheless, the relationship between androgens and macrophages in lung cancer has not been explored. In this context, Padgett et al. found that physiological concentrations of DHEA abolish the secretion of TNF-α, IL-1, and IL-6 in murine macrophages [635]. Another study proved that TES at physiological and supraphysiological concentrations suppresses the expression and release of TNF-α from human macrophages [636]. These results point out that androgens possess anti-inflammatory properties that modulate macrophage activity. In LC, IL-10, mainly produced by TAMs, has pro-tumoral qualities and correlates with non-response to anti-tumoral therapy [600, 624, 637]. In 2012, Wang et al. proposed that this interleukin promotes tumor malignancy by stimulating T-cell apoptosis and tumor cell survival in the lung [620]. Interestingly, IL-10 transgenic mice injected with Lewis lung (3LL) carcinoma cells develop larger tumors than control mice, pointing out that this interleukin may prevent an adequate response against tumor cells [638]. Moreover, patients with NSCLS show increased IL-10 serum levels (compared to healthy subjects), and this increase is associated with reduced survival [639]. In this regard, the androgen effects on IL-10 have been studied in tissues different from lung cancer cells. It was demonstrated that TES at physiological concentrations acts as an inducer of IL-10 synthesis in mice monocyte macrophages [205]. Furthermore, TES replacement therapy in men with symptomatic androgen defi-

ciency resulted in increased serum concentrations of IL-10 and decreased levels of TNF-α and IL-1 [204]. These studies suggest that probably androgens favor the pro-tumoral IL-10 effects, but specific studies in lung cancer cells are needed to confirm this assumption.

Another target of androgen regulation is the transmembrane prostate androgen-induced protein (TMEPAI). The synthesis of TMEPAI is elicited mainly by androgens, but also by TGF-β1 [640–642]. This oncogenic protein is expressed in lung cancer cell lines. Moreover, TMEPAI plays a vital role in TGF-β1-induced EMT by modulating ROS signaling and causing changes in epithelial cells, including migration, invasion, and proliferation of the tumor [643]. Furthermore, a research group found in lung cancer cells that the ablation of TMEPAI prevents cell proliferation, migration, and invasion. Also, they observed that the expression of TMEPAI in nude mice facilitates tumorigenesis. Finally, they proposed that the activation of TGF-β pathway induces TMEPAI expression and, subsequently, TMEPAI downregulates TGF-β signaling by promoting lysosomal degradation of the TGF-β receptor [644].

In conclusion, the evidence suggests that androgens might positively or negatively regulate lung cancer development. On the one hand, androgens may stimulate the growth of lung cancer cells and might play a role in the EMT by inducing the synthesis of TMEPAI. On the other hand, androgens may interfere with the signaling of NF-κB, leading to the downregulation of inflammatory cytokines.

15.5.3.6 Estrogens' Effects on Inflammation in Lung Cancer

Histological subtypes of lung cancer differ between men and women, being adenocarcinoma and bronchioloalveolar carcinoma, the most common subtypes in women [645, 646]. Estrogens have an essential role in lung carcinogenesis and can be locally synthesized by lung cancer cells [647, 648]. Moreover, ERs are found in lung cancer cells, suggesting that local production of estrogens is a response to a process of carcinogenesis [649, 650]. Increasing evidence shows that estrogens are involved in lung cancer proliferation and progression and most human lung tumors express the ERβ subtype and the aromatase enzyme [651–653]. In patients with NSCLC, elevated aromatase and ERβ expression are associated with poorer survival [653, 654]. Furthermore, Stabile et al., using a murine model of lung cancer induced by tobacco carcinogen (NKK, 4-(methylnitrosamino)-1-(3-pyridyl)-1-butanone), demonstrated that aromatase is expressed in TAMs, whereas ERβ is found in both macrophages and lung tumor cells. Interestingly, they also observed that the combination of anastrozole (aromatase inhibitor) and fulvestrant (ER antagonist) inhibits tobacco carcinogen-induced lung tumorigenesis [655]. Moreover, the increased ERα expression has been linked to macrophage infiltration into the tumor microenvironment [656]. The infiltration of macrophages is favored by CCL2 and its receptor (CCR2), i.e., the stimulation of ERα by E2, may activate the CCR2 signaling, leading to macrophage infiltration, MMP9 production, tumor progression, growth, and metastasis [656, 657].

During the inflammatory response, IL-6 acts as a regulator of neutrophil trafficking [658]. Some authors have focused on IL-6 and its role as a biomarker of ongoing inflammation. It has been reported that patients with LC have increased serum levels of this interleukin [659, 660]. Another research group observed that ovariectomized mice have decreased total neutrophils and this condition was recovered by E2 replacement therapy [661]. Human NSCLC cell lines and more important human NSCLC samples show tumor expression of IL-6 and its receptor. This interleukin can stimulate and enhance pathways in tumorigenesis, including signal transducer and activator of transcription 3 (STAT3), which regulates cell cycle progression, apoptosis, and tumor angiogenesis [662, 663]. In 2018, Huang et al. confirmed that E2 induces the activation of ERβ and promotes IL-6 expression via the MAPK/ERK and the PI3K/AKT signaling pathways in lung cancer cells. Moreover, they proposed that ERβ/IL6 signaling pathway could be a target for

therapeutic intervention [664]. Furthermore, the estrogen-related receptor alpha (ERRα), a protein with a similar structure to ERα, has been implicated in LC. This receptor is expressed in LC cells and can regulate cell proliferation and migration [665, 666]. In 2018, Zhang et al. demonstrated that ERRα was significantly elevated in NSCLC cell lines compared with a normal bronchial epithelial cell line. They also reported that the overexpression and activation of ERRα increase the expression of IL-6 and the inhibition of NF-κB eliminates the ERRα effect in IL-6 synthesis [667]. The role of estrogen action on the inflammatory response in lung cancer is much better understood than the role of androgens. Moreover, it has been proposed that aromatase inhibitors may serve as potential drugs in lung cancer therapy [515, 516]. In this regard, further research on estrogen-targeted therapies that could improve patient survival and reduce tumor invasion is needed.

15.5.3.7 Coronavirus Disease 2019

The coronavirus disease 2019 (COVID-19) is caused by a virus named severe acute respiratory syndrome coronavirus 2 (SARS-CoV-2). According to the World Health Organization, by September 17, 2020, SARS-CoV-2 had infected more than 30,055,710 people worldwide and killed more than 943,433. This disease is characterized by respiratory symptoms and is transmitted from human to human through respiratory secretions and saliva [668, 669]. Patients with COVID-19 exhibit fever, dry cough, difficulty in breathing, myalgia, headache, diarrhea, and nausea [670–672]. Also, clinical data indicate decreased oxygen saturation, blood gas deviation, and abnormalities observed by chest X-rays, lymphopenia, and an increase of C-reactive protein [673].

Severe COVID-19 cases progress to acute respiratory distress syndrome (ARDS), around 8–9 days after symptom onset, and may lead to respiratory failure [674–676]. Essentially, SARS-CoV-2 binds to host cells such as airway epithelial cells, alveolar cells, vascular endothelial cells, and macrophages in the lung by the angiotensin-converting enzyme 2 (ACE2) host target receptor [677–680]. After the SARS-CoV-2 infection, a reduction in the ACE2 function that is associated with acute lung injury occurs [681–683]. Furthermore, ACE2 may be cleaved by transmembrane serine protease-2 (TMPRSS2), leading to an enhancement in the entry of the virus [684, 685]. On the other hand, there is an association between the gender of COVID-19 patients and fatality rates. In this context, data from the WHO show that a lower percentage of women (1.7%) infected with the virus will die in comparison with men (2.8%). Other investigations reported that less female patients with the severe form of the disease require intensive care or die compared to male patients [60, 686]. In this context, it has been insinuated that there may be alleles that confer resistance to this disease, since ACE2 gene is located on the X chromosome [687]. Interestingly, it also has been demonstrated that E2 downregulated the expression of ACE2 in differentiated normal human bronchial epithelial (NHBE) cells [688], which might explain the lower fatality rate in females.

The severity of COVID-19 is due to the host response featured by an uncontrolled inflammation associated with high levels of circulating cytokines, lymphopenia, and mononuclear cell infiltration in the lungs [676, 689]. When SARS-CoV-2 infects epithelial cells expressing the surface receptors ACE2 and TMPRSS2 in the airway, it activates a local immune response that in most cases resolves the infection [690]. Alveolar endothelial cells and macrophages detect the released PAMPs such as viral RNA and trigger the generation of pro-inflammatory cytokines and chemokines including IL-6, IFN-γ-induced protein 10 (IP-10/CXCL10), macrophage inflammatory protein 1α (MIP1α), and MIP1β in order to recruit immune cells [674, 676, 691]. The infiltration of monocytes, macrophages, and T cells to the site of infection promotes further inflammation [689, 692, 693] and may explain the lymphopenia seen in patients with this disease [670, 694]. In addition, it has been suggested that high exhaustion and a decreased functional diversity of T cells in peripheral blood may be an indicator of developing severe acute respiratory syndrome in patients with COVID-19 [695].

Moreover, higher levels of cytokines and chemokines have been related to the severity of the disease and eventually death [674, 694, 696]. In a study, a research group measured plasma levels of diverse cytokines in patients with severe COVID-19, founding increased levels of IL-2, IL-7, IL-10, G-CSF, IP-10, MCP-1, MIP1, and TNF [674]. Interestingly, Zhou et al. showed that IL-6 levels were elevated in non-survivors compared with survivors [697]. This cytokine storm, along with the cell infiltration, provokes lung damage by the excessive secretion of proteases and ROS [689, 692]. In addition, neutralizing antibodies produced by B cells can block viral infection, a situation that may occur around 1 week following symptoms onset [698, 699]. However, it has been suggested that some patients may not develop long-lasting antibodies to this virus and is unknown whether they are susceptible to reinfection. In this context, Elizaldi et al. conducted a study focused in CD4 T follicular helper (T_{fh}) cells (entities with high importance in the generation of long-lasting and specific humoral protection against viral infections) and reported that following infection with SARS-CoV-2, adult rhesus macaques exhibited transient accumulation of activated proliferating T_{fh} cells toward a Th1 response. They also proposed that a vaccine promoting Th1-type T_{fh} responses that target the S protein of the virus may lead to protective immunity [700]. Since the disease continues to spread worldwide, several immunosuppressive therapies have been used. Clinical trials have been focused on targeting pro-inflammatory mediators such as IL-6 and GM-CSF (clinical trials: ChiCTR2000029765).

15.5.3.8 Androgens' Effects on Inflammation in Coronavirus Disease 2019

More male patients with COVID-19 have higher mortality and develop the severe form of the disease than women [60, 701]. The difference in the number of cases reported by gender augments progressively in favor of male patients [702]. Studies have proposed that androgens modulate the immune system response and may predispose men to different clinical course and prognosis of COVID-19. Since men with severe COVID-19 are ≥60 years old, decreased TES levels during aging may be involved in a pro-inflammatory condition [703–705]. Furthermore, it is well established that TES concentration in plasma is reduced by comorbidities like obesity, diabetes, and COPD, which are prevalent in COVID-19 patients [706–711]. In this regard, it has been shown that hypogonadism is highly prevalent (22–69%) in male patients with COPD [712]. Meanwhile, testosterone treatment improved the exercise capacity, muscle strength, and oxygen consumption in men [713].

It has been established that ACE2 mediates the cell entry of SARS-CoV-2, insinuating a protective role of this receptor against the viral infection [702]. Remarkably, this enzyme is selectively expressed by adult Leydig cells [714], pointing out a possible role of testicular secretion of TES in COVID-19 patients [702]. Also, it has been exhibited that patients with SARS-CoV-2 pneumonia, who were transferred to the intensive care unit (ICU) or died in respiratory ICU (RICU), had lower amounts of total TES and calculated free TES, compared to patients who were transferred to the internal medicine unit or were at a stable condition in RICU [715]. Alveolar endothelial cells and macrophages detect the SARS-CoV-2 and trigger the generation of pro-inflammatory cytokines and chemokines. In this regard, it has been demonstrated a correlation between hypogonadism and augmented pro-inflammatory cytokines and that testosterone treatment reduces IL-1β, IL-6, and TNF-α [704]. Furthermore, a study reported that low serum TES concentrations were significantly associated with elevated levels of TNF-α, MIP1α (CCL3), and MIP1β (CCL4) [716]. In the healthy immune response, these and other cytokines promote the infiltration of monocytes, macrophages, and T lymphocytes to the site of infection, leading to a pro-inflammatory feedback loop [691]. Moreover, it was reported that DHEA (at physiological concentrations) eliminates the release of TNF-α, IL-1, and IL-6 in murine macrophages [635]. Another study proved that TES at physiological

and supraphysiological concentrations suppresses the expression and secretion of TNF-α from human macrophages [636]. In relation to T cells, Olsen et al. reported that an increase in peripheral T cells was reversed by androgen replacement [717]. Other research group found that castration of post-pubertal male mice increases T-cell numbers in peripheral lymphoid tissues [718]. In peripheral blood cells, TES treatment reduces the relative number of monocytes and increases CD8+ T-cell number [246]. Therefore, it is possible that androgen actions on CD8+ T cells favor the recognition and elimination of infected cells with SARS-CoV-2.

On the other hand, some critically ill patients have been treated with convalescent plasma, and growing evidence indicates positive results [719–721]. In an observational study, male patients with COVID-19 respond with a lower generation of effective SARS-CoV-2 IgG antibodies compared to women, confirming that a reduced antibody response in men is associated with worse prognosis [722]. This finding suggests a sexual hormone influence on B-cell proliferation. In this regard, androgen/AR actions on B lymphocytes have been studied by Altuwaijri et al. They reported that the lack of AR in B cells in different strains of mice results in increased B cells in the blood and bone marrow [243]. This insight supports the hypothesis that androgen-mediated B-cell maturation is AR dependent.

As previously mentioned, TMPRSS2 is a critical protease for the pathogenesis and spread of SARS-CoV-2 [723, 724]. The gene transcription of TMPRSS2 depends on the activity of the AR. It has been suggested that TES may promote higher expression of this protease in the lung of males, which might improve the ability of SARS-CoV-2 to enter cells [725–727]. The regulation of TMPRSS2 by TES has been suggested to influence on the male predominance displayed in COVID-19 infection [726]. Moreover, the hyperandrogenic condition could explain the severe COVID-19 cases in young males [702]. Also, it has been proposed a role for TMPRSS2 variants and their expression levels in the regulating the severity of COVID-19; however, further experimental research and a hypothesis that fosters validation on large cohorts of patients with different clinical manifestations are required [727]. Given the fact that TRPMSS2 is found in the lung, the use of inhibitors of this protease (currently employed for cancer prostate) against COVID-19 pneumonia seems a promissory therapeutic tool. Additionally, the assessment of potential drugs that interfere with androgen activity, such as androgen receptor inhibitors, steroidogenesis inhibitors, and 5-alpha reductase inhibitors, has been suggested [702].

Finally, it is well established that androgens play inhibitory roles in the inflammatory response, and the evidence indicates that reduced TES levels associated with age or comorbidities may increase the pro-inflammatory response in men, contributing to the development of a severe form of COVID-19. In this context, the quantification of TES levels may be considered when a COVID-19-positive patient is identified. Moreover, if the values are low, the use of testosterone has been remarkably proposed to reduce the associated pulmonary syndrome, thus preventing the progression to severe COVID-19 disease [702].

15.5.3.9 Estrogens' Effects on Inflammation in Coronavirus Disease 2019

Emerging studies have suggested that women are less susceptible to COVID-19 and exhibit lower mortality than men [670, 671], which may be explained by a potential protective role of estrogens. X chromosome encoding the greatest density of genes related to immune response [728] supposes the immunological advantage of women over males. Interestingly, Channappanavar et al. examined the gender-dependent difference outcomes of the infection by SARS-CoV. They demonstrated that the estrogen depletion by ovariectomy or the use of an ER antagonist increases the morbidity and mortality in SARS-CoV-infected female mice [729]. In addition, they suggested that infected female mice have a sex-specific protection during their reproductive period. These findings strengthen the crucial hypothesis about the protective role of estrogen and the ER signaling against the respiratory virus. Another research group observed that, in an age group of 40–60 years, the transcriptomic profile of female lung tissue has more similarities to that evoked upon SARS-CoV-2 infection compared to male tissue [730]. In this regard, characterizing the most activated intracellular pathways during the viral infection may provide a molecular explanation of the lower incidence of COVID-19 in females.

The estrogen/ER signaling regulates the development of immune cells and the pathways of innate and adaptive immune system [731–733]. It appears that the development and severity of COVID-19 depend on individual propensities for the massive release of pro-inflammatory mediators [674, 676]. Data have pointed out that high doses of E2 may inhibit the production of inflammatory cytokines (IL1, IL6, and TNF-α), whereas stimulation with low doses of E2 enhances the production [734–736]. It was proved that, in the early phase of antiviral immune response against SARS-CoV infection, pro-inflammatory mediators (IL-6, CCL2, and CXCL1) display similar increased levels in both sexes and 72 h post-infection these cytokines are upregulated in the lung of male mice compared with females [729]. To understand this finding, it has been suggested that monocyte-macrophage recruitment is suppressed by estrogens, which favor the downregulation of CCL2/MCP-1 expression and inhibit NF-κB activation in macrophages [737, 738]. Similarly, reduced TNF-α and CCL2 levels are observed in gonadectomized mice treated with estrogen, which protect them from influenza virus infection [739, 740]. Furthermore, in SARS-CoV-infected mice, the predominant sources of these pro-inflammatory mediators are the inflammatory monocyte macrophages (IMMs) [741]. In male mice, these cells are increased in numbers and also produced more mediators compared with female mice [729]. Additionally, increased numbers of IMMs in ovariectomized mice compared with intact female mice suggest that estrogen signaling in females abolishes the accumulation and function of IMMs in the lung [729]. Moreover, it has been reported that the downregulation of IL-6 gene expression by E2 is induced via the interaction of the estrogen receptor with NF-κB [742].

Zheng et al. showed in an observational study that, the humoral response in seriously ill male patients exposes a delayed peak of antibody response with a lower generation of effective IgG compared to women [722]. Females exhibit a predominant Th2 cytokine profile, which could be involved in immune responses characterized principally by the secretion of antibodies [743]. Estrogen significantly enhances the generation of a Th2 response [233, 744], corresponding to the findings that this hormone increases the frequency of antibody secreting B cells in the follicular phase (when the levels of estrogen are high) of the menstrual cycle in rhesus macaques. Furthermore, the stimulation of a CD8 + -enriched cell population induced the expression of IFN-γ and IL-12 [745]. These findings are actually attributable to estrogen influence on B-cell proliferation, activation, and maturation by upregulating the expression of CD22, Src homology region 2 domain-containing phosphatase-1 (SHP-1), and B-cell lymphoma 2 (Bcl-2) [733, 746]. Finally, it has been hypothesized that estrogens could protect women from the most serious complications of COVID-19, especially women before the menopause due to high serum estrogen levels [747].

15.6   Conclusions

Inflammation is a complex biological process that involves multiple immune mechanisms. The literature confirms the influence of sex hormones on the incidence and severity of the inflammatory response in lung diseases. The effects of sex steroids have been observed in pathophysiological conditions of the lung related to diseases such as asthma, COPD, lung fibrosis, lung cancer, and COVID-19. This chapter highlights the importance of sex-specific research taking into account the hormonal status of the patients. Moreover, sex steroid actions depend on very particular circumstances such as the hormone concentration, duration of the stimulus, genomic or non-genomic pathway, and interaction between male and female sex hormones. The evidence indicates that sex hormone actions on the inflammatory response can be beneficial or detrimental. Generally, male sex steroids have beneficial anti-inflammatory properties in asthma, COPD, and other lung diseases. On the other hand, E2 displays anti-inflammatory and pro-inflammatory properties on lung diseases. It is crucial to consider that the effects of sex hormones on the inflammatory responses in lung diseases need to be further explored in order to find novel therapeutic approaches and pursue an individualized medicine in the future.

Acknowledgments Jorge Reyes-García is grateful to Posgrado en Ciencias Biológicas UNAM and CONACYT for the support to obtain a postdoctoral fellowship (EPE 2019). Abril Carbajal-García is grateful to the Programa de Doctorado en Ciencias Biomédicas, Universidad Nacional Autónoma de México, for the instruction received during her studies to get a Ph.D. degree. She received fellowship from the Consejo Nacional de Ciencia y Tecnología, México (application # 2018-000068-02NACF-17950; CVU 826027).

Declaration of Interest Figures in this book chapter were created with BioRender.com. The authors declare that they do not have a conflict of interest or financial relationship that may have influenced the work.

Funding This study was partly supported by grants from Dirección General de Asuntos del Personal Académico (DGAPA), Universidad Nacional Autónoma de México IN204319, and CONACYT 137725 to LM Montaño.

References

1. Aghasafari P, George U, Pidaparti R. A review of inflammatory mechanism in airway diseases. Inflamm Res. 2019;68(1):59–74.
2. Germolec DR, Shipkowski KA, Frawley RP, Evans E. Markers of inflammation. Methods Mol Biol. 1803;2018:57–79.
3. Guo H, Callaway JB, Ting JP. Inflammasomes: mechanism of action, role in disease, and therapeutics. Nat Med. 2015;21(7):677–87.
4. Wong J, Magun BE, Wood LJ. Lung inflammation caused by inhaled toxicants: a review. Int J Chron Obstruct Pulmon Dis. 2016;11:1391–401.

5. Chen L, Deng H, Cui H, Fang J, Zuo Z, Deng J, et al. Inflammatory responses and inflammation-associated diseases in organs. Oncotarget. 2018;9(6):7204–18.

6. Dallacasagrande V, Hajjar KA. Annexin A2 in inflammation and host defense. Cell. 2020;9(6)

7. Costela-Ruiz VJ, Illescas-Montes R, Puerta-Puerta JM, Ruiz C, Melguizo-Rodriguez L. SARS-CoV-2 infection: The role of cytokines in COVID-19 disease. Cytokine Growth Factor Rev. 2020;

8. Lee IT, Yang CM. Inflammatory signalings involved in airway and pulmonary diseases. Mediat Inflamm. 2013;2013:791231.

9. Netea MG, Balkwill F, Chonchol M, Cominelli F, Donath MY, Giamarellos-Bourboulis EJ, et al. A guiding map for inflammation. Nat Immunol. 2017;18(8):826–31.

10. Moldoveanu B, Otmishi P, Jani P, Walker J, Sarmiento X, Guardiola J, et al. Inflammatory mechanisms in the lung. J Inflamm Res. 2009;2:1–11.

11. Zhang Q, Zhu S, Cheng X, Lu C, Tao W, Zhang Y, et al. Euphorbia factor L2 alleviates lipopolysaccharide-induced acute lung injury and inflammation in mice through the suppression of NF-kappaB activation. Biochem Pharmacol. 2018;155:444–54.

12. Hagimoto N, Kuwano K, Kawasaki M, Yoshimi M, Kaneko Y, Kunitake R, et al. Induction of interleukin-8 secretion and apoptosis in bronchiolar epithelial cells by Fas ligation. Am J Respir Cell Mol Biol. 1999;21(3):436–45.

13. Peteranderl C, Sznajder JI, Herold S, Lecuona E. Inflammatory responses regulating alveolar ion transport during pulmonary infections. Front Immunol. 2017;8:446.

14. Kirkpatrick CT, Wang Y, Leiva Juarez MM, Shivshankar P, Pantaleon Garcia J, Plumer AK, et al. Inducible lung epithelial resistance requires multisource reactive oxygen species generation to protect against viral infections. MBio. 2018;9(3)

15. Shimbashi R, Chang B, Tanabe Y, Takeda H, Watanabe H, Kubota T, et al. Epidemiological and clinical features of invasive pneumococcal disease caused by serotype 12F in adults. Japan PLoS One. 2019;14(2):e0212418.

16. Vrolyk V, Wobeser BK, Al-Dissi AN, Carr A, Singh B. Lung inflammation associated with clinical acute necrotizing pancreatitis in dogs. Vet Pathol. 2017;54(1):129–40.

17. Nagashima R, Kosai H, Masuo M, Izumiyama K, Noshikawaji T, Morimoto M, et al. Nrf2 suppresses allergic lung inflammation by attenuating the type 2 innate lymphoid cell response. J Immunol. 2019;202(5):1331–9.

18. McCall AL, Salemi J, Bhanap P, Strickland LM, Elmallah MK. The impact of Pompe disease on smooth muscle: a review. J Smooth Muscle Res. 2018;54:100–18.

19. Weidner J, Jogdand P, Jarenback L, Aberg I, Helihel D, Ankerst J, et al. Expression, activity and localiza-tion of lysosomal sulfatases in chronic obstructive pulmonary disease. Sci Rep. 2019;9(1):1991.

20. Yang L, Wang L, Zhang Y. Immunotherapy for lung cancer: advances and prospects. Am J Clin Exp Immunol. 2016;5(1):1–20.

21. Gomes M, Teixeira AL, Coelho A, Araujo A, Medeiros R. The role of inflammation in lung cancer. Adv Exp Med Biol. 2014;816:1-23.

22. Charavaryamath C, Janardhan KS, Townsend HG, Willson P, Singh B. Multiple exposures to swine barn air induce lung inflammation and airway hyper-responsiveness. Respir Res. 2005;6:50.

23. Wang H, Song L, Ju W, Wang X, Dong L, Zhang Y, et al. The acute airway inflammation induced by PM2.5 exposure and the treatment of essential oils in Balb/c mice. Sci Rep. 2017;7:44256.

24. Suratt BT, Parsons PE. Mechanisms of acute lung injury/acute respiratory distress syndrome. Clin Chest Med. 2006;27(4):579–89; abstract viii.

25. Fuentes N, Silveyra P. Endocrine regulation of lung disease and inflammation. Exp Biol Med (Maywood). 2018;243(17–18):1313–22.

26. Carey MA, Card JW, Voltz JW, Arbes SJ Jr, Germolec DR, Korach KS, et al. It's all about sex: gender, lung development and lung disease. Trends Endocrinol Metab. 2007;18(8):308–13.

27. Carey MA, Card JW, Voltz JW, Germolec DR, Korach KS, Zeldin DC. The impact of sex and sex hormones on lung physiology and disease: lessons from animal studies. Am J Physiol Lung Cell Mol Physiol. 2007;293(2):L272–8.

28. DeBoer MD, Phillips BR, Mauger DT, Zein J, Erzurum SC, Fitzpatrick AM, et al. Effects of endogenous sex hormones on lung function and symptom control in adolescents with asthma. BMC Pulm Med. 2018;18(1):58.

29. Han MK, Arteaga-Solis E, Blenis J, Bourjeily G, Clegg DJ, DeMeo D, et al. Female sex and gender in lung/sleep health and disease. Increased understanding of basic biological, pathophysiological, and behavioral mechanisms leading to better health for female patients with lung disease. Am J Respir Crit Care Med. 2018;198(7):850–8.

30. Seaborn T, Simard M, Provost PR, Piedboeuf B, Tremblay Y. Sex hormone metabolism in lung development and maturation. Trends Endocrinol Metab. 2010;21(12):729–38.

31. Townsend EA, Miller VM, Prakash YS. Sex differences and sex steroids in lung health and disease. Endocr Rev. 2012;33(1):1–47.

32. Traglia M, Bseiso D, Gusev A, Adviento B, Park DS, Mefford JA, et al. Genetic mechanisms leading to sex differences across common diseases and anthropometric traits. Genetics. 2017;205(2):979–92.

33. Barnes PJ. Sex differences in chronic obstructive pulmonary disease mechanisms. Am J Respir Crit Care Med. 2016;193(8):813–4.

34. Gonzalez AV, Suissa S, Ernst P. Gender differences in survival following hospitalisation for COPD. Thorax. 2011;66(1):38–42.

35. Raherison C, Tillie-Leblond I, Prudhomme A, Taille C, Biron E, Nocent-Ejnaini C, et al. Clinical characteristics and quality of life in women with COPD: an observational study. BMC Womens Health. 2014;14(1):31.

36. Tam A, Morrish D, Wadsworth S, Dorscheid D, Man SF, Sin DD. The role of female hormones on lung function in chronic lung diseases. BMC Womens Health. 2011;11:24.

37. Tam A, Tanabe N, Churg A, Wright JL, Hogg JC, Sin DD. Sex differences in lymphoid follicles in COPD airways. Respir Res. 2020;21(1):46.

38. Fuseini H, Newcomb DC. Mechanisms Driving Gender Differences in Asthma. Curr Allergy Asthma Rep. 2017;17(3):19.

39. Holguin F. Sex hormones and asthma. Am J Respir Crit Care Med. 2020;201(2):127–8.

40. Naeem A, Silveyra P. Sex differences in Paediatric and adult asthma. Eur Med J (Chelmsf). 2019;4(2):27–35.

41. Shah R, Newcomb DC. Sex Bias in asthma prevalence and pathogenesis. Front Immunol. 2018;9:2997.

42. Yung JA, Fuseini H, Newcomb DC. Hormones, sex, and asthma. Ann Allergy Asthma Immunol. 2018;120(5):488–94.

43. Baldacara RP, Silva I. Association between asthma and female sex hormones. Sao Paulo Med J. 2017;135(1):4–14.

44. Han YY, Forno E, Celedon JC. Sex steroid hormones and asthma in a Nationwide study of U.S. adults. Am J Respir Crit Care Med. 2020;201(2):158–66.

45. Meltzer EB, Noble PW. Idiopathic pulmonary fibrosis. Orphanet J Rare Dis. 2008;3:8.

46. Salisbury ML, Xia M, Zhou Y, Murray S, Tayob N, Brown KK, et al. Idiopathic pulmonary fibrosis: gender-age-physiology index stage for predicting future lung function decline. Chest. 2016;149(2):491–8.

47. Zaman T, Moua T, Vittinghoff E, Ryu JH, Collard HR, Lee JS. Differences in clinical characteristics and outcomes between men and women with idiopathic pulmonary fibrosis: a multicenter retrospective cohort study. Chest. 2020;

48. Gharaee-Kermani M, Hatano K, Nozaki Y, Phan SH. Gender-based differences in bleomycin-induced pulmonary fibrosis. Am J Pathol. 2005;166(6):1593–606.

49. Voltz JW, Card JW, Carey MA, Degraff LM, Ferguson CD, Flake GP, et al. Male sex hormones exacerbate lung function impairment after bleomycin-induced pulmonary fibrosis. Am J Respir Cell Mol Biol. 2008;39(1):45–52.

50. Harness-Brumley CL, Elliott AC, Rosenbluth DB, Raghavan D, Jain R. Gender differences in outcomes of patients with cystic fibrosis. J Womens Health (Larchmt). 2014;23(12):1012–20.

51. Johannesson M, Ludviksdottir D, Janson C. Lung function changes in relation to menstrual cycle in females with cystic fibrosis. Respir Med. 2000;94(11):1043–6.

52. Batton KA, Austin CO, Bruno KA, Burger CD, Shapiro BP, Fairweather D. Sex differences in pulmonary arterial hypertension: role of infection and autoimmunity in the pathogenesis of disease. Biol Sex Differ. 2018;9(1):15.

53. Lahm T, Tuder RM, Petrache I. Progress in solving the sex hormone paradox in pulmonary hypertension. Am J Physiol Lung Cell Mol Physiol. 2014;307(1):L7–26.

54. Li C, Zhang Z, Xu Q, Wu T, Shi R. Potential mechanisms and serum biomarkers involved in sex differences in pulmonary arterial hypertension. Medicine (Baltimore). 2020;99(13):e19612.

55. Mair KM, Johansen AK, Wright AF, Wallace E, MacLean MR. Pulmonary arterial hypertension: basis of sex differences in incidence and treatment response. Br J Pharmacol. 2014;171(3):567–79.

56. Lahm T, Kawut SM. Inhibiting oestrogen signalling in pulmonary arterial hypertension: sex, drugs and research. Eur Respir J. 2017;50(2)

57. Wright AF, Ewart MA, Mair K, Nilsen M, Dempsie Y, Loughlin L, et al. Oestrogen receptor alpha in pulmonary hypertension. Cardiovasc Res. 2015;106(2):206–16.

58. Dou M, Zhu K, Fan Z, Zhang Y, Chen X, Zhou X, et al. Reproductive hormones and their receptors may affect lung Cancer. Cell Physiol Biochem. 2017;44(4):1425–34.

59. Slowikowski BK, Lianeri M, Jagodzinski PP. Exploring estrogenic activity in lung cancer. Mol Biol Rep. 2017;44(1):35–50.

60. Jin JM, Bai P, He W, Wu F, Liu XF, Han DM, et al. Gender differences in patients with COVID-19: focus on severity and mortality. Front Public Health. 2020;8:152.

61. Uwamino Y, Nishimura T, Sato Y, Tamizu E, Asakura T, Uno S, et al. Low serum estradiol levels are related to Mycobacterium avium complex lung disease: a cross-sectional study. BMC Infect Dis. 2019;19(1):1055.

62. Cole TJ, Short KL, Hooper SB. The science of steroids. Semin Fetal Neonatal Med. 2019;24(3):170–5.

63. Miller WL, Auchus RJ. The molecular biology, biochemistry, and physiology of human steroidogenesis and its disorders. Endocr Rev. 2011;32(1):81–151.

64. Taraborrelli S. Physiology, production and action of progesterone. Acta Obstet Gynecol Scand. 2015;94 Suppl 161:8–16.

65. Wang Y, Li H, Zhu Q, Li X, Lin Z, Ge RS. The cross talk of adrenal and Leydig cell steroids in Leydig cells. J Steroid Biochem Mol Biol. 2019;192:105386.

66. Barakat R, Oakley O, Kim H, Jin J, Ko CJ. Extragonadal sites of estrogen biosynthesis and function. BMB Rep. 2016;49(9):488–96.

67. Baburski AZ, Andric SA, Kostic TS. Luteinizing hormone signaling is involved in synchronization of Leydig cell's clock and is crucial for rhythm robust-

ness of testosterone productiondagger. Biol Reprod. 2019;100(5):1406–15.

68. Talamillo A, Ajuria L, Grillo M, Barroso-Gomila O, Mayor U, Barrio R. SUMOylation in the control of cholesterol homeostasis. Open Biol. 2020;10(5):200054.

69. Brown MS, Radhakrishnan A, Goldstein JL. Retrospective on cholesterol homeostasis: The central role of Scap. Annu Rev Biochem. 2018;87:783–807.

70. Chen H, Li Z, Dong L, Wu Y, Shen H, Chen Z. Lipid metabolism in chronic obstructive pulmonary disease. Int J Chron Obstruct Pulmon Dis. 2019;14:1009–18.

71. Spann NJ, Glass CK. Sterols and oxysterols in immune cell function. Nat Immunol. 2013;14(9):893–900.

72. Papadopoulos V, Miller WL. Role of mitochondria in steroidogenesis. Best Pract Res Clin Endocrinol Metab. 2012;26(6):771–90.

73. Hu J, Zhang Z, Shen WJ, Azhar S. Cellular cholesterol delivery, intracellular processing and utilization for biosynthesis of steroid hormones. Nutr Metab (Lond). 2010;7:47.

74. Venugopal S, Martinez-Arguelles DB, Chebbi S, Hullin-Matsuda F, Kobayashi T, Papadopoulos V. Plasma membrane origin of the Steroidogenic Pool of cholesterol used in hormone-induced acute steroid formation in Leydig cells. J Biol Chem. 2016;291(50):26109–25.

75. Dufau ML, Catt KJ. Gonadotropin receptors and regulation of steroidogenesis in the testis and ovary. Vitam Horm. 1978;36:461–592.

76. Montaño LM, Flores-Soto E, Sommer B, Solis-Chagoyan H, Perusquia M. Androgens are effective bronchodilators with anti-inflammatory properties: a potential alternative for asthma therapy. Steroids. 2020;153:108509.

77. van Rooyen D, Yadav R, Scott EE, Swart AC. CYP17A1 exhibits 17alphahydroxylase/17,20-lyase activity towards 11beta-hydroxyprogesterone and 11-ketoprogesterone metabolites in the C11-oxy backdoor pathway. J Steroid Biochem Mol Biol. 2020;199:105614.

78. Zirkin BR, Papadopoulos V. Leydig cells: formation, function, and regulation. Biol Reprod. 2018;99(1):101–11.

79. Faienza MF, Baldinotti F, Marrocco G, TyuTyusheva N, Peroni D, Baroncelli GI, et al. 17beta-hydroxysteroid dehydrogenase type 3 deficiency: female sex assignment and follow-up. J Endocrinol Investig. 2020;

80. Vinklarova L, Schmidt M, Benek O, Kuca K, Gunn-Moore F, Musilek K. Friend or enemy? Review of 17beta-HSD10 and its role in human health or disease. J Neurochem. 2020;

81. Endo S, Morikawa Y, Kudo Y, Suenami K, Matsunaga T, Ikari A, et al. Human dehydrogenase/reductase SDR family member 11 (DHRS11) and aldo-keto reductase 1C isoforms in comparison:

substrate and reaction specificity in the reduction of 11-keto-C19-steroids. J Steroid Biochem Mol Biol. 2020;199:105586.

82. Swerdloff RS, Dudley RE, Page ST, Wang C, Salameh WA. Dihydrotestosterone: biochemistry, physiology, and clinical implications of elevated blood levels. Endocr Rev. 2017;38(3):220–54.

83. Traish AM. Negative impact of testosterone deficiency and 5alpha-reductase inhibitors therapy on metabolic and sexual function in men. Adv Exp Med Biol. 2017;1043:473–526.

84. DeWitt NA, Whirledge S, Kallen AN. Updates on molecular and environmental determinants of luteal progesterone production. Mol Cell Endocrinol. 2020;110930

85. Cui J, Shen Y, Li R. Estrogen synthesis and signaling pathways during aging: from periphery to brain. Trends Mol Med. 2013;19(3):197–209.

86. Labrie F. Extragonadal synthesis of sex steroids: intracrinology. Ann Endocrinol (Paris). 2003;64(2):95–107.

87. Labrie F. Each tissue becomes master of its sex steroid environment at menopause. Climacteric. 2015;18(5):764–5.

88. Luu-The V, Labrie F. The intracrine sex steroid biosynthesis pathways. Prog Brain Res. 2010;181:177–92.

89. Guercio G, Saraco N, Costanzo M, Marino R, Ramirez P, Berensztein E, et al. Estrogens in human male gonadotropin secretion and testicular physiology from infancy to late puberty. Front Endocrinol (Lausanne). 2020;11:72.

90. Tian Y, Shen W, Lai Z, Shi L, Yang S, Ding T, et al. Isolation and identification of ovarian theca-interstitial cells and granulose cells of immature female mice. Cell Biol Int. 2015;39(5):584–90.

91. Havelock JC, Rainey WE, Carr BR. Ovarian granulosa cell lines. Mol Cell Endocrinol. 2004;228(1–2):67–78.

92. Leung PC, Steele GL. Intracellular signaling in the gonads. Endocr Rev. 1992;13(3):476–98.

93. Wood JR, Strauss JF 3rd. Multiple signal transduction pathways regulate ovarian steroidogenesis. Rev Endocr Metab Disord. 2002;3(1):33–46.

94. Andersen CY, Ezcurra D. Human steroidogenesis: implications for controlled ovarian stimulation with exogenous gonadotropins. Reprod Biol Endocrinol. 2014;12:128.

95. Sathish V, Martin YN, Prakash YS. Sex steroid signaling: implications for lung diseases. Pharmacol Ther. 2015;150:94–108.

96. Carreau S, Bouraima-Lelong H, Delalande C. Estrogens in male germ cells. Spermatogenesis. 2011;1(2):90–4.

97. Hammes SR, Levin ER. Impact of estrogens in males and androgens in females. J Clin Invest. 2019;129(5):1818–26.

98. Petra PH. The plasma sex steroid binding protein (SBP or SHBG). A critical review of recent developments on the structure, molecular biol-

ogy and function. J Steroid Biochem Mol Biol. 1991;40(4–6):735–53.

99. Ramachandran S, Hackett GI, Strange RC. Sex hormone binding globulin: a review of its interactions with testosterone and age, and its impact on mortality in men with type 2 diabetes. Sex Med Rev. 2019;7(4):669–78.

100. Grishkovskaya I, Avvakumov GV, Sklenar G, Dales D, Hammond GL, Muller YA. Crystal structure of human sex hormone-binding globulin: steroid transport by a laminin G-like domain. EMBO J. 2000;19(4):504–12.

101. Dunn JF, Nisula BC, Rodbard D. Transport of steroid hormones: binding of 21 endogenous steroids to both testosterone-binding globulin and corticosteroid-binding globulin in human plasma. J Clin Endocrinol Metab. 1981;53(1):58–68.

102. Heinrich-Balard L, Zeinyeh W, Dechaud H, Rivory P, Roux A, Pugeat M, et al. Inverse relationship between hSHBG affinity for testosterone and hSHBG concentration revealed by surface plasmon resonance. Mol Cell Endocrinol. 2015;399:201–7.

103. de Ronde W, van der Schouw YT, Pierik FH, Pols HA, Muller M, Grobbee DE, et al. Serum levels of sex hormone-binding globulin (SHBG) are not associated with lower levels of non-SHBG-bound testosterone in male newborns and healthy adult men. Clin Endocrinol. 2005;62(4):498–503.

104. Mendel CM. The free hormone hypothesis. Distinction from the free hormone transport hypothesis. J Androl. 1992;13(2):107–16.

105. de Marco R, Locatelli F, Sunyer J, Burney P. Differences in incidence of reported asthma related to age in men and women. A retrospective analysis of the data of the European respiratory health survey. Am J Respir Crit Care Med. 2000;162(1):68–74.

106. Laffont S, Blanquart E, Guery JC. Sex differences in asthma: a key role of androgen-signaling in group 2 innate lymphoid cells. Front Immunol. 2017;8:1069.

107. Traish A, Bolanos J, Nair S, Saad F, Morgentaler A. Do androgens modulate the pathophysiological pathways of inflammation? Appraising the contemporary evidence. J Clin Med. 2018;7(12)

108. Guiochon-Mantel A. Regulation of the differentiation and proliferation of smooth muscle cells by the sex hormones. Rev Mal Respir. 2000;17(2 Pt 2):604–8.

109. Hanley K, Rassner U, Jiang Y, Vansomphone D, Crumrine D, Komuves L, et al. Hormonal basis for the gender difference in epidermal barrier formation in the fetal rat. Acceleration by estrogen and delay by testosterone. J Clin Invest. 1996;97(11):2576–84.

110. Kimura Y, Suzuki T, Kaneko C, Darnel AD, Akahira J, Ebina M, et al. Expression of androgen receptor and 5alpha-reductase types 1 and 2 in early gestation fetal lung: a possible correlation with branching morphogenesis. Clin Sci (Lond). 2003;105(6):709–13.

111. Zarazua A, Gonzalez-Arenas A, Ramirez-Velez G, Bazan-Perkins B, Guerra-Araiza C, Campos-Lara MG. Sexual dimorphism in the regulation of estro-gen, progesterone, and androgen receptors by sex steroids in the rat airway smooth muscle cells. Int J Endocrinol. 2016;2016:8423192.

112. Liu QH, Zheng YM, Wang YX. Two distinct signaling pathways for regulation of spontaneous local Ca2+ release by phospholipase C in airway smooth muscle cells. Pflugers Arch. 2007;453(4):531–41.

113. Mikkonen L, Pihlajamaa P, Sahu B, Zhang FP, Janne OA. Androgen receptor and androgen-dependent gene expression in lung. Mol Cell Endocrinol. 2010;317(1–2):14–24.

114. Butts CL, Bowers E, Horn JC, Shukair SA, Belyavskaya E, Tonelli L, et al. Inhibitory effects of progesterone differ in dendritic cells from female and male rodents. Gend Med. 2008;5(4):434–47.

115. Rubinow KB, Houston B, Wang S, Goodspeed L, Ogimoto K, Morton GJ, et al. Androgen receptor deficiency in monocytes/macrophages does not alter adiposity or glucose homeostasis in male mice. Asian J Androl. 2018;20(3):276–83.

116. Trigunaite A, Dimo J, Jorgensen TN. Suppressive effects of androgens on the immune system. Cell Immunol. 2015;294(2):87–94.

117. Mantalaris A, Panoskaltsis N, Sakai Y, Bourne P, Chang C, Messing EM, et al. Localization of androgen receptor expression in human bone marrow. J Pathol. 2001;193(3):361–6.

118. Viselli SM, Olsen NJ, Shults K, Steizer G, Kovacs WJ. Immunochemical and flow cytometric analysis of androgen receptor expression in thymocytes. Mol Cell Endocrinol. 1995;109(1):19–26.

119. Heinlein CA, Chang C. The roles of androgen receptors and androgen-binding proteins in nongenomic androgen actions. Mol Endocrinol. 2002;16(10):2181–7.

120. Heinlein CA, Chang C. Androgen receptor (AR) coregulators: an overview. Endocr Rev. 2002;23(2):175–200.

121. Fang H, Tong W, Branham WS, Moland CL, Dial SL, Hong H, et al. Study of 202 natural, synthetic, and environmental chemicals for binding to the androgen receptor. Chem Res Toxicol. 2003;16(10):1338–58.

122. Berg AH, Rice CD, Rahman MS, Dong J, Thomas P. Identification and characterization of membrane androgen receptors in the ZIP9 zinc transporter subfamily: I. discovery in female Atlantic croaker and evidence ZIP9 mediates testosterone-induced apoptosis of ovarian follicle cells. Endocrinology. 2014;155(11):4237–49.

123. Thomas P, Converse A, Berg HA. ZIP9, a novel membrane androgen receptor and zinc transporter protein. Gen Comp Endocrinol. 2018;257:130–6.

124. Jorgensen CV, Brauner-Osborne H. Pharmacology and physiological function of the orphan GPRC6A receptor. Basic Clin Pharmacol Toxicol. 2020;126 Suppl 6:77–87.

125. Lucas-Herald AK, Alves-Lopes R, Montezano AC, Ahmed SF, Touyz RM. Genomic and nongenomic effects of androgens in the cardiovascu-

lar system: clinical implications. Clin Sci (Lond). 2017;131(13):1405–18.

126. Clemmensen C, Smajilovic S, Wellendorph P, Brauner-Osborne H. The GPCR, class C, group 6, subtype a (GPRC6A) receptor: from cloning to physiological function. Br J Pharmacol. 2014;171(5):1129–41.

127. Pi M, Kapoor K, Wu Y, Ye R, Senogles SE, Nishimoto SK, et al. Structural and functional evidence for testosterone activation of GPRC6A in peripheral tissues. Mol Endocrinol. 2015;29(12):1759–73.

128. Thomas P, Pang Y, Dong J, Berg AH. Identification and characterization of membrane androgen receptors in the ZIP9 zinc transporter subfamily: II. Role of human ZIP9 in testosterone-induced prostate and breast cancer cell apoptosis. Endocrinology. 2014;155(11):4250–65.

129. Filardo EJ, Quinn JA, Frackelton AR Jr, Bland KI. Estrogen action via the G protein-coupled receptor, GPR30: stimulation of adenylyl cyclase and cAMP-mediated attenuation of the epidermal growth factor receptor-to-MAPK signaling axis. Mol Endocrinol. 2002;16(1):70–84.

130. Paterni I, Granchi C, Katzenellenbogen JA, Minutolo F. Estrogen receptors alpha (ERalpha) and beta (ERbeta): subtype-selective ligands and clinical potential. Steroids. 2014;90:13–29.

131. Prossnitz ER, Arterburn JB, Smith HO, Oprea TI, Sklar LA, Hathaway HJ. Estrogen signaling through the transmembrane G protein-coupled receptor GPR30. Annu Rev Physiol. 2008;70:165–90.

132. Prossnitz ER, Barton M. The G-protein-coupled estrogen receptor GPER in health and disease. Nat Rev Endocrinol. 2011;7(12):715–26.

133. Heldring N, Pike A, Andersson S, Matthews J, Cheng G, Hartman J, et al. Estrogen receptors: how do they signal and what are their targets. Physiol Rev. 2007;87(3):905–31.

134. Lai YJ, Yu D, Zhang JH, Chen GJ. Cooperation of genomic and rapid nongenomic actions of estrogens in synaptic plasticity. Mol Neurobiol. 2017;54(6):4113–26.

135. Saczko J, Michel O, Chwilkowska A, Sawicka E, Maczynska J, Kulbacka J. Estrogen receptors in cell membranes: regulation and signaling. Adv Anat Embryol Cell Biol. 2017;227:93–105.

136. Bjornstrom L, Sjoberg M. Estrogen receptor-dependent activation of AP-1 via non-genomic signalling. Nucl Recept. 2004;2(1):3.

137. Bjornstrom L, Sjoberg M. Mechanisms of estrogen receptor signaling: convergence of genomic and nongenomic actions on target genes. Mol Endocrinol. 2005;19(4):833–42.

138. Madak-Erdogan Z, Lupien M, Stossi F, Brown M, Katzenellenbogen BS. Genomic collaboration of estrogen receptor alpha and extracellular signal-regulated kinase 2 in regulating gene and proliferation programs. Mol Cell Biol. 2011;31(1):226–36.

139. Yu L, Moore AB, Castro L, Gao X, Huynh HL, Klippel M, et al. Estrogen regulates MAPK-related genes through genomic and nongenomic interactions between IGF-I receptor tyrosine kinase and estrogen receptor-alpha signaling pathways in human uterine leiomyoma cells. J Signal Transduct. 2012;2012:204236.

140. Adamski J, Benveniste EN. 17beta-estradiol activation of the c-Jun N-terminal kinase pathway leads to down-regulation of class II major histocompatibility complex expression. Mol Endocrinol. 2005;19(1):113–24.

141. Ambhore NS, Kalidhindi RSR, Loganathan J, Sathish V. Role of differential estrogen receptor activation in airway Hyperreactivity and remodeling in a murine model of asthma. Am J Respir Cell Mol Biol. 2019;61(4):469–80.

142. Ambhore NS, Kalidhindi RSR, Pabelick CM, Hawse JR, Prakash YS, Sathish V. Differential estrogen-receptor activation regulates extracellular matrix deposition in human airway smooth muscle remodeling via NF-kappaB pathway. FASEB J. 2019;33(12):13935–50.

143. Cunningham M, Gilkeson G. Estrogen receptors in immunity and autoimmunity. Clin Rev Allergy Immunol. 2011;40(1):66–73.

144. Ivanova MM, Mazhawidza W, Dougherty SM, Minna JD, Klinge CM. Activity and intracellular location of estrogen receptors alpha and beta in human bronchial epithelial cells. Mol Cell Endocrinol. 2009;305(1–2):12–21.

145. Kalidhindi RSR, Ambhore NS, Bhallamudi S, Loganathan J, Sathish V. Role of estrogen receptors alpha and beta in a murine model of asthma: exacerbated airway Hyperresponsiveness and remodeling in ERbeta knockout mice. Front Pharmacol. 2019;10:1499.

146. Ladikou EE, Kassi E. The emerging role of estrogen in B cell malignancies. Leuk Lymphoma. 2017;58(3):528–39.

147. Beyer C, Kuppers E, Karolczak M, Trotter A. Ontogenetic expression of estrogen and progesterone receptors in the mouse lung. Biol Neonate. 2003;84(1):59–63.

148. Massaro D, Massaro GD. Estrogen receptor regulation of pulmonary alveolar dimensions: alveolar sexual dimorphism in mice. Am J Physiol Lung Cell Mol Physiol. 2006;290(5):L866–70.

149. Massaro GD, Mortola JP, Massaro D. Estrogen modulates the dimensions of the lung's gas-exchange surface area and alveoli in female rats. Am J Phys. 1996;270(1 Pt 1):L110–4.

150. Flouriot G, Brand H, Denger S, Metivier R, Kos M, Reid G, et al. Identification of a new isoform of the human estrogen receptor-alpha (hER-alpha) that is encoded by distinct transcripts and that is able to repress hER-alpha activation function 1. EMBO J. 2000;19(17):4688–700.

151. Wang Z, Zhang X, Shen P, Loggie BW, Chang Y, Deuel TF. Identification, cloning, and expression of human estrogen receptor-alpha36, a novel variant of

human estrogen receptor-alpha66. Biochem Biophys Res Commun. 2005;336(4):1023–7.

152. Acconcia F, Ascenzi P, Bocedi A, Spisni E, Tomasi V, Trentalance A, et al. Palmitoylation-dependent estrogen receptor alpha membrane localization: regulation by 17beta-estradiol. Mol Biol Cell. 2005;16(1):231–7.

153. Acconcia F, Ascenzi P, Fabozzi G, Visca P, Marino M. S-palmitoylation modulates human estrogen receptor-alpha functions. Biochem Biophys Res Commun. 2004;316(3):878–83.

154. Marino M, Ascenzi P, Acconcia F. S-palmitoylation modulates estrogen receptor alpha localization and functions. Steroids. 2006;71(4):298–303.

155. Razandi M, Oh P, Pedram A, Schnitzer J, Levin ER. ERs associate with and regulate the production of caveolin: implications for signaling and cellular actions. Mol Endocrinol. 2002;16(1):100–15.

156. Billon-Gales A, Fontaine C, Filipe C, Douin-Echinard V, Fouque MJ, Flouriot G, et al. The transactivating function 1 of estrogen receptor alpha is dispensable for the vasculoprotective actions of 17beta-estradiol. Proc Natl Acad Sci U S A. 2009;106(6):2053–8.

157. Warnmark A, Treuter E, Wright AP, Gustafsson JA. Activation functions 1 and 2 of nuclear receptors: molecular strategies for transcriptional activation. Mol Endocrinol. 2003;17(10):1901–9.

158. Li L, Haynes MP, Bender JR. Plasma membrane localization and function of the estrogen receptor alpha variant (ER46) in human endothelial cells. Proc Natl Acad Sci U S A. 2003;100(8):4807–12.

159. Wang Z, Zhang X, Shen P, Loggie BW, Chang Y, Deuel TF. A variant of estrogen receptor-{alpha}, hER-{alpha}36: transduction of estrogen- and antiestrogen-dependent membrane-initiated mitogenic signaling. Proc Natl Acad Sci U S A. 2006;103(24):9063–8.

160. Lin AH, Li RW, Ho EY, Leung GP, Leung SW, Vanhoutte PM, et al. Differential ligand binding affinities of human estrogen receptor-alpha isoforms. PLoS One. 2013;8(4):e63199.

161. Aravamudan B, Goorhouse KJ, Unnikrishnan G, Thompson MA, Pabelick CM, Hawse JR, et al. Differential expression of estrogen receptor variants in response to inflammation signals in human airway smooth muscle. J Cell Physiol. 2017;232(7):1754–60.

162. Arias-Pulido H, Royce M, Gong Y, Joste N, Lomo L, Lee SJ, et al. GPR30 and estrogen receptor expression: new insights into hormone dependence of inflammatory breast cancer. Breast Cancer Res Treat. 2010;123(1):51–8.

163. Prossnitz ER, Maggiolini M. Mechanisms of estrogen signaling and gene expression via GPR30. Mol Cell Endocrinol. 2009;308(1–2):32–8.

164. Prossnitz ER, Oprea TI, Sklar LA, Arterburn JB. The ins and outs of GPR30: a transmembrane estrogen receptor. J Steroid Biochem Mol Biol. 2008;109(3–5):350–3.

165. Thomas P, Pang Y, Filardo EJ, Dong J. Identity of an estrogen membrane receptor coupled to a G protein in human breast cancer cells. Endocrinology. 2005;146(2):624–32.

166. Jala VR, Radde BN, Haribabu B, Klinge CM. Enhanced expression of G-protein coupled estrogen receptor (GPER/GPR30) in lung cancer. BMC Cancer. 2012;12:624.

167. Townsend EA, Thompson MA, Pabelick CM, Prakash YS. Rapid effects of estrogen on intracellular Ca2+ regulation in human airway smooth muscle. Am J Physiol Lung Cell Mol Physiol. 2010;298(4):L521–30.

168. Merlino AA, Welsh TN, Tan H, Yi LJ, Cannon V, Mercer BM, et al. Nuclear progesterone receptors in the human pregnancy myometrium: evidence that parturition involves functional progesterone withdrawal mediated by increased expression of progesterone receptor-a. J Clin Endocrinol Metab. 2007;92(5):1927–33.

169. Piette PCM. The pharmacodynamics and safety of progesterone. Best Pract Res Clin Obstet Gynaecol. 2020;

170. Shao R, Egecioglu E, Weijdegard B, Ljungstrom K, Ling C, Fernandez-Rodriguez J, et al. Developmental and hormonal regulation of progesterone receptor A-form expression in female mouse lung in vivo: interaction with glucocorticoid receptors. J Endocrinol. 2006;190(3):857–70.

171. Shao R, Weijdegard B, Ljungstrom K, Friberg A, Zhu C, Wang X, et al. Nuclear progesterone receptor a and B isoforms in mouse fallopian tube and uterus: implications for expression, regulation, and cellular function. Am J Physiol Endocrinol Metab. 2006;291(1):E59–72.

172. Giangrande PH, DP MD. The a and B isoforms of the human progesterone receptor: two functionally different transcription factors encoded by a single gene. Recent Prog Horm Res. 1999;54:291–313; discussion -4.

173. Tetel MJ, Giangrande PH, Leonhardt SA, McDonnell DP, Edwards DP. Hormone-dependent interaction between the amino- and carboxyl-terminal domains of progesterone receptor in vitro and in vivo. Mol Endocrinol. 1999;13(6):910–24.

174. McKenna NJ, Lanz RB, O'Malley BW. Nuclear receptor coregulators: cellular and molecular biology. Endocr Rev. 1999;20(3):321–44.

175. Onate SA, Tsai SY, Tsai MJ, O'Malley BW. Sequence and characterization of a coactivator for the steroid hormone receptor superfamily. Science. 1995;270(5240):1354–7.

176. Migliaccio A, Piccolo D, Castoria G, Di Domenico M, Bilancio A, Lombardi M, et al. Activation of the Src/p21ras/Erk pathway by progesterone receptor via cross-talk with estrogen receptor. EMBO J. 1998;17(7):2008–18.

177. Jain R, Ray JM, Pan JH, Brody SL. Sex hormone-dependent regulation of cilia beat frequency in

airway epithelium. Am J Respir Cell Mol Biol. 2012;46(4):446–53.

178. Barberis MC, Veronese S, Bauer D, De Juli E, Harari S. Immunocytochemical detection of progesterone receptors. A study in a patient with primary pulmonary hypertension. Chest. 1995;107(3):869–72.

179. Welter BH, Hansen EL, Saner KJ, Wei Y, Price TM. Membrane-bound progesterone receptor expression in human aortic endothelial cells. J Histochem Cytochem. 2003;51(8):1049–55.

180. Gilliver SC. Sex steroids as inflammatory regulators. J Steroid Biochem Mol Biol. 2010;120(2–3):105–15.

181. Dressing GE, Goldberg JE, Charles NJ, Schwertfeger KL, Lange CA. Membrane progesterone receptor expression in mammalian tissues: a review of regulation and physiological implications. Steroids. 2011;76(1–2):11–7.

182. Shah NM, Lai PF, Imami N, Johnson MR. Progesterone-related immune modulation of pregnancy and labor. Front Endocrinol (Lausanne). 2019;10:198.

183. Zhu Y, Rice CD, Pang Y, Pace M, Thomas P. Cloning, expression, and characterization of a membrane progestin receptor and evidence it is an intermediary in meiotic maturation of fish oocytes. Proc Natl Acad Sci U S A. 2003;100(5):2231–6.

184. Tokumoto T, Hossain MB, Wang J. Establishment of procedures for studying mPR-interacting agents and physiological roles of mPR. Steroids. 2016;111:79–83.

185. Zhu Y, Bond J, Thomas P. Identification, classification, and partial characterization of genes in humans and other vertebrates homologous to a fish membrane progestin receptor. Proc Natl Acad Sci U S A. 2003;100(5):2237–42.

186. Areia A, Vale-Pereira S, Alves V, Rodrigues-Santos P, Moura P, Mota-Pinto A. Membrane progesterone receptors in human regulatory T cells: a reality in pregnancy. BJOG. 2015;122(11):1544–50.

187. Areia A, Vale-Pereira S, Alves V, Rodrigues-Santos P, Santos-Rosa M, Moura P, et al. Can membrane progesterone receptor alpha on T regulatory cells explain the ensuing human labour? J Reprod Immunol. 2016;113:22–6.

188. Dosiou C, Hamilton AE, Pang Y, Overgaard MT, Tulac S, Dong J, et al. Expression of membrane progesterone receptors on human T lymphocytes and Jurkat cells and activation of G-proteins by progesterone. J Endocrinol. 2008;196(1):67–77.

189. Feng L, Ransom CE, Nazzal MK, Allen TK, Li YJ, Truong T, et al. The role of progesterone and a novel progesterone receptor, progesterone receptor membrane component 1, in the inflammatory response of fetal membranes to Ureaplasma parvum infection. PLoS One. 2016;11(12):e0168102.

190. Ndiaye K, Poole DH, Walusimbi S, Cannon MJ, Toyokawa K, Maalouf SW, et al. Progesterone effects on lymphocytes may be mediated by membrane progesterone receptors. J Reprod Immunol. 2012;95(1–2):15–26.

191. Jasper AE, McIver WJ, Sapey E, Walton GM. Understanding the role of neutrophils in chronic inflammatory airway disease. F1000Res. 2019;8.

192. Gordon S, Taylor PR. Monocyte and macrophage heterogeneity. Nat Rev Immunol. 2005;5(12):953–64.

193. Chuang KH, Altuwaijri S, Li G, Lai JJ, Chu CY, Lai KP, et al. Neutropenia with impaired host defense against microbial infection in mice lacking androgen receptor. J Exp Med. 2009;206(5):1181–99.

194. Marin DP, Bolin AP, dos Santos RC, Curi R, Otton R. Testosterone suppresses oxidative stress in human neutrophils. Cell Biochem Funct. 2010;28(5):394–402.

195. Cabello N, Mishra V, Sinha U, DiAngelo SL, Chroneos ZC, Ekpa NA, et al. Sex differences in the expression of lung inflammatory mediators in response to ozone. Am J Physiol Lung Cell Mol Physiol. 2015;309(10):L1150–63.

196. Hu G, Christman JW. Editorial: alveolar macrophages in lung inflammation and resolution. Front Immunol. 2019;10:2275.

197. Yamasaki K, Eeden SFV. Lung macrophage phenotypes and functional responses: role in the pathogenesis of COPD. Int J Mol Sci. 2018;19(2)

198. Kapellos TS, Bassler K, Aschenbrenner AC, Fujii W, Schultze JL. Dysregulated functions of lung macrophage populations in COPD. J Immunol Res. 2018;2018:2349045.

199. Schyns J, Bureau F, Marichal T. Lung interstitial macrophages: past, present, and future. J Immunol Res. 2018;2018:5160794.

200. Olsen HH, Grunewald J, Tornling G, Skold CM, Eklund A. Bronchoalveolar lavage results are independent of season, age, gender and collection site. PLoS One. 2012;7(8):e43644.

201. Guilliams M, De Kleer I, Henri S, Post S, Vanhoutte L, De Prijck S, et al. Alveolar macrophages develop from fetal monocytes that differentiate into long-lived cells in the first week of life via GM-CSF. J Exp Med. 2013;210(10):1977–92.

202. Gordon S, Martinez FO. Alternative activation of macrophages: mechanism and functions. Immunity. 2010;32(5):593–604.

203. Rettew JA, Huet-Hudson YM, Marriott I. Testosterone reduces macrophage expression in the mouse of toll-like receptor 4, a trigger for inflammation and innate immunity. Biol Reprod. 2008;78(3):432–7.

204. Malkin CJ, Pugh PJ, Jones RD, Kapoor D, Channer KS, Jones TH. The effect of testosterone replacement on endogenous inflammatory cytokines and lipid profiles in hypogonadal men. J Clin Endocrinol Metab. 2004;89(7):3313–8.

205. D'Agostino P, Milano S, Barbera C, Di Bella G, La Rosa M, Ferlazzo V, et al. Sex hormones modulate inflammatory mediators produced by macrophages. Ann N Y Acad Sci. 1999;876:426–9.

206. Melgert BN, Oriss TB, Qi Z, Dixon-McCarthy B, Geerlings M, Hylkema MN, et al. Macrophages:

regulators of sex differences in asthma? Am J Respir Cell Mol Biol. 2010;42(5):595–603.

207. Villa A, Rizzi N, Vegeto E, Ciana P, Maggi A. Estrogen accelerates the resolution of inflammation in macrophagic cells. Sci Rep. 2015;5:15224.

208. Pisetsky DS, Spencer DM. Effects of progesterone and estradiol sex hormones on the release of microparticles by RAW 264.7 macrophages stimulated by poly(I:C). Clin Vaccine Immunol. 2011;18(9):1420–6.

209. Nelson RK, Bush A, Stokes J, Nair P, Akuthota P. Eosinophilic asthma. J Allergy Clin Immunol Pract. 2020;8(2):465–73.

210. Hogan SP, Rosenberg HF, Moqbel R, Phipps S, Foster PS, Lacy P, et al. Eosinophils: biological properties and role in health and disease. Clin Exp Allergy. 2008;38(5):709–50.

211. Hallstrand TS, Henderson WR Jr. An update on the role of leukotrienes in asthma. Curr Opin Allergy Clin Immunol. 2010;10(1):60–6.

212. Nakanishi H, Horii Y, Fujita K. Effect of testosterone on the eosinophil response of C57BL/6 mice to infection with Brugia pahangi. Immunopharmacology. 1992;23(2):75–9.

213. Hamano N, Terada N, Maesako K, Numata T, Konno A. Effect of sex hormones on eosinophilic inflammation in nasal mucosa. Allergy Asthma Proc. 1998;19(5):263–9.

214. Riffo-Vasquez Y, Ligeiro de Oliveira AP, Page CP, Spina D, Tavares-de-Lima W. Role of sex hormones in allergic inflammation in mice. Clin Exp Allergy. 2007;37(3):459–70.

215. Hellings PW, Vandekerckhove P, Claeys R, Billen J, Kasran A, Ceuppens JL. Progesterone increases airway eosinophilia and hyper-responsiveness in a murine model of allergic asthma. Clin Exp Allergy. 2003;33(10):1457–63.

216. Amin K. The role of mast cells in allergic inflammation. Respir Med. 2012;106(1):9–14.

217. da Silva EZ, Jamur MC, Oliver C. Mast cell function: a new vision of an old cell. J Histochem Cytochem. 2014;62(10):698–738.

218. Chen W, Beck I, Schober W, Brockow K, Effner R, Buters JT, et al. Human mast cells express androgen receptors but treatment with testosterone exerts no influence on IgE-independent mast cell degranulation elicited by neuromuscular blocking agents. Exp Dermatol. 2010;19(3):302–4.

219. Sinha-Hikim I, Taylor WE, Gonzalez-Cadavid NF, Zheng W, Bhasin S. Androgen receptor in human skeletal muscle and cultured muscle satellite cells: up-regulation by androgen treatment. J Clin Endocrinol Metab. 2004;89(10):5245–55.

220. Guhl S, Artuc M, Zuberbier T, Babina M. Testosterone exerts selective anti-inflammatory effects on human skin mast cells in a cell subset dependent manner. Exp Dermatol. 2012;21(11):878–80.

221. Russi AE, Ebel ME, Yang Y, Brown MA. Male-specific IL-33 expression regulates sex-dimorphic EAE susceptibility. Proc Natl Acad Sci U S A. 2018;115(7):E1520–E9.

222. Boberg E, Johansson K, Malmhall C, Weidner J, Radinger M. House dust mite induces bone marrow IL-33-responsive ILC2s and TH cells. Int J Mol Sci. 2020;21(11)

223. Vellutini M, Viegi G, Parrini D, Pedreschi M, Baldacci S, Modena P, et al. Serum immunoglobulins E are related to menstrual cycle. Eur J Epidemiol. 1997;13(8):931–5.

224. Narita S, Goldblum RM, Watson CS, Brooks EG, Estes DM, Curran EM, et al. Environmental estrogens induce mast cell degranulation and enhance IgE-mediated release of allergic mediators. Environ Health Perspect. 2007;115(1):48–52.

225. Vliagoftis H. Thrombin induces mast cell adhesion to fibronectin: evidence for involvement of protease-activated receptor-1. J Immunol. 2002;169(8):4551–8.

226. Zaitsu M, Narita S, Lambert KC, Grady JJ, Estes DM, Curran EM, et al. Estradiol activates mast cells via a non-genomic estrogen receptor-alpha and calcium influx. Mol Immunol. 2007;44(8):1977–85.

227. Belot MP, Abdennebi-Najar L, Gaudin F, Lieberherr M, Godot V, Taieb J, et al. Progesterone reduces the migration of mast cells toward the chemokine stromal cell-derived factor-1/CXCL12 with an accompanying decrease in CXCR4 receptors. Am J Physiol Endocrinol Metab. 2007;292(5):E1410–7.

228. Vasiadi M, Kempuraj D, Boucher W, Kalogeromitros D, Theoharides TC. Progesterone inhibits mast cell secretion. Int J Immunopathol Pharmacol. 2006;19(4):787–94.

229. Geissmann F, Manz MG, Jung S, Sieweke MH, Merad M, Ley K. Development of monocytes, macrophages, and dendritic cells. Science. 2010;327(5966):656–61.

230. O'Keeffe M, Mok WH, Radford KJ. Human dendritic cell subsets and function in health and disease. Cell Mol Life Sci. 2015;72(22):4309–25.

231. Patente TA, Pinho MP, Oliveira AA, Evangelista GCM, Bergami-Santos PC, Barbuto JAM. Human dendritic cells: their heterogeneity and clinical application potential in Cancer immunotherapy. Front Immunol. 2018;9:3176.

232. Tai Y, Wang Q, Korner H, Zhang L, Wei W. Molecular mechanisms of T cells activation by dendritic cells in autoimmune diseases. Front Pharmacol. 2018;9:642.

233. Hepworth MR, Hardman MJ, Grencis RK. The role of sex hormones in the development of Th2 immunity in a gender-biased model of Trichuris muris infection. Eur J Immunol. 2010;40(2):406–16.

234. Paharkova-Vatchkova V, Maldonado R, Kovats S. Estrogen preferentially promotes the differentiation of CD11c+ CD11b(intermediate) dendritic cells from bone marrow precursors. J Immunol. 2004;172(3):1426–36.

235. Laffont S, Seillet C, Guery JC. Estrogen receptor-dependent regulation of dendritic cell development and function. Front Immunol. 2017;8:108.
236. Chaplin DD. Overview of the immune response. J Allergy Clin Immunol. 2010;125(2 Suppl 2):S3–23.
237. Hoffman W, Lakkis FG, Chalasani G. B cells, antibodies, and more. Clin J Am Soc Nephrol. 2016;11(1):137–54.
238. Nicholson LB. The immune system. Essays Biochem. 2016;60(3):275–301.
239. Famili F, Wiekmeijer AS, Staal FJ. The development of T cells from stem cells in mice and humans. Future Sci OA. 2017;3(3):FSO186.
240. Zhu J, Yamane H, Paul WE. Differentiation of effector CD4 T cell populations (*). Annu Rev Immunol. 2010;28:445–89.
241. Fitzpatrick F, Lepault F, Homo-Delarche F, Bach JF, Dardenne M. Influence of castration, alone or combined with thymectomy, on the development of diabetes in the nonobese diabetic mouse. Endocrinology. 1991;129(3):1382–90.
242. Viselli SM, Reese KR, Fan J, Kovacs WJ, Olsen NJ. Androgens alter B cell development in normal male mice. Cell Immunol. 1997;182(2):99–104.
243. Altuwaijri S, Chuang KH, Lai KP, Lai JJ, Lin HY, Young FM, et al. Susceptibility to autoimmunity and B cell resistance to apoptosis in mice lacking androgen receptor in B cells. Mol Endocrinol. 2009;23(4):444–53.
244. Erlandsson MC, Jonsson CA, Islander U, Ohlsson C, Carlsten H. Oestrogen receptor specificity in oestradiol-mediated effects on B lymphopoiesis and immunoglobulin production in male mice. Immunology. 2003;108(3):346–51.
245. Fu Y, Li L, Liu X, Ma C, Zhang J, Jiao Y, et al. Estrogen promotes B cell activation in vitro through down-regulating CD80 molecule expression. Gynecol Endocrinol. 2011;27(8):593–6.
246. Yao G, Liang J, Han X, Hou Y. In vivo modulation of the circulating lymphocyte subsets and monocytes by androgen. Int Immunopharmacol. 2003;3(13–14):1853–60.
247. Bebo BF Jr, Schuster JC, Vandenbark AA, Offner H. Androgens alter the cytokine profile and reduce encephalitogenicity of myelin-reactive T cells. J Immunol. 1999;162(1):35–40.
248. Liva SM, Voskuhl RR. Testosterone acts directly on CD4+ T lymphocytes to increase IL-10 production. J Immunol. 2001;167(4):2060–7.
249. Freeman BM, Mountain DJ, Brock TC, Chapman JR, Kirkpatrick SS, Freeman MB, et al. Low testosterone elevates interleukin family cytokines in a rodent model: a possible mechanism for the potentiation of vascular disease in androgen-deficient males. J Surg Res. 2014;190(1):319–27.
250. Lelu K, Laffont S, Delpy L, Paulet PE, Perinat T, Tschanz SA, et al. Estrogen receptor alpha signaling in T lymphocytes is required for estradiol-mediated inhibition of Th1 and Th17 cell differentiation and protection against experimental autoimmune encephalomyelitis. J Immunol. 2011;187(5):2386–93.
251. Priyanka HP, Krishnan HC, Singh RV, Hima L, Thyagarajan S. Estrogen modulates in vitro T cell responses in a concentration- and receptor-dependent manner: effects on intracellular molecular targets and antioxidant enzymes. Mol Immunol. 2013;56(4):328–39.
252. Lambert KC, Curran EM, Judy BM, Milligan GN, Lubahn DB, Estes DM. Estrogen receptor alpha (ERalpha) deficiency in macrophages results in increased stimulation of CD4+ T cells while 17beta-estradiol acts through ERalpha to increase IL-4 and GATA-3 expression in CD4+ T cells independent of antigen presentation. J Immunol. 2005;175(9):5716–23.
253. Fuseini H, Yung JA, Cephus JY, Zhang J, Goleniewska K, Polosukhin VV, et al. Testosterone decreases house dust mite-induced type 2 and IL-17A-mediated airway inflammation. J Immunol. 2018;201(7):1843–54.
254. Wang Y, Cela E, Gagnon S, Sweezey NB. Estrogen aggravates inflammation in Pseudomonas aeruginosa pneumonia in cystic fibrosis mice. Respir Res. 2010;11:166.
255. Chang J, Mosenifar Z. Differentiating COPD from asthma in clinical practice. J Intensive Care Med. 2007;22(5):300–9.
256. Peltola L, Patsi H, Harju T. COPD comorbidities predict high mortality – asthma-COPD-overlap has better prognosis. COPD. 2020:1–7.
257. Tommola M, Won HK, Ilmarinen P, Jung H, Tuomisto LE, Lehtimaki L, et al. Relationship between age and bronchodilator response at diagnosis in adult-onset asthma. Respir Res. 2020;21(1):179.
258. Zhou A, Luo L, Liu N, Zhang C, Chen Y, Yin Y, et al. Prospective development of practical screening strategies for diagnosis of asthma-COPD overlap. Respirology. 2020;25(7):735–42.
259. Ober C, Yao TC. The genetics of asthma and allergic disease: a 21st century perspective. Immunol Rev. 2011;242(1):10–30.
260. Smit LA, Lenters V, Hoyer BB, Lindh CH, Pedersen HS, Liermontova I, et al. Prenatal exposure to environmental chemical contaminants and asthma and eczema in school-age children. Allergy. 2015;70(6):653–60.
261. DeChristopher LR. Excess free fructose and childhood asthma. Eur J Clin Nutr. 2015;69(12):1371.
262. Chhabra SK. Clinical application of spirometry in asthma: why, when and how often? Lung India. 2015;32(6):635–7.
263. Langley RJ, Dryden C, Westwood J, Anderson E, Thompson A, Urquhart D. Once daily combined inhaled steroid and ultra long-acting bronchodilator prescribing in pediatric asthma: a dual center retrospective cohort study. J Asthma. 2019:1–2.
264. Vink NM, Postma DS, Schouten JP, Rosmalen JG, Boezen HM. Gender differences in asthma development and remission during transition through

puberty: the TRacking Adolescents' individual lives survey (TRAILS) study. J Allergy Clin Immunol. 2010;126(3):498–504 e1–6.

265. Melgert BN, Ray A, Hylkema MN, Timens W, Postma DS. Are there reasons why adult asthma is more common in females? Curr Allergy Asthma Rep. 2007;7(2):143–50.

266. Holgate ST, Wenzel S, Postma DS, Weiss ST, Renz H, Sly PD. Asthma. Nat Rev Dis Primers. 2015;1:15025.

267. de Oliveira AP, Peron JP, Damazo AS, Franco AL, Domingos HV, Oliani SM, et al. Female sex hormones mediate the allergic lung reaction by regulating the release of inflammatory mediators and the expression of lung E-selectin in rats. Respir Res. 2010;11:115.

268. Stanford KI, Mickleborough TD, Ray S, Lindley MR, Koceja DM, Stager JM. Influence of menstrual cycle phase on pulmonary function in asthmatic athletes. Eur J Appl Physiol. 2006;96(6):703–10.

269. Vrieze A, Postma DS, Kerstjens HA. Perimenstrual asthma: a syndrome without known cause or cure. J Allergy Clin Immunol. 2003;112(2):271–82.

270. Chhabra SK. Premenstrual asthma. Indian J Chest Dis Allied Sci. 2005;47(2):109–16.

271. Farha S, Asosingh K, Laskowski D, Hammel J, Dweik RA, Wiedemann HP, et al. Effects of the menstrual cycle on lung function variables in women with asthma. Am J Respir Crit Care Med. 2009;180(4):304–10.

272. Eid RC, Palumbo ML, Cahill KN. Perimenstrual asthma in aspirin-exacerbated respiratory disease. J Allergy Clin Immunol Pract. 2020;8(2):573–8. e4

273. Graziottin A, Serafini A. Perimenstrual asthma: from pathophysiology to treatment strategies. Multidiscip Respir Med. 2016;11:30.

274. Marques-Mejias MA, Barranco P, Laorden D, Romero D, Quirce S. Worsening of severe asthma due to menstruation and sensitization to albumins. J Investig Allergol Clin Immunol. 2018;28(5):330–2.

275. Semik-Orzech A, Skoczynski S, Pierzchala W. Serum estradiol concentration, estradiol-to-progesterone ratio and sputum IL-5 and IL-8 concentrations are increased in luteal phase of the menstrual cycle in perimenstrual asthma patients. Eur Ann Allergy Clin Immunol. 2017;49(4):161–70.

276. Koper I, Hufnagl K, Ehmann R. Gender aspects and influence of hormones on bronchial asthma - secondary publication and update. World Allergy Organ J. 2017;10(1):46.

277. Rao CK, Moore CG, Bleecker E, Busse WW, Calhoun W, Castro M, et al. Characteristics of perimenstrual asthma and its relation to asthma severity and control: data from the severe asthma research program. Chest. 2013;143(4):984–92.

278. Forbes L, Jarvis D, Burney P. Do hormonal contraceptives influence asthma severity? Eur Respir J. 1999;14(5):1028–33.

279. Dratva J, Schindler C, Curjuric I, Stolz D, Macsali F, Gomez FR, et al. Perimenstrual increase in bronchial

hyperreactivity in premenopausal women: results from the population-based SAPALDIA 2 cohort. J Allergy Clin Immunol. 2010;125(4):823–9.

280. Kwon HL, Belanger K, Bracken MB. Asthma prevalence among pregnant and childbearing-aged women in the United States: estimates from national health surveys. Ann Epidemiol. 2003;13(5):317–24.

281. Namazy JA, Schatz M. Update in the treatment of asthma during pregnancy. Clin Rev Allergy Immunol. 2004;26(3):139–48.

282. Virchow JC. Asthma and pregnancy. Semin Respir Crit Care Med. 2012;33(6):630–44.

283. Schatz M, Dombrowski MP. Clinical practice. Asthma in pregnancy. N Engl J Med. 2009;360(18):1862–9.

284. Balzano G, Fuschillo S, Melillo G, Bonini S. Asthma and sex hormones. Allergy. 2001;56(1):13–20.

285. Scioscia G, Carpagnano GE, Lacedonia D, Soccio P, Quarato CMI, Trabace L, et al. The role of airways 17beta-estradiol as a biomarker of severity in postmenopausal asthma: a pilot study. J Clin Med. 2020;9(7)

286. Balzano G, Fuschillo S, De Angelis E, Gaudiosi C, Mancini A, Caputi M. Persistent airway inflammation and high exacerbation rate in asthma that starts at menopause. Monaldi Arch Chest Dis. 2007;67(3):135–41.

287. Farina F, Colombi S, Cantone R, Pastore M, Centanni S, Galimberti M. Study of hypophyseal and gonadal hormones and cases of postmenopausal occurrence of bronchial asthma. Minerva Med. 1986;77(7–8):243–7.

288. Troisi RJ, Speizer FE, Willett WC, Trichopoulos D, Rosner B. Menopause, postmenopausal estrogen preparations, and the risk of adult-onset asthma. A prospective cohort study. Am J Respir Crit Care Med. 1995;152(4 Pt 1):1183–8.

289. Romieu I, Fabre A, Fournier A, Kauffmann F, Varraso R, Mesrine S, et al. Postmenopausal hormone therapy and asthma onset in the E3N cohort. Thorax. 2010;65(4):292–7.

290. Hoffstein V. Relationship between lung volume, maximal expiratory flow, forced expiratory volume in one second, and tracheal area in normal men and women. Am Rev Respir Dis. 1986;134(5):956–61.

291. Pagtakhan RD, Bjelland JC, Landau LI, Loughlin G, Kaltenborn W, Seeley G, et al. Sex differences in growth patterns of the airways and lung parenchyma in children. J Appl Physiol Respir Environ Exerc Physiol. 1984;56(5):1204–10.

292. Tantisira KG, Colvin R, Tonascia J, Strunk RC, Weiss ST, Fuhlbrigge AL, et al. Airway responsiveness in mild to moderate childhood asthma: sex influences on the natural history. Am J Respir Crit Care Med. 2008;178(4):325–31.

293. Carbajal-Garcia A, Reyes-Garcia J, Casas-Hernandez MF, Flores-Soto E, Diaz-Hernandez V, Solis-Chagoyan H, et al. Testosterone augments beta2 adrenergic receptor genomic transcription

increasing salbutamol relaxation in airway smooth muscle. Mol Cell Endocrinol. 2020;110801

294. Espinoza J, Montaño LM, Perusquia M. Nongenomic bronchodilating action elicited by dehydroepiandrosterone (DHEA) in a Guinea pig asthma model. J Steroid Biochem Mol Biol. 2013;138:174–82.

295. Flores-Soto E, Reyes-Garcia J, Carbajal-Garcia A, Campuzano-Gonzalez E, Perusquia M, Sommer B, et al. Sex steroids effects on Guinea pig airway smooth muscle tone and intracellular ca(2+) basal levels. Mol Cell Endocrinol. 2017;439:444–56.

296. Montaño LM, Flores-Soto E, Reyes-Garcia J, Diaz-Hernandez V, Carbajal-Garcia A, Campuzano-Gonzalez E, et al. Testosterone induces hyporesponsiveness by interfering with IP3 receptors in Guinea pig airway smooth muscle. Mol Cell Endocrinol. 2018;473:17–30.

297. Perusquia M, Flores-Soto E, Sommer B, Campuzano-Gonzalez E, Martinez-Villa I, Martinez-Banderas AI, et al. Testosterone-induced relaxation involves L-type and store-operated Ca2+ channels blockade, and PGE 2 in Guinea pig airway smooth muscle. Pflugers Arch. 2015;467(4):767–77.

298. Cephus JY, Stier MT, Fuseini H, Yung JA, Toki S, Bloodworth MH, et al. Testosterone attenuates group 2 innate lymphoid cell-mediated airway inflammation. Cell Rep. 2017;21(9):2487–99.

299. Laffont S, Blanquart E, Savignac M, Cenac C, Laverny G, Metzger D, et al. Androgen signaling negatively controls group 2 innate lymphoid cells. J Exp Med. 2017;214(6):1581–92.

300. Kjellman B, Gustafsson PM. Asthma from childhood to adulthood: asthma severity, allergies, sensitization, living conditions, gender influence and social consequences. Respir Med. 2000;94(5):454–65.

301. Zannolli R, Morgese G. Does puberty interfere with asthma? Med Hypotheses. 1997;48(1):27–32.

302. Mileva Z, Maleeva A. The serum testosterone level of patients with bronchial asthma treated with corticosteroids and untreated. Vutr Boles. 1988;27(4):29–32.

303. Kwon HL, Belanger K, Holford TR, Bracken MB. Effect of fetal sex on airway lability in pregnant women with asthma. Am J Epidemiol. 2006;163(3):217–21.

304. Guiedem E, Ikomey GM, Nkenfou C, Walter PE, Mesembe M, Chegou NN, et al. Chronic obstructive pulmonary disease (COPD): neutrophils, macrophages and lymphocytes in patients with anterior tuberculosis compared to tobacco related COPD. BMC Res Notes. 2018;11(1):192.

305. Vogelmeier CF, Criner GJ, Martinez FJ, Anzueto A, Barnes PJ, Bourbeau J, et al. Global strategy for the diagnosis, management, and prevention of chronic obstructive lung disease 2017 report. GOLD executive summary. Am J Respir Crit Care Med. 2017;195(5):557–82.

306. Davis RM, Novotny TE. The epidemiology of cigarette smoking and its impact on chronic obstructive pulmonary disease. Am Rev Respir Dis. 1989;140(3 Pt 2):S82–4.

307. Vozoris NT. Opioid utility for dyspnea in chronic obstructive pulmonary disease: a complicated and controversial story. Ann Palliat Med. 2020;9(2):571–8.

308. Varmaghani M, Dehghani M, Heidari E, Sharifi F, Moghaddam SS, Farzadfar F. Global prevalence of chronic obstructive pulmonary disease: systematic review and meta-analysis. East Mediterr Health J. 2019;25(1):47–57.

309. Chapman KR. Chronic obstructive pulmonary disease: are women more susceptible than men? Clin Chest Med. 2004;25(2):331–41.

310. Afonso AS, Verhamme KM, Sturkenboom MC, Brusselle GG. COPD in the general population: prevalence, incidence and survival. Respir Med. 2011;105(12):1872–84.

311. Artyukhov IP, Arshukova IL, Dobretsova EA, Dugina TA, Shulmin AV, Demko IV. Epidemiology of chronic obstructive pulmonary disease: a population-based study in Krasnoyarsk region, Russia. Int J Chron Obstruct Pulmon Dis. 2015;10:1781–6.

312. Ng M, Freeman MK, Fleming TD, Robinson M, Dwyer-Lindgren L, Thomson B, et al. Smoking prevalence and cigarette consumption in 187 countries, 1980-2012. JAMA. 2014;311(2):183–92.

313. Barnes PJ, Burney PG, Silverman EK, Celli BR, Vestbo J, Wedzicha JA, et al. Chronic obstructive pulmonary disease. Nat Rev Dis Primers. 2015;1:15076.

314. Birring SS, Brightling CE, Bradding P, Entwisle JJ, Vara DD, Grigg J, et al. Clinical, radiologic, and induced sputum features of chronic obstructive pulmonary disease in nonsmokers: a descriptive study. Am J Respir Crit Care Med. 2002;166(8):1078–83.

315. Silverman EK, Weiss ST, Drazen JM, Chapman HA, Carey V, Campbell EJ, et al. Gender-related differences in severe, early-onset chronic obstructive pulmonary disease. Am J Respir Crit Care Med. 2000;162(6):2152–8.

316. Gan WQ, Man SF, Postma DS, Camp P, Sin DD. Female smokers beyond the perimenopausal period are at increased risk of chronic obstructive pulmonary disease: a systematic review and meta-analysis. Respir Res. 2006;7:52.

317. Terzikhan N, Verhamme KM, Hofman A, Stricker BH, Brusselle GG, Lahousse L. Prevalence and incidence of COPD in smokers and non-smokers: the Rotterdam study. Eur J Epidemiol. 2016;31(8):785–92.

318. Camp PG, Coxson HO, Levy RD, Pillai SG, Anderson W, Vestbo J, et al. Sex differences in emphysema and airway disease in smokers. Chest. 2009;136(6):1480–8.

319. Martinez FJ, Curtis JL, Sciurba F, Mumford J, Giardino ND, Weinmann G, et al. Sex differences in severe pulmonary emphysema. Am J Respir Crit Care Med. 2007;176(3):243–52.

320. Moll M, Regan EA, Hokanson JE, Lutz SM, Silverman EK, Crapo JD, et al. The Association of Multiparity with lung function and chronic obstructive pulmonary disease-related phenotypes. Chronic Obstr Pulm Dis. 2020;7(2):86–98.

321. Agache I, Beltran J, Akdis C, Akdis M, Canelo-Aybar C, Canonica GW, et al. Efficacy and safety of treatment with biologicals (benralizumab, dupilumab, mepolizumab, omalizumab and reslizumab) for severe eosinophilic asthma. A systematic review for the EAACI guidelines - recommendations on the use of biologicals in severe asthma. Allergy. 2020;75(5):1023–42.

322. Barnes PJ. Targeting cytokines to treat asthma and chronic obstructive pulmonary disease. Nat Rev Immunol. 2018;18(7):454–66.

323. Fahy JV. Type 2 inflammation in asthma--present in most, absent in many. Nat Rev Immunol. 2015;15(1):57–65.

324. Lambrecht BN, Hammad H. The immunology of asthma. Nat Immunol. 2015;16(1):45–56.

325. Nagata M, Nakagome K, Soma T. Mechanisms of eosinophilic inflammation. Asia Pac Allergy. 2020;10(2):e14.

326. Upham JW, Xi Y. Dendritic cells in human lung disease: recent advances. Chest. 2017;151(3):668–73.

327. Scanlon ST, McKenzie AN. Type 2 innate lymphoid cells: new players in asthma and allergy. Curr Opin Immunol. 2012;24(6):707–12.

328. Rosenberg HF, Dyer KD, Foster PS. Eosinophils: changing perspectives in health and disease. Nat Rev Immunol. 2013;13(1):9–22.

329. Taskar VS, Coultas DB. Is idiopathic pulmonary fibrosis an environmental disease? Proc Am Thorac Soc. 2006;3(4):293–8.

330. Galli SJ, Tsai M. IgE and mast cells in allergic disease. Nat Med. 2012;18(5):693–704.

331. Brightling CE, Bradding P, Symon FA, Holgate ST, Wardlaw AJ, Pavord ID. Mast-cell infiltration of airway smooth muscle in asthma. N Engl J Med. 2002;346(22):1699–705.

332. Arthur G, Bradding P. New developments in mast cell biology: clinical implications. Chest. 2016;150(3):680–93.

333. Barnes PJ. The cytokine network in asthma and chronic obstructive pulmonary disease. J Clin Invest. 2008;118(11):3546–56.

334. Brightling C, Berry M, Amrani Y. Targeting TNF-alpha: a novel therapeutic approach for asthma. J Allergy Clin Immunol. 2008;121(1):5–10. quiz 1-2

335. Amrani Y, Panettieri RA Jr, Frossard N, Bronner C. Activation of the TNF alpha-p55 receptor induces myocyte proliferation and modulates agonist-evoked calcium transients in cultured human tracheal smooth muscle cells. Am J Respir Cell Mol Biol. 1996;15(1):55–63.

336. Guedes AG, Jude JA, Paulin J, Kita H, Lund FE, Kannan MS. Role of CD38 in TNF-alpha-induced airway hyperresponsiveness. Am J Physiol Lung Cell Mol Physiol. 2008;294(2):L290–9.

337. Makwana R, Gozzard N, Spina D, Page C. TNF-alpha-induces airway hyperresponsiveness to cholinergic stimulation in Guinea pig airways. Br J Pharmacol. 2012;165(6):1978–91.

338. Sathish V, Thompson MA, Bailey JP, Pabelick CM, Prakash YS, Sieck GC. Effect of proinflammatory cytokines on regulation of sarcoplasmic reticulum Ca2+ reuptake in human airway smooth muscle. Am J Physiol Lung Cell Mol Physiol. 2009;297(1):L26–34.

339. Stober VP, Johnson CG, Majors A, Lauer ME, Cali V, Midura RJ, et al. TNF-stimulated gene 6 promotes formation of hyaluronan-inter-alpha-inhibitor heavy chain complexes necessary for ozone-induced airway hyperresponsiveness. J Biol Chem. 2017;292(51):20845–58.

340. Nagase H, Ueki S, Fujieda S. The roles of IL-5 and anti-IL-5 treatment in eosinophilic diseases: asthma, eosinophilic granulomatosis with polyangiitis, and eosinophilic chronic rhinosinusitis. Allergol Int. 2020;69(2):178–86.

341. Brode S, Farahi N, Cowburn AS, Juss JK, Condliffe AM, Chilvers ER. Interleukin-5 inhibits glucocorticoid-mediated apoptosis in human eosinophils. Thorax. 2010;65(12):1116–7.

342. Wallen N, Kita H, Weiler D, Gleich GJ. Glucocorticoids inhibit cytokine-mediated eosinophil survival. J Immunol. 1991;147(10):3490–5.

343. Halim TY, Steer CA, Matha L, Gold MJ, Martinez-Gonzalez I, McNagny KM, et al. Group 2 innate lymphoid cells are critical for the initiation of adaptive T helper 2 cell-mediated allergic lung inflammation. Immunity. 2014;40(3):425–35.

344. Laffont S, Guery JC. Deconstructing the sex bias in allergy and autoimmunity: from sex hormones and beyond. Adv Immunol. 2019;142:35–64.

345. Marone G, Granata F, Pucino V, Pecoraro A, Heffler E, Loffredo S, et al. The intriguing role of interleukin 13 in the pathophysiology of asthma. Front Pharmacol. 2019;10:1387.

346. Rael EL, Lockey RF. Interleukin-13 signaling and its role in asthma. World Allergy Organ J. 2011;4(3):54–64.

347. Townley RG, Sapkota M, Sapkota K. IL-13 and its genetic variants: effect on current asthma treatments. Discov Med. 2011;12(67):513–23.

348. Corren J, Lemanske RF, Hanania NA, Korenblat PE, Parsey MV, Arron JR, et al. Lebrikizumab treatment in adults with asthma. N Engl J Med. 2011;365(12):1088–98.

349. Zhou T, Huang X, Zhou Y, Ma J, Zhou M, Liu Y, et al. Associations between Th17-related inflammatory cytokines and asthma in adults: a case-control study. Sci Rep. 2017;7(1):15502.

350. Newcomb DC, Peebles RS Jr. Th17-mediated inflammation in asthma. Curr Opin Immunol. 2013;25(6):755–60.

351. Zhao ST, Wang CZ. Regulatory T cells and asthma. J Zhejiang Univ Sci B. 2018;19(9):663–73.

352. Aujla SJ, Alcorn JF. T(H)17 cells in asthma and inflammation. Biochim Biophys Acta. 2011;1810(11):1066–79.

353. Seys SF, Lokwani R, Simpson JL, Bullens DMA. New insights in neutrophilic asthma. Curr Opin Pulm Med. 2019;25(1):113–20.

354. McKinley L, Alcorn JF, Peterson A, Dupont RB, Kapadia S, Logar A, et al. TH17 cells mediate steroid-resistant airway inflammation and airway hyperresponsiveness in mice. J Immunol. 2008;181(6):4089–97.

355. Ling SH, McDonough JE, Gosselink JV, Elliott WM, Hayashi S, Hogg JC, et al. Patterns of retention of particulate matter in lung tissues of patients with COPD: potential role in disease progression. Chest. 2011;140(6):1540–9.

356. Lange P, Celli B, Agusti A. Lung-function trajectories and chronic obstructive pulmonary disease. N Engl J Med. 2015;373(16):1575.

357. Barnes PJ. Inflammatory mechanisms in patients with chronic obstructive pulmonary disease. J Allergy Clin Immunol. 2016;138(1):16–27.

358. Caramori G, Adcock IM, Di Stefano A, Chung KF. Cytokine inhibition in the treatment of COPD. Int J Chron Obstruct Pulmon Dis. 2014;9:397–412.

359. Forsslund H, Yang M, Mikko M, Karimi R, Nyren S, Engvall B, et al. Gender differences in the T-cell profiles of the airways in COPD patients associated with clinical phenotypes. Int J Chron Obstruct Pulmon Dis. 2017;12:35–48.

360. Paats MS, Bergen IM, Hoogsteden HC, van der Eerden MM, Hendriks RW. Systemic CD4+ and CD8+ T-cell cytokine profiles correlate with GOLD stage in stable COPD. Eur Respir J. 2012;40(2):330–7.

361. Di Stefano A, Caramori G, Gnemmi I, Contoli M, Vicari C, Capelli A, et al. T helper type 17-related cytokine expression is increased in the bronchial mucosa of stable chronic obstructive pulmonary disease patients. Clin Exp Immunol. 2009;157(2):316–24.

362. Pridgeon C, Bugeon L, Donnelly L, Straschil U, Tudhope SJ, Fenwick P, et al. Regulation of IL-17 in chronic inflammation in the human lung. Clin Sci (Lond). 2011;120(12):515–24.

363. Hiemstra PS. Altered macrophage function in chronic obstructive pulmonary disease. Ann Am Thorac Soc. 2013;10(Suppl):S180–5.

364. Barnes PJ. Alveolar macrophages in chronic obstructive pulmonary disease (COPD). Cell Mol Biol (Noisy-le-Grand). 2004;50 Online Pub:OL627–37.

365. Culpitt SV, Rogers DF, Shah P, De Matos C, Russell RE, Donnelly LE, et al. Impaired inhibition by dexamethasone of cytokine release by alveolar macrophages from patients with chronic obstructive pulmonary disease. Am J Respir Crit Care Med. 2003;167(1):24–31.

366. Barnes PJ. Cellular and molecular mechanisms of asthma and COPD. Clin Sci (Lond). 2017;131(13):1541–58.

367. Vlahos R, Bozinovski S, Chan SP, Ivanov S, Linden A, Hamilton JA, et al. Neutralizing granulocyte/macrophage colony-stimulating factor inhibits cigarette smoke-induced lung inflammation. Am J Respir Crit Care Med. 2010;182(1):34–40.

368. Saetta M, Mariani M, Panina-Bordignon P, Turato G, Buonsanti C, Baraldo S, et al. Increased expression of the chemokine receptor CXCR3 and its ligand CXCL10 in peripheral airways of smokers with chronic obstructive pulmonary disease. Am J Respir Crit Care Med. 2002;165(10):1404–9.

369. Majo J, Ghezzo H, Cosio MG. Lymphocyte population and apoptosis in the lungs of smokers and their relation to emphysema. Eur Respir J. 2001;17(5):946–53.

370. Churg A, Zhou S, Wright JL. Series "matrix metalloproteinases in lung health and disease": matrix metalloproteinases in COPD. Eur Respir J. 2012;39(1):197–209.

371. Papi A, Bellettato CM, Braccioni F, Romagnoli M, Casolari P, Caramori G, et al. Infections and airway inflammation in chronic obstructive pulmonary disease severe exacerbations. Am J Respir Crit Care Med. 2006;173(10):1114–21.

372. Byers DE, Alexander-Brett J, Patel AC, Agapov E, Dang-Vu G, Jin X, et al. Long-term IL-33-producing epithelial progenitor cells in chronic obstructive lung disease. J Clin Invest. 2013;123(9):3967–82.

373. Bafadhel M, Saha S, Siva R, McCormick M, Monteiro W, Rugman P, et al. Sputum IL-5 concentration is associated with a sputum eosinophilia and attenuated by corticosteroid therapy in COPD. Respiration. 2009;78(3):256–62.

374. Reyes-Garcia J, Flores-Soto E, Carbajal-Garcia A, Sommer B, Montaño LM. Maintenance of intracellular Ca2+ basal concentration in airway smooth muscle (review). Int J Mol Med. 2018;42(6):2998–3008.

375. An SS, Bai TR, Bates JH, Black JL, Brown RH, Brusasco V, et al. Airway smooth muscle dynamics: a common pathway of airway obstruction in asthma. Eur Respir J. 2007;29(5):834–60.

376. Cazzola M, Rogliani P, Matera MG. The future of bronchodilation: looking for new classes of bronchodilators. Eur Respir Rev. 2019;28(154)

377. Kouloumenta V, Hatziefthimiou A, Paraskeva E, Gourgoulianis K, Molyvdas PA. Non-genomic effect of testosterone on airway smooth muscle. Br J Pharmacol. 2006;149(8):1083–91.

378. Bordallo J, de Boto MJ, Meana C, Velasco L, Bordallo C, Suarez L, et al. Modulatory role of endogenous androgens on airway smooth muscle tone in isolated Guinea-pig and bovine trachea; involvement of beta2-adrenoceptors, the polyamine system and external calcium. Eur J Pharmacol. 2008;601(1–3):154–62.

379. Montaño LM, Espinoza J, Flores-Soto E, Chavez J, Perusquia M. Androgens are bronchoactive drugs that act by relaxing airway smooth muscle and preventing bronchospasm. J Endocrinol. 2014;222(1):1–13.

380. Carbajal-Garcia A, Reyes-Garcia J, Casas-Hernandez MF, Flores-Soto E, Diaz-Hernandez V, Solis-Chagoyan H, et al. Testosterone augments beta2 adrenergic receptor genomic transcription increasing salbutamol relaxation in airway smooth muscle. Mol Cell Endocrinol. 2020;510:110801.

381. Verthelyi D, Klinman DM. Sex hormone levels correlate with the activity of cytokine-secreting cells in vivo. Immunology. 2000;100(3):384–90.

382. Dunn PJ, Mahood CB, Speed JF, Jury DR. Dehydroepiandrosterone sulphate concentrations in asthmatic patients: pilot study. N Z Med J. 1984;97(768):805–8.

383. Kasperska-Zajac A. Asthma and dehydroepiandrosterone (DHEA): facts and hypotheses. Inflammation. 2010;33(5):320–4.

384. Weinstein RE, Lobocki CA, Gravett S, Hum H, Negrich R, Herbst J, et al. Decreased adrenal sex steroid levels in the absence of glucocorticoid suppression in postmenopausal asthmatic women. J Allergy Clin Immunol. 1996;97(1 Pt 1):1–8.

385. Choi IS, Cui Y, Koh YA, Lee HC, Cho YB, Won YH. Effects of dehydroepiandrosterone on Th2 cytokine production in peripheral blood mononuclear cells from asthmatics. Korean J Intern Med. 2008;23(4):176–81.

386. Koziol-White CJ, Goncharova EA, Cao G, Johnson M, Krymskaya VP, Panettieri RA Jr. DHEA-S inhibits human neutrophil and human airway smooth muscle migration. Biochim Biophys Acta. 2012;1822(10):1638–42.

387. Wenzel SE, Robinson CB, Leonard JM, Panettieri RA Jr. Nebulized dehydroepiandrosterone-3-sulfate improves asthma control in the moderate-to-severe asthma results of a 6-week, randomized, double-blind, placebo-controlled study. Allergy Asthma Proc. 2010;31(6):461–71.

388. Marozkina N, Zein J, DeBoer MD, Logan L, Veri L, Ross K, et al. Dehydroepiandrosterone supplementation may benefit women with asthma who have low androgen levels: a pilot study. Pulm Ther. 2019;5(2):213–20.

389. Asano T, Kanemitsu Y, Takemura M, Fukumitsu K, Kurokawa R, Inoue Y, et al. Small airway inflammation is associated with residual airway hyperresponsiveness in Th2-high asthma. J Asthma. 2019:1–9.

390. Doherty TA, Broide DH. Airway innate lymphoid cells in the induction and regulation of allergy. Allergol Int. 2019;68(1):9–16.

391. Neill DR, Wong SH, Bellosi A, Flynn RJ, Daly M, Langford TK, et al. Nuocytes represent a new innate effector leukocyte that mediates type-2 immunity. Nature. 2010;464(7293):1367–70.

392. Warren KJ, Sweeter JM, Pavlik JA, Nelson AJ, Devasure JM, Dickinson JD, et al. Sex differences in activation of lung-related type 2 innate lymphoid cells in experimental asthma. Ann Allergy Asthma Immunol. 2017;118(2):233–4.

393. Bartemes KR, Kephart GM, Fox SJ, Kita H. Enhanced innate type 2 immune response in peripheral blood from patients with asthma. J Allergy Clin Immunol. 2014;134(3):671–8. e4

394. Christianson CA, Goplen NP, Zafar I, Irvin C, Good JT Jr, Rollins DR, et al. Persistence of asthma requires multiple feedback circuits involving type 2 innate lymphoid cells and IL-33. J Allergy Clin Immunol. 2015;136(1):59–68. e14

395. Nagakumar P, Denney L, Fleming L, Bush A, Lloyd CM, Saglani S. Type 2 innate lymphoid cells in induced sputum from children with severe asthma. J Allergy Clin Immunol. 2016;137(2):624–6. e6

396. Smith SG, Chen R, Kjarsgaard M, Huang C, Oliveria JP, O'Byrne PM, et al. Increased numbers of activated group 2 innate lymphoid cells in the airways of patients with severe asthma and persistent airway eosinophilia. J Allergy Clin Immunol. 2016;137(1):75–86. e8

397. Yu QN, Tan WP, Fan XL, Guo YB, Qin ZL, Li CL, et al. Increased group 2 innate lymphoid cells are correlated with eosinophilic granulocytes in patients with allergic airway inflammation. Int Arch Allergy Immunol. 2018;176(2):124–32.

398. Stadhouders R, Li BWS, de Bruijn MJW, Gomez A, Rao TN, Fehling HJ, et al. Epigenome analysis links gene regulatory elements in group 2 innate lymphocytes to asthma susceptibility. J Allergy Clin Immunol. 2018;142(6):1793–807.

399. Yu CK, Liu YH, Chen CL. Dehydroepiandrosterone attenuates allergic airway inflammation in Dermatophagoides farinae-sensitized mice. J Microbiol Immunol Infect. 2002;35(3):199–202.

400. Liou CJ, Huang WC. Dehydroepiandrosterone suppresses eosinophil infiltration and airway hyperresponsiveness via modulation of chemokines and Th2 cytokines in ovalbumin-sensitized mice. J Clin Immunol. 2011;31(4):656–65.

401. Hayashi T, Adachi Y, Hasegawa K, Morimoto M. Less sensitivity for late airway inflammation in males than females in BALB/c mice. Scand J Immunol. 2003;57(6):562–7.

402. Cerqua I, Terlizzi M, Bilancia R, Riemma MA, Citi V, Martelli A, et al. 5alpha-dihydrotestosterone abrogates sex bias in asthma like features in the mouse. Pharmacol Res. 2020;158:104905.

403. Lund SJ, Portillo A, Cavagnero K, Baum RE, Naji LH, Badrani JH, et al. Leukotriene C4 potentiates IL-33-induced group 2 innate lymphoid cell activation and lung inflammation. J Immunol. 2017;199(3):1096–104.

404. Laidlaw TM, Boyce JA. Cysteinyl leukotriene receptors, old and new; implications for asthma. Clin Exp Allergy. 2012;42(9):1313–20.

405. Ueno H, Koya T, Takeuchi H, Tsukioka K, Saito A, Kimura Y, et al. Cysteinyl leukotriene synthesis via phospholipase A2 group IV mediates exercise-induced bronchoconstriction and airway remodeling. Am J Respir Cell Mol Biol. 2020;63(1):57–66.

406. Kanaoka Y, Boyce JA. Cysteinyl leukotrienes and their receptors; emerging concepts. Allergy Asthma Immunol Res. 2014;6(4):288–95.

407. Pergola C, Rogge A, Dodt G, Northoff H, Weinigel C, Barz D, et al. Testosterone suppresses phospholipase D, causing sex differences in leukotriene biosynthesis in human monocytes. FASEB J. 2011;25(10):3377–87.
408. Bair AM, Turman MV, Vaine CA, Panettieri RA Jr, Soberman RJ. The nuclear membrane leukotriene synthetic complex is a signal integrator and transducer. Mol Biol Cell. 2012;23(22):4456–64.
409. Radmark O, Werz O, Steinhilber D, Samuelsson B. 5-lipoxygenase, a key enzyme for leukotriene biosynthesis in health and disease. Biochim Biophys Acta. 2015;1851(4):331–9.
410. Pergola C, Dodt G, Rossi A, Neunhoeffer E, Lawrenz B, Northoff H, et al. ERK-mediated regulation of leukotriene biosynthesis by androgens: a molecular basis for gender differences in inflammation and asthma. Proc Natl Acad Sci U S A. 2008;105(50):19881–6.
411. Pace S, Pergola C, Dehm F, Rossi A, Gerstmeier J, Troisi F, et al. Androgen-mediated sex bias impairs efficiency of leukotriene biosynthesis inhibitors in males. J Clin Invest. 2017;127(8):3167–76.
412. Johnston NW, Mandhane PJ, Dai J, Duncan JM, Greene JM, Lambert K, et al. Attenuation of the September epidemic of asthma exacerbations in children: a randomized, controlled trial of montelukast added to usual therapy. Pediatrics. 2007;120(3):e702–12.
413. Kalidhindi RSR, Katragadda R, Beauchamp KL, Pabelick CM, Prakash YS, Sathish V. Androgen receptor-mediated regulation of intracellular calcium in human airway smooth muscle cells. Cell Physiol Biochem. 2019;53(1):215–28.
414. Jovanovic DV, Di Battista JA, Martel-Pelletier J, Jolicoeur FC, He Y, Zhang M, et al. IL-17 stimulates the production and expression of proinflammatory cytokines, IL-beta and TNF-alpha, by human macrophages. J Immunol. 1998;160(7):3513–21.
415. Vazquez-Tello A, Halwani R, Hamid Q, Al-Muhsen S. Glucocorticoid receptor-beta up-regulation and steroid resistance induction by IL-17 and IL-23 cytokine stimulation in peripheral mononuclear cells. J Clin Immunol. 2013;33(2):466–78.
416. Cato AC, Wade E. Molecular mechanisms of anti-inflammatory action of glucocorticoids. BioEssays. 1996;18(5):371–8.
417. Bamberger CM, Bamberger AM, de Castro M, Chrousos GP. Glucocorticoid receptor beta, a potential endogenous inhibitor of glucocorticoid action in humans. J Clin Invest. 1995;95(6):2435–41.
418. Sumi Y, Hamid Q. Airway remodeling in asthma. Allergol Int. 2007;56(4):341–8.
419. Evasovic JM, Singer CA. Regulation of IL-17A and implications for TGF-beta1 comodulation of airway smooth muscle remodeling in severe asthma. Am J Physiol Lung Cell Mol Physiol. 2019;316(5):L843–L68.
420. Miettinen PJ, Ebner R, Lopez AR, Derynck R. TGF-beta induced transdifferentiation of mammary epithelial cells to mesenchymal cells: involvement of type I receptors. J Cell Biol. 1994;127(6 Pt 2):2021–36.
421. Chakir J, Shannon J, Molet S, Fukakusa M, Elias J, Laviolette M, et al. Airway remodeling-associated mediators in moderate to severe asthma: effect of steroids on TGF-beta, IL-11, IL-17, and type I and type III collagen expression. J Allergy Clin Immunol. 2003;111(6):1293–8.
422. Xu L, Xiang X, Ji X, Wang W, Luo M, Luo S, et al. Effects and mechanism of dehydroepiandrosterone on epithelial-mesenchymal transition in bronchial epithelial cells. Exp Lung Res. 2014;40(5):211–21.
423. Atlantis E, Martin SA, Haren MT, O'Loughlin PD, Taylor AW, Anand-Ivell R, et al. Demographic, physical and lifestyle factors associated with androgen status: the Florey Adelaide male ageing study (FAMAS). Clin Endocrinol. 2009;71(2):261–72.
424. Jankowska EA, Filippatos G, Ponikowska B, Borodulin-Nadzieja L, Anker SD, Banasiak W, et al. Reduction in circulating testosterone relates to exercise capacity in men with chronic heart failure. J Card Fail. 2009;15(5):442–50.
425. Atlantis E, Fahey P, Cochrane B, Wittert G, Smith S. Endogenous testosterone level and testosterone supplementation therapy in chronic obstructive pulmonary disease (COPD): a systematic review and meta-analysis. BMJ Open. 2013;3(8)
426. Casaburi R, Bhasin S, Cosentino L, Porszasz J, Somfay A, Lewis MI, et al. Effects of testosterone and resistance training in men with chronic obstructive pulmonary disease. Am J Respir Crit Care Med. 2004;170(8):870–8.
427. Creutzberg EC, Casaburi R. Endocrinological disturbances in chronic obstructive pulmonary disease. Eur Respir J Suppl. 2003;46:76s–80s.
428. Kamischke A, Kemper DE, Castel MA, Luthke M, Rolf C, Behre HM, et al. Testosterone levels in men with chronic obstructive pulmonary disease with or without glucocorticoid therapy. Eur Respir J. 1998;11(1):41–5.
429. Semple PD, Beastall GH, Watson WS, Hume R. Serum testosterone depression associated with hypoxia in respiratory failure. Clin Sci (Lond). 1980;58(1):105–6.
430. Svartberg J. Androgens and chronic obstructive pulmonary disease. Curr Opin Endocrinol Diabetes Obes. 2010;17(3):257–61.
431. Karadag F, Ozcan H, Karul AB, Yilmaz M, Cildag O. Sex hormone alterations and systemic inflammation in chronic obstructive pulmonary disease. Int J Clin Pract. 2009;63(2):275–81.
432. Baillargeon J, Urban RJ, Zhang W, Zaiden MF, Javed Z, Sheffield-Moore M, et al. Testosterone replacement therapy and hospitalization rates in men with COPD. Chron Respir Dis. 2019;16:1479972318793004.
433. Ponce-Gallegos MA, Ramirez-Venegas A, Falfan-Valencia R. Th17 profile in COPD exacerbations. Int J Chron Obstruct Pulmon Dis. 2017;12:1857–65.

434. Barin JG, Baldeviano GC, Talor MV, Wu L, Ong S, Quader F, et al. Macrophages participate in IL-17-mediated inflammation. Eur J Immunol. 2012;42(3):726–36.

435. Finkelstein R, Fraser RS, Ghezzo H, Cosio MG. Alveolar inflammation and its relation to emphysema in smokers. Am J Respir Crit Care Med. 1995;152(5 Pt 1):1666–72.

436. Hautamaki RD, Kobayashi DK, Senior RM, Shapiro SD. Requirement for macrophage elastase for cigarette smoke-induced emphysema in mice. Science. 1997;277(5334):2002–4.

437. Balhara J, Gounni AS. The alveolar macrophages in asthma: a double-edged sword. Mucosal Immunol. 2012;5(6):605–9.

438. Shaykhiev R, Krause A, Salit J, Strulovici-Barel Y, Harvey BG, O'Connor TP, et al. Smoking-dependent reprogramming of alveolar macrophage polarization: implication for pathogenesis of chronic obstructive pulmonary disease. J Immunol. 2009;183(4):2867–83.

439. Moreira AP, Hogaboam CM. Macrophages in allergic asthma: fine-tuning their pro- and anti-inflammatory actions for disease resolution. J Interf Cytokine Res. 2011;31(6):485–91.

440. Becerra-Diaz M, Strickland AB, Keselman A, Heller NM. Androgen and androgen receptor as enhancers of M2 macrophage polarization in allergic lung inflammation. J Immunol. 2018;201(10):2923–33.

441. Fiandalo MV, Stocking JJ, Pop EA, Wilton JH, Mantione KM, Li Y, et al. Inhibition of dihydrotestosterone synthesis in prostate cancer by combined frontdoor and backdoor pathway blockade. Oncotarget. 2018;9(13):11227–42.

442. Zein JG, Erzurum SC. Asthma is Different in Women. Curr Allergy Asthma Rep. 2015;15(6):28.

443. Ambhore NS, Katragadda R, Raju Kalidhindi RS, Thompson MA, Pabelick CM, Prakash YS, et al. Estrogen receptor beta signaling inhibits PDGF induced human airway smooth muscle proliferation. Mol Cell Endocrinol. 2018;476:37–47.

444. Fuentes N, Cabello N, Nicoleau M, Chroneos ZC, Silveyra P. Modulation of the lung inflammatory response to ozone by the estrous cycle. Physiol Rep. 2019;7(5):e14026.

445. Sathish V, Freeman MR, Long E, Thompson MA, Pabelick CM, Prakash YS. Cigarette smoke and estrogen signaling in human airway smooth muscle. Cell Physiol Biochem. 2015;36(3):1101–15.

446. Wang SY, Freeman MR, Sathish V, Thompson MA, Pabelick CM, Prakash YS. Sex steroids influence brain-derived neurotropic factor secretion from human airway smooth muscle cells. J Cell Physiol. 2016;231(7):1586–92.

447. Corteling R, Trifilieff A. Gender comparison in a murine model of allergen-driven airway inflammation and the response to budesonide treatment. BMC Pharmacol. 2004;4:4.

448. Sakazaki F, Ueno H, Nakamuro K. 17beta-estradiol enhances expression of inflammatory cytokines and inducible nitric oxide synthase in mouse contact hypersensitivity. Int Immunopharmacol. 2008;8(5):654–60.

449. Seymour BW, Friebertshauser KE, Peake JL, Pinkerton KE, Coffman RL, Gershwin LJ. Gender differences in the allergic response of mice neonatally exposed to environmental tobacco smoke. Dev Immunol. 2002;9(1):47–54.

450. Haggerty CL, Ness RB, Kelsey S, Waterer GW. The impact of estrogen and progesterone on asthma. Ann Allergy Asthma Immunol. 2003;90(3):284–91; quiz 91-3, 347.

451. Intapad S, Dimitropoulou C, Snead C, Piyachaturawat P, Catravas JD. Regulation of asthmatic airway relaxation by estrogen and heat shock protein 90. J Cell Physiol. 2012;227(8):3036–43.

452. Myers JR, Sherman CB. Should supplemental estrogens be used as steroid-sparing agents in asthmatic women? Chest. 1994;106(1):318–9.

453. Townsend EA, Sathish V, Thompson MA, Pabelick CM, Prakash YS. Estrogen effects on human airway smooth muscle involve cAMP and protein kinase a. Am J Physiol Lung Cell Mol Physiol. 2012;303(10):L923–8.

454. Foster PS, Goldie RG, Paterson JW. Effect of steroids on beta-adrenoceptor-mediated relaxation of pig bronchus. Br J Pharmacol. 1983;78(2):441–5.

455. Pang JJ, Xu XB, Li HF, Zhang XY, Zheng TZ, Qu SY. Inhibition of beta-estradiol on trachea smooth muscle contraction in vitro and in vivo. Acta Pharmacol Sin. 2002;23(3):273–7.

456. Teoh H, Man RY. Enhanced relaxation of porcine coronary arteries after acute exposure to a physiological level of 17beta-estradiol involves non-genomic mechanisms and the cyclic AMP cascade. Br J Pharmacol. 2000;129(8):1739–47.

457. Dimitropoulou C, White RE, Ownby DR, Catravas JD. Estrogen reduces carbachol-induced constriction of asthmatic airways by stimulating large-conductance voltage and calcium-dependent potassium channels. Am J Respir Cell Mol Biol. 2005;32(3):239–47.

458. Nunez FJ, Johnstone TB, Corpuz ML, Kazarian AG, Mohajer NN, Tliba O, et al. Glucocorticoids rapidly activate cAMP production via Galphas to initiate non-genomic signaling that contributes to one-third of their canonical genomic effects. FASEB J. 2020;34(2):2882–95.

459. Townsend EA, Meuchel LW, Thompson MA, Pabelick CM, Prakash YS. Estrogen increases nitric-oxide production in human bronchial epithelium. J Pharmacol Exp Ther. 2011;339(3):815–24.

460. Perusquia M, Hernandez R, Montaño LM, Villalon CM, Campos MG. Inhibitory effect of sex steroids on Guinea-pig airway smooth muscle contractions. Comp Biochem Physiol C Pharmacol Toxicol Endocrinol. 1997;118(1):5–10.

461. Carey MA, Card JW, Bradbury JA, Moorman MP, Haykal-Coates N, Gavett SH, et al. Spontaneous airway hyperresponsiveness in estrogen receptor-

alpha-deficient mice. Am J Respir Crit Care Med. 2007;175(2):126–35.

462. Matsubara S, Swasey CH, Loader JE, Dakhama A, Joetham A, Ohnishi H, et al. Estrogen determines sex differences in airway responsiveness after allergen exposure. Am J Respir Cell Mol Biol. 2008;38(5):501–8.

463. Dimitropoulou C, Drakopanagiotakis F, Chatterjee A, Snead C, Catravas JD. Estrogen replacement therapy prevents airway dysfunction in a murine model of allergen-induced asthma. Lung. 2009;187(2):116–27.

464. Bhallamudi S, Connell J, Pabelick CM, Prakash YS, Sathish V. Estrogen receptors differentially regulate intracellular calcium handling in human nonasthmatic and asthmatic airway smooth muscle cells. Am J Physiol Lung Cell Mol Physiol. 2020;318(1):L112–L24.

465. Dunican EM, Fahy JV. The role of type 2 inflammation in the pathogenesis of asthma exacerbations. Ann Am Thorac Soc. 2015;12 Suppl 2:S144–9.

466. Haselkorn T, Szefler SJ, Simons FE, Zeiger RS, Mink DR, Chipps BE, et al. Allergy, total serum immunoglobulin E, and airflow in children and adolescents in TENOR. Pediatr Allergy Immunol. 2010;21(8):1157–65.

467. Bengtsson AK, Ryan EJ, Giordano D, Magaletti DM, Clark EA. 17beta-estradiol (E2) modulates cytokine and chemokine expression in human monocyte-derived dendritic cells. Blood. 2004;104(5):1404–10.

468. Uemura Y, Liu TY, Narita Y, Suzuki M, Matsushita S. 17 Beta-estradiol (E2) plus tumor necrosis factor-alpha induces a distorted maturation of human monocyte-derived dendritic cells and promotes their capacity to initiate T-helper 2 responses. Hum Immunol. 2008;69(3):149–57.

469. Melgert BN, ten Hacken NH, Rutgers B, Timens W, Postma DS, Hylkema MN. More alternative activation of macrophages in lungs of asthmatic patients. J Allergy Clin Immunol. 2011;127(3):831–3.

470. Girodet PO, Nguyen D, Mancini JD, Hundal M, Zhou X, Israel E, et al. Alternative macrophage activation is increased in asthma. Am J Respir Cell Mol Biol. 2016;55(4):467–75.

471. Blacquiere MJ, Hylkema MN, Postma DS, Geerlings M, Timens W, Melgert BN. Airway inflammation and remodeling in two mouse models of asthma: comparison of males and females. Int Arch Allergy Immunol. 2010;153(2):173–81.

472. Melgert BN, Postma DS, Kuipers I, Geerlings M, Luinge MA, van der Strate BW, et al. Female mice are more susceptible to the development of allergic airway inflammation than male mice. Clin Exp Allergy. 2005;35(11):1496–503.

473. Keselman A, Fang X, White PB, Heller NM. Estrogen signaling contributes to sex differences in macrophage polarization during asthma. J Immunol. 2017;199(5):1573–83.

474. Newcomb DC, Cephus JY, Boswell MG, Fahrenholz JM, Langley EW, Feldman AS, et al. Estrogen and progesterone decrease let-7f microRNA expression and increase IL-23/IL-23 receptor signaling and IL-17A production in patients with severe asthma. J Allergy Clin Immunol. 2015;136(4):1025–34. e11

475. McGeachy MJ, Bak-Jensen KS, Chen Y, Tato CM, Blumenschein W, McClanahan T, et al. TGF-beta and IL-6 drive the production of IL-17 and IL-10 by T cells and restrain T(H)-17 cell-mediated pathology. Nat Immunol. 2007;8(12):1390–7.

476. McGeachy MJ, Chen Y, Tato CM, Laurence A, Joyce-Shaikh B, Blumenschein WM, et al. The interleukin 23 receptor is essential for the terminal differentiation of interleukin 17-producing effector T helper cells in vivo. Nat Immunol. 2009;10(3):314–24.

477. Ahmadi-Vasmehjani A, Baharlou R, Atashzar MR, Raofi R, Jafari M, Razavi FS. Regulatory effects of estradiol on peripheral blood mononuclear cells activation in patients with asthma. Iran J Allergy Asthma Immunol. 2018;17(1):9–17.

478. Molnar I, Bohaty I, Somogyine-Vari E. High prevalence of increased interleukin-17A serum levels in postmenopausal estrogen deficiency. Menopause. 2014;21(7):749–52.

479. Schatz M, Dombrowski MP, Wise R, Thom EA, Landon M, Mabie W, et al. Asthma morbidity during pregnancy can be predicted by severity classification. J Allergy Clin Immunol. 2003;112(2):283–8.

480. Juniper EF, Daniel EE, Roberts RS, Kline PA, Hargreave FE, Newhouse MT. Improvement in airway responsiveness and asthma severity during pregnancy. A prospective study. Am Rev Respir Dis. 1989;140(4):924–31.

481. Straub RH. The complex role of estrogens in inflammation. Endocr Rev. 2007;28(5):521–74.

482. Gilmore W, Weiner LP, Correale J. Effect of estradiol on cytokine secretion by proteolipid protein-specific T cell clones isolated from multiple sclerosis patients and normal control subjects. J Immunol. 1997;158(1):446–51.

483. Cenci S, Weitzmann MN, Roggia C, Namba N, Novack D, Woodring J, et al. Estrogen deficiency induces bone loss by enhancing T-cell production of TNF-alpha. J Clin Invest. 2000;106(10):1229–37.

484. Triebner K, Johannessen A, Puggini L, Benediktsdottir B, Bertelsen RJ, Bifulco E, et al. Menopause as a predictor of new-onset asthma: a longitudinal northern European population study. J Allergy Clin Immunol. 2016;137(1):50–7. e6

485. Triebner K, Matulonga B, Johannessen A, Suske S, Benediktsdottir B, Demoly P, et al. Menopause is associated with accelerated lung function decline. Am J Respir Crit Care Med. 2017;195(8):1058–65.

486. Fait T. Menopause hormone therapy: latest developments and clinical practice. Drugs Context. 2019;8:212551.

487. van Hylckama VA, Helmerhorst FM, Vandenbroucke JP, Doggen CJ, Rosendaal FR. The venous thrombotic risk of oral contraceptives, effects of oestrogen

dose and progestogen type: results of the MEGA case-control study. BMJ. 2009;339:b2921.

488. van Rooijen M, Silveira A, Thomassen S, Hansson LO, Rosing J, Hamsten A, et al. Rapid activation of haemostasis after hormonal emergency contraception. Thromb Haemost. 2007;97(1):15–20.

489. Jaimez R, Cooney A, Jackson K, Lemus AE, Lemini C, Cardenas M, et al. In vivo estrogen bioactivities and in vitro estrogen receptor binding and transcriptional activities of anticoagulant synthetic 17beta-aminoestrogens. J Steroid Biochem Mol Biol. 2000;73(1–2):59–66.

490. Lemini C, Rubio-Poo C, Franco Y, Jaimez R, Avila ME, Medina M, et al. In vivo profile of the anticoagulant effect of 17ss-amino-1,3,5(10)estratrien-3-ol. Eur J Pharmacol. 2013;700(1–3):210–6.

491. Lemini C, Rubio-Poo C, Silva G, Garcia-Mondragon J, Zavala E, Mendoza-Patino N, et al. Anticoagulant and estrogenic effects of two new 17 beta-aminoestrogens, butolame [17 beta-(4-hydroxy-1-butylamino)-1,3,5(10)-estratrien-3-ol] and pentolame [17 beta-(5-hydroxy-1-pentylamino)-1,3,5(10)-estratrien-3-ol]. Steroids. 1993;58(10):457–61.

492. Flores-Soto E, Martinez-Villa I, Solis-Chagoyan H, Sommer B, Lemini C, Montaño LM. 17beta-Aminoestrogens induce Guinea pig airway smooth muscle hyperresponsiveness through L-type ca(2+) channels activation. Steroids. 2015;101:64–70.

493. Stamatiou R, Paraskeva E, Papagianni M, Molyvdas PA, Hatziefthimiou A. The mitogenic effect of testosterone and 17beta-estradiol on airway smooth muscle cells. Steroids. 2011;76(4):400–8.

494. Cheng B, Song J, Zou Y, Wang Q, Lei Y, Zhu C, et al. Responses of vascular smooth muscle cells to estrogen are dependent on balance between ERK and p38 MAPK pathway activities. Int J Cardiol. 2009;134(3):356–65.

495. Li H, Cheng Y, Simoncini T, Xu S. 17beta-estradiol inhibits TNF-alpha-induced proliferation and migration of vascular smooth muscle cells via suppression of TRAIL. Gynecol Endocrinol. 2016;32(7):581–6.

496. Zheng S, Chen X, Hong S, Long L, Xu Y, Simoncini T, et al. 17beta-estradiol inhibits vascular smooth muscle cell migration via up-regulation of striatin protein. Gynecol Endocrinol. 2015;31(8):618–24.

497. Chakir J, Haj-Salem I, Gras D, Joubert P, Beaudoin EL, Biardel S, et al. Effects of bronchial Thermoplasty on airway smooth muscle and collagen deposition in asthma. Ann Am Thorac Soc. 2015;12(11):1612–8.

498. Mostaco-Guidolin LB, Osei ET, Ullah J, Hajimohammadi S, Fouadi M, Li X, et al. Defective Fibrillar collagen organization by fibroblasts contributes to airway remodeling in asthma. Am J Respir Crit Care Med. 2019;200(4):431–43.

499. Royce SG, Tan L, Koek AA, Tang ML. Effect of extracellular matrix composition on airway epithelial cell and fibroblast structure: implications for airway remodeling in asthma. Ann Allergy Asthma Immunol. 2009;102(3):238–46.

500. Chakrabarti S, Patel KD. Matrix metalloproteinase-2 (MMP-2) and MMP-9 in pulmonary pathology. Exp Lung Res. 2005;31(6):599–621.

501. Chung FT, Huang HY, Lo CY, Huang YC, Lin CW, He CC, et al. Increased ratio of matrix Metalloproteinase-9 (MMP-9)/tissue inhibitor Metalloproteinase-1 from alveolar macrophages in chronic asthma with a fast decline in FEV1 at 5-year follow-up. J Clin Med. 2019;8(9)

502. Ohbayashi H, Shimokata K. Matrix metalloproteinase-9 and airway remodeling in asthma. Curr Drug Targets Inflamm Allergy. 2005;4(2):177–81.

503. Barr RG, Camargo CA Jr. Hormone replacement therapy and obstructive airway diseases. Treat Respir Med. 2004;3(1):1–7.

504. Konings GFJ, Reynaert NL, Delvoux B, Verhamme FM, Bracke KR, Brusselle GG, et al. Increased levels of enzymes involved in local estradiol synthesis in chronic obstructive pulmonary disease. Mol Cell Endocrinol. 2017;443:23–31.

505. Zienolddiny S, Campa D, Lind H, Ryberg D, Skaug V, Stangeland LB, et al. A comprehensive analysis of phase I and phase II metabolism gene polymorphisms and risk of non-small cell lung cancer in smokers. Carcinogenesis. 2008;29(6):1164–9.

506. Van Winkle LS, Gunderson AD, Shimizu JA, Baker GL, Brown CD. Gender differences in naphthalene metabolism and naphthalene-induced acute lung injury. Am J Physiol Lung Cell Mol Physiol. 2002;282(5):L1122–34.

507. Chichester CH, Buckpitt AR, Chang A, Plopper CG. Metabolism and cytotoxicity of naphthalene and its metabolites in isolated murine Clara cells. Mol Pharmacol. 1994;45(4):664–72.

508. Han W, Pentecost BT, Pietropaolo RL, Fasco MJ, Spivack SD. Estrogen receptor alpha increases basal and cigarette smoke extract-induced expression of CYP1A1 and CYP1B1, but not GSTP1, in normal human bronchial epithelial cells. Mol Carcinog. 2005;44(3):202–11.

509. Lee AJ, Cai MX, Thomas PE, Conney AH, Zhu BT. Characterization of the oxidative metabolites of 17beta-estradiol and estrone formed by 15 selectively expressed human cytochrome p450 isoforms. Endocrinology. 2003;144(8):3382–98.

510. Osborne MP, Bradlow HL, Wong GY, Telang NT. Upregulation of estradiol C16 alpha-hydroxylation in human breast tissue: a potential biomarker of breast cancer risk. J Natl Cancer Inst. 1993;85(23):1917–20.

511. Schneider J, Kinne D, Fracchia A, Pierce V, Anderson KE, Bradlow HL, et al. Abnormal oxidative metabolism of estradiol in women with breast cancer. Proc Natl Acad Sci U S A. 1982;79(9):3047–51.

512. Lu LJ, Cree M, Josyula S, Nagamani M, Grady JJ, Anderson KE. Increased urinary excretion of 2-hydroxyestrone but not 16alpha-hydroxyestrone in

premenopausal women during a soya diet containing isoflavones. Cancer Res. 2000;60(5):1299–305.

513. Michnovicz JJ, Hershcopf RJ, Naganuma H, Bradlow HL, Fishman J. Increased 2-hydroxylation of estradiol as a possible mechanism for the anti-estrogenic effect of cigarette smoking. N Engl J Med. 1986;315(21):1305–9.

514. Spivack SD, Hurteau GJ, Fasco MJ, Kaminsky LS. Phase I and II carcinogen metabolism gene expression in human lung tissue and tumors. Clin Cancer Res. 2003;9(16 Pt 1):6002–11.

515. Weinberg OK, Marquez-Garban DC, Fishbein MC, Goodglick L, Garban HJ, Dubinett SM, et al. Aromatase inhibitors in human lung cancer therapy. Cancer Res. 2005;65(24):11287–91.

516. Young PA, Pietras RJ. Aromatase inhibitors combined with aspirin to prevent lung cancer in preclinical models. Transl Lung Cancer Res. 2018;7(Suppl 4):S373–S6.

517. Travis WD, Costabel U, Hansell DM, King TE Jr, Lynch DA, Nicholson AG, et al. An official American Thoracic Society/European Respiratory Society statement: update of the international multidisciplinary classification of the idiopathic interstitial pneumonias. Am J Respir Crit Care Med. 2013;188(6):733–48.

518. Raghu G, Collard HR, Egan JJ, Martinez FJ, Behr J, Brown KK, et al. An official ATS/ERS/JRS/ALAT statement: idiopathic pulmonary fibrosis: evidence-based guidelines for diagnosis and management. Am J Respir Crit Care Med. 2011;183(6):788–824.

519. Raghu G, Chen SY, Hou Q, Yeh WS, Collard HR. Incidence and prevalence of idiopathic pulmonary fibrosis in US adults 18-64 years old. Eur Respir J. 2016;48(1):179–86.

520. Raghu G, Chen SY, Yeh WS, Maroni B, Li Q, Lee YC, et al. Idiopathic pulmonary fibrosis in US Medicare beneficiaries aged 65 years and older: incidence, prevalence, and survival, 2001-11. Lancet Respir Med. 2014;2(7):566–72.

521. Glassberg MK. Overview of idiopathic pulmonary fibrosis, evidence-based guidelines, and recent developments in the treatment landscape. Am J Manag Care. 2019;25(11 Suppl):S195–203.

522. Venkataraman T, Frieman MB. The role of epidermal growth factor receptor (EGFR) signaling in SARS coronavirus-induced pulmonary fibrosis. Antivir Res. 2017;143:142–50.

523. Wang Q, Pan S, Zhang S, Shen G, Huang M, Wu M. Lung transplantation in pulmonary fibrosis secondary to influenza a pneumonia. Ann Thorac Surg. 2019;108(4):e233–e5.

524. Wang L, Cheng W, Zhang Z. Respiratory syncytial virus infection accelerates lung fibrosis through the unfolded protein response in a bleomycin-induced pulmonary fibrosis animal model. Mol Med Rep. 2017;16(1):310–6.

525. Paolocci G, Folletti I, Toren K, Ekstrom M, Dell'Omo M, Muzi G, et al. Occupational risk factors for idio-

pathic pulmonary fibrosis in southern Europe: a case-control study. BMC Pulm Med. 2018;18(1):75.

526. Hu Y, Wang LS, Li Y, Li QH, Li CL, Chen JM, et al. Effects of particulate matter from straw burning on lung fibrosis in mice. Environ Toxicol Pharmacol. 2017;56:249–58.

527. Fang L, Cheng Q, Zhao F, Cheng H, Luo Y, Bao X, et al. Cigarette smoke exposure combined with lipopolysaccharides induced pulmonary fibrosis in mice. Respir Physiol Neurobiol. 2019;266:9–17.

528. Pan M, Zheng Z, Chen Y, Sun N, Zheng B, Yang Q, et al. Angiotensin-(1-7) attenuated cigarette smoking-related pulmonary fibrosis via improving the impaired autophagy caused by Nicotinamide adenine dinucleotide phosphate reduced oxidase 4-dependent reactive oxygen species. Am J Respir Cell Mol Biol. 2018;59(3):306–19.

529. Gille T, Didier M, Boubaya M, Moya L, Sutton A, Carton Z, et al. Obstructive sleep apnoea and related comorbidities in incident idiopathic pulmonary fibrosis. Eur Respir J. 2017;49(6)

530. Gribbin J, Hubbard R, Smith C. Role of diabetes mellitus and gastro-oesophageal reflux in the aetiology of idiopathic pulmonary fibrosis. Respir Med. 2009;103(6):927–31.

531. Hutchinson J, Fogarty A, Hubbard R, McKeever T. Global incidence and mortality of idiopathic pulmonary fibrosis: a systematic review. Eur Respir J. 2015;46(3):795–806.

532. Agusti AG, Roca J, Gea J, Wagner PD, Xaubet A, Rodriguez-Roisin R. Mechanisms of gas-exchange impairment in idiopathic pulmonary fibrosis. Am Rev Respir Dis. 1991;143(2):219–25.

533. Robinson HC. Respiratory conditions update: restrictive lung disease. FP Essent. 2016;448:29–34.

534. Fernandez IE, Eickelberg O. New cellular and molecular mechanisms of lung injury and fibrosis in idiopathic pulmonary fibrosis. Lancet. 2012;380(9842):680–8.

535. Wynn TA. Integrating mechanisms of pulmonary fibrosis. J Exp Med. 2011;208(7):1339–50.

536. Salton F, Volpe MC, Confalonieri M. Epithelial(−) mesenchymal transition in the pathogenesis of idiopathic pulmonary fibrosis. Medicina (Kaunas). 2019;55(4)

537. Kim KK, Kugler MC, Wolters PJ, Robillard L, Galvez MG, Brumwell AN, et al. Alveolar epithelial cell mesenchymal transition develops in vivo during pulmonary fibrosis and is regulated by the extracellular matrix. Proc Natl Acad Sci U S A. 2006;103(35):13180–5.

538. Willis BC, Liebler JM, Luby-Phelps K, Nicholson AG, Crandall ED, du Bois RM, et al. Induction of epithelial-mesenchymal transition in alveolar epithelial cells by transforming growth factor-beta1: potential role in idiopathic pulmonary fibrosis. Am J Pathol. 2005;166(5):1321–32.

539. Ji Y, Dou YN, Zhao QW, Zhang JZ, Yang Y, Wang T, et al. Paeoniflorin suppresses TGF-beta mediated epithelial-mesenchymal transition in pulmonary

fibrosis through a Smad-dependent pathway. Acta Pharmacol Sin. 2016;37(6):794–804.

540. Cho N, Razipour SE, McCain ML. Featured article: TGF-beta1 dominates extracellular matrix rigidity for inducing differentiation of human cardiac fibroblasts to myofibroblasts. Exp Biol Med (Maywood). 2018;243(7):601–12.

541. Desmouliere A, Geinoz A, Gabbiani F, Gabbiani G. Transforming growth factor-beta 1 induces alpha-smooth muscle actin expression in granulation tissue myofibroblasts and in quiescent and growing cultured fibroblasts. J Cell Biol. 1993;122(1):103–11.

542. Nakagome K, Dohi M, Okunishi K, Tanaka R, Miyazaki J, Yamamoto K. In vivo IL-10 gene delivery attenuates bleomycin induced pulmonary fibrosis by inhibiting the production and activation of TGF-beta in the lung. Thorax. 2006;61(10):886–94.

543. Saito A, Okazaki H, Sugawara I, Yamamoto K, Takizawa H. Potential action of IL-4 and IL-13 as fibrogenic factors on lung fibroblasts in vitro. Int Arch Allergy Immunol. 2003;132(2):168–76.

544. Upparahalli Venkateshaiah S, Niranjan R, Manohar M, Verma AK, Kandikattu HK, Lasky JA, et al. Attenuation of allergen-, IL-13-, and TGF-alpha-induced lung fibrosis after the treatment of rIL-15 in mice. Am J Respir Cell Mol Biol. 2019;61(1):97–109.

545. Xiao L, Li ZH, Hou XM, Yu RJ. Evaluation of interleukin-13 in the serum and bronchoalveolar lavage fluid of patients with idiopathic pulmonary fibrosis. Zhonghua Jie He He Hu Xi Za Zhi. 2003;26(11):686–8.

546. Hancock A, Armstrong L, Gama R, Millar A. Production of interleukin 13 by alveolar macrophages from normal and fibrotic lung. Am J Respir Cell Mol Biol. 1998;18(1):60–5.

547. Park SW, Ahn MH, Jang HK, Jang AS, Kim DJ, Koh ES, et al. Interleukin-13 and its receptors in idiopathic interstitial pneumonia: clinical implications for lung function. J Korean Med Sci. 2009;24(4):614–20.

548. Khosravi AR, Alheidary S, Nikaein D, Asghari N. Aspergillus fumigatus conidia stimulate lung epithelial cells (TC-1 JHU-1) to produce IL-12, IFNgamma, IL-13 and IL-17 cytokines: modulatory effect of propolis extract. J Mycol Med. 2018;28(4):594–8.

549. Jia Y, Fang X, Zhu X, Bai C, Zhu L, Jin M, et al. IL-13(+) type 2 innate lymphoid cells correlate with asthma control status and treatment response. Am J Respir Cell Mol Biol. 2016;55(5):675–83.

550. Nie Y, Hu Y, Yu K, Zhang D, Shi Y, Li Y, et al. Akt1 regulates pulmonary fibrosis via modulating IL-13 expression in macrophages. Innate Immun. 2019;25(7):451–61.

551. Fichtner-Feigl S, Strober W, Kawakami K, Puri RK, Kitani A. IL-13 signaling through the IL-13alpha2 receptor is involved in induction of TGF-beta1 production and fibrosis. Nat Med. 2006;12(1):99–106.

552. Ingram JL, Rice AB, Geisenhoffer K, Madtes DK, Bonner JC. IL-13 and IL-1beta promote lung fibroblast growth through coordinated up-regulation of PDGF-AA and PDGF-Ralpha. FASEB J. 2004;18(10):1132–4.

553. Fanny M, Nascimento M, Baron L, Schricke C, Maillet I, Akbal M, et al. The IL-33 receptor ST2 regulates pulmonary inflammation and fibrosis to Bleomycin. Front Immunol. 2018;9:1476.

554. Zhang C, Cai R, Lazerson A, Delcroix G, Wangpaichitr M, Mirsaeidi M, et al. Growth hormone-releasing hormone receptor antagonist modulates lung inflammation and fibrosis due to Bleomycin. Lung. 2019;197(5):541–9.

555. Wilson MS, Madala SK, Ramalingam TR, Gochuico BR, Rosas IO, Cheever AW, et al. Bleomycin and IL-1beta-mediated pulmonary fibrosis is IL-17A dependent. J Exp Med. 2010;207(3):535–52.

556. Liu T, Jin H, Ullenbruch M, Hu B, Hashimoto N, Moore B, et al. Regulation of found in inflammatory zone 1 expression in bleomycin-induced lung fibrosis: role of IL-4/IL-13 and mediation via STAT-6. J Immunol. 2004;173(5):3425–31.

557. Mi S, Li Z, Yang HZ, Liu H, Wang JP, Ma YG, et al. Blocking IL-17A promotes the resolution of pulmonary inflammation and fibrosis via TGF-beta1-dependent and -independent mechanisms. J Immunol. 2011;187(6):3003–14.

558. Wang T, Liu Y, Zou JF, Cheng ZS. Interleukin-17 induces human alveolar epithelial to mesenchymal cell transition via the TGF-beta1 mediated Smad2/3 and ERK1/2 activation. PLoS One. 2017;12(9):e0183972.

559. Simonian PL, Roark CL, Wehrmann F, Lanham AK. Diaz del Valle F, born WK, et al. Th17-polarized immune response in a murine model of hypersensitivity pneumonitis and lung fibrosis. J Immunol. 2009;182(1):657–65.

560. Chen Y, Li C, Weng D, Song L, Tang W, Dai W, et al. Neutralization of interleukin-17A delays progression of silica-induced lung inflammation and fibrosis in C57BL/6 mice. Toxicol Appl Pharmacol. 2014;275(1):62–72.

561. Harari S, Madotto F, Caminati A, Conti S, Cesana G. Epidemiology of idiopathic pulmonary fibrosis in northern Italy. PLoS One. 2016;11(2):e0147072.

562. Richeldi L, Rubin AS, Avdeev S, Udwadia ZF, Xu ZJ. Idiopathic pulmonary fibrosis in BRIC countries: the cases of Brazil, Russia, India, and China. BMC Med. 2015;13:237.

563. American Thoracic Society. Idiopathic pulmonary fibrosis: diagnosis and treatment. International consensus statement. American Thoracic Society (ATS), and the European Respiratory Society (ERS). Am J Respir Crit Care Med. 2000;161(2 Pt 1):646–64.

564. Olson AL, Gifford AH, Inase N, Fernandez Perez ER, Suda T. The epidemiology of idiopathic pulmonary fibrosis and interstitial lung diseases at risk of a progressive-fibrosing phenotype. Eur Respir Rev. 2018;27(150)

565. Redente EF, Jacobsen KM, Solomon JJ, Lara AR, Faubel S, Keith RC, et al. Age and sex dimorphisms

contribute to the severity of bleomycin-induced lung injury and fibrosis. Am J Physiol Lung Cell Mol Physiol. 2011;301(4):L510–8.

566. Khalil N, Bereznay O, Sporn M, Greenberg AH. Macrophage production of transforming growth factor beta and fibroblast collagen synthesis in chronic pulmonary inflammation. J Exp Med. 1989;170(3):727–37.

567. Buford TW, Willoughby DS. Impact of DHEA(S) and cortisol on immune function in aging: a brief review. Appl Physiol Nutr Metab. 2008;33(3):429–33.

568. Mendoza-Milla C, Valero Jimenez A, Rangel C, Lozano A, Morales V, Becerril C, et al. Dehydroepiandrosterone has strong antifibrotic effects and is decreased in idiopathic pulmonary fibrosis. Eur Respir J. 2013;42(5):1309–21.

569. Lekgabe ED, Royce SG, Hewitson TD, Tang ML, Zhao C, Moore XL, et al. The effects of relaxin and estrogen deficiency on collagen deposition and hypertrophy of nonreproductive organs. Endocrinology. 2006;147(12):5575–83.

570. Tofovic SP, Zhang X, Jackson EK, Zhu H, Petrusevska G. 2-methoxyestradiol attenuates bleomycin-induced pulmonary hypertension and fibrosis in estrogen-deficient rats. Vasc Pharmacol. 2009;51(2–3):190–7.

571. Pan LC, Wilson DW, Segall HJ. Strain differences in the response of Fischer 344 and Sprague-Dawley rats to monocrotaline induced pulmonary vascular disease. Toxicology. 1993;79(1):21–35.

572. Hirano S, Furutama D, Hanafusa T. Physiologically high concentrations of 17beta-estradiol enhance NF-kappaB activity in human T cells. Am J Physiol Regul Integr Comp Physiol. 2007;292(4):R1465–71.

573. Speyer CL, Rancilio NJ, McClintock SD, Crawford JD, Gao H, Sarma JV, et al. Regulatory effects of estrogen on acute lung inflammation in mice. Am J Physiol Cell Physiol. 2005;288(4):C881–90.

574. Frazier-Jessen MR, Kovacs EJ. Estrogen modulation of JE/monocyte chemoattractant protein-1 mRNA expression in murine macrophages. J Immunol. 1995;154(4):1838–45.

575. Kurokawa A, Azuma K, Mita T, Toyofuku Y, Fujitani Y, Hirose T, et al. 2-Methoxyestradiol reduces monocyte adhesion to aortic endothelial cells in ovariectomized rats. Endocr J. 2007;54(6):1027–31.

576. Barron L, Wynn TA. Fibrosis is regulated by Th2 and Th17 responses and by dynamic interactions between fibroblasts and macrophages. Am J Physiol Gastrointest Liver Physiol. 2011;300(5):G723–8.

577. Bonner JC. Regulation of PDGF and its receptors in fibrotic diseases. Cytokine Growth Factor Rev. 2004;15(4):255–73.

578. Antoniades HN, Bravo MA, Avila RE, Galanopoulos T, Neville-Golden J, Maxwell M, et al. Platelet-derived growth factor in idiopathic pulmonary fibrosis. J Clin Invest. 1990;86(4):1055–64.

579. Yoshida M, Sakuma J, Hayashi S, Abe K, Saito I, Harada S, et al. A histologically distinctive interstitial pneumonia induced by overexpression of the interleukin 6, transforming growth factor beta 1, or platelet-derived growth factor B gene. Proc Natl Acad Sci U S A. 1995;92(21):9570–4.

580. Mukherjee S, Duan F, Kolb MR, Janssen LJ. Platelet derived growth factor-evoked Ca2+ wave and matrix gene expression through phospholipase C in human pulmonary fibroblast. Int J Biochem Cell Biol. 2013;45(7):1516–24.

581. Chen G, Qiao Y, Xiao X, Zheng S, Chen L. Effects of estrogen on lung development in a rat model of diaphragmatic hernia. J Pediatr Surg. 2010;45(12):2340–5.

582. Bentzen SM, Skoczylas JZ, Overgaard M, Overgaard J. Radiotherapy-related lung fibrosis enhanced by tamoxifen. J Natl Cancer Inst. 1996;88(13):918–22.

583. Smith LC, Moreno S, Robertson L, Robinson S, Gant K, Bryant AJ, et al. Transforming growth factor beta1 targets estrogen receptor signaling in bronchial epithelial cells. Respir Res. 2018;19(1):160.

584. Elliot S, Periera-Simon S, Xia X, Catanuto P, Rubio G, Shahzeidi S, et al. MicroRNA let-7 Downregulates ligand-independent estrogen receptor-mediated male-predominant pulmonary fibrosis. Am J Respir Crit Care Med. 2019;200(10):1246–57.

585. Spitz MR, Wei Q, Dong Q, Amos CI, Wu X. Genetic susceptibility to lung cancer: the role of DNA damage and repair. Cancer Epidemiol Biomark Prev. 2003;12(8):689–98.

586. Jonsson S, Thorsteinsdottir U, Gudbjartsson DF, Jonsson HH, Kristjansson K, Arnason S, et al. Familial risk of lung carcinoma in the Icelandic population. JAMA. 2004;292(24):2977–83.

587. Remen T, Pintos J, Abrahamowicz M, Siemiatycki J. Risk of lung cancer in relation to various metrics of smoking history: a case-control study in Montreal. BMC Cancer. 2018;18(1):1275.

588. Siegel RL, Miller KD, Jemal A. Cancer statistics, 2019. CA Cancer J Clin. 2019;69(1):7–34.

589. Jeon J, Holford TR, Levy DT, Feuer EJ, Cao P, Tam J, et al. Smoking and lung Cancer mortality in the United States from 2015 to 2065: a comparative modeling approach. Ann Intern Med. 2018;169(10):684–93.

590. Lortet-Tielent J, Renteria E, Sharp L, Weiderpass E, Comber H, Baas P, et al. Convergence of decreasing male and increasing female incidence rates in major tobacco-related cancers in Europe in 1988-2010. Eur J Cancer. 2015;51(9):1144–63.

591. Harichand-Herdt S, Ramalingam SS. Gender-associated differences in lung cancer: clinical characteristics and treatment outcomes in women. Semin Oncol. 2009;36(6):572–80.

592. Kabat GC. Aspects of the epidemiology of lung cancer in smokers and nonsmokers in the United States. Lung Cancer. 1996;15(1):1–20.

593. Rivera GA, Wakelee H. Lung Cancer in never smokers. Adv Exp Med Biol. 2016;893:43–57.

594. Donington JS, Colson YL. Sex and gender differences in non-small cell lung cancer. Semin Thorac Cardiovasc Surg. 2011;23(2):137–45.

595. Kligerman S, White C. Epidemiology of lung cancer in women: risk factors, survival, and screening. AJR Am J Roentgenol. 2011;196(2):287–95.

596. Radkiewicz C, Dickman PW, Johansson ALV, Wagenius G, Edgren G, Lambe M. Sex and survival in non-small cell lung cancer: a nationwide cohort study. PLoS One. 2019;14(6):e0219206.

597. Inamura K. Lung Cancer: understanding its molecular pathology and the 2015 WHO classification. Front Oncol. 2017;7:193.

598. Martey CA, Pollock SJ, Turner CK, O'Reilly KM, Baglole CJ, Phipps RP, et al. Cigarette smoke induces cyclooxygenase-2 and microsomal prostaglandin E2 synthase in human lung fibroblasts: implications for lung inflammation and cancer. Am J Physiol Lung Cell Mol Physiol. 2004;287(5):L981–91.

599. D'Anna C, Cigna D, Costanzo G, Ferraro M, Siena L, Vitulo P, et al. Cigarette smoke alters cell cycle and induces inflammation in lung fibroblasts. Life Sci. 2015;126:10–8.

600. Montuenga LM, Pio R. Tumour-associated macrophages in nonsmall cell lung cancer: the role of interleukin-10. Eur Respir J. 2007;30(4):608–10.

601. Ding M, He SJ, Yang J. MCP-1/CCL2 mediated by autocrine loop of PDGF-BB promotes invasion of lung Cancer cell by recruitment of macrophages via CCL2-CCR2 Axis. J Interf Cytokine Res. 2019;39(4):224–32.

602. Cai Z, Chen Q, Chen J, Lu Y, Xiao G, Wu Z, et al. Monocyte chemotactic protein 1 promotes lung cancer-induced bone resorptive lesions in vivo. Neoplasia. 2009;11(3):228–36.

603. Wang N, Liang H, Zen K. Molecular mechanisms that influence the macrophage m1-m2 polarization balance. Front Immunol. 2014;5:614.

604. Mantovani A, Sica A, Sozzani S, Allavena P, Vecchi A, Locati M. The chemokine system in diverse forms of macrophage activation and polarization. Trends Immunol. 2004;25(12):677–86.

605. Chittezhath M, Dhillon MK, Lim JY, Laoui D, Shalova IN, Teo YL, et al. Molecular profiling reveals a tumor-promoting phenotype of monocytes and macrophages in human cancer progression. Immunity. 2014;41(5):815–29.

606. Anderson CF, Mosser DM. A novel phenotype for an activated macrophage: the type 2 activated macrophage. J Leukoc Biol. 2002;72(1):101–6.

607. Hu P, Shen M, Zhang P, Zheng C, Pang Z, Zhu L, et al. Intratumoral neutrophil granulocytes contribute to epithelial-mesenchymal transition in lung adenocarcinoma cells. Tumour Biol. 2015;36(10):7789–96.

608. Pham CT. Neutrophil serine proteases: specific regulators of inflammation. Nat Rev Immunol. 2006;6(7):541–50.

609. Canli O, Nicolas AM, Gupta J, Finkelmeier F, Goncharova O, Pesic M, et al. Myeloid cell-derived reactive oxygen species induce epithelial mutagenesis. Cancer Cell. 2017;32(6):869–83. e5

610. Hiraoka K, Miyamoto M, Cho Y, Suzuoki M, Oshikiri T, Nakakubo Y, et al. Concurrent infiltration

611. Dai F, Liu L, Che G, Yu N, Pu Q, Zhang S, et al. The number and microlocalization of tumor-associated immune cells are associated with patient's survival time in non-small cell lung cancer. BMC Cancer. 2010;10:220.

612. Miotto D, Lo Cascio N, Stendardo M, Querzoli P, Pedriali M, De Rosa E, et al. CD8+ T cells expressing IL-10 are associated with a favourable prognosis in lung cancer. Lung Cancer. 2010;69(3):355–60.

613. Neurath MF, Becker C, Barbulescu K. Role of NF-kappaB in immune and inflammatory responses in the gut. Gut. 1998;43(6):856–60.

614. Ben-Neriah Y, Karin M. Inflammation meets cancer, with NF-kappaB as the matchmaker. Nat Immunol. 2011;12(8):715–23.

615. Chen W, Li Z, Bai L, Lin Y. NF-kappaB in lung cancer, a carcinogenesis mediator and a prevention and therapy target. Front Biosci (Landmark Ed). 2011;16:1172–85.

616. Luo X, Ding Q, Wang M, Li Z, Mao K, Sun B, et al. In vivo disruption of TGF-beta signaling by Smad7 in airway epithelium alleviates allergic asthma but aggravates lung carcinogenesis in mouse. PLoS One. 2010;5(4):e10149.

617. Bruno A, Focaccetti C, Pagani A, Imperatori AS, Spagnoletti M, Rotolo N, et al. The proangiogenic phenotype of natural killer cells in patients with non-small cell lung cancer. Neoplasia. 2013;15(2):133–42.

618. Siegel PM, Massague J. Cytostatic and apoptotic actions of TGF-beta in homeostasis and cancer. Nat Rev Cancer. 2003;3(11):807–21.

619. Saito A, Horie M, Micke P, Nagase T. The role of TGF-beta signaling in lung Cancer associated with idiopathic pulmonary fibrosis. Int J Mol Sci. 2018;19(11)

620. Wang YC, Sung WW, Wu TC, Wang L, Chien WP, Cheng YW, et al. Interleukin-10 haplotype may predict survival and relapse in resected non-small cell lung cancer. PLoS One. 2012;7(7):e39525.

621. Lan X, Lan T, Faxiang Q. Interleukin-10 promoter polymorphism and susceptibility to lung cancer: a systematic review and meta-analysis. Int J Clin Exp Med. 2015;8(9):15317–28.

622. Chow MT, Moller A, Smyth MJ. Inflammation and immune surveillance in cancer. Semin Cancer Biol. 2012;22(1):23–32.

623. Zeng L, O'Connor C, Zhang J, Kaplan AM, Cohen DA. IL-10 promotes resistance to apoptosis and metastatic potential in lung tumor cell lines. Cytokine. 2010;49(3):294–302.

624. Wang R, Lu M, Zhang J, Chen S, Luo X, Qin Y, et al. Increased IL-10 mRNA expression in tumor-associated macrophage correlated with late stage of lung cancer. J Exp Clin Cancer Res. 2011;30:62.

625. Beattie CW, Hansen NW, Thomas PA. Steroid receptors in human lung cancer. Cancer Res. 1985;45(9):4206–14.

626. Kaiser U, Hofmann J, Schilli M, Wegmann B, Klotz U, Wedel S, et al. Steroid-hormone receptors in cell lines and tumor biopsies of human lung cancer. Int J Cancer. 1996;67(3):357–64.

627. Maasberg M, Rotsch M, Jaques G, Enderle-Schmidt U, Weehle R, Havemann K. Androgen receptors, androgen-dependent proliferation, and 5 alpha-reductase activity of small-cell lung cancer cell lines. Int J Cancer. 1989;43(4):685–91.

628. Jeong Y, Xie Y, Xiao G, Behrens C, Girard L, Wistuba II, et al. Nuclear receptor expression defines a set of prognostic biomarkers for lung cancer. PLoS Med. 2010;7(12):e1000378.

629. Raso MG, Behrens C, Herynk MH, Liu S, Prudkin L, Ozburn NC, et al. Immunohistochemical expression of estrogen and progesterone receptors identifies a subset of NSCLCs and correlates with EGFR mutation. Clin Cancer Res. 2009;15(17):5359–68.

630. Jeong Y, Xie Y, Lee W, Bookout AL, Girard L, Raso G, et al. Research resource: diagnostic and therapeutic potential of nuclear receptor expression in lung cancer. Mol Endocrinol. 2012;26(8):1443–54.

631. Hyde Z, Flicker L, McCaul KA, Almeida OP, Hankey GJ, Chubb SA, et al. Associations between testosterone levels and incident prostate, lung, and colorectal cancer. A population-based study. Cancer Epidemiol Biomark Prev. 2012;21(8):1319–29.

632. Harlos C, Musto G, Lambert P, Ahmed R, Pitz MW. Androgen pathway manipulation and survival in patients with lung cancer. Horm Cancer. 2015;6(2–3):120–7.

633. Kissick HT, Sanda MG, Dunn LK, Pellegrini KL, On ST, Noel JK, et al. Androgens alter T-cell immunity by inhibiting T-helper 1 differentiation. Proc Natl Acad Sci U S A. 2014;111(27):9887–92.

634. Wu CT. Chen WC, Lin PY, Liao SK. Chen MF Androgen deprivation modulates the inflammatory response induced by irradiation BMC Cancer. 2009;9:92.

635. Padgett DA, Loria RM. Endocrine regulation of murine macrophage function: effects of dehydroepiandrosterone, androstenediol, and androstenetriol. J Neuroimmunol. 1998;84(1):61–8.

636. Corcoran MP, Meydani M, Lichtenstein AH, Schaefer EJ, Dillard A, Lamon-Fava S. Sex hormone modulation of proinflammatory cytokine and C-reactive protein expression in macrophages from older men and postmenopausal women. J Endocrinol. 2010;206(2):217–24.

637. Zeni E, Mazzetti L, Miotto D, Lo Cascio N, Maestrelli P, Querzoli P, et al. Macrophage expression of interleukin-10 is a prognostic factor in nonsmall cell lung cancer. Eur Respir J. 2007;30(4):627–32.

638. Hagenbaugh A, Sharma S, Dubinett SM, Wei SH, Aranda R, Cheroutre H, et al. Altered immune responses in interleukin 10 transgenic mice. J Exp Med. 1997;185(12):2101–10.

639. De Vita F, Orditura M, Galizia G, Romano C, Roscigno A, Lieto E, et al. Serum interleukin-10 levels as a prognostic factor in advanced non-small cell lung cancer patients. Chest. 2000;117(2):365–73.

640. Xu LL, Shanmugam N, Segawa T, Sesterhenn IA, McLeod DG, Moul JW, et al. A novel androgen-regulated gene, PMEPA1, located on chromosome 20q13 exhibits high level expression in prostate. Genomics. 2000;66(3):257–63.

641. Brunschwig EB, Wilson K, Mack D, Dawson D, Lawrence E, Willson JK, et al. PMEPA1, a transforming growth factor-beta-induced marker of terminal colonocyte differentiation whose expression is maintained in primary and metastatic colon cancer. Cancer Res. 2003;63(7):1568–75.

642. Itoh S, Thorikay M, Kowanetz M, Moustakas A, Itoh F, Heldin CH, et al. Elucidation of Smad requirement in transforming growth factor-beta type I receptor-induced responses. J Biol Chem. 2003;278(6):3751–61.

643. Hu Y, He K, Wang D, Yuan X, Liu Y, Ji H, et al. TMEPAI regulates EMT in lung cancer cells by modulating the ROS and IRS-1 signaling pathways. Carcinogenesis. 2013;34(8):1764–72.

644. Bai X, Jing L, Li Y, Li Y, Luo S, Wang S, et al. TMEPAI inhibits TGF-beta signaling by promoting lysosome degradation of TGF-beta receptor and contributes to lung cancer development. Cell Signal. 2014;26(9):2030–9.

645. Fu JB, Kau TY, Severson RK, Kalemkerian GP. Lung cancer in women: analysis of the national surveillance, epidemiology, and end results database. Chest. 2005;127(3):768–77.

646. North CM, Christiani DC. Women and lung cancer: what is new? Semin Thorac Cardiovasc Surg. 2013;25(2):87–94.

647. Niikawa H, Suzuki T, Miki Y, Suzuki S, Nagasaki S, Akahira J, et al. Intratumoral estrogens and estrogen receptors in human non-small cell lung carcinoma. Clin Cancer Res. 2008;14(14):4417–26.

648. Ikeda K, Shiraishi K, Yoshida A, Shinchi Y, Sanada M, Motooka Y, et al. Synchronous multiple lung adenocarcinomas: estrogen concentration in peripheral lung. PLoS One. 2016;11(8):e0160910.

649. Zhang G, Liu X, Farkas AM, Parwani AV, Lathrop KL, Lenzner D, et al. Estrogen receptor beta functions through nongenomic mechanisms in lung cancer cells. Mol Endocrinol. 2009;23(2):146–56.

650. Schwartz AG, Prysak GM, Murphy V, Lonardo F, Pass H, Schwartz J, et al. Nuclear estrogen receptor beta in lung cancer: expression and survival differences by sex. Clin Cancer Res. 2005;11(20):7280–7.

651. Hershberger PA, Stabile LP, Kanterewicz B, Rothstein ME, Gubish CT, Land S, et al. Estrogen receptor beta (ERbeta) subtype-specific ligands increase transcription, p44/p42 mitogen activated protein kinase (MAPK) activation and growth in human non-small cell lung cancer cells. J Steroid Biochem Mol Biol. 2009;116(1–2):102–9.

652. Stabile LP, Davis AL, Gubish CT, Hopkins TM, Luketich JD, Christie N, et al. Human non-small cell lung tumors and cells derived from normal lung express both estrogen receptor alpha and beta and show biological responses to estrogen. Cancer Res. 2002;62(7):2141–50.

653. Mah V, Seligson DB, Li A, Marquez DC, Wistuba II, Elshimali Y, et al. Aromatase expression predicts survival in women with early-stage non small cell lung cancer. Cancer Res. 2007;67(21):10484–90.

654. Stabile LP, Dacic S, Land SR, Lenzner DE, Dhir R, Acquafondata M, et al. Combined analysis of estrogen receptor beta-1 and progesterone receptor expression identifies lung cancer patients with poor outcome. Clin Cancer Res. 2011;17(1):154–64.

655. Stabile LP, Rothstein ME, Cunningham DE, Land SR, Dacic S, Keohavong P, et al. Prevention of tobacco carcinogen-induced lung cancer in female mice using antiestrogens. Carcinogenesis. 2012;33(11):2181–9.

656. He M, Yu W, Chang C, Miyamoto H, Liu X, Jiang K, et al. Estrogen receptor alpha promotes lung cancer cell invasion via increase of and cross-talk with infiltrated macrophages through the CCL2/CCR2/MMP9 and CXCL12/CXCR4 signaling pathways. Mol Oncol. 2020;

657. O'Connor T, Borsig L, Heikenwalder M. CCL2-CCR2 signaling in disease pathogenesis. Endocr Metab Immune Disord Drug Targets. 2015;15(2):105–18.

658. Fielding CA, McLoughlin RM, McLeod L, Colmont CS, Najdovska M, Grail D, et al. IL-6 regulates neutrophil trafficking during acute inflammation via STAT3. J Immunol. 2008;181(3):2189–95.

659. Yanagawa H, Sone S, Takahashi Y, Haku T, Yano S, Shinohara T, et al. Serum levels of interleukin 6 in patients with lung cancer. Br J Cancer. 1995;71(5):1095–8.

660. Pine SR, Mechanic LE, Enewold L, Chaturvedi AK, Katki HA, Zheng YL, et al. Increased levels of circulating interleukin 6, interleukin 8, C-reactive protein, and risk of lung cancer. J Natl Cancer Inst. 2011;103(14):1112–22.

661. Fuentes N, Nicoleau M, Cabello N, Montes D, Zomorodi N, Chroneos ZC, et al. 17beta-estradiol affects lung function and inflammation following ozone exposure in a sex-specific manner. Am J Physiol Lung Cell Mol Physiol. 2019;317(5):L702–L16.

662. Haura EB, Livingston S, Coppola D. Autocrine interleukin-6/interleukin-6 receptor stimulation in non-small-cell lung cancer. Clin Lung Cancer. 2006;7(4):273–5.

663. Song L, Turkson J, Karras JG, Jove R, Haura EB. Activation of Stat3 by receptor tyrosine kinases and cytokines regulates survival in human non-small cell lung carcinoma cells. Oncogene. 2003;22(27):4150–65.

664. Huang Q, Zhang Z, Liao Y, Liu C, Fan S, Wei X, et al. 17beta-estradiol upregulates IL6 expression through the ERbeta pathway to promote lung adenocarcinoma progression. J Exp Clin Cancer Res. 2018;37(1):133.

665. Wang J, Wang Y, Wong C. Oestrogen-related receptor alpha inverse agonist XCT-790 arrests A549 lung cancer cell population growth by inducing mitochondrial reactive oxygen species production. Cell Prolif. 2010;43(2):103–13.

666. Huang JW, Guan BZ, Yin LH, Liu FN, Hu B, Zheng QY, et al. Effects of estrogen-related receptor alpha (ERRalpha) on proliferation and metastasis of human lung cancer A549 cells. J Huazhong Univ Sci Technolog Med Sci. 2014;34(6):875–81.

667. Zhang J, Guan X, Liang N, Li S. Estrogen-related receptor alpha triggers the proliferation and migration of human non-small cell lung cancer via interleukin-6. Cell Biochem Funct. 2018;36(5):255–62.

668. Li Q, Guan X, Wu P, Wang X, Zhou L, Tong Y, et al. Early transmission dynamics in Wuhan, China, of novel coronavirus-infected pneumonia. N Engl J Med. 2020;382(13):1199–207.

669. Miller R, Englund K. Transmission and risk factors of OF COVID-19. Cleve Clin J Med. 2020;

670. Guan WJ, Ni ZY, Hu Y, Liang WH, Ou CQ, He JX, et al. Clinical characteristics of coronavirus disease 2019 in China. N Engl J Med. 2020;382(18):1708–20.

671. Chen N, Zhou M, Dong X, Qu J, Gong F, Han Y, et al. Epidemiological and clinical characteristics of 99 cases of 2019 novel coronavirus pneumonia in Wuhan, China: a descriptive study. Lancet. 2020;395(10223):507–13.

672. Liu Y, Yang Y, Zhang C, Huang F, Wang F, Yuan J, et al. Clinical and biochemical indexes from 2019-nCoV infected patients linked to viral loads and lung injury. Sci China Life Sci. 2020;63(3):364–74.

673. Velavan TP, Meyer CG. The COVID-19 epidemic. Tropical Med Int Health. 2020;25(3):278–80.

674. Huang C, Wang Y, Li X, Ren L, Zhao J, Hu Y, et al. Clinical features of patients infected with 2019 novel coronavirus in Wuhan. China Lancet. 2020;395(10223):497–506.

675. Wang D, Hu B, Hu C, Zhu F, Liu X, Zhang J, et al. Clinical characteristics of 138 hospitalized patients with 2019 novel coronavirus-infected pneumonia in Wuhan, China. JAMA. 2020;

676. Zhang B, Zhou X, Qiu Y, Song Y, Feng F, Feng J, et al. Clinical characteristics of 82 cases of death from COVID-19. PLoS One. 2020;15(7):e0235458.

677. Zhou P, Yang XL, Wang XG, Hu B, Zhang L, Zhang W, et al. A pneumonia outbreak associated with a new coronavirus of probable bat origin. Nature. 2020;579(7798):270–3.

678. Zhao Y, Zhao Z, Wang Y, Zhou Y, Ma Y, Zuo W. Single-cell RNA expression profiling of ACE2, The receptor of SARS-CoV-2. Am J Respir Crit Care Med. 2020;

679. Walls AC, Park YJ, Tortorici MA, Wall A, McGuire AT, Veesler D. Structure, function, and antigenicity of the SARS-CoV-2 spike glycoprotein. Cell. 2020;181(2):281–92. e6

680. Xu X, Chen P, Wang J, Feng J, Zhou H, Li X, et al. Evolution of the novel coronavirus from the ongoing Wuhan outbreak and modeling of its spike protein for risk of human transmission. Sci China Life Sci. 2020;63(3):457–60.

681. Imai Y, Kuba K, Rao S, Huan Y, Guo F, Guan B, et al. Angiotensin-converting enzyme 2 protects from severe acute lung failure. Nature. 2005;436(7047):112–6.

682. Imai Y, Kuba K, Penninger JM. The discovery of angiotensin-converting enzyme 2 and its role in acute lung injury in mice. Exp Physiol. 2008;93(5):543–8.

683. Kuba K, Imai Y, Rao S, Gao H, Guo F, Guan B, et al. A crucial role of angiotensin converting enzyme 2 (ACE2) in SARS coronavirus-induced lung injury. Nat Med. 2005;11(8):875–9.

684. Heurich A, Hofmann-Winkler H, Gierer S, Liepold T, Jahn O, Pohlmann S. TMPRSS2 and ADAM17 cleave ACE2 differentially and only proteolysis by TMPRSS2 augments entry driven by the severe acute respiratory syndrome coronavirus spike protein. J Virol. 2014;88(2):1293–307.

685. Lukassen S, Chua RL, Trefzer T, Kahn NC, Schneider MA, Muley T, et al. SARS-CoV-2 receptor ACE2 and TMPRSS2 are primarily expressed in bronchial transient secretory cells. EMBO J. 2020;39(10):e105114.

686. Meng Y, Wu P, Lu W, Liu K, Ma K, Huang L, et al. Sex-specific clinical characteristics and prognosis of coronavirus disease-19 infection in Wuhan, China: a retrospective study of 168 severe patients. PLoS Pathog. 2020;16(4):e1008520.

687. Gemmati D, Bramanti B, Serino ML, Secchiero P, Zauli G, Tisato V. COVID-19 and individual genetic susceptibility/receptivity: role of ACE1/ACE2 genes, immunity, inflammation and coagulation. Might the double X-chromosome in females be protective against SARS-CoV-2 compared to the single X-chromosome in males? Int J Mol Sci. 2020;21(10)

688. Stelzig KE, Canepa-Escaro F, Schiliro M, Berdnikovs S, Prakash YS, Chiarella SE. Estrogen regulates the expression of SARS-CoV-2 receptor ACE2 in differentiated airway epithelial cells. Am J Physiol Lung Cell Mol Physiol. 2020;318(6):L1280–L1.

689. Xu Z, Shi L, Wang Y, Zhang J, Huang L, Zhang C, et al. Pathological findings of COVID-19 associated with acute respiratory distress syndrome. Lancet Respir Med. 2020;8(4):420–2.

690. Zhang H, Zhou P, Wei Y, Yue H, Wang Y, Hu M, et al. Histopathologic changes and SARS-CoV-2 Immunostaining in the lung of a patient with COVID-19. Ann Intern Med. 2020;172(9):629–32.

691. Tay MZ, Poh CM, Renia L, MacAry PA, Ng LFP. The trinity of COVID-19: immunity, inflammation and intervention. Nat Rev Immunol. 2020;20(6):363–74.

692. Tian S, Hu W, Niu L, Liu H, Xu H, Xiao SY. Pulmonary pathology of early-phase 2019 novel coronavirus (COVID-19) pneumonia in two patients with lung Cancer. J Thorac Oncol. 2020;15(5):700–4.

693. Huang KJ, Su IJ, Theron M, Wu YC, Lai SK, Liu CC, et al. An interferon-gamma-related cytokine storm in SARS patients. J Med Virol. 2005;75(2):185–94.

694. Qin C, Zhou L, Hu Z, Zhang S, Yang S, Tao Y, et al. Dysregulation of immune response in patients with COVID-19 in Wuhan. Clin Infect Dis: China; 2020.

695. Zheng HY, Zhang M, Yang CX, Zhang N, Wang XC, Yang XP, et al. Elevated exhaustion levels and reduced functional diversity of T cells in peripheral blood may predict severe progression in COVID-19 patients. Cell Mol Immunol. 2020;17(5):541–3.

696. Chen G, Wu D, Guo W, Cao Y, Huang D, Wang H, et al. Clinical and immunological features of severe and moderate coronavirus disease 2019. J Clin Invest. 2020;130(5):2620–9.

697. Zhou F, Yu T, Du R, Fan G, Liu Y, Liu Z, et al. Clinical course and risk factors for mortality of adult inpatients with COVID-19 in Wuhan, China: a retrospective cohort study. Lancet. 2020;395(10229):1054–62.

698. Nie Y, Wang G, Shi X, Zhang H, Qiu Y, He Z, et al. Neutralizing antibodies in patients with severe acute respiratory syndrome-associated coronavirus infection. J Infect Dis. 2004;190(6):1119–26.

699. Thevarajan I, Nguyen THO, Koutsakos M, Druce J, Caly L, van de Sandt CE, et al. Breadth of concomitant immune responses prior to patient recovery: a case report of non-severe COVID-19. Nat Med. 2020;26(4):453–5.

700. Elizaldi SR, Lakshmanappa YS, Roh JW, Schmidt BA, Carroll TD, Weaver KD, et al. SARS-CoV-2 infection induces germinal center responses with robust stimulation of CD4 T follicular helper cells in rhesus macaques. bioRxiv. 2020.

701. Qin L, Li X, Shi J, Yu M, Wang K, Tao Y, et al. Gendered effects on inflammation reaction and outcome of COVID-19 patients in Wuhan. J Med Virol. 2020;

702. Pozzilli P, Lenzi A. Commentary: testosterone, a key hormone in the context of COVID-19 pandemic. Metabolism. 2020;108:154252.

703. Maggio M, Basaria S, Ceda GP, Ble A, Ling SM, Bandinelli S, et al. The relationship between testosterone and molecular markers of inflammation in older men. J Endocrinol Invest. 2005;28(11 Suppl Proceedings):116–9.

704. Mohamad NV, Wong SK, Wan Hasan WN, Jolly JJ, Nur-Farhana MF, Ima-Nirwana S, et al. The relationship between circulating testosterone and inflammatory cytokines in men. Aging Male. 2019;22(2):129–40.

705. Zhao M, Wang M, Zhang J, Gu J, Zhang P, Xu Y, et al. Comparison of clinical characteristics and outcomes of patients with coronavirus disease 2019 at different ages. Aging (Albany NY). 2020;12(11):10070–86.

706. Bhasin S, Brito JP, Cunningham GR, Hayes FJ, Hodis HN, Matsumoto AM, et al. Testosterone therapy in men with hypogonadism: An Endocrine Society clinical practice guideline. J Clin Endocrinol Metab. 2018;103(5):1715–44.

707. Van Vliet M, Spruit MA, Verleden G, Kasran A, Van Herck E, Pitta F, et al. Hypogonadism, quadriceps weakness, and exercise intolerance in chronic obstructive pulmonary disease. Am J Respir Crit Care Med. 2005;172(9):1105–11.

708. Richardson S, Hirsch JS, Narasimhan M, Crawford JM, McGinn T, Davidson KW, et al. Presenting characteristics, comorbidities, and outcomes among 5700 patients hospitalized with COVID-19 in the new York City area. JAMA. 2020;

709. Alqahtani JS, Oyelade T, Aldhahir AM, Alghamdi SM, Almehmadi M, Alqahtani AS, et al. Prevalence, severity and mortality associated with COPD and smoking in patients with COVID-19: a rapid systematic review and meta-analysis. PLoS One. 2020;15(5):e0233147.

710. Kapoor D, Aldred H, Clark S, Channer KS, Jones TH. Clinical and biochemical assessment of hypogonadism in men with type 2 diabetes: correlations with bioavailable testosterone and visceral adiposity. Diabetes Care. 2007;30(4):911–7.

711. Svartberg J, von Muhlen D, Sundsfjord J, Jorde R. Waist circumference and testosterone levels in community dwelling men. The Tromso study Eur J Epidemiol. 2004;19(7):657–63.

712. Balasubramanian V, Naing S. Hypogonadism in chronic obstructive pulmonary disease: incidence and effects. Curr Opin Pulm Med. 2012;18(2):112–7.

713. Caminiti G, Volterrani M, Iellamo F, Marazzi G, Massaro R, Miceli M, et al. Effect of long-acting testosterone treatment on functional exercise capacity, skeletal muscle performance, insulin resistance, and baroreflex sensitivity in elderly patients with chronic heart failure a double-blind, placebo-controlled, randomized study. J Am Coll Cardiol. 2009;54(10):919–27.

714. Douglas GC, O'Bryan MK, Hedger MP, Lee DK, Yarski MA, Smith AI, et al. The novel angiotensin-converting enzyme (ACE) homolog, ACE2, is selectively expressed by adult Leydig cells of the testis. Endocrinology. 2004;145(10):4703–11.

715. Rastrelli G, Di Stasi V, Inglese F, Beccaria M, Garuti M, Di Costanzo D, et al. Low testosterone levels predict clinical adverse outcomes in SARS-CoV-2 pneumonia patients. Andrology. 2020;

716. Bobjer J, Katrinaki M, Tsatsanis C, Lundberg Giwercman Y, Giwercman A. Negative association between testosterone concentration and inflammatory markers in young men: a nested cross-sectional study. PLoS One. 2013;8(4):e61466.

717. Olsen NJ, Kovacs WJ. Evidence that androgens modulate human thymic T cell output. J Investig Med. 2011;59(1):32–5.

718. Roden AC, Moser MT, Tri SD, Mercader M, Kuntz SM, Dong H, et al. Augmentation of T cell levels and responses induced by androgen deprivation. J Immunol. 2004;173(10):6098–108.

719. Shen C, Wang Z, Zhao F, Yang Y, Li J, Yuan J, et al. Treatment of 5 critically ill patients with COVID-19 with convalescent plasma. JAMA. 2020;

720. Valk SJ, Piechotta V, Chai KL, Doree C, Monsef I, Wood EM, et al. Convalescent plasma or hyperimmune immunoglobulin for people with COVID-19: a rapid review. Cochrane Database Syst Rev. 2020;5:CD013600.

721. Sheridan C. Convalescent serum lines up as first-choice treatment for coronavirus. Nat Biotechnol. 2020;38(6):655–8.

722. Zeng F, Dai C, Cai P, Wang J, Xu L, Li J, et al. A comparison study of SARS-CoV-2 IgG antibody between male and female COVID-19 patients: a possible reason underlying different outcome between sex. J Med Virol. 2020;

723. Glowacka I, Bertram S, Muller MA, Allen P, Soilleux E, Pfefferle S, et al. Evidence that TMPRSS2 activates the severe acute respiratory syndrome coronavirus spike protein for membrane fusion and reduces viral control by the humoral immune response. J Virol. 2011;85(9):4122–34.

724. Hoffmann M, Kleine-Weber H, Schroeder S, Kruger N, Herrler T, Erichsen S, et al. SARS-CoV-2 cell entry depends on ACE2 and TMPRSS2 and is blocked by a clinically proven protease inhibitor. Cell. 2020;181(2):271–80. e8

725. Lucas JM, Heinlein C, Kim T, Hernandez SA, Malik MS, True LD, et al. The androgen-regulated protease TMPRSS2 activates a proteolytic cascade involving components of the tumor microenvironment and promotes prostate cancer metastasis. Cancer Discov. 2014;4(11):1310–25.

726. Stopsack KH, Mucci LA, Antonarakis ES, Nelson PS, Kantoff PW. TMPRSS2 and COVID-19: serendipity or opportunity for intervention? Cancer Discov. 2020;10(6):779–82.

727. Asselta R, Paraboschi EM, Mantovani A, Duga S. ACE2 and TMPRSS2 variants and expression as candidates to sex and country differences in COVID-19 severity in Italy. Aging (Albany NY). 2020;12(11):10087–98.

728. Ross MT, Grafham DV, Coffey AJ, Scherer S, McLay K, Muzny D, et al. The DNA sequence of the human X chromosome. Nature. 2005;434(7031):325–37.

729. Channappanavar R, Fett C, Mack M, Ten Eyck PP, Meyerholz DK, Perlman S. Sex-based differences in susceptibility to severe acute respiratory syndrome coronavirus infection. J Immunol. 2017;198(10):4046–53.

730. Fagone P, Ciurleo R, Lombardo SD, Iacobello C, Palermo CI, Shoenfeld Y, et al. Transcriptional landscape of SARS-CoV-2 infection dismantles pathogenic pathways activated by the virus, proposes unique sex-specific differences and predicts tailored therapeutic strategies. Autoimmun Rev. 2020;19(7):102571.

731. Laffont S, Rouquie N, Azar P, Seillet C, Plumas J, Aspord C, et al. X-chromosome complement and estrogen receptor signaling independently contribute to the enhanced TLR7-mediated IFN-alpha production of plasmacytoid dendritic cells from women. J Immunol. 2014;193(11):5444–52.

732. Phiel KL, Henderson RA, Adelman SJ, Elloso MM. Differential estrogen receptor gene expression in human peripheral blood mononuclear cell populations. Immunol Lett. 2005;97(1):107–13.
733. Grimaldi CM, Cleary J, Dagtas AS, Moussai D, Diamond B. Estrogen alters thresholds for B cell apoptosis and activation. J Clin Invest. 2002;109(12):1625–33.
734. Brannstrom M, Friden BE, Jasper M, Norman RJ. Variations in peripheral blood levels of immunoreactive tumor necrosis factor alpha (TNFalpha) throughout the menstrual cycle and secretion of TNFalpha from the human corpus luteum. Eur J Obstet Gynecol Reprod Biol. 1999;83(2):213–7.
735. Rachon D, Mysliwska J, Suchecka-Rachon K, Wieckiewicz J, Mysliwski A. Effects of oestrogen deprivation on interleukin-6 production by peripheral blood mononuclear cells of postmenopausal women. J Endocrinol. 2002;172(2):387–95.
736. Berg G, Ekerfelt C, Hammar M, Lindgren R, Matthiesen L, Ernerudh J. Cytokine changes in postmenopausal women treated with estrogens: a placebo-controlled study. Am J Reprod Immunol. 2002;48(2):63–9.
737. Murphy AJ, Guyre PM, Pioli PA. Estradiol suppresses NF-kappa B activation through coordinated regulation of let-7a and miR-125b in primary human macrophages. J Immunol. 2010;184(9):5029–37.
738. Zhang X, Wang L, Zhang H, Guo D, Qiao Z, Qiao J. Estrogen inhibits lipopolysaccharide-induced tumor necrosis factor-alpha release from murine macrophages. Methods Find Exp Clin Pharmacol. 2001;23(4):169–73.
739. Robinson DP, Lorenzo ME, Jian W, Klein SL. Elevated 17beta-estradiol protects females from influenza a virus pathogenesis by suppressing inflammatory responses. PLoS Pathog. 2011;7(7):e1002149.
740. Robinson DP, Hall OJ, Nilles TL, Bream JH, Klein SL. 17beta-estradiol protects females against influenza by recruiting neutrophils and increasing virus-specific CD8 T cell responses in the lungs. J Virol. 2014;88(9):4711–20.
741. Channappanavar R, Fehr AR, Vijay R, Mack M, Zhao J, Meyerholz DK, et al. Dysregulated type I interferon and inflammatory monocyte-macrophage responses cause lethal pneumonia in SARS-CoV-infected mice. Cell Host Microbe. 2016;19(2):181–93.
742. Stein B, Yang MX. Repression of the interleukin-6 promoter by estrogen receptor is mediated by NF-kappa B and C/EBP beta. Mol Cell Biol. 1995;15(9):4971–9.
743. Giron-Gonzalez JA, Moral FJ, Elvira J, Garcia-Gil D, Guerrero F, Gavilan I, et al. Consistent production of a higher TH1:TH2 cytokine ratio by stimulated T cells in men compared with women. Eur J Endocrinol. 2000;143(1):31–6.
744. Faas M, Bouman A, Moesa H, Heineman MJ, de Leij L, Schuiling G. The immune response during the luteal phase of the ovarian cycle: a Th2-type response? Fertil Steril. 2000;74(5):1008–13.
745. Lu FX, Abel K, Ma Z, Rourke T, Lu D, Torten J, et al. The strength of B cell immunity in female rhesus macaques is controlled by CD8+ T cells under the influence of ovarian steroid hormones. Clin Exp Immunol. 2002;128(1):10–20.
746. Kanda N, Tamaki K. Estrogen enhances immunoglobulin production by human PBMCs. J Allergy Clin Immunol. 1999;103(2 Pt 1):282–8.
747. Grandi G, Facchinetti F, Bitzer J. The gendered impact of coronavirus disease (COVID-19): do estrogens play a role? Eur J Contracept Reprod Health Care. 2020;25(3):233–4.

Synopsis of Clinical Acute Respiratory Distress Syndrome (ARDS)

16

Archana Mane and Naldine Isaac

Abstract

The entity of acute respiratory distress syndrome (ARDS) is an acute inflammatory lung condition associated with lung damage and increased vascular permeability. In the ICU, ARDS was reported to be the cause of 10.4% of admissions. The syndrome is associated with conditions such as sepsis, burns, trauma, and many others. The Berlin Definition which is the most up-to-date definition defines ARDS as an early onset of severe and refractory hypoxemia, PaO2/FiO2 ratio less than 300 mmHg, bilateral infiltrates on chest x-ray, and alveolar edema not explained by a cardiogenic cause or fluid overload.

The entity of ARDS and its treatment have been studied for many years to better understand it and help find therapies. However, the mainstay of medical management is supportive with specific strategies for mechanical ventilation. No specific drug therapy is available at present.

In this chapter, the history, clinical picture, and therapeutic approaches to ARDS will be discussed. We include lung-protective ventilation, prone positioning, use of neuromuscular blockade, corticosteroids, as well as discussion of studies done on this important clinical and morbid condition. We emphasize that there are ongoing trials and research being done to better identify patients earlier in their clinical course so that supportive care with lung-protective ventilation and a conservative fluid approach can be implemented. We also mention promising therapies such as cell-based therapies which would help in decreasing lung inflammation.

Keywords

Acute respiratory distress syndrome (ARDS) · Definition · Diagnostic criteria · Inciting events · Ventilator strategies · Future therapies

16.1 Introduction

Acute respiratory distress syndrome (ARDS) is a severe respiratory response to damage of the lungs leading to acute respiratory failure and often multi-organ failure. Medical management of this common and clinically morbid condition is constantly evolving. With multiple therapies under evaluation and ongoing research, this field of lung injury remains challenging and imperative, as it carries a high rate of mortality [1].

A. Mane (✉) · N. Isaac
Department of Anesthesiology, Albany Medical Center, Albany, NY, USA
e-mail: ManeA@amc.edu

© The Author(s), under exclusive license to Springer Nature Switzerland AG 2021
Y. -X. Wang (ed.), *Lung Inflammation in Health and Disease, Volume II*, Advances in Experimental Medicine and Biology 1304, https://doi.org/10.1007/978-3-030-68748-9_16

Part of the challenge in identifying effective therapeutic modalities can be attributed to the complex pathogenesis of this syndrome along with the insensitive and nonspecific diagnostic criteria used to diagnose ARDS. (These may also contribute to the under-detection of ARDS by clinicians.) The definition of ARDS has been reported to have a relatively low specificity of 51% [2].

Up until the 1990s, the mortality rate for ARDS was reported to be as high as 40–70%. A better understanding of the disease etiology, recognition of its management by specific mechanical ventilation protocols, and early treatment has contributed to the decline in mortality [3]. More recently, mortality has been correlated with increases in disease severity. According to Bellani et al., unadjusted ICU and hospital mortality rates were reported to be 35% among those with mild ARDS, 40% for moderate disease, and 46% for severe ARDS. The cause of early mortality is most commonly due to the underlying source of the disease. In contrast, nosocomial pneumonia and sepsis are the most common causes of death among patients who die later in their clinical course. It is less common for patients to die from the respiratory failure alone [4].

16.2   The Evolving Definition of ARDS

The historical identification and description of the syndrome of ARDS are interesting, dating back to the work of Laennec in 1821. He first documented pulmonary infiltrates in the lungs of trauma patients which he called "idiopathic pulmonary edema" or "shock lung" [5]. In 1964, Ashbaugh and his colleagues made some noteworthy observations in 12 civilian patients who were victims of trauma and other conditions such as hemorrhagic pancreatitis. Over a period of 24 hours, these patients developed respiratory distress with severe hypoxemia, fluffy infiltrates on x-ray, and the need for increased ventilatory pressures [6].

They published their findings in *The Lancet* in 1967 which was read by military surgeons who were able to identify these findings in Southeast Asia during the Vietnam War when previously healthy young servicemen developed respiratory failure after trauma that did not respond to oxygen therapy [7].

More recently in 1994, the American Thoracic Society and the European Society of Intensive Critical Medicine established a task force to address discrepancies in identifying the mechanism, incidence, outcomes, and preventative strategies for ARDS globally. The definition for ARDS included "arterial oxygen tension/fractional inspired oxygen (PaO2/FIO2 ≤200 mmHg) regardless of the level of positive end-expiratory pressure (PEEP), bilateral pulmonary infiltrates, and no evidence of left heart failure as measured by pulmonary wedge pressure" [8, 9].

In 2011, the definition was updated to its current definition due to ongoing concerns over the reliability of the definition. The reliability of the chest radiographic criteria of ARDS by this definition has been demonstrated to be moderate, with substantial interobserver variability. In addition, the hypoxemia criterion (i.e., PaO2/FiO2 <200 mmHg) was in question as it could be markedly affected by the patient's ventilator settings, especially the PEEP level used. Finally, the wedge pressure was considered difficult to interpret, and if a patient with ARDS develops a high wedge pressure, that should not preclude diagnosing that patient as having ARDS [10, 11].

16.3   Today's Definition of ARDS

The Berlin Definition declared a new classification of ARDS in 2011. All four of the following criteria must be met to diagnose the condition:

1. Acute onset: respiratory failure or significant worsening of respiratory status <1 week of a predisposing factor.
2. Imaging: bilateral chest opacities on chest x-ray or CT.

3. Respiratory failure must not be explained by heart failure or fluid overload.
4. Hypoxemia:
    (a) Mild: PaO2/FiO2 ratio ≤300 and >200
    (b) Moderate: PaO2/FiO2 ratio 100–200
    (c) Severe: PaO2/FiO2 ratio <100 [12]

Of note, a lesser form of respiratory failure is known as acute lung injury (ALI). While ARDS defines hypoxemia as a PaO2/FiO2 < 200 mmHg, ALI is defined as a PaO2/FiO2 < 300 mmHg [13].

16.4   Etiology

There are a multitude of factors that can lead to the development of ARDS. These possibilities can be divided into two categories: pulmonary

Table 16.1 Pulmonary ARDS vs extrapulmonary ARDS

	Pulmonary ARDS	Extrapulmonary ARDS
Etiology	Direct insult to the lung (trauma, contusion, pneumonia, aspiration)	Sepsis, smoke inhalation, intra-abdominal infections, pancreatitis
Site of damage	Alveolar epithelium	Vascular endothelium
Ventilator duration	~28 days	~28 days
Chest x-ray	Alveolar consolidation, intra-alveolar damage	Ground-glass appearance from interstitial edema
Treatment with PEEP, lung recruitment, prone positioning	Not as effective	Effective
CT	Extensive consolidation with equal amounts of normal lung and ground-glass opacification (Fig. 16.1)	Predominantly ground-glass opacification (Fig. 16.1)

ARDS or extrapulmonary ARDS [14] (Table 16.1, Fig. 16.1).

16.5   Pathophysiology of ARDS

In pulmonary ARDS, injury originates at the thin alveolar membrane which is largely composed of Type 1 pneumocytes responsible for gas exchange. In extrapulmonary ARDS, the site of injury is at the vascular endothelium [1]. Both result in damage that occurs in three stages.

1. Exudative Stage: Cellular damage results in the release of pro-inflammatory cytokines such as tumor necrosis factor (TNF), IL-1, and IL-6. Neutrophils are activated and release additional toxic mediators. This leads to the movement of fluid from the vasculature that contains plasma proteins, red blood cells, and leukocytes. The fluid enters the alveoli resulting in pulmonary edema, preventing the adequate exchange of oxygen across the membrane. Additionally, damage to the surfactant-producing alveolar Type 2 cells causes alveolar collapse.
2. Proliferative Stage: In response to injury, alveolar Type 2 cells proliferate to regenerate the endothelium, and remodeling occurs. This leads to decreased lung compliance.
3. Fibrotic Stage: Some patients progress to an irreversible fibrotic stage. The inflammatory exudates result in the scarring of the lung tissue leading to fibrosis, cysts, and permanent changes to the architecture of the lung [15].

16.6   Risk Factors for ARDS

16.6.1 Age

With increasing age, the incidence of ARDS rises. After the age of 74, the incidence has been reported to be as high as 73.9 cases per 100,000 individuals [5].

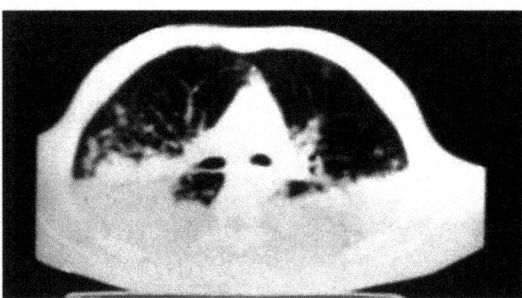 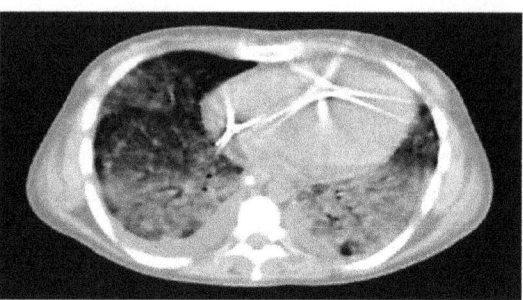

Fig. 16.1 The computed tomography of the lungs showing ground-glass opacities (especially in the posterior gravity-dependent portions) with some normal lung tissue (*left*) and more diffusely dispersed bilateral ground-glass opacities (*right*). (The images were taken from a previous publication by Pelosi et al. [14])

16.6.2 Burns

Incidence of ARDS in burn patients depends on the severity and ranges between 22% and 56%, making it one of the leading causes of death among burn patients [16]. Inhalation burns and full-thickness burns covering over 20% of the total body surface area were determined as risk factors for the development of ARDS [17].

16.6.3 The Obesity Paradox

Obesity carries an increased likelihood of developing ARDS due to increased inflammatory cytokines and impaired pulmonary vascular homeostasis. Obese patients have altered pulmonary mechanics and increased incidence of atelectasis and pulmonary mismatch. However, the mortality risk is decreased. The reason for this paradox is unclear, but may be related to the increase in metabolic reserve found in adipose tissue [18].

16.6.4 Alcohol

A meta-analysis by Simou et al. reported that there was an increased incidence of ARDS arising from sepsis in those who abused alcohol. Although the mechanism for this is not well understood, the study postulates that the effects of alcohol on the function of macro-phages and the depletion of glutathione may play a role [19].

16.6.5 Diabetes

Diabetes is associated with a decreased incidence of ARDS as shown in many but not all studies. This reduced incidence is seen in both Type I and Type II diabetes. Diabetes may reduce development of ARDS through a compromised immune system and attenuation of cytokine release and impairment of neutrophil function [20].

16.7 Clinical Presentation

Generally, the clinical presentation of ARDS develops within 24 hours. Respiratory symptoms such an acute onset of tachypnea, cough and decreased oxygen saturation by pulse oximetry develop. Lung auscultation may be normal, but as the syndrome progresses, rales are often heard. The patient may complain of chest pain and experience tachycardia. As illustrated in Fig. 16.2, the chest radiography shows patchy infiltrates that progress to diffuse opacities [15].

Analysis of arterial blood gases plays a critical role in the evaluation and management of ARDS. More specifically, it allows a provider to assess oxygenation as a marker of the severity of ARDS. Additionally, the alveolar-arterial gradient can detect the degree of hypoxemia [21].

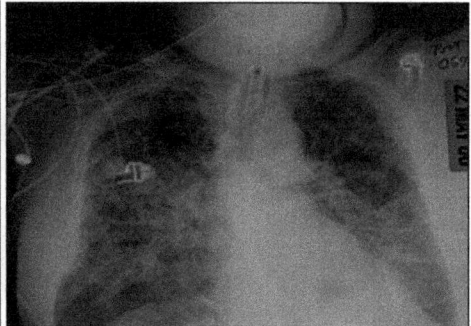

Fig. 16.2 Chest x-ray taken during early ARDS, showing that the opacities are patchy and less dense than those observed in later stages (*left*), and during late ARDS, illustrating the consolidation and bilateral infiltrates that are dispersed throughout the lung tissue (*right*). (The images were adapted from a previous study [15])

16.8 Management of ARDS

The variety of clinical presentations makes it challenging to treat. Additionally, patients can deteriorate rapidly over the span of hours leading to refractory hypoxemia. The mainstay of treatment is mechanical ventilation and supportive care.

Other possible methods include prone positioning, neuromuscular blockade, sedation, fluid management, corticosteroid therapy, and management of stress ulcers.

GRADE (Grading of Recommendations Assessment, Development and Evaluation) Method: The quality of evidence from various studies is analyzed and given a score that determines the reliability of the recommendation for therapy. GRADE 1 demonstrates a high level of validity, while a GRADE 4 signifies lower validity of evidence [22], as shown in Fig. 16.3. GRADE 1 recommendations include maintaining low tidal volume, limiting plateau pressure, placing patients in the prone position for at least 12 hours a day, and avoiding the use of oscillatory ventilation. GRADE 2 recommendations supported the utilization of PEEP in moderate and severe ARDS for the recruitment of the lungs, as well as the use of muscle relaxants and extracorporeal membrane oxygenation (ECMO) [22–24].

16.8.1 Mechanical Ventilation

The utilization of low tidal volumes via endotracheal intubation (~6 ml/kg) has been shown to reduce mortality when compared to higher tidal volume (~12 ml/kg) therapy from 39.8% to 31%. Lower tidal volumes reduce the overdistension of the alveoli that would otherwise lead to volutrauma and barotrauma. PEEP is employed to help prevent lung damage by inhibiting the cyclic opening and closing of alveoli and reducing shear stress [25].

16.8.2 Non-invasive Methods of Ventilation and Oxygenation

Ventilatory modes such as pressure support, bilevel positive airway pressure (BiPAP), and continuous positive airway pressure (CPAP) may be used with devices such as a venturi mask or a high-flow nasal cannula. This method is reserved for patients presenting with mild ARDS who are able to maintain their own airway. It is not sufficient for patients with a PaO2/FiO2 lower than 150 mmHg, as it is associated with higher ICU mortality in this population. However, it is important to note that there is not yet enough research on these methods of ventilation to support their use [26].

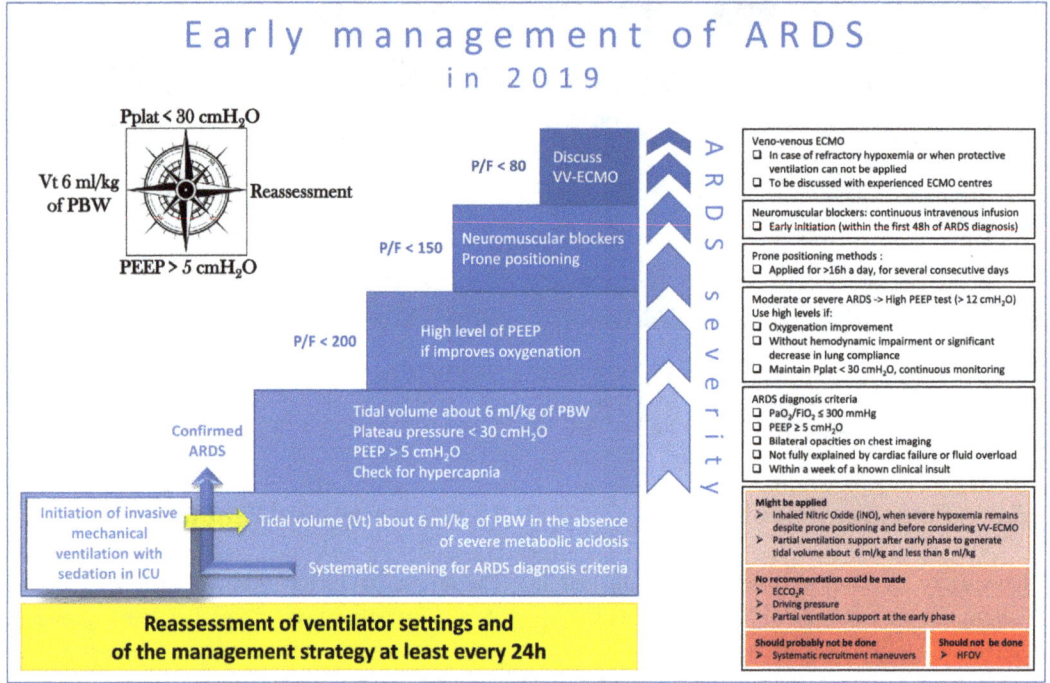

Fig. 16.3 Formal guidelines: management of acute respiratory distress syndrome. (Taken from a previous publication [22])

When all else fails, ECMO may be a possibility, although several studies have not provided enough evidence to support its use. In a study by Combes et al., 60-day mortality was not significantly lower with ECMO when compared to conventional mechanical ventilation that included ECMO as rescue therapy. It still remains an option for severe refractory ARDS [27].

16.8.3 Weaning from Ventilator Therapy

Weaning a patient off of ventilator support depends on the individual's condition. It often takes 24–48 hours, but can be longer.

It involves gradually decreasing FiO2 and PEEP and then shifting to either a spontaneous breathing trial or reduction in the amount of ventilatory support.

16.8.4 Prone Positioning

Prone positioning is a strategy used in order to recruit the dorsal regions of the lungs. This provides better distribution of air throughout the lungs and improves perfusion to the lung tissue. The PROSEVA (Proning Severe ARDS Patients), a large randomized clinical trial, demonstrated a major decrease in mortality rate at 28 and 90 days in patients who were treated with prone positioning within 48 hours of ARDS diagnosis. This treatment involved remaining prone 12 hours a day. At 28 days, prone patients had a mortality rate of 16% compared to 32.8% in supine patients. At 90 days, the mortality rate was 23.6% for prone patients versus 41.0% in the supine group. In addition, prone patients had better outcomes with successful extubation and were weaned from ventilatory support earlier than their supine counterparts [28].

16.8.5 Neuromuscular Blockade Agents (NMDAs) and Sedation in ARDS Management

Invasively ventilated patients are often given neuromuscular blockade requiring sedation to prevent recall. Commonly used sedating agents include dexmedetomidine, propofol, and midazolam as continuous infusions. This combined therapy prevents patient-ventilator asynchrony which is key to reducing the incidence of ventilator-induced lung injury. In severe ARDS, NMBA therapy had a 90-day mortality benefit of up to 9.1% according to ARDS et Curarization Systematique (ACURASYS) trial, an international study. NMBAs assist with lung recruitment which increases oxygenation. Furthermore, NMBAs allow patients to tolerate the endotracheal tube, decrease systemic inflammation, and improve V/Q mismatch [29, 30].

16.8.6 Fluid Management in ARDS

Studies have supported the benefits of negative fluid balance in improving patient outcomes. Limiting the administration of fluids is recommended due to the increased permeability of the pulmonary membranes. Colloid therapy with albumin has been shown to improve oxygenation, although more research is needed. Oftentimes, diuretics are also administered to help with fluid balance [31, 32].

16.8.7 Corticosteroids

Dexamethasone is another potential therapy worth exploring as a means of decreasing the duration of mechanical ventilation and decreasing mortality [33].

16.9 Potential Complications of ARDS

A potential complication of ARDS treatment is ventilator-related injury. Several mechanisms can result in this type of injury:

1. Volutrauma is caused by high tidal volumes and is the most harmful of the VILI.
2. Atelectrauma is caused by repetitive opening and closing of the airway.
3. The interaction of collapsed or fluid-filled alveoli with surrounding alveoli.
4. The endotracheal tube can cause laryngeal edema.
5. A patient may become dependent on the ventilator requiring a tracheostomy.

Other sources of complication include nosocomial infections and antibiotic resistance. Immobility can cause conditions such as DVT and muscle weakness [34, 35].

16.10 Future Directions

Further exploration is being done on pharmacologic therapies such as vitamin C, vitamin D, thiamine, corticosteroids, and many others [36].

Finelli et al. claimed that those being treated with a low tidal volume ventilation of 6 ml/kg may still suffer from VILI. He proposed that an even lower tidal volume ventilation of 4 ml/kg may provide more benefit. A barrier preventing this decrease in ventilation is the potential accumulation of CO_2. His group is therefore now studying extracorporeal carbon dioxide removal [37].

Another potential therapy targets the FLT1 gene which encodes vascular endothelial growth factor receptor 1 (VEGFR-1). The goal is to use VEGFR-1 to repair the vascular endothelium [38].

References

1. Pierrakos C, Karanikolas M, Scolletta S, et al. Acute respiratory distress syndrome: pathophysiology and therapeutic options. J Clin Med Res. 2012;4(1):7–16. https://doi.org/10.4021/jocmr761w.

2. Fanelli V, Vlachou A, Ghannadian S, et al. Acute respiratory distress syndrome: new definition, current and future therapeutic options. J Thorac Dis. 2013;5:326–34. https://doi.org/10.3978/j.issn.2072-1439.2013.04.05.

3. Harman EM. Acute respiratory distress syndrome (ARDS): background, pathophysiology, etiology. https://emedicine.medscape.com/article/165139-overview. Accessed 4 Sept 2020.

4. Bellani G, Laffey JG, Pham T, et al. Epidemiology, patterns of care, and mortality for patients with acute respiratory distress syndrome in intensive care units in 50 countries. JAMA. 2016;315(8):788–800. https://doi.org/10.1001/jama.2016.0291.

5. Rezoagli E, Fumagalli R, Bellani G. Definition and epidemiology of acute respiratory distress syndrome. Ann Transl Med. 2017;5:282. https://doi.org/10.21037/atm.2017.06.62.

6. Ashbaugh D. David Ashbaugh reminisces. Lancet Respir Med. 2017;5:474. https://doi.org/10.1016/S2213-2600(17)30182-0.

7. Petty TL. In the cards was ARDS: (how we discovered the acute respiratory distress syndrome). Am J Respir Crit Care Med. 2001;163:602–3. https://doi.org/10.1164/ajrccm.163.3.16331.

8. Bernard GR, Artigas A, Brigham KL, et al. Report of the American-European consensus conference on ARDS: definitions, mechanisms, relevant outcomes and clinical trial coordination. Intensive Care Med. 1994;20(3):225–32. https://doi.org/10.1007/BF01704707.

9. Villar J, Pérez-Méndez L, Kacmarek RM. Current definitions of acute lung injury and the acute respiratory distress syndrome do not reflect their true severity and outcome. Intensive Care Med. 1999;25(9):930–5. https://doi.org/10.1007/s001340050984.

10. Zompatori M, Ciccarese F, Fasano L. Overview of current lung imaging in acute respiratory distress syndrome. Eur Respir Rev. 2014;23(134):519–30. https://doi.org/10.1183/09059180.00001314.

11. Amin Z, Amanda AP. Comparison of new ARDS criteria (Berlin) with old criteria (AECC) and its application in country with limited facilities. J Gen Emerg Med. 2017;2(1):2–4.

12. Ranieri VM, Rubenfeld GD, Thompson BT, et al. Acute respiratory distress syndrome: the Berlin definition. JAMA. 2012;307(23):2526–33. https://doi.org/10.1001/jama.2012.5669.

13. Laycock H, Rajah A. Acute lung injury and acute respiratory distress syndrome: a review article. Br J Med Pract. 2010;3:324.

14. Pelosi P, D'Onofrio D, Chiumello D, et al. Pulmonary and extrapulmonary acute respiratory distress syndrome are different. Eur Respir J Suppl. 2003;42:48s–56s. https://doi.org/10.1183/09031936.03.00420803.

15. Udobi KF, Childs E, Touijer K. Acute respiratory distress syndrome. Am Fam Physician. 2003;67(2):315–22.

16. Lam NN, Hung TD, Hung DK. Acute respiratory distress syndrome among severe burn patients in a developing country: application result of the berlin definition. Ann Burns Fire Disasters. 2018;31(1):9–12.

17. Lam NN, Hung TD. ARDS among cutaneous burn patients combined with inhalation injury: early onset and bad outcome. Ann Burns Fire Disasters. 2019;32(1):37–42.

18. Guo Z, Wang X, Wang Y, Xing G, Liu S. "Obesity paradox" in acute respiratory distress syndrome: a systematic review and meta-analysis. PLoS One. 2016;11(9):e0163677. https://doi.org/10.1371/journal.pone.0163677.

19. Simou E, Leonardi-Bee J, Britton J. The effect of alcohol consumption on the risk of ARDS: a systematic review and meta-analysis. Chest. 2018;154(1):58–68. https://doi.org/10.1016/j.chest.2017.11.041.

20. Ji M, Chen M, Hong X, Chen T, Zhang N. The effect of diabetes on the risk and mortality of acute lung injury/acute respiratory distress syndrome: a meta-analysis. Medicine (Baltimore). 2019;98(13):e15095. https://doi.org/10.1097/MD.0000000000015095.

21. Wilmott RW, Bush A, Deterding R, et al. Kendig's disorders of the respiratory tract in children. Philadelphia: Elsevier; 2018.

22. Papazian L, Aubron C, Brochard L, et al. Formal guidelines: management of acute respiratory distress syndrome. Ann Intensive Care. 2019;9:69. https://doi.org/10.1186/s13613-019-0540-9.

23. Walkey AJ, Goligher EC, Del Sorbo L, et al. Low tidal volume versus non-volume-limited strategies for patients with acute respiratory distress syndrome: a systematic review and meta-analysis. Ann Am Thorac Soc. 2017;14:S271–9. https://doi.org/10.1513/AnnalsATS.201704-337OT.

24. Fan E, Del Sorbo L, Goligher EC, et al. An official American Thoracic Society/European Society of intensive care medicine/society of critical care medicine clinical practice guideline: mechanical ventilation in adult patients with acute respiratory distress syndrome. Am J Respir Crit Care Med. 2017;195(9):1253–63. https://doi.org/10.1164/rccm.201703-0548ST.

25. Carrasco Loza R, Villamizar Rodríguez G, Medel Fernández N. Ventilator-induced lung injury (VILI) in acute respiratory distress syndrome (ARDS): volutrauma and molecular effects. Open Respir Med J. 2015;9(1):112–9. https://doi.org/10.2174/1874306401509010112.

26. Bellani G, Laffey JG, Pham T, et al. Noninvasive ventilation of patients with acute respiratory distress syndrome: insights from the LUNG SAFE study. Am J Respir Crit Care Med. 2017;195(1):67–77. https://doi.org/10.1164/rccm.201606-1306OC.

27. Combes A, Hajage D, Capellier G, et al. Extracorporeal membrane oxygenation for severe acute respiratory distress syndrome. N Engl J Med. 2018;378(21):1965–75. https://doi.org/10.1056/NEJMoa1800385.

28. Guérin C, Reignier J, Richard JC, et al. Prone positioning in severe acute respiratory distress syndrome. N Engl J Med. 2013;368(23):2159–68. https://doi.org/10.1056/NEJMoa1214103.

29. Bourenne J, Hraiech S, Roch A, et al. Sedation and neuromuscular blocking agents in acute respiratory distress syndrome. Ann Transl Med. 2017;5:291. https://doi.org/10.21037/atm.2017.07.19.

30. Torbic H, Krishnan S, Duggal A. Neuromuscular blocking agents for acute respiratory distress syndrome: how did we get conflicting results? Crit Care. 2019;23(1):305. https://doi.org/10.1186/s13054-019-2586-3.

31. Silversides JA, Fitzgerald E, Manickavasagam US, et al. Deresuscitation of patients with iatrogenic fluid overload is associated with reduced mortality in critical illness*. Crit Care Med. 2018;46(10):1600–7. https://doi.org/10.1097/CCM.0000000000003276.

32. Uhlig C, Silva PL, Deckert S, Schmitt J, De Abreu MG. Albumin versus crystalloid solutions in patients with the acute respiratory distress syndrome: a systematic review and meta-analysis. Crit Care. 2014;18(1):R10. https://doi.org/10.1186/cc13187.

33. Villar J, Ferrando C, Martínez D, et al. Dexamethasone treatment for the acute respiratory distress syndrome: a multicentre, randomised controlled trial. Lancet Respir Med. 2020;8(3):267–76. https://doi.org/10.1016/S2213-2600(19)30417-5.

34. Cruz FF, Ball L, Rocco PRM, Pelosi P. Ventilator-induced lung injury during controlled ventilation in patients with acute respiratory distress syndrome: less is probably better. Expert Rev Respir Med. 2018;12:403–14. https://doi.org/10.1080/17476348.2018.1457954.

35. Beitler JR, Malhotra A, Thompson BT. Ventilator-induced lung injury. Clin Chest Med. 2016;37:633–46. https://doi.org/10.1016/j.ccm.2016.07.004.

36. Nanchal RS, Truwit JD. Recent advances in understanding and treating acute respiratory distress syndrome [version 1; referees: 2 approved]. F1000Res. 2018;7:F1000. https://doi.org/10.12688/f1000research.15493.1.

37. Fanelli V, Ranieri MV, Mancebo J, et al. Feasibility and safety of low-flow extracorporeal carbon dioxide removal to facilitate ultra-protective ventilation in patients with moderate acute respiratory distress syndrome. Crit Care. 2016;20(1):36. https://doi.org/10.1186/s13054-016-1211-y.

38. Wilson JG, Calfee CS. ARDS subphenotypes: understanding a heterogeneous syndrome. Crit Care. 2020;24:102. https://doi.org/10.1186/s13054-02.

Redox and Inflammatory Signaling, the Unfolded Protein Response, and the Pathogenesis of Pulmonary Hypertension

Adiya Katseff, Raed Alhawaj, and Michael S. Wolin

Abstract

Protein folding overload and oxidative stress disrupt endoplasmic reticulum (ER) homeostasis, generating reactive oxygen species (ROS) and activating the unfolded protein response (UPR). The altered ER redox state induces further ROS production through UPR signaling that balances the cell fates of survival and apoptosis, contributing to pulmonary microvascular inflammation and dysfunction and driving the development of pulmonary hypertension (PH). UPR-induced ROS production through ER calcium release along with NADPH oxidase activity results in endothelial injury and smooth muscle cell (SMC) proliferation. ROS and calcium signaling also promote endothelial nitric oxide (NO) synthase (eNOS) uncoupling, decreasing NO production and increasing vascular resistance through persistent vasoconstriction and SMC proliferation. C/EBP-homologous protein further inhibits eNOS, interfering with endothelial function. UPR-induced NF-κB activity regulates inflammatory processes in lung tissue and contributes to pulmonary vascular remodeling. Conversely, UPR-activated nuclear factor erythroid 2-related factor 2-mediated antioxidant signaling through heme oxygenase 1 attenuates inflammatory cytokine levels and protects against vascular SMC proliferation. A mutation in the bone morphogenic protein type 2 receptor (BMPR2) gene causes misfolded BMPR2 protein accumulation in the ER, implicating the UPR in familial pulmonary arterial hypertension pathogenesis. Altogether, there is substantial evidence that redox and inflammatory signaling associated with UPR activation is critical in PH pathogenesis.

A. Katseff
Department of Microbiology and Immunology,
New York Medical College, Valhalla, NY, USA

R. Alhawaj
Department of Physiology, New York Medical
College, Valhalla, NY, USA

Department of Physiology, Faculty of Medicine,
Kuwait University, Safat, Kuwait

M. S. Wolin (✉)
Department of Physiology, New York Medical
College, Valhalla, NY, USA
e-mail: mike_wolin@nymc.edu

Keywords

Antioxidant signaling · Cytokines ·
Endothelial dysfunction · Endothelial injury ·
NADPH oxidase · Oxidative stress · Protein
folding · Pulmonary vascular remodeling ·
Vasoconstriction

Y. -X. Wang (ed.), *Lung Inflammation in Health and Disease, Volume II*, Advances in Experimental
Medicine and Biology 1304, https://doi.org/10.1007/978-3-030-68748-9_17

Abbreviations

Ang II	Angiotensin II
AP-1	Activator protein 1
APx	Ascorbate peroxidase
ARE	Antioxidant response element
ASC	Apoptosis-associated speck-like protein containing a CARD
ASK1	Apoptosis signal-regulating kinase 1
ATF	Activating transcription factor
ATFS-1	Activating transcription factor associated with stress 1
Bach1	BTB and CNC homology 1
BAK	BCL-2 homologous antagonist/killer
BAX	BCL-2-associated X protein
BBF2H7	Box B-binding factor 2 human homolog on chromosome 7
BCL-2	B cell lymphoma 2
BH3	BCL-2 homology 3
BH4	Tetrahydrobiopterin
BIM	BCL-2-interacting mediator of cell death
BiP	Immunoglobulin binding protein
BMPR2	Bone morphogenic protein type 2 receptor
BPA	Bovine pulmonary artery
bZIP	Basic leucine zipper
C/EBP	CCAAT/enhancer-binding protein
CaMKII	Calcium/calmodulin-dependent kinase II
CHOP	C/EBP-homologous protein
CO	Carbon monoxide
COMP	Cartilage oligomeric matrix protein
COPII	Coat protein II
CRE	Cyclic AMP response element
DAMP	Damage-associated molecular pattern
EC	Endothelial cell
ECM	Extracellular matrix
eIF2α	Eukaryotic translation initiation factor 2α
EndMT	Endothelial-mesenchymal transition
eNOS	Endothelial nitric oxide synthase
EPC	Endothelial progenitor cell
ER	Endoplasmic reticulum
ERAD	ER-associated degradation
ERK	Extracellular signal-regulated kinase
ERO1	ER oxidoreductase 1
ERSE	ER stress response element
ET-1	Endothelin-1
FAD	Flavin adenine dinucleotide
FPAH	Familial pulmonary arterial hypertension
GADD34	Growth arrest and DNA damage-inducible 34
GAG	Glycosaminoglycan
GCLC	Glutamate-cysteine ligase catalytic subunit
GCN2	General control nonderepressible 2
GM-CSF	Granulocyte-macrophage colony-stimulating factor
GPx7/8	Glutathione peroxidase 7 and 8
GR	Glutathione reductase
GSH/GSSG	Glutathione (reduced/oxidized)
GSH1	γ-glutamylcysteine synthetase
GST	Glutathione S-transferase
H/R	Hypoxia/ischemia and reoxygenation
H_2O_2	Hydrogen peroxide
HA	Hyaluronan
HDAC4	Histone deacetylase 4
HIFα	Hypoxia-inducible factor α
HIV-PAH	HIV-induced pulmonary arterial hypertension
HO-1	Heme oxygenase 1
HOCl	Hypochlorous acid
HPAEC	Human pulmonary arterial endothelial cell
HPMEC	Human pulmonary microvascular endothelial cell
HRE	Hypoxia response element
HSP47	Heat shock protein 47
IKK	Inhibitor of nuclear factor-κB (IκB) kinase
IL	Interleukin
IP3R	Inositol 1,4,5-triphosphate receptor
IRE1	Inositol-requiring protein 1
ISR	Integrated stress response
IκB	Inhibitor of nuclear factor-κB
JNK	JUN N-terminal kinase
KEAP1	Kelch-like ECH-associated protein 1
LC20	20-kDa regulatory light chain of myosin II
LPS	Lipopolysaccharide

MA	Methamphetamine	QSOX	Quiescin sulfhydryl oxidase
Maf	Musculoaponeurotic fibrosarcoma	RHAMM	Receptor for HA-mediated motility
MAM	Mitochondria-associated membrane	RIDD	Regulated IRE1-dependent decay
MAPK	Mitogen-activated protein kinase	ROS	Reactive oxygen species
MCP	Monocyte chemoattractant protein	RVSP	Right ventricle systolic pressure
MCT	Monocrotaline	S1P	Site 1 protease
Mdm2	Mouse double minute 2 homolog	S2P	Site 2 protease
MEF	Myocyte enhancer factor	SFN	Sulforaphane
MIP	Macrophage inflammatory protein	SMC	Smooth muscle cell
MLCK	Myosin light chain kinase	SOD	Superoxide dismutase
MMP	Matrix metalloproteinase	SRXN-1	Sulfiredoxin-1
NAD	Nicotine adenine dinucleotide	Tat	Trans-activator of transcription
NF-Y	Nuclear transcription factor Y	TGFβ	Transforming growth factor β
NF-κB	Nuclear factor kappa-light-chain-enhancer of activated B cells	TIM	Translocase of the inner membrane
		TLR	Toll-like receptor
NIK	NF-κB inducing kinase	TNF-α	Tumor necrosis factor α
NLRP3	Nucleotide-binding oligomerization domain (NOD)-like receptor (NLR) family, leucine-rich repeat (LRR), and pyrin domain (PYD)-containing protein-3	TOM	Translocase of the outer membrane
		TRAF	Tumor necrosis factor receptor-associated factor
		Treg	Regulatory T cell
		TRx	Thioredoxin
NO	Nitric oxide	TXNIP	TRx interacting protein
NOX	NADPH oxidase	UGT	UDP glucuronosyl transferase
NQO1	NAD(P)H: quinone oxidoreductase 1	UPR	Unfolded protein response
NRF2	Nuclear factor erythroid 2-related factor 2	UPR^{am}	UPR activated by protein mistargeting
O_2	Oxygen	UPR^{mt}	Mitochondrial UPR
OASIS	Old astrocyte specifically induced substance	UPS	Ubiquitin-proteasome system
		VDCC	Voltage-dependent calcium channel
PAEC	Pulmonary arterial endothelial cell	VHL	Von Hippel-Lindau
PAH	Pulmonary arterial hypertension	VKOR	Vitamin K epoxide reductase
PAMP	Pathogen-associated molecular pattern	VPO1	Vascular peroxidase 1
		XBP1	X-box binding protein 1
PASMC	Pulmonary artery smooth muscle cell		
PCNA	Proliferating cell nuclear antigen		
PDI	Protein disulfide isomerase		
PERK	Protein kinase RNA-like endoplasmic reticulum kinase		
PH	Pulmonary hypertension		
PHD	Proline hydroxylase		
PI3K	Phosphoinositide-3 kinase		
PKB	Protein kinase B		
PKC	Protein kinase C		
PP1C	Protein phosphatase 1C		
PRxIV	Peroxiredoxin IV		
PTP	Permeability transition pore		
PUMA	p53 upregulated modulator of apoptosis		

17.1 Introduction

17.1.1 Relevance of Redox Modulation of the Unfolded Protein Response to Pulmonary Hypertension

Within a cell, the processes of protein translation and post-translational modification are highly regulated. All secretory proteins, resident proteins of the secretory pathway organelles, and membrane surface proteins are co-translationally inserted into the endoplasmic reticulum (ER) lumen for processing [204]. The ER handles approximately 30% of proteins that are folded in a typical cell, with an even higher percentage [83] in specialized secretory cells.

Secretory proteins are translated directly into the ER lumen, where they mature to the proper conformation by acquiring post-translational modifications, including the introduction of disulfide bonds between cysteine residues [22]. This folding process involves oxidation and reduction as disulfide bonds are added, removed, and shuffled around within the protein as it acquires its shape [94]. Proper protein folding and maturation requires the integration of multiple signals and feedback mechanisms [83].

Protein overexpression can overload the folding capacity of the ER, during which proteins are translated but cannot be folded and exported fast enough to accommodate for the influx of nascent polypeptides [83]. Misfolded proteins accumulate and oxidative stress increases within the ER lumen to activate the unfolded protein response (UPR), which is conserved in eukaryotes from yeast through humans and controls the fate of the stressed cell [184]. First, there is an adaptive response aimed at restoring homeostasis. If the adaptive phase is insufficient and the stress is prolonged, the UPR promotes apoptosis [255]. The UPR regulates several cellular processes including energy homeostasis, inflammation, and cell differentiation [197]. Importantly, altered redox homeostasis is involved in the activation of the UPR as well as downstream of the UPR, and may lead to the development of several diseases [212].

Pulmonary hypertension (PH) is a rare and incurable life-threatening disease in which mean pulmonary arterial pressure is greater than 25 mm Hg at rest as measured by right heart catheterization [87]. PH is characterized by pulmonary vascular remodeling at the different layers of the vascular wall: proliferation and dysfunction of endothelial cells of the intima as well as the pulmonary artery smooth muscle cells (PASMCs) of the media [34, 219]. The adventitia lacks precise boundaries in the human lung, making it difficult to measure remodeling, but the adventitia does play a role in PH as a hub for signaling interactions

between local fibroblasts and arriving macrophages [190, 221]. Nevertheless, pulmonary vascular remodeling in PH results in a thickening of at least two levels of the vascular compartment [236].

Expansion of the extracellular matrix (ECM) is involved in vascular remodeling at all layers [230]. Excess production of secreted ECM proteins may overload the ER of pulmonary vascular cells to trigger UPR signaling and PH progression. Interestingly, PH patients have upregulated UPR genes [129], and several recent studies have shown UPR activation in PH models using rodents and cultured PASMCs [29, 60, 264, 265]. Conditions leading to PH including hypoxia and endothelin-1 (ET-1) production induce elevated pulmonary vascular pressure and exacerbate oxidant formation [35, 141, 250], which may stimulate UPR signaling to further PH pathogenesis in a feed-forward manner. However, little is known about the precise pathways connecting UPR activation with the development of PH.

This chapter will present information indicating that altered ER redox homeostasis serves to activate the UPR as well as signaling downstream of the UPR, potentially leading to PH. First will be a definition of the UPR and a brief discussion of the redox-related aspects of the adaptive and apoptotic phases following activation of each UPR sensor. Next will be a discussion of the role of redox in protein folding at the ER and how these thiol redox mechanisms activate the UPR and may play a role in PH. Finally, other redox mechanisms both upstream and downstream of the UPR will be discussed along with their implications in PH.

17.2 The UPR Is Redox-Regulated and Promotes Redox Signaling

17.2.1 Chronological Order of the Stress Response

The UPR is a response to stress that consists of several complementary adaptive mechanisms, both transcriptional and non-transcriptional. This has been well-covered in several other reviews [83, 84, 184, 204, 220, 255], but we will include a description here of the events relevant to redox signaling and PH. The collective function of these responses is to reduce the unfolded protein load at the ER and increase its folding capacity. The ER has several sensors that send information about the status of the ER lumen to the nucleus and cytosol where the UPR is carried out.

The most immediate adaptive action after activation of the UPR sensors is the phosphorylation of eukaryotic translation initiation factor 2α (eIF2α), which deactivates it [78]. This significantly slows the translation of new proteins, because eIF2α is necessary for the initiation of translation. At the same time, degradation of misfolded proteins begins through the process of macroautophagy, including ER-phagy, in which portions of the ER including misfolded proteins are degraded in lysosomes [83].

The slower actions of the UPR involve the activation of transcription factors that promote gene expression to relieve ER stress, which takes more time than phosphorylation. These UPR-activated genes are involved in adaptive processes such as expanding the ER membrane [218], ER-associated degradation (ERAD) to eliminate misfolded proteins via the proteasome [234], translating more folding chaperones [244], and increasing quality control by preventing certain proteins from entering the ER [101]. The redox balance of both the cytosol and the ER is affected by antioxidant genes that are upregulated via the UPR-activated protein activating transcription factor 4 (ATF4) and others [83].

Finally, the UPR shifts to become pro-apoptotic if the ER stress or cellular damage is prolonged or severe. Excess reactive oxygen species (ROS) contribute to mitochondrial damage and apoptotic signaling, resulting in cell death [255].

17.2.2 Specific Actions of Each UPR Sensor

Sensors of ER stress that carry out the UPR reside in the ER membrane. They are all activated by ER luminal stress but primarily act on the cyto-

solic side of the membrane [204]. One sensor, inositol-requiring protein 1 (IRE1), is also sensitive to cytoplasmic stress [91]. Each sensor can be affected by altered redox homeostasis and may also activate downstream mechanisms to either correct or potentiate the redox state, depending on whether UPR activity is adaptive or pro-apoptotic.

17.2.2.1 IRE1

IRE1 is the most evolutionarily conserved sensor of the UPR [197]. Mammalian cells express both IRE1α and IRE1β, transmembrane proteins with kinase and endoribonuclease activity [43, 171, 233]. During non-stress conditions, IRE1α exists as an inactive monomer in the ER membrane, bound to the chaperone immunoglobulin binding protein (BiP; also called GRP78 and HSPA5). During ER stress, BiP dissociates to interact with unfolded proteins [18], allowing the IRE1α monomers to dimerize or oligomerize and then autotransphosphorylate (Fig. 17.1). There is some evidence showing that phosphorylation is not necessary to activate IRE1, but only a conformational change [196]. Kinase activity may instead attenuate IRE1 endoribonuclease activity, inactivating IRE1 by separating it into monomers [196].

Once IRE1 is activated, the adaptive response begins through its endoribonuclease activity, in which it splices X-box binding protein 1 unspliced (XBP1u) mRNA to the spliced form (XBP1s) [267]. Once XBP1s is translated, it acts as a transcription factor, interacting with the general nuclear transcription factor Y (NF-Y) at gene promoters containing an X-box element or the ER stress response element (ERSE) to regulate expression [204, 267]. The newly expressed genes aid in several processes, including protein folding through the expression of chaperones such as BiP [43].

Another process promoted by XBP1s is the ubiquitination and proteasomal degradation of misfolded ER proteins, known as ERAD [234]. The ubiquitin-proteasome system (UPS) is itself redox-controlled [238]. The 20S proteasome specifically degrades oxidized proteins and is upregulated by elevated ROS and mito-

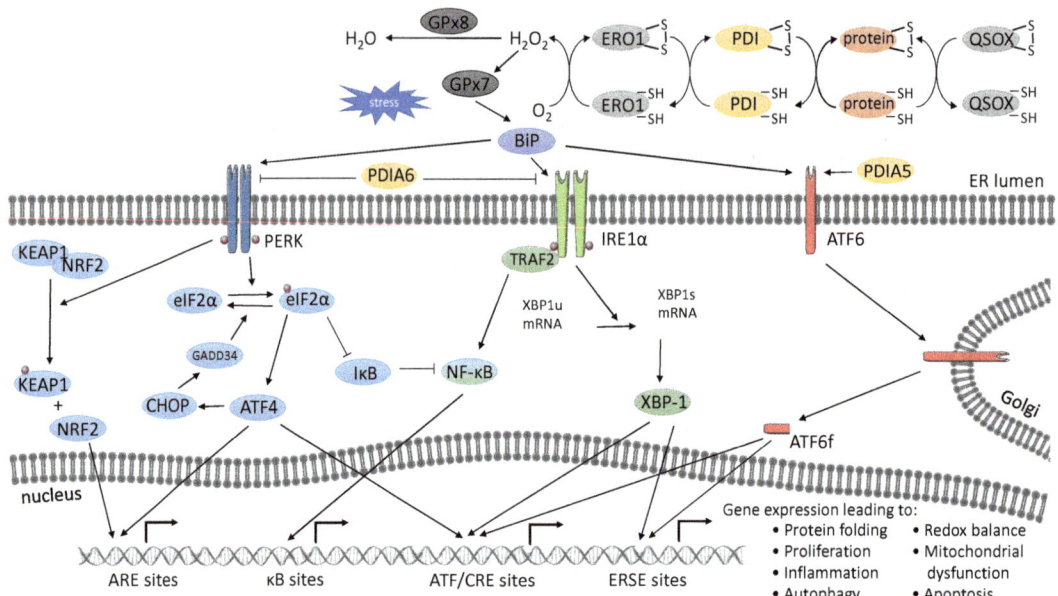

Fig. 17.1 Protein-folding overload activates adaptive and apoptotic UPR signaling to potentially promote PH pathogenesis. Disulfide bond formation in protein substrates occurs though electron relay involving enzymes including protein disulfide isomerases (PDIs), endoplasmic reticulum (ER) oxidoreductase 1 (ERO1), and quiescin sulfhydryl oxidase (QSOX). Excess protein-folding load can overwhelm chaperones, leading to thiol redox-mediated activation of the unfolded protein response (UPR) due to disrupted redox homeostasis. ERO1α has two regulatory disulfide bonds. Disrupting these bonds deregulates the enzyme, resulting in the hyper-oxidation of substrates and production of excess reactive oxygen species (ROS; denoted H_2O_2). Glutathione peroxidase 8 (GPx8) is induced by ER stress and binds to ERO1, converting H_2O_2 to water (H_2O). Some ROS avoids this and oxidizes GPx7, forming a disulfide bond which then oxidizes BiP, allowing BiP to interact more strongly with protein substrates and dissociate from the UPR sensors, promoting UPR activation. If redox homeostasis is restored, UPR signaling ceases. However, if the redox state is not restored, sustained UPR signaling can lead to adaptation or apoptosis. PDIs are active during protein-folding overload, forming and isomerizing disulfide bonds to result in hyperoxidized and misfolded substrates. Specific PDIs activate and inhibit specific UPR sensors to promote and prevent adaptive and apoptotic signaling. The UPR sensors protein kinase RNA-like endoplasmic reticulum kinase (PERK) and inositol-requiring protein 1 (IRE1) are activated by oligomerization associated with disulfide bond formation between monomers after the dissociation of BiP as well as phosphorylation (dark red circles). PERK phosphorylates eIF2α, promoting the transcription of activating transcription factor 4 (ATF4) which regulates transcription in the nucleus at ARE and ATF/cyclic AMP response element (CRE) sites. Activated nuclear factor erythroid 2-related factor 2 (NF-κB) translocates to the nucleus and increases the expression of genes regulating inflammation and autophagy. PERK also phosphorylates the Kelch-like ECH-associated protein 1 (KEAP1)/nuclear factor erythroid 2-related factor 2 (NRF2) complex in the cytosol, allowing the transcription factor NRF2 to dissociate and translocate to the nucleus to upregulate gene expression at ARE sites. IRE1 also acts as an endonuclease to splice mRNA encoding transcription factor X-box binding protein 1 (XBP-1). After translation and nuclear entry, it promotes gene expression at ATF/CRE and ER stress response element (ERSE) sites. ATF6 activity begins at the ER membrane, where disulfide bonds are reduced to produce ATF6 monomers, which travel to the Golgi for processing. The mature ATF6 fragment then acts in the nucleus as a transcription factor at ATF/CRE and ERSE sites. Sustained adaptive and apoptotic signaling from UPR sensors alters gene expression to affect cellular processes in pulmonary artery smooth muscle cells (PASMCs) and endothelial cells (ECs), leading to pulmonary hypertension (PH). TRAF2 tumor necrosis factor receptor-associated factor 2

chondrial dysfunction. It is also redox-activated by glutathionylation during oxidizing conditions. However, this shift to the 20S may decrease 26S proteasome activity, promoting the accumulation of non-oxidized misfolded proteins and possibly a further shift away from homeostasis. There is some evidence that the high ratio of oxidized to reduced nicotine adenine dinucleotide (NAD$^+$:NADH) in the cytoplasm during oxidizing conditions may open and activate the 26S proteasome [238]. While the UPS is clearly subject to redox control, the precise nature of this control is still under debate.

Other gene products upregulated by XBP1s act to deny ER entry to certain nascent polypeptides based on their signal sequences, preventing their translation within the ER and resulting in protein quality control [101]. Another process promotes phospholipid synthesis at the ER, expanding the ER membrane and increasing the volume of the lumen to make more space for folding proteins [218]. These activities all promote a decrease in ER stress through increasing ER folding capacity or decreasing protein folding load at the ER.

Aside from XBP1s, IRE1 also activates the regulated IRE1-dependent decay (RIDD) pathway to degrade mRNA [89]. RIDD decreases protein folding load at the ER by making mRNA unavailable for translation, thus abrogating the need to fold any resulting proteins.

Tumor necrosis factor receptor-associated factor 2 (TRAF2), an adaptor protein, can bind to activated IRE1, which then activates apoptosis signal-regulating kinase 1 (ASK1) and further targets in a signaling cascade eventually leading to "alarm stress pathways" that activate JUN N-terminal kinase (JNK), which is pro-apoptotic and promotes macroautophagy. Other distinct adaptor proteins interact with IRE1 to activate p38, extracellular signal-regulated kinase (ERK), and the pro-inflammatory transcription factor nuclear factor kappa-light-chain-enhancer of activated B cells (NF-κB) [83, 100, 179, 237]. These "alarm stress" pathways promote the reduction of ER stress through increased protein folding or autophagy of misfolded or damaged proteins and organelles. In addition to these adaptive processes, IRE1 may be involved in the apoptosis phase under some conditions, but the mechanism is unclear [83].

17.2.2.2 Protein Kinase RNA-Like Endoplasmic Reticulum Kinase

Protein kinase RNA-like endoplasmic reticulum kinase (PERK) exists as an inactive monomer in the ER membrane during non-stress conditions. During ER stress, PERK monomers are activated by dimerization and autotransphosphorylation (Fig. 17.1). Once phosphorylated, PERK dimers are active kinases [77]. PERK activity is perhaps most important during the early stages of the UPR.

The adaptive response begins with the phosphorylation of eIF2α, a protein required for the initiation of translation at a ribosome [77]. Phosphorylation of eIF2α inactivates most translation initiation [79]. Importantly, phosphorylated eIF2α represses translation of the NF-κB inhibitor IκB, thus activating NF-κB, a redox-sensitive transcription factor that regulates gene expression at κB sites in the promoters of genes that regulate inflammatory processes [51, 112, 173]. However, phosphorylated eIF2α does allow for the selective translation of ATF4, which is expressed at low levels during unstressed conditions [78]. The ATF4 protein is a transcription factor for genes involved in adaptive processes including autophagy, amino acid metabolism, and the antioxidant response [204].

PERK signaling also activates nuclear factor erythroid 2-related factor 2 (NRF2), a transcription factor that regulates redox and antioxidant metabolism along with ATF4 [178]. NRF2 is kept inactive in the cytoplasm by associating with Kelch-like ECH-associated protein 1 (KEAP1), which targets NRF2 for proteasomal degradation. KEAP1 contains several redox-sensitive cysteine residues that may affect its ability to interact with NRF2 [88]. PERK phosphorylates this complex to dissociate NRF2, allowing its nuclear import and regulation of gene expression [45]. Additionally, oxidation-dependent aggregation of p62 both stimulates autophagy and allows p62 to occupy the NRF2-binding site in KEAP1, further stabilizing NRF2 [30, 114, 238].

If the ER stress fails to resolve, apoptosis signaling begins, regulated by the B cell lymphoma 2 (BCL-2) family of proteins. Pro-apoptotic BCL-2 homology 3 (BH3)-only proteins such as BCL-2-interacting mediator of cell death (BIM) and p53 upregulated modulator of apoptosis (PUMA) are upregulated, which then activate BCL-2-associated X protein (BAX) and/or BCL-2 homologous antagonist/killer (BAK), promoting permeabilization of the outer mitochondrial membrane and further pro-apoptotic signaling [255].

During the apoptosis phase, ATF4 upregulates C/EBP-homologous protein (CHOP, also known as GADD153) [78, 204]. CHOP both upregulates the expression of BIM and downregulates the expression of BCL-2, an anti-apoptotic BH3-containing protein. Additionally, ATF4 directly upregulates BH3-only proteins including PUMA [255]. ATF4 and CHOP also promote growth arrest and DNA damage-inducible 34 (GADD34) expression. GADD34 then upregulates a phosphatase complex including protein phosphatase 1C (PP1C), which dephosphorylates eIF2α and promotes the resumption of translation [180]. This causes further oxidative stress at the ER, generating ROS and ultimately leading to apoptosis.

17.2.2.3   Activating Transcription Factor 6 (ATF6)

ATF6 is different from IRE1α and PERK, which have enzymatic activity and are inactive as monomers. ATF6 represents a group of structurally similar basic leucine zipper (bZIP) transcription factors in the ATF6α/β and old astrocyte specifically induced substance (OASIS) families [266]. They are specialized in their activation, tissue distribution, and response element binding [5]. Interestingly, the OASIS family member box B-binding factor 2 human homolog on chromosome 7 (BBF2H7) is strongly expressed during ER stress in the lungs, as well as the long bones, spleen, gonads, and nervous system [115]. ATF6 is kept inactive in non-stress conditions through glycosylation as well as oligomerization via intra- and inter-peptide disulfide bridges, which trap it at the ER membrane. During ER stress,

ATF6 is under-glycosylated, reduced, and monomeric [90, 220]. It is then free to be sent to the Golgi for additional processing and activation. Some OASIS family members lack both sites for BiP interaction and a Golgi-localization sequence, and hence they are activated by an alternate mechanism [5, 175].

The adaptive response to ER stress begins with the transport of monomeric ATF6 from the ER to the Golgi apparatus via coat protein II (COPII)-coated vesicles [36, 203]. Once at the Golgi membrane, site 1 protease (S1P) and site 2 protease (S2P) cut ATF6 at specific sites to release a cytosolic fragment (ATF6f) that can travel to the nucleus [263]. ATF6f interacts with DNA, acting as a transcription factor at the ER stress-response element (ERSE) to upregulate the expression of genes for components of ERAD as well as XBP1 [266]. XBP1 itself, as described earlier, promotes the transcription of genes involved in several adaptive processes, including ERAD, protein folding, and quality control. ATF6f, ATF4, and XBP1s all partially overlap in the target genes whose expression they regulate (Fig. 17.1). Both the type of stress stimulus and the cell type affect the activation of the UPR and the gene products that are ultimately expressed to produce the adaptive response [83].

17.2.3   Activation of the UPR Sensors

The mechanism for activation of the UPR has been studied most extensively in IRE1α, which probably does not directly recognize unfolded proteins in mammals [83]. There is evidence that PERK interacts directly with misfolded proteins, causing its oligomerization and subsequent activation [246]. The individual UPR sensors respond to stress signals; here there will be a focus on the signals caused by misfolded proteins that are related to the redox state of the ER.

During non-stress conditions, the ER chaperone BiP binds to the inactive IRE1α and PERK monomers, suppressing their oligomerization or dimerization. During ER stress, BiP is required to fold excess unfolded or misfolded proteins. BiP must dissociate from the UPR sensors to interact with these misfolded proteins [18]. This allows the free UPR sensors IRE1α and PERK to dimerize or oligomerize and begin activation through autotransphosphorylation. However, the activation of each of the individual UPR sensors is dependent on slightly different regulatory mechanisms. For example, IRE1α may specifically require heat shock protein 47 (HSP47) to help displace BiP [208]. Also, deleting the BiP-binding site on IRE1 does not alter the induction of ER stress, indicating that other factors govern the activation of the UPR [110].

BiP also binds ATF6 family members during normal conditions, blocking the Golgi-localization signal and retaining them at the ER [209]. In non-stress conditions, ATF6 family members are sufficiently glycosylated, and the lectin-like ER chaperone calreticulin may interact with them to further retain them at the ER. ER stress conditions result in under-glycosylated ATF6, which is unable to interact with calreticulin [90]. The disulfide bridges that oligomerize ATF6 must be also be reduced before ATF6 can travel to the Golgi as monomers.

The UPR is highly regulated by complex feedback mechanisms, which allows for responses that are dynamic and diverse that may be either transient or sustained. The UPR pathways are not linear or parallel. There is overlap among the UPR pathways: for example, ATF6 and XBP1s can form a heterodimer, promoting transcription of the same genes [210]. Redox metabolism and inflammation pathways also interact both upstream and downstream of the UPR sensors, which together play a role in PH.

17.2.4 Mitochondrial UPR Redox Signaling Activates the UPR

The mitochondrial genome encodes 13 essential proteins that are localized in the mitochondria. The remainder of the proteins that reside in the mitochondria are nuclear-encoded and are trans-lated on cytosolic ribosomes [161]. They must be imported through both mitochondrial membranes via the translocase of the outer membrane (TOM) and the translocase of the inner membrane (TIM) [31]. Once inside the mitochondria, these polypeptides must be folded with the help of chaperones to achieve their proper conformations. This echoes the protein folding that takes place in the ER lumen, and oxidative stress at the mitochondria can similarly disturb protein folding [161].

Mitochondrial stress, including increased ROS due to respiratory chain dysfunction, can activate the cytosolic kinase general control nonderepressible 2 (GCN2) as part of the integrated stress response (ISR), which, like PERK, phosphorylates eIF2α to slow general protein synthesis and alleviate oxidative stress [8, 80]. During mitochondrial stress, there is a reduced ability to import nascent polypeptides, resulting in the accumulation of misfolded proteins in the cytosol. The UPR activated by protein mistargeting (UPR^{am}) is one response that promotes decreased protein synthesis and the proteasomal degradation of these potentially toxic proteins [257]. There is also an adaptive mitochondrial UPR (UPR^{mt}) coordinated by the activating transcription factor associated with stress 1 (ATFS-1) that promotes the expression of mitochondrial chaperones, proteases, protein import components, and ROS detoxification enzymes [177].

The exchange of calcium between the ER and the mitochondria at the mitochondria-associated membrane (MAM) ties mitochondrial stress to ER stress [214]. Additionally, ROS produced during mitochondrial metabolic stress enhances UPR activation of XBP1 and CHOP associated with inflammation marked by interleukin (IL)-23 and IL-6 expression [168]. If the UPR is insufficient to resolve oxidative stress, severely damaged mitochondria are degraded via mitophagy as a last resort [6, 161]. A prolonged UPR^{mt} might delay or impair mitophagy. Interestingly, there is evidence of increased mitochondrial fragmentation, indicating inadequate mitophagy, in PASMCs of PH patients [153].

17.3 Thiol Redox Dysregulation During ER Protein Folding Activates the UPR, Leading to PH

17.3.1 Thiol Redox and Protein Folding Overload Cause ER Stress, Leading to UPR Activation

Protein folding at the ER is stabilized by the formation of disulfide bonds [204]. Disulfide bond formation between cysteine residues in a protein is an oxidative process that can help maintain stability, allowing the protein to remain functional in the potentially harsh extracellular environment [22]. This process essentially requires both a source of oxidizing equivalents and an enzyme to catalyze the electron transfer. An understanding of the process of disulfide bond formation during protein folding illustrates the role of dysfunctional thiol redox homeostasis in the activation of the UPR, potentially leading to the development of PH.

17.3.2 The Role of Glutathione in Protein Folding

Glutathione is abundant in animal cells, so it is considered the major cellular redox buffer and was previously considered to be the source of oxidizing equivalents for disulfide bond formation [151]. Glutathione in its reduced form (GSH) is a tripeptide consisting of glutamate with a gamma peptide linkage to glycine, which has a peptide linkage to cysteine. Oxidized glutathione disulfide (GSSG) contains a disulfide bond between the cysteines of two glutathione molecules. Due to its abundance, the redox state of the local cellular environment is indicated by the redox state, or ratio, of the two forms of glutathione [10].

More than half of the glutathione in the ER is in the form of mixed disulfides with proteins [14]. These mixed disulfide bonds are likely to be with both nascent polypeptides that are in the process of folding and exposed free thiols on mature resident ER proteins, both membrane-bound and within the ER lumen. It is not known whether these mixed disulfides are formed with specific proteins or a broad range, but there are a few potential consequences [14]. First, glutathionylation of proteins can affect their function if the active site requires a free thiol group [14]. Additionally, any bound glutathione may contribute even further to the redox buffering capacity of the ER lumen. Importantly, the fact that glutathione forms bonds with ER proteins shows that it plays a role in protein folding, even if it is not required as the major contributor of oxidizing equivalents [14]. The ER environment must be tightly regulated, as a change in the redox state in either direction can affect ER function [10].

The concentration of glutathione in both the cytoplasm and the ER is about 1–10 mM or higher [94, 95]. However, the redox ratio in each compartment differs. In the cytoplasm, the ratio of reduced glutathione to glutathione disulfide ranges from 30:1 to 100:1 but in the secretory pathway, which includes the ER, the ratio is closer to 3:1 [61]. This indicates that the ER is a more oxidizing environment than the cytoplasm. More recent studies using fluorescent probes to report directly from the ER lumen in live cells suggest that the ER may be more reducing than originally thought, perhaps with a GSH:GSSG of 35:1 [94]. However, the ER is still considered to be a more oxidizing environment than the cytoplasm.

One reason for this oxidizing nature is that reduced glutathione, which is synthesized in the cytoplasm, is capable of slow, mediated transport across the ER membrane, but oxidized glutathione disulfide is trapped inside the ER [9, 49]. Furthermore, there is very little glutathione reductase (GR) in the ER lumen, so glutathione remains in its oxidized form [186]. This trapped glutathione may then slowly convert into mixed disulfides with ER proteins.

There has been some debate over whether glutathione acts as an oxidant, a reductant, or that it simply reflects the oxidative ER milieu [49]. Previously, glutathione was thought to provide oxidizing equivalents for protein folding [204]. However, *Saccharomyces cerevisiae* (budding yeast) mutants lacking the γ-glutamylcysteine synthetase (GSH1) gene, which is required for glutathione biosynthesis, can form disulfide bonds at their normal rate, indicating that GSSG is not required as an oxidant [67].

The most accepted current model suggests that glutathione is not a source of oxidizing equivalents, but that GSH acts as a reductant, or electron donor, for PDI and its protein substrates, allowing disulfide bonds to be isomerized until the protein achieves its final conformation [10, 151]. This is in parallel with another thiol reducing pathway involving thioredoxin (TRx) and thioredoxin reductase [49]. Fortunately, GSH is less likely to reduce native disulfides on proteins due to the oxidizing nature of the ER lumen [22].

17.3.3 Disulfide Bond Formation as an Electron Relay

Protein disulfide formation is catalyzed by an electron relay system independent of glutathione in which oxygen (O_2) is the ultimate electron acceptor (Fig. 17.1). The process is mediated by thiol-oxidoreductase enzymes including ER oxidoreductase 1 (ERO1) and the protein disulfide isomerase (PDI) family as the main ER protein folding chaperones [10]. ERO1 shuttles oxidizing equivalents via its flavin adenine dinucleotide (FAD) cofactor to PDI. There is evidence in vitro showing that ERO1 oxidizes PDI. However, there is conflicting evidence regarding whether ERO1 is required to oxidize glutathione in yeast [46, 235].

17.3.4 Formation of Disulfide Bonds in Protein Substrates Occurs Through a Thiol Redox Mechanism

PDI and other thiol-oxidoreductases help to catalyze thiol-disulfide exchange with protein substrates [33]. Two reduced thiols in the protein substrate are exchanged with an oxidized disulfide in a thiol-oxidoreductase enzyme such as PDI. The thiol-oxidoreductase, now in the reduced thiol form, must subsequently be re-oxidized to the disulfide by a thiol oxidase such as ERO1 to regain its oxidative activity. The protein substrate contains a new disulfide bond; disulfide bonds are often formed due to the spatial proximity of cysteine residues, even if the bond is not part of the final conformation of the protein substrate. Disulfide bonds can be rearranged until the substrate achieves the final conformation [118]. Other ER chaperones such as BiP detect exposed hydrophobic regions of the protein, which often indicate an incorrect or unstable conformation [61, 204]. The disulfide bonds are progressively rearranged until the hydrophobic regions are no longer exposed.

PDI catalyzes protein folding in a variety of substrates, acting as both an oxidase to introduce new disulfide bonds and an isomerase to rearrange incorrect disulfide bonds. PDI consists of four TRx-like domains: two A domains and two B domains [275]. The A domains are catalytically active and carry out the oxidase activity. The B domains are redox-inactive, but the full protein is required to perform the isomerase activity. PDI introduces disulfide bonds co-translationally, acting as a placeholder that is removed when the two cysteine thiol groups of the protein substrate are paired [118]. The enzymes in the TRx superfamily that catalyze thiol-disulfide exchange share a structural fold and a highly conserved Cys-X-X-Cys motif in the active site [69, 118].

The reaction mechanism for both oxidase and isomerase activity of PDI involves the formation of a mixed disulfide between the N-terminal cysteine in the PDI active site and a cysteine in the protein substrate [69, 118]. The difference arises in the initial redox state of the PDI active site and the protein substrate. For oxidase activity, the PDI active site cysteines are initially oxidized in the form of a disulfide bond. After the oxidized PDI active site disulfide forms a mixed disulfide with a reduced cysteine in the protein substrate, another reduced cysteine in the substrate attacks the mixed disulfide to release reduced PDI, resulting in a new disulfide bond in the substrate [69, 118]. For isomerase activity, the PDI active site cysteines are initially in the reduced thiol form. The reduced N-terminal PDI active site cysteine forms a mixed disulfide with an existing disulfide bond in the substrate. The C-terminal cysteine in the PDI active site then reacts with the mixed disulfide to release oxidized PDI and a reduced protein substrate with free cysteine residues [69, 118].

17.3.5 Thiol Oxidases Regenerate Oxidizing Equivalents After Substrate Oxidation by PDI

After a disulfide bond is introduced into the protein substrate through PDI oxidase activity as described above, the PDI active site cysteine thiol groups are reduced and must be re-oxidized to the disulfide form before another substrate can be oxidized. This requires a continuous influx of oxidizing equivalents that are usually provided by ERO1, which directly oxidizes PDI [69]. In a yeast model with a conditional loss-of-function Ero1p mutation, protein substrates that usually contain oxidized disulfide bonds remain reduced [68]. Conversely, overexpression of functional Ero1p confers resistance to the reducing agent DTT by promoting oxidation [187].

Some mammalian thiol oxidases for the PDI family, including ERO1α and β and the quiescin sulfhydryl oxidase family (QSOX), use O_2 as the electron acceptor and generate ROS in the form of hydrogen peroxide (H_2O_2) [227]. Other mammalian thiol oxidases in the ER include peroxiredoxin IV (PRxIV), glutathione peroxidase 7 and 8 (GPx7/8), and ascorbate peroxidase (APx), which are ROS scavengers and consume H_2O_2 as the electron acceptor to generate water [272]. These thiol oxidases become reduced when they oxidize their substrates, so they transfer electrons to O_2 or H_2O_2 to regain their oxidative activity. Altogether, this electron relay system maintains the redox homeostasis that is critical to maintaining the oxidative protein folding environment of the ER. However, the use of O_2 as the primary electron acceptor means that excess protein folding activity results in excess ROS, causing ER stress that can activate the UPR.

ERO1, the primary oxidase of PDI, is tightly associated with its FAD cofactor that assists with electron transfer starting at the cysteine thiols of substrate proteins and continuing through PDI to the ERO1 shuttle disulfides to the ERO1 active site to FAD to O_2 [227]. ERO1 activity, as well as overall protein folding, is highly sensitive to levels of free FAD. FAD is rapidly equilibrated between the cytosol and the ER lumen with a robust transport system. This dependence on FAD links protein folding to the metabolic status of the cell [235]. Additionally, mutations in the FAD-binding site cause ERO1 to lose stability as well as its ability to oxidize PDI [53]. O_2 entry and H_2O_2 exit from the flavin cofactor at the ERO1α active site is regulated by a pair of cysteines, Cys208-Cys241, that block the cofactor and are unlocked by forming a mixed sulfide complex with PDI [194]. ERO1 dysfunction and FAD displacement disrupt protein folding, promoting UPR signaling [21].

Alternatively, the QSOX family is capable of directly oxidizing protein substrates without the involvement of PDI (Fig. 17.1). QSOX can fold proteins efficiently as the sole oxidant in conditions where PDI is reduced and is only capable of isomerase activity [94]. QSOX, like ERO1, uses FAD as a cofactor to aid in electron transfer from thiols to O_2. Also like ERO1, QSOX uses a shuttle disulfide to mediate electron transfer from the protein substrate to the active site [113]. In yeast with a deletion of the ERO1 gene, overexpression of the QSOX family member hQSOX1a restores disulfide bond formation, suppressing lethality [32].

Reduced glutathione can compete as a substrate for any of these enzymes. While the oxidation of glutathione is a by-product of protein folding, it also allows glutathione to act as a buffer preventing the overoxidation of true protein substrates [46, 61]. Glutathione may further protect against hyperoxidizing conditions driven by ERO1 in the ER by consuming excess oxidizing equivalents in the place of PDI or protein substrates, preventing oxidative stress and subsequent UPR activation [46].

17.3.6 Thiol Redox Dysregulation Disrupts Protein Folding and Activates the UPR

The presence of misfolded proteins in the ER induces the expression of folding chaperones including BiP and PDI [55, 120]. Additionally, overexpression of GPX1 and PDI1 in yeast rescues correct protein folding [50]. However, if these enzymes are unable to compensate for the protein load, misfolded proteins can accumulate in the ER lumen, generating ROS through fruitless disulfide bond formation and activating PERK and IRE1 signaling. UPR signaling can be attenuated by antioxidants [147], but if left unchecked, the altered redox state may potentially lead to pathophysiological conditions such as PH.

ERO1α activity converts one molecule of O_2 to one molecule of H_2O_2 for each disulfide bond formed, and increased ERO1 activity promotes ER hyperoxidation and stress. ERO1 contains several cysteine residues that can form regulatory disulfide bonds, which might constrain the flexible loop containing the shuttle disulfide and prevent it from migrating to the active site [227]. ERO1α contains two: Cys94-Cys131 and Cys99-Cys104. Deregulating ERO1α by changing cysteine to alanine (C104A/C131A) is hyperoxidizing, highlighting the importance of ER thiol redox balance for regulating thiol oxidase activity to ultimately regulate ER redox balance [75]. Similarly, ERO1β contains two regulatory cysteine pairs: Cys90-Cys130 and Cys95-Cys100. ERO1β mutants lacking these residues or changing cysteine to alanine (C100A/C130A) also increases ERO1β activity, resulting in the hyperoxidation of substrates, protein misfolding, and UPR activation [76]. Indeed, ERO1β was initially characterized as inducible during treatments to specifically elicit the UPR. Additionally, ERO1α is upregulated later in the UPR downstream of CHOP, potentially contributing to apoptosis signaling [227].

The H_2O_2 produced by ERO1 may be cleared by GPx8, which forms a complex with ERO1 (Fig. 17.1). This peroxidase activity is induced by ER stress to protect against ERO1-mediated hyperoxidation [193]. In contrast, PRxIV is not induced in response to ER stress [228]. ERO1 and PDI may operate at increased levels in a futile attempt to fold misfolded protein substrates during high ER load, resulting in excess ROS production [199]. Unresolved ROS production and ER stress due to ERO1α activity is a factor contributing to UPR signaling, including increased expression of BiP [75].

GPx7 (also called NPGPx) acts as a ROS sensor, transmitting oxidative stress signals through thiol redox. ROS promote the oxidized form of GPx7, which has a disulfide bond between Cys57 and Cys86 [192, 249]. Cys86 then binds to Cys41 or Cys420 of BiP, promoting the disulfide bond formation between Cys41 and Cys420. This bond enhances the interaction of BiP with misfolded proteins. Cells lacking GPx7 have impaired BiP activity and are sensitive to oxidative stress [249].

Thiol redox alterations at the specific UPR sensors can affect their activity, altering downstream events. For example, sulfenylation of IRE1 by ROS at Cys715, which is in the kinase activation loop, attenuates canonical UPR signaling by impairing IRE1 kinase activity but promotes the antioxidant response [91].

Members of the PDI family have been shown to directly interact with the ER luminal domains of specific UPR sensors (Fig. 17.1), affecting their activity level and duration of signaling [57]. The PDI family member PDIA6 interacts with Cys148 of IRE1α, preventing Cys148 from forming an interchain disulfide bond during IRE1α oligomerization. This limits IRE1α activation and helps return it to the inactive monomeric form [58]. The depletion of PDIA6 prolongs and increases the amplitude of both IRE1 and PERK signaling in mammalian cells [57, 58]. Additionally, PDIA5 activates ATF6α, reducing the intermolecular disulfide bonds that retain it at the ER and allowing it to be transported to the Golgi for activation after dissociation from BiP [85]. Upregulation of PDIA5 and PDIA6 promotes cell survival, possibly though this redox regulation that both promotes adaptive signaling through ATF6 and inhibits sustained apoptotic signaling of IRE1 and PERK [58]. Additionally, oxidized PDIA1 activates PERK signaling during ER stress in colon carcinoma cells [121].

17.3.7 Alternative Pathways in the ER for Thiol Oxidation Leading to UPR Activation

There are also alternative minor redox pathways involving molecules that may re-oxidize the protein-folding ER enzymes by assisting in the transfer of electrons from reduced thiols to O_2. Dehydroascorbate, the oxidized form of ascorbate (vitamin C), is transported into the ER via facilitated diffusion in species that are unable to synthesize it [176]. It can accept electrons from PDI and other protein thiols, supporting non-enzymatic oxidative disulfide bond formation while itself reducing to ascorbate [10, 94, 176, 252]. Ascorbate can then act at the ER membrane to regenerate tocopherol (vitamin E) from its radical form, allowing it to scavenge radicals and other ROS to prevent lipid peroxidation and ROS accumulation [44]. Also, vitamin K-dependent proteins in the ER require reduced vitamin K as a cofactor during γ-carboxylation. After its oxidation during this process to vitamin K epoxide, reduced vitamin K can be regenerated by accepting electrons from PDI via vitamin K epoxide reductase (VKOR) [94, 243].

Dysregulation of these alternative pathways can lead to UPR signaling. Elevated levels of dehydroascorbate upregulate UPR signaling through BiP, CHOP, and XBP1s in neuroblastoma cells [231]. Treatment of liver cancer cells with menadione, a vitamin K precursor and oxidizing agent, when combined with NADPH depletion, can trigger UPR pro-apoptotic signaling through induction of CHOP and activation of ATF6 and procaspase-4. Interestingly, this signaling does not induce apoptosis via the activation of the effector caspases but instead promotes autophagy [225].

17.3.8 Altered ER Redox Homeostasis and Subsequent UPR Activation Lead to PH

ER protein folding is easily disrupted because of the complexity of the process, which involves a careful balance of folding chaperones, calcium ions, and redox signaling molecules including ROS, glutathione, and FAD; they work together to help fold protein substrates into their final conformations and introduce post-translational modifications such as N-linked glycosylations and the disulfide bonds discussed above [204]. Disulfide bond formation demonstrates how altered thiol redox states of protein-folding chaperones and their substrates can activate specific signaling pathways of the UPR or alter the length or strength of signaling, ultimately affecting cell function and fate. Thiol redox dysfunction within the ER, along with additional redox and antioxidant mechanisms generating ROS and inflammation both upstream and downstream of the UPR sensors, shifts the cell toward either adaptive, potentially dysfunctional, survival; or apoptosis, both of which may contribute to the development or attenuation of PH. For example, ROS including superoxide and H_2O_2 stimulate the proliferation of fetal PASMCs, but antioxidants such as ascorbate slow cell proliferation and even promote PASMC apoptosis [248].

17.4 Hypoxia-Generated ROS Modulates UPR Signaling and PH Development

17.4.1 ROS-Induced UPR Activation Induces Gene Expression to Modulate ROS

The loss of redox homeostasis in the cell promotes oxidative stress, which contributes to the pathogenesis of a vast number of diseases [251]. ROS are generated at very low levels as a by-product of normal oxidative protein folding in the ER, catalyzed by PDI and ERO1α [270]. Increased protein-folding load may cause an accumulation of ROS, which may, in turn, disrupt disulfide bond formation in the ER and interfere with proper protein folding, causing a buildup in misfolded and unfolded proteins in the ER lumen leading to ER stress and UPR activation [188, 204]. ROS-induced ER stress activates the PERK-eIF2α-ATF4 axis, which suppresses global cellular translation and, among other functions, promotes the expression of several antioxidant genes that modulate ROS [17, 45, 80]. The IRE1α axis is also activated in response to ROS-induced ER stress to splice XBP1 mRNA, allowing the translated XBP1 protein to increase the expression of proteins involved in disulfide bond formation, which attenuates excess ROS generation [58, 198, 267]. The ATF6 axis is also activated in response to ROS-induced ER stress, but it does not appear to be directly involved in modulating ROS.

17.4.2 Hypoxia-Inducible Factor Plays a Key Role in the Cellular Response to Hypoxia

Mammals express three isoforms of the transcription factor subunit hypoxia-inducible factor α (HIFα): HIF1α, HIF2α, and HIF3α. HIF1α is ubiquitously expressed in all cells, while HIF2α and HIF3α are expressed in certain cell types such as vascular endothelial cells, renal interstitial cells, liver parenchymal cells, and type II pneumocytes [20, 146].

Under normoxic conditions, the HIF1α subunit is produced by the cell and marked for degradation by O_2-sensitive proline hydroxylases (PHD). This post-translational regulation is achieved through the hydroxylation of proline residues at the O_2-dependent domain of HIFα [19, 146]. PHD-modified HIFα is then recognized by Von Hippel-Lindau (VHL) E3 ubiquitin ligase, resulting in HIFα ubiquitination and subsequent 26S proteasomal degradation to minimize HIFα half-life in the cell [19, 98].

During hypoxia, decreased O_2 levels inhibit PHD, preventing PHD modification of HIFα and allowing HIFα to escape degradation and accumulate in the cell. HIFα then forms a heterodimer with the constitutively produced HIFβ subunit, yielding the functionally active HIFα/β transcription factor [13, 98]. HIF binds to hypoxia response elements (HREs) in the promoters of key genes,

modulating their transcription and translation to result in a cell-wide adaptive response to hypoxia.

17.4.2.1 HIF1α Promotes Redox Homeostasis Through a Metabolic Shift from Oxidative Phosphorylation to Non-oxidative Glycolysis

HIF1α reprograms cellular metabolism by shifting the flow of carbon atoms away from oxidative phosphorylation and toward non-oxidative glycolysis. This metabolic shift lessens O_2 dependence during ATP generation, allowing for minimal ROS production [72]. This shift is achieved through HIF1α-mediated upregulation of glucose transporters and glycolytic enzymes [72]. HIF1α also upregulates pyruvate dehydrogenase kinase 1, which represses the conversion of pyruvate to acetyl-CoA, attenuating the citric acid cycle and by extension oxidative phosphorylation [109]. Moreover, HIF1α upregulates lactate dehydrogenase A, which converts pyruvate to lactate and concomitantly converts NADH to NAD^+, which is required to sustain glycolysis [72].

17.4.2.2 HIF2α Promotes Redox Homeostasis Through the Upregulation of Key Antioxidant Genes

HIF2α contributes to redox homeostasis under hypoxic conditions through the targeted upregulation of key enzymes such as pyruvate dehydrogenase kinase 4, which inhibits mitochondrial utilization of glucose-derived carbon to attenuate oxidative phosphorylation [92]. HIF2α upregulates the expression of mitochondrial superoxide dismutase 2 (SOD2) and the antioxidant heme oxygenase 1 (HO-1) [20, 72]. HIF1α and HIF2α promote increased electron transfer efficiency from cytochrome c oxidase to O_2, and by extension increase the efficiency of the electron transport chain under hypoxia [72].

17.4.3 Hypoxia-Induced UPR Activation

Despite the mitigating role played by HIF, hypoxia can result in the deregulation of many mechanisms critical to cellular and ER homeostasis. This can cause unfolded proteins to accumulate in the ER lumen, resulting in ER stress [29]. Hypoxia-induced ER stress activates the PERK, ATF6, and IRE1α UPR sensors [128]. Furthermore, hypoxia induces HIF1α-independent transcription and splicing of XBP1 mRNA [128]. XBP1s, once translated, creates a transcriptional complex with HIF1α and recruits RNA polymerase II. The assembled XBP1s-HIF1α complex regulates the HIF1α transcriptional program and eventually the expression of HIF1α target genes, which ultimately augments and sustains HIF activity and the adaptive response to hypoxia [37].

17.4.3.1 Hypoxia- and ROS-Induced UPR^{mt}

Hypoxia-induced ROS production in mitochondria may interfere with mitochondrial protein folding, causing the accumulation of misfolded or unfolded mitochondrial proteins. This leads to further ROS imbalance and mitochondrial stress, resulting in the UPR^{mt} [70, 125, 136, 161, 174, 177, 211]. The major sensor of mitochondrial stress that triggers the UPR^{mt} is ATFS-1, which under normal conditions is trafficked from the nucleus to the mitochondrial matrix where it is degraded. This import process is disrupted under mitochondrial stress, and ATFS-1 accumulates in the nucleus where it triggers the expression of genes associated with restoring mitochondrial homeostasis, including mitochondrial antioxidant, protease, chaperone, and import machinery [177]. Mitochondrial stress also promotes the phosphorylation of elF2α, inhibiting cellular translation while augmenting translation of the transcription factors CHOP, ATF4, and ATF5 [13, 256]. All three transcription factors upregulate the expression of UPR^{mt}-associated genes. Additionally, CHOP and ATF4 further induce transcription of ATF5 [8, 62, 125, 161, 174, 191].

17.4.4 The UPR Differentially Employs NOX2 and NOX4 to Mediate Pro-Apoptotic and Pro-Survival Responses with Implications in PH Pathogenesis

Hypoxia and other ER stressors [60, 128] may increase ROS generation through several sources in the cell, including NADPH oxidases (NOXs) [132, 185]. Elevated ROS can activate the UPR to promote both pro-oxidant and antioxidant mechanisms described in detail later in this chapter [181, 205].

UPR Pro-Oxidant Response ROS-activated UPR signaling may mediate a pro-oxidant response that may either confer cell death or survival [80, 82, 181]. A UPR pro-oxidant response may involve a host of ROS sources, including, for example, NOX2 and NOX4. NOX2 is subject to induction through the CHOP-ERO1α-inositol 1,4,5-triphosphate receptor (IP3R)-calcium/calmodulin-dependent kinase II (CaMKII)-NOX2 signaling axis or through JNK-mediated upregulation [131]. NOX2 may contribute to mitochondrial-driven pro-apoptotic signaling pathways, resulting in cell death [82, 131–133, 150, 205]. On the other hand, a UPR pro-oxidant response may also induce NOX4 through the IRE1α-JNK signaling pathway [185]. NOX4 activity is associated with H_2O_2 generation in addition to superoxide, which mediates adaptive signaling and cell survival.

UPR Antioxidant Response Conversely, ROS-activated UPR signaling may promote an antioxidant response through, for example, the upregulation of NRF2. NRF2 attenuates ROS levels through negative feedback, contributing to redox and cellular homeostasis [97, 229].

17.4.4.1 NOX2 and NOX4 Signaling in Pulmonary Vascular Endothelial Cells

Chronic exposure to hypoxia may result in endothelial injury and dysfunction, which is the initial step in PH development and progression.

Prominent mechanisms contributing to endothelial injury include the upregulation of ROS-producing NOX2 and NOX4 to the detriment of the underlying smooth and adventitial cell layers (Fig. 17.2) [2].

NOX2-Mediated Endothelial Cell Apoptosis in HIV-Induced Pulmonary Arterial Hypertension

During the early phases of the HIV-induced pulmonary arterial hypertension (HIV-PAH) disease model, human pulmonary microvascular endothelial cells (HPMECs) upregulate NOX2, which is native to the plasma membrane. NOX2 upregulation is associated with induced augmentation of Ras-Raf-ERK1/2 signaling, resulting in the disruption of endothelial tight junctions to elevate endothelial permeability and dysfunction and contribute to ROS-mediated apoptosis [2].

NOX4-Mediated Endothelial Cell Survival and Proliferation in HIV-PAH

As the disease progresses, HIV-PAH HPMECs upregulate NOX4, which is localized to

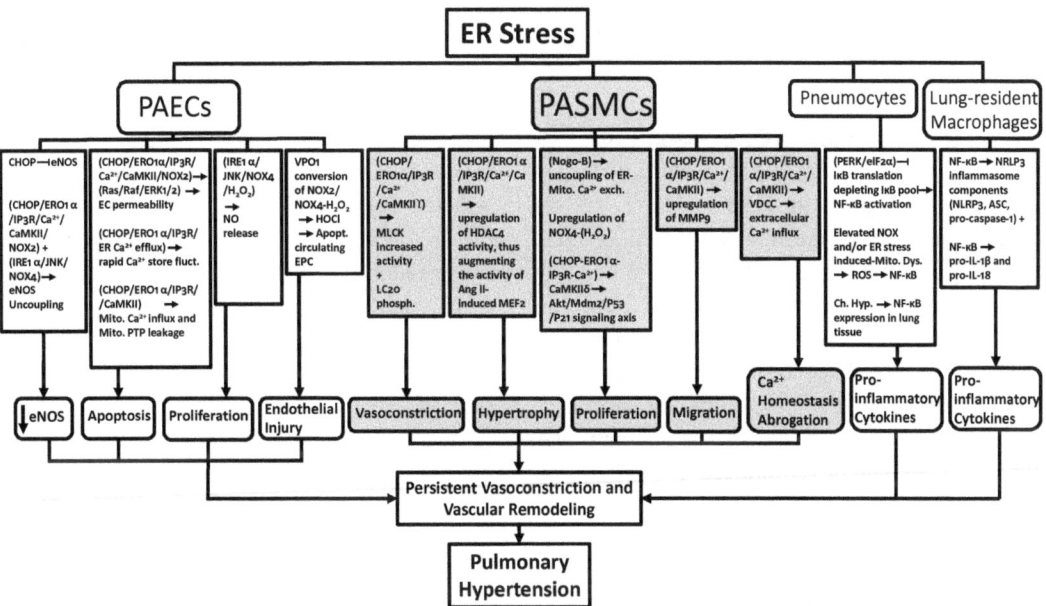

Fig. 17.2 Potential roles for redox-modulated UPR in PH pathogenesis. Increased endoplasmic reticulum (ER) stress initiates a cascade of signaling events in pulmonary arterial endothelial cells (PAECs), pulmonary artery smooth muscle cells (PASMCs), pneumocytes, and lung-resident macrophages. In PAECs, redox-modulated unfolded protein response (UPR) signaling can result in decreased nitric oxide (NO) production, apoptosis, maladaptive proliferation, and injury. In PASMCs, redox-modulated UPR signaling cascades can result in vasoconstriction, hypertrophy, proliferation, migration, and increased cytosolic calcium. In pneumocytes and lung-resident macrophages, these signaling events may result in elevated production of pro-inflammatory cytokines. All together, these signaling pathways converge and culminate in persistent vasoconstriction and vascular remodeling, contributing to the pathogenesis of pulmonary hypertension (PH). Ang II angiotensin II, Apopt. apoptosis, ASC apoptosis-associated speck-like protein containing a CARD, CaMKII Ca2+/calmodulin-dependent protein kinase II, Ch. Hyp. chronic hypoxia, CHOP C/EBP-homologous protein, eNOS endothelial nitric oxide (NO) synthase, ERO1α endoplasmic reticulum oxidoreductin-1α, HDAC4 histone deacetylase 4, IL-1β/IL-18 interleukin-1β/18, IP3R inositol 1,4,5-trisphosphate receptor, IRE1α inositol-requiring protein 1α, JNK c-Jun N-terminal kinase, LC20 20-kDa regulatory light chain of myosin II, Mdm2 mouse double minute 2 homolog, MEF2 myocyte enhancer factor-2, Mito. mitochondria, MLCK myosin light-chain kinase, MMP9 matrix metallopeptidase 9, NOX2/4 NADPH oxidase-2/4, PTP permeability transition pore, IκB inhibitor of κB, NF-κB nuclear factor kappa-light-chain-enhancer of activated B cells, NLRP3 NOD-, LRR- and pyrin domain-containing protein 3, ROS reactive oxygen species, VDCC voltage-dependent calcium channel, VPO1 vascular peroxidase 1

intracellular membranes including the ER and mitochondria. NOX4 upregulation in the late stages of HIV-PAH is associated with ROS-mediated autophagy, which switches the cells to an apoptosis-resistant hyper-proliferative state conferring cell survival and protection. NOX4 upregulation also promotes vascular structural integrity due to the release of NO secondary to NOX4-produced H_2O_2 [2, 181, 205]. NOX4 upregulation eventually results in ROS-induced maladaptive vascular remodeling that has been observed in the monocrotaline- and chronic hypoxia-induced models of PAH, as indicated by the proliferation of adventitial cells and SMCs, elevated right ventricle systolic pressure (RVSP), and right ventricle hypertrophy [12].

17.4.4.2 NOX2 and NOX4 Signaling in Circulating Endothelial Progenitor Cells

Endothelial injuries are mended by either local endothelial cell (EC) replication or by circulating endothelial progenitor cells (EPCs). NOX2 and NOX4 upregulation is not confined to ECs upon exposure to pro-PH conditions such as hypoxia; it is extended to the EPCs that are normally tasked with endothelial repair and homeostasis [28, 247]. In a hypoxia-induced PH rat model, vascular peroxidase 1 (VPO1) mediates the conversion of NOX2- and NOX4-generated H_2O_2 to hypochlorous acid (HOCl), a stronger and more destructive oxidant than H_2O_2 [247]. This HOCl-induced oxidative stress may induce EPC apoptosis and dysfunction interfering with EPC-mediated endothelial homeostasis, further compounding endothelial injury and dysfunction in hypoxia-induced PH.

17.4.4.3 NOX2 and NOX4 Signaling in Pulmonary Vascular Smooth Muscle Cells

Hypoxia, through the activation of ATF6, upregulates the ER structural protein Nogo-B in PASMCs in vitro. Nogo-B promotes the uncoupling of ER-mitochondria calcium exchange, attenuating mitochondrial ROS-mediated pro-apoptotic pathways to confer cell survival and proliferation [224]. Additionally, PASMCs

showed elevated NOX4 expression in chronic hypoxia-induced PH animal models [166]. NOX4 knockdown augmented the IRE1α pro-apoptotic effectors JNK-ASK1 and attenuated PASMC proliferation in vitro [166, 185, 200]. Increased NOX4 expression was accompanied by increased ROS production. Interestingly, increased ROS generation was attenuated through GADD34 plasmid transfection, which interfered with UPR signal transduction (Fig. 17.2) [200]. Indeed, NOX4-generated ROS appears to be associated with promoting a pro-survival, and in the case of PH-PASMCs, maladaptive, UPR response, the mechanism of which is not yet clear.

Furthermore, chronic hypoxia may deplete the cartilage oligomeric matrix protein (COMP), which may result in the loss of the bone morphogenic protein type 2 receptor (BMPR2) in isolated bovine pulmonary arteries (BPAs) [268]. Depleted COMP and BMPR2 loss are associated with NOX2 and NOX4 upregulation that contributes to redox imbalance in isolated BPAs [268]. UPR signaling, BMPR2 loss, and other signaling pathways appear to converge at the NOX2 and NOX4 level to drive pro-PH signaling mechanisms in the pulmonary vasculature.

17.5 UPR Signaling Induces Pro-PH Calcium Signaling

17.5.1 Redox Modulation of Calcium Release from the Stressed ER

The ER is the major site of calcium storage in the cell, and ER calcium stores play a significant role in ER stress-response mechanisms. Most ER calcium ions are bound to foldases, chaperones, and other luminal proteins and may aid in protein folding [204]. ER stress leading to prolonged UPR signaling through CHOP induces ERO1α-mediated hyper-oxidation of the ER lumen and may promote IP3R-mediated calcium efflux to the cytosol [86, 150]. This altered ER redox state contributes to rapid fluctuations in calcium stores that may interfere with calcium-mediated protein folding [253], inhibit normal chaperone function, disrupt protein folding, alter protein conformation, and affect protein-protein interactions in the ER [270]. This promotes further UPR signaling leading to apoptosis in ECs [52, 103, 111, 131, 144, 240, 277] and proliferation in SMCs (Fig. 17.2).

17.5.1.1 ER Stress-Induced Calcium Signaling and ROS Generation Mediate Apoptosis in Pulmonary Vascular ECs

Calcium released from the stressed ER through the CHOP-ERO1α-IP3R signaling pathway as described above [86, 150] binds to CaMKII, which then forms a node for several pro-apoptotic signaling mechanisms [52, 63, 126, 149, 240, 277]. One example is CaMKII-mediated mitochondrial calcium influx, which abolishes the mitochondrial inner membrane potential, promotes mitochondrial ROS generation, and increases mitochondrial permeability through the opening of mitochondrial permeability transition pores (PTPs) [52]. Open PTPs allow leakage of pro-apoptotic mediators such as cytochrome c into the cytosol [52, 73, 124, 232]. CaMKII activity is sustained through autophosphorylation and by calcium-CaMKII-independent ROS-mediated oxidation of Met281/282 [59]. Mitochondrial dysfunction-induced ROS can be viewed as part of a positive feedback loop amplifying CaMKII-mediated apoptosis triggered by the stressed ER [232]. In ECs, this aberrant calcium flux-induced apoptosis may result in vascular injury and arteriolar remodeling, contributing to the progression of PH (Fig. 17.2).

17.5.1.2 ER Stress-Induced Calcium Signaling Mediates Remodeling in Pulmonary Vascular SMCs

Contrary to its pro-apoptotic role in ECs, CHOP-ERO1α-IP3R-calcium-activated CaMKII may assume a pro-remodeling role in PASMCs. CaMKII promotes extracellular calcium influx into the cytosol through its modulation of the β3 subunit of voltage-dependent calcium channels [189]. CaMKIIδ also stimulates Akt activation [130] which results in high mouse double minute 2 homolog (Mdm2) phosphorylation at the

Akt-specific Ser166 site [273]. This process results in p53 degradation and decreased p21 expression to prevent apoptosis, ultimately conferring vascular SMC proliferation [134, 142]. Moreover, CaMKII may contribute to increased vasoconstriction through the activation of myosin light chain kinase (MLCK) and subsequent phosphorylation of 20-kDa regulatory light chain of myosin II (LC20) [108]. CaMKII also modulates histone deacetylase 4 (HDAC4) activity, contributing to myocyte enhancer factor 2 (MEF2) activation by angiotensin II to promote SMC hypertrophy [133]. Furthermore, CaMKII regulates matrix metalloproteinase 9 (MMP9) expression, which aids in ECM degradation that may allow for PASMC migration [71, 206, 271]. Taken together, ER stress-activated CaMKII assumes divergent roles in ECs and PASMCs, both of which ultimately result in pulmonary arteriolar remodeling contributing to the pathogenesis of PH (Fig. 17.2).

17.6 UPR and ROS Signaling Modulate eNOS, Contributing to PH Pathogenesis

Decreased nitric oxide (NO) in pulmonary arterial ECs promotes endothelial dysfunction, causing persistent vasoconstriction and PASMC proliferation [24, 188]. This increases pulmonary vascular resistance and contributes to PH pathogenesis. NO bioavailability in ECs is regulated by endothelial NO synthase (eNOS) activity, expression, and uncoupling [7, 48, 188, 242].

17.6.1 Calcium-Dependent Regulation of eNOS May Contribute to PH Pathogenesis

eNOS activity is regulated by several factors, including CaMKII. Activated CaMKII mediates Ser1177 phosphorylation in eNOS, increasing NO release [116]. Additionally, calcium-bound calmodulin binds to its specific domain in eNOS, aligning the eNOS oxygenase and reductase domains and inhibiting protein kinase C (PKC)-mediated phosphorylation of Thr495, both of which promote NO synthesis [116]. Additionally, UPR signaling and ROS generation may negatively modulate eNOS through several mechanisms.

17.6.2 CHOP Inhibits eNOS Expression, Contributing to PH

The ER stress-induced PERK effector CHOP regulates cellular responses including immune and inflammatory responses, cellular differentiation and proliferation, and apoptosis [140]. Under normal physiological conditions, CHOP is minimally expressed, but transcriptional upregulation during ER stress mediates ischemia and hypoxia-induced apoptosis [259]. CHOP inhibits eNOS transcription in ECs through the postulated binding to an optimal CHOP-responsive element at the eNOS promoter [140], directly inhibiting eNOS transcription to confer a wide range of antiangiogenic effects. Inhibited eNOS expression interferes with normal endothelial cell growth, migration, vasodilation, and bone marrow-derived cell-related functions [140]. ER stress-induced CHOP inhibition of eNOS expression underscores its role in modulating postnatal vessel formation and maturation [140] and potential contribution to PH pathogenesis.

17.6.3 ROS Induces eNOS Uncoupling in Endothelia During Chronic Hypoxia and Reoxygenation Injury

In addition to CHOP, chronic exposure to hypoxia and hypoxia/ischemia and reoxygenation (H/R) injury interfere with eNOS function in ECs [48, 140, 259]. Chronic hypoxia and H/R injury elevate oxidative stress in ECs through NOX2 and NOX4, mitochondrial ROS leakage, and ischemia-induced conversion of xanthine dehydrogenase to xanthine oxidase. This conversion

hinders the FAD-binding site and allows O_2 to act as an alternative electron acceptor to NAD^+, resulting in superoxide generation [48].

Chronic hypoxia and H/R-mediated ROS generation also result in eNOS uncoupling in which eNOS generates superoxide instead of NO causing endothelial dysfunction and contributing to a variety of cardiovascular diseases [48, 65]. eNOS uncoupling is mediated by ROS through two mechanisms. One is ROS-mediated oxidation of the eNOS cofactor tetrahydrobiopterin (BH_4) to dihydrobiopterin (BH_2), resulting in both forms of biopterin competing for a single eNOS-binding site. The second mechanism is ROS-mediated S-glutathionylation of eNOS, in which elevated ROS disrupts the ratio of reduced to oxidized glutathione in the cytosol. This shift toward GSSG promotes disulfide exchange between GSSG and Cys689 and Cys908 in the eNOS reductase domain, resulting in eNOS S-glutathionylation and uncoupling [48].

In the chronic hypoxia-induced pulmonary hypertension animal model, it has been shown that eNOS is subject to uncoupling [7, 48, 65]. Moreover, eNOS uncoupling exacerbates and further compounds existing oxidative stress in endothelial cells through its aberrant generation of superoxide [65]. Decreased NO bioavailability and increased ROS production may ultimately result in persistent vasoconstriction and SMC proliferation in small pulmonary arteries and arterioles, increasing pulmonary vascular resistance and ultimately contributing to PH pathogenesis [48, 247].

17.7 ROS and NF-κB Signaling Regulate Inflammatory Processes Leading to PH

17.7.1 NF-κB Mediates ER-Based Signals and Is Subject to ROS-Induced Activation

NF-κB is a redox-sensitive transcription factor that regulates the expression of a multitude of genes implicated in inflammatory and immune responses, cell proliferation, and tumorigenesis [81, 245, 260]. NF-κB can be activated through

IκB kinase (IKK)-mediated phosphorylation and subsequent disassociation and degradation of the inhibitory IκB subunit in a polyubiquitin-dependent manner. Active NF-κB subsequently dimerizes and undergoes nuclear translocation [162, 183]. NF-κB can be activated by elevated ROS, which may occur through NOX signaling as well as ER stress induced-mitochondrial dysfunction [183]. NF-κB also promotes increased ROS production [170], enabling a feed-forward loop. Importantly, ER stress and UPR signaling activate NF-κB [181]. PERK phosphorylation of elF2α inhibits IκB translation, depleting the IκB pool to promote NF-κB activation [258].

Prolonged exposure to hypoxia upregulates the expression of NF-κB in lung tissue [138]. In the nucleus of lung cells, NF-κB regulates the expression of several pro-inflammatory mediator-encoding genes, contributing to pulmonary vascular remodeling and PAH disease state [207, 270]. Moreover, NF-κB regulates inflammatory processes through inflammasome activation [74, 157].

17.7.2 NF-κB Is Critical for Inflammasome Activation to Regulate Inflammation Leading to PH

In response to intra- and extracellular stress signals, a number of multi-protein complexes termed inflammasomes mediate macrophage-driven immune responses that culminate in the production of the pro-inflammatory cytokines IL-1β and IL-18, contributing to the regulation of inflammatory processes [122, 155]. The nucleotide-binding oligomerization domain (NOD)-like receptor (NLR) family, leucine-rich repeat (LRR), and pyrin domain (PYD)-containing protein-3 (NLRP3) inflammasome has been implicated in the progression of several diseases, including PH [123]. NLRP3 activation requires two steps, priming and activation initiation, which attests to the highly regulated nature of the inflammasome response in macrophages.

NF-κB is critical for the priming phase, which occurs in response to extracellular pathogen-associated molecular patterns (PAMPs) recognition by cell surface toll-like receptors (TLRs) and pro-inflammatory cytokine receptors on antigen-presenting cells [56, 66, 96, 148] and/or in response to intracellular damage-associated molecular patterns (DAMPs) such as tumor necrosis factor α (TNF-α) or monosodium urate and calcium pyrophosphate dihydrate crystals [41, 156]. NF-κB mediates upregulation of the NRLP3 inflammasome protein components NLRP3, an apoptosis-associated speck-like protein containing a CARD (ASC) and pro-caspase-1 as well as upregulation of pro-IL-1β and pro-IL-18 [16, 123].

Elevated ROS mediates NLRP3 inflammasome activation at both phases: indirectly during priming, through ROS-induced NF-κB activation [16], and directly during activation initiation, through ROS-induced activation of NLRP3 [241, 275, 276]. Interestingly, all known NLRP3 activators induce ROS production [56, 157, 241, 275]. ROS-induced TRx interacting protein (TXNIP) dissociation from the redox-sensitive domain of TRx allows free TXNIP to bind and activate NLRP3 [275]. In the macrophage-like cell line THP1, NLRP3 activators MSU and R-837 generate ROS and mediate TXNIP dissociation from TRX, which can be prevented by the ROS inhibitor APDC [56]. Elevated H_2O_2 mediates the association of TXNIP with NRLP3, further affirming the role of ROS in NLRP3 inflammasome activation [56, 275].

Upon activation, NLRP3 oligomerizes, then recruits, and activates pro-caspase-1 through ASC [148, 202]. Activated caspase-1 cleaves pro-IL-1β and pro-IL-18 [54], activating them to promote the release of several downstream cytokines that propagate the inflammasome-mediated inflammatory response [15].

17.7.3 NLRP3 Itself Promotes NF-κB Activation Through IL-1β in a Feed-Forward Manner

IL-1β is a pro-inflammatory cytokine activated by the NLRP3 inflammasome that elicits an expansive array of secondary cytokines to transduce inflammatory signaling [239]. IL-1β mediates the feed-forward activation of NF-κB through two known pathways [154, 254]. First, through the activation of the phosphoinositide-3 kinase (PI3K)-Akt/protein kinase B (PKB) signaling axis, IKKα is activated to dissociate IκB from NF-κB as described earlier [25, 154, 254]. Second, IL-1β associates with TNFα and transforming growth factor β (TGFβ) to induce NF-κB inducing kinase (NIK), which in turn activates IKKα and NF-κB.

17.7.4 ROS-Induced Hyaluronan Fragmentation Plays a Major Role in Inflammasome Activation in Chronic Hypoxia-Induced PH

As mentioned, all known NLRP3 inflammasome activators generate ROS, which mediates NLRP3 activation in chronic hypoxia-induced PH [56, 241]. Exposure to hypoxia elevates ROS production, rendering the pulmonary vasculature, including the ECM in the outer adventitial layer, susceptible to oxidative stress [99, 137, 241]. Elevated extracellular ROS may promote the degradation of critical components of ECM such as glycosaminoglycan (GAG) and hyaluronan (HA). The degradation and biosynthesis of HA are implicated in rapid matrix remodeling during processes such as inflammation and tumorigenesis [26, 27, 164]. Chronic hypoxia-mediated ROS generation may cause HA fragmentation, generating biologically active HA fragments that form ligands for the macrophage cell surface receptors CD44, the receptor for HA-mediated motility (RHAMM), and TLR4 [167, 216, 226, 241]. Ligand-activated RHAMM induces macrophage recruitment [165, 241] and TLR4 promotes NF-κB activation and the eventual upregulation of pro-inflammatory cytokines [107, 165]. Interestingly, extracellular ROS-induced HA fragments bound to CD44 receptors trigger NLRP3 inflammasome activation to play a major role in chronic hypoxia-induced PH. In a chronic hypoxia-induced PH model, adminis-

tration of the superoxide dismutase (SOD) mimetic MnTE-2PyP to reduce ROS production decreases NLRP3 activation, reducing vascular remodeling and attenuating PH development [241].

17.7.5 The Inflammasome Activates Pro-Inflammatory Cytokines Evident in PH

IL-6 is a pro-inflammatory cytokine that can be induced by several upstream pro-inflammatory mediators including IL-1β, TNFα, and TGFβ [25]. IL-1β promotes IL-6 transcription through the PI3K-Akt/PKB signaling axis and activator protein 1 (AP-1) [3, 40, 47, 152, 195], as well as the activation of IKKα-NF-κB and AP-1 to promote further IL-6 expression [25]. Long-term upregulation of IL-6 has been observed during PH [261].

CD4+CD25+ regulatory T cells (Tregs) impart anti-inflammatory effects in PAH and other diseases. Tregs suppress the expression of the pro-inflammatory cytokines IL-1β, IL-6, and monocyte chemoattractant protein 1 (MCP-1) while upregulating the expression of IL-10 in chronic hypoxia-induced PH [42].

17.8 Redox Signaling Via the UPR Sensors Drives PH Progression or Attenuation

17.8.1 UPR Signaling Modulates Oxidative Stress Through NRF2 Activation

NRF2 is an anti-inflammatory and antioxidant transcription factor that is targeted for degradation under non-stress conditions while bound to the ubiquitin ligase adaptor KEAP1 in the cytoplasm [188, 229]. Cellular oxidative stress promotes PERK phosphorylation of this complex to disassociate NRF2 and allow it to translocate to the nucleus [178], where it heterodimerizes to other bZIP proteins at the antioxidant response element (ARE), a specific cis-acting regulatory element located in the promoter region of several antioxidant and anti-inflammatory genes, to upregulate transcription of target genes [145, 188, 229] including as glutathione S-transferase (GST) A1 and A2 subunits, heme oxygenase 1 (HO-1), sulfiredoxin-1 (SRXN-1), glutamate-cysteine ligase catalytic subunit (GCLC), UDP glucuronosyl transferase (UGT), and NAD(P)H: quinone oxidoreductase 1 (NQO1) [127, 159, 269]. NRF2 is the most important protein for inducing ARE-mediated transcription to counteract oxidative stress and activated inflammatory pathways in the cell (Fig. 17.3) [127, 178].

17.8.1.1 NRF2 Balances Activation and Repression of Antioxidant Gene Expression

Upon heterodimerization with other bZIP proteins, NRF2 and its binding partner form a transcriptional activator or repressor. Transcriptional activators are formed with c-Jun, Jun-B, Jun-D, and ATF4 [178]. Activator heterodimers may compete with transcriptional repressor protein complexes such NRF2-MafK or Maf homodimers for binding to the ARE, exerting transcriptional regulation over antioxidant gene expression [105, 106, 172, 178].

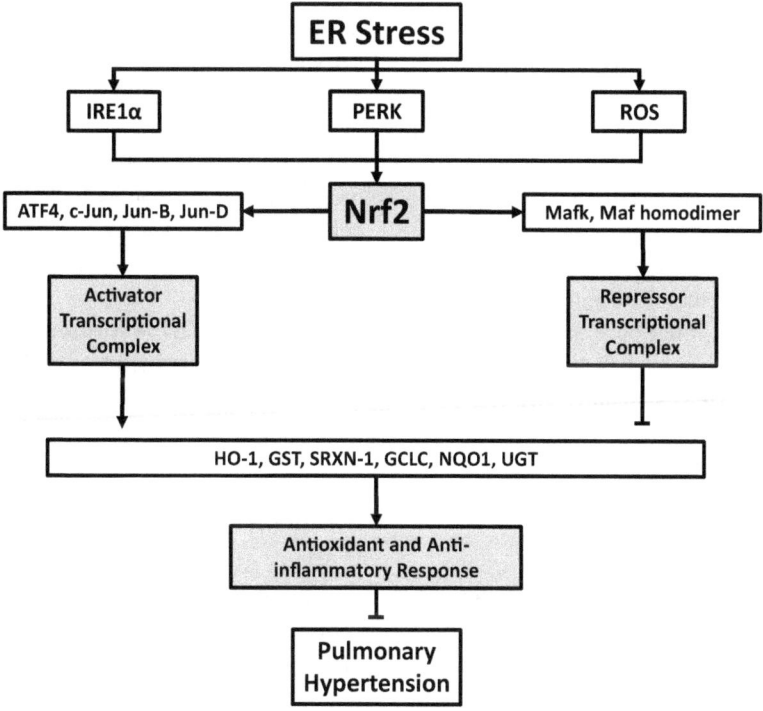

Fig. 17.3 ER stress potentially induces an antioxidant and anti-inflammatory UPR that attenuates the development of PH. Endoplasmic reticulum (ER) stress initiates inositol-requiring protein 1α (IRE1α)- and protein kinase RNA-like endoplasmic reticulum kinase (PERK)-mediated induction of nuclear factor erythroid 2-related factor 2 (NRF2), which can also be activated through elevated cytoplasmic reactive oxygen species (ROS) secondary to ER stress. Activated NRF2 undergoes nuclear translocation where it may form an activator transcriptional complex that binds to the antioxidant response element (ARE) at which it upregulates the transcription of antioxidant and anti-inflammatory genes. This signaling cascade is modulated through the formation of NRF2 transcriptional complex. NRF2-driven antioxidant and anti-inflammatory gene transcriptional upregulation attenuates oxidant stress and triggered inflammatory processes triggered by the initial ER stress, mitigating the pathogenesis of pulmonary hypertension (PH). ATF4 activating transcription factor 4, GCLC glutamate-cysteine ligase catalytic subunit, GST glutathione S-transferase, HO-1 heme oxygenase-1, Maf musculoaponeurotic fibrosarcoma, MafK musculoaponeurotic fibrosarcoma K, NQO1 NAD(P)H: quinone oxidoreductase 1, SRXN-1 sulfiredoxin-1, UGT UDP glucuronosyl transferase

17.8.1.2 NRF2 Activation Is Redox-Mediated

ARE gene expression may additionally be subject to redox regulation. Altered redox status during ER stress may promote ROS-induced modification of cysteine residues on KEAP1, allowing NRF2 dissociation and participation in ARE gene transcription [97, 178]. ARE-expressed proteins such as glutathione-S-transferase and thioredoxin may participate in a negative feedback mechanism. Both proteins are reactive cysteine-based inhibitors of ASK, which activates the JNK and p38 mitogen-activated protein kinase (MAPK) signaling pathways to induce ARE gene expression [1, 39]. Taken together, thiol redox modulation plays a pivotal role in regulating ARE gene expression in response to cellular oxidative stress.

The NRF2 antioxidant response is also promoted through another mechanism mediated by IRE1α. Elevated cytosolic ROS sulfenylate Cys715 at the IRE1α kinase activation loop [91, 117] and attenuate IRE1α kinase activity and canonical UPR signaling. However, sulfenylation also initiates the NRF2 antioxidant response (Fig. 17.3) [91]. The UPR and antioxidant response functions of IRE1α are mutually exclusive, such that elevated ER stress triggers IRE1α UPR signaling, while cytosolic ROS-mediated cytoplasmic stress initiates the IRE1α antioxidant response [91].

17.8.2 IRE1α Regulates and Is Regulated by Innate Immune Pathways

The IRE1α-XBP1 axis contributes to innate immunity by mediating lipopolysaccharide (LPS)-triggered TLR4 signaling in macrophages [158]. Both TLR2 and TLR4 activate NOX2 through the adaptor protein TRAF6, elevating ROS generation to promote IRE1α activation and XBP1 splicing. XBP1s translocates to the nucleus and binds to the promoter regions of pro-inflammatory genes such as IL-6, TNF, and IFN-β, upregulating their transcription [158]. IRE1α may also promote ROS production through NOX and the mitochondria during ER stress [158]. This ROS-IRE1α-ROS feed-forward loop may exacerbate UPR-mediated inflammatory pathways in the cell, contributing to metabolic deterioration and eventually disease [270].

17.8.3 PERK Signaling Contributes to ROS Production Through CHOP

17.8.3.1 CHOP Downregulates BCL-2 Transcription Via GSH Depletion to Promote ROS

Prolonged UPR signaling can result in overexpression of the PERK effector CHOP, which can complex with other transcription factors at the BCL-2 promoter to suppress its transcription and attenuate BCL-2-mediated anti-apoptotic signaling [140, 217]. Increased CHOP levels are also associated with depletion of GSH while GSSG remains unchanged, shifting this ratio to disrupt the cellular redox state and elevating intracellular ROS [160]. The mechanism by which overexpressed CHOP or suppressed BCL-2 may deplete cellular GSH remains unclear, but this ultimately promotes ER stress-induced apoptosis [160].

17.8.3.2 GADD34 Induces ROS Generation and Misfolded Protein Overload

GADD34, which is upregulated by the CHOP-ATF4 signaling axis, activates a PP1C complex to dephosphorylate eIF2α. This attenuates the initial global translation inhibition imposed by PERK in the early stages of the UPR [23, 181]. Resuming translation contributes to increased cellular ROS along with misfolded and unfolded protein overload in the ER, while CHOP signaling promotes calcium efflux and apoptosis.

17.9 NRF2 Antioxidant and Anti-Inflammatory Signaling Attenuates PH

17.9.1 The Anti-Inflammatory Role of the NRF2-Induced Heme Degradation Pathway

As mentioned earlier, NRF2 promotes the expression of several antioxidant genes including HO-1, the inducible HO isoform that degrades free heme, a pro-oxidant, into equimolar amounts of biliverdin, carbon monoxide (CO), and ferrous iron. Ferritin is co-induced with HO-1 to sequester HO-1-generated ferrous iron [93, 104].

Biliverdin reductase converts heme-derived biliverdin to bilirubin. Both biliverdin and bilirubin have antioxidant properties and can interfere with inflammatory cascades by altering the expression of endothelial adhesion molecules and blocking leukocyte adhesion to endothelial cells in the vasculature [11, 139, 222].

HO-1-generated carbon monoxide (CO) confers anti-inflammatory properties by interfering with AP-1 binding to the IL-6 promoter in LPS-activated macrophages through JNK-mediated signaling [104]. In LPS-induced systemic inflammation, CO inhibits the expression of pro-inflammatory IL1-β, macrophage inflammatory protein 1β (MIP-1β), and TNFα. CO also increases the expression of the antioxidant macrophage cytokine IL-10 [163]. Moreover, CO exhibits anti-inflammatory properties during vascular injury [182, 215]. CO, biliverdin, and bilirubin activate MAPK signaling, inhibiting SMC proliferation [274].

17.9.2 HO-1 Activity Attenuates PH Progression

Chronic hypoxia increases the expression of pro-inflammatory cytokines and chemokines in the lungs within the first 2–5 days of exposure, increasing vascular permeability to target ECs, epithelial cells, and monocytes. This leads to leukocyte recruitment, further releasing proteolytic and pro-oxidant mediators promoting vasoconstriction [4, 223, 262]. Chronic hypoxia-induced pro-inflammatory cytokines may induce proliferation in pulmonary vascular smooth cells contributing to vascular remodeling, eventually leading to PH [163]. Moreover, increased HO-1 activity promotes Th-2 cytokine expression, as opposed to Th-1, suggesting a key role for HO-1 modulation of lymphocyte maturation [104]. Interestingly, the lungs of HO-1 overexpressing mice exposed to chronic hypoxia showed attenuated levels of pro-inflammatory cytokines such as monocyte chemoattractant protein (MCP)-1, IL-1β, IL-6, and macrophage inflammatory protein (MIP)-2 [104]. Furthermore, in a monocrotaline (MCT)-induced PAH model, the immunosuppressant agent rapamycin mediates its vascular SMC anti-proliferative effects through HO-1 induction [274].

17.9.3 NRF2 Activation Is Protective Against PAH

NRF2 activation through sulforaphane (SFN) attenuates pulmonary vascular inflammation, remodeling, and fibrosis as well as preventing right ventricle hypertrophy and fibrosis in SU5416- and chronic hypoxia-induced PAH animal models [102]. These effects are associated with NRF2-mediated upregulation of NQO1 and downregulation of NLRP3 [102]. TGF1β-mediated endothelial-mesenchymal transition (EndMT), which contributes to vascular remodeling in PAH, is partially driven by oxidative stress to promote TGFβ1 and TGFβ2 expression and secretion [169]. The NRF2 activator salvianolic acid A may attenuate EndMT and reduce oxidative stress in the pulmonary vasculature through the NRF2 and HO-1 signaling mechanism in PAH [38].

17.9.4 NRF2 Deregulation in PAH Pathogenesis

17.9.4.1 Xenobiotic-Induced NRF2 Deregulation

Abuse of the inhaled form of methamphetamine (MA) may contribute to PAH by preventing NRF2 nuclear translocation, contributing to oxidative stress. This correlates with MA-induced PASMC proliferation, mediated by downregulation of the pro-apoptotic mediators BAX and caspase-3 and upregulation of the anti-apoptotic mediators BCL-2 and proliferating cell nuclear antigen (PCNA) [135]. Taken together, excessive oxidative stress-induced NRF2 deregulation may contribute to pulmonary arterial remodeling in chronic MA-induced PAH.

17.9.4.2 Viral Infection-Induced NRF2 Deregulation

NRF2-ARE activity and target genes are repressed in HIV-PAH concomitant with elevated oxidative stress evident in increased ROS production in human primary arterial endothelial cells (HPAECs) [213]. NRF2 levels were not altered in these HPAECs, but NRF2-ARE-regulated genes were transcriptionally repressed, suggesting differential regulation of NRF2-ARE through other mechanisms. It is possible that the pro-oxidant and pro-inflammatory HIV transactivator of transcription (Tat) can bind to the enhancer element or to AP-1 sequences proximal to the ARE in the promoter region, subverting the transcriptional expression of ARE-regulated genes [213]. Small Maf proteins (sMaf), co-transcriptional factors that normally heterodimerize with NRF2 or with BTB and CNC homology 1 (Bach1), the transcriptional repressor of ARE-regulated genes, contribute to the transcriptional regulation of ARE-driven genes [106]. Tat may modulate sMaf heterodimerization with NRF2 as well as Bach1, mediating the transcriptional repression of ARE-driven genes. This repressed antioxidant gene expression contributes to the impairment of redox homeostasis and elevated oxidative stress in HIV-PAH HPAECs, resulting in endothelial dysfunction, imbalance of endothelial proliferation and apoptosis, and pulmonary arterial remodeling, contributing to HIV-PAH development and progression [213].

17.10 Misfolded BMPR2 Implicates UPR Signaling in Familial PAH Pathogenesis

17.10.1 BMPR2 Mutation or Reduced Expression in FPAH and PAH Pathogenesis

Familial pulmonary arterial hypertension (FPAH), caused by an inherited autosomal dominant mutation, accounts for 6% of PAH cases. A clear association has been established between a heterozygous mutation or reduced expression of a gene that encodes for BMPR2 and with FPAH and PAH, respectively [143]. Interestingly, attenuated expression of the BMPR2 gene is closely linked with vascular inflammation [201]. BMPR2 is part of the TGF-β receptor superfamily, which through SMAD signaling modulates many critical cellular processes including cell differentiation, proliferation, migration, and apoptosis, as well as secretion and deposition of the ECM. BMPR2 is composed of three domains: a ligand-binding domain, a kinase domain, and a cytoplasmic tail [143]. Upon BMPR2 mutation, highly conserved cysteine residues in the ligand-binding domain of BMPR2 are prone to alteration, causing the aberrant protein to be retained and possibly build up in the ER. This accumulation of mutant proteins in the ER may trigger the activation of UPR, implicating it in the pathogenesis of FPAH [265].

17.10.2 Attenuated BMPR2 Expression Promotes Inflammatory Cell Recruitment and Vascular Remodeling in PH

Reduced BMPR2 expression in HPAECs prolongs p-p38-MAPK signaling in response to TNF stimulation. Augmented p-p38 signaling activates the GADD34-PP1 complex, dephosphorylating eIF2α to disrupt stress granule formation. Translation resumes, amplifying the synthesis of granulocyte-macrophage colony-stimulating factor (GM-CSF), a powerful chemokine that stimulates stem cell production of granulocytes and macrophages [201]. GM-CSF also mediates macrophage polarization and increased production of the pro-inflammatory cytokines IL-6, IL-8, IL-12, TNF, and leukotriene B4 [64, 119]. Moreover, augmented GM-CSF mRNA translation in HPAECs may induce GM-CSFRα expressing HPAECs to express inflammatory cell adhesion molecules. Uninhibited GM-CSF mRNA translation in HPAECs may promote the recruitment of GM-CSFRα-expressing inflammatory cells to the pulmonary vasculature, mediating enhanced macrophage production of inflammatory cytokines and increased MMP activity, all of which contribute to vascular remodeling [201]. Altogether, attenuated BMPR2 expression mediates p-p38-dependent deactivation of eIF2α stress granule formation, allowing translation of GM-CSF mRNA in HPAECs to promote inflammatory cell recruitment, vascular remodeling, and ultimately the development of PAH.

17.11 Summary and Conclusion

Despite the mitigating role played by HIF, hypoxia can result in the deregulation of a number of mechanisms critical to cellular and ER homeostasis, resulting in ER stress and the onset of UPR. Elevated ER and mitochondrial ROS secondary to hypoxia trigger UPR and UPR^{mt}, respectively. ROS-activated UPR may mediate a pro-oxidant response that may either confer cell death through NOX2 as seen in EPCs in chronic hypoxia-PH and endothelial cells in early HIV-PAH or survival through NOX4 as observed in PASMCs in chronic hypoxia-PH and endothelial cells in late HIV-PAH. UPR-associated NOX4 may also mediate pro-proliferative signaling pathways in PASMCs. On the other hand, ROS-activated UPR may promote an antioxidant response through NRF2, for instance, contributing to redox and cellular homeostasis.

Prolonged ER stress may result in CHOP-induced ERO1α-mediated hyper-oxidation of the ER lumen resulting in IP3R-mediated ER calcium efflux. Rapid fluctuation of calcium stores may contribute to apoptosis. Elevated cytosolic calcium triggers CaMKII-mediated mitochondrial calcium influx contributing to apoptosis in ECs. On the other hand, calcium-triggered CaMKII can initiate an array of signaling cascades that culminate in proliferation and hypertrophy in PASMCs (Fig. 17.4).

Redox-sensitive NF-κB regulates the expression of a multitude of genes implicated in inflammatory/immune response, cell proliferation, and tumorigenesis. NF-κB mediates ER-based signals and is subject to ROS-induced activation. NF-κB promotes ROS production and regulates the expression of several pro-inflammatory mediator-encoding genes in lung cells, contributing to pulmonary vascular remodeling and PAH.

In response to intra-/extracellular stress signals, inflammasomes such as NLRP3 mediate macrophage-driven immune responses that culminate in the production of the pro-inflammatory cytokines IL-1β and IL-18, contributing to the regulation of inflammatory processes. NLRP3 has been implicated in the progression of several diseases, including PH. All known NLRP3 activators induce increased ROS production. Elevated ROS mediates NLRP3 inflammasome activation at both phases: indirectly during priming, through ROS-induced NF-κB activation and directly during activation initiation, through ROS-induced activation of NLRP3. NF-κB-activated NLRP3 promotes NF-κB activation through IL-1β in a feed-forward manner. IL-1β also promotes IL-6 transcription which if sustained may contribute to PH.

Chronic hypoxia-mediated ROS generation may cause HA fragmentation in pulmonary vascular adventitial ECM, generating HA fragment ligands that bind to the following macrophage receptors: RHAMM inducing macrophage recruitment, TLR4 promoting NF-κB activation and the eventual upregulation of pro-inflammatory cytokines, and CD44 triggering NLRP3 activation contributing to the development of chronic hypoxia-PH.

The anti-inflammatory and antioxidant transcription factor NRF2 is targeted for degradation under non-stress conditions. This process is prevented under cellular oxidative stress through PERK-mediated phosphorylation of Nrf2 ubiquitin ligase adaptor, allowing for NRF2 binding to ARE mediating the transcriptional upregulation of many antioxidant and anti-inflammatory genes such as glutathione S-transferase A1/A2 subunits, HO-1, ϒ-glutamyl cysteine synthetase, UDP glucuronosyl transferase, and NAD(P)H: quinone oxidoreductase. NRF2 antioxidant response is also promoted through another mechanism mediated by IRE1α, in which elevated cytosolic ROS sulfenylate Cys715 at the IRE1α kinase activation loop, initiating the NRF2 antioxidant response. HO-1, the inducible HO iso-

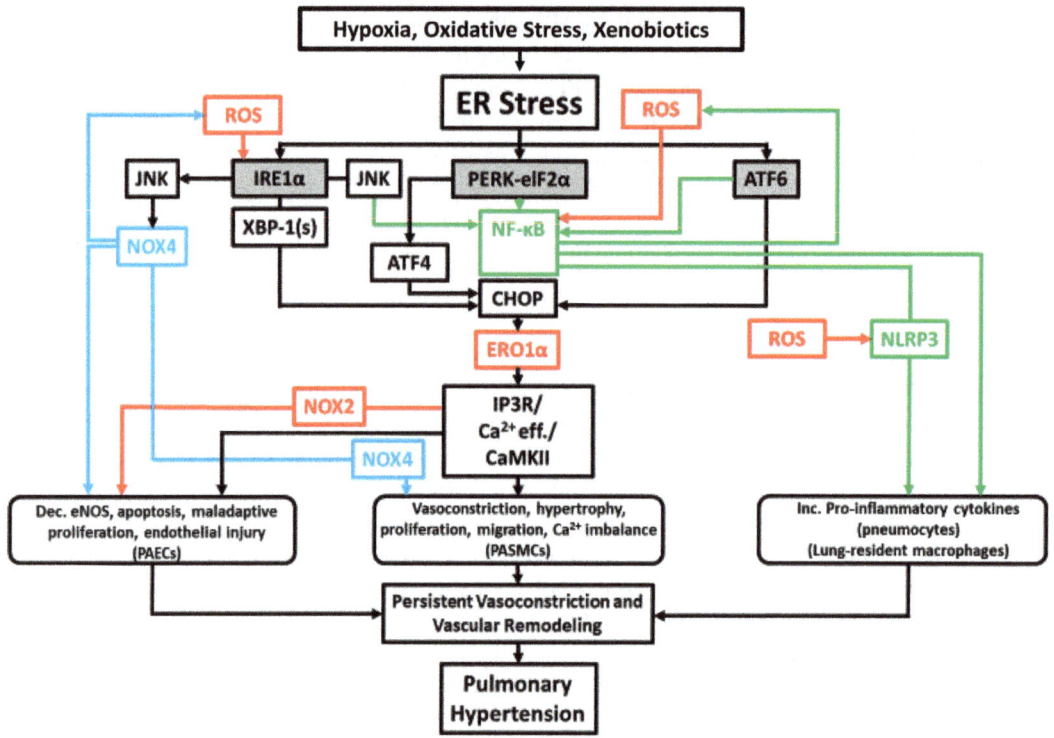

Fig. 17.4 Interplay among redox, inflammatory, and UPR signaling potentially contributes to PH pathogenesis. Hypoxia, oxidative stress, xenobiotics (chemicals and antigens), and other inducers of endoplasmic reticulum (ER) stress trigger the three arms of unfolded protein response (UPR), contributing to elevated reactive oxygen species (ROS) production through NADPH oxidases 2 and 4 (NOX2, NOX4) and the mitochondria, along with elevated cytosolic calcium and nuclear factor kappa-light-chain-enhancer of activated B cells (NF-κB)/nucleotide-binding oligomerization domain (NOD)-like receptor (NLR) family, leucine-rich repeat (LRR), and pyrin domain (PYD)-containing protein-3 (NLRP3)-mediated inflammasome activation. These signaling events result in decreased nitric oxide (NO) production, apoptosis, and maladaptive proliferation and injury in pulmonary arterial endothelial cells (PAECs), while potentially augmenting vasoconstriction, hypertrophy, proliferation, migration, and further cytosolic calcium imbalance in pulmonary artery smooth muscle cells (PASMCs). In pneumocytes and lung-resident macrophages, triggered UPR and elevated cytosolic calcium and ROS converge to activate NF-κB/NLRP3 which contribute to the upregulation of pro-inflammatory cytokines. These signaling events culminate to cause persistent vasoconstriction and vascular remodeling in pulmonary small arteries and arterioles, elevating pulmonary vascular resistance and contributing to the development of pulmonary hypertension (PH). Redox signaling mediators are represented in red and blue boxes, while inflammatory signaling mediators are shown in green boxes. ATF 4 and 6 activating transcription factor 4 and 6, CaMKII calcium-/calmodulin-dependent kinase II, CHOP C/EBP-homologous protein, ERO1α ER oxidoreductase 1α, IP3R inositol 1,4,5-triphosphate receptor, IRE1α inositol-requiring protein 1α, JNK JUN N-terminal kinase, PERK protein kinase RNA-like endoplasmic reticulum kinase, XBP1 X-box binding protein 1

form, is a key NRF2/ARE-driven antioxidant gene that degrades free heme, a pro-oxidant, into equimolar amounts of biliverdin, CO, and ferrous iron, all three of which have antioxidant and anti-inflammatory properties that attenuate the progression of PH. The NRF2 activator, salvianolic acid A, may attenuate EndMT and reduce oxidative stress in the pulmonary vasculature through the NRF2/HO-1 signaling mechanism in PAH. NRF2 mediates upregulation of NQO1 and downregulation of NLRP3, which are associated with attenuated pulmonary vascular inflammation, remodeling, and fibrosis in SU5416/chronic hypoxia-PAH. Furthermore, oxidative stress-induced NRF2 deregulation may contribute to pulmonary arterial remodeling in chronic

MA-induced PAH. NRF2/ARE activity and target genes are repressed in HIV-induced PAH concomitant with increased ROS production in HPAECs. This may result in endothelial dysfunction, endothelial proliferation/apoptosis imbalance, and pulmonary arterial remodeling contributing to HIV-induced PAH.

This repressed antioxidant gene expression contributes to the impairment of redox homeostasis and elevated oxidative stress in HIV-induced PAH PAECs, resulting in endothelial dysfunction, endothelial proliferation/apoptosis imbalance, and pulmonary arterial remodeling contributing to HIV-induced PAH development and progression.

ARE gene expression may additionally be subject to redox regulation. Altered redox status during ER stress may promote modification of cysteine residues on Keap1 and allow the dissociation of NRF2 to participate in ARE gene transcription. ARE-expressed proteins such as glutathione-S-transferase and thioredoxin may participate in a negative feedback mechanism. Both proteins are reactive cysteine-based inhibitors of ASK, an inducer of ARE gene expression.

LPS-triggered TLR2/TLR4 through TRAF6-NOX2-ROS signaling promote IRE1α-XBP1 UPR axis, upregulating transcription of proinflammatory genes such as IL-6, TNF, and IFN-β in macrophages. ROS-activated IRE1α may also promote further ROS production through NOX and the mitochondria in a feed-forward (ROS-IRE1α-ROS) loop which may exacerbate UPR-mediated inflammatory pathways in the cell.

ER-stress induction of PERK-CHOP downregulates the transcription of the pro-survival bcl2 gene via GSH depletion promoting elevated cellular ROS, which ultimately promotes ER stress-induced apoptosis. CHOP-ATF4 upregulation of GADD34 induces ROS generation through attenuating PERK-induced elF2α phosphorylation/inactivation, an event that may also result in overloading the ER with misfolded/unfolded proteins. Elevated ROS and ER overload promote cell death. ER stress-induced CHOP has been shown to inhibit eNOS transcrip-

tion in ECs through the postulated binding to an optimal CHOP-responsive element at the eNOS promoter potentially contributing to PH pathogenesis.

Chronic hypoxia and H/R-mediated ROS generation results in eNOS uncoupling through oxidation of biopterin and eNOS S-glutathionylation. Decreased endothelial NO bioavailability and increased ROS production may result in persistent vasoconstriction and PASMC proliferation ultimately contributing to PH pathogenesis.

Upon BMPR2 mutation, highly conserved cysteine residues in the ligand-binding domain of BMPR2 are prone to alteration, causing the aberrant protein to be retained in the ER triggering UPR that may contribute to the pathogenesis of FPAH. Furthermore, attenuated BMPR2 expression deactivates elF2α stress granule formation, allowing translation of GM-CSF mRNA in HPAECs to promote inflammatory cell recruitment and vascular remodeling, contributing to FPAH.

Altered metabolic-redox states due to factors promoting PH within various cells at the different levels of the pulmonary vasculature induce UPR signaling to control processes such as autophagy, proliferation, feed-forward roles for inflammatory factors, and apoptosis. This signaling further activates NOX 2 and 4, promoting EC apoptosis and PASMC proliferation; uncouples eNOS to promote SMC proliferation; and promotes inflammation and IL-6 production as well as apoptosis – all of which may contribute to pulmonary vascular remodeling and PH development (Fig. 17.4). PH progression may be further exacerbated by dysregulation of NRF2 antioxidant signaling, causing EC and SMC dysfunction, as well as BMPR2 downregulation that promotes inflammatory cell recruitment and vascular remodeling. Interestingly, FPAH is caused by a BMPR2 mutation that may cause the misfolded BMPR2 protein to accumulate in the ER, triggering UPR signaling. In conclusion, there are several mechanisms through which cellular redox processes are able to modulate the UPR during ER stress, allowing these redox processes to have a major role in directing the subsequent expansive signaling events that participate as an integral part of remodeling and PH pathogenesis. While many aspects of the role of UPR signaling in PH need to be better defined, it appears that multiple redox processes appear to be sensors for controlling the balance between UPR signaling mechanisms and the processes they influence in the progression of PH.

Acknowledgment Recent studies from our lab have been funded by NIH grants R01HL115124, R01129797, and R01HL151187.

References

1. Adler V, Yin Z, Fuchs SY, et al. Regulation of JNK signaling by GSTp. EMBO J. 1999;18(5):1321–34. https://doi.org/10.1093/emboj/18.5.1321.
2. Agarwal S, Sharma H, Chen L, et al. NADPH oxidase mediated endothelial injury in HIV and opioid in duced pulmonary 2 arterial hypertension. Am J Physiol Lung Cell Mol Physiol. 2020;318(5):1097. https://doi.org/10.1152/ajplung.00480.2019.
3. Alessi DR, Cohen P. Mechanism of activation and function of protein kinase B. Curr Opin Genet Dev. 1998;8(1):55–62. https://doi.org/10.1016/S0959-437X(98)80062-2.
4. Ali MH, Schlidt SA, Chandel NS, et al. Endothelial permeability and IL-6 production during hypoxia: role of ROS in signal transduction. Am J Physiol Lung Cell Mol Physiol. 1999;277(5):1057. https://doi.org/10.1152/ajplung.1999.277.5.l1057.
5. Asada R, Kanemoto S, Kondo S, et al. The signalling from endoplasmic reticulum-resident bZIP transcription factors involved in diverse cellular physiology. J Biochem. 2011;149(5):507–18. https://doi.org/10.1093/jb/mvr041.
6. Ashrafi G, Schwarz TL. The pathways of mitophagy for quality control and clearance of mitochondria. Cell Death Differ. 2013;20(1):31–42. https://doi.org/10.1038/cdd.2012.81.
7. Badran M, Abuyassin B, Golbidi S, et al. Uncoupling of vascular nitric oxide synthase caused by intermittent hypoxia. Oxidative Med Cell Longev. 2016;2016:2354870. https://doi.org/10.1155/2016/2354870.
8. Baker BM, Nargund AM, Sun T, et al. Protective coupling of mitochondrial function and protein synthesis via the eIF2α kinase GCN-2. PLoS Genet. 2012;8(6):e1002760. https://doi.org/10.1371/journal.pgen.1002760.
9. Bánhegyi G, Lusini L, Puskás F, et al. Preferential transport of glutathione versus glutathione disulfide in rat liver microsomal vesicles. J Biol Chem. 1999;274(18):12213–6.
10. Bánhegyi G, Benedetti A, Csala M, et al. Stress on redox. FEBS Lett. 2007;581(19):3634–40. https://doi.org/10.1016/j.febslet.2007.04.028.
11. Barañano DE, Rao M, Ferris CD, et al. Biliverdin reductase: a major physiologic cytoprotectant. Proc Natl Acad Sci. 2002;99(25):16093–8. https://doi.org/10.1073/pnas.252626999.
12. Barman SA, Chen F, Su Y, et al. NADPH oxidase 4 is expressed in pulmonary artery adventitia and contributes to hypertensive vascular remodeling. Arterioscler Thromb Vasc Biol. 2014;34(8):1704–15. https://doi.org/10.1161/ATVBAHA.114.303848.
13. Bartoszewska S, Collawn JF. Unfolded protein response (UPR) integrated signaling networks determine cell fate during hypoxia. Cell Mol Biol Lett. 2020;25:18. https://doi.org/10.1186/s11658-020-00212-1.
14. Bass R, Ruddock LW, Klappa P, et al. A major fraction of endoplasmic reticulum-located glutathione is present as mixed disulfides with protein. J Biol Chem. 2004;279(7):5257–62. https://doi.org/10.1074/jbc.M304951200.
15. Basset C, Holton J, O'Mahony R, et al. Innate immunity and pathogen-host interaction. Vaccine. 2003;21(Suppl 2):12. https://doi.org/10.1016/S0264-410X(03)00195-6.
16. Bauernfeind FG, Horvath G, Stutz A, et al. Cutting edge: NF-kappaB activating pattern recognition and cytokine receptors license NLRP3 inflammasome activation by regulating NLRP3 expression. J Immunol. 2009;183(2):787–91. https://doi.org/10.4049/jimmunol.0901363.
17. B'Chir W, Maurin AC, Carraro V, et al. The eIF2α/ATF4 pathway is essential for stress-induced autophagy gene expression. Nucleic Acids Res. 2013;41(16):7683–99. https://doi.org/10.1093/nar/gkt563.
18. Bertolotti A, Zhang Y, Hendershot LM, et al. Dynamic interaction of BiP and ER stress transducers in the unfolded-protein response. Nat Cell Biol. 2000;2(6):326–32. https://doi.org/10.1038/35014014.
19. Bertout JA, Patel SA, Simon MC. The impact of O2 availability on human cancer. Nat Rev Cancer. 2008;8(12):967–75. https://doi.org/10.1038/nrc2540.
20. Bertout JA, Majmundar AJ, Gordan JD, et al. HIF2α inhibition promotes p53 pathway activity, tumor cell death, and radiation responses. Proc Natl Acad Sci. 2009;106(34):14391–6. https://doi.org/10.1073/pnas.0907357106.
21. Blais JD, Chin K, Zito E, et al. A small molecule inhibitor of endoplasmic reticulum oxidation 1 (ERO1) with selectively reversible thiol reactivity. J Biol Chem. 2010;285(27):20993–1003. https://doi.org/10.1074/jbc.M110.126599.
22. Braakman I, Bulleid NJ. Protein folding and modification in the mammalian endoplasmic reticulum. Annu Rev Biochem. 2011;80:71–99. https://doi.org/10.1146/annurev-biochem-062209-093836.

23. Brush MH, Weiser DC, Shenolikar S. Growth arrest and DNA damage-inducible protein GADD34 targets protein phosphatase 1 alpha to the endoplasmic reticulum and promotes dephosphorylation of the alpha subunit of eukaryotic translation initiation factor 2. Mol Cell Biol. 2003;23(4):1292–303. https://doi.org/10.1128/MCB.23.4.1292-1303.2003.

24. Budhiraja R, Tuder RM, Hassoun PM. Endothelial dysfunction in pulmonary hypertension. Circulation. 2004;109(2):159–65. https://doi.org/10.1161/01.CIR.0000102381.57477.50.

25. Cahill CM, Rogers JT. Interleukin (IL) 1β induction of IL-6 is mediated by a novel phosphatidylinositol 3-kinase-dependent AKT/IκB kinase α pathway targeting activator protein-1. J Biol Chem. 2008;283(38):25900–12. https://doi.org/10.1074/jbc.M707692200.

26. Camenisch TD, Spicer AP, Brehm-Gibson T, et al. Disruption of hyaluronan synthase-2 abrogates normal cardiac morphogenesis and hyaluronan-mediated transformation of epithelium to mesenchyme. J Clin Invest. 2000;106(3):349–60. https://doi.org/10.1172/JCI10272.

27. Camenisch TD, Schroeder JA, Bradley J, et al. Heart-valve mesenchyme formation is dependent on hyaluronan-augmented activation of ErbB2-ErbB3 receptors. Nat Med. 2002;8(8):850–5. https://doi.org/10.1038/nm742.

28. Cao JP, He XY, Xu HT, et al. Autologous transplantation of peripheral blood-derived circulating endothelial progenitor cells attenuates endotoxin-induced acute lung injury in rabbits by direct endothelial repair and indirect immunomodulation. Anesthesiology. 2012;116(6):1278–87. https://doi.org/10.1097/ALN.0b013e3182567f84.

29. Cao X, He Y, Li X, et al. The IRE1α-XBP1 pathway function in hypoxia-induced pulmonary vascular remodeling, is upregulated by quercetin, inhibits apoptosis and partially reverses the effect of quercetin in PASMCs. Am J Transl Res. 2019;11(2):641–54.

30. Carroll B, Otten EG, Manni D, et al. Oxidation of SQSTM1/p62 mediates the link between redox state and protein homeostasis. Nat Commun. 2018;9(1):256. https://doi.org/10.1038/s41467-017-02746-z.

31. Chacinska A, Koehler CM, Milenkovic D, et al. Importing mitochondrial proteins: machineries and mechanisms. Cell. 2009;138(4):628–44. https://doi.org/10.1016/j.cell.2009.08.005.

32. Chakravarthi S, Jessop CE, Willer M, et al. Intracellular catalysis of disulfide bond formation by the human sulfhydryl oxidase, QSOX1. Biochem J. 2007;404(3):403–11. https://doi.org/10.1042/BJ20061510.

33. Chamberlain N, Anathy V. Pathological consequences of the unfolded protein response and downstream protein disulphide isomerases in pulmonary viral infection and disease. J Biochem.

2020;167(2):173–84. https://doi.org/10.1093/jb/mvz101.

34. Chazova I, Loyd JE, Zhdanov VS, et al. Pulmonary artery adventitial changes and venous involvement in primary pulmonary hypertension. Am J Pathol. 1995;146(2):389–97.

35. Chen YF, Oparil S. Endothelin and pulmonary hypertension. J Cardiovasc Pharmacol. 2000;35(4 Suppl 2):49. https://doi.org/10.1097/00005344-200000002-00012.

36. Chen X, Shen J, Prywes R. The luminal domain of ATF6 senses endoplasmic reticulum (ER) stress and causes translocation of ATF6 from the ER to the Golgi. J Biol Chem. 2002;277(15):13045–52. https://doi.org/10.1074/jbc.M110636200.

37. Chen X, Iliopoulos D, Zhang Q, et al. XBP1 promotes triple-negative breast cancer by controlling the HIF1α pathway. Nature. 2014;508(7494):103–7. https://doi.org/10.1038/nature13119.

38. Chen Y, Yuan T, Zhang H, et al. Activation of Nrf2 attenuates pulmonary vascular remodeling via inhibiting endothelial-to-mesenchymal transition: an insight from a plant polyphenol. Int J Biol Sci. 2017;13(8):1067–81. https://doi.org/10.7150/ijbs.20316.

39. Cho S, Lee YH, Park H, et al. Glutathione S-transferase mu modulates the stress-activated signals by suppressing apoptosis signal-regulating kinase 1. J Biol Chem. 2001;276(16):12749–55. https://doi.org/10.1074/jbc.M005561200.

40. Chou C, Wei L, Kuo M, et al. Up-regulation of interleukin-6 in human ovarian cancer cell via a Gi/PI3K-Akt/NF-κB pathway by lysophosphatidic acid, an ovarian cancer-activating factor. Carcinogenesis. 2005;26(1):45–52. https://doi.org/10.1093/carcin/bgh301.

41. Chow MT, Duret H, Andrews DM, et al. Type I NKT-cell-mediated TNF-α is a positive regulator of NLRP3 inflammasome priming. Eur J Immunol. 2014;44(7):2111–20. https://doi.org/10.1002/eji.201344329.

42. Chu Y, Xiangli X, Xiao W. Regulatory T cells protect against hypoxia-induced pulmonary arterial hypertension in mice. Mol Med Rep. 2015;11(4):3181–7. https://doi.org/10.3892/mmr.2014.3106.

43. Cox JS, Shamu CE, Walter P. Transcriptional induction of genes encoding endoplasmic reticulum resident proteins requires a transmembrane protein kinase. Cell. 1993;73(6):1197–206. https://doi.org/10.1016/0092-8674(93)90648-a.

44. Csala M, Szarka A, Margittai É, et al. Role of vitamin E in ascorbate-dependent protein thiol oxidation in rat liver endoplasmic reticulum. Arch Biochem Biophys. 2001;388(1):55–9. https://doi.org/10.1006/abbi.2000.2260.

45. Cullinan SB, Zhang D, Hannink M, et al. Nrf2 is a direct PERK substrate and effector of PERK-dependent cell survival. Mol Cell Biol.

2003;23(20):7198–209. https://doi.org/10.1128/MCB.23.20.7198-7209.2003.

46. Cuozzo JW, Kaiser CA. Competition between glutathione and protein thiols for disulphide-bond formation. Nat Cell Biol. 1999;1(3):130–5. https://doi.org/10.1038/11047.

47. Dahle MK, Øverland G, Myhre AE, et al. The phosphatidylinositol 3-kinase/protein kinase B signaling pathway is activated by lipoteichoic acid and plays a role in Kupffer cell production of interleukin-6 (IL-6) and IL-10. Infect Immun. 2004;72(10):5704–11. https://doi.org/10.1128/IAI.72.10.5704-5711.2004.

48. De Pascali F, Hemann C, Samons K, et al. Hypoxia and reoxygenation induce endothelial nitric oxide synthase uncoupling in endothelial cells through tetrahydrobiopterin depletion and S-glutathionylation. Biochemistry. 2014;53(22):3679–88. https://doi.org/10.1021/bi500076r.

49. Delaunay-Moisan A, Ponsero A, Toledano MB. Reexamining the function of glutathione in oxidative protein folding and secretion. Antioxid Redox Signal. 2017;27(15):1178–99. https://doi.org/10.1089/ars.2017.7148.

50. Delic M, Rebnegger C, Wanka F, et al. Oxidative protein folding and unfolded protein response elicit differing redox regulation in endoplasmic reticulum and cytosol of yeast. Free Radic Biol Med. 2012;52(9):2000–12. https://doi.org/10.1016/j.freeradbiomed.2012.02.048.

51. Deng J, Lu PD, Zhang Y, et al. Translational repression mediates activation of nuclear factor kappa B by phosphorylated translation initiation factor 2. Mol Cell Biol. 2004;24(23):10161–8. https://doi.org/10.1128/MCB.24.23.10161-10168.2004.

52. Deniaud A, Sharaf el Dein O, Maillier E, et al. Endoplasmic reticulum stress induces calcium-dependent permeability transition, mitochondrial outer membrane permeabilization and apoptosis. Oncogene. 2008;27(3):285–99. https://doi.org/10.1038/sj.onc.1210638.

53. Dias-Gunasekara S, van Lith M, Williams JAG, et al. Mutations in the FAD binding domain cause stress-induced misoxidation of the endoplasmic reticulum oxidoreductase Ero1β. J Biol Chem. 2006;281(35):25018–25. https://doi.org/10.1074/jbc.M602354200.

54. Dinarello CA. The IL-1 family and inflammatory diseases. Clin Exp Rheumatol. 2002;20(5 Suppl 27):1.

55. Dorner AJ, Wasley LC, Raney P, et al. The stress response in Chinese hamster ovary cells. Regulation of ERp72 and protein disulfide isomerase expression and secretion. J Biol Chem. 1990;265(35):22029–34.

56. Dostert C, Pétrilli V, Van Bruggen R, et al. Innate immune activation through Nalp3 inflammasome sensing of asbestos and silica. Science. 2008;320(5876):674–7. https://doi.org/10.1126/science.1156995.

57. Eletto D, Chevet E, Argon Y, et al. Redox controls UPR to control redox. J Cell Sci. 2014;127(Pt 17):3649–58. https://doi.org/10.1242/jcs.153643.

58. Eletto D, Eletto D, Dersh D, et al. Protein disulfide isomerase A6 controls the decay of IRE1α signaling via disulfide-dependent association. Mol Cell. 2014;53(4):562–76. https://doi.org/10.1016/j.molcel.2014.01.004.

59. Erickson JR, Joiner MA, Guan X, et al. A dynamic pathway for calcium-independent activation of CaMKII by methionine oxidation. Cell. 2008;133(3):462–74. https://doi.org/10.1016/j.cell.2008.02.048.

60. Federti E, Matté A, Ghigo A, et al. Peroxiredoxin-2 plays a pivotal role as multimodal cytoprotector in the early phase of pulmonary hypertension. Free Radic Biol Med. 2017;112:376–86. https://doi.org/10.1016/j.freeradbiomed.2017.08.004.

61. Fewell SW, Travers KJ, Weissman JS, et al. The action of molecular chaperones in the early secretory pathway. Annu Rev Genet. 2001;35:149–91. https://doi.org/10.1146/annurev.genet.35.102401.090313.

62. Fiorese CJ, Schulz AM, Lin YF, et al. The transcription factor ATF5 mediates a mammalian mitochondrial UPR. Curr Biol. 2016;26(15):2037–43. https://doi.org/10.1016/j.cub.2016.06.002.

63. Fladmark KE, Brustugun OT, Mellgren G, et al. Ca2+/calmodulin-dependent protein kinase II is required for microcystin-induced apoptosis. J Biol Chem. 2002;277(4):2804–11. https://doi.org/10.1074/jbc.M109049200.

64. Fleetwood AJ, Lawrence T, Hamilton JA, et al. Granulocyte-macrophage colony-stimulating factor (CSF) and macrophage CSF-dependent macrophage phenotypes display differences in cytokine profiles and transcription factor activities: implications for CSF blockade in inflammation. J Immunol. 2007;178(8):5245–52. https://doi.org/10.4049/jimmunol.178.8.5245.

65. Förstermann U, Münzel T. Endothelial nitric oxide synthase in vascular disease: from marvel to menace. Circulation. 2006;113(13):1708–14. https://doi.org/10.1161/CIRCULATIONAHA.105.602532.

66. Franchi L, Eigenbrod T, Muñoz-Planillo R, et al. The inflammasome: a caspase-1-activation platform that regulates immune responses and disease pathogenesis. Nat Immunol. 2009;10(3):241–7. https://doi.org/10.1038/ni.1703.

67. Frand AR, Kaiser CA. The ERO1 gene of yeast is required for oxidation of protein dithiols in the endoplasmic reticulum. Mol Cell. 1998;1(2):161–70. https://doi.org/10.1016/S1097-2765(00)80017-9.

68. Frand AR, Kaiser CA. Ero1p oxidizes protein disulfide isomerase in a pathway for disulfide bond formation in the endoplasmic reticulum. Mol Cell. 1999;4(4):469–77. https://doi.org/10.1016/S1097-2765(00)80198-7.

69. Fujimoto T, Inaba K, Kadokura H. Methods to identify the substrates of thiol-disulfide oxidoreduc-

tases. Protein Sci. 2019;28(1):30–40. https://doi.org/10.1002/pro.3530.

70. Fulda S. Alternative cell death pathways and cell metabolism. Int J Cell Biol. 2013;2013:463637. https://doi.org/10.1155/2013/463637.

71. George J, D'Armiento J. Transgenic expression of human matrix metalloproteinase-9 augments monocrotaline-induced pulmonary arterial hypertension in mice. J Hypertens. 2011;29(2):299–308. https://doi.org/10.1097/HJH.0b013e328340a0e4.

72. Gordan JD, Thompson CB, Simon MC. HIF and c-Myc: sibling rivals for control of cancer cell metabolism and proliferation. Cancer Cell. 2007;12(2):108–13. https://doi.org/10.1016/j.ccr.2007.07.006.

73. Görlach A, Klappa P, Kietzmann T. The endoplasmic reticulum: folding, calcium homeostasis, signaling, and redox control. Antioxid Redox Signal. 2006;8(9–10):1391–418. https://doi.org/10.1089/ars.2006.8.1391.

74. Greten FR, Arkan MC, Bollrath J, et al. NF-κB is a negative regulator of IL-1β secretion as revealed by genetic and pharmacological inhibition of IKKβ. Cell. 2007;130(5):918–31. https://doi.org/10.1016/j.cell.2007.07.009.

75. Hansen HG, Schmidt JD, Søltoft CL, et al. Hyperactivity of the Ero1α oxidase elicits endoplasmic reticulum stress but no broad antioxidant response. J Biol Chem. 2012;287(47):39513–23. https://doi.org/10.1074/jbc.M112.405050.

76. Hansen HG, Søltoft CL, Schmidt JD, et al. Biochemical evidence that regulation of Ero1β activity in human cells does not involve the isoform-specific cysteine 262. Biosci Rep. 2014;34(2):e00103. https://doi.org/10.1042/BSR20130124.

77. Harding HP, Zhang Y, Ron D. Protein translation and folding are coupled by an endoplasmic-reticulum-resident kinase. Nature. 1999;397(6716):271–4. https://doi.org/10.1038/16729.

78. Harding HP, Novoa I, Zhang Y, et al. Regulated translation initiation controls stress-induced gene expression in mammalian cells. Mol Cell. 2000;6(5):1099–108. https://doi.org/10.1016/S1097-2765(00)00108-8.

79. Harding HP, Zhang Y, Bertolotti A, et al. Perk is essential for translational regulation and cell survival during the unfolded protein response. Mol Cell. 2000;5(5):897–904. https://doi.org/10.1016/S1097-2765(00)80330-5.

80. Harding HP, Zhang Y, Zeng H, et al. An integrated stress response regulates amino acid metabolism and resistance to oxidative stress. Mol Cell. 2003;11(3):619–33. https://doi.org/10.1016/S1097-2765(03)00105-9.

81. Hayden MS, Ghosh S. Regulation of NF-κB by TNF family cytokines. Semin Immunol. 2014;26(3):253–66. https://doi.org/10.1016/j.smim.2014.05.004.

82. Haynes CM, Titus EA, Cooper AA. Degradation of misfolded proteins prevents ER-derived oxidative stress and cell death. Mol Cell. 2004;15(5):767–76. https://doi.org/10.1016/j.molcel.2004.08.025.

83. Hetz C. The unfolded protein response: controlling cell fate decisions under ER stress and beyond. Nat Rev Mol Cell Biol. 2012;13(2):89–102. https://doi.org/10.1038/nrm3270.

84. Hetz C, Martinon F, Rodriguez D, et al. The unfolded protein response: integrating stress signals through the stress sensor IRE1α. Physiol Rev. 2011;91(4):1219–43. https://doi.org/10.1152/physrev.00001.2011.

85. Higa A, Taouji S, Lhomond S, et al. Endoplasmic reticulum stress-activated transcription factor ATF6α requires the disulfide isomerase PDIA5 to modulate chemoresistance. Mol Cell Biol. 2014;34(10):1839–49. https://doi.org/10.1128/MCB.01484-13.

86. Higo T, Hattori M, Nakamura T, et al. Subtype-specific and ER lumenal environment-dependent regulation of inositol 1,4,5-trisphosphate receptor type 1 by ERp44. Cell. 2005;120(1):85–98. https://doi.org/10.1016/j.cell.2004.11.048.

87. Hoeper MM, Bogaard HJ, Condliffe R, et al. Definitions and diagnosis of pulmonary hypertension. J Am Coll Cardiol. 2013;62(25 Suppl):42. https://doi.org/10.1016/j.jacc.2013.10.032.

88. Holland R, Fishbein JC. Chemistry of the cysteine sensors in Kelch-like ECH-associated protein 1. Antioxid Redox Signal. 2010;13(11):1749–61. https://doi.org/10.1089/ars.2010.3273.

89. Hollien J, Weissman JS. Decay of endoplasmic reticulum-localized mRNAs during the unfolded protein response. Science. 2006;313(5783):104–7. https://doi.org/10.1126/science.1129631.

90. Hong M, Luo S, Baumeister P, et al. Underglycosylation of ATF6 as a novel sensing mechanism for activation of the unfolded protein response. J Biol Chem. 2004;279(12):11354–63. https://doi.org/10.1074/jbc.M309804200.

91. Hourihan J, Moronetti Mazzeo L, Fernández-Cárdenas L, et al. Cysteine sulfenylation directs IRE-1 to activate the SKN-1/Nrf2 antioxidant response. Mol Cell. 2016;63(4):553–66. https://doi.org/10.1016/j.molcel.2016.07.019.

92. Huang B, Wu P, Bowker-Kinley MM, et al. Regulation of pyruvate dehydrogenase kinase expression by peroxisome proliferator-activated receptor-alpha ligands, glucocorticoids, and insulin. Diabetes. 2002;51(2):276–83. https://doi.org/10.2337/diabetes.51.2.276.

93. Huang Y, Li W, Su Z, et al. The complexity of the Nrf2 pathway: beyond the antioxidant response. J Nutr Biochem. 2015;26(12):1401–13. https://doi.org/10.1016/j.jnutbio.2015.08.001.

94. Hudson DA, Gannon SA, Thorpe C. Oxidative protein folding: from thiol–disulfide exchange reactions to the redox poise of the endoplasmic reticulum.

Free Radic Biol Med. 2015;80:171–82. https://doi.org/10.1016/j.freeradbiomed.2014.07.037.

95. Hwang C, Sinskey AJ, Lodish HF. Oxidized redox state of glutathione in the endoplasmic reticulum. Science. 1992;257(5076):1496–502. https://doi.org/10.1126/science.1523409.

96. Ishii M, Hogaboam CM, Joshi A, et al. CC chemokine receptor 4 modulates Toll-like receptor 9-mediated innate immunity and signaling. Eur J Immunol. 2008;38(8):2290–302. https://doi.org/10.1002/eji.200838360.

97. Itoh K, Wakabayashi N, Katoh Y, et al. Keap1 represses nuclear activation of antioxidant responsive elements by Nrf2 through binding to the amino-terminal Neh2 domain. Genes Dev. 1999;13(1):76–86. https://doi.org/10.1101/gad.13.1.76.

98. Ivanova IG, Park CV, Yemm AI, et al. PERK/eIF2α signaling inhibits HIF-induced gene expression during the unfolded protein response via YB1-dependent regulation of HIF1α translation. Nucleic Acids Res. 2018;46(8):3878–90. https://doi.org/10.1093/nar/gky127.

99. Jernigan NL, Naik JS, Weise-Cross L, et al. Contribution of reactive oxygen species to the pathogenesis of pulmonary arterial hypertension. PLoS One. 2017;12(6):e0180455. https://doi.org/10.1371/journal.pone.0180455.

100. Kaneko M, Niinuma Y, Nomura Y. Activation signal of nuclear factor-κB in response to endoplasmic reticulum stress is transduced via IRE1 and tumor necrosis factor receptor-associated factor 2. Biol Pharm Bull. 2003;26(7):931–5. https://doi.org/10.1248/bpb.26.931.

101. Kang S, Rane NS, Kim SJ, et al. Substrate-specific translocational attenuation during ER stress defines a pre-emptive quality control pathway. Cell. 2006;127(5):999–1013. https://doi.org/10.1016/j.cell.2006.10.032.

102. Kang Y, Zhang G, Huang EC, et al. Sulforaphane prevents right ventricular injury and reduces pulmonary vascular remodeling in pulmonary arterial hypertension. Am J Physiol Heart Circ Physiol. 2020;318(4):853. https://doi.org/10.1152/ajpheart.00321.2019.

103. Kaplin AI, Ferris CD, Voglmaier SM, et al. Purified reconstituted inositol 1,4,5-trisphosphate receptors. Thiol reagents act directly on receptor protein. J Biol Chem. 1994;269(46):28972–8.

104. Kapturczak MH, Wasserfall C, Brusko T, et al. Heme oxygenase-1 modulates early inflammatory responses: evidence from the heme oxygenase-1-deficient mouse. Am J Pathol. 2004;165(3):1045–53. https://doi.org/10.1016/S0002-9440(10)63365-2.

105. Kataoka K, Igarashi K, Itoh K, et al. Small Maf proteins heterodimerize with Fos and may act as competitive repressors of the NF-E2 transcription factor. Mol Cell Biol. 1995;15(4):2180–90. https://doi.org/10.1128/mcb.15.4.2180.

106. Katsuoka F, Yamamoto M. Small Maf proteins (MafF, MafG, MafK): history, structure and function. Gene. 2016;586(2):197–205. https://doi.org/10.1016/j.gene.2016.03.058.

107. Kawai T, Akira S. TLR signaling. Semin Immunol. 2007;19(1):24–32. https://doi.org/10.1016/j.smim.2006.12.004.

108. Kim I, Je HD, Gallant C, et al. Ca2+-calmodulin-dependent protein kinase II-dependent activation of contractility in ferret aorta. J Physiol. 2000;526(2):367–74. https://doi.org/10.1111/j.1469-7793.2000.00367.x.

109. Kim J, Tchernyshyov I, Semenza GL, et al. HIF-1-mediated expression of pyruvate dehydrogenase kinase: a metabolic switch required for cellular adaptation to hypoxia. Cell Metab. 2006;3(3):177–85. https://doi.org/10.1016/j.cmet.2006.02.002.

110. Kimata Y, Oikawa D, Shimizu Y, et al. A role for BiP as an adjustor for the endoplasmic reticulum stress-sensing protein Ire1. J Cell Biol. 2004;167(3):445–56. https://doi.org/10.1083/jcb.200405153.

111. Kiselyov K, Xu X, Mozhayeva G, et al. Functional interaction between InsP3 receptors and store-operated Htrp3 channels. Nature. 1998;396(6710):478–82. https://doi.org/10.1038/24890.

112. Kitamura M. Biphasic, bidirectional regulation of NF-κB by endoplasmic reticulum stress. Antioxid Redox Signal. 2009;11(9):2353–64. https://doi.org/10.1089/ars.2008.2391.

113. Kodali VK, Thorpe C. Oxidative protein folding and the Quiescin–sulfhydryl oxidase family of flavoproteins. Antioxid Redox Signal. 2010;13(8):1217–30. https://doi.org/10.1089/ars.2010.3098.

114. Komatsu M, Kurokawa H, Waguri S, et al. The selective autophagy substrate p62 activates the stress responsive transcription factor Nrf2 through inactivation of Keap1. Nat Cell Biol. 2010;12(3):213–23. https://doi.org/10.1038/ncb2021.

115. Kondo S, Saito A, Hino S, et al. BBF2H7, a novel transmembrane bZIP transcription factor, is a new type of endoplasmic reticulum stress transducer. Mol Cell Biol. 2007;27(5):1716–29. https://doi.org/10.1128/MCB.01552-06.

116. Koo B, Hwang H, Yi B, et al. Arginase II contributes to the Ca2+/CaMKII/eNOS axis by regulating Ca2+ concentration between the cytosol and mitochondria in a p32-dependent manner. J Am Heart Assoc. 2018;7(18):e009579. https://doi.org/10.1161/JAHA.118.009579.

117. Kornmann B. The molecular hug between the ER and the mitochondria. Curr Opin Cell Biol. 2013;25(4):443–8. https://doi.org/10.1016/j.ceb.2013.02.010.

118. Kosuri P, Alegre-Cebollada J, Feng J, et al. Protein folding drives disulfide formation. Cell. 2012;151(4):794–806. https://doi.org/10.1016/j.cell.2012.09.036.

119. Kotlyarov A, Neininger A, Schubert C, et al. MAPKAP kinase 2 is essential for LPS-induced

TNF-α biosynthesis. Nat Cell Biol. 1999;1(2):94–7. https://doi.org/10.1038/10061.

120. Kozutsumi Y, Segal M, Normington K, et al. The presence of malfolded proteins in the endoplasmic reticulum signals the induction of glucose-regulated proteins. Nature. 1988;332(6163):462–4. https://doi.org/10.1038/332462a0.

121. Kranz P, Neumann F, Wolf A, et al. PDI is an essential redox-sensitive activator of PERK during the unfolded protein response (UPR). Cell Death Dis. 2017;8(8):e2986. https://doi.org/10.1038/cddis.2017.369.

122. Krishnan SM, Sobey CG, Latz E, et al. IL-1β and IL-18: inflammatory markers or mediators of hypertension? Br J Pharmacol. 2014;171(24):5589–602. https://doi.org/10.1111/bph.12876.

123. Krishnan SM, Dowling JK, Ling YH, et al. Inflammasome activity is essential for one kidney/deoxycorticosterone acetate/salt-induced hypertension in mice. Br J Pharmacol. 2016;173(4):752–65. https://doi.org/10.1111/bph.13230.

124. Kroemer G, Galluzzi L, Brenner C. Mitochondrial membrane permeabilization in cell death. Physiol Rev. 2007;87(1):99–163. https://doi.org/10.1152/physrev.00013.2006.

125. Kueh HY, Niethammer P, Mitchison TJ. Maintenance of mitochondrial oxygen homeostasis by cosubstrate compensation. Biophys J. 2013;104(6):1338–48. https://doi.org/10.1016/j.bpj.2013.01.030.

126. Laabich A, Li G, Cooper NG. Characterization of apoptosis-genes associated with NMDA mediated cell death in the adult rat retina. Brain Res Mol Brain Res. 2001;91(1–2):34–42. https://doi.org/10.1016/s0169-328x(01)00116-4.

127. Lee C. Collaborative power of Nrf2 and PPARγ activators against metabolic and drug-induced oxidative injury. Oxidative Med Cell Longev. 2017;2017:1378175. https://doi.org/10.1155/2017/1378175.

128. Lee P, Chandel NS, Simon MC. Cellular adaptation to hypoxia through hypoxia inducible factors and beyond. Nat Rev Mol Cell Biol. 2020;21(5):268–83. https://doi.org/10.1038/s41580-020-0227-y.

129. Lenna S, Farina AG, Martyanov V, et al. Increased expression of endoplasmic reticulum stress and unfolded protein response genes in peripheral blood mononuclear cells from patients with limited cutaneous systemic sclerosis and pulmonary arterial hypertension. Arthritis Rheum. 2013;65(5):1357–66. https://doi.org/10.1002/art.37891.

130. Li F, Malik KU. Angiotensin II-induced Akt activation is mediated by metabolites of arachidonic acid generated by CaMKII-stimulated Ca2(+)-dependent phospholipase A2. Am J Physiol Heart Circ Physiol. 2005;288(5):2306. https://doi.org/10.1152/ajpheart.00571.2004.

131. Li G, Mongillo M, Chin K, et al. Role of ERO1-alpha-mediated stimulation of inositol 1,4,5-triphosphate receptor activity in endoplasmic reticulum stress-induced apoptosis. J Cell Biol. 2009;186(6):783–92. https://doi.org/10.1083/jcb.200904060.

132. Li G, Scull C, Ozcan L, et al. NADPH oxidase links endoplasmic reticulum stress, oxidative stress, and PKR activation to induce apoptosis. J Cell Biol. 2010;191(6):1113–25. https://doi.org/10.1083/jcb.201006121.

133. Li H, Li W, Gupta AK, et al. Calmodulin kinase II is required for angiotensin II-mediated vascular smooth muscle hypertrophy. Am J Physiol Heart Circ Physiol. 2010;298(2):688. https://doi.org/10.1152/ajpheart.01014.2009.

134. Li W, Li H, Sanders PN, et al. The multifunctional Ca2+/calmodulin-dependent kinase II delta (CaMKIIdelta) controls neointima formation after carotid ligation and vascular smooth muscle cell proliferation through cell cycle regulation by p21. J Biol Chem. 2011;286(10):7990–9. https://doi.org/10.1074/jbc.M110.163006.

135. Liang L, Wang M, Liu M, et al. Chronic toxicity of methamphetamine: oxidative remodeling of pulmonary arteries. Toxicol In Vitro. 2020;62:104668. https://doi.org/10.1016/j.tiv.2019.104668.

136. Lin W, Harding HP, Ron D, et al. Endoplasmic reticulum stress modulates the response of myelinating oligodendrocytes to the immune cytokine interferon-γ. J Cell Biol. 2005;169(4):603–12. https://doi.org/10.1083/jcb.200502086.

137. Liu JQ, Zelko IN, Erbynn EM, et al. Hypoxic pulmonary hypertension: role of superoxide and NADPH oxidase (gp91phox). Am J Physiol Lung Cell Mol Physiol. 2006;290(1):2. https://doi.org/10.1152/ajplung.00135.2005.

138. Liu W, Wang L, Lai Y. Hepcidin protects pulmonary artery hypertension in rats by activating NF-κB/TNF-α pathway. Eur Rev Med Pharmacol Sci. 2019;23(17):7573–81. https://doi.org/10.26355/eurrev_201909_18878.

139. Llesuy SF, Tomaro ML. Heme oxygenase and oxidative stress. Evidence of involvement of bilirubin as physiological protector against oxidative damage. Biochim Biophys Acta. 1994;1(1):9–14. https://doi.org/10.1016/0167-4889(94)90067-1.

140. Loinard C, Zouggari Y, Rueda P, et al. C/EBP homologous protein-10 (CHOP-10) limits postnatal neovascularization through control of endothelial nitric oxide synthase gene expression. Circulation. 2012;125(8):1014–26. https://doi.org/10.1161/CIRCULATIONAHA.111.041830.

141. Luke T, Maylor J, Undem C, et al. Kinase-dependent activation of voltage-gated Ca2+ channels by ET-1 in pulmonary arterial myocytes during chronic hypoxia. Am J Physiol Lung Cell Mol Physiol. 2012;302(10):1128. https://doi.org/10.1152/ajplung.00396.2011.

142. Luo Q, Wang X, Liu R, et al. alpha1A-adrenoceptor is involved in norepinephrine-induced proliferation of pulmonary artery smooth muscle cells via CaMKII signaling. J Cell Biochem. 2019;120(6):9345–55. https://doi.org/10.1002/jcb.28210.

143. Ma L, Chung WK. The role of genetics in pulmonary arterial hypertension. J Pathol. 2017;241(2):273–80. https://doi.org/10.1002/path.4833.

144. Ma HT, Patterson RL, van Rossum DB, et al. Requirement of the inositol trisphosphate receptor for activation of store-operated Ca2+ channels. Science. 2000;287(5458):1647–51. https://doi.org/10.1126/science.287.5458.1647.

145. Magesh S, Chen Y, Hu L. Small molecule modulators of Keap1-Nrf2-ARE pathway as potential preventive and therapeutic agents. Med Res Rev. 2012;32(4):687–726. https://doi.org/10.1002/med.21257.

146. Majmundar AJ, Wong WJ, Simon MC. Hypoxia inducible factors and the response to hypoxic stress. Mol Cell. 2010;40(2):294–309. https://doi.org/10.1016/j.molcel.2010.09.022.

147. Malhotra JD, Hongzhi M, Zhang K, et al. Antioxidants reduce endoplasmic reticulum stress and improve protein secretion. Proc Natl Acad Sci. 2008;105(47):18525–30. https://doi.org/10.1073/pnas.0809677105.

148. Malik A, Kanneganti T. Inflammasome activation and assembly at a glance. J Cell Sci. 2017;130(23):3955–63. https://doi.org/10.1242/jcs.207365.

149. Mao W, Fukuoka S, Iwai C, et al. Cardiomyocyte apoptosis in autoimmune cardiomyopathy: mediated via endoplasmic reticulum stress and exaggerated by norepinephrine. Am J Physiol Heart Circ Physiol. 2007;293(3):1636. https://doi.org/10.1152/ajpheart.01377.2006.

150. Marciniak SJ, Yun CY, Oyadomari S, et al. CHOP induces death by promoting protein synthesis and oxidation in the stressed endoplasmic reticulum. Genes Dev. 2004;18(24):3066–77. https://doi.org/10.1101/gad.1250704.

151. Margittai É, Enyedi B, Csala M, et al. Composition of the redox environment of the endoplasmic reticulum and sources of hydrogen peroxide. Free Radic Biol Med. 2015;83:331–40. https://doi.org/10.1016/j.freeradbiomed.2015.01.032.

152. Marmiroli S, Bavelloni A, Faenza I, et al. Phosphatidylinositol 3-kinase is recruited to a specific site in the activated IL-1 receptor I. FEBS Lett. 1998;438(1–2):49–54. https://doi.org/10.1016/S0014-5793(98)01270-8.

153. Marsboom G, Toth PT, Ryan JJ, et al. Dynamin-related protein 1–mediated mitochondrial mitotic fission permits hyperproliferation of vascular smooth muscle cells and offers a novel therapeutic target in pulmonary hypertension. Circ Res. 2012;110(11):1484–97. https://doi.org/10.1161/CIRCRESAHA.111.263848.

154. Martin MU, Wesche H. Summary and comparison of the signaling mechanisms of the Toll/interleukin-1 receptor family. Biochim Biophys Acta. 2002;1592(3):265–80. https://doi.org/10.1016/s0167-4889(02)00320-8.

155. Martinon F, Burns K, Tschopp J. The inflammasome: a molecular platform triggering activation of inflammatory caspases and processing of proIL-beta. Mol Cell. 2002;10(2):417–26. https://doi.org/10.1016/S1097-2765(02)00599-3.

156. Martinon F, Pétrilli V, Mayor A, et al. Gout-associated uric acid crystals activate the NALP3 inflammasome. Nature. 2006;440(7081):237–41. https://doi.org/10.1038/nature04516.

157. Martinon F, Mayor A, Tschopp J. The inflammasomes: guardians of the body. Annu Rev Immunol. 2009;27:229–65. https://doi.org/10.1146/annurev.immunol.021908.132715.

158. Martinon F, Chen X, Lee A, et al. TLR activation of the transcription factor XBP1 regulates innate immune responses in macrophages. Nat Immunol. 2010;11(5):411–8. https://doi.org/10.1038/ni.1857.

159. Mathers J, Fraser JA, McMahon M, et al. Antioxidant and cytoprotective responses to redox stress. Biochem Soc Symp. 2004;71:157–76. https://doi.org/10.1042/bss0710157.

160. McCullough KD, Martindale JL, Klotz L, et al. Gadd153 sensitizes cells to endoplasmic reticulum stress by down-regulating Bcl2 and perturbing the cellular redox state. Mol Cell Biol. 2001;21(4):1249–59. https://doi.org/10.1128/MCB.21.4.1249-1259.2001.

161. Melber A, Haynes CM. UPRmt regulation and output: a stress response mediated by mitochondrial-nuclear communication. Cell Res. 2018;28(3):281–95. https://doi.org/10.1038/cr.2018.16.

162. Meyer M, Caselmann WH, Schlüter V, et al. Hepatitis B virus transactivator MHBst: activation of NF-kappa B, selective inhibition by antioxidants and integral membrane localization. EMBO J. 1992;11(8):2991–3001. https://doi.org/10.1002/j.1460-2075.1992.tb05369.x.

163. Minamino T, Christou H, Hsieh C, et al. Targeted expression of heme oxygenase-1 prevents the pulmonary inflammatory and vascular responses to hypoxia. Proc Natl Acad Sci. 2001;98(15):8798–803. https://doi.org/10.1073/pnas.161272598.

164. Misra S, Heldin P, Hascall VC, et al. Hyaluronan-CD44 interactions as potential targets for cancer therapy. FEBS J. 2011;278(9):1429–43. https://doi.org/10.1111/j.1742-4658.2011.08071.x.

165. Misra S, Hascall VC, Markwald RR, et al. Interactions between hyaluronan and its receptors (CD44, RHAMM) regulate the activities of inflammation and cancer. Front Immunol. 2015;6:201. https://doi.org/10.3389/fimmu.2015.00201.

166. Mittal M, Roth M, König P, et al. Hypoxia-dependent regulation of nonphagocytic NADPH oxidase subunit NOX4 in the pulmonary vasculature. Circ Res. 2007;101(3):258–67. https://doi.org/10.1161/CIRCRESAHA.107.148015.

167. Miyake K, Underhill CB, Lesley J, et al. Hyaluronate can function as a cell adhesion molecule and CD44 participates in hyaluronate recognition. J Exp

Med. 1990;172(1):69–75. https://doi.org/10.1084/jem.172.1.69.

168. Mogilenko DA, Haas JT, L'homme L, et al. Metabolic and innate immune cues merge into a specific inflammatory response via the UPR. Cell. 2019;177(5):1201–1216.e19. https://doi.org/10.1016/j.cell.2019.03.018.

169. Montorfano I, Becerra A, Cerro R, et al. Oxidative stress mediates the conversion of endothelial cells into myofibroblasts via a TGF-β1 and TGF-β2-dependent pathway. Lab Investig. 2014;94(10):1068–82. https://doi.org/10.1038/labinvest.2014.100.

170. Morgan MJ, Liu ZG. Crosstalk of reactive oxygen species and NF-κB signaling. Cell Res. 2011;21(1):103–15. https://doi.org/10.1038/cr.2010.178.

171. Mori K, Ma W, Gething M, et al. A transmembrane protein with a cdc2+CDC28-related kinase activity is required for signaling from the ER to the nucleus. Cell. 1993;74(4):743–56. https://doi.org/10.1016/0092-8674(93)90521-Q.

172. Motohashi H, Shavit JA, Igarashi K, et al. The world according to Maf. Nucleic Acids Res. 1997;25(15):2953–9. https://doi.org/10.1093/nar/25.15.2953.

173. Mulero MC, Wang VY, Huxford T, et al. Genome reading by the NF-κB transcription factors. Nucleic Acids Res. 2019;47(19):9967–89. https://doi.org/10.1093/nar/gkz739.

174. Münch C. The different axes of the mammalian mitochondrial unfolded protein response. BMC Biol. 2018;16(1):81. https://doi.org/10.1186/s12915-018-0548-x.

175. Murakami T, Kondo S, Ogata M, et al. Cleavage of the membrane-bound transcription factor OASIS in response to endoplasmic reticulum stress. J Neurochem. 2006;96(4):1090–100. https://doi.org/10.1111/j.1471-4159.2005.03596.x.

176. Nardai G, Braun L, Csala M, et al. Protein-disulfide isomerase- and protein thiol-dependent dehydroascorbate reduction and ascorbate accumulation in the lumen of the endoplasmic reticulum. J Biol Chem. 2001;276(12):8825–8. https://doi.org/10.1074/jbc.M010563200.

177. Nargund AM, Pellegrino MW, Fiorese CJ, et al. Mitochondrial import efficiency of ATFS-1 regulates mitochondrial UPR activation. Science. 2012;337(6094):587–90. https://doi.org/10.1126/science.1223560.

178. Nguyen T, Sherratt PJ, Pickett CB. Regulatory mechanisms controlling gene expression mediated by the antioxidant response element. Annu Rev Pharmacol Toxicol. 2003;43:233–60. https://doi.org/10.1146/annurev.pharmtox.43.100901.140229.

179. Nishitoh H, Matsuzawa A, Tobiume K, et al. ASK1 is essential for endoplasmic reticulum stress-induced neuronal cell death triggered by expanded polyglutamine repeats. Genes Dev. 2002;16(11):1345–55. https://doi.org/10.1101/gad.992302.

180. Novoa I, Zeng H, Harding HP, et al. Feedback inhibition of the unfolded protein response by GADD34-mediated dephosphorylation of eIF2α. J Cell Biol. 2001;153(5):1011–22. https://doi.org/10.1083/jcb.153.5.1011.

181. Ochoa CD, Wu RF, Terada LS. ROS signaling and ER stress in cardiovascular disease. Mol Asp Med. 2018;63:18–29. https://doi.org/10.1016/j.mam.2018.03.002.

182. Otterbein LE, Bach FH, Alam J, et al. Carbon monoxide has anti-inflammatory effects involving the mitogen-activated protein kinase pathway. Nat Med. 2000;6(4):422–8. https://doi.org/10.1038/74680.

183. Pahl HL, Baeuerle PA. A novel signal transduction pathway from the endoplasmic reticulum to the nucleus is mediated by transcription factor NF-kappa B. EMBO J. 1995;14(11):2580–8. https://doi.org/10.1002/j.1460-2075.1995.tb07256.x.

184. Patil C, Walter P. Intracellular signaling from the endoplasmic reticulum to the nucleus: the unfolded protein response in yeast and mammals. Curr Opin Cell Biol. 2001;13(3):349–55. https://doi.org/10.1016/S0955-0674(00)00219-2.

185. Pedruzzi E, Guichard C, Ollivier V, et al. NAD(P)H oxidase Nox-4 mediates 7-ketocholesterol-induced endoplasmic reticulum stress and apoptosis in human aortic smooth muscle cells. Mol Cell Biol. 2004;24(24):10703–17. https://doi.org/10.1128/MCB.24.24.10703-10717.2004.

186. Piccirella S, Czegle I, Lizák B, et al. Uncoupled redox systems in the lumen of the endoplasmic reticulum. Pyridine nucleotides stay reduced in an oxidative environment. J Biol Chem. 2006;281(8):4671–7. https://doi.org/10.1074/jbc.M509406200.

187. Pollard MG, Travers KJ, Weissman JS. Ero1p: a novel and ubiquitous protein with an essential role in oxidative protein folding in the endoplasmic reticulum. Mol Cell. 1998;1(2):171–82. https://doi.org/10.1016/S1097-2765(00)80018-0.

188. Polverino F, Celli BR, Owen CA. COPD as an endothelial disorder: endothelial injury linking lesions in the lungs and other organs? Pulm Circ. 2018;8(1):2045894018758528. https://doi.org/10.1177/2045894018758528.

189. Prasad AM, Nuno DW, Koval OM, et al. Differential control of calcium homeostasis and vascular reactivity by Ca2+/calmodulin-dependent kinase II. Hypertension. 2013;62(2):434–41. https://doi.org/10.1161/HYPERTENSIONAHA.113.01508.

190. Pugliese SC, Poth JM, Fini MA, et al. The role of inflammation in hypoxic pulmonary hypertension: from cellular mechanisms to clinical phenotypes. Am J Physiol Lung Cell Mol Physiol. 2015;308(3):229. https://doi.org/10.1152/ajplung.00238.2014.

191. Quirós PM, Prado MA, Zamboni N, et al. Multiomics analysis identifies ATF4 as a key regulator of the mitochondrial stress response in mammals. J Cell Biol. 2017;216(7):2027–45. https://doi.org/10.1083/jcb.201702058.

192. Ramming T, Appenzeller-Herzog C. Destroy and exploit: catalyzed removal of hydroperoxides from the endoplasmic reticulum. Int J

Cell Biol. 2013;2013:180906–13. https://doi.org/10.1155/2013/180906.

193. Ramming T, Hansen HG, Nagata K, et al. GPx8 peroxidase prevents leakage of H2O2 from the endoplasmic reticulum. Free Radic Biol Med. 2014;70:106–16. https://doi.org/10.1016/j.freeradbiomed.2014.01.018.

194. Ramming T, Okumura M, Kanemura S, et al. A PDI-catalyzed thiol–disulfide switch regulates the production of hydrogen peroxide by human Ero1. Free Radic Biol Med. 2015;83:361–72. https://doi.org/10.1016/j.freeradbiomed.2015.02.011.

195. Reddy SAG, Huang JH, Liao WS. Phosphatidylinositol 3-kinase in interleukin 1 signaling. Physical interaction with the interleukin 1 receptor and requirement in NFkappaB and AP-1 activation. J Biol Chem. 1997;272(46):29167–73. https://doi.org/10.1074/jbc.272.46.29167.

196. Rubio C, Pincus D, Korennykh A, et al. Homeostatic adaptation to endoplasmic reticulum stress depends on Ire1 kinase activity. J Cell Biol. 2011;193(1):171–84. https://doi.org/10.1083/jcb.201007077.

197. Rutkowski DT, Hegde RS. Regulation of basal cellular physiology by the homeostatic unfolded protein response. J Cell Biol. 2010;189(5):783–94. https://doi.org/10.1083/jcb.201003138.

198. Sacerdote P, Massi P, Panerai AE, et al. In vivo and in vitro treatment with the synthetic cannabinoid CP55, 940 decreases the in vitro migration of macrophages in the rat: involvement of both CB1 and CB2 receptors. J Neuroimmunol. 2000;109(2):155–63. https://doi.org/10.1016/S0165-5728(00)00307-6.

199. Santos CXC, Tanaka LY, Wosniak J, et al. Mechanisms and implications of reactive oxygen species generation during the unfolded protein response: roles of endoplasmic reticulum oxidoreductases, mitochondrial electron transport, and NADPH oxidase. Antioxid Redox Signal. 2009;11(10):2409–27. https://doi.org/10.1089/ars.2009.2625.

200. Santos CXC, Nabeebaccus AA, Shah AM, et al. Endoplasmic reticulum stress and Nox-mediated reactive oxygen species signaling in the peripheral vasculature: potential role in hypertension. Antioxid Redox Signal. 2014;20(1):121–34. https://doi.org/10.1089/ars.2013.5262.

201. Sawada H, Saito T, Nickel NP, et al. Reduced BMPR2 expression induces GM-CSF translation and macrophage recruitment in humans and mice to exacerbate pulmonary hypertension. J Exp Med. 2014;211(2):263–80. https://doi.org/10.1084/jem.20111741.

202. Sborgi L, Ravotti F, Dandey VP, et al. Structure and assembly of the mouse ASC inflammasome by combined NMR spectroscopy and cryo-electron microscopy. Proc Natl Acad Sci. 2015;112(43):13237–42. https://doi.org/10.1073/pnas.1507579112.

203. Schindler AJ, Schekman R. In vitro reconstitution of ER-stress induced ATF6 transport in COPII ves-

icles. Proc Natl Acad Sci. 2009;106(42):17775–80. https://doi.org/10.1073/pnas.0910342106.

204. Schröder M, Kaufman RJ. The mammalian unfolded protein response. Annu Rev Biochem. 2005;74:739–89. https://doi.org/10.1146/annurev.biochem.73.011303.074134.

205. Sciarretta S, Zhai P, Shao D, et al. Activation of NADPH oxidase 4 in the endoplasmic reticulum promotes cardiomyocyte autophagy and survival during energy stress through the protein kinase RNA-activated-like endoplasmic reticulum kinase/eukaryotic initiation factor 2α/activating transcription factor 4 pathway. Circ Res. 2013;113(11):1253–64. https://doi.org/10.1161/CIRCRESAHA.113.301787.

206. Scott JA, Xie L, Li H, et al. The multifunctional Ca2+/calmodulin-dependent kinase II regulates vascular smooth muscle migration through matrix metalloproteinase 9. Am J Physiol Heart Circ Physiol. 2012;302(10):1953. https://doi.org/10.1152/ajpheart.00978.2011.

207. Scott TE, Kemp-Harper BK, Hobbs AJ. Inflammasomes: a novel therapeutic target in pulmonary hypertension? Br J Pharmacol. 2019;176(12):1880–96. https://doi.org/10.1111/bph.14375.

208. Sepulveda D, Rojas-Rivera D, Rodríguez DA, et al. Interactome screening identifies the ER luminal chaperone Hsp47 as a regulator of the unfolded protein response transducer IRE1α. Mol Cell. 2018;69(2):238–252.e7. https://doi.org/10.1016/j.molcel.2017.12.028.

209. Shen J, Chen X, Hendershot L, et al. ER stress regulation of ATF6 localization by dissociation of BiP/GRP78 binding and unmasking of Golgi localization signals. Dev Cell. 2002;3(1):99–111. https://doi.org/10.1016/S1534-5807(02)00203-4.

210. Shoulders MD, Ryno LM, Genereux JC, et al. Stress-independent activation of XBP1s and/or ATF6 reveals three functionally diverse ER proteostasis environments. Cell Rep. 2013;3(4):1279–92. https://doi.org/10.1016/j.celrep.2013.03.024.

211. Shpilka T, Haynes CM. The mitochondrial UPR: mechanisms, physiological functions and implications in ageing. Nat Rev Mol Cell Biol. 2018;19(2):109–20. https://doi.org/10.1038/nrm.2017.110.

212. Sicari D, Delaunay-Moisan A, Combettes L, et al. A guide to assessing endoplasmic reticulum homeostasis and stress in mammalian systems. FEBS J. 2020;287(1):27–42. https://doi.org/10.1111/febs.15107.

213. Simenauer A, Assefa B, Rios-Ochoa J, et al. Repression of Nrf2/ARE regulated antioxidant genes and dysregulation of the cellular redox environment by the HIV transactivator of transcription. Free Radic Biol Med. 2019;141:244–52. https://doi.org/10.1016/j.freeradbiomed.2019.06.015.

214. Simmen T, Lynes EM, Gesson K, et al. Oxidative protein folding in the endoplasmic reticulum: tight links to the mitochondria-associated membrane (MAM).

Biochim Biophys Acta. 2010;1798(8):1465–73. https://doi.org/10.1016/j.bbamem.2010.04.009.

215. Soares MP, Lin Y, Anrather J, et al. Expression of heme oxygenase-1 can determine cardiac xenograft survival. Nat Med. 1998;4(9):1073–7. https://doi.org/10.1038/2063.

216. Sohara Y, Ishiguro N, Machida K, et al. Hyaluronan activates cell motility of v-Src-transformed cells via Ras-mitogen-activated protein kinase and phosphoinositide 3-kinase-Akt in a tumor-specific manner. Mol Biol Cell. 2001;12(6):1859–68. https://doi.org/10.1091/mbc.12.6.1859.

217. Song B, Scheuner D, Ron D, et al. Chop deletion reduces oxidative stress, improves β cell function, and promotes cell survival in multiple mouse models of diabetes. J Clin Invest. 2008;118(10):3378–89. https://doi.org/10.1172/JCI34587.

218. Sriburi R, Jackowski S, Mori K, et al. XBP1: a link between the unfolded protein response, lipid biosynthesis, and biogenesis of the endoplasmic reticulum. J Cell Biol. 2004;167(1):35–41. https://doi.org/10.1083/jcb.200406136.

219. Stacher E, Graham BB, Hunt JM, et al. Modern age pathology of pulmonary arterial hypertension. Am J Respir Crit Care Med. 2012;186(3):261–72. https://doi.org/10.1164/rccm.201201-0164OC.

220. Stauffer WT, Arrieta A, Blackwood EA, et al. Sledgehammer to scalpel: broad challenges to the heart and other tissues yield specific cellular responses via transcriptional regulation of the ER-stress master regulator ATF6α. Int J Mol Sci. 2020;21(3):1134. https://doi.org/10.3390/ijms21031134.

221. Stenmark KR, Tuder RM, El Kasmi KC. Metabolic reprogramming and inflammation act in concert to control vascular remodeling in hypoxic pulmonary hypertension. J Appl Physiol. 2015;119(10):1164–72. https://doi.org/10.1152/japplphysiol.00283.2015.

222. Stocker R, Yamamoto Y, McDonagh AF, et al. Bilirubin is an antioxidant of possible physiological importance. Science. 1987;235(4792):1043–6. https://doi.org/10.1126/science.3029864.

223. Strieter RM, Kunkel SL, Keane MP, et al. Chemokines in lung injury: Thomas A. Neff Lecture. Chest. 1999;116(1 Suppl):103S–10S. https://doi.org/10.1378/chest.116.suppl_1.103s.

224. Sutendra G, Dromparis P, Wright P, et al. The role of Nogo and the mitochondria-endoplasmic reticulum unit in pulmonary hypertension. Sci Transl Med. 2011;3(88):88ra55. https://doi.org/10.1126/scitranslmed.3002194.

225. Száraz P, Bánhegyi G, Benedetti A. Altered redox state of luminal pyridine nucleotides facilitates the sensitivity towards oxidative injury and leads to endoplasmic reticulum stress dependent autophagy in HepG2 cells. Int J Biochem Cell Biol. 2010;42(1):157–66. https://doi.org/10.1016/j.biocel.2009.10.004.

226. Tammi MI, Day AJ, Turley EA. Hyaluronan and homeostasis: a balancing act. J Biol Chem. 2002;277(7):4581–4. https://doi.org/10.1074/jbc.R100037200.

227. Tavender TJ, Bulleid NJ. Molecular mechanisms regulating oxidative activity of the Ero1 family in the endoplasmic reticulum. Antioxid Redox Signal. 2010;13(8):1177–87. https://doi.org/10.1089/ars.2010.3230.

228. Tavender TJ, Sheppard AM, Bulleid NJ. Peroxiredoxin IV is an endoplasmic reticulum-localized enzyme forming oligomeric complexes in human cells. Biochem J. 2008;411(1):191–9. https://doi.org/10.1042/BJ20071428.

229. Tebay LE, Robertson H, Durant ST, et al. Mechanisms of activation of the transcription factor Nrf2 by redox stressors, nutrient cues, and energy status and the pathways through which it attenuates degenerative disease. Free Radic Biol Med. 2015;88(Pt B):108–46. https://doi.org/10.1016/j.freeradbiomed.2015.06.021.

230. Thenappan T, Chan SY, Weir EK. Role of extracellular matrix in the pathogenesis of pulmonary arterial hypertension. Am J Physiol Heart Circ Physiol. 2018;315(5):1322. https://doi.org/10.1152/ajpheart.00136.2018.

231. Thon M, Hosoi T, Ozawa K. Dehydroascorbic acid-induced endoplasmic reticulum stress and leptin resistance in neuronal cells. Biochem Biophys Res Commun. 2016;478(2):716–20. https://doi.org/10.1016/j.bbrc.2016.08.013.

232. Timmins JM, Ozcan L, Seimon TA, et al. Calcium/calmodulin-dependent protein kinase II links ER stress with Fas and mitochondrial apoptosis pathways. J Clin Invest. 2009;119(10):2925–41. https://doi.org/10.1172/JCI38857.

233. Tirasophon W, Welihinda AA, Kaufman RJ. A stress response pathway from the endoplasmic reticulum to the nucleus requires a novel bifunctional protein kinase/endoribonuclease (Ire1p) in mammalian cells. Genes Dev. 1998;12(12):1812–24. https://doi.org/10.1101/gad.12.12.1812.

234. Travers KJ, Patil CK, Wodicka L, et al. Functional and genomic analyses reveal an essential coordination between the unfolded protein response and ER-associated degradation. Cell. 2000;101(3):249–58. https://doi.org/10.1016/s0092-8674(00)80835-1.

235. Tu BP, Ho-Schleyer SC, Travers KJ, et al. Biochemical basis of oxidative protein folding in the endoplasmic reticulum. Science. 2000;290(5496):1571–4. https://doi.org/10.1126/science.290.5496.1571.

236. Tuder RM. Pulmonary vascular remodeling in pulmonary hypertension. Cell Tissue Res. 2017;367(3):643–9. https://doi.org/10.1007/s00441-016-2539-y.

237. Urano F, Wang X, Bertolotti A, et al. Coupling of stress in the ER to activation of JNK protein kinases by transmembrane protein kinase IRE1. Science.

2000;287(5453):664–6. https://doi.org/10.1126/science.287.5453.664.

238. van Dam L, Dansen TB. Cross-talk between redox signalling and protein aggregation. Biochem Soc Trans. 2020;48(2):379–97. https://doi.org/10.1042/BST20190054.

239. van Vliet AR, Verfaillie T, Agostinis P. New functions of mitochondria associated membranes in cellular signaling. Biochim Biophys Acta. 2014;1843(10):2253–62. https://doi.org/10.1016/j.bbamcr.2014.03.009.

240. Vila-Petroff M, Salas MA, Said M, et al. CaMKII inhibition protects against necrosis and apoptosis in irreversible ischemia-reperfusion injury. Cardiovasc Res. 2007;73(4):689–98. https://doi.org/10.1016/j.cardiores.2006.12.003.

241. Villegas LR, Kluck D, Field C, et al. Superoxide dismutase mimetic, MnTE-2-PyP, attenuates chronic hypoxia-induced pulmonary hypertension, pulmonary vascular remodeling, and activation of the NALP3 inflammasome. Antioxid Redox Signal. 2013;18(14):1753–64. https://doi.org/10.1089/ars.2012.4799.

242. Wagner L, Laczy B, Tamaskó M, et al. Cigarette smoke-induced alterations in endothelial nitric oxide synthase phosphorylation: role of protein kinase C. Endothelium. 2007;14(4–5):245–55. https://doi.org/10.1080/10623320701606707.

243. Wajih N, Hutson SM, Wallin R. Disulfide-dependent protein folding is linked to operation of the vitamin K cycle in the endoplasmic reticulum. J Biol Chem. 2007;282(4):2626–35. https://doi.org/10.1074/jbc.M608954200.

244. Wang X, Zhang Y, Jolicoeur EM, et al. Cloning of mammalian Ire1 reveals diversity in the ER stress responses. EMBO J. 1998;17(19):5708–17. https://doi.org/10.1093/emboj/17.19.5708.

245. Wang Q, Zuo X, Wang Y, et al. Monocrotaline-induced pulmonary arterial hypertension is attenuated by TNF-α antagonists via the suppression of TNF-α expression and NF-κB pathway in rats. Vasc Pharmacol. 2013;58(1–2):71–7. https://doi.org/10.1016/j.vph.2012.07.006.

246. Wang P, Li J, Tao J, et al. The luminal domain of the ER stress sensor protein PERK binds misfolded proteins and thereby triggers PERK oligomerization. J Biol Chem. 2018;293(11):4110–21. https://doi.org/10.1074/jbc.RA117.001294.

247. Wang E, Jia M, Luo F, et al. Coordination between NADPH oxidase and vascular peroxidase 1 promotes dysfunctions of endothelial progenitor cells in hypoxia-induced pulmonary hypertensive rats. Eur J Pharmacol. 2019;857:172459. https://doi.org/10.1016/j.ejphar.2019.172459.

248. Wedgwood S, Dettman RW, Black SM. ET-1 stimulates pulmonary arterial smooth muscle cell proliferation via induction of reactive oxygen species. Am J Physiol Lung Cell Mol Physiol. 2001;281(5):1058. https://doi.org/10.1152/ajplung.2001.281.5.L1058.

249. Wei P, Hsieh Y, Su M, et al. Loss of the oxidative stress sensor NPGPx compromises GRP78 chaperone activity and induces systemic disease. Mol Cell. 2012;48(5):747–59. https://doi.org/10.1016/j.molcel.2012.10.007.

250. Weise-Cross L, Sands MA, Sheak JR, et al. Actin polymerization contributes to enhanced pulmonary vasoconstrictor reactivity after chronic hypoxia. Am J Physiol Heart Circ Physiol. 2018;314(5):1011. https://doi.org/10.1152/ajpheart.00664.2017.

251. Weise-Cross L, Resta TC, Jernigan NL. Redox regulation of ion channels and receptors in pulmonary hypertension. Antioxid Redox Signal. 2019;31(12):898–915. https://doi.org/10.1089/ars.2018.7699.

252. Wells WW, Xu DP, Yang YF, et al. Mammalian thioltransferase (glutaredoxin) and protein disulfide isomerase have dehydroascorbate reductase activity. J Biol Chem. 1990;265(26):15361–4.

253. Wetmore DR, Hardman KD. Roles of the propeptide and metal ions in the folding and stability of the catalytic domain of stromelysin (matrix metalloproteinase 3). Biochemistry. 1996;35(21):6549–58. https://doi.org/10.1021/bi9530752.

254. Wietek C, O'Neill LAJ. Diversity and regulation in the NF-κB system. Trends Biochem Sci. 2007;32(7):311–9. https://doi.org/10.1016/j.tibs.2007.05.003.

255. Woehlbier U, Hetz C. Modulating stress responses by the UPRosome: a matter of life and death. Trends Biochem Sci. 2011;36(6):329–37. https://doi.org/10.1016/j.tibs.2011.03.001.

256. Woo CW, Cui D, Arellano J, et al. Adaptive suppression of the ATF4–CHOP branch of the unfolded protein response by toll-like receptor signalling. Nat Cell Biol. 2009;11(12):1473–80. https://doi.org/10.1038/ncb1996.

257. Wrobel L, Topf U, Bragoszewski P, et al. Mistargeted mitochondrial proteins activate a proteostatic response in the cytosol. Nature. 2015;524(7566):485–8. https://doi.org/10.1038/nature14951.

258. Wu S, Tan M, Hu Y, et al. Ultraviolet light activates NFκB through translational inhibition of IκBα synthesis. J Biol Chem. 2004;279(33):34898–902. https://doi.org/10.1074/jbc.M405616200.

259. Wu Y, Adi D, Long M, et al. 4-Phenylbutyric acid induces protection against pulmonary arterial hypertension in rats. PLoS One. 2016;11(6):e0157538. https://doi.org/10.1371/journal.pone.0157538.

260. Xia Y, Shen S, Verma IM. NF-κB, an active player in human cancers. Cancer Immunol Res. 2014;2(9):823–30. https://doi.org/10.1158/2326-6066.CIR-14-0112.

261. Xing Z, Gauldie J, Cox G, et al. IL-6 is an antiinflammatory cytokine required for controlling local or systemic acute inflammatory responses. J Clin Invest. 1998;101(2):311–20. https://doi.org/10.1172/JCI1368.

262. Xing Z, Jordana M, Gauldie J, et al. Cytokines and pulmonary inflammatory and immune diseases. Histol Histopathol. 1999;14(1):185–201. https://doi.org/10.14670/HH-14.185.

263. Ye J, Rawson RB, Komuro R, et al. ER stress induces cleavage of membrane-bound ATF6 by the same proteases that process SREBPs. Mol Cell. 2000;6(6):1355–64. https://doi.org/10.1016/S1097-2765(00)00133-7.

264. Yeager ME, Belchenko DD, Nguyen CM, et al. Endothelin-1, the unfolded protein response, and persistent inflammation: role of pulmonary artery smooth muscle cells. Am J Respir Cell Mol Biol. 2012;46(1):14–22. https://doi.org/10.1165/rcmb.2010-0506OC.

265. Yeager ME, Reddy MB, Nguyen CM, et al. Activation of the unfolded protein response is associated with pulmonary hypertension. Pulm Circ. 2012;2(2):229–40. https://doi.org/10.4103/2045-8932.97613.

266. Yoshida H, Haze K, Yanagi H, et al. Identification of the cis-acting endoplasmic reticulum stress response element responsible for transcriptional induction of mammalian glucose-regulated proteins. Involvement of basic leucine zipper transcription factors. J Biol Chem. 1998;273(50):33741–9. https://doi.org/10.1074/jbc.273.50.33741.

267. Yoshida H, Matsui T, Yamamoto A, et al. XBP1 mRNA is induced by ATF6 and spliced by IRE1 in response to ER stress to produce a highly active transcription factor. Cell. 2001;107(7):881–91. https://doi.org/10.1016/S0092-8674(01)00611-0.

268. Yu H, Alruwaili N, Hu B, et al. Potential role of cartilage oligomeric matrix protein in the modulation of pulmonary arterial smooth muscle superoxide by hypoxia. Am J Physiol Lung Cell Mol Physiol. 2019;317(5):L569–77. https://doi.org/10.1152/ajplung.00080.2018.

269. Zhang DD. Mechanistic studies of the Nrf2-Keap1 signaling pathway. Drug Metab Rev. 2006;38(4):769–89. https://doi.org/10.1080/03602530600971974.

270. Zhang K, Kaufman RJ. From endoplasmic-reticulum stress to the inflammatory response. Nature. 2008;454(7203):455–62. https://doi.org/10.1038/nature07203.

271. Zhang W, Chen D, Qi F, et al. Inhibition of calcium-calmodulin-dependent kinase II suppresses cardiac fibroblast proliferation and extracellular matrix secretion. J Cardiovasc Physiol. 2010;55(1):96–105. https://doi.org/10.1097/FJC.0b013e3181c9548b.

272. Zhang Z, Zhang L, Zhou L, et al. Redox signaling and unfolded protein response coordinate cell fate decisions under ER stress. Redox Biol. 2019;25:101047. https://doi.org/10.1016/j.redox.2018.11.005.

273. Zhou BP, Liao Y, Xia W, et al. HER-2/neu induces p53 ubiquitination via Akt-mediated MDM2 phosphorylation. Nat Cell Biol. 2001;3(11):973–82. https://doi.org/10.1038/ncb1101-973.

274. Zhou H, Liu H, Porvasnik SL, et al. Heme oxygenase-1 mediates the protective effects of rapamycin in monocrotaline-induced pulmonary hypertension. Lab Investig. 2006;86(1):62–71. https://doi.org/10.1038/labinvest.3700361.

275. Zhou R, Tardivel A, Thorens B, et al. Thioredoxin-interacting protein links oxidative stress to inflammasome activation. Nat Immunol. 2010;11(2):136–40. https://doi.org/10.1038/ni.1831.

276. Zhou R, Yazdi AS, Menu P, et al. A role for mitochondria in NLRP3 inflammasome activation. Nature. 2011;469(7329):221–6. https://doi.org/10.1038/nature09663.

277. Zhu W, Woo AYH, Yang D, et al. Activation of CaMKIIδC is a common intermediate of diverse death stimuli-induced heart muscle cell apoptosis. J Biol Chem. 2007;282(14):10833–9. https://doi.org/10.1074/jbc.M611507200.

Index

A

Absent in melanoma 2 (AIM2), 60
Activating transcription factor 4 (ATF4), 338
Activating transcription factor 6 (ATF6), 340
Activator protein-1 (AP-1), 190
Acute lung injury (ALI), 96, 188, 190, 325
  and ARDS, 96, 103
  in a Caucasian population, 102
  coagulation and inflammation, 98
  cytokine stimuli, 98
  development, 101
  endothelial barrier, 97
  genome-wide significance, 98
  lung endothelial and epithelial barriers, 97
  molecular marker, 98
  pathophysiology, 98
  treatment, 100
Acute myeloid leukemia (AML), 182
Acute respiratory distress syndrome (ARDS), 96, 194,
     290, 323
  ALI, 325
  classification, 324
  clinical presentation, 326
  definition, 324
  etiology, 325
  extrapulmonary ARDS, 325
  management, 327
    corticosteroids, 329
    fluid management, 329
    mechanical ventilation, 327
    NMBA therapy, 329
    prone positioning, 328
    ventilator therapy, 328
    ventilatory modes, 327
  medical management, 323
  mortality rate, 324
  pathophysiology, 325
  pharmacologic therapies, 329
  potential complication, 329
  pulmonary ARDS, 325
  risk factors
    age, 325
    alcohol, 326

    burns, 326
    diabetes, 326
    obesity paradox, 326
Adaptive immune response, 270
Adaptive RV hypertrophy, 240
Adenylyl cyclase (AC), 262
Airway hyperresponsiveness (AHR), 4, 151
  allergic asthma, 209
  Sema3E-Fc Ig effect, 209
  TNF and IL-1β, 209
Airway inflammation, 207
Airway inflammatory disorders, 208
Airway parasympathetic ganglia, 114
Airway remodeling, 209
  HDM allergen, 209
  Sema3E-deficient mice, 209
Airway smooth muscle (ASM), 2, 110, 191, 275
Airway smooth muscle cells (ASMCs), 151
  hyperplasia, 151
  hyperresponsiveness, 151
  migration, 151
  remodeling, 151
Allergic asthma, 191
Alpha-1 antitrypsin (AATD), 150
Alveolar macrophages, 41–43, 46, 97, 191
Androgen biosynthetic pathway
  in Leydig cells, 262
Androgen receptor (AR), 266
Angio-obliterative PH, 29
Angiotensin-converting enzyme (ACE), 30
Angiotensin-converting enzyme 2
    (ACE2), 219
Anti-apoptotic pathways, 198
Anti-citrullinated protein antibodies (ACPA), 84
Anti-inflammatory agents, 150
Antioxidant response element (ARE), 354
Antioxidants, 195
Antioxidant signaling, 361
Anti-PAR2 antibodies, 9
Antiplatelet therapy (APT), 195
Apoptosis signal-regulating kinase 1 (ASK1), 339
Apparent diffusion coefficient (ADC), 129
Aprotinin, 222

ARDS et Curarization Systematique (ACURASYS)
    trial, 329
Arterial spin labelling (ASL), 135
Ascorbate peroxidase (APx), 344
ASM cells (ASMCs), 278, 280–283, 287
Asthma, 61, 148, 149, 205, 232
    AHR, 280
    airway inflammation, 207
    airway mucus, 2
    allergen-induced, 283
    allergic, 276
    allergic asthma, 208
    androgen and estrogen effects, 277
    androgens, 273
    androgens' effects on inflammation, 275, 276, 278,
        279
    antioxidant therapies, 192
    chronic airway inflammatory disease, 271
    clinical aspects, 192
    and COPD, 274
    corticosteroids, 5
    cytokines, 282
    DHEA-S, 276
    estrogens' effects on inflammation, 279–284
    gender differences, 271
    IL-17A, 278
    IL-17-mediated neutrophil inflammatory response,
        274
    lung, 3
    menopausal-onset, 273
    neutrophilic inflammation, 208
    outcomes, 206
    oxidant production, 192
    persistent asthma, 278
    PMA, 272
    prevalence, 205
    semaphorins and plexins, 205
    severe asthma symptoms, 282
    sex steroid, 284
    TES levels, 273
    TLRs, 192
    TNF-α, 274
Asthma and Allergy Foundation of America
        (AAFA), 191
Asthmatic airway, 116
Asthmatic symptoms, 148
Asthma symptoms, 261
Autoimmune mechanisms, 41

B
Bacterial infection, 181
Basic leucine zipper (bZIP) transcription factors, 340
BCL-2-associated X protein (BAX), 340
Beta-agonists, 4, 9
Biased agonism pharmacology, 13
Bleomycin-induced PF, 261
Bone marrow-derived macrophage (BMDM), 101
Bone morphogenic protein type 2 receptor
        (BMPR2), 350

Brain-derived neurotrophic factor (BDNF), 112
    application, 115
    ASM layer, 115
    asthma, 115
    cellular responses, 113
    production, 111, 115
    protein synthesis, 111
    regulation, 111
    treatment, 114
Bromhexine, 222
Bronchial airway, 109
Bronchial airway disease, 81
Bronchial asthma, 191
Bronchial hyperreactivity (BHR), 61
Bronchiolitis obliterans syndrome (BOS), 133, 219
Bronchoalveolar lavage (BAL), 192
Bronchoalveolar lavage fluid (BALF), 152, 158
Bronchoconstriction, 4
Bronchodilator (BD), 3, 5, 8, 150
Bronchopulmonary dysplasia (BPD), 135, 229, 231, 244
    antenatal corticosteroids, 231
    disease diagnosis, 231
    lung disease, 231
    multiple clinical studies, 231
    newborn, 230
    sexual dimorphism, 231

C
Calcium-sensing receptor (CaSR), 7
Calcium signaling, 351
Calmodulin, 158
Camostat, 222
Carbon monoxide (CO), 357
Cartesian sampling methods, 135
Cartilage oligomeric matrix protein (COMP), 350
Ca^{2+} signaling, 155
    NF-κB signaling, 158
    RyR channels, 158
C/EBP-homologous protein (CHOP), 356
Cellular apoptosis, 195
Cellular oxidative stress, 354
Cellular stimulation, 261
Chemokines, 45
Chest tomographic analysis, 79
Childhood asthma, 232
Chitotriosidase, 44, 45
Cholesterol, 262
Chronic cough, 232
Chronic hypoxia, 350, 352, 357
Chronic inflammation, 154
Chronic obstructive pulmonary disease (COPD), 61, 84,
        100, 109, 193, 218, 234, 235, 271
    airways, 149
    androgens' effects on inflammation, 275, 276, 278, 279
    asthma, 150
    cardiac manifestations, 149
    corticosteroids, 150
    development and progression, 149
    diagnosis, 273

estrogens' effects on inflammation, 279–284
factors, 149
IFN response, 61, 62
IL-17A and IL-22, 275
inflammasome activation, 61, 62
inflammation, 62
lung disease, 273
morbidity and mortality, 194
neutrophils and macrophages, 152
NF-κB activation, 62
pathogenesis, 61
PH, 149, 150
pharmacological treatments, 150
prevalence, 273
pulmonary and extrapulmonary, 149
remodeling, 152
risk factors and causes, 61
ROS levels, 193
severity, 61
treatments, 150
Chymotrypsin-like (CTL), 216
Cigarette smoke (CS), 61
Collagen synthesis, 154
Computed tomography (CT), 124
Connective tissue disease (CTD), 73, 74
Conventional respiratory diagnostics, 124
Coronavirus disease 2019 (COVID-19), 62, 196
    ACE2, 196
    androgens' effects on inflammation, 291–293
    ARDS, 290
    chromosomal differences, 242
    cytokines and chemokines, 291
    demographic and clinical data, 242
    estrogens' effects on inflammation, 293, 294
    gender factors, 243
    male sex hormones, 243
    mitochondrial ROS functions, 197
    morbidity and mortality, 242
    peroxynitrite anions, 196
    public health crisis, 242
    SARS-CoV, 196
    screening, 197
    serum ACE2, 243
    severity, 291
    sex differences, 242, 243
    sex-disaggregated data, 242
    sex-specific immune responses, 242
    treatment, 196
Corticosteroids (CS), 3
Crohn disease (CD), 41
CS/nicotine inhalation, 155
CT pulmonary angiography (CTPA), 137
CXC chemokine receptor 4 (CXCR4)), 189
Cysteine residues, 336
Cystic fibrosis (CF), 235, 236
Cystic fibrosis transmembrane conductance regulator
    (CFTR) gene, 235
Cytochrome c (CytC), 26, 28
Cytokine release syndrome (CRS), 196
Cytokines, 353, 357–360

D
Damage-associated molecular patterns (DAMPs), 55, 353
Dehydroepiandrosterone (DHEA), 240, 262, 263, 275,
    276, 278, 279, 286, 289, 292
Dendritic cells (DCs), 208, 270
    conventional, 208
    pulmonary, 209
    role, 208
    subsets, 208
Depalmitoylation, 182
Dexamethasone, 329
Diffuse alveolar damage (DAD), 82, 86
Diffusion capacity of carbon monoxide (DLCO), 79, 231
Diffusion-weighted imaging (DWI), 129, 131
Dissolved phase ^{129}Xe MR, 131
DNA methylation, 100
DNA methyltransferase inhibitor (DNMTi), 100
Drug repurposing, 222
Dynamic contrast-enhanced (DCE), 136

E
Elastase-like (EL), 216
Electron relay (ER), 343
Emphysema, 193
Endoplasmic reticulum (ER), 336, 355
Endothelial cells (ECs), 338
Endothelial dysfunction, 351, 352, 358, 361
Endothelial injury, 348, 350
Endothelial-mesenchymal transition (EndMT), 357
Endothelial progenitor cells (EPCs), 350
Eosinophilic airway inflammation, 191
Eosinophils, 269
EP receptor family, 7
Epidermal growth factor (EGF), 28
Epigenetic mechanisms, 100
Epigenetic processes, 99
Epigenome-wide association study (EWAS), 100
Epithelial-mesenchymal transition (EMT), 190
ER-associated degradation (ERAD), 337
ER oxidoreductase 1 (ERO1), 343
ER stress response element (ERSE), 337, 338, 340
17β-Estradiol (E2), 263, 280, 283
Erythropoietin (EPO), 27
Estrogen receptors (ERs), 238, 267
Estrogen response elements (EREs), 267
Evolutionarily bitter taste receptor signaling, 6
Exercise-induced bronchospasm (EIB), 233, 234
Extracellular matrix (ECM), 110, 116, 283, 336
Extracellular signal-regulated kinase (ERK), 339
Extrapulmonary ARDS, 325

F
Familial pulmonary arterial hypertension (FPAH), 358
Farnesylation, 176
Female sex hormones, 228
Fibrinolytic process, 98
Fibrinous variant, 85
Fibrogenesis, 284

Fibrosing lung disease, 74
F2-isoprostane, 193
Flavin adenine dinucleotide (FAD), 343

G
G protein-coupled receptors (GPCRs), 2, 178
   agonists, 3, 6
   airway and asthma biology, 3, 6
   ASM, 5
   biased ligand pharmacology, 11
   biology, 10
   endogenous levels, 4
   inflammatory agents, 4
   LABAs/LAMAs, 6
   ligands, 13
   limitations, 12
   m3mAChR antagonists, 4
   properties, 13
   pro-relaxant signaling, 3
   transmembrane, 6
Gadolinium, 136
Gadolinium enhancement, 137
GDNF family receptor (GFR) isoforms, 111
Gender, 228
Gender differences
   COVID-19, 243
   in EIB, 233
   in lung ailments, 262
   in respiratory disease, 228 (*see also* Sex differences)
Gene markers, 41
General control nonderepressible 2 (GCN2), 341
Genome-wide association studies (GWAS), 41–42, 99
Genomic approaches, 99
Glial-derived neurotrophic factor (GDNF), 113
   ASM, 116
   functional perspective, 116
   functionality, 114
   protein, 112
   secretion, 112
   signaling, 114
   synthesize and secrete, 111
Glucose transporters, 347
Glutathione (GSH), 197, 342, 343
Glutathione reductase (GR), 342
Glycolytic enzymes, 347
Glycosaminoglycan (GAG), 354
Golgi membrane interface, 178
Gonadotropin-releasing hormone (GnRH),
      262–265
GRADE (Grading of Recommendations Assessment,
      Development and Evaluation) Method, 327
Gradient recall echo (GRE), 136
Granulocyte-macrophage colony-stimulating factor
      (GM-CSF), 358
Granuloma formation, 43
Granuloma structure, 44
Granulomatous pulmonary disease, 40
Granulosa cells (GCs), 263

H
HA-mediated motility (RHAMM), 354
Heat shock protein 47 (HSP47), 341
High-resolution chest tomography (HRCT), 74
Histone acetylation, 101, 102
Histone acetyltransferase (HAT) inhibitor, 102
Histone deacetylase 1 (HDAC1), 198
Histopathology analysis, 84
HIV-induced pulmonary arterial hypertension
      (HIV-PAH), 349
Hormone replacement therapy (HRT), 238
Human leukocyte antigen (HLA) gene patterns, 41
Human lung development, 229, 230
Human neutrophil elastase (HNE), 217
Human pulmonary artery endothelial cells (HPAEC),
      102, 358
Human pulmonary microvascular endothelial cells
      (HPMECs), 349
Human umbilical endothelial vein cells (HUVECs), 29
Hyaluronan (HA), 354
Hyperpolarization techniques, 128
Hyperpolarized gas, 128
Hyperproliferation, 159
Hyperresponsiveness, 159
Hypoxia, 154, 190, 347
Hypoxia-inducible factors (HIFs), 26, 190, 347
   gene products, 26
   normoxic conditions, 190
Hypoxia response elements (HREs), 347
Hypoxic pulmonary vasoconstriction (HPV), 154

I
Icatibant, 219
Idiopathic inflammatory muscle disease, 75
Idiopathic interstitial pneumonias (IIPs), 74
Idiopathic pulmonary edema, 324
Idiopathic pulmonary fibrosis (IPF), 130, 236, 237
IFN-I receptor (IFNAR), 58
Immune cells
   DCs, 270
   eosinophils, 269
   macrophages, 269
   mast cells, 270, 274, 276, 281
   neutrophils, 268
   T and B lymphocytes, 270, 271
Immune-related genes, 243
Inflammasome, 59
Inflammation, 124, 189, 216, 217, 260
   acute phase, 260
   chronic phase, 261
   and coagulation, 216
   infectious origin, 40
   inorganic and organic substances, 40
Inflammatory cytokine signaling, 97
Inflammatory responses, 156
Influenza, 241, 242
Innate immune pathways, 356
Innate immune reactivity, 42

Innate immune responses
 acute respiratory disease, 62
 COVID-19, 62
 dysfunctions, 63
 inflammasome, 63
 SARS-CoV-2, 63
Innate immune system, 55
 NLRP1, 56
 NLRs, 56
 RLRs, 55
 TLRs, 55
Innate immunity, 40
Inositol-requiring protein 1 (IRE1), 337, 338
 alarm stress pathways, 339
 ATF6, 340
 chaperone immunoglobulin binding protein, 337
 endoribonuclease activity, 337
 ERAD, 337
 gene products, 339
 kinase activity, 337
 mammalian cells, 337
 nicotine adenine dinucleotide, 338–339
 PERK, 339, 340
 phospholipid synthesis, 339
 phosphorylation, 337
Integrated stress response (ISR), 341
Interferon gamma (IFN-γ) responses, 41
Interstitial lung disease (ILD), 74
 causes, 74
 clinical features, 75
 CTD, 75
 CTD-ILD, 74
 evaluation, 75, 77
 history, 75
 nailfold capillaroscopy, 77
 nature, 85
 pathogenic mechanisms, 74
 respiratory symptoms, 75
 serology, 84
Interstitial pneumonia with autoimmune
   features (IPAF), 87
 clinical domain, 87
 diagnostic criteria, 89
 morphological domain, 87
 serological domain, 87
 utility, 89
Intracellular nucleic acid sensors, 56

J
JUN N-terminal kinase (JNK), 339

K
Kallikrein-related peptidases, 220
 cellular and tissue localization, 220
 DX-2300, 220
 human kallikrein KLK1, 220
Kallistatin, 220
Keap1-Nrf-ARE signaling, 153

Kelch-like ECH-associated protein 1 (KEAP1), 339
Kinin-kallikrein system, 220
K-space data, 126

L
Leucine-rich repeat (LRR), 353
Leukocytes, 189, 194
Leydig cells, 262, 263, 265, 292
Lipid peroxidation, 195
Lipopolysaccharide (LPS), 356
Long-acting beta-agonist (LABA), 2
L-type voltage-gated Ca^{2+} channels (LTCCs), 155
Lung cancer (LC), 237–239, 287, 288
 androgens' effects on inflammation, 288, 289
 environmental risk factor, 197
 estrogens' effects on inflammation, 289, 290
 metastasis, 198
 NOX inhibitors, 198
 NSCLC, 197
 ROS, 197, 198
 treatments, 198
Lung clearance index (LCI), 139
Lung conditions, 227
Lung diseases
 adenocarcinoma, 228
 epithelial cells, 217
 men and women, 228
 pediatric and adult
  asthma, 232
  CF, 235, 236
  COPD, 234, 235
  DLCO, 231
  EIB, 233, 234
  IPF, 236, 237
  LAM, 239
  lung cancer, 237–239
  OSA, 239
  PAH, 240, 241
  respiratory infection, 241
  sexual dimorphism, 231
Lung infiltration, 98
Lung inflammation, 217, 222
 cellular mechanisms, 261
 resident macrophages, 261
Lung injury, 102
Lung neutrophilia, 208
Lung-resident macrophages, 349
Lung volume reduction (LVR), 139
Lymphangioleiomyomatosis (LAM), 131, 228, 239
Lymphocytes, 40, 42, 45, 47
Lymphocytic interstitial pneumonia (LIP), 81, 86

M
Macrophages, 45, 46, 269
Magnetic resonance imaging (MRI), 124
 anoxic mixture, 128
 coils transmit, 126
 computer algorithms, 126

Magnetic resonance imaging (MRI) (*cont.*)
  conventional, 126–128
  DCE MRI, 137
  function, 125
  gadolinium, 127
  hyperpolarization techniques, 128
  hyperpolarized, 128
  in vivo human imaging, 126
  lungs, 127
  physics, 126
  principles, 125
  resonance element, 126
  RF pulse, 126
  scanner, 126
  signals, 125–127
  structural imaging, 127
  thoracic, 127, 138
  xenon, 130
Male infants, 229
Male lung maturation, 229
Male sex steroids, 266
Mammalian target of rapamycin (mTOR), 43
Mast cells, 270, 274, 276, 281
Matrix metalloproteinase 9 (MMP9), 351
Mechanical stress signals, 97
Metabolic reprogramming, 28
Methamphetamine (MA), 357
Microarray data, 100
MicroRNAs (miRNAs), 231
Misfolded proteins, 336
Mitochondria, 23
Mitochondria-associated membrane (MAM), 341
Mitochondrial Ca^{2+}, 198
Mitochondrial dysfunction, 23
Mitochondrial membrane structure, 101
Mitogen-activated protein kinase (MAPK) signaling
      pathways, 356
MR angiography (MRA), 137
Multiorgan dysfunction syndrome (MODS), 96
Murine lung epithelial cell line (MLE-12), 102
Mycobacterial ligands, 43
Myocyte enhancer factor 2 (MEF2), 351
Myofibroblast, 237

N
NADPH oxidase (NOX), 348
  oxidase family, 25
  Nox4, 26
  transverse aortic constrictions, 26
Nafamostat, 222
Nafamostat mesylate, 222
NAMPT transcriptional regulation, 102
Neonatal intensive care, 135
Neonatal lung disease, 228
Neonatal Research Network (NRN), 227
Nerve growth factor (NGF), 110
Neuromuscular Blockade Agents (NMDAs), 329
Neurotrophin, 110
  airways, 114
  BDNF gene, 111

  classical, 110
  environmental, 110
  expression, 116
  fibroblasts, 110
  GDNF, 111
  non-neuronal systems, 110
  regulatory pathways, 110
  resident airway cell function, 110
  signaling, 110
  smooth muscles, 114
Neurotrophin signaling
  BDNF, 113
  GDNF family member ligands, 114
  heterogeneity, 113
  p75NTR receptor, 114
  TrkB gene, 113
Neutrophil elastase (HNE), 217, 219
  ALI/ARDS, 219
  inhibition, 219
  inhibitors, 219
  NETs, 219
Neutrophils, 158, 217, 268
NF-κB dimers, 57
NF-κB signaling, 57, 58, 63, 156, 158
  airways, 158
  COPD, 158
  inflammation, 158
  pathway, 153, 159
  reulation, 57
Nicotine, 155
Nicotinic receptors (nAChRs), 155
N-linked glycosylation, 112
NLR Inflammasome Network, 43
NLRP3 inflammasome, 60
NOD-like receptors (NLRs), 56
Nonallergic/nonatopic asthma, 2
Noncanonical NF-κB pathway, 157
Non-interstitial pneumonia (NSIP), 81
Non-small cell lung cancer (NSCLC), 197
Nonspecific interstitial pneumonia (NSIP), 85
Nrp-Plexin complexes, 207
Nuclear factor (NF)-κB
  activation, 189
  activators, 189
  cytoplasm, 189
  selectivity, 189
  thioredoxin, 189
Nuclear factor erythroid 2-related factor 2
      (NRF2), 339
Nucleic acid sensors, 56
Nucleotide-binding oligomerization domain (NOD), 353

O
Obesity, 150
Obstructive lung diseases (OLDs), 2
Obstructive sleep apnea (OSA), 239
Old astrocyte specifically induced substance
      (OASIS), 340
Organizing pneumonia (OP), 79, 81, 85
Oxidant/antioxidant balance, 194

Oxidative stress (OS), 153, 188, 336, 340, 341, 344–346, 350, 352, 354, 357, 358, 360
  bronchial asthma, 191
  nonallergic asthma, 191
  TLRs, 191
Oxygen-enhanced MRI (OE-MRI)
  gadolinium-based perfusion imaging, 135
  hyperpolarized gases, 132
  standard clinical MRI scanners, 133
  T1 maps, 133
  UTE images, 135
  VDP, 133
Oxygen transfer function (OTF), 133

P
Palmitoylation, 166
Pathogen-associated molecular pattern (PAMP), 42, 55, 192, 353
Pattern recognition receptors (PRRs), 42, 55, 157
Perimenstrual asthma (PMA), 234, 272
Permeability transition pores (PTPs), 351
Peroxiredoxin IV (PRxIV), 344
Persistent asthma, 278
Pharmacological treatments, 150
Plasma kallikrein, 219
Plexins, 206
Pneumocytes, 349
Positron emission tomography (PET), 124
Pregnenolone, 262–265
Progesterone (P4), 263, 268, 280–282
Progesterone receptors (PRs), 268
Pro-inflammatory cytokines, 235, 242
Pro-inflammatory genes, 195
Proline hydroxylases (PHD), 347
Prone positioning, 328
Protease-activated receptor 2 (PAR2), 9
Protein depalmitoylation, 182
Protein disulfide isomerase (PDI), 338, 343, 344
Protein interactions, 180
Protein kinase A (PKA) signaling, 262, 264
Protein kinase C (PKC), 352
Protein kinase RNA-like endoplasmic reticulum kinase (PERK), 338–340
Protein modifications, 197
Protein secretion
  CCN3, 179
  CCR5, 178
  embryos, 179
  S-palmitoylation, 178
  Wnts, 179
Protein stability
  zDHHC-9 knockout, 179
Protumorigenic inflammatory responses, 190
Puberty, 228
Pulmonary ARDS, 325
Pulmonary arterial endothelial cells (PAECs), 349
Pulmonary arterial hypertension (PAH), 240, 241, 261
Pulmonary arterial smooth muscle cells (PASMCs), 151, 336, 338, 349

Pulmonary diseases
  COPD, 61
Pulmonary embolism (PE), 137
Pulmonary fibrosis (PF), 261, 284, 286
  androgens effect on inflammation, 286
  estrogens effect on inflammation, 286, 287
Pulmonary function tests (PFTs), 75, 79, 124
Pulmonary granuloma formation, 43
Pulmonary hypertension (PH), 22, 194
  activation, UPR sensors, 340, 341
  anti-inflammatory role, 356, 357
  antioxidant gene expression, 361
  antioxidant response, 345, 348
  antioxidants, 345
  apoptosis, 361
  biopterin, 361
  BMPR2 mutation, 358, 359, 361
  calcium release, 350
  calcium signaling, 351
  CHOP, 356
  CHOP inhibits, 352
  chronic hypoxia, 352
  chronic hypoxia-mediated ROS generation, 359
  colon carcinoma cells, 345
  disulfide bonds, 343
  endothelial cells, 359
  eNOS activity, 351
  enzymes, 345
  EPCs, 350
  ER enzymes, 346
  ER redox homeostasis, 346
  ER stress, 355, 359
  ERO1α activity, 345
  glutathione, 342, 343
  GPx7, 345
  GPx8, 345
  HIF1α, 347
  HIF2α, 347
  HIV-PAH, 349
  HO-1 activity, 357
  hypoxia, 359
  hypoxia-induced UPR activation, 348
  hypoxia-inducible factor, 347
  inflammasome activates, 354
  inflammatory factors, 361
  innate immune pathways, 356
  interplay, 360
  intra-/extracellular stress signals, 359
  IRE1α oligomerization, 345
  misfolded proteins, 345
  mitochondrial genome, 341
  NF-κB, 352, 353
  NLRP3, 353, 354
  NLRP3 inflammasome activation, 359
  NOX2-mediated endothelial cell apoptosis, 349, 350
  NRF2, 359
  NRF2 activation, 354–357
  NRF2 deregulation, 357, 358
  PDI, 344
  pro-inflammatory genes, 361

Pulmonary hypertension (PH) (*cont.*)
    protein folding, 338, 342
    pulmonary vascular remodeling, 361
    pulmonary vascular smooth muscle cells, 350
    redox and cellular homeostasis, 359
    redox modulation, 335, 336
    redox-sensitive NF-κB, 359
    reoxygenation injury, 352
    ROS-induced hyaluronan fragmentation, 354
    ROS-induced UPR activation, 346, 347
    stress response, 336, 337
    subsequent UPR activation, 346
    thiol redox, 342
    unfolded protein response, 335, 336
    UPR pro-oxidant response, 348
    UPR sensors, 337, 338
Pulmonary Inflammation, 40
Pulmonary surfactant, 229, 230
Pulmonary vascular remodeling (PVR), 22, 336, 353, 359, 361
Pulmonary vascular smooth muscle cells, 350
Pulse sequence, 126
Pyruvate dehydrogenase (PDH), 28

Q
Quiescin sulfhydryl oxidase (QSOX), 338, 344

R
Radiofrequency (RF) pulses, 125
Radiological patterns, 79
Raynaud's phenomenon, 77
Reactive nitrogen species (RNS), 192
Reactive oxygen species (ROS), 23, 97, 187, 337
    biological systems, 187
    cell signaling pathways, 153
    cellular, 188
    cellular injury, 195
    cigarette smoke, 188
    COPD patients, 154
    COVID-19, 196
    DNA oxidation, 196
    exacerbated, 194
    factors, 188
    free radicals, 23
    generation, 23–25, 152, 194
    inflammatory signaling pathways, 157
    leakage, 24
    lipid, 195
    mitochondria, 152
    mitochondrial electron transport chain, 23
    NF-κB signaling, 157
    nonradical derivatives, 23
    NOX regulation, 153
    overproduction, 25, 153
    in PASMCs, 156
    physiological role, 188
    preconditioning adaptive response, 25
    reactive forms, 23

        signaling, 154
        signaling molecules, 23
        in vascular homeostasis, 194
Redox modulation, 335, 336
Regulatory T cells (Tregs), 48
Rel homology domain (RHD), 158
Relative enhancement ratio (RER), 133
Renin-angiotensin system (RAS), 30, 196
Reoxygenation injury, 352
Respiratory burst, 189
Respiratory diseases, 222, 229
Respiratory distress syndrome (RDS)
    female neonates, 229
    pathophysiology, 230
    premature born, 230
    preventative and treatment options, 230
Respiratory infection, 241
Respiratory syncytial virus (RSV), 241
Retinoic acid-inducible gene I (RIG-I)-like receptors, 55
Rheumatoid arthritis (RA), 75
Rieske iron-sulfur protein (RISP), 152
Rregulated IRE1-dependent decay (RIDD) pathway, 339

S
Saccharomyces cerevisiae, 342
Sarcoidosis, 40, 45, 48
    autoimmune, 41
    chemokines, 45
    chitotriosidase, 44, 45
    fibrosis, 48
    gene markers, 41
    genetic studies, 41
    granuloma structure, 44
    immune factors, 42
    Inflammation, 40
    macrophage activation, 46
    macrophages, 46
    MMP12 expression, 45
    MTOR pathways, 43
    mycobacteria, 40
    PPARγ, 46
    PRRs, 42
    SAA expression, 44
    sIL-2R assay, 45
    TLR2 expression, 43
Secretory proteins, 335, 336
Selectin P ligand (SELPLG) gene, 99
Sema3E, 206
Sema3E pathway, 208
Sema3E/plexinD1 axis, 207, 209
Sema3E/plexinD1 interaction, 209
Sema3E/plexinD1 pathway, 206
Sema3e-deficient mice, 208
Sema3E-plexinD1 function, 207
Semaphorins, 206
    morphogenesis, 206
    plexinD1 axis, 206, 207
    plexins, 206
    Sema3E, 206

Serine proteases, 215, 218
  inflammation responses, 216
  inhibitors, 222
  N-terminus, 217
  protein/peptide, 216
  structure, 216
Serology screening method, 78
Serum amyloid A (SAA), 44
Severe acute respiratory syndrome coronavirus 2
    (SARS-CoV-2), 290
Sex, 228
Sex differences
  in asthma, 232
  in BPD, 231
  in COVID-19, 243
  in influenza severity, 241
  in IPF, 237
  in lung and airway development, 231
  in lung development, 229
  in lung disease progression, 244
  in lung inflammatory diseases, 261
  in neonatal, pediatric and adult lung disease
    prevalence, 229
  in respiratory disease, 228
  in RSV infection and bronchiolitis, 241
Sex hormone binding globulin (SHBG), 266
Sex hormone receptors
  AR, 266
  ER, 267
  PRs, 268
  sex steroids, 266
  SHBG, 266
Sex hormones, 228, 229, 232, 235, 236, 238, 239
  cholesterol, 262
  classification, 262
  GnRH, 262
  in inflammatory lung pathologies
    asthma (see Asthma)
Sex steroids, 266
Sex-related differences, 227
Sex-specific immune responses, 242
Sexual dimorphism, 229, 231
Short-acting beta-agonists (SABAs), 2, 4
Single nucleotide polymorphisms (SNPs), 99, 232
Site 1 protease (S1P), 340
Site 2 protease (S2P), 340
Sjogren's syndrome, 75, 80, 82
Sledgehammer effect, 5
S-palmitoylation, 176, 179, 180
  agonists, 183
  CCR5, 183
  cycle rates, 182
  α1D AR, 181
  and depalmitoylation, 183
  exploitation, 183
  GLUT1, 183
  Gαq, 182
  inflammatory lung diseases, 180
  inhibitors, 183
  LpdA targets, 181
  MYD88, 181

  protein, 182
  protein trafficking, 183
  pulmonary blood vessels, 181
  zDHHC-7, 181
S-palmitoylation regulating protein
  BMP signaling, 178
  CD61, 175
  cellular differentiation, 175
  characteristics, 177
  protein modification, 176
  PSD-95, 175, 176, 178
  regulatory effect, 176
  transmembrane, 178
  zDHHC, 175, 177
  zDHHC-3, 176
  zDHHC-6, 176
  zDHHC-9, 175
  zDHHC-18, 176
Speckle-type POZ protein (SPOP), 57
Static ventilation imaging, 128
Static ventilation studies, 131
Steroid hormone biosynthesis pathways, 265
Steroidogenic acute regulatory protein (STAR), 262, 263
Steroid-resistant asthmatics, 5
Sulforaphane (SFN), 357
Superoxide dismutase 2 (SOD2), 347
Surfactant protein B (SFTPB) expression, 102
Systemic lupus erythematosus (SLE), 217
Systemic sclerosis, 77

T
T lymphocyte activation, 45
Tachykinin-expressing sensory fibers, 114
Th17 cells, 47
Theca (TCs), 263
Thiol antioxidants, 192
Thioredoxin (TRx), 343
Thoracic MRI research, 138
Thoracic radiographic images, 79
Thromboembolic pulmonary hypertension, 137
Thrombomodulin (TM), 98
Thymic stromal lymphopoietin (TSLP) gene, 232
Tissue factor (TF) inhibitor, 98
Tissue remodeling, 205
TMPRSS2, 221
TNF meta-analysis, 41
Toll-like receptors (TLRs), 42, 55, 180, 191
  functions, 55
  PAMPs and DAMPs, 55
Transforming growth factor β (TGFβ), 354
Trans-Golgi network, 180
Transient receptor potential (TRP) channels, 156
Translocase of the inner membrane (TIM), 341
Translocase of the outer membrane (TOM), 341
Translocator protein (TSPO), 262, 264
Transmembrane domains (TMDs), 177
Transmembrane protease serine type 2 (TMPRSS2), 220
  AC2, 221
  COVID-19, 221
  C-terminal domain, 220

Transmembrane protease serine type 2
        (TMPRSS2) (*cont.*)
    extracellular region, 221
    genetic variations, 221
    metastatic prostate cancer, 222
    TMPRSS2 inhibitors, 222
TRx interacting protein (TXNIP), 353
Tumor necrosis factor receptor-associated factor 2
        (TRAF2), 339

U
Ubiquitin-proteasome system (UPS), 337
Ultrashort echo time (UTE), 135
Undifferentiated connective tissue disease (UCTD), 80
Unfolded protein response (UPR), 335, 336, 338
Usual interstitial pneumonia (UIP), 80, 85
UTE MRI
    axial CT, 136
    bronchiectasis, 135
    COPD, 135
    CT-like images, 135
    functional lung MRI, 136
    OE-MRI, 136
    short-term reproducibility, 135

V
Valproic acid (VPA), 101
Vascular damage
    endothelium, 98
    endothelium forms, 97

mechanical ventilation, 97
    oxidative stress, 97
    pathologic mechanisms, 97
Vascular endothelial growth factor (VEGF),
        26, 190
Vascular hyperresponsiveness, 159
Vasculature, 99
Vasoconstriction, 152, 349, 351, 352, 357, 360, 361
VEGF receptor 1 (VEGFR1), 28
Ventilation defect percentage (VDP), 128
Ventilation-perfusion (V/Q) imaging, 124
Ventilator-associated pneumonia (VAP), 102
Ventilator-induced lung injury (VILI), 97
Vitamin K epoxide reductase (VKOR), 346
Voltage-dependent anion channels, 31
Voltage-gated K^+ (K_V) channels, 155
Von Willebrand factor (VWF), 98

W
Wheezing, 148
Wntless (WLS), 179

X
^{129}Xe isotope, 130
^{129}Xe static ventilation imaging, 130

Z
Zero echo time (ZTE) sequences, 135
ZIP9 (zinc transporter from the ZIP family), 266

Lightning Source UK Ltd.
Milton Keynes UK
UKHW052149270522
403612UK00008B/26

9 783030 687502